Encyclopaedia of
APPLIED GEO
ENVIRONMENTAL
STUDIES

Encyclopaedia of APPLIED GEO ENVIRONMENTAL STUDIES

Volume I: Ecology and Applied Environmental Science

Contributors

Mohammad Ali Zare Chahouki, Paulo Cesar Fernandes da Silva, and Bizhan Shirdel et al.

AURIS Reference

www.aurisreference.com

Encyclopaedia of Applied Geo Environmental Studies

Volume I: Ecology and Applied Environmental Science

Contributors: Mohammad Ali Zare Chahouki, Paulo Cesar Fernandes da Silva, and Bizhan Shirdel et al.

Published by Auris Reference Limited

www.aurisreference.com

United Kingdom

Printed in 2017 for Sale in the Indian Subcontinent

Ecology and Applied Environmental Science

ISBN: 978-1-78154-980-3 (Set of 3 Volumes)

British Library Cataloguing in Publication Data
A CIP record for this book is available from the British Library

Printed in the United Kingdom

Exclusively distributed by CBS Publishers & Distributors Pvt. Ltd.

Contents

List of Abbreviations

AFOLU	Agriculture, forest management, and other land uses
ALF	Animal Liberation Front
AR	Animal rights
AW	Animal welfare
AWE	Animal welfare education
BES LTER	Baltimore Ecosystem Study Long-Term Ecological Research Program
BHC	Building Healthy Communities
BD	Bulk density
CCA	Canonical correspondence analysis
CCA	Canonical Correspondence Analysis
CC	Carbonate carbon
CAM	Computer Aided Manufacturing
CIM	Computer Integrated Manufacturing
CAFOs	Concentrated animal feeding operations
CRW	correlated random walk
DE	Deep ecology
DI	Deprivation index
ELF	Earth Liberation Front
ENA	Ecological network analysis
ESD	Education for sustainable development
EC	Elemental carbon
ERP	Enterprise Resource Planning
EE	Environmental education
EFA	Environmental flow assessment
EIA	Environmental Impact Assessment
EO	Environmental objective
ETP	European Technology Platforms
FC	Field capacity
FDE	Foundation for Deep Ecology
IE	Industrial Ecology
ICT	Information Communication Technology
IWRM	Integrated Water Resources Management
IEA	International Energy Agency
IFAW	International Fund for Animal Welfare
JIT	Just-in-time
KLT	Karhunen–Loève transform
LM	Lean Management
LP	Lean Production
MAF	Mean annual flow
MRPP	Multi-response Permutation Procedure
NDRC	National Development and Reform Commission
NPCSs	National Priority Contaminated Sites
NSF	National Science Foundation
NMS	Non-metric Multidimensional Scaling
OC	Organic carbon
PTFs	Pedo transfer functions

PETA	People for the Ethical Treatment of Animals
PWP	Permanent wilting point
PIOT	Physical input-output table
PAHs	Polycyclic aromatic hydrocarbons
PEU	population energy use
PCA	Principal component analysis
PCoA	Principal Coordinate Analysis
POD	Proper orthogonal decomposition
RDA	Redundancy analysis
RSPCA	Royal Society for the Prevention of Cruelty to Animals
SCEU	size class energy use
SEP	Socioeconomic position level
SOM	Soil organic matter
SDRs	Standardized death rates
SIR	Standardized incidence ratio
SMR	Standardized mortality ratios
TC	Total carbon
TOVR	Total Ordinary Volume Regime
TQM	Total quality management
TVFR	Total Volume for Flood Regime
TEF	Toxic equivalent factor
TPS	Toyota Production System
UDWG	Urban Design Working Group

List of Contributors

Mohammad Ali Zare Chahouki
Department of Rehabilitation of Arid and Mountainous Regions, University of Tehran, Iran

Khageshwar Singh Patel
School of Studies in Chemistry/Environmental Science, Pt. Ravishankar Shukla University, Raipur, India

Shobhana Ramteke
School of Studies in Chemistry/Environmental Science, Pt. Ravishankar Shukla University, Raipur, India

Yogita Naik[1],
School of Studies in Chemistry/Environmental Science, Pt. Ravishankar Shukla University, Raipur, India

Bharat Lal Sahu
School of Studies in Chemistry/Environmental Science, Pt. Ravishankar Shukla University, Raipur, India

Saroj Sharma
Department of Chemistry, Devi Rathi Mahila Mahavidhaya, Rajnandagaon, India

Jutta Lintelmann
GSF-Forschungszentrum für Umwelt und Gesundheit, Institut für Ökologische Chemie, Neuherberg, Germany

Matuschek Georg
GSF-Forschungszentrum für Umwelt und Gesundheit, Institut für Ökologische Chemie, Neuherberg, Germany

Roberta Pirastu
Department of Biology and Biotechnologies Charles Darwin, Sapienza Rome University, Piazzale Aldo Moro 5, 00185 Rome, Italy

Pietro Comba
Department of Environment and Primary Prevention, National Health Insti-

tute, Viale Regina Elena 299, 00161 Rome, Italy

Ivano Iavarone
Department of Environment and Primary Prevention, National Health Institute, Viale Regina Elena 299, 00161 Rome, Italy

Amerigo Zona
Department of Environment and Primary Prevention, National Health Institute, Viale Regina Elena 299, 00161 Rome, Italy

Susanna Conti
Unit of Statistics of the National Health Institute, National Center for Epidemiology, Surveillance and Health Promotion, Viale Regina Elena 299, 00161 Rome, Italy

Giada Minelli
Unit of Statistics of the National Health Institute, National Center for Epidemiology, Surveillance and Health Promotion, Viale Regina Elena 299, 00161 Rome, Italy

Valerio Manno
Unit of Statistics of the National Health Institute, National Center for Epidemiology, Surveillance and Health Promotion, Viale Regina Elena 299, 00161 Rome, Italy

Antonia Mincuzzi
Taranto Local Health Unit, Epidemiological and Statistical Unit, Viale Virgilio 31, 74121 Taranto, Italy

Sante Minerba
Taranto Local Health Unit, Epidemiological and Statistical Unit, Viale Virgilio 31, 74121 Taranto, Italy

Francesco Forastiere
Department of Epidemiology, Lazio Regional Health Service, Via di Santa Costanza 53, 00198 Rome, Italy

Francesca Mataloni
Department of Epidemiology, Lazio Regional Health Service, Via di Santa Costanza 53, 00198 Rome, Italy

Annibale Biggeri
Biostatistics Unit, ISPO Cancer Research and Prevention Institute, Via Cosimo il Vecchio 2, 50139 Florence, Italy
Annibale Biggeri Department of Statistics "G. Parenti", University of Florence, Viale Morgagni 59, 50134 Firenze, Italy

María Antonieta Gómez-Balandra,
Hydrobiology and Environmental Assessment Department, Mexican Institute of Water Technology, Jiutepec, México

María del Pilar Saldaña-Fabela,
Hydrobiology and Environmental Assessment Department, Mexican Institute of Water Technology, Jiutepec, México

Maricela Martínez-Jiménez
Hydrobiology and Environmental Assessment Department, Mexican Institute of Water Technology, Jiutepec, México

Fereshte Haghighi
Soil Conservation and Watershed Management Institute, Tehran

Mirmasoud Kheirkhah
Soil Conservation and Watershed Management Research Institute, Tehran, Iran

Bahram Saghafian
Soil Conservation and Watershed Management Research Institute, Tehran, Iran

Kevin Lo
Department of Resource Management and Geography, University of Melbourne, Melbourne, VIC 3010, Australia

Helen Kopnina
Institute Cultural Anthropology and Development Sociology, Social and Behavioural Sciences, Leiden University, Wassenaarseweg 52, Leiden 2300 RB, The Netherlands

Brett Cherniak
47 Jones Street, Hamilton, Ontario, ON L8R 1X9, Canada

Daniel L. Childers
School of Sustainability, Arizona State University, Tempe, AZ 85287, USA

Mary L. Cadenasso
Department of Plant Sciences, University of California, Davis, CA 95616, USA

J. Morgan Grove
U.S. Forest Service, Baltimore, MD 21201, USA

Victoria Marshall
Parsons The New School for Design, New York, NY 10011, USA

Brian McGrath
Parsons The New School for Design, New York, NY 10011, USA

Steward T. A. Pickett
Cary Institute of Ecosystem Studies, Millbrook, NY 12545, USA

Giuseppe Ioppolo
Department of Economics, Business, Environment and Quantitative Methods (SEAM), University of Study of Messina, P.zza Puglatti 1, Messina 98122, Italy

Stefano Cucurachi
Department CML-Industrial Ecology, Leiden University, 2300 RA Leiden, The Netherlands

Roberta Salomone
Department of Economics, Business, Environment and Quantitative Methods (SEAM), University of Study of Messina, P.zza Puglatti 1, Messina 98122, Italy

Giuseppe Saija
Department of Economics, Business, Environment and Quantitative Methods (SEAM), University of Study of Messina, P.zza Puglatti 1, Messina 98122, Italy

Luigi Ciraolos
Department of Economics, Business, Environment and Quantitative Methods (SEAM), University of Study of Messina, P.zza Puglatti 1, Messina 98122, Italy

Christian Mulder
National Institute for Public Health and the Environment (RIVM)

Yajun Song
School of Information Science & Technology, Beijing Forestry University
Library of Beijing International Studies University

Tianzhong Zhao
School of Information Science & Technology, Beijing Forestry University

Brent J. Sewall
Department of Biology, Temple University, Philadelphia, Pennsylvania, United States of America
Department of Wildlife, Fish, and Conservation Biology, University of California Davis, Davis, California, United States of America

Amy L. Freestone
Department of Biology, Temple University, Philadelphia, Pennsylvania, United States of America

Joseph E. Hawes
School of Environmental Sciences, University of East Anglia, Norwich Research Park, Norwich, United Kingdom

Ernest Andriamanarina
De´partement des Sciences de la Nature et de l'Environnement, Universite´ d'Antsiranana, Antsiranana, Madagascar

Jure Kokošin
Sweco Norge AS, Dronningensgate 52/54, 8509 Narvik, Norway

Andrej Gosar
Faculty of Natural Sciences and Engineering, University of Ljubljana, Aškerčeva 12, SI-1000 Ljubljana, Slovenia
Slovenian Environment Agency, Seismology and Geology Office, Dunajska 47, SI-1000 Ljubljana, Slovenia

Ichiro Ken Shimatani
The Institute of Statistical Mathematics, Tokyo, Japan

Ken Yoda
Graduate School of Environmental Studies, Nagoya University, Furo, Chikusa, Nagoya, Japan

Nobuhiro Katsumata
International Coastal Research Center, Atmosphere and Ocean Research Institute, The University of Tokyo, Kashiwa, Chiba, Japan

Katsufumi Sato
International Coastal Research Center, Atmosphere and Ocean Research Institute, The University of Tokyo, Kashiwa, Chiba, Japan

Preface

Encyclopaedia of Applied Geo Environmental Studies presents a variety of scientific ecological issues and uses these to examine a range of environmental problems while considering potential engineering, scientific, and managerial solutions. Ecology is the scientific analysis and study of interactions among organisms and their environment. The text *Ecology and Applied Environmental Science* addresses the impact of contemporary environmental problems by using the main principles of scientific ecology. It presents a variety of scientific ecological issues and examine a range of environmental problems while considering potential engineering, scientific, and managerial solutions. Multivariate analysis techniques in environmental science have been presented in first chapter. In second chapter, we discuss the contamination of environment with polycyclic aromatic hydrocarbons in India. Third chapter presents an epidemiological profile of Taranto NPCS residents analyzing different health indicators available at municipality level. In fourth chapter, an analysis of the environmental flow standard and examples of the suggested hydrological methods are presented. In fifth chapter, we review soil water content, pedotransfer functions, and some infiltration models applicability for two land-use types. Sixth chapter identifies three types of model environmental cities in China and examines their levels of energy-related carbon emissions using a bottom-up accounting system. Seventh chapter discusses how to cultivate a value for non-human interests through the convergence of animal welfare, animal rights, and deep ecology in environmental education. In eighth chapter, we present a transformative model that merges urban design and ecology into an inclusive, creative, knowledge-to-action process. The aim of ninth chapter is to highlight the key factors relevant to their integration in an environmental lean management system, both of positive terms (lights) and possible barrier (shadows). Tenth chapter focuses on ecology and eScience. A bibliometric analysis of global forest ecology research during 2002–2011 has been proposed in eleventh chapter. Size-energy relationships in ecological communities have been established in twelfth chapter. Last chapter represents a contribution to the evaluation of the seismic hazard in Breginjski kot, which is among the most endangered seismic areas in Slovenia.

Chapter 1

MULTIVARIATE ANALYSIS TECHNIQUES IN ENVIRONMENTAL SCIENCE

Mohammad Ali Zare Chahouki

Department of Rehabilitation of Arid and Mountainous Regions, University of Tehran, Iran

INTRODUCTION

One of the characteristics of environmental data, many of them and the complex relationships between them. To reduce the number variables, different statistical methods exist. Multivariate statistics is used extensively in environmental science. It helps ecologists discover structure and previous relatively objective summary of the primary features of the data for easier comprehension. However, it is complicated in theorical structure and in operational methodology.

In this chapter some important statistical methods such as Principal component analysis (PCA), Canonical correspondence analysis (CCA), Redundancy analysis (RDA), Cluster analysis, and Discriminate function analysis will be explained briefly. This chapter too cover the statistical analysis of assemblage data (species by samples matrices of abundance, area cover etc) and/or multi variable environmental data which arise in a wide range of applications in ecology and environmental science, from basic ecological studies (e.g. of dietary composition or population size-structure), through community-based field studies, environmental impact assessments and monitoring of largescale biodiversity change, to purely physical or chemical analyses.

The use of multivariate analysis has been extended much more widely over the past 20 years. Much more is included on techniques such as Canonical Correspondence Analysis (CCA) and Non-metric Multidimensional Scaling (NMS) and another technique to include organisms and organism-environment relationships other than vegetation. Spatially constrained data analysis will be introduced and the importance of accounting for spatial autocorrelation will be emphasized. Use of the methods within ecology and in environmental reconstruction will also be covered. A study and review of the application of multivariate analysis in biogeography and ecology is provided in: Kent, M. (2006).

LANDSCAPE ECOLOGY

Landscape is simply an area of land (at any scale) containing an interesting pattern that affects and is affected by an ecological process of interest. Landscape ecology, then, involves the study of these landscape patterns, the interactions among the elements of this pattern, and how these patterns and interactions change over time. In addition, landscape ecology involves the application of these principles in the formulation and solving of real-world problems (Turner et al, 2001).

Landscape ecology is perhaps best distinguished by its focus on: 1) spatial heterogeneity, 2) broader spatial extents than those traditionally studied in ecology, and 3) the role of humans in creating and affecting landscape patterns and process (Turner et al, 2001).

In effect the role of ecology, and especially that of vegetation science, has been mainly restricted to the evaluation of the landscape with respect to particular demands: either evaluation as an assessment of the qualities of the ecosystem or evaluation as a socioeconomic procedure intended to estimate the functions the natural environment fulfills for human societ (Van der Ploeg and Vlijm, 1978). Landscape ecology theory stresses the role of human impacts on landscape structures and functions. It also proposes ways for restoring degraded landscapes. Landscape ecology explicitly includes humans as entities that cause functional changes on the landscape (Mielke& Berry, 2001).

Landscape ecology theory includes the landscape stability principle, which emphasizes the importance of landscape structural heterogeneity in developing resistance to disturbances, recovery from disturbances, and promoting total system stability (Mielke& Berry, 2001). This principle is a major contribution to general ecological theories which highlight the importance of relationships among the various components of the landscape. Integrity of landscape components helps maintain resistance to external threats, including development and land transformation by human activity. Analysis of land use change has included a strongly geographical approach which has led to the acceptance of the idea of multifunctional properties of landscapes (Mielke et al, 1976). There are still calls for a more unified theory of landscape ecology due to differences in professional opinion among ecologists and its interdisciplinary approach (Bastian 2001).

Landscape ecology is distinguished by its focus on broader spatial extents than those traditionally studied in ecology. This stems from the anthropocentric origins of the discipline. Despite early attention to the effects of sample area on measurements, such as species-area relationships, the importance of scale was not widely recognized until the 1980's. Recognition that pattern-process relationships vary with scale demanded that ecologist give explicit consideration to scale in designing experiments and interpreting results.

It became evident that different problems require different scales of study, and that most problems require multiple scales of study. The theory of scale and hierarchy emerged as a framework for dealing with scale. The emergence of scale and hierarchy theory provided a partial theoretical framework for understanding pattern-process relationships, which became the basis for the emergence of landscape ecology as a discipline (Turner MG. 1989).

SAMPLING METHODS

Sampling design varies considerably with habitat type and specific taxonomic groups. Sampling design begins with a clear statement of the question(s) being asked. This may be the most difficult part of the procedure because the quality of the results is dependent on the nature of the original design.

If the sampling is for densities of organisms then at least five replicate samples per sampling site are needed because many statistical tests require that minimal number. Better yet, consider 20 replicates per sampling site and in some cases 50 or more. If sample replicates are less than five then bootstrapping techniques can be used to analyze the data (name). Some type of random sampling should be attempted (e.g., stratified random sampling) or a line intercept method used to estimate densities (e.g., Strong Method).

Measurements of important physical–chemical variables should be made (e.g.,temperature, salinity, sediment grain size, etc...). Field experiments need to be carried out with carefully designed controls. The correct spatial scale needs to be considered when planning experiments (Stiling, 2002). If vegetation is correlated with geomorphologic landforms, for this reason stratified random sampling method is better and this method employ (Greig-Smith, 1983; Ludwig and Reynold, 1988)

Environmental impact assessments ideally attempt to compare before and after studies. There were differences in sampling sites, sampling dates, effort, replication, taxonomic categories, and recovery data. (Ferson et al., 1986). Gotelli and Ellison (2004) and Odum and Barrett (2005) studied on sampling design. Diserud and Aagaard (2002) present a method that tests for changes in community structure based on repeated sampling. This may be the plot method of sampling generally consists of three major types: (1) Simple random or random sampling without replacement, (2) stratified random, and (3) systematic (Cochran, 1977)

Simple random sampling with replacement is inherently less efficient than simple random sampling without replacement (Thompson, 2002). It is important not to have to determine whether any unit in the data is included more than once. Simple random sampling consists of using a grid or a series of coordinate lines (transects) and a table of random numbers to select several plots (quadrats), the size depending on the dimensions and densities of the organisms present.

CLASSIFICATION METHODS

Measures of Similarity and Difference (Similarity and Dissimilarity)

Dissimilarity or distance measures can be categorized as metric, semi-metric, or nonmetric (McCune et al., 2002). Symmetric distances are extremely useful in community ecology. There are different Similarity coefficients based on binary data. Two methods that are simple and give good results are Jaccard and Czekanowski (Sorenson) coefficients range 0 to 1.o. Following equations are for Jaccard (1901) and Sorenson similarity coefficients

$$J = \frac{c}{a+b+c}$$

Where J= Jcacard coeffiecient, a= number of occurrences of species a alone, b= number of occurrences of b alone, c= number of co-occurences of two species (a and b).

$$C = \frac{2c}{2a+b+c}$$

Where: (same as in Jaccard)

There are different Dissimilarity coefficients based on meristic or metric data. Coefficient that are widely used in ecology studies are Bray-Curtis (1975) dissimilarity coefficient and Morisita's Index.

Bray-Curtis is recommended by Clarke and Warwick (2001) and others as the most appropriate dissimilarity coefficient for community studies.

$$BC = \frac{\sum_{j=1}^{n} \left| X_{1j} - X_{2j} \right|}{(X_{1j} + X_{2j})}$$

Where

Σ= sum (from 1 to n)
X1j= # organisms of species j (attribute) collected at site 1 (entity).
X2j= # organisms of species j collected at site 2.
BC = Bray - Curtis coefficient of distance
| | = absolute value

J= 1 to n

N= number of species

Krebs (1999) considers Morisita's index as the best similarity index, as follows:

$$C_\lambda = \frac{2\sum X_{ij}X_{ik}}{(\lambda_1 + \lambda_2)N_jN_k}$$

$$\lambda_1 = \frac{\sum\left[X_{ij}(X_{ij}-1)\right]}{N_i(N_i-1)}$$

$$\lambda_2 = \frac{\sum\left[X_{ik}(X_{ik}-1)\right]}{N_k(N_k-1)}$$

Where

Σ= sum

Cm= Morisita's Index of Similarity

Xij= No. individuals of species I in sample j

Xik= No. individuals of species I in sample k

Nj= Total No. individuals in sample j

Nk= Total No. individuals in sample k

And Coefficients of Association are two types, ranging from -1 to +1 (e.g., Pearson Correlation Coefficient) and from 0 to X (e.g., χ^2) and applicable to binary and continuous data. Pearson coefficient moment correlation uses conjoint absences, the use of which is inappropriate for comparing sites and appropriate for comparing species (Clarke and Warwick, 2001)

Euclidian Distance Euclidean distance is another measure of distance that can be applied to a site by species matrix. It has been widely used in the past because it is compatible with virtually all cluster techniques.

Cluster Analysis

Clustering is a straightforward method to show association data, however, the confidence of the nodes are highly dependent on data quality, and levels of similarity for cluster nodes is dependent on the similarity index used. Krebs

(1989) shows that mean linkage is superior to single and complete linkage methods for ecological purposes because the other two are extremes, either producing long or tight, compact clusters respectively. There are, however, no guidelines as to which mean-linkage method is the best (Swan, 1970).

The purpose of two-way clustering (also known as biclustering) is to graphically expose the relationship between cluster analyses and your individual data points. The resulting graph makes it easy to see similarities and differences between rows in the same group, rows in different groups, columns in the same group, and columns in different groups. You can see graphically how groups of rows and columns relate to each other. Two-way clustering refers to doing a cluster analysis on both the rows and columns of your matrix, followed by graphing the two dendrograms simultaneously, adjacent to a representation of your main matrix. Rows and columns of your main matrix are re-ordered to match the order of items in your dendrogram.

Table 1: Major types of hierarchical, agglomerative, polythetic clustering strategies

Group Linkage Methods	
1. Nearest Neighbor	5. Centroid
2. Farthest Neighbor	6. Ward's Method
3. Median	7. Flexible Beta
4. Group Average	8. McQuitty's Method

Ward's is also know as Orloci's and Minimum Variance Method

Cluster analysis can be performed using either presence–absence or quantitative data. Each pair of sites is evaluated on the degree of similarity, and then combined sequentially into clusters to form a dendrogram with the branching point representing the measure of similarity.

TWINSPAN

The TWINSPAN method (from Two Way Indicator Species Analysis, Hill 1979; Hill et al. 1975) is a very popular method among community ecologists and it was partially inspired by the classificatory methods of classical phytosociology (use of indicators for the definition of vegetation types). Two popular agglomerative polythetic techniques are Group Average and Flexible. McCune et al. (2002) recommend Ward's method in addition. Gauch (1982) preferred to use divisive polythetic techniques such as TWINSPAN.

This method works with qualitative data only. In order not to lose the information about the species abundances, the concepts of pseudo-species and pseudo-species cut levels were introduced. Each species can be represented by several pseudo-species, depending on its quantity in the sample. A pseudo-species is present if the species quantity exceeds the corresponding cut level. TWINSPAN is a program for classifying species and samples, producing an ordered two-way table of their occurrence. The process of classification is hierarchical; samples are successively divided into categories, and species are then divided into categories on the basis of the sample classification. TWINSPAN, like DECORANA, has been widely used by ecologists.

Indicator Species Analysis

Indicator species analysis is a divisive polythetic method of numerical classification applicable to large sets of qualitative or quantitative data. This method provides a simple, intuitive solution to the problem of evaluating species associated with groups of sample units Dufrêne and Legendre's (1997). It combines information on the concentration of species abundance in a particular group and the faithfulness of occurrence of a species in a particular group. This method produces indicator values for each species in each group. These are tested for statistical significance using a Monte Carlo technique. It requires data from two or more sample units. Indicator values (range 0 for no indication to 100 for perfect indication) are presented for each species. The statistical significance of the maximum indicator value recorded for a given species is generated by a Monte Carlo test. There are many types of indicator species ranging from individual species (Dufrêne & Legendre's ,1997).

The identification of characteristic or indicator species is traditional in ecology and biogeography. Field studies describing sites or habitats usually mention one or several species that characterize each habitat. There is clearly a need for the identification of characteristic or indicator species in the fields of monitoring, conservation, and management. Indicator species add ecological meaning to groups of sites discovered by clustering, they can compare typologies derived from data analysis, identify where to stop dividing clusters into subsets, and point out the main levels in a hierarchical classification of sites.

Indicator species differ from species associations in that they are indicative of particular groups of sites. Good indicator species should be found mostly in a single group of a typology and be present at most of the sites belonging to that group (Hill, 1975). With Classical problem in community ecology and biogeography, species are the best indicators we have for particular environmental conditions. And in long-term environmental follow-up, conservation, ecological management, researchers are looking for bioindicators of habitat types to preserve or rehabilitate.

McGeoch & Chown (1998) found the indicator value method important to conservation ecosystem because it is conceptually straightforward and allows researchers to identify bioindicators for any combination of habitat types or areas of interest, e.g. existing conservation areas, or groups of sites based on the outcome of a classification procedure. Indicator species are species that, due to their niche preferences, can be used as ecological indicators of community types, habitat conditions, or environmental changes (McGeoch 1998, Carignan and Villard 2002, Niemi and McDonald 2004). They are usually determined using an analysis of the relationship between the observed species presence–absence or abundance values in a set of sampled sites and a classification of the same sites (Dufrêne and Legendre 1997).

Finally this method may represent the groups of sites in the classification, in different qualitative characteristics of the ecosystem, such as habitat or community types, environmental or succession states, or the levels of controlled experimental designs. Since indicator species analysis relates two elements, the species and the groups of sites, it can be used for gaining information on either or both. Indeed, indicator species analysis allows the characterization of the qualitative environmental preferences of the target species (for instance, when the groups are habitat types), and identifyes indicators of particular groups of sites, which can be used in further surveys.

The applications of indicator species analysis are many, including conservation, land management, landscape mapping, or design of natural reserves. Indicator species are commonly referred to as 'diagnostic species' in vegetation studies (Chytr et al. 2002).

Multi-response Permutation Procedures (MRPP)

Multi-response Permutation Procedure (MRPP) was introduced by Mielke, Berry, and Johnson (1976) as a technique for detecting the difference between a priori classified groups.

It turned out to be an extremely versatile data-analytic framework from which a number of applications fall out, such as the measurement of agreement, multivariate correlation and association coefcients, and the detection of autocorrelation (see Mielke & Berry, 2001 for a complete coverage of applications of the MRPP framework). MRPP is a non-parametric method for testing the hypothesis of no difference between two or more groups of entities. The groups must be a priori. For example, one could compare species composition between burned and unburned plots to test the hypothesis of no treatment effect. Discriminant analysis is a parametric method that can be used on the same general class of questions. However, MRPP has the advantage of not requiring assumptions (such as multivariate normality and homogeneity of variances) that are seldom met with ecological community data (Bakus, 2007).

Multiple Response Permutation Procedure (MRPP) provides a test of whether there is a significant difference between two or more groups of sampling units. This difference may be one of location (differences in mean) or one of spread (differences in within-group distance). Function MRPP operates on a data frame matrix where rows are observations and responses data matrix. The response(s) may be uni or multivariate. The method is mathematically allied with analysis of variance, in that it compares dissimilarities within and among groups. If two groups of sampling units are really different (e.g. in their species composition), then average of the within-group compositional dissimilarities ought to be less than the average of the dissimilarities between two random collection of sampling units drawn from the entire population.

The MRPP method is simply the overall weighted mean of within-group means of the pairwise dissimilarities among sampling units. The MRPP algorithm first calculates all pairwise distances in the entire dataset, then calculates delta. It then permutes the sampling units and their associated pairwise distances, and recalculates a delta based on the permuted data. It repeats the permutation step permutations times.

The function also calculates the change-corrected within-group agreement. And it also calculates classification strength which is defined as the difference between average between group dissimilarities and within group dissimilarities (Van Sickle 1997). If the first argument data can be interpreted as dissimilarities, they will be used directly. In other cases the function treats datas observations, and uses vegdist to find the dissimilarities. The default distance is Euclidean as in the traditional use of the method, but other dissimilarities in vegdist also are available.

Function meandist calculates a matrix of mean within-cluster dissimilarities (diagonal) and between-cluster dissimilarities (off-diagonal elements), and an attribute n of grouping counts. Function summary finds the within-class, between-class and overall means of these dissimilarities, and the MRPP statistics with all weight type options and the classification strength. The MRPP a robust alternative to the traditional normal theory based parametric tests, such as the t test and the analysis of variance. However, MRPP is not widely known to researchers, and part of the reason is that it has not been incorporated into major statistical packages.

Mantel Test

The Mantel test evaluates the null hypothesis of no relationship between two dissimilarity (distance) or similarity matrices. The Mantel test is an alternative to regressing distance matrices that circumvents the problem of partial dependence in these matrices.

For example evaluating the correspondence between two groups of organisms from the same set of sample units or comparing community structure before and after a disturbance is evaluate by mantel test. In Consequency, Two methods are available in PC-ORD: Mantel's asymptotic approximation and a randomization (Monte Carlo) method.

The Mantel Test (Mantel, 1967) compares two dissimilarity (distance) matrices using Pearson correlation. The matrices that use in this test must be of the same size (i.e., same number of rows). It would appear that the Mantel Test could be used to evaluate the internal structure in two sets of samples by comparing the two (dis)similarity matrices. McCune and Mefford (1999) have a computer program for performing the Mantel test. After the Mantel statistic has been calculated, the statistical significance of the relationship is tested by a permutation test or by using an asymptotic t-approximation test. The minimum number of permutations (randomizations) recommended by Manly (1997) is 1000. The Mantel and ANOSIM procedures produce similar probabilities (Legendre and Legendre, 1998).

Some of the problems associated with the Mantel test include (1) weakness in detecting spatial autocorrelation where the spatial pattern is complex and not easily modeled with distance matrices, (2) a larger number of data points may be needed for field experiments than is usually obtained, and (3) multivariate data are summarized into a single distance or dissimilarity and it is not possible to identify which variable(s) contributed the most (Fortin and Gurevitch, 2001). The use of Partial Mantel tests can distinguish the relative contributions of the factors of a third matrix considered as covariables. Mantel and partial Mantel tests can produce complementary information that other methods such ANOVA cannot provide and may do a better job than other method, in detecting block effects (Fortin and Gurevitch, 2001).

Thus, it is possible to distinguish the effects of spatial pattern from those of experimentally imposed treatment effects with these Mantel tests (Bakus, 2007).

The Best Clustering Strategies

The similarity or dissimilarity measures, forming the backbone of most multivariate clustering and ordination techniques, which are favored by many ecologists, are Jaccard and Bray–Curtis. The most popular clustering techniques in ecological studies are Group Average and Flexible group average clustering. For many of these techniques, it is impossible to choose a "best method" because of the heuristic nature of the methods (Jongman et al., 1995; Anderson, 2001).

The choice of an index must be made based on the investigator's experience, the type of data collected, and the ecological question to be answered. When comparisons are made, the Jaccard index is among the least sensitive of the similarity (or dissimilarity) indices (Jongman et al. 1995). Similarly, hierarchical agglomerative techniques (e.g., particularly Group Average) have proven to be very useful to ecologists in the construction of dendrograms. As part of a study is needed information on plant communities and their distribution. Then the objective of best method of multivariate analysis is to:

1. Identify and describe plant communities on the ecosystem
2. Map these communities to provide a tool for ecological studies and monitoring of ecological community;
3. Characterize the interrelationships between plant communities, soil particle size, moisture availability, grazing pressure and elevation.

ORDINATION METHODS

Ordination is a collective term for multivariate techniques which adapt a multi-dimensional swarm of data points in such a way that when it is projected onto a two dimensional space any intrinsic pattern the data may possess becomes apparent upon visual inspection (Pielou, 1984). As mentioned in the introduction, ordination is the arrangement of samples along gradients. Indeed, ordination can be considered a synonym for multivariate gradient analysis. Basically, ordination serves to summarize community data by producing a low-dimensional ordination space in which similar species and samples are plotted close together, and dissimilar species and samples are placed far apart.

Indirect gradient analysis

- Distance-based approaches
- Polar ordination, PO (Bray-Curtis ordination)
- Principal Coordinates Analysis, PCoA (Metric multidimensional scaling)
- Nonmetric Multidimensional Scaling, NMDS

Eigenanalysis-based approaches

- Linear model
- Principal Components Analysis, PCA
- Unimodal model
- Correspondence Analysis, CA (Reciprocal Averaging)
- Detrended Correspondence Analysis, DCA

Direct gradient analysis

- Linear model
- Redundancy Analysis, RDA
- Unimodal model
- Canonical Correspondence Analysis, CCA
- Detrended Canonical Correspondence Analysis, DCCA

Polar Ordination (Bray- Curtis)

(Polar Ordination) arranges samples with respect to "poles" (also termed end points or reference points) according to a distance matrix (Bray and Curtis 1957). These endpoints are two samples with the highest ecological distance between them (objective approach), OR two samples suspected of being at opposite ends of an important gradient (subjective approach). This procedure is especially useful for investigating ecological change (e.g., succession, recovery).

Advantages of this method is Ideal for evaluating problems with discrete endpoints. Polar Ordination ideal for testing specific hypotheses (e.g., reference condition or experimental design) by subjectively selecting the end points Disadvantages: This technique does not provide a general-purpose description of the community (perspective is biased) Very sensitive to outliers (by definition – "end points") Select a distance measure (usually Sorensen Index) and calculate matrix of distances (D) between all pairs of points (Beals, 1984). In the earliest versions of PO, these endpoints were the two samples with the highest ecological distance between them, or two samples which are suspected of being at opposite ends of an important gradient (thus introducing a degree of subjectivity).

Beals (1984) extended Bray-Curtis ordination and discussed its variants, and is thus a useful reference. The polar ordination, simplest method is to choose the pair of samples, not including the previous endpoints, with the maximum distance of separation.

Principal Component Analysis

PCA was invented in 1901 by Karl Pearson (Dunn, et al, 1987) Now it is mostly used as a tool in exploratory data analysis and for making predictive models. Principal Components Analysis is a method that reduces data dimensionality by performing a covariance analysis between factors.

It can be done by eigenvalue decomposition of a data covariance matrix or singular value decomposition of a data matrix, usually after mean centering the data for each attribute. The results of a PCA are usually discussed in terms of component scores (the transformed variable values corresponding to a particular case in the data) and loadings (the weight by which each standarized original variable should be multiplied to get the component score (Feoli and Orl¢ci. 1992). Principal components analysis is the basic eigenanalysis technique. It maximizes the variance explained by each successive axis.

PCA was one of the earliest ordination techniques applied to ecological data. PCA uses a rigid rotation to derive orthogonal axes, which maximize the variance in the data set. Both species and sample ordinations result from a single analysis. Computationally, PCA is basically an eigenanalysis. The sum of the eigenvalues will equal the sum of the variance of all variables in the data set. PCA is relatively objective and provides a reasonable but crude indication of relationships.

This method is a mathematical procedure that uses an orthogonal transformation to convert a set of observations of possibly correlated variables into a set of values of uncorrelated variables called principal components.

The number of principal components is less than or equal to the number of original variables. This transformation is defined in such a way that the first principal component has as high a variance as possible (that is, accounts for as much of the variability in the data as possible), and each succeeding component in turn has the highest variance possible under the constraint that it be orthogonal to (uncorrelated with) the preceding components.

Principal components are guaranteed to be independent only if the data set is jointly normally distributed. It is sensitive to the relative scaling of the original variables. Depending on the field of application, it is also named the discrete Karhunen–Loève transform (KLT), the Hotelling transform or proper orthogonal decomposition (POD). Broken-stick eigenvalues are provided to help evaluating statistical significance. Principal component analysis (PCA) (ter Braak and Sˇ milauer, 1998) was used to determine the association between plant communities and environmental variables, i.e. in an indirect noncanonical way (ter Braak and Loomans, 1987).

While PCA finds the mathematically optimal method (as in minimizing the squared error), it is sensitive to outliers in the data that produce large errors PCA tries to avoid. It therefore is common practice to remove outliers before computing PCA. However, in some contexts, outliers can be difficult to identify. For example in data mining algorithms like correlation clustering, the assignment of points to clusters and outliers is not known beforehand. A recently proposed generalization of PCA based on Weighted PCA increases robustness by assigning different weights to data objects based on their estimated relevancy.

Although it has severe faults with many community data sets, it is probably the best technique to use when a data set approximates multivariate normality. PCA is usually a poor method for community data, but it is the best method for many other kinds of multivariate (Bakus, 2007).

Principal Coordinate Analysis (PCoA)

Principal Coordinate Analysis (PCoA) is a method to represent on a 2 or 3 dimensional chart objects described by a square matrix containing that contains resemblance indices between these objects. This method is due to Gower (1966). It is sometimes called metric MDS (MDS: Mutidimensional scaling) as opposed to the MDS (or non-metric MDS). Both methods have the same objective and produce similar results if the similarity matrix f square distances are metric and if the dimensionality is sufficient.

Principle coordinates are similar to principal components in concept. The advantage of PCoA is that it may be used with all types of variables (Legendre and Legendre, 1998). Because of this, PCoA is an ordination method of considerable interest to ecologists. Most metric (PCoA) and nonmetric MDS plots are very similar or even identical, provided that a similar distance measure is used. The occurrence of negative eigenvalues, lack of emphasis on distance preservation, and other problems are discussed in detail by Legendre and Legendre (1998) and Clarke and Warwick (2001).

One of the biggest differences between PCA and PCoA is that the variables (i.e. species) representing the original axes are projected as biplot arrows. In the bryophyte communities, these biplot arrows greatly aid in interpretation (Bakus, 2007). In most applications of PCA (e.g. as a factor analysis technique), variables are often measured in different units. For example, PCA of taxonomic data may include measures of size, shape, color, age, numbers, and chemical concentrations. For such data, the data must be standardized to zero mean and unit variance (the typical default for most computer programs).

For ordination of ecological communities, however, all species are measured in the same units, and data should not be standardized. In matrix algebra terms, most PCAs are eigenanalyses of the correlation matrix, but for ordination they should be PCAs of the covariance matrix. In contrast to Correspondence Analysis and related methods, species are represented by arrows. This implies that the abundance of the species is continuously increasing in the direction of the arrow, and decreasing in the opposite direction. Thus PCA is a 'linear method'.

Although the discussion above implies that PCA is distinctly different from PCoA, the two techniques end up being identical, if the distance metric is Euclidean. Unfortunately, this linear assumption causes PCA to suffer from a serious problem, the horseshoe effect, which makes it unsuitable for most ecological data sets (Gauch 1982). Principal Coordinates Analysis (PCoA, = Multidimensional scaling, MDS) is a method to explore and to visualize imilarities or dissimilarities of data. It starts with a similarity matrix or dissimilarity matrix (= distance matrix) and assigns for each item a location in a lowdimensional space, e.g. as a 3D graphics. PCOA tries to find the main axes through a matrix. It is a kind of eigenanalysis (sometimes referred as "singular value decomposition") and calculates a series of eigenvalues and eigenvectors. Each eigenvalue has an eigenvector, and there are as many eigenvectors and eigenvalues as there are rows in the initial matrix. Eigenvalues are usually ranked from the greatest to the least. The first eigenvalue is often called the "dominant" or leading" eigenvalue. Using the eigenvectors we can visualize the main axes through the initial distance matrix. Eigenvalues are also often called "latent values". The result is a rotation of the data matrix: it does not change the positions of points relative to each other but it just changes the coordinate systems!

By using PCoA we can visualize individual and/or group differences. Individual differences can be used to show outliers. There is also a method called 'Principal Component Analysis' (PCA, sometimes also misleadingly abbreviated as 'PCoA') which is different from PCOA.

- Principal Coordinates Analysis (Principal Coordinates Analysis (PCoA PCoA or PCO) or PCO)
- Maximizes the linear correlation between distance measures and distance in the ordination.
- useful if one has only a distance (or similarity) matrix
- The underlying model is that there a fixed number of explanatoryoriginal variables. In contrast, PCA, RA, and DCA assume that there are potentially many variables, but of declining importance.
- One cannot easily put new points in a PCoA.
- For Euclidean distance, PCoA= PCA
- PCoA can be expressed as an eigenanalysis.

Factor Analysis (FA)

FA and PCA (principal components analysis) are methods of data reduction. Take many variables and explain them with a few "factors" or "components". Correlated variables are grouped together and separated from other variables with low or no correlation. Patterns of correlations are identified and either used as descriptives (PCA) or as indicative of underlying theory (FA). It process of providing an operational definition for latent construct (through regression equation). FA and PCA are not much different than canonical correlation in terms of generating canonical variates from linear combinations of variables. Although there are now no "sides" of the equation.

For calcuting this method we should:

1. Selecting and measuring a set of variables in a given domain
2. Data screening in order to prepare the correlation matrix
3. Factor Extraction
4. Factor Rotation to increase interpretability
5. Interpretation

Factor analysis is seldom used in ecology. Several statisticians state that it should not be used because it is based on a special statistical model. Estimating the unique variance is the most difficult and ambiguous task in FA (McGarigal et al., 2000).

Redundancy Analysis

RDA is a linear method and since it is a linear method, species as well as environmental variables are represented by arrows. In most cases, it is best to represent the two sets of arrows in two figures for ease of display. Thus, if have a gradient along which all species are positively correlated, RDA will detect such a gradient while CCA will not. RDA can use 'species' that are measured in different units. If so, the data must be centered and standardized. But in fact, as an ordination technique, the species should not be standardized. Redundancy analysis is the linear method of direct ordination. It also goes under the name of principal components analysis with respect to instrumental variables, which in our case are environmental variables (Sabatier et al., 1989). It is also called least-squares reduced rank regression so as to emphasize its link with multivariate regression (ter Braak & Prentice, 1988; ter Braak & Looman, 1994). In statistical textbooks on multivariate analysis, redundancy analysis is usually neglected.

RDA is useful when gradients are short. In particular, RDA may be the method of choice in a short-term experimental study. In such cases, the treatments are the explanatory variables (and are usually dummy variables). The sample ID or block might be a covariable in a partial RDA, if one wishes to factor out local effects (Bakus, 2007). 'variance explained' is actually a variance explained, and not merely inertia. Thus, variance partitioning, and interpretation of eigenvalues, are more straightforward than for CCA (Lepš & Šmilauer, 1999).

The relatively short space devoted to RDA should not be given as an indication that it is less valuable than, or inferior to, other method. It is simply used for different purposes. Consequently, the linear method of direct ordination, redundancy analysis, can often efficiently display the interesting effects, and there is no need for methods that also work when the range of community variation is larger (such as canonical correspondence analysis) nor for unconstrained nonmetric multidimensional scaling.

Correspondence Analysis (CA) or Reciprocal Averaging (RA)

Reciprocal averaging is also known as correspondence analysis (CA) because one algorithm for finding the solution involves the repeated averaging of sample scores and species scores (citations). It is a graphical display ordination technique which simultaneously displays the rows (sites) and columns (species) of a data matrix in low dimensional space (Gittins, 1985). Row identifiers (species) plotted close together are similar in their relative profiles, and column identifiers plotted close together are correlated, enabling one to interpret not only which of the taxa are clustered, but also why they are clustered (Bakus,2007).

Reciprocal averaging (RA) yields both normal and transpose ordinations automatically. Like DCA, RA ordinates both species and samples simultaneously. Instead of maximizing 'variance explained', CA maximizes the correspondence between species scores and sample scores. If species scores are standardized to zero mean and unit variance, the eigenvalues also represent the variance in the sample scores (but not, as is often misunderstood, the variance in species abundance).

Since CA is a unimodal model, species are represented by a point rather than an arrow. This is (under some choices of scaling; see ter Braak and Šmilauer 1998) the weighted average of the samples in which that species occurs. With some simplifying assumptions (ter Braak and Looman 1986), the species score can be considered an estimate of the location of the peak of the species response curve.

The CA distortion is called the arch effect, which is not as serious as the horseshoe effect of PCA because the ends of the gradients are not incurved. Nevertheless, the distortion is prominent enough to seriously impair ecological interpretation (Bakus, 2007). In other words, the spacing of samples along an axis may not affect true differences in species composition. Gradient compression can be quite blatant in simulated data sets. The problems of gradient compression and the arch effect led to the development of Detrended Correspondence Analysis (Bakus, 2007).

Detrended Correspondence Analysis (DCA)

Detrended Correspondence Analysis (DCA) eliminates the arch effect by detrending (Hill and Gauch 1982). It's a series of rules that are used to reshape data to make it friendlier for analysis. Once again, primarily used for ecological data, but can be extended to anything (data simply can't contain negative values). The reason that this technique is used is to overcome the arch effect (the horseshoe effect). Found in data whenever "PCA or other distance conserving ordination techniques are applied to data which follow a continuous gradient, along which there is a progressive turnover of dominant variables." Such as in ecological succession.

After ordination by a distance conserving technique and the first two axes are plotted against each other, one would find an arch shape. DCA is another eigenanalysis ordination technique that based on reciprocal averaging (RA; Hill 1979). DCA is geared to ecological sets, is based on samples and species. DCA ordinates both species and samples simultaneously. There are two basic approaches to detrending: by polynomials and by segments (ter Braak and Šmilauer 1998). Detrending by polynomials is the more elegant of the two: a regression is performed in which the second axis is a polynomial function of the first axis, after which the second axis is replaced by the residuals from this regression. Similar procedures are followed for the third and higher axes.

The compression of the ends of the gradients is corrected by nonlinear rescaling. Rescaling shifts sample scores along each axis such that the average width is equal to 1. Rescaling has a beneficial consequence: the axes are scaled in units of beta diversity (SD units, or units of species standard deviations). Thus if the underlying gradient is important well known, it is possible to plot the DCA scores as a function of the gradient, and there by determine whether the species 'perceive' the gradient differently than we measure it. The shape of the species response curves may change if axes are rescaled. Thus, skewness and kurtosis are largely artifacts of the units of measurement for which we choose to measure the environment.

Nonmetric multimentional scaling (MDS, NMDS, NMS, NM-MDS)

Nonmetric Multidimensional Scaling (NMDS) rectifies this by maximizing the rank order correlation. For this proceeds at first the user selects the number of dimensions (N) for the solution, and chooses an appropriate distance metric and then The distance matrix is calculated. And initial configuration of samples in N dimensions is selected. This configuration can be random, though the chances of reaching the correct solution are enhanced if the configuration is derived from another ordination method. And finally, the final configuration of points represents your ordination solution. The configuration is dependent on the number of dimensions selected; e.g. the first two axes of a 3-dimensional solution does not necessarily resemble a 2-dimensional solution. The stress will typically decrease as a function of the number of dimensions chosen; this function can aid in the selection of the results (Bakus, 2007).

This is why it is sometimes useful to rotate the solution (such as by the Varimax method) – although there is no theory that states that the final solution will represent a 'gradient' Other problems and advantages of NMDS will be discussed later, when comparing it to Detrended Correspondence Analysis (Bakus, 2007).

MANOVA AND MANCOVA

A factorial MANOVA may be used to determine whether or not two or more categorical grouping variables (and their interactions) significantly affect optimally weighted linear combinations of two or more normally distributed outcome variables. These parametric multivariate techniques (Multivariate Analysis of Variance and Multivariate Analysis of Covariance) are similar to ANOVA and ANCOVA MANOVA (Wilks' Lambda) and ANCOVA are advantageous in that performing multiple univariate tests can inflate the a value, leading to false conclusions (Scheiner, 2001).

MANOVA seeks differences in the dependent variables among the groups (McCune et al, 2002). Assumptions of MANOVA include multivariate normality (error effects included), independent observations, and equality of variance-covariance matrices (Paukert and Wittig, 2002). Because of these assumptions, among others, MANOVA is not often used in ecology although its use in increasing. The power of traditional MANOVA declines with an increase in the number of response variables (Scheiner, 2001). Unequal sample sizes are not a large problem for MANOVA, but may bias the results for factorial or nested designs. Before ANCOVA is run, tests of the assumption of homogeneity of slopes need to be performed (Petratis et al., 2001).

Early attempts to develop nonparametric multivariate analysis include those of Mantel and Valand (1970). They are more complex as they handle three or more variables simultaneously. They are frequently used with the analysis of experimental studies, especially laboratory experiments. More recently, the Analysis of Similarities was developed to compare communities or changes in communities because of pollution (Clarke, 1993). For typical species abundance matrices, an Analysis of Similarities (ANOSIM) permutation procedure is recommended over MANOVA.

Multiple analysis of variance (MANOVA) is used to see the main and interaction effects of categorical variables on multiple dependent interval variables. MANOVA uses one or more categorical independents as predictors, like ANOVA, but unlike ANOVA, there is more than one dependent variable. Where ANOVA tests the differences in means of the interval dependent for various categories of the independent(s), MANOVA tests the differences in the centroid (vector) of means of the multiple interval dependents, for various categories of the independent(s). One may also perform planned comparison or post hoc comparisons to see which values of a factor contribute most to the explanation of the dependents. There are multiple potential purposes for MANOVA.

To compare groups formed by categorical independent variables on group differences in a set of interval dependent variables. To use lack of difference for a set of dependent variables as a criterion for reducing a set of independent variables to a smaller, more easily modeled number of variables. Multiple analysis of covariance (MANCOVA) is similar to MANOVA, but interval independents may be added as "covariates." These covariates serve as control variables for the independent factors, serving to reduce the error term in the model. Like other control procedures, MANCOVA can be seen as a form of "what if" analysis, asking what would happen if all cases scored equally on the covariates, so that the effect of the factors over and beyond the covariates can be isolated. The discussion of concepts in the ANOVA section also applies, including the discussion of assumptions.

Discriminate Analysis

Discriminate Analysis (DA) is a powerful tool that can be used with both clusters of species data and environmental variables. It determines which variables discriminate between two or more groups, that is, independent variables are used as predictors of group membership (McCune et al., 2002). It is very similar to MANOVA and multiple regression analysis (Statsoft, Inc., 1995; McGarigal et al., 2000). Clusters can be identified by several methods using raw data: (1) constructing a dendrogram, (2) using PCA (even if you have field data) for initial visual identification of clusters, and (3) point rotation in space by rotating ordinations (i.e., rotating axes – see McCune and Mefford, 1999). If any method indicates groups or clusters of data then DA can be used. However, the number of groups is set before the DA analysis. DA finds a transform for the minimum ratio of difference between pairs of multivariate means and variances in which two clusters are separated the most and inflated the least. DA produces two functions: (1) classification function consisting of 2 groups or clusters of points (this information can be used for prediction with probabilities) and (2) discriminate function containing environmental variables that can be used to discriminate differences among the groups.

DA differs from PCA and Factor Analysis in that no standardization of data is needed (PCA and FA need standardization because of scaling problems) and the position of the axes distinguishes the maximum distance between clusters. (Davis, 1986). DA assumes a multivariate normal distribution, homogeneity of variances, and independent samples (Paukert and Wittig, 2002). Violations of normality are usually not fatal (i.e., somewhat non-normal data can be used). A description of the procedure to use DA with Statistica is given in the Appendix. Multiple Discriminate Analysis (MDA) is the term often used when three or more clusters of data are processed simultaneously. MDA is particularly susceptible to rounding error. Calculations in double precision for at least the eigenvalueeigenvector routines are advisable (Green, 1979). Limitations of Discriminate Analysis are discussed by McGarigal et al. (2000) Dytham (1999).

Canonical Correspondence Analysis (CCA)

In ecology studies, the ordination of samples and species is constrained by their relationships to environmental variables. When species responses are unimodal (humpshaped), and by measuring the important underlying environmental variables, CCA is most likely to be useful. It was used to examine the relationships between the measured variables and the distribution of plant communities (Ter Braak, 1986). CCA expresses species relationships as linear combinations of environmental variables and combines the features of CA with canonical correlation analysis (Green, 1989). This provides a graphical representation of the relationships between species and environmental factors.

Canonical Correlation Analysis is presented as the standard method to relate two sets of variables (Gittins, 1985). However, the latter method is useless if there are many species compared to sites, as in many ecological studies, because its ordination axes are very unstable in such cases.

The best weight for CCA describes environment variables with the first axis shows. Species information structure using a reply CCA Nonlinear with the linear combination of variables will consider environmental characteristics of acceptable behavior characteristics of species with environment shows. CCA analysis combined with non-linear species and environmental factors shows the most important environmental variable in connection with the axes shows. In Canonical Correspondence Analysis, the sample scores are constrained to be linear combinations of explanatory variables. CCA focuses more on species composition, i.e. relative abundance.

When a combination of environmental variables is highly related to species composition, this method, will create an axis from these variables that makes the species response curves most distinct. The second and higher axes will also maximize the dispersion of species, subject to the constraints that these higher axes are linear combinations of the explanatory variables, and that they are orthogonal to all previous axis. Monte Carlo permutation tests were subsequently used within canonical correspondence analysis (CCA) to determine the significance of relations between species composition and environmental variables (ter Braak, 1987)

The outcome of CCA is highly dependent on the scaling of the explanatory variables. Unfortunately, we cannot know a priori what the best transformation of the data will be, and it would be arrogant to assume that our measurement scale is the same scale used by plants and animals. Nevertheless, we must make intelligent guesses (Bakus, 2007). In CCA possible that patterns result from the combination of several explanatory variables. And many extensions of multiple regression (e.g. stepwise analysis and partial analysis) also apply to CCA. It is possible to test hypotheses (though in CCA, hypothesis testing is based on randomization procedures rather than distributional assumptions). Explanatory variables can be of many types (e.g. continuous, ratio scale, nominal) and do not need to meet distributional assumptions.

Another advantage of CCA lies in the intuitive nature of its ordination diagram, or triplot. It is called a triplot because it simultaneously displays three pieces of information: samples as points, species as points, and environmental variables as arrows (or points). If data sets are few, CCA triplots can get very crowded then should be separate the parts of the triplot into biplots or scatterplots (e.g. plotting the arrows in a different panel of the same figure) or rescaling the arrows so that the species and sample scores are more spread out. And we can only plotting the most abundant species (but by all means, keep the rare species in the analysis).

Noise in the species abundance data set is not much of a problem for CCA (Palmer, 1988). However, it has been argued that noise in the environmental data can be a problem (McCune 1999). It is not at all surprising that noise in the predictor variables will cause noise in the sample scores, since the latter are linear combinations of the former. It is probably obvious that the choice of variables in CCA is crucial for the output. Meaningless variables will produce meaningless results. However, a meaningful variable that is not necessarily related to the most important gradient may still yield meaningful results (Palmer, 1988).

If many variables are included in an analysis, much of the inertia becomes 'explained'. Any linear transformation of variables (e.g. kilograms to grams, meters to inches, Fahrenheit to Centigrade) will not affect the outcome of CCA whatsoever. There are as many constrained axes as there are explanatory variables. The total 'explained inertia' is the sum of the eigenvalues of the constrained axes. The remaining axes are unconstrained, and can be considered 'residual'. The total inertia in the species data is the sum of eigenvalues of the constrained and the unconstrained axes, and is equivalent to the sum of eigenvalues, or total inertia, of CA. Thus, explained inertia, compared to total inertia, can be used as a measure of how well species composition is explained by the variables. Unfortunately, a strict measure of 'goodness of fit' for CCA is elusive, because the arch effect itself has some inertia associated with it (Bakus, 2007).

The ordination diagrams of canonical correlation analysis and redundancy analysis display the same data tables; the difference lies in the precise weighing of the species (ter Braak & Looman, 1994; Van der Myer, 1991) One of limitations to CCA is that correlation does not imply causation, and a variable that appears to be strong may merely be related to an unmeasured but 'true' gradient. As with any technique, results should be interpreted in light of these limitations (McCune, 1999).

Multiple Regression (MR) (Multiple Linear Regression)

The mechanics of testing the "significance" of a multiple regression model is basically the same as testing the significance of a simple regression model, we will consider an F-test, a ttest (multiple t's) and R-sqrd. However, unlike simple regression where the F & t tests tested the same hypothesis, in multiple regression these two tests have different purposes. R-sqrd is still the percent of variance explained but is no longer the correlation squared (as it was with in simple linear regression) and we will also introduce adjusted R-sqrd. When considering a multiple regression (MR) model the most common order to interpret things consists of first looking at the R-sqrd, then testing the entire model by looking at the F-test, and finally looking at each individual coefficient individually using the t-tests. NOTE: The term "significance" is a nice convenience but is very ambiguous in definition if not properly specified. Thus when taking this class you should avoid simply saying something is significant without explaining (1) how you made that determination, and (2) what that specifically means in this case. You will see from the examples that those two things are always done. If you cannot do that then any time you use the word "significant" you are potentially hurting yourself in two ways; (1) you won't do well on the quizzes or exams where you have to be able to be more explicit than simply throwing out the word "significant", and (2) you will look like a fool in the business world when somebody asks you to explain what you mean by "significant" and you are stumped. Remember if you can't explain your results in managerial terms than you do not really understand what you are doing.

In order to show the relationship between biotic (principal component axes) and abiotic factors (environmental factors), a multiple regression type analysis is used. Multiple regression solves simultaneously normal equations and produces partial regression coefficients. Partial regression coefficients each give the rate of change or slope in the dependent variable for a unit of change in a particular independent variable, assuming all other independent variables are held constant. MR is not considered by some statisticians as a multivariate procedure because it includes only one dependent variable (Paukert and Wittig, 2002).

The objective of multiple regression is to determine the influence of independent variables on a dependent variable, for example, the effect of depth, sediment grain size, salinity, temperature, and predator density on the population density of species. The parameters are estimated by the least-squares method that is, minimizing the sum of squares of the differences between the observed and expected response (Jongman et al., 1995)

Normally component loadings (e.g., scores in PCA) suggest which variables are most important. However, with species abundances the only variable in some ordinations, one must use other techniques to attempt to suggest what may have produced the gradients for Axis 1, 2, 3, and so forth. Univariate analyses such as the Spearman rank correlation coefficient are not as ecologically realistic as multivariate analyses such as multiple regression because some variables are correlated and there are interaction effects between variables (Jongman et al., 1995). The highest standardized partial regression coefficients (positive or negative) suggest the most important factors (e.g., sediment size, predator density, etc.) in controlling the population density of species X. Significance tests and standard errors then can be calculated from the data.

Multiple regression has many potential problems such as the type of response curve, error distribution, and outliers that may unduly influence the results (Jongman et al., 1995). Multiple regression variables may be highly correlated, therefore, examine the correlation coefficients first (i.e., run a multiple correlation analysis between variables) to exclude some of them before doing multiple regressions. Multiple regression generally should employ a maximum of 6 variables. Legendre and Legendre (1998) suggest a stepwise procedure for reducing numerous variables. This involves a process of alternating between forward selection and backward elimination (Kutner et al., 1996). Lee and Sampson (2000) took ordination scores, representing a gradient in fish communities, and regressed them against a group of environmental variables and time. Many of scientist study on multiple regression and computer program (Such as Davis, 1986; and Sokal and Rohlf, 1995)

In multiple regression, it is typical to include quadratic terms for explanatory variables. For example, if you expect a response variable to reach a maximum at an intermediate value of an explanatory variable, including this explanatory variable AND the square of the explanatory variable may allow a concave-down parabola to provide a reasonable fit. This is an analogous situation to multiple regression: the multiple r2 or 'variance explained' increases as a function of the number of variables included. Both multiple regression and CCA find the best linear combination of explanatory variables, they are not guaranteed to find the true underlying gradient (which may be related to unmeasured or unmeasurable factors), nor are they guaranteed to explain a large portion of variation in the data.

Path Analysis

In statistics, path analysis is used to describe the directed dependencies among a set of variables. This includes models equivalent to any form of multiple regression analysis, factor analysis, canonical correlation analysis, discriminate analysis, as well as more general families of models in the multivariate analysis of variance and covariance analyses (MANOVA, ANOVA, ANCOVA).

Path analysis is a straightforward extension of multiple regression. Its aim is to provide estimates of the magnitude and significance of hypothesised causal connections between sets of variables. This is best explained by considering a path diagram. Path analysis is an extension of multiple linear regression, allowing interpretation of linear relationships among a small number of descriptors (Legendre and Legendre, 1998). This method was originally developed by Sewall Wright in which he introduced the concept of a path diagram. It handles more than one dependent variable and the effects of dependent variables on one another (Mitchell, 2001). Path analysis assumes a normal distribution of residuals, additive and linear effects, inclusion of all important variables, that residual errors are uncorrelated, and that there is no measurement error.

Path analysis was developed as a method of decomposing correlations into different pieces for interpretation of effects (e.g., how does parental education influence children's income 40 years later?). Path analysis is closely related to multiple regression; you might say that regression is a special case of path analysis. Some people call this stuff (path analysis and related techniques) causal modeling. The reason for this name is that the techniques allow us to test theoretical propositions about cause and effect without manipulating variables.

However, the causal in causal modeling refers to an assumption of the model rather than a property of the output or consequence of the technique. That is, people assume some variables are causally related, and test propositions about them using the techniques.

Canonical Correlation Analysis (CVA)

Canonical variate analysis (CVA) is a widely used method for analyzing group structure in multivariate data. It is mathematically equivalent to a one-way multivariate analysis of variance and often goes by the name of canonical discriminate analysis. Change over time is a central feature of many phenomena of interest to researchers. This dissertation extends CVA to longitudinal data. It develops models whose purpose is to determine what is changing and what is not changing in the group structure. Three approaches are taken: a maximum likelihood approach, a least squares approach, and a covariance structure analysis approach. All methods have in common that they hypothesize canonical variates which are stable over time.

The maximum likelihood approach models the positions of the group means in the subspace of the canonical variates. It also requires modeling the structure of the within-groups covariance matrix, which is assumed to be constant or proportional over time. In addition to hypothesizing stable variates over time, one can also hypothesize canonical variates that change over time. Hypothesis tests and confidence intervals are developed. The least squares methods are exploratory. They are based on three-mode PCA methods such as the Tucker2 and parallel factor analysis. Graphical methods are developed to display the relationships between the variables over time.

Stable variates over time imply a particular structure for the between-groups covariance matrix. This structure is modeled using covariance structure analysis, which is available in the SAS package Proc Calis. Canonical Variate Analysis is a special case of Canonical Correlation Analysis (Jongman et al., 1995). It is also described as a type of linear discriminate analysis (McGarigal et al., 2000). The set of environmental variables consists of a single nominal variable defining the classes.

CVA is usable only if the number of sites is much greater than the number of species and the number of classes. Many ecological data sets cannot be analyzed by CVA without dropping many species, thus CVA is not used much in ecology. Instead, they usually give the impression that there is only one such hypothesis, and therefore only one statistical technique is needed-Hotelling's canonical variate analysis (CVA). Most discussions of CVA are restricted almost entirely to a description of the underlying mathematical theory, computing directions, and perhaps an example of the computations.

Very little is usually said about the logic of the method, so that the reader is unable to judge for himself whether the method described can actually be used to test the hypothesis of interest to him.

Even when CVA is appropriate, several prominent sources have recommended misleading interpretations of the statistics computed in CVA. Perhaps because of these deficiencies, many behavioral scientists have concluded incorrectly that CVA has few or no valid and important uses in the behavioral sciences. The use originally proposed by Hotelling has been rejected by most behavioral scientists.

ORDINATION AND CLASSIFICATION METHODS IN VARIOUS ECOLOGY STUDIES

The more applied an ecological study is, the more the emphasis is on the effects on ecological communities of particular environmental factors, for example pollutants, management regimes, and other human-induced changes in the environments. (ter Braak, 1994). Correspondingly, the statistical analysis should not 'just' show the major variation in the species assemblage, but focus on the effects on the variables of prime interest. Applied studies thus call for direct methods of ordination, typically with a very limited number of (qualitative or quantitative) environmental variables (ter Braak, 1994) The range of community variation in an applied study tends to be quite small compared to that in the early ordination studies (e.g. Whittaker, 1965; Hill & Gauch, 1980).

Determining which factors control the distribution patterns of plant communities remains a central goal in ecology Classification assumes from the outset that the species assemblages fall into discontinuous groups, whereas ordination starts from the idea that such assemblages vary gradually. Ordination compares sites on their degree of similarity, and then plots them in Euclidian space, with the distance between points representing their degree of Similarity. Ordination techniques include principal components analysis (PCA), detrended correspondence analyses (DCA), and nonmetric multidimensional scaling (NMS)

Ordination (or inertia) methods, like principal component and correspondence analysis, and clustering and classification methods are currently used in many ecological studies (Zare Chahouki et al., 2009, Anderson, 2002; Gauch et al., 1977; Orloci, 1975; Whittaker (ed.), 1967; Legendre & Legendre, 1988). Ordination methods can be divided in two main groups, direct and indirect methods. Direct methods use species and environment data in a single, integrated analysis. Indirect methods use the species data only (Jongman et al.,

1987). In contrast, if a unimodal response model is assumed, the relationships are unimodal. Unimodal relations are usefully summarized by their modes or - more conveniently - weighted averages (ter Braak & Looman, 1986), so that a sensible coefficient for the species × environment table is the weighted average. The other way round, if a correlation coefficient is chosen, the implied response model is linear or approximately linear and if the chosen coefficient is the weighted average than the implied response model is unimodal (i.e. if the true model is bimodal, the ordination will fail, and if the true model is linear, the ordination will be inefficient). Assumptions about the response model and the choice of coefficients to use in secondary tables are thus interrelated.

New methods of exploring differences among groups include the nonparametric, recursive classification, and regression tree (CART). It is used to classify habitats or vegetation types and their environmental variables (McCune et al., 2002). It produces a top-to-bottom visual classification tree that undergoes a "pruning" or optimization process. CART is used to generate community maps, wildlife habitats, and land cover types in conjunction with a GIS (see p. 195 in Chapter 3). Another multivariate technique is Structural Equation Modeling (unfortunately termed SEM), a merger of factor analysis and path analysis (McCune et al. (2002). It is a method of evaluating complex hypotheses (e.g., effects of abiotic factors on plant species richness).

With multiple causal pathways among variables. It requires the initial development of a path diagram. It is an analysis of covariance relationships, effectively limited to about 10 variables. See Shipley (2000), McCune et al. (2002), Pugesek et al. (2002) and Bakus (2007).

REFERENCES

1. Anderson, M.J. (2001). A new method for non-parametric multivariate analysis of variance. Austral. Ecol. 26:32-46.
2. Alisauskas, R.T. (1998). Winter range expansion and relationships between landscape and morphometrics of midcontinent Lesser Snow Geese. Auk 115(4):851-862.
3. Anderson, C.W., Barnett, V., Chatwin, P.C. and El-Shaarawi A.H. (2002). Quantitative Methods for Current EnvironmentalIssues. Springer-Verlag, New York.
4. C.T. Bastian, S.R. Koontz and Menkhaus D.J. (2001). The Impact of Forward Contract Information on the Fed Cattle Market: An Experimental Investigation into Mandatory Price Reporting, UW Agricultural and Applied Economics Seminar. August 31, 2001. (Presented by Bastian).
5. Bakus Gerald J. (2007). Quantitative Analysis of Marine Biological

Communities Field Biology and Environment. WILEY-INTERSCIENCE, A John Wiley & Sons, Inc., Publication, 453p.

6. Beals, E.W. (1973). Ordination: Mathematical elegance and ecological naivete. J. Ecol. 61:23– 35.
7. Bray, J.R. and Curtis J.T. (1957). An ordination of the upland forest communities of southern Wisconsin.Ecol. Monogr. 27:325-349.
8. Carignan, V. and Villard M. (2002). Selecting indicator species to monitor ecological integrity: a review. Environ. Monitor. Assess. 78: 45-61.
9. Chytr, M. (2002). Determination of diagnostic species with statistical fidelity measures. J. Veg. Sci. 13: 79–90.
10. Clarke, K.R. (1993). Non-parametric multivariate analyses of changes in community structure. Aust. J. Ecol.18:117-143.
11. Clarke, K.R. and Warwick R.M. (2001). Change in Marine Communities: An Approach to Statistical Analysis and Interpretation. 2nd edition. PRIMER- E, Plymouth Marine Laboratory, Plymouth, U.K.
12. Clifford, H.T. and Stephenson W. (1975). An introduction to numerical classification. Academic Press, New York, pp. 229.
13. Cochran, W.G. (1977). Sampling Techniques. Wiley, New York.
14. Davis, J.C. (1986). Statistics and Data Analysis in Geology. Wiley, New York.
15. Diserud, O.H. and Aagaard K. (2002). Testing for changes in community structure based on repeated sampling. Ecology, 83(8): 2271–2277.
16. Dufrene, M. and Legendre P. (1997). Species assemblages and indicator species: the need for a flexible asymmetrical approach. Ecol. Monogr. 67:345–366.
17. Dunn, C. P., and Stearns F. (1987). Relationship of vegetation layers to soils in southeastern Wisconsin forested wetlands. Am. Midl. Nat. 118:366-74.
18. Dytham, C. (1999). Choosing and Using Statistics: A Biologist's Guide. Blackwell Publishing, Williston, VT.
19. Ferson, S., Downey P., Klerks P., Weissburg M., Kroot S.I., S. Jacquez O., Ssemakula J., Malenky R. and Anderson K. (1986). Competing reviews, or why do Connell and Schoener disagree? Am. Nat., 127: 571–576.
20. Feoli, E., and Orl¢ci L. (1992). Thre properties and interpretation of observations in vegetation study. Coenoses 6:61-70.
21. Fortin, M-J. and Gurevitch J. (2001). Mantel tests: Spatial structure in field experiments. pp. 308-326 in: Scheiner S.M. and J. Gurevitch (eds.).

Design and Analysis of Ecological Experiments. Oxford University Press, Oxford.

22. Gauch, H.G. (1977). ORDIFLIX—A flexible computer program for four ordination techniques: weighted averages, polar ordination, principal components analysis, and reciprocal averaging. In: Ecology and Systematics, Cornell University, Ithaca, N.Y.
23. Gauch, H.G., J. (1982). Multivariate Analysis in Community Ecology. Cambridge University Press, New York.
24. Gittins, R. (1985). Canonical analysis: a review with applications in ecology. SpringerVerlag, Berlin.
25. Gotelli, N.J. and Ellison A.M. (2004). A Primer of Ecological Statistics. Sinauer Associates, Sunderland, Maine
26. Gower, J. C. (1966). Some Distance Properties of Latent Root and Vector Methods used in Multivariate Analysis. Biometrika 53, 325-338.
27. Green, R.H. (1979). Sampling Design and Statistical Methods for Environmental Biologists. John Wiley and Sons, New York.
28. Greig-Smith, P. (1983). Quantitative Plant Ecology, 3rd Edition. Blackwell Scientific Publications, London, 359 pp.
29. Green, R.H. 1989. Power analysis and practical strategies for environmental monitoring. Environ. Res. 50:195–205.
30. Hill, M.O., Bunce R.G.H. & Shaw M.V. (1975). Indicator species analysis, a divisive polythetic method of classification, and its application to survey of native pinewoods in Scotland. Journal of Ecology, 63: 597–613
31. Hill, M.O. (1979). TWINSPAN – a FORTRAN Program for Arranging Multivariate Data in an Ordered Two-way Table by Classification of the Individuals and Attributes. Ithaca: Section of Ecology and Systematic, Cornell University.
32. Hill M.O. & Gauch H.G. (1980). Detrended correspondence analysiss, an improved ordination technique. Vegetatio, 42: 47-58
33. Jaccard, P. (1901): Etude comparative de la distribution florale dans une portion des Alpes et du Jura. Bulletin de la Socie´ te´ Vaudoisedes SciencesNaturelles, 37: 547–579.
34. Jongman, R.H.G., ter Braak C.J.F. and Van Tongeren O.F.R. (1987). Data analysis in community and landscape ecology. Cambridge University Press, Cambridge, UK.
35. Jongman, R.H.G., Ter Braak, C.J.F. and van Tongeren O.F.R. (eds.) (1995). Data Analysis in Community and Landscape Ecology. Cambridge University Press, Cambridge.

36. Kent, M. (2006). Numerical classification and ordination methods in biogeography. Progress in Physical Geography, 30: 399-408
37. Krebs, C.J. (1989). Ecol. Method. Harper Collins, New York.
38. Krebs, C.J. (1999). Ecological Methodology. Harper & Row, New York.
39. Kutner, M.H., Nachtscheim, C.J., Wasserman, W. and Neter J. (1996). Applied linear statistical models. WCB/McGraw-Hill, New York.
40. Lee, Y.W. and Sampson D.B. (2000). Spatial and temporal stability of commercial ground fish assemblages off Oregon andWashington as inferred from Oregon travel logbooks. Canadian J. Fish. Aqua. Sci., 57:2443-2454.
41. Legendre, P. & Gallagher E.D. (2001). Ecologically meaningful transformations for ordination of species data. Oecologia, 129: 271–280
42. Legendre, P. and Legendre L. (1998). Numerical Ecology. 2nd Edition. Elsevier, Amsterdam.
43. Lepš, Jan & Šmilauer P. (1999). Multivariate Analysis of Ecological Data Faculty of Biological Sciences, University of South Bohemia Ceské Budejovice,110pp
44. Ludwig, J.A., Reynold, J.F. (1988). Statistical Ecology. Wiley, New York, 337pp
45. Niemi, G.J. and McDonald M.E. (2004). Application of ecological indicators. – Annu. Rev. Ecol. Evol. Syst. 35: 89–111.
46. Manly, B.F.G. (1997). Randomization, Bootstrap and Monte Carlo Methods in Biology. Chapman and Hall, London.
47. Mantel, N. 1967. The detection of disease clustering and generalized regression approach. Cancer Res. 27:209-220.
48. Mantel, N. and Valand R.S. (1970). A technique of nonparametric multivariate analysis. Biometrics 26:547–558.
49. McCune, B. and Mefford M.J. (1999). PCORD, Multivariate Analysis of Ecological Data, Version 4. MjM Software Design, Gleneden Beach, Oregon, USA.
50. McCune, B., Grace J.B. and Urban D.L. (2002). Analysis of Ecological Communities. MjM Software Design, Gleneden Beach, Oregon.
51. McGarigal, K., Cushman, S. and Stafford S. (2000). Multivariate Statistics for Wildlife and Ecology Research. Springer-Verlag, New York
52. McGeogh, M.A. (1998). The selection, testing and application of terrestrial insects as bioindicators. Biol. Rev. 73: 181–201.
53. McGeoch, M.A. and Chown. S.L. (1998). Scaling up the value of

bioindicators. Trends Ecol. Evol. 13: 46-47.

54. Mielke, P.W. & Berry K.J. (2001). Permutation methods: A distance function approach. New York: Springer-Verlag.
55. Mielke, P. W., Berry K.J. & Johnson E. S. (1976). Multi-response permutation procedures for a priori classications. Communications in Statistics- Theory and Methods, 5: 1409- 1424
56. Mitchell, R.J. (2001). Path analysis: Pollination. pp. 217–234 in: Scheiner S.M. and J. Gurevitch (eds.). Design and Analysis of Ecological Experiments. Oxford University Press.
57. Petratis, P.S., Beaupre S.J. and Dunham A.E. (2001). ANCOVA: Nonparametric and Randomization Approaches. pp. 116–133 in: Scheiner S.M. and Gurevitch (eds.). Design and Analysis of Ecological Experiments. Oxford University Press, Oxford.
58. Pugesek, B., Tomer A. and von Eye A. (eds.) (2002). Structural Equations Modeling: Applications in Ecological and Evolutionary Biology Research. Cambridge University Press, Cambridge, U.K.
59. Paukert, C.P. and Wittig T.A. (2002). Applications of multivariate statistical methods in fi sheries. Fisheries 27(9):16-22
60. Palmer, M.W. (1988). Fractal geometry: a tool for describing spatial patterns of plant communities. Vegetatio 75:91–102.
61. Odum, E.P. and Barrett G.W. (2005). Fundamentals of Ecology. Thomas Brooks/Cole, Belmont, CA.
62. Orloci, L. (1975). Multivariate Analysis inVegetation Research. Junk, The Hague.
63. Rohlf, F.J. (1995). BIOM: A Package of Statistical Programs to Accompany the Text Biometry. Exeter Software, Setauket, New York.
64. Scheiner, S.M. (2001). Theories, hypotheses, and statistics. pp. 3-13 in: Scheiner S.M. and J. Gurevitch (eds.). Design and Analysis of Ecological Experiments. Oxford University Press, Oxford.
65. Shipley, B. (2000). Cause and Correlation in Biology. Cambridge University Press, Cambridge, U.K.
66. Statsoft, Inc. 1995. STATISTICA for Windows. 2nd edition. Tulsa, OK.
67. Stiling, P. (2002). Ecology: Theories and Applications. Prentice Hall, Upper Saddle River, N.J.
68. Swan, J.M.A. (1970). An examination of some ordination problems by use of simulatedvegetational data. Ecology 51: 89–102.
69. Ter Braak C.J.F. & Looman C.W.N. (1986). Weighted averaging, logistic

regression and the Gaussian response model. Vegetatio, 65: 3-11

70. Ter Braak, C.J.F. (1987). The analysis of vegetation-environment relationships by canonical correspondence analysis. Vegetatio, 69:69-77.
71. Ter Braak, C. J. F. (1994). Canonical community ordination. Part I: Basic theory and linear methods. Ecoscience 1 (2), 127-140.
72. Ter Braak C.J.F. & Šmilauer P. (1998). CANOCO Reference Manual and User›s Guide to Canoco for Windows. Microcomputer Power, Ithaca, USA. 352 pp.
73. Thompson, S.K. (2002). Sampling. John Wiley & Sons, New York. Second Edition.
74. Turner MG. (1989). Landscape ecology: the effect of pattern on process. Ann. Rev. Ecol. Syst. 20:171-197.
75. Turner MG, RH Gardner and RV O'Neill (2001). Landscape Ecology in Theory and Practice: Pattern and Process. Springer, New York.
76. van der Meer, J. Heip C.H., Herman P.J.M., Moens T. and van Oevelen D. (2005). Measuring the Flow of Energy and Matter in Marine Benthic Animal Populations. pp. 326–408 in: Eleftheriou. A. and A. McIntyre (eds.). 2005. Methods for Study of Marine Benthos. Blackwell Science Ltd., Oxford, UK.
77. Van der Plogeg, S.W.F. & Vlijm L. (1978). Ecological evaluation, nature conservation and land use planning with particular reference to the methods used in the Netherlands.Biol.Consero.14:197-221.
78. Van Sickle, J. (1997). Using mean similarity dendrograms to evaluate classifications. Journal of Agricultural, Biological, and Environmental Statistics, 2:370-388.
79. Whittaker R.H. (1965). Dominance and diversity in land plant communities. Science 147: 250–260.
80. Whittaker, R.H. (1967). Gradient analysis of vegetation. Biol. Rev. 42:207-264.
81. Zare Chahouki, M. A. Azarnivand H., Jafari M. & Tavili A. (2009). Multivariate Statistical Methods as a Tool for Model_Based Prediction of Vegetation Types, Russian Journal of Ecology, 41(1): 84-94.
82. Zare Chahouki, M.A. (2006). Modeling the spatial distribution of plant species in arid and semi-arid rangelands. PhD Thesis in Range management, Faculty of Natural Resources, University of Tehran, 180 p. (In Persian).
83. Schluter. D. and Grant P.R. (1982). The distribution of Geospiza difficilis on Galapagos islands: test of three hypotheses. Evolution 36:1213-1226

Chapter 2

CONTAMINATION OF ENVIRONMENT WITH POLYCYCLIC AROMATIC HYDROCARBONS IN INDIA

Khageshwar Singh Patel[1], Shobhana Ramteke[1], Yogita Naik[1], Bharat Lal Sahu[1], Saroj Sharma[2], Jutta Lintelmann[3], Matuschek Georg[3]

[1] School of Studies in Chemistry/Environmental Science, Pt. Ravishankar Shukla University, Raipur, India

[2] Department of Chemistry, Devi Rathi Mahila Mahavidhaya, Rajnandagaon, India

[3] GSF-Forschungszentrum für Umwelt und Gesundheit, Institut für Ökologische Chemie, Neuherberg, Germany

ABSTRACT

In India is contaminated with polycyclic aromatic hydrocarbons (PAHs) due to occurring of large anthropogenic activities i.e. fuel combustion, mineral roasting and biomass burning. Hence, contamination of 13 toxic PAHs: phenanthrene, anthracene, fluoranthene, pyrene, benz (a) anthracene, benzo (b) fluoranthene, benzo (k) fluoranthene, benzo (a) pyrene, benzo (ghi) perylene, dibenz (ah) anthracene, indeno1,2,3-(cd) pyrene, coronene and coronene in the environment (i.e. ambient particulate matter, road dust, sludge and sewage) of the most industrialized area: Raipur city, India is described. The $\sum PAH_{13}$ concentration in the 16 environment materials was ranged from 7980 - 1,051,300 μg/kg with mean value of 172,613 ± 154,726 μg/kg. The concentration variations, toxicities and sources of the PAHs in various environmental compartments are discussed.

INTRODUCTION

Polycyclic aromatic hydrocarbons (PAHs) are a large group of chemical compounds, I, with a similar structure comprising two or more joined aromatic carbon rings [1]. The compounds are formed by combustion of fuels, biomass and waste materials [2]. Polycyclic aromatic compounds are carcinogenic and mutagenic compounds, causing irreversible changes in the structure and functioning of living organisms [3]. There are thousands of PAH compounds in the environment but 13 compounds i.e. naphthalene, phenanthrene, anthracene, fluoranthene, pyrene, chrysene, benz (a) anthracene, benzo (b) fluoranthene, benzo (k) fluoranthene, benzo (a) pyrene, benzo (ghi) perylene, dibenz (ah) anthracene, indeno1,2,3-(cd) pyrene and coronene of the increased environmental and health interests. These compounds differ substantially in their physical, chemical and toxicological properties and therefore, their quantification in the environment is needed. The most potent carcinogens have been shown to be benzo [a] anthracene, benzo [a] pyrene and dibenz [ah] anthracene. They are multimedia contaminants, reported at elevated levels in several environmental samples i.e. dust, particulate matter, sludge and sewage of various region of the World [4]-[26].

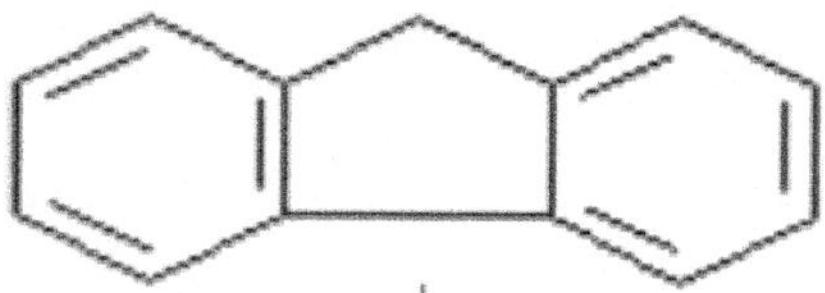

I

METHODS AND MATERIALS

Selection of Sampling Sites

Raipur (21°23'N, 81°63'E) is the capital city of the Chhattisgarh state with population of ≈2 million. The Raipur city is now becoming an important regional commercial and industrial destination for the coal, power, steel and aluminum industries. Several steel rolling mills, sponge iron plants, steel plants, agro-industries, thermal power plants and vehicles ($>1.0 \times 10^5$) are emitting effluents in and around the city.

Collection of Samples

The road dust, sludge and sewage samples were collected using a stainless-steel scoop from 13 locations of Raipur city in February 2010, Figure 1. They were kept in 250-mL glass bottle and dried at 30°C in an oven for overnight. The samples were crushed into fine particles by mortar and sieved out the particles of mesh size < 0.1 mm. The samples were stored in aluminum foil.

The coarse particulate matter (PM_{10}) and fine particulate matter ($PM_{2.5}$) were collected by using Partisol Model 2300 Sequential speciation air sampler. The sampler was installed at the roof of the building, ≈10 m above from the ground level at residential site: Dagania, Raipur. Both $PM2_{.5}$ and PM_{10} were collected simultaneously over 47 mm quartz fiber filters housed in molded filter cassette. The sampler was run for 24 hrs (6.00 am - 6.00 am) at flow rate of 10 L/min. One sample blank was used for collection of both PM_{10} and $PM_{2.5}$. The loaded filters were dismounted, brought to laboratory, and heated up to 30°C for 6 hrs to remove the moisture contents. The filters were transferred into the desiccator, and finally weighted to record the particulate contents.

Analysis of Carbons

The CHNSO-IRMS Analyzer by SV Instruments Analytica Pvt Ltd. was used for analysis of the total carbon (TC). Three carbons i.e. elemental carbon (EC), organic carbon (OC) and carbonate carbon (CC) were analyzed in the samples. The total carbn (TC) sample was oxidized with O_2 at 1020°C with constant helium flow by measuring the resulting CO_2 with thermal conductivity detector. The CC content was analyzed by treating the sample with HCl acid in the CO_2 free atmosphere. The resulting CO_2 was measured by coulometric titration method. The OC content was analyzed by titration method using $K_2Cr_2O_7$ as oxidant, and the excess of $K_2Cr_2O_7$ was determined by titration with the $FeSO_4 \cdot 7H_2O$ solution. The EC content was evaluated by using following equation.

$$EC = TC - (OC + CC)$$

Analysis of PAHs

The PAH samples were analyzed by capillary gas chromatography (Varian STAR 3400 CX) using temperature programmable splitless injection, a fused silica RTX5-MS column and ion trap mass spectrometric detection [27].

RESULTS AND DISCUSSION

Carbon Concentration

All samples are colored, ranging from brown to black, depending on the EC content. The content of EC, OC and CC in the 16 environmental shown in Table 1. Relatively high content of EC in all samples was achieved, ranging from 6.5% - 13.5% with mean value of 8.4% ± 1.1%. Very low content of OC and CC was observed in the dust, sludge and sewage samples unlikely to PM samples may be due to their degradation and water solubility, Figure 1. The EC content with the OC and CC had good relation (r = 0.94 - 0.96), indicating origin from the similar sources.

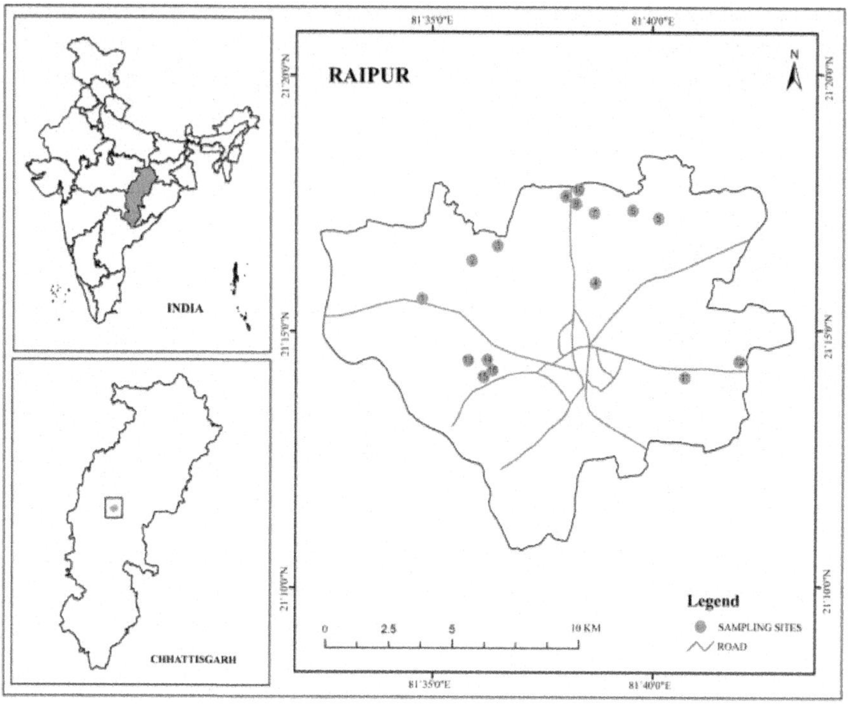

Figure 1: Representation of sampling locations.

PAHs Concentration

The chemical characteristics of 13 PAHs i.e. phenanthrene (Phe), anthracene (Ant), fluoranthene (Fla), pyrene (Pyr), benz [a] anthracene (Baa), chrysene (Cry), benzo [b] fluoranthene (Bbf), benzo [k] fluoranthene (Bkf), benzo [a] pyrene (Bap), benzo [ghi] perylene (Bgh), dibenz [a,h] anthracene (Dba), indeno [1,2,3-cd] pyrene (Ind) and coronene is summarized in Table 2. The content of 13 PAHs in 16 environmental samples is presented in Table 3. The sum of total concentration of PAHs (ΣPAH13) in the road dust of Raipur city (n = 8) was ranged from 10,427 - 26,031 μg/kg with mean value of 15,282 ± 3377 μg/kg. The highest concentration of the ΣPAH_{13} was observed at site no 5 (i.e. Birgaon) due to higher industrial and traffic emissions, Figure 2. Similarly, the concentration of ΣPAH_{13} in the SL, MW, AW and TPPW was found to be 7980, 9669, 10,570 and 8326 μg/kg, respectively. No signal for Cor was detected in the environmental samples i.e. RD, SL, MW, AW and TPPW samples. The major fraction of PAHs in the RD, AW and TPPW samples was contributed by three compounds i.e. Phe, Fla and Pyr, Figure 3. A different distribution pattern of PAHs in the SL and MW samples was observed, dominated by Pyr and Bgh contents, Figure 3. The concentration of ΣPAHs in the PM_{10} and $PM_{2.5}$ was strongly enriched, >25-folds higher than the road dust with appearing of strong Cor signal. The $PM_{2.5}$ sample was dominated by higher PAHs i.e. Bbf, Bgh and Ind, Figure 4.

Table 1: Chemical characteristics of PAHs

S. No.	PAHs	No. of rings	Abbreviation	Formula
1	Phenanthrene	3	Phe	$C_{14}H_{10}$
2	Anthracene	3	Ant	$C_{14}H_{10}$
3	Fluoranthene	4	Fla	$C_{16}H_{10}$
4	Pyrene	4	Pyr	$C_{16}H_{10}$
5	Benz [a] anthracene	4	Baa	$C_{18}H_{12}$
6	Chrysene	4	Cry	$C_{18}H_{12}$
7	Benzo [b] fluoranthene	5	Bbf	$C_{20}H_{12}$
8	Benzo [k] fluoranthene	5	Bkf	$C_{20}H_{12}$
9	Benzo [a] pyrene	5	Bap	$C_{20}H_{12}$
10	Dibenz [ah] anthracene	5	Dba	$C_{22}H_{14}$
11	Benzo [ghi] perylene	6	Bgh	$C_{22}H_{12}$
12	Indeno [1,2,3-cd] pyrene	6	Ind	$C_{22}H_{12}$
13	Coronene	6	Cor	$C_{24}H_{12}$

Table 2: Concentration of carbons and polycyclic aromatic hydrocarbons in environmental samples

S. No.	Sample type	Location	BC, %	OC, %	CC, %	ΣPAHs, mg/kg
1	RD1	Tatibandh	6.5	0.32	0.45	10.4
2	RD2	Hirapur	6.7	0.38	0.47	11.2
3	RD3	Sarora	6.9	0.34	0.41	13.1
4	RD4	Khamtarai	6.8	0.36	0.47	14.1
5	RD5	Birgaon	10.4	0.53	0.68	26.0
6	RD6	Urla	7.2	0.39	0.52	14.2
7	RD7	Sankra	8.2	0.42	0.56	16.7
8	RD8	Siltara	8.2	0.44	0.52	16.5
9	SL	Siltara	7.5	0.03	0.12	8.0
10	MW	Siltara	7.6	0.03	0.11	9.7
11	TPPW	Monate, Urla	8.9	0.04	0.17	8.3
12	AW	IGAU	9.0	0.05	0.19	10.6
13	$(PM_{10})1$	Dagania	6.8	5.8	4.6	505
14	$(PM_{10})2$	Dagania	6.7	5.6	4.2	347
15	$(PM_{2.5})1$	Dagania	13.5	8.3	5.1	1051
16	$(PM_{2.5})2$	Dagania	13.2	8.1	5.0	700

RD, SL, MW, TPPW, AW, PM_{10} and $PM_{2.5}$ represent road dust, sludge, municipal/sewage waste, thermal power plant waste, agricultural waste, coarse particulate matter and fine particulate matter, respectively.

Whereas, the PM10 sample was dominated by PAHs i.e. Fla, Pyr, Bgh and Ind, Figure 4. The PAHs content in the dust was negatively and fairly correlated with particle size (r = −0.89), Figure 5. The concentration of the PAHs in the environmental samples of studied area was found to be comparable to the other parts of the country and World [4]-[25].

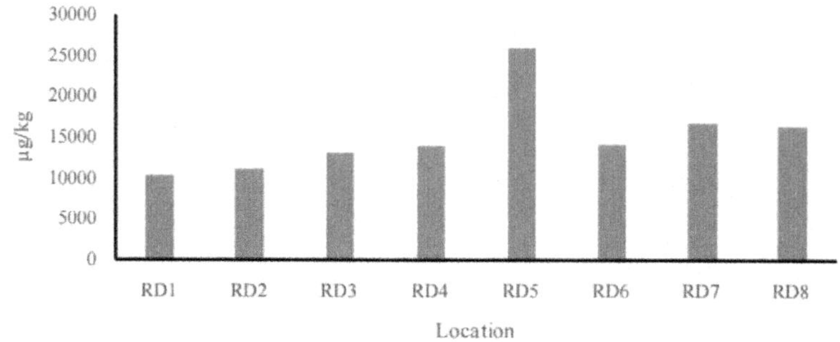

Figure 2: Spatial distribution of PAHs in the road dust.

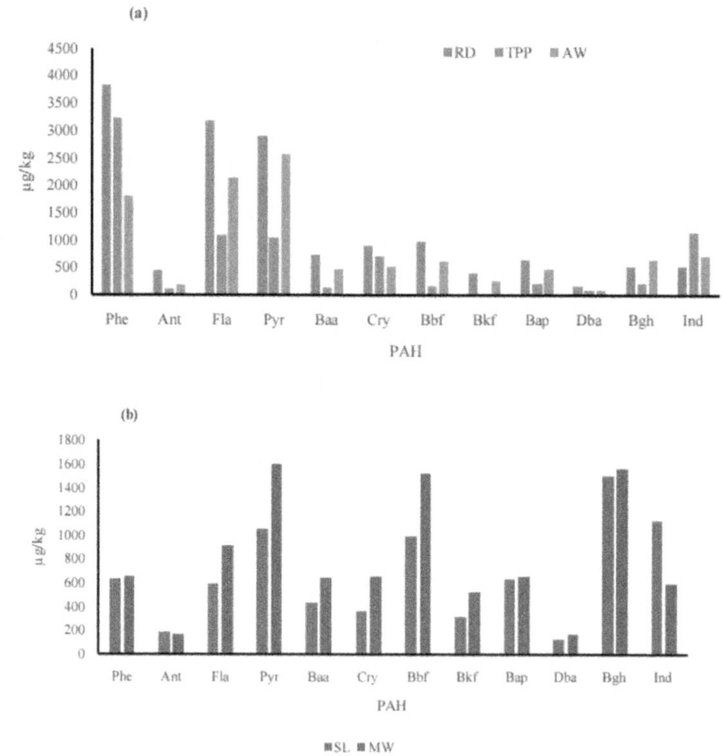

Figure 3: Distribution of PAHs in dust (RD), thermal power plant ash (TPPA), agricul-

tural waste (AW), sludge (SL) and municipal waste (MW).

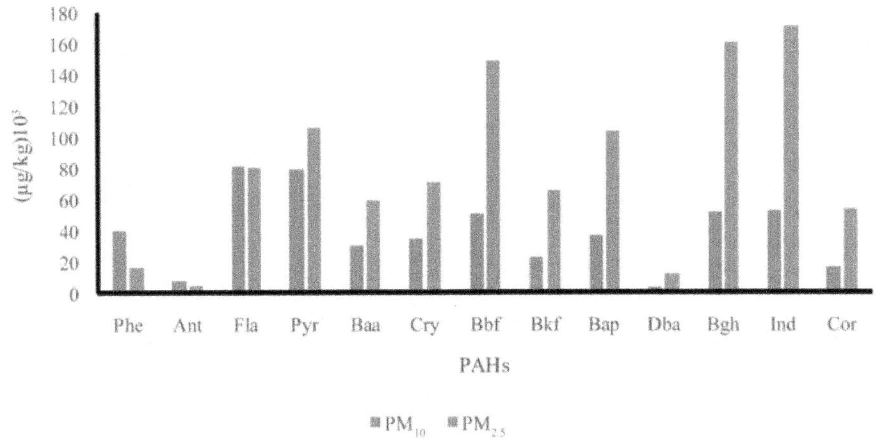

Figure 4: Distribution of PAHs in the PM_{10} and $PM_{2.5}$.

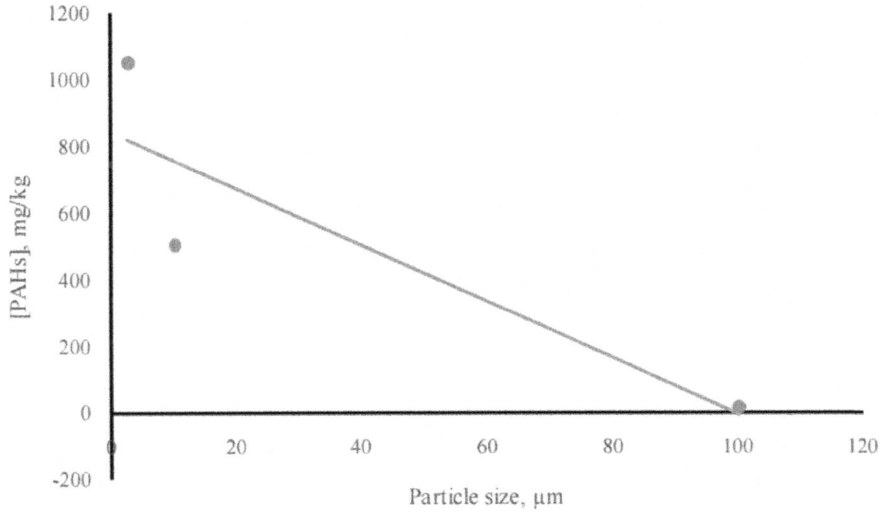

Figure 5: Correlation of PAHs content in the sample with particle size.

Table 3: Concentration of PAHs in environmental samples, μg/kg

S. No.	Phe	Ant	Fla	Pyr	Baa	Cry	Bbf	Bkf	Bap	Dba	Bgh	Ind	Cor
RD1	3168	301	1921	2222	394	622	586	227	341	18	373	254	0
RD2	2549	218	2669	2494	408	869	629	295	358	151	262	334	0
RD3	3402	320	2874	2240	600	902	866	343	439	97	602	430	0
RD4	3484	252	2208	2978	797	895	1078	442	710	28	582	632	0
RD5	5022	763	4759	5258	1886	1595	1986	880	1636	103	1019	1124	0
RD6	2636	178	3073	2254	781	1135	1358	480	732	200	670	677	0
RD7	5688	797	4655	2928	414	490	468	175	275	496	156	194	0
RD8	4622	732	3240	2974	713	826	931	398	720	226	521	550	0
SL	632	186	598	1056	439	365	990	318	637	132	1505	1122	0
MW	653	168	914	1598	642	658	1525	526	653	167	1564	601	0
TPPW	3240	120	1104	1056	154	708	168	145	204	84	204	1139	0
AW	1829	191	2149	2574	475	515	630	256	486	101	644	720	0
(PM10)1*	39.3	7.7	81.7	79.8	29.6	34.6	50.5	22.1	35.9	3.5	51.5	52.7	15.9
(PM10)2*	24.7	5.2	39.9	39.6	18.5	23.3	43.5	19.0	22.2	2.7	45.4	47.8	15.6
(PM2.5)1*	16.1	3.8	79.9	106.1	59.1	70.6	149.1	65.4	104.4	11.3	160.3	171.1	54.1
(PM2.5)2*	15.9	4.4	37.7	47.3	30.7	42.4	115.6	51.1	53.3	6.8	119.5	129.2	46.3

* = 10^3.

Vertical Distribution of PAHs

The vertical distribution of the PAHs from 0 - 30 cm in the sludge samples was studied, and presented in Figure 6. The ∑PAHs content was strongly increased with increase of the sludge depth profile from 0 - 30 cm, may be due to their poor adsorption with the geo-media. Among them, extremely high vertical distribution of compounds i.e. Fla, Pyr, Bbf and Bap was observed.

Toxicities

The toxicities of PAHs increases as the mass number increases, and seven PAHs (i.e. Pyr, Baa, Bbf, Bkf, Bap, Dba and Ind) are considered to be more toxic, may bedue to higher thermal stability and delocalization of π-electrons. The carcinogenic potentiality of Pyr, Baa, Bbf, Bkf, Bpa, Dba and Ind reported was 0.01, 0.1, 0.1, 0.1, 0.1, 1.0 and 0.1, respectively [28].

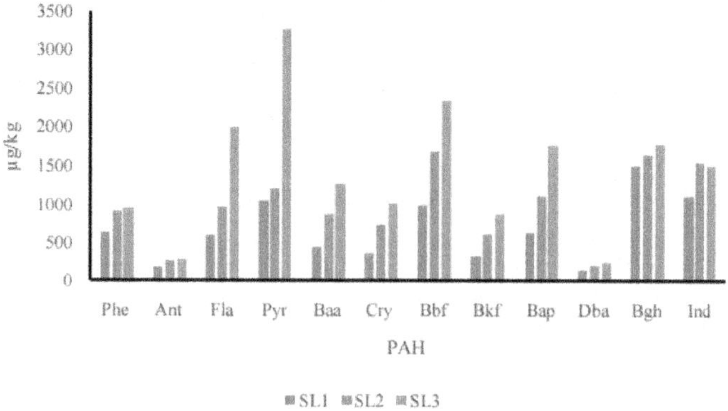

Figure 6: Vertical distribution of PAHs in the sludge, SL1 = 0 - 10 cm, SL2 = 10 - 20

cm, SL3 = 20 - 30 cm.

The benzo [a] pyrene equivalent (BapE) value was computed by using the following equation:

$$\text{Total BapE} = [\Sigma_i C_i] \times [TEF_i]$$

where, C_i and TEF_i are the concentration and the corresponding toxic equivalent factor (TEF) value of PAHs. The BapE value for RD, SL, MW, AW, TPPW, PM_{10} and $PM_{2.5}$ was found to be 1108, 1135, 1237, 846, 475, 52,000 and 138,500 µg/kg in the term of Bap. The carcinogenic fraction of PAHs in RD, SL, MW, AW, TPPW, PM_{10} and $PM2_{.5}$ samples was ranged from 5.7% - 15.7% with significantly higher value for SL, MW and PM samples, Figure 7. The concentration of PAHs in the environmental samples was found to be several folds higher than recommended value of 1000 µg/kg [29].

Correlation and Sources

The correlation matrix of the carbons and PAHs are summarized in Table 4. The PAHs had fair correlation with the BC, OC and CC contents (r = 0.70 - 0.96), indicating their origin from the burning processes. The lower PAHs (i.e. Phe, Ant, Fla and Pyr) among themselves had fair correlation, may be due to existence of their larger fractions in the gaseous forms, Table 4. The higher PAHs (i.e. Baa, Cry, Bbf, Bkf, Bap, Bgh and Ind except Dba) among themselves had good correlation, indicating origin from the burning processes, Table 4.

The diagnosis ratios: Phe/Antand [Fla]/[Fla + Pyr] were used to find out the sources of PAHs in the studied samples [30] [31]. The Phe/Ant ratio for TPPW, RD, AW, SL, MW, PM_{10} and $PM_{2.5}$ was found to be 27, 10.2, 9.6, 3.4, 3.9, 5.0 and 3.9, respectively, suggesting the domination of petrogenic PAHs in the TPPW, RD and AW samples, Figure 8. The Fla./Fla + Pyr. ratio of >0.5, 0.5 - >0.4 and <0.4 was used as signature for PAHs emission from combustion of grass, wood/ coal, petroleum and diesel, respectively [31]. The Fla./Fla + Pyr. ratio was ranged from 0.36 - 0.52, indicating domination of biomass or coal origin PAHs in the RD, TPPW and PM_{10} samples, Figure 8.

CONCLUSION

The light PAHs (3 - 4 ring) was found to be dominated in the RD, TPPW, AW and PM_{10} samples unlikely to SL, MW and $PM_{2.5}$ samples. Their origins were largely pyrogenic, emitted by combustion of biomass, coal and diesel. The higher PAHs (5 - 6 ring) was found to extremely enrich in the $PM_{2.5}$ sample due to origin by the combustion processes.

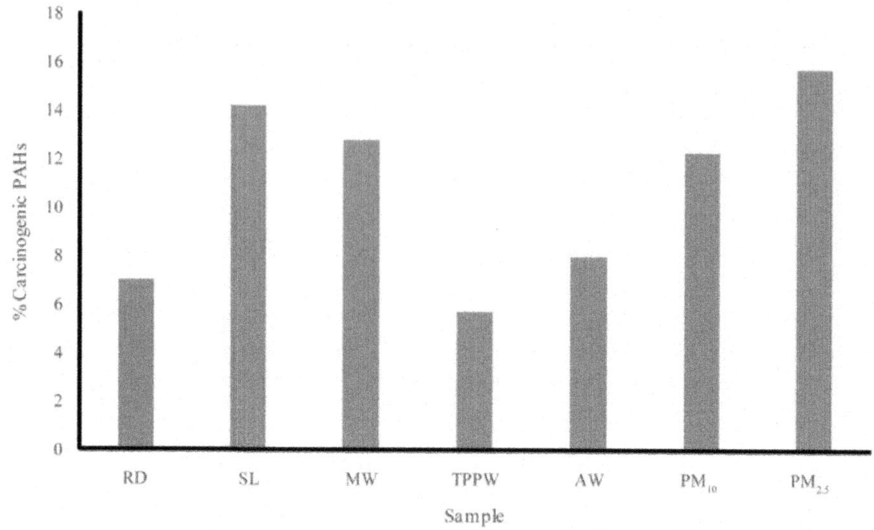

Figure 7: Percentage of carcinogenic PAHs in environmental samples.

The PHAs concentration was remarkably increased vertically and might be due to poor adsorption by the geo-media. The PAHs content in the environmental samples of the studied area was found to be several folds higher than recommended value of 1000 μg/kg.

(a)

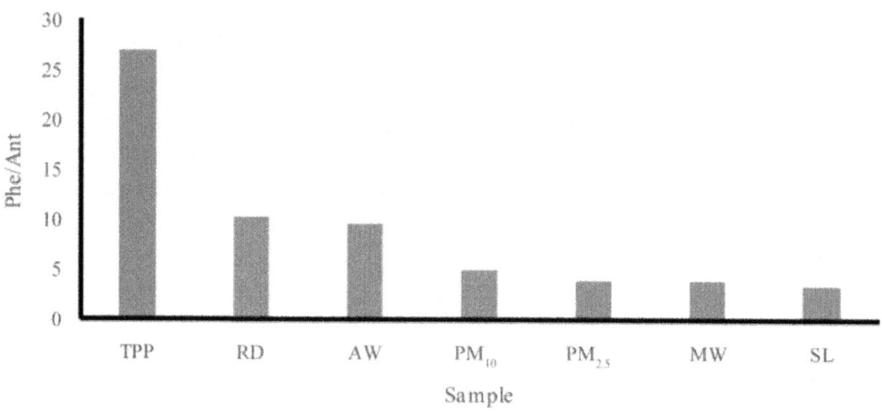

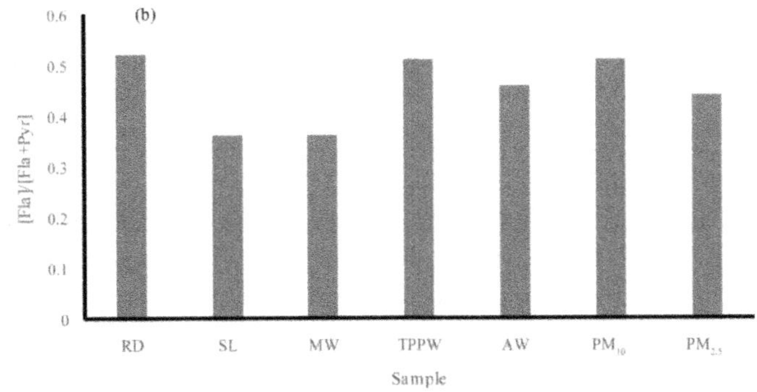

Figure 8: Diagnostic ratio for source determination of PAHs.

Table 4: Correlation matrix of PAHs in the road dust

	Phe	Ant	Fla	Pyr	Baa	Cry	Bbf	Bkf	Bap	Dba	Bgh	Ind
Phe	1											
Ant	0.96	1										
Fla	0.80	0.81	1									
Pyr	0.61	0.64	0.69	1								
Baa	0.35	0.39	0.54	0.91	1							
Cry	−0.02	0.06	0.34	0.71	0.91	1						
Bbf	0.12	0.17	0.39	0.75	0.95	0.96	1					
Bkf	0.18	0.24	0.42	0.83	0.98	0.97	0.98	1				
Bap	0.31	0.37	0.48	0.89	0.99	0.92	0.96	0.99	1			
Dba	0.60	0.57	0.66	−0.01	−0.23	−0.37	−0.31	−0.33	−0.26	1		
Bgh	0.05	0.10	0.24	0.65	0.90	0.92	0.96	0.94	0.91	−0.47	1	
Ind	0.15	0.20	0.37	0.78	0.96	0.95	0.99	0.99	0.98	−0.33	0.95	1

ACKNOWLEDGEMENTS

We are thankful the CCOST, Raipur for granting research support for doing this work.

REFERENCES

1. Bjorseth, A. and Dennis, A. (1980) Proceedings of International Polynuclear Aromatic Hydrocarbon Symposia, PAHIV: Polynuclear Aromatic Hydrocarbons, Chemistry and Biological Effects. Battelle Press, Columbus, OH, 1125 p.
2. Finlayson-Pitts, B.J. and Pitts, J.N. (2000) Chemistry of the Upper and Lower Atmosphere. Academic Press, San Diego.
3. Pickering, R.W. (1999) A Toxicological Review of Polycyclic Aromatic Hydrocarbons. Journal of Toxicology—Cutaneous and Ocular

Toxicology, 18, 101-135. http://dx.doi.org/10.3109/15569529909037562

4. Baran, S. and Oleszczuk, P. (2003) The Concentration of Polycyclic Aromatic Hydrocarbons in Sewage Sludge in Relation to the Amount and Origin of Purified Sewage. Polish Journal of Environmental Studies, 12, 523-529.
5. Pérez, S., Farré, M., Jesús Garcí a, M. and Barceló, D. (2001) Occurrence of Polycyclic Aromatic Hydrocarbons in Sewage Sludge and their Contribution to its Toxicity in the ToxAlert® 100 Bioassay. Chemosphere, 45, 705-712. http://dx.doi.org/10.1016/S0045-6535(01)00152-7
6. Blanchard, M., Teil, M.J., Ollivon, D., Legenti, L. and Chevreuil, M. (2004) Polycyclic Aromatic Hydrocarbons and Polychlorobiphenyls in Wastewaters and Sewage Sludges from the Paris Area (France). Environmental Research, 95, 184-197. http://dx.doi.org/10.1016/j.envres.2003.07.003
7. Harrison, E.Z., Oakes, S.R., Hysell, M. and Hay, A. (2006) Organic Chemicals in Sewage Sludge. Science of Total Environment, 367, 481-497. http://dx.doi.org/10.1016/j.scitotenv.2006.04.002
8. Cai, Q.Y., Mo, C.H., Wu, Q.T., Zeng, Q.Y. and Katsoyiannis, A. (2007) Occurrence of Organic Contaminants in Sewage Sludges from Eleven Wastewater Treatment Plants, China. Chemosphere, 68, 1751-1762. http://dx.doi.org/10.1016/j.chemosphere.2007.03.041
9. Lorenzi, D., Entwistle, J.A., Cave, M. and Dean, J.R. (2011) Determination of Polycyclic Aromatic Hydrocarbons in Urban Street Dust: Implications for Human Health. Chemosphere, 83, 970-977. http://dx.doi.org/10.1016/j.chemosphere.2011.02.020
10. Tang, X., Shen, C., Cheema, S.A., Chen, L., Xiao, X., Zhang, C., Liu, W., Li, F. and Chen, Y. (2010) Levels and Distributions of Polycyclic Aromatic Hydrocarbons in Agricultural Soils in an Emerging e-waste Recycling town in Taizhou Area, China. Journal of Environmental Science and Health Part A, Toxic/Hazardous Substances and Environmental Engineering, 45, 1076-1084. http://dx.doi.org/10.1080/10934529.2010.486336
11. Amuda, O.S. and Adelowo-Imeokparia, F.E. (2007) Polycyclic Aromatic Hydrocarbons in Municipal Waste Ashes from Three Waste Dumps in Lagos, Nigeria. Bulletin of the Chemical Society of Ethiopia, 21, 141-144. http://dx.doi.org/10.4314/bcse.v21i1.61402
12. Pengchai, P., Nakajima, F. and Furumai, H. (2005) Estimation of Origins of Polycyclic Aromatic Hydrocarbons in Size-Fractionated Road Dust in Tokyo with Multivariate Analysis. Water Science and Technology, 51,

169-175.

13. Boonyatumanond, R., Murakami, M., Wattayakorn, G., Togo, A. and Takada, H. (2007) Sources of Polycyclic Aromatic Hydrocarbons (PAHs) in Street Dust in a Tropical Asian Mega-City, Bangkok, Thailand. Science of the Total Environment, 384, 420-432. http://dx.doi.org/10.1016/j.scitotenv.2007.06.046
14. Naspinski, C., Lingenfelter, R., Cizmas, L., Naufal, Z., He, L.Y., Islamzadeh, A., Li, Z.W., Li, Z., McDonald, T. and Donnelly, K.C. (2008) A Comparison of Concentrations of Polycyclic Aromatic Compounds Detected in Dust Samples from Various Regions of the World. Environment International, 34, 988-993. http://dx.doi.org/10.1016/j.envint.2008.03.008
15. Hassanien, M.A. and Abdel-Latif, N.M. (2008) Polycyclic Aromatic Hydrocarbons in Road Dust over Greater Cairo, Egypt. Journal of Hazardous Materials, 151, 247-254. http://dx.doi.org/10.1016/j.jhazmat.2007.05.079
16. Han, B., Bai, Z.P., Guo, G.H., Wang, F., Li, F., Liu, Q.X., Ji, Y.Q., Li, X. and Hu, Y.D. (2009) Characterization of PM10 Fraction of Road Dust for Polycyclic Aromatic Hydrocarbons (PAHs) from Anshan, China. Journal of Hazardous Materials, 170, 934-940. http://dx.doi.org/10.1016/j.jhazmat.2009.05.059
17. Wang, W., Huang, M.J., Kang, Y., Wang, H.S., Leung, A.O., Cheung, K.C. and Wong, M.H. (2011) Polycyclic Aromatic Hydrocarbons (PAHs) in Urban Surface Dust of Guangzhou, China: Status, Sources and Human Health Risk Assessment. Science of the Total Environment, 409, 4519-4527. http://dx.doi.org/10.1016/j.scitotenv.2011.07.030
18. Yu, B., Xie, X., Ma, L.Q., Kan, H. and Zhou, Q. (2013) Source, Distribution, and Health Risk Assessment of Polycyclic Aromatic Hydrocarbons in Urban Street Dust from Tianjin, China. Environmental Science and Pollution Research, 21, 2817-2825. http://dx.doi.org/10.1007/s11356-013-2190-z
19. Mohd Radzi, N.A.S., Bakar, N.K.A., Emenike, C.U. and Abas, M.R. (2015) Polycyclic Aromatic Hydrocarbons (PAHs): Contamination Level and Risk Assessment in Urban Areas, Kuala Lumpur, Malaysia. Desalination and Water Treatment. http://dx.doi.org/10.1080/19443994.2015.1021103
20. Martuzevicius, D., Kliucininkas, L., Prasauskas, T., Krugly, E., Kauneliene, V. and Strandberg, B. (2011) Resuspension of Particulate Matter and PAHs from Street Dust. Atmospheric Environment, 45, 310-

317. http://dx.doi.org/10.1016/j.atmosenv.2010.10.026

21. Sharma, H., Jain, V.K. and Khan, Z.H. (2007) Characterization and Source Identification of Polycyclic Aromatic Hydrocarbons (PAHs) in the Urban Environment of Delhi. Chemosphere, 66, 302-310. http://dx.doi.org/10.1016/j.chemosphere.2006.05.003
22. Hassan, Y. and Sarma, H.P. (2013) Distribution of Polycyclic Aromatic Hydrocarbons (PAHS) in Roadside Soils from Industrial and High Traffic Area of Greater Guwahati City, Assam, India. Archives of Applied Science Research, 5, 85-93.
23. Ray, S., Khillare, P.S., Kim, K.H. and Brown, R.J. (2012) Distribution, Sources, and Association of Polycyclic Aromatic Hydrocarbons, Black Carbon, and Total Organic Carbon in Size-Segregated Soil Samples Along a BackgroundUrban-Rural Transect. Environmental Engineering Science, 29, 1008-1019. http://dx.doi.org/10.1089/ees.2011.0323
24. Cheng, H.R., Deng, Z.M., Chakraborty, P., Liu, D., Zhang, R.J., Xu, Y., Luo, C.L., Zhang, G. and Li, J. (2013) A Comparison Study of Atmospheric Polycyclic Aromatic Hydrocarbons in Three Indian Cities Using PUF Disk Passive Air Samplers. Atmospheric Environment, 73, 16-21. http://dx.doi.org/10.1016/j.atmosenv.2013.03.001
25. Fu, F.F., Tian, B.Y., Lin, G.S., Chen, Y.Q. and Zhang, J.H. (2010) Chemical Characterization and Source Identification of Polycyclic Aromatic Hydrocarbons in Aerosols Originating from Different Sources. Journal of the Air and Waste Management Association, 60, 1309-1314. http://dx.doi.org/10.3155/1047-3289.60.11.1309
26. Wingfors, H., Hägglund, L. and Magnusson, R. (2011) Characterization of the Size-Distribution of Aerosols and Particle-Bound Content of Oxygenated PAHs, PAHs, and n-Alkanes in Urban Environments in Afghanistan. Atmospheric Environment, 45, 4360-4369. http://dx.doi.org/10.1016/j.atmosenv.2011.05.049
27. Lintelmann, J., Fischer, K., Karg, E. and Schroppel, A. (2005) Determination of Selected Polycyclic Aromatic Hydrocarbons and Oxygenated Polycyclic Aromatic Hydrocarbons in Aerosol Samples by High-Performance Liquid Chromatography and Liquid Chromatography-Tandem Mass Spectrometry. Analytical and Bioanalytical Chemistry, 381, 508-519. http://dx.doi.org/10.1007/s00216-004-2883-8
28. EPA (2005) Air Toxics Hot Spots Program Risk Assessment Guidelines, Part II Technical Support Document for Describing Available Cancer Potency Factors. Office of Environmental Health Hazard Assessment, California Environmental Protection Agency, Sacramento.

29. CCME (Canadian Council of Ministers of the Environment) (2008) Canadian Soil Quality Guidelines for Carcinogenic and Other Polycyclic Aromatic Hydrocarbons (PAHs). Environmental and Human Health Effects, Scientific Supporting Document, 218 p.

30. Magi, E., Bianco, R., Ianni, C. and Carro, M.D. (2000) Distribution of Polycyclic Aromatic Hydrocarbons in the Sedi-ments of the Adriatic Sea. Environmental Pollution, 119, 91-98. http://dx.doi.org/10.1016/S0269-7491(01)00321-9

31. Li, G.C., Xia, X.H., Yang, Z.F., Wang, R. and Voulvoulis, N. (2006) Distribution and Sources of Polycyclic Aromatic Hydrocarbons in the Middle and Lower Reaches of the Yellow River, China. Environmental Pollution, 144, 985-993. http://dx.doi.org/10.1016/j.envpol.2006.01.047

Chapter 3

ENVIRONMENT AND HEALTH IN CONTAMINATED SITES: THE CASE OF TARANTO, ITALY

Roberta Pirastu,[1] Pietro Comba,[2] Ivano Iavarone,[2] Amerigo Zona,[2] Susanna Conti,[3] Giada Minelli,[3] Valerio Manno,[3] Antonia Mincuzzi,[4] Sante Minerba,[4] Francesco Forastiere,[5] Francesca Mataloni,[5] and Annibale Biggeri[6,7]

[1]Department of Biology and Biotechnologies Charles Darwin, Sapienza Rome University, Piazzale Aldo Moro 5, 00185 Rome, Italy

[2]Department of Environment and Primary Prevention, National Health Institute, Viale Regina Elena 299, 00161 Rome, Italy

[3]Unit of Statistics of the National Health Institute, National Center for Epidemiology, Surveillance and Health Promotion, Viale Regina Elena 299, 00161 Rome, Italy

[4]Taranto Local Health Unit, Epidemiological and Statistical Unit, Viale Virgilio 31, 74121 Taranto, Italy

[5]Department of Epidemiology, Lazio Regional Health Service, Via di Santa Costanza 53, 00198 Rome, Italy

[6]Biostatistics Unit, ISPO Cancer Research and Prevention Institute, Via Cosimo il Vecchio 2, 50139 Florence, Italy

[7]Annibale Biggeri Department of Statistics "G. Parenti", University of Florence, Viale Morgagni 59, 50134 Firenze, Italy

ABSTRACT

The National Environmental Remediation programme in Italy includes sites with documented contamination and associated potential health impacts (National Priority Contaminated Sites—NPCSs). SENTIERI Project, an extensive investigation of mortality in 44 NPCSs, considered the area of Taranto, a NPCS where a number of polluting sources are present. Health indicators available at municipality level were analyzed, that is, mortality (2003–2009), mortality time trend (1980–2008), and cancer incidence (2006-2007). In addition, the cohort of individuals living in the area was followed up to evaluate mortality (1998–2008) and morbidity (1998–2010) by district of residence. The results of the study consistently showed excess risks for a number of causes of death in both genders, among them: all causes, all cancers, lung cancer, and cardiovascular and respiratory diseases, both acute and chronic. An increased infant mortality was also observed from the time trends analysis. Mortality/morbidity excesses were detected in residents living in districts near the industrial area, for several disorders including cancer, cardiovascular, and respiratory diseases. These coherent findings from different epidemiological approaches corroborate the need to promptly proceed with environmental cleanup interventions. Most diseases showing an increase in Taranto NPCS have a multifactorial etiology, and preventive measures of proven efficacy (e.g., smoking cessation and cardiovascular risk reduction programs, breast cancer screening) should be planned. The study results and public health actions are to be communicated objectively and transparently so that a climate of confidence and trust between citizens and public institutions is maintained.

INTRODUCTION

Contaminated sites are extensively present in Europe where approximately 250,000 sites require cleanup interventions, as listed by the European Environment Agency [1]. Several thousands of these sites are located in Italy, and a total of 57 sites, defined in 2009 as National Priority Contaminated Sites (NPCSs), qualify for remediation because of contamination documented in qualitative and/or quantitative terms, and because of a potential health impact.

The site of Taranto, located in Apulia region (southern Italy), includes two municipalities and 216,618 inhabitants at 2001 Census. This site is of interest because of several polluting sources, such as a large steel plant, a refinery, the harbor, and both controlled and illegal waste dumps.

Previous environmental and epidemiological investigations in the area have provided evidence of environmental contamination [2–9]. These studies have documented severe air pollution originating mainly from the steel industry, that is, particulate matter, heavy metals, polycyclic aromatic hydrocarbons, and organ-halogenated compounds.

Epidemiological studies showed an increased mortality/morbidity from respiratory, cardiovascular diseases and several cancer sites [10–13].

SENTIERI Project (epidemiological study of residents in National Priority Contaminated Sites—NPCSs) funded by the Italian Ministry of Health studied mortality for 63 causes among residents of 44 NPCSs included in the "National Environmental Remediation Programme" [14].

The distinguishing feature of SENTIERI Project is that the epidemiological evidence evaluation has been carried out before the study to minimize the risk for researchers to be data-driven when performing and interpreting the specific epidemiologic investigation. SENTIERI dealt with the complexity of the relation between area contamination and health effects by examining, for each combination of causes of death/environmental exposures, the epidemiological evidence (1998–2009) and then building a matrix of the a priori evaluation of the strength of the causal association. The environmental exposures were classified as chemicals, petrochemicals and refineries, steel plants, power plants, mines and/or quarries, harbour areas, asbestos or other mineral fibers, landfills, and incinerators—labelled on the basis of the legislative decrees defining the sites' boundaries. A standardized procedure was set up to collect the available epidemiological literature, which was reviewed on the basis of explicit criteria and led to classify each cause of death/environmental exposures combination in terms of strength of causal association. The evaluation was categorized as sufficient to infer the presence of a causal association (S), limited to infer the presence of a causal association (L), and inadequate to infer the presence or the absence of a causal association (I). The rationale, scope, methods, and details of the a priori evidence evaluation can be found in [15], and the procedures and results of the evidence evaluation have been published in Italian [16].

With specific reference to the environmental exposures in Taranto NPCS, this procedure led to classify the presence of a steel industry, a refinery, a harbour area, and a number of landfills and waste dump sites as associated, with limited evidence, with an increased risk of lung cancer, pleural mesothelioma, nonmalignant respiratory diseases, congenital malformations, and perinatal conditions. In this context, residence near steel industry was evaluated as specifically associated with the occurrence of both acute and chronic respiratory diseases in adults and children, based on studies by [17–23]. It should be underlined, however, that air pollution from particulate matter has been causally linked by WHO with several health effects, including all-cause mortality and cardiovascular and respiratory morbidity [24].

SENTIERI analyzed mortality at municipality level in the period 1995–2002, computing standardized mortality ratios (SMR) both crude and adjusted for a deprivation index [25]. While SENTIERI strengths are the a prioriepidemiological evidence evaluation and the mortality analysis of all NPCSs adopting the same analytical approach and adjusting for deprivation, there are several limitations that should be noted, such as its ecological design and the use of mortality data at municipal level for a short period of time.

With the aim of overcoming the above limitations, this paper presents an epidemiological profile of Taranto NPCS residents analyzing different health indicators available at municipality level, that is, cause-specific mortality (2003–2009), mortality time trend (1980–2008), and cancer incidence (2006-

2007). A cohort study of the resident population examined mortality (1998–2008) and morbidity (1998–2010) in the districts close to the steel plant.

MATERIAL AND METHODS

Data Source

Details on the codes used during the study period for the 9th and 10th revisions of the International Classification of Diseases (ICD-9 and ICD-10) and on the demographic data of the two municipalities included in Taranto NPCS are presented in Appendix A.

Mortality 2003–2009

Mortality in Taranto NPCS residents was initially studied for the period 1995–2002 [12] and then updated for the years 2003–2009 (note that the period 2004-2005 was not available from ISTAT). The analysis considered 63 single or grouped causes (all ages, both genders); 0-1 and 0–14 age classes were also analyzed for a selection of causes (both genders combined). Standardized mortality ratios both crude (SMRs) and adjusted for deprivation together with 90% confidence intervals (90% CIs) were computed using regional rates for comparison [14]. In SENTIERI Project the deprivation index (DI) was constructed using the 2001 national census variables representing the following socioeconomic domains: education, unemployment, dwelling ownership, and overcrowding. The strengths and weaknesses of SENTIERI ID, its correlation with 2001 national deprivation index, its efficacy in representing deprivation in different categories of demographic dimensions, together with suggestions about the use of socioeconomic indices in small area studies of environment and health are discussed in Pasetto et al. 2011 [26].

Mortality Time Trend 1980–2008

Mortality was analyzed for a twenty-seven-year period: 1980–2008. The analysis was performed for the population 0–99 years separately for men and women; directly standardized death rates (SDRs) per 100,000 were calculated (standard: Italian population at 2001 Census) together with their 90% CI. The population size of Taranto NPCS is relatively small; therefore the study period was divided into three-year periods to obtain stable values of the indicators. The overall mortality was analyzed together with specific diseases, selected on the basis of the a priori evidence evaluation of their link with environmental exposures in Taranto NPCS. The selected diseases were all cancers (in particular lung cancer), circulatory diseases (in particular ischemic heart disease), and respiratory diseases (in particular, the acute, and chronic ones). We also analyzed the overall infant mortality (i.e. mortality from all causes during the first year of life) without gender distinction, as not informative in this age group.

Cancer Incidence 2006-2007

For cancer incidence (2006-2007), standardized incidence ratio (SIR) and 90% CI were calculated for both genders; the incidence rates of Italian South and Islands Cancer Registries macroarea (2005–2007) and of Taranto Province, excluding NPCSs municipalities (2006-2007), were used for comparison.

Mortality (1998–2008) and Morbidity (1998–2010) of the Residential Cohort

A cohort study design was applied to evaluate cause-specific mortality and hospitalization in relation to residence in specific districts close to the industrial sites. A cohort of residents (all subjects living in Taranto, Massafra, and Statte as January 1, 1998, and subsequently entered in these municipalities up to 2010) was enrolled from the municipal register. Individual follow-up for vital status assessment at 31.01.2010 was performed using municipality data. This cohort population is different from the base population analyzed for mortality in 2003–2009 (see previous paragraph for details). The socioeconomic position level (SEP) of the census block of residence and the district of residence were assigned to each participant (five categories from low to high SEP), on the basis of the addresses geocoded at the beginning of the follow-up. Occupational history for all cohort members was traced through the national insurance company (INPS) database (people employed in 1974 and subsequently), and the subcohort of individuals employed in industries located in the area was identified. Mortality/morbidity information was retrieved from Regional Health Databases (1998–2008 for mortality, 1998–2010 for hospital admissions). The associations of district of residence with mortality/morbidity were estimated by calculating mortality and morbidity hazard ratios (HR, CI 95%) using the proportional Cox models. All models considered age (temporal axis), calendar period, and area-based socioeconomic status [25], and they were calculated separately for men and women [13].

RESULTS

Mortality 2003–2009-SENTIERI

Tables 1–3 show mortality results in the periods 1995–2002 and 2003–2009.

Table 1: SENTIERI—Taranto NPCS. Mortality for the main causes of death. Number of observed cases (Obs), standardized mortality ratio crude (SMR), and adjusted for deprivation (SMR ID); 90% IC: 90% confidence interval; regional references: 1995–2002 and 2003–2009. Males and females

Causes of death (ICD IX)	1995-2002						2003-2009*					
	Males			Females			Males			Females		
	OBS	SMR (90% CI)	SMR ID (90% CI)	OBS	SMR (90% CI)	SMR ID (90% CI)	OBS	SMR (90% CI)	SMR ID (90% CI)	OBS	SMR (90% CI)	SMR ID (90% CI)
All causes (1-999.9)	7585	109 (107-111)	107 (105-109)	7104	107 (105-109)	107 (105-109)	4936	114 (111-117)	111 (108-113)	4847	108 (105-110)	107 (104-109)
All neoplasms (140.0-239.9)	2529	115 (112-119)	113 (109-116)	1716	113 (108-117)	112 (108-117)	1650	114 (110-119)	111 (106-115)	1208	113 (108-118)	111 (106-116)
Diseases of the circulatory system (401.0-405.9)	2654	105 (102-108)	103 (99-106)	3118	101 (98-104)	100 (97-103)	1645	114 (109-119)	109 (105-114)	1968	104 (100-108)	103 (99-107)
Diseases of the respiratory system (460.0-519.9)	666	107 (100-114)	107 (100-114)	406	113 (104-123)	111 (102-120)	447	117 (108-126)	112 (103-121)	268	104 (94-115)	105 (95-116)
Diseases of the digestive system (520.0-579.9)	442	114 (105-123)	114 (106-124)	472	142 (132-153)	141 (131-153)	283	147 (133-162)	136 (123-150)	233	119 (106-132)	117 (104-130)
Diseases of the genitourinary system (580.0-629.9)	101	92 (78-109)	97 (82-115)	107	89 (75-104)	91 (77-108)	71	94 (77-115)	101 (82-123)	85	89 (74-107)	87 (72-104)

*2004-2005 not available from ISTAT.

Table 2: SENTIERI—Taranto NPCS. Mortality for causes of death with limited evidence of association with environmental exposures in Taranto NPCS. Number of observed cases (Obs), standardized mortality ratio crude (SMR), and adjusted for deprivation (SMR ID); 90% IC: 90% confidence interval; regional references: 1995–2002 and 2003–2009. Males and females

Causes of death (ICD IX)	1995-2002						2003-2009*					
	Males			Females			Males			Females		
	OBS	SMR (90% CI)	SMR ID (90% CI)	OBS	SMR (90% CI)	SMR ID (90% CI)	OBS	SMR (90% CI)	SMR ID (90% CI)	OBS	SMR (90% CI)	SMR ID (90% CI)
Malignant neoplasm of trachea bronchus and lung (162.0-162.9)	840	130 (122-137)	119 (112-126)	121	135 (115-157)	130 (111-151)	516	133 (124-143)	123 (114-132)	97	130 (109-153)	121 (101-143)
Malignant pleural neoplasm (163.0-163.9)	83	521 (430-625)	293 (242-352)	14	242 (147-379)	190 (115-297)	44	519 (397-667)	272 (208-350)	12	311 (180-505)	210 (121-340)
Diseases of the respiratory system (460.0-519.9)	666	107 (100-114)	107 (100-114)	406	113 (104-123)	111 (102-120)	447	117 (108-126)	112 (103-121)	268	104 (94-115)	105 (95-116)
Acute diseases of the respiratory system (460.0-466.9, 480.0-487.9)	125	156 (134-181)	149 (127-173)	135	145 (125-167)	138 (119-159)	50	136 (106-172)	137 (107-174)	58	112 (89-140)	115 (91-143)
Chronic diseases of the respiratory system (491.0-492.9, 494.0-496.9)	388	96 (88-105)	97 (89-105)	151	92 (80-105)	92 (80-105)	322	116 (106-127)	110 (100-121)	149	104 (90-119)	100 (87-114)
Asthma (493.0-493.9)	9	41 (22-72)	42 (22-73)	11	73 (41-121)	68 (38-113)	0			1	25 (1-118)	29 (1-137)

*2004-2005 not available from ISTAT.

Table 3: SENTIERI—Taranto NPCS. Mortality for causes of death with limited evidence of association with environmental exposures in Taranto NPCS. Number of observed cases (Obs), standardized mortality ratio crude (SMR), and adjusted for deprivation (SMR ID); 90% IC: 90% confidence interval; regional references: 1995–2002 and 2003–2009. Males and females combined

Causes of death (age classes) (ICD IX)	1995–2002 Total			2003–2009* Total		
	OBS	SMR (90% CI)	SMR ID (90% CI)	OBS	SMR (90% CI)	SMR ID (90% CI)
Congenital anomalies (all ages) (740.0–759.9)	59	115 (91–142)	117 (93–145)	20	82 (54–119)	93 (62–135)
Certain conditions originating in the perinatal period (0-1) (760.0–779.9)	79	135 (111–162)	121 (100–146)	37	165 (123–218)	147 (110–193)
Acute diseases of the respiratory system (0–14) (460.0–466.9, 480.0–487.9)	4	96 (33–219)	95 (33–219)	<3	—	—
Asthma (0–14) (493.0–493.9)	<3	—	—	<3	—	—

*2004-2005 not available from ISTAT.

In Table 1, mortality, from the main causes of death, is displayed for descriptive purposes; Tables 2 and 3present the results for the causes selected on the basis of the a priori evidence evaluation, the distinguishing feature of SENTIERI Project [15].

Table 1 shows that in both periods, for both genders, for all causes and all neoplasms, there was an excess of mortality ranging between 7% and 15%; adjustment for deprivation did not substantially change SMRs values. In both periods, among males and females, the observed mortality was above expected for circulatory, respiratory, and digestive systems diseases; also in these cases, accounting for socioeconomic factors did not essentially change the study results. For diseases of the genitourinary system, the observed mortality was similar to the expected one.

Table 2 presents the results for the causes of death for which SENTIERI classified the epidemiological evidence of causal association with the environmental exposures in Taranto NPCS as "Limited". From now on, reference will only be made to results adjusted for deprivation. Among males, lung cancer showed a 20% excess in the first period, confirmed in the second one; among females the excesses were, respectively, about 30% (1995–2002) and 20%. Correspondingly, in the two periods excesses for pleural tumors were 193% and 167% among males, 90% and 103% among females. Excesses for acute respiratory diseases among males were 49% (1995–2002) and 37% (2003–2009), and for females 38% and 14%, respectively. The observed mortality for chronic respiratory diseases in 1995–2002 was as expected in both genders, while in 2003–2009 a 10% excess was present for males. Asthma mortality was not increased, but the observed number of death is small.

Table 3 displays the results combined for males and females, again for causes with limited epidemiological evidence of causal association with the environmental exposures in Taranto NPCS. The mortality from congenital anomalies showed a 17% excess in 1995–2002, while in 2003–2009 it was below expectation. For mortality from perinatal conditions, an heterogeneous group of diseases affecting fetus or newborn spanning from pregnancy and delivery complications to digestive or hematological disorders, the excess was 21% and 47%, in the first and second periods, respectively. In the age class 0–14 less than 3 deaths from acute respiratory diseases and asthma were observed.

Some noteworthy results should be considered (results not shown in Tables). Among males, the observed deaths were above expected in 1995–2002 and 2003–2009 for dementia (resp., 105 and 102 deaths), hypertensive diseases (resp., 307 and 287 deaths), ischemic heart diseases (resp., 1032 and 679 deaths), and cirrhosis (resp., 266 and 156 deaths). In 2003–2009 excesses were reported for melanoma (50%, 26 deaths), non-Hodgkin lymphoma (34%, 45 deaths), and myeloid leukemia (35%, 37 deaths).

Mortality Time Trend Analysis 1980–2008

Time trends for mortality from all causes and selected causes among adults and overall infant mortality are presented in Figures 1–17 and in Appendix B (see Tables S2, S3, and S4 in Supplementary Material available online at http://dx.doi.org/10.1155/2013/753719).

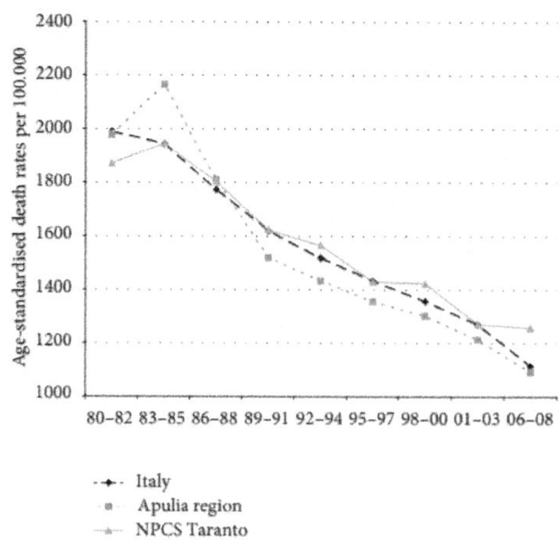

Figure 1: Overall mortality. Trends in age-standardised (Italian Census 2001) death rates per 100.000 (1980–2008) (2004-2005 data were not available). All ages. Men.

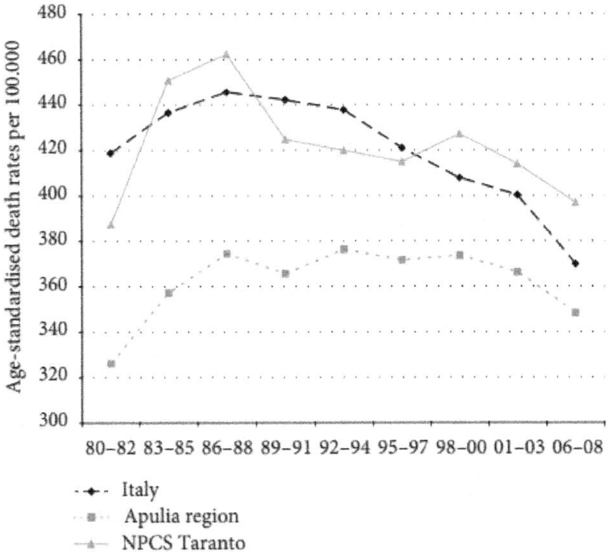

Figure 2: All cancers. Trends in age-standardised (Italian Census 2001) death rates per 100.000, from selected causes of death (1980–2008) (2004-2005 data were not available). All ages. Men.

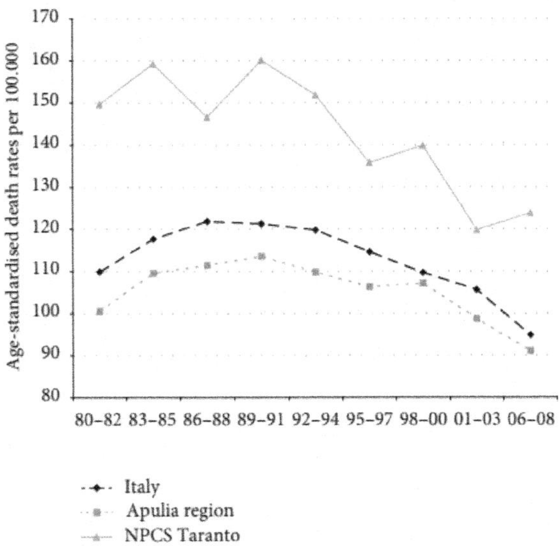

Figure 3: Lung cancer. Trends in age-standardised (Italian Census 2001) death rates per 100.000, from selected causes of death (1980–2008) (2004-2005 data were not available). All ages. Men.

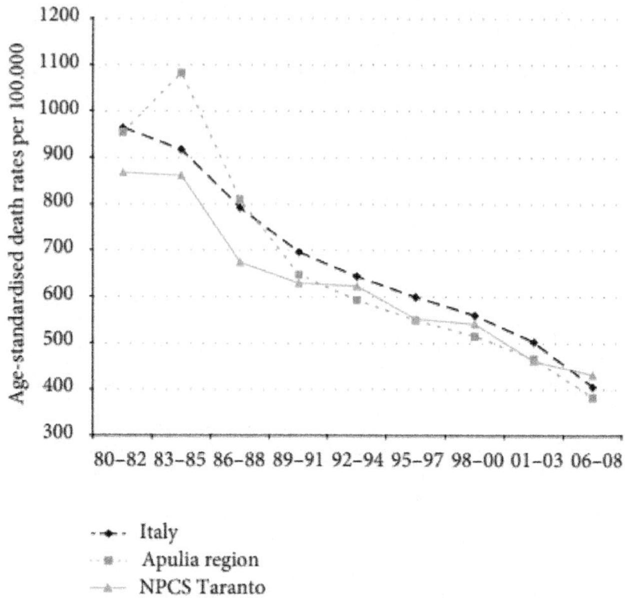

Figure 4: Circulatory diseases. Trends in age-standardised (Italian Census 2001) death rates per 100.000, from selected causes of death (1980–2008) (2004-2005 data were not available). All ages. Men.

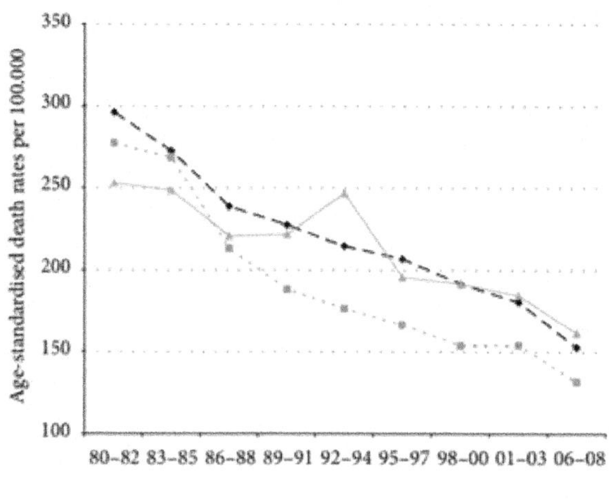

Figure 5: Ischaemic heart diseases. Trends in age-standardised (Italian Census 2001) death rates per 100.000, from selected causes of death (1980–2008) (2004-2005 data were not available). All ages. Men.

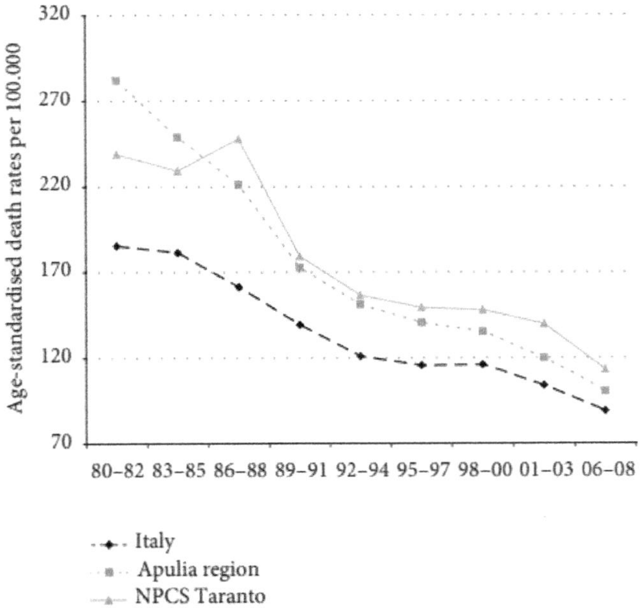

Figure 6: Respiratory diseases. Trends in age-standardised (Italian Census 2001) death rates per 100.000, from selected causes of death (1980–2008) (2004-2005 data were not available). All ages. Men.

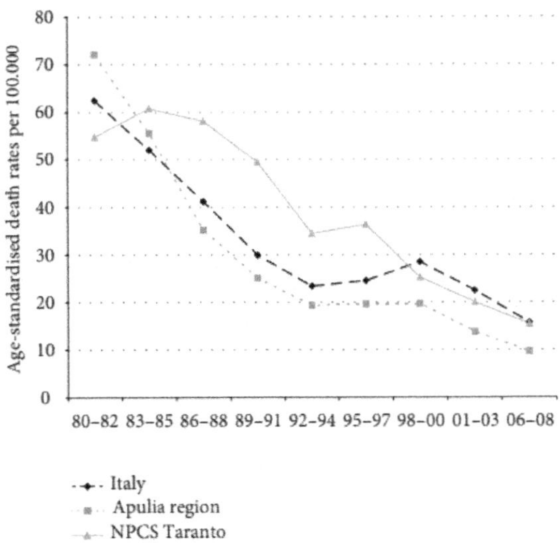

Figure 7: Acute respiratory diseases. Trends in age-standardised (Italian Census 2001) death rates per 100.000, from selected causes of death (1980–2008) (2004-2005 data were not available). All ages. Men.

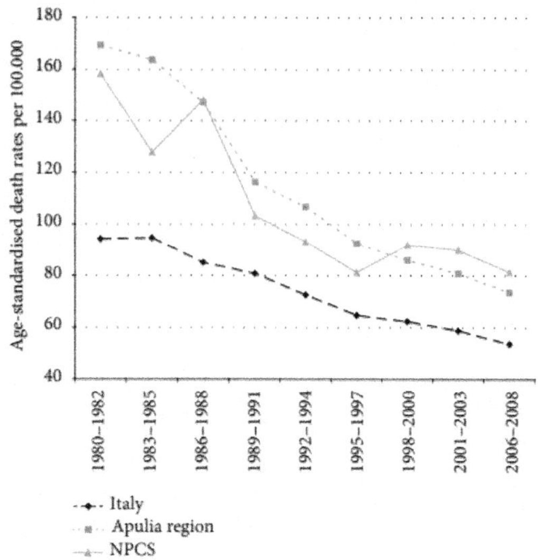

Figure 8: Chronic respiratory diseases. Trends in age-standardised (Italian Census 2001) death rates per 100.000, from selected causes of death (1980–2008) (2004-2005 data were not available). All ages. Men.

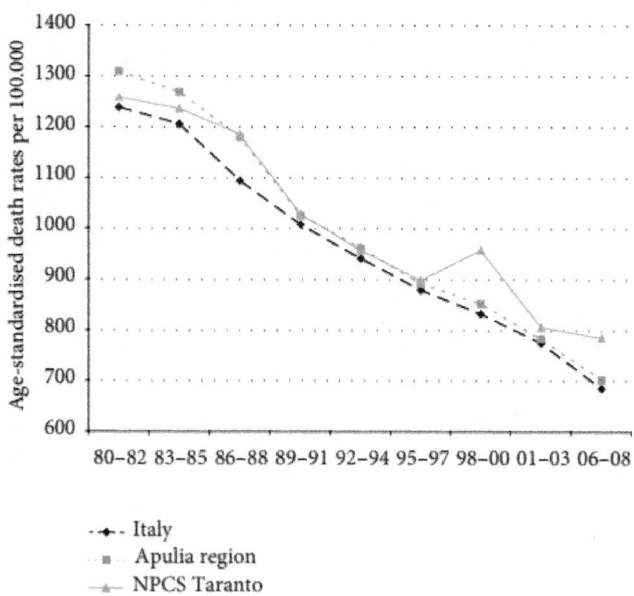

Figure 9: Overall mortality. Trends in age-standardised (Italian Census 2001) death rates per 100.000 (1980–2008) (2004-2005 data were not available). All ages. Women.

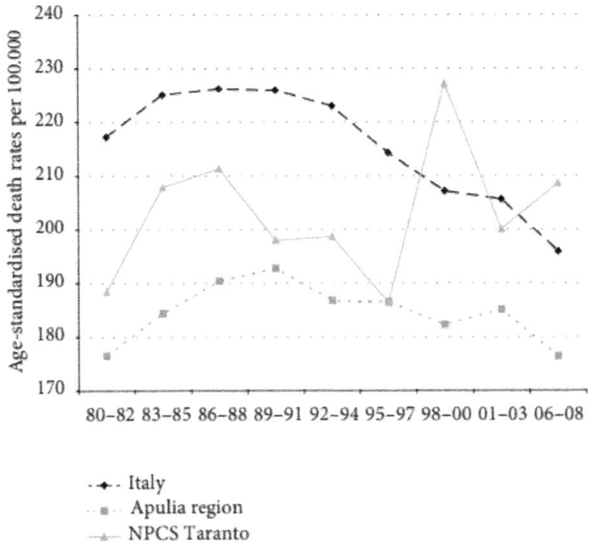

Figure 10: All cancer. Trends in age-standardised (Italian Census 2001) death rates per 100.000, from selected causes of death (1980–2008) (2004-2005 data were not available). All ages. Women.

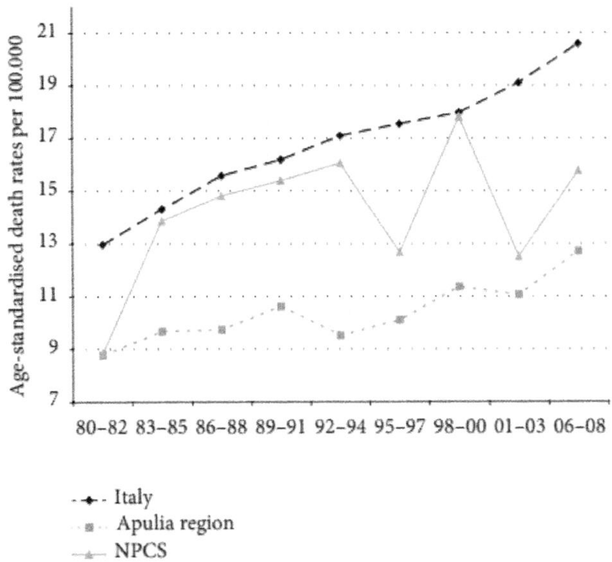

Figure 11: Lung cancers. Trends in age-standardised (Italian Census 2001) death rates per 100.000, from selected causes of death (1980–2008) (2004-2005 data were not available). All ages. Women.

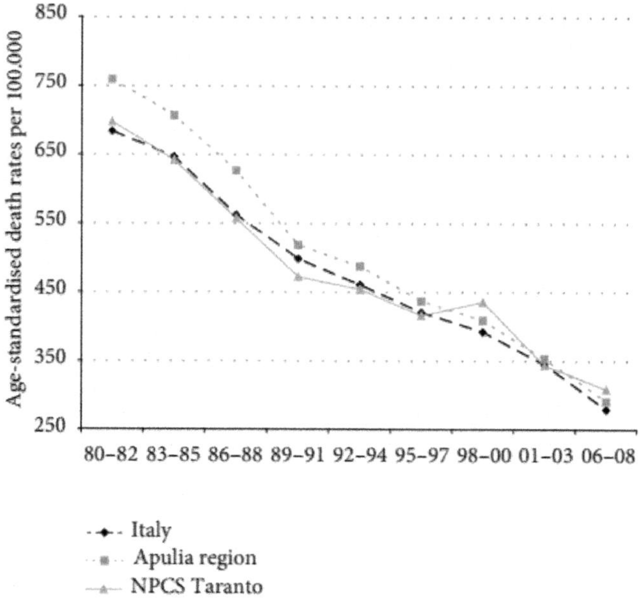

Figure 12: Circulatory diseases. Trends in age-standardised (Italian Census 2001) death rates per 100.000, from selected causes of death (1980–2008) (2004-2005 data were not available). All ages. Women.

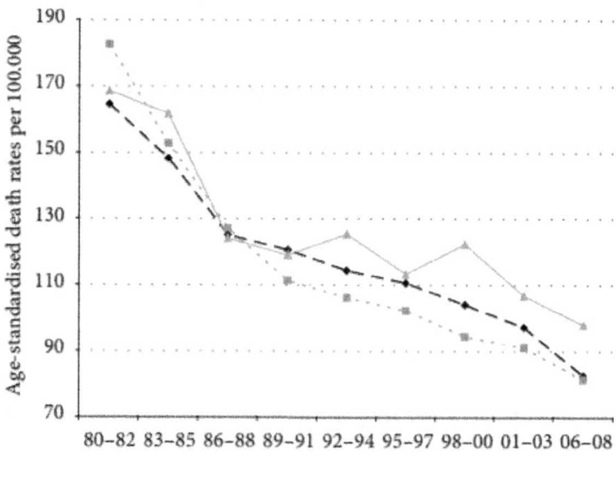

Figure 13: Ischaemic heart diseases. Trends in age-standardised (Italian Census 2001) death rates per 100.000, from selected causes of death (1980–2008) (2004-2005 data were not available). All ages. Women.

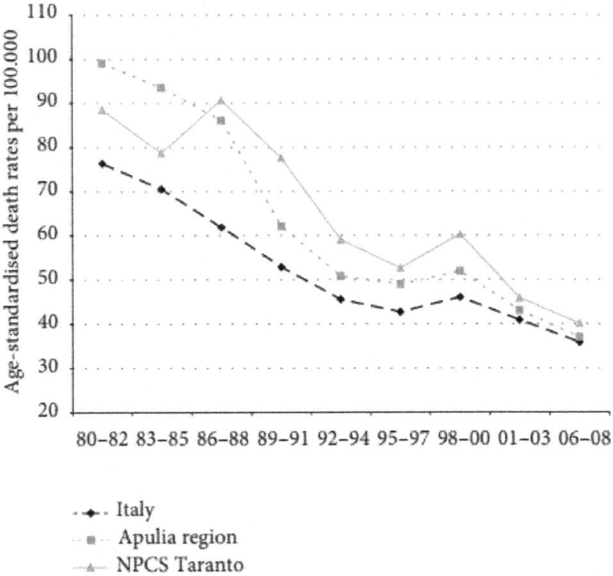

Figure 14: Respiratory diseases. Trends in age-standardised (Italian Census 2001) death rates per 100.000, from selected causes of death (1980–2008) (2004-2005 data were not available). All ages. Women.

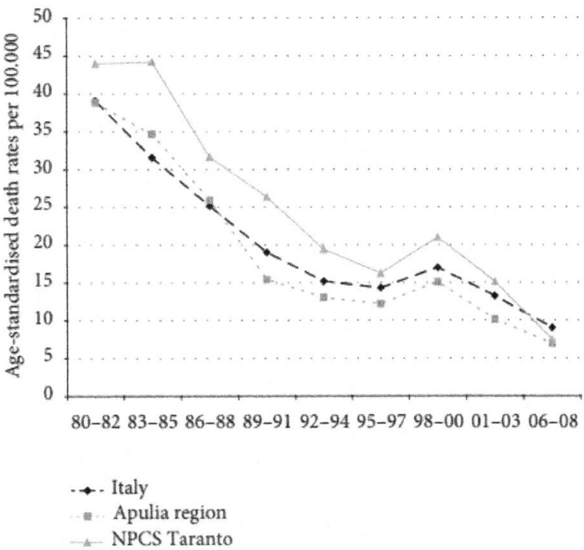

Figure 15: Acute respiratory diseases. Trends in age-standardised (Italian Census 2001) death rates per 100.000, from selected causes of death (1980–2008) (2004-2005 data were not available). All ages. Women.

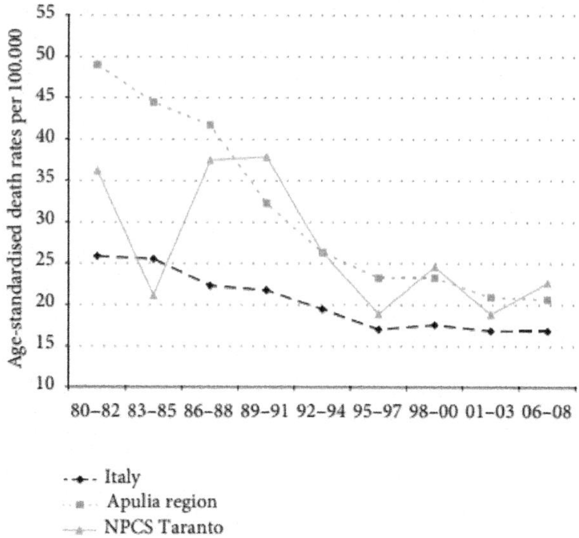

Figure 16: Chronic respiratory diseases. Trends in age-standardised (Italian Census 2001) death rates per 100.000, from selected causes of death (1980–2008) (2004-2005 data were not available). All ages. Women.

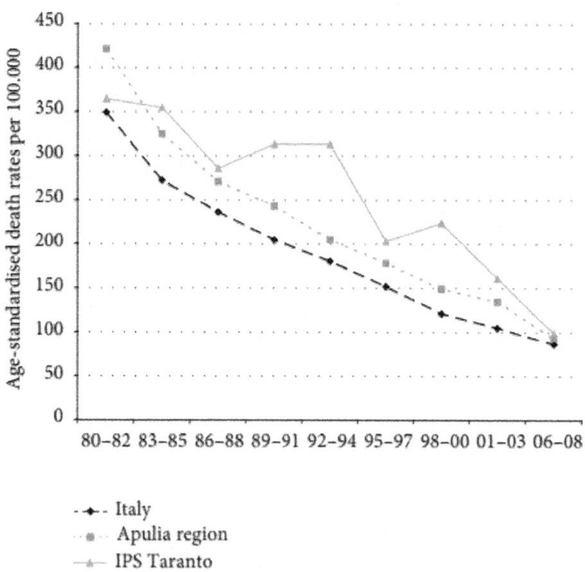

Figure 17: Overall mortality. Trends in age-standardised (Italian Census 2001) death rates per 100.000 (1980–2008) (2004-2005 data were not available). Infant Mortality (0 yrs).

Men

Since many decades, overall mortality in Italy and in Apulia has been declining (resp., 44% and 45%). This favorable trend was also observed in Taranto NPCS, where the regular fall showed a slackening in the last three-year period. Since the early 90s, the SDRs observed in Taranto NPCS were higher than those observed in Apulia, which in turn were lower than the Italian ones; in the most recent three-year period the SDR observed in Taranto site was higher than that in Apulia and Italy (Figure 1). In Italy, mortality from all neoplasms has been falling throughout the study period, while in Taranto NPCS and in Apulia, the trend has been moderately increasing for the same study period. In Taranto NPCS SDRs tend to be higher than in Apulia, which in turn are lower than in Italy (Figure 2).

Mortality from lung cancer has been declining in Italy, Apulia, and Taranto NPCS throughout the years 1980–2008; SDRs in the study site are higher than in Italy and Apulia, since 1995–1997, this differential shows a reduction (Figure 3). Mortality from circulatory diseases in Italy, Apulia, and Taranto site decreased, and the mortality rates almost halved during the study period; yet, in the most recent time, Taranto's mortality rates are higher than in Italy (Figure 4).

Mortality trends from ischemic heart disease have been declining in Italy, Apulia, and Taranto site; yet, since the end of the 1980s, Taranto's SDRs are higher than in Apulia (Figure 5). Mortality trends from respiratory diseases (overall, acute, and chronic) showed a decrease in Italy, Apulia, and Taranto. The rates in Taranto site are higher than those in the Italian ones over the whole study period, the exception being the last decade, when only mortality due to acute respiratory was higher in Taranto (Figures 6,7, and 8).

Women

Overall mortality among women in Italy, Apulia, and Taranto site showed a long-term decreasing trend: from 1980 to 2008 the decrease was respectively 45%, 46%, and 38%; yet, since the beginning of the 2000s, mortality rates in Taranto are higher compared to those in Apulia and Italy (Figure 9).

Mortality from cancer (all sites) has been decreasing in Italy, remaining stable in Apulia, and increasing in Taranto site (Figure 10).

In contrast with the overall cancer mortality, lung cancer mortality has been rising steadily; in the study period the increase was 59% in Italy, 44% in Apulia, and 78% in Taranto NPCS, where SDRs were higher than in Apulia (Figure 11).

Mortality from circulatory diseases showed a declining trend in Italy, Apulia, and Taranto site (Figure 12). Mortality from ischemic heart disease declined as well, but the rates observed in Taranto NPCS were higher than in Apulia and Italy throughout the study period (Figure 13).

Finally, mortality from respiratory diseases (overall, acute, and chronic) showed a decline in Italy, Apulia, and Taranto site, but the rates in Taranto are higher than in Italy (Figures 14, 15, and 16). The results indicate that some differential between Taranto and other areas are emerging in recent years; it may be attributed to the presence for many years of pollutants in the environment, given that no remediation has been performed in the Taranto area.

Infant Mortality

Infant mortality showed a steady decline both in Italy and Apulia; SDRs in Taranto were decreasing, but they stayed higher in Taranto than in Apulia and Italy (Table S4 and Figure 17).

Cancer Incidence 2006-2007

In Taranto NPCS (Table 4) excesses were observed among both males and females when using for comparison both the rates of South and Islands macroarea and of the province (Taranto and Statte excluded) for all tumors and a number of tumor sites (stomach, colon-rectum, liver, pancreas, lung, mesothelioma, skin melanoma, kidney and other unspecified genitourinary organs, and leukemia). Among males an increase was present for prostate cancer and among females for breast cancer. For most sites, the excesses were confirmed when province rates (Taranto and Statte excluded) were used for reference, thus supporting a major health impact among residents in Taranto NPCS.

Table 4: Cancer incidence Taranto NPCS. Number of observed cases (Obs), standardized incidence ratio crude (SIR); 90% IC: 90% confidence interval; Reference SIR: macroarea South and Islands 2005–2007; Reference SIR: Taranto province excluding NPCS municipalities (TA-NPCS) 2006-2007

Cancer site	Males			Females		
	OBS	SIR-macroarea South and Islands (90% CI)	SIR (TA-NPCS) (90% CI)	OBS	SIR-macro-area South and Islands (90% CI)	SIR (TA-NPCS) (90% CI)
Head and neck	52	98 (77–123)	131 (103–165)	13	96 (57–153)	134 (79–213)
Stomach	51	130 (102–164)	117 (91–148)	43	167 (127–215)	224 (171–289)
Colon and rectum	145	112 (97–129)	122 (106–140)	131	115 (99–133)	121 (104–140)
Liver	64	110 (88–135)	140 (113–172)	36	110 (82–145)	175 (130–231)
Pancreas	28	100 (71–137)	135 (96–185)	31	113 (82–153)	129 (93–174)
Lung	245	144 (129–160)	150 (135–167)	47	117 (90–149)	148 (114–189)
Skin melanoma	35	214 (158–284)	193 (143–256)	23	143 (98–203)	120 (82–170)
Mesothelioma	21	429 (287–618)	256 (172–369)	3	197 (53–509)	81 (22–209)
Breast	—	—	—	317	130 (118–143)	124 (113–136)
Prostate	204	129 (112–148)	121 (105–139)	—	—	—
Testis	12	109 (63–177)	79 (46–128)	—	—	—
Uterus, cervix	—	—	—	14	93 (56–145)	88 (53–138)
Uterus, body	—	—	—	69	134 (109–164)	188 (152–230)
Ovary	—	—	—	35	119 (88–158)	81 (60–107)
Kidney and other unspecified urinary organs	51	164 (128–207)	201 (157–254)	18	119 (77–176)	114 (74–169)
Bladder	188	141 (125–159)	136 (120–153)	23	62 (42–88)	92 (63–130)
Brain and CNS (malignant)	16	86 (54–131)	88 (55–134)	12	78 (45–126)	65 (37–105)
Thyroid	23	169 (115–239)	126 (86–179)	71	152 (124–185)	94 (76–115)
Hodgkin lymphoma	6	88 (38–174)	63 (27–124)	8	131 (65–236)	70 (35–126)
Non-Hodgkin lymphoma	42	119 (90–154)	160 (122–207)	28	88 (63–121)	143 (102–196)
Myeloma	18	140 (90–208)	135 (87–200)	15	107 (66–165)	97 (60–149)
Leukemia	30	108 (78–146)	82 (59–111)	37	164 (122–216)	103 (77–136)
All tumors excluding skin, nonmalignant brain, and CNS	1338	131 (125–137)	130 (124–136)	1084	126 (120–132)	121 (115–127)

Mortality 1998–2008 and Morbidity 1998–2010 of the Residential Cohort

The study area was divided into 9 districts in Taranto city and 2 municipalities (Massafra and Statte). Figure 18shows the districts investigated and the location of the industrial area. We considered that the districts located close to the industrial zone were the most areas affected by environmental pollution, especially considering the prevailing winds from northwest. The exposed districts were (1) Tamburi (we also included in this category the small districts of Isola, Porta Napoli, and Lido Azzurro), (2) Borgo, (3) Paolo VI, and (4) Statte. All the other districts were considered the reference zone

(Italia-Montegragnano, San Vito-Lama-Carelli, Salinella, Solito, Corvisea, Talsano, Tre Carrare-Battisti, and Massafra).

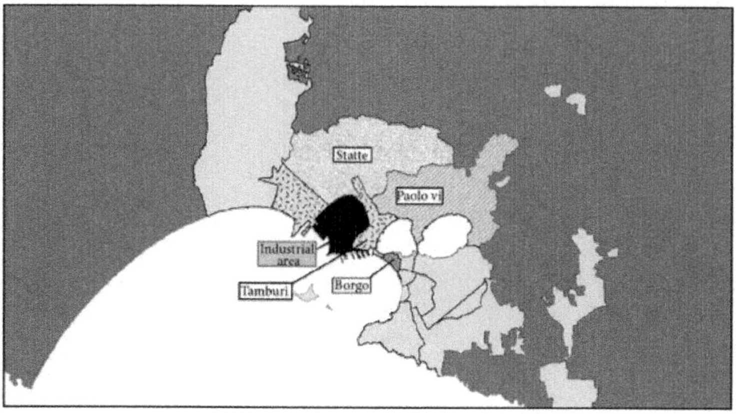

Figure 18: Taranto study area, districts.

A total of 321,356 people (157,031 males and 164,325 females) were enrolled in the cohort. At the time of the enrolment (January 1, 1998), 84.9% of the subjects were already resident in the study area, and 39.1% of them had been residing at the same address for more than 20 years. As far as socioeconomic status is concerned, 35% of the cohort members were in the low SEP category and 21.4% in the high SEP category. The social distribution in the different districts was heterogeneous, with elevated proportion of high social level (62.2%) in some districts in the reference area (San Vito, Lama, Carelli) and low social level in Tamburi (69.4%) and Paolo VI (64.3%). In the Tamburi and Paolo VI districts there was a higher proportion of subjects with previous employment at the steel industry than in other areas.

A total of 3,384,302 person years were accumulated for the cohort. At the end of follow-up (December 31, 2010), 76.6% of the cohort members were alive and resident in the study area, 14.6% had moved, and 28,171 individuals (8.8%) had died. The cause of death was known only for 23,004 individuals deceased by 2008. Only 2.3% of the study subjects were born abroad, and most of the cohort members were born in Taranto (81.6%) and in southern Italy (93.5%).

An analysis of the mortality differences by socioeconomic status (low versus high category) shows (data not in Tables) in both genders a higher mortality for all causes (Hazard ratio—HR 1.25, 95% CI 1.19–1.31 among males and HR 1.18, 95% CI 1.13–1.24 among females), cardiovascular diseases (HR 1.14, 95% CI 1.04–1.25 among males and HR 1.21, 95% CI 1.11–1.32 among females), respiratory diseases (HR 1.89, 95% CI 1.55–2.29 among males and HR 1.38, 95% CI 1.11–1.72 among females), and digestive diseases (HR 1.46, 95% CI 1.19–1.79 among males and HR 1.56, 95% CI 1.24–1.95 among females). Socioeconomic differences in mortality were observed among males for all neoplasms (HR 1.26, 95% CI 1.15–1.37) and cancer of the stomach (HR 1.69, 95% CI 1.10–2.59), larynx (HR 3.32, 95% CI 1.55–7.09), lung (HR 1.40, 95% CI 1.20–1.64), and bladder (HR 1.55, 95% CI 1.08–2.23). The results for hospitalizations confirm the increased risks among subjects in the low socioeconomic category when compared with those in the highest socioeconomic group.

Tables 5 and 6 show, respectively, for males and females, cause-specific mortality in the exposed subareas, that is, Tamburi, Borgo, Paolo VI, and Statte, compared with mortality observed in the reference ones. After adjusting for socioeconomic status, the exposed subareas (Tamburi, Borgo, Paolo VI, and Statte) have a higher mortality for all causes in comparison with the reference in males (in particular, Paolo VI and Tamburi). The most notable increases in mortality among males were in Paolo VI, with 42% excess for all malignant neoplasms (especially lung cancer, +76%), diseases of the cardiovascular (+28%), respiratory (+64%) and digestive (+47%) systems. In Tamburi, an excess was observed among males for all malignant neoplasms (+11%) and cardiovascular diseases (+10%), specifically ischemic heart diseases (+20%). Among females, in Paolo VI, excesses are present for all cancers (+23%), in particular lung, pleural and liver cancer, cardiovascular diseases (+18%), chronic obstructive pulmonary disease (COPD), and digestive system diseases. In Tamburi, excesses were present among females for cardiovascular diseases (+15%), in particular ischemic heart diseases, COPD (+39%), and renal diseases (+57%).

Table 5: Association between district and cause-specific mortality (HR, 90% CI) (males, 1998–2008). Hazard Ratio (HR) from the Cox model stratified by calendar period and adjusted for age (underlying time) and socioeconomic position

Cause of Death (ICD-9-CM)	Reference districts n = 107,909	Tamburi n = 14,067				Borgo n = 16,312				Paolo VI n = 10,097				Statte n = 8,283			
	n	n	HR	90% CI		n	HR	90% CI		n	HR	90% CI		n	HR	90% CI	
All causes (001–999)*	9.378	1.470	1,12	1,07	1,18	1.973	1,07	1,03	1,12	684	1,27	1,19	1,36	654	1,08	1,01	1,15
Malignant cancers (140–208)	2.650	400	1,11	1,01	1,21	505	1,00	0,93	1,09	223	1,42	1,26	1,59	178	1,05	0,93	1,20
Stomach (151)	126	22	1,24	0,83	1,84	28	1,20	0,85	1,71	12	1,62	0,97	2,69	7	0,85	0,45	1,61
Colorectal (153-154)	220	19	0,62	0,41	0,92	48	1,13	0,87	1,48	12	1,07	0,65	1,75	11	0,79	0,48	1,32
Trachea, bronchus, and lung (162)	829	127	1,09	0,93	1,28	150	0,97	0,84	1,13	94	1,76	1,46	2,11	61	1,12	0,90	1,39
Pleura (163)	80	12	1,09	0,64	1,85	16	1,08	0,68	1,71	6	1,19	0,59	2,42	2	0,39	0,12	1,27
Prostate (185)	244	45	1,42	1,07	1,89	53	0,98	0,76	1,26	8	0,84	0,46	1,52	12	0,85	0,52	1,39
Bladder (188)	189	34	1,20	0,87	1,66	29	0,73	0,53	1,02	13	1,45	0,90	2,35	13	1,17	0,73	1,87
Kidney (189)	21	5	2,23	0,93	5,33	5	1,30	0,56	2,98	2	1,85	0,53	6,39	5	3,69	1,62	8,45
Brain and other parts of CNS (191-192; 225)	83	15	1,37	0,85	2,23	18	1,23	0,80	1,90	10	1,64	0,93	2,88	6	1,07	0,53	2,14
Lymphatic and hematopoietic tissue (200–208)	212	27	1,05	0,74	1,49	41	1,07	0,81	1,43	14	1,01	0,64	1,60	9	0,69	0,39	1,20
Neurological diseases (330–349)	190	28	1,09	0,77	1,54	30	0,76	0,54	1,05	7	0,72	0,38	1,36	15	1,32	0,85	2,06
Cardiovascular diseases (390–459)	2.442	378	1,10	1,00	1,21	551	1,02	0,94	1,10	147	1,28	1,11	1,47	137	0,93	0,81	1,08
Cardiac diseases (390–429)	1.688	260	1,09	0,98	1,23	387	1,03	0,94	1,13	106	1,27	1,08	1,51	84	0,82	0,68	0,99
Ischemic heart diseases (410–414)	733	116	1,20	1,01	1,42	152	1,04	0,89	1,20	56	1,37	1,09	1,73	30	0,66	0,49	0,90
Cerebrovascular diseases (430–438)	551	86	1,06	0,87	1,29	109	0,87	0,73	1,04	25	1,07	0,76	1,50	43	1,34	1,04	1,75
Respiratory diseases (460–519)	697	122	1,08	0,91	1,27	177	1,05	0,91	1,20	48	1,64	1,28	2,11	59	1,46	1,17	1,82
COPD (490–492, 494, 496)	469	93	1,17	0,96	1,41	110	0,94	0,79	1,12	32	1,70	1,25	2,32	39	1,44	1,10	1,90
Diseases of the digestive system (520–579)	527	81	1,06	0,86	1,30	111	1,07	0,90	1,27	47	1,47	1,14	1,90	27	0,79	0,57	1,09
Renal diseases (580–599)	146	27	1,36	0,94	1,95	37	1,13	0,83	1,54	3	0,50	0,19	1,30	9	1,00	0,57	1,76

*Referring to the period 1998–2010.

Table 6: Association between district and cause-specific mortality (HR, 90% CI) (females, 1998–2008). Hazard Ratio (HR) from the Cox model stratified by calendar period and adjusted for age (underlying time) and socioeconomic position

Cause of death (ICD-9-CM)	Reference districts *n* = 112,897	Tamburi *n* = 14,625				Borgo *n* = 18,528				Paolo VI *n* = 9,714				Statte *n* = 8,271			
	n	*n*	HR	90% CI		*n*	HR	90% CI		*n*	HR	90% CI		*n*	HR	90% CI	
All causes (001–999)*	9.015	1.479	1,09	1,04	1,15	2.482	1,01	0,97	1,05	489	1,28	1,18	1,38	547	1,06	0,98	1,14
Malignant cancers (140–208)	1.900	230	0,84	0,75	0,95	434	0,95	0,87	1,04	126	1,23	1,06	1.44	102	0,92	0,78	1,08
Stomach (151)	96	20	1,52	0,99	2,34	24	1,01	0,69	1,48	7	1,47	0,76	2,83	7	1,31	0,69	2,51
Colorectal (153-154)	226	23	0,62	0,43	0,90	45	0,78	0,59	1,02	16	1,35	0,87	2,08	7	0,54	0,29	1,02
Trachea, bronchus and lung (162)	144	15	0,76	0,48	1,20	34	1,06	0,77	1,46	13	1,71	1,05	2,79	6	0,68	0,34	1,34
Pleura (163)	20	2	0,66	0,19	2,31	6	1,16	0,53	2,52	3	2,95	1,03	8,46	0			
Breast (174)	349	41	0,92	0,69	1,22	89	1,18	0,96	1,44	28	1,29	0,93	1,79	22	1,04	0,72	1,49
Bladder (188)	33	7	1,23	0,61	2,50	12	1,13	0,64	1,97	2	1,29	0,39	4,34	1	0,58	0,11	3,06
Kidney (189)	17	0				3	0,87	0,30	2,49	0	0,00	0,00	0,00	1	1,06	0,19	5,79
Brain and other parts of CNS (191-192; 225)	90	6	0,48	0,24	0,97	17	0,85	0,55	1,33	4	0,67	0,29	1,57	7	1,30	0,68	2,48
Lymphatic and hematopoietic tissue (200–208)	202	22	0,74	0,50	1,08	33	0,65	0,47	0,88	11	0,98	0,59	1,65	11	0,99	0,60	1,66
Neurological diseases (330–349)	216	35	1,08	0,79	1,47	50	0,83	0,64	1,08	13	1,68	1,04	2,71	11	0,87	0,52	1,45
Cardiovascular diseases (390–459)	2.945	529	1,15	1,06	1,24	876	0,93	0,88	1.00	125	1,18	1,01	1,37	166	0,98	0,86	1,12
Cardiac diseases (390–429)	1.910	371	1,24	1,12	1,37	623	1,04	0,96	1,12	84	1,22	1,01	1,47	90	0,81	0,68	0,97
Ischemic heart diseases (410–414)	565	124	1,46	1,23	1,73	171	1,02	0.88	1,18	24	1,15	0,81	1,63	27	0,86	0.62	1,19
Cerebrovascular diseases (430–438)	820	122	0,93	0,79	1,10	207	0,77	0,68	0,88	35	1,19	0,90	1,59	62	1,38	1,11	1,71
Respiratory diseases (460–519)	476	82	1,09	0,89	1,34	169	1,09	0,94	1,26	22	1,26	0,88	1,82	34	1,28	0,95	1,71
COPD (490–492, 494, 496)	220	49	1,39	1,06	1,83	70	0,97	0,77	1,21	16	2,14	1,38	3,29	14	1,16	0,74	1,83
Diseases of the digestive system (520–579)	484	77	0,95	0,77	1,16	119	0,88	0,74	1,04	29	1,43	1,04	1,97	30	1,13	0,83	1,54
Renal diseases (580–599)	166	38	1,57	1,15	2,14	49	1,01	0,77	1,33	10	1,68	0,98	2,90	11	1,12	0,67	1,87

*Referring to the period 1998–2010.

Tables 7 and 8 display the results for hospital admissions. The hospitalization analysis confirms the mortality results, documenting the major health impact on residents in Tamburi and Paolo VI areas, where excesses were observed among males for a number of causes such as lung cancer (29% and 61% in Tamburi and Paolo VI, resp.), neurological (26% and 43% in Tamburi and Paolo VI, resp.), cardiovascular (18% and 32% in Tamburi and Paolo VI, resp.), respiratory (36% and 52% in Tamburi and Paolo VI, resp.), and renal diseases (35% both in Tamburi and Paolo VI). Among females, similar excesses were observed for cardiovascular diseases (15 and 31% in Tamburi and Paolo VI, resp.), respiratory diseases (28% and 39% in Tamburi and Paolo VI, resp.), digestive diseases (18% and 25% in Tamburi and Paolo VI, resp.), and renal diseases (47% and 35% among males in Tamburi and Paolo VI, resp.). In Paolo VI, pleural (235%) and breast cancer (33%) were also in excess among females.

Table 7: Association between district and cause-specific hospitalization (HR, 90% CI) (males, 1998–2010). Hazard Ratio (HR) from the Cox model stratified by calendar period and adjusted for age (underlying time) and socioeconomic position

Diagnosis (ICD-9-CM)	Reference districts *n* = 108,272	Tamburi *n* = 14,067				Borgo *n* = 16,312				Paolo VI *n* = 10,097				Statte *n* = 8,283			
	n	*n*	HR	90% CI		*n*	HR	90% CI		*n*	HR	90% CI		*n*	HR	90% CI	
Malignant cancers (140–208)	4.818	685	1,12	1,05	1,21	861	1,06	0,99	1,12	444	1,31	1,21	1,43	354	1,06	0,97	1,16
Stomach (151)	166	27	1,21	0,85	1,73	30	1,05	0,75	1,46	19	1,63	1,09	2,46	15	1,29	0,83	2,02
Colorectal (153-154)	520	52	0,86	0,67	1,10	73	0,87	0,70	1,07	28	0,82	0,59	1,13	39	1,08	0,82	1,41
Trachea, bronchus, and lung (162)	866	149	1,29	1,10	1,50	156	1,06	0,92	1,22	101	1,61	1,34	1,92	60	0,98	0,78	1,22
Pleura (163)	75	18	1,80	1,14	2,84	18	1,38	0,89	2,15	7	1,44	0,74	2,79	4	0,77	0,33	1,79
Connective and other soft tissue (171)	32	4	0,91	0,37	2,24	10	1,80	0,98	3,31	5	1,66	0,74	3,72	3	1,30	0,48	3,52
Prostate (185)	639	80	1,10	0,90	1,35	113	1,04	0,88	1,24	40	0,98	0,75	1,29	51	1,22	0,96	1,56
Testis (186)	49	3	0,42	0,15	1,13	9	1,23	0,67	2,25	2	0,40	0,12	1,34	5	1,31	0,60	2,84
Bladder (188)	787	106	1,12	0,94	1,34	121	0,90	0,77	1,06	84	1,62	1,34	1,97	50	0,95	0,74	1,20
Kidney (189)	149	23	1,26	0,85	1,85	37	1,47	1,08	1,99	15	1,41	0,89	2,22	14	1,37	0,86	2,17
Brain and other parts of CNS (191-192; 225)	187	23	0,98	0,67	1,43	30	1,01	0,73	1,41	14	0,89	0,56	1,41	16	1,19	0,77	1,83
Lymphatic and hematopoietic tissue (200–208)	400	58	1,20	0,94	1,53	80	1,26	1,03	1,55	35	1,13	0,84	1,52	24	0,84	0,60	1,19
Neurological diseases (330–349)	1.850	329	1,26	1,14	1,40	337	1,11	1,00	1,22	226	1,43	1,27	1,61	140	1,04	0,90	1,20
Cardiovascular diseases (390–459)	14.504	2.078	1,18	1,13	1,23	2.457	1,03	1,00	1,07	1.388	1,32	1,26	1,39	1.076	1,06	1,01	1,12
Cardiac diseases (390–429)	9.866	1.411	1,20	1,15	1,27	1.699	1,04	0,99	1,09	957	1,41	1,33	1,49	752	1,10	1,04	1,17
Acute coronary events (410-411)	2.328	310	1,13	1,02	1,26	396	1,09	0,99	1,19	241	1,39	1,24	1,56	167	1,00	0,87	1,14
Heart failure (428)	1.878	317	1,21	1,09	1,34	375	1,03	0,94	1,14	180	1,54	1,35	1,76	119	0,97	0,83	1,13
Cerebrovascular diseases (430–438)	3.124	525	1,30	1,19	1,41	581	1,04	0,96	1,12	211	1,01	0,89	1,13	226	1,09	0,97	1,22
Respiratory diseases (460–519)	8.906	1.836	1,36	1,30	1,43	1.493	1,01	0,96	1,06	1.255	1,52	1,44	1,60	752	1,12	1,05	1,19
Acute respiratory infections (460–466, 480–487)	3.769	961	1,51	1,42	1,61	746	1,19	1,11	1,27	625	1,54	1,44	1,66	315	1,05	0,95	1,15
COPD (490–492, 494, 496)	2.350	450	1,31	1,20	1,43	379	0,85	0,78	0,93	252	1,60	1.43	1,79	203	1,29	1,14	1,46
Diseases of the digestive system (520–579)	15.628	2.465	1,20	1,15	1,24	2.301	0,94	0,91	0,98	1.682	1,23	1,18	1,29	1.229	1,06	1,01	1,12
Renal diseases (580–599)	3.252	562	1,35	1,25	1,46	521	0,98	0,90	1,06	331	1,35	1,22	1,49	282	1,22	1,10	1,36

Table 8: Association between district and cause-specific hospitalization (HR, 90% CI) (females, 1998–2010). Hazard Ratio (HR) from the Cox model stratified by calendar period and adjusted for age (underlying time) and socioeconomic position

Diagnosis (ICD-9-CM)	Reference districts	Tamburi				Borgo				Paolo VI				Statte			
	n = 113,187	*n* = 14,625				*n* = 18,528				*n* = 9,714				*n* = 8,271			
	n	*n*	HR	90% CI		*n*	HR	90% CI		*n*	HR	90% CI		*n*	HR	90% CI	
Malignant cancers (140–208)	3.878	530	1,03	0,95	1,12	713	0,91	0,85	0,97	295	1,17	1,06	1,29	225	0,91	0,81	1,02
Stomach (151)	129	23	1,26	0,85	1,85	24	0,89	0.61	1,28	7	1,06	0,55	2,02	9	1,10	0,62	1,94
Colorectal (153-154)	483	60	0,93	0,74	1,18	62	0,62	0,50	0,78	35	1,21	0,90	1,62	17	0,58	0,39	0,88
Trachea, bronchus, and lung (162)	149	17	0,89	0,57	1,37	30	0,94	0,67	1,33	10	1,04	0,60	1,80	5	0,53	0,25	1,12
Pleura (163)	23	3	0,87	0,31	2,44	2	0,44	0,13	1,48	4	3,35	1,31	8,54	2	1,25	0,37	4,22
Breast (174)	990	127	1,03	0,87	1,21	179	0.98	0,86	1,12	94	1,33	1,11	1,59	59	0,89	0,72	1,11
Vescica (188)	146	22	1,12	0,76	1,66	37	1,16	0,85	1,58	5	0,61	0,29	1,29	5	0,56	0,27	1,19
Kidney (189)	78	10	0,95	0,53	1,68	13	0,84	0,51	1,39	5	1,06	0,49	2,28	10	2,10	1,21	3,66
Brain and other parts of CNS (191-192; 225)	204	25	0,94	0,65	1,35	30	0,77	0,55	1,06	17	1,18	0,77	1,80	17	1,27	0,84	1,93
Lymphatic and hematopoietic tissue (200–208)	398	55	1,03	0,80	1,31	58	0,72	0,57	0.91	26	1,00	0,71	1,40	24	0,95	0,67	1,35
Neurological diseases (330–349)	2.151	351	1,11	1,01	1,23	378	0,89	0,81	0,98	168	1,06	0,93	1,21	141	0,98	0,85	1,14
Cardiovascular diseases (390–459)	13.500	2.072	1,15	1,11	1,20	2.611	0,90	0,87	0.93	1.059	1,31	1,25	1,39	888	1,05	0,99	1,11
Cardiac diseases (390–429)	9.366	1.478	1,17	1,12	1,23	1.897	0,93	0,89	0,97	752	1,40	1,31	1,49	632	1,09	1,02	1,17
Acute coronary events (410-411)	1.060	197	1,32	1,16	1,51	254	1,08	0,96	1,22	88	1,42	1,18	1,72	64	1,02	0,82	1,26
Heart failure (428)	2.449	368	0,95	0,87	1,05	628	1,02	0,94	1,09	154	1,32	1,15	1,52	106	0,77	0,65	0,90
Cerebrovascular diseases (430–438)	3.595	600	1,15	1,06	1,24	728	0,84	0,79	0,90	220	1,21	1,07	1,35	212	1,01	0,90	1,13
Respiratory diseases (460–519)	6.673	1.336	1,28	1,21	1,34	1.273	1,00	0,95	1,05	813	1,39	1,31	1,49	514	1,07	0,99	1,16
Acute respiratory infections (460–466, 480–487)	3.020	705	1,39	1,29	1,49	599	1,08	1,00	1,17	422	1,37	1,25	1,49	228	0,98	0,88	1,10
COPD (490–492, 494, 496)	1.433	262	1,19	1,06	1,33	325	0,94	0.85	1,04	126	1,62	1,39	1,89	91	1,09	0,91	1,30
Diseases of the digestive system (520–579)	12.952	2.067	1,18	1,13	1,22	2.038	0,89	0,85	0,92	1.288	1,25	1,19	1,31	905	1,00	0,95	1,06
Renal diseases (580–599)	3.187	662	1,47	1,37	1,59	609	0,99	0,92	1,07	320	1,35	1,22	1,49	248	1,17	1,05	1,31

DISCUSSION

The health impact of residence in Taranto NPCS was investigated using different epidemiological approaches: geographical (mortality and cancer incidence at municipality level), historical (mortality time trends at municipality level), and residential cohort studies (mortality and hospital discharge records at individual level). We adopted an analysis at small-area scale which includes a mix of small area and individual based data. The design issues relevant to this type of investigation have been thoroughly examined by Elliott and Savitz [27], some of them are briefly examined in the following paragraphs.

In environmental epidemiology, exposure ascertainment is a key phase because the exposure/s affecting the study population should ideally be described in detail, while in most instances the available exposure information is indirect and qualitative. In ecological investigations, the exposure/s can be a single event from a point emission source of some contaminants; more often the contaminants are heterogeneous mixture progressively polluting different matrices in the area. For example, in SENTIERI Project the sources ofenvironmental exposures were abstracted from the legislative decrees defining sites' boundaries and fixed on the basis of the possible sources of contamination (e.g., chemical industry, steel plants, and landfills). A further limitation lies in the implicit assumption that all residents in the area under investigation experience the same exposures, while exposure variability is likely to be substantial, due to many factors (e.g., concentration of contaminants and their diffusion to soil and water, distance of residence from polluting sources). The possible consequences of such nondifferential exposure misclassification are complex, and the direction of the resulting bias is not predictable [28]. In addition, information exposure source/s with possible health impact, such as concurrent air pollution from road traffic and exposures in the occupational setting, are often not available. Finally, vital statistics are accessible for a given administrative area whose boundaries hardly correspond to the distribution of environmental pollutants, so that the misclassification of exposure (and loss of statistical power) is common. A more detailed description of these limitations of ecological study design is available [15, 28, 29].

Exposure ascertainment is a critical issue in ecological investigations as well as in studies based on individuals, as cohort of residents are. In this case residential history, overtime information on exposure/s in different residences and different environmental matrices, daily variability, and seasonal variability should be available to classify subjects in different exposure categories. Obtaining individual-based measures of exposure as described above is clearly infeasible, and modeling of exposures, ranging from simple measures such as distance from a point source or distance to nearest road to more complex estimation, for example, dispersion modeling around a point source, is used. The possible exposure misclassification in such approaches has been discussed in [27].

As far as outcome measures are concerned, many studies of environmental healthin polluted areas consider mortality, based on death records. However, the analysis of hospital discharge records, ad hoc registry data of specific pathologies (e.g., cancer, congenital malformations) can give a better picture of the health profile of residents in NPCSs. The key issue is that, whatever the outcome under investigation, databases need to be validated for use in epidemiological studies.

Reporting the event of death is usually exhaustive; therefore the overall mortality can be analyzed with confidence [30]. In Italy, validity of cause of death certification has been documented for specific diseases [31–34]. The validity of hospital discharge records (HDRs), indicators of hospital activity, has not been systematically evaluated, although in Italy some critical aspects of this novel utilization of HDRs have been examined [35].

Another crucial aspect in environmental health studies is that factors such as socioeconomic status, occupational exposures/s, and individual lifestyles can have an etiologic role on the health effects under study thus possibly confounding the exposure-disease relationships.

For a review of adjustment for socioeconomic status using census data in ecological studies of environment and health refer to Pasetto et al, 2010 [36]. To account for deprivation in SENTIERI Project, mortality data both crude and adjusted were analyzed using an ad hoc built deprivation index [25]. Also in individual-based studies, the analyses can be adjusted for socioeconomic factors, usually with an aggregate indicator based on residence address. In the present cohort of residents in Taranto NPCS SEP was assigned to each participant on the basis of the geocoded addresses at the beginning of the follow-up.

Occupational exposure/s are also potential confounders in environmental health studies. The ecological components of the present investigation are affected by this limitation, while for the cohort study individual occupational history was traced through the national insurance company (INPS) database, and the subcohort of individuals employed in industries located in the area was identified indicating a high proportion of past employment at the steel plant among residents in Tamburi and Paolo VI.

Again on this topic, with specific reference to pleural mesothelioma, it should be noted that the Italian National Mesothelioma Registry analysis of residential asbestos exposure showed that steel mills and iron foundries were the second most frequent sources of asbestos in the neighborhood (after asbestos-cement factories), ranking equal to asbestos textiles production [37]. Furthermore a case-control study in the area of Taranto (HDR, 1998–2002) adjusted for occupational exposures [11], showed an increase in malignant pleural neoplasms among residents close to the steel mill (or 1.62, 95% CI 0.37–7.10, 11 cases) and the coke plant (or 2.18, 95% CI 0.31–15.31, 9 cases).

Another aspect to consider in environmental epidemiology is that large populations are needed to study many of the health concerns of greatest interest. In ecological investigations, the reference populations should be selected balancing the need for comparability of study and reference populations for factors other than the environmental exposure/s with possible health impact (socioeconomic status and lifestyle factors as diet and tobacco use) and the requirement of sufficiently numerous populations to have stable reference rates also for rare diseases. These needs are satisfied in the present investigation where national, macroregional, and regional populations were used for comparison in the mortality, time trend, and cancer incidence analysis. In the cohort study an internal comparison was carried out, and the reference population was composed of residents in districts distant from the industrial area.

For chronic diseases including most cancers, latency effects are important, such that exposures experienced many years previously, or accumulated exposures, may be crucial. In Taranto NPCS, the mortality (1995–2009) and the time trends analyses (1980–2008) show consistent results and cover a time span which should encompass latency effects. Analogously in the cohort of residents, most subjects were present at enrolment in 1998 (85%), and half of them had a residence duration of 10 or more years.

A brief comment on the use of 90% CI in our analyses is needed. In this respect we refer to Sterne and Smith [38], who affirm that confidence intervals for the main results should always be included, but 90% rather than 95% levels should be used. CI should not be used as a surrogate means of examining significance at the conventional 5% level. Interpretation of CI should focus on the implications (clinical importance) of the range of values in the interval.

CONCLUSIONS

In Taranto NPCS, mortality data at municipality level analyzed, in the context of SENTIERI Project, time trend analysis and cancer incidence results coherently showed, in both genders, excess risks for a number of causes of death, among them: all causes, all cancers, lung cancer, cardiovascular, and respiratory diseases, both acute and chronic. For these causes, an etiologic role of environmental exposure present in Taranto NPCS can be supported on the basis of a priori evaluation of the epidemiological evidence completed in SENTIERI. In the cohort study among residents in the districts nearer to the industrial area, excess mortality/morbidity risks were shown for natural cause, cancers, cardiovascular, and respiratory diseases. These excesses were also observed in low socioeconomic position groups compared to high ones, some of them could be explained on the basis of previous employment of residents in industries active in the study area.

As discussed above the present results for Taranto NPCS are based on sound study design and valid data, which make a low potential for bias and help to strengthen etiologic inference. The present findings further corroborate the need to promptly proceed with environmental cleanup interventions.

It should not be disregarded the fact that most diseases showing an increased risk have multifactorial etiology, therefore interventions of proven efficacy, such as smoking cessation, food education, measures for cardiovascular risk reduction, and breast cancer screening programs, should be planned. To build a climate of confidence and trust between citizens and public institutions, study results and public health actions are to be communicated objectively and transparently.

APPENDICES

A. International Classification of Diseases to Code Mortality: From ICD-9 to ICD-10

Mortality data are coded according to the International Classification of Disease (ICD) which has been revised approximately every 10 years; the purpose of the revision is to stay abreast of medical advances in terms of disease nomenclature and etiology. In Italy, deaths have been coded according to the Ninth Revision (ICD-9) until 2002 [39]; since 2003, the Tenth Revision (ICD-10) has been adopted [40].

ICD-10 differs from ICD-9 in several respects: ICD-10 is far more detailed than ICD-9, with about 12,000 categories versus about 5,000 categories; ICD-10 uses alphanumeric codes compared with numeric codes in ICD-9; some additions and modifications were made to the chapters in the ICD; and some of the coding rules and rules for selecting the underlying cause of death have been changed.

Because of these modifications, comparability studies, also called bridge-coding studies, were carried out to measure the effects of a new revision of the ICD on the comparability with the previous revision of mortality statistics by cause of death. These studies involve the dual classification of a single-year mortality data, that is, classifying the underlying cause of death on mortality records by both the new revision and the previous revision. The key element of a comparability study is the "comparability ratio", which is derived from the dual classification.

Operationally, the comparability ratio for the cause of death i (C_i) is calculated as follows:

$$Ci = \frac{D_{i,\text{ICD-10}}}{D_{i,\text{ICD-9}}}, \qquad \text{(A.1)}$$

where $D_{i,\text{ICD-10}}$ is the number of deaths due to cause i classified by ICD-10. $D_{i,\text{ICD-9}}$, is the number of deaths due to cause i classified by ICD-9.

A comparability ratio of 1.00 indicates that the same number of deaths was assigned to cause i under both ICD-9 and ICD-10.

Comparability studies between ICD-9 and ICD-10 have been conducted in the USA, in Europe (by Eurostat), and also in Italy (by the Italian Census Bureau, ISTAT) [41–44]. The Italian study [44] documented that comparability ratios of the main causes of death are close to the value of one.

Table S1 presents the comparability study results that referred to some causes investigated in the time trend analysis.

In Italy the definition of municipalities has varied over time: some municipalities have been suppressed, others have been modified, and others have been created.

Generally, these modifications are not implemented at the same time by the office that registers deaths (Civil Status Office) and by the office that registers the population (General Registry office).

Therefore, it may occur that a "new" municipality (created by the subdivision of another municipality) creates first the General Registry, beginning to register its own populations, while the deaths continue to be registered by the Civil Status Office of the "old" municipality [45, 46].

In the context of SENTIERI Project the situation of all of the municipalities constituting the NPCSs has been studied [47].

As far as Taranto and Statte municipalities, that constitute Taranto NPCS, the description is as follows: until 1993, Statte was a district of Taranto; in 1994 it became a new municipality and created immediately its own General Registry Office; on the other hand, its own Civil Status office was set up some years later (1998).

To "synchronize" data regarding deaths and populations of both Statte and Taranto, we decided to attribute the population and the deaths of Statte to Taranto until 1997; since 1998, they have been attributed to Statte itself.

B. Mortality Trends from Selected Causes in Taranto NPCS, 1980–2008

See Supplementary Tables S2, S3, and S4.

ACKNOWLEDGMENTS

SENTIERI Project was funded, for the years 2008–2010, by the Italian Ministry of Health (Strategic Programme Environment and Health) and for the years 2009–2013 by the Ministry of Health's Project CCM 2009 "Epidemiological surveillance of populations living in contaminated sites". The authors wish to thank Letizia Sampaolo for the editorial assistance and language revision.

REFERENCES

1. European Environment Agency, "Progress in management of contaminated sites (CSI 015): assessment published Aug 2007," http://www.eea.europa.eu/data-and-maps/indicators/progress-in-management-of-contaminated-sites/progress-in-management-of-contaminated-1.
2. L. Liberti, M. Notarnicola, R. Primerano, and G. Vitucci, "Air pollution from a large steel factory: toxic contaminants from coke-ovenplants," in Proceedings of the 12th international conference on modelling, monitoring and management of air pollution, pp. 485–494, WIT Press, Southampton, UK, 2004.
3. M. M. Storelli and G. O. Marcotrigiano, "Bioindicator organisms: heavy metal pollution evaluation in the Ionian Sea (Mediterranean Sea, Italy)," Environmental Monitoring and Assessment, vol. 102, no. 1–3, pp. 159–166, 2005.
4. L. Liberti, M. Notarnicola, R. Primerano, and P. Zannetti, "Air pollution from a large steel factory: polycyclic aromatic hydrocarbon emissions from coke-oven batteries," Journal of the Air and Waste Management Association, vol. 56, no. 3, pp. 255–260, 2006.
5. C. Gariazzo, V. Papaleo, A. Pelliccioni, G. Calori, P. Radice, and G. Tinarelli, "Application of a Lagrangian particle model to assess the impact of harbour, industrial and urban activities on air quality in the Taranto area, Italy," Atmospheric Environment, vol. 41, no. 30, pp. 6432–6444, 2007.
6. L. Bisceglia, R. Giua, A. Morabito et al., "Source apportionment of benzo(a)pyrene in Taranto and carcinogenic risk estimate in general population," Giornale Italiano di Medicina del Lavoro ed Ergonomia, vol. 32, no. 4, supplement, pp. 355–356, 2010 (Italian).
7. A. Di Leo, N. Cardellicchio, S. Giandomenico, and L. Spada, "Mercury and methylmercury contamination in Mytilus galloprovincialis from Taranto Gulf (Ionian Sea, Southern Italy): risk evaluation for consumers," Food and Chemical Toxicology, vol. 48, no. 11, pp. 3131–3136, 2010.
8. S. Giandomenico, L. Spada, C. Annicchiarico, G. Assennato, N. Cardellicchio, N. Ungaro, et al., "Chlorinated compounds and polybrominated diphenyl ethers (PBDEs) in mussels (Mytilus

galloprovincialis) collected from Apulia Region coasts," Marine Pollution Bulletin, vol. 73, no. 1, pp. 243–251, 2013.

9. I. Iavarone, E. De Felip, A. M. Ingelido, N. Iacovella, A. Abballe, S. Valentini, et al., "Exploratory biomonitoring study among workers of livestock farms of the Taranto Province," Epidemiologia and Prevenzione, vol. 36, no. 6, pp. 321–331, 2012 (Italian).
10. M. Martuzzi, F. Mitis, A. Biggeri, B. Terracini, and R. Bertollini, "Environment and health status of the population in areas with high risk of environmental crisis in Italy," Epidemiologia and prevenzione, vol. 26, no. 6, supplement, pp. 1–53, 2002 (Italian).
11. A. Marinaccio, S. Belli, A. Binazzi et al., "Residential proximity to industrial sites in the area of Taranto (Southern Italy): a case-control cancer incidence study," Annali dell›Istituto Superiore di Sanita, vol. 47, no. 2, pp. 192–199, 2011.
12. P. Comba, R. Pirastu, S. Conti, M. De Santis, I. Iavarone, G. Marsili, et al., "Environment and health in Taranto, Southern Italy: epidemiological studies and public health recommendations," Epidemiologia and Prevenzione, vol. 36, no. 6, pp. 305–320, 2012 (Italian).
13. F. Mataloni, M. Stafoggia, E. Alessandrini, M. Triassi, A. Biggeri, and F. Forastiere, "A cohort study on mortality and morbidity in the area of Taranto, Southern Italy," Epidemiologia and Prevenzione, vol. 36, no. 5, pp. 237–252, 2012 (Italian).
14. R. Pirastu, C. Ancona, I. Iavarone et al., "SENTIERI Project. Mortality study of residents in Italian polluted sites: evaluation of the epidemiological evidence," Epidemiologia and prevenzione, vol. 34, no. 5-6, pp. 1–2, 2010 (Italian).
15. R. Pirastu, R. Pasetto, A. Zona, C. Ancona, I. Iavarone, M. Martuzzi, et al., "The health profile of population living in contaminated sites: SENTIERI approach," Journal of Environmental and Public Health, vol. 2013, Article ID 939267, 13 pages, 2013.
16. R. Pirastu, C. Ancona, I. Iavarone et al., "SENTIERI Project. Mortality study of residents in Italian polluted sites: evaluation of the epidemiological evidence," Epidemiologia and prevenzione, vol. 34, supplement 3, no. 5-6, pp. 1–2, 2010 (Italian).
17. C. A. Pope III, "Respiratory hospital admissions associated with PM10 pollution in Utah, Salt Lake, and Cache Valleys," Archives of Environmental Health, vol. 46, no. 2, pp. 90–97, 1991.
18. R. S. Bhopal, S. Moffatt, T. Pless-Mulloli et al., "Does living near a constellation of petrochemical, steel, and other industries impair

health?" Occupational and Environmental Medicine, vol. 55, no. 12, pp. 812–822, 1998.

19. P. R. Lewis, M. J. Hensley, J. Wlodarczyk et al., "Outdoor air pollution and children›s respiratory symptoms in the steel cities of New South Wales," Medical Journal of Australia, vol. 169, no. 9, pp. 459–463, 1998.
20. J. Petrela, V. M. Câmara, G. Kennedy, B. Bouyahi, and J. Zayed, "Health effects of residential exposure to aluminum plant air pollution," Archives of Environmental Health, vol. 56, no. 5, pp. 456–460, 2001.
21. M. Wilhelm, G. Eberwein, J. Hölzer et al., "Influence of industrial sources on children›s health: hot spot studies in North Rhine Westphalia, Germany," International Journal of Hygiene and Environmental Health, vol. 210, no. 5, pp. 591–599, 2007.
22. A. C. Câra, F. Buntinx, M. van den Akker, G.-J. Dinant, and C. Manolovici, "Industrial air pollution and children›s respiratory health: a natural experiment in Călărasi," European Journal of General Practice, vol. 13, no. 3, pp. 135–143, 2007.
23. A. C. Cra, J. Degryse, M. van den Akker, G.-J. Dinant, C. Manolovici, and F. Buntinx, "Impact of early childhood air pollution on respiratory status of school children," European Journal of General Practice, vol. 16, no. 3, pp. 133–138, 2010.
24. World Health, Air Quality Guidelines. Global Update 2005. Particulate Matter Ozone, Nitrogen Dioxide and Sulfur Dioxide, World Health Organization, Copenhagen, Denmark, 2006,http://www.euro.who.int/__data/assets/pdf_file/0005/78638/E90038.pdf.
25. M. De Santis, R. Pasetto, G. Minelli, and S. Conti, "Methods for mortality analysis in SENTIERI Project,"Epidemiologia and prevenzione, vol. 35, supplement 4, no. 5-6, pp. 24–28, 2011 (Italian).
26. R. Pasetto, N. Caranci, and R. Pirastu, "Deprivation indices in small-area studies of environment and health in Italy," Epidemiologia e prevenzione, vol. 35, supplement 4, no. 5-6, pp. 174–180, 2011 (Italian).
27. P. Elliott and D. A. Savitz, "Design issues in small-area studies of environment and health,"Environmental Health Perspectives, vol. 116, no. 8, pp. 1098–1104, 2008.
28. J. Wakefield, "Ecologic studies revisited," Annual Review of Public Health, vol. 29, pp. 75–90, 2008.
29. D. A. Savitz, "Commentary: a niche for ecologic studies in environmental epidemiology," Epidemiology, vol. 23, no. 1, pp. 53–54, 2012.
30. B. Terracini and R. Pirastu, "General guidance to the interpretation of

vital statistics in polluted areas," inHuman Health in Areas with Local Industrial Contamination. Challenges and Perspectives, with Examples From Sicily, WHO Europe, 2013.

31. A. Barchielli, R. De Angelis, and L. Frova, "Mortality data for the study of digestive cancers in Italy: characteristics and quality," Annali dell›Istituto Superiore di Sanita, vol. 32, no. 4, pp. 433–442, 1996 (Italian).
32. C. Bruno, P. Comba, P. Maiozzi, and T. Vetrugno, "Accuracy of death certification of pleural mesothelioma in Italy," European Journal of Epidemiology, vol. 12, no. 4, pp. 421–423, 1996.
33. S. Conti, M. Masocco, V. Toccaceli et al., "Mortality from human transmissible spongiform encephalopathies: a record linkage study," Neuroepidemiology, vol. 24, no. 4, pp. 214–220, 2005.
34. P. Comba, F. Bianchi, S. Conti et al., "SENTIERI Project: discussion and conclusions," Epidemiologia e prevenzione, vol. 35, no. 5-6, pp. 163–171, 2011 (Italian).
35. A. Cernigliaro, A. Marras, S. Pollina Addario, and S. Scondotto, Stato di Salute Della Popolazione Residente Nelle Aree a Rischio Ambientale e nei siti di Interesse Nazionale per le Bonifiche in Sicilia. Analisi dei dati ReNCam (anni 2004–2011) e dei Ricoveri Ospedalieri (anni 2007–2011), Supplemento Monografico Notiziario Osservatorio Epidemiologico, 2013.
36. R. Pasetto, L. Sampaolo, and R. Pirastu, "Measures of material and social circumstances to adjust for deprivation in small-area studies of environment and health: review and perspectives," Annali dell›Istituto Superiore di Sanita, vol. 46, no. 2, pp. 185–197, 2010.
37. D. Mirabelli, D. Cavone, E. Merler et al., "Non-occupational exposure to asbestos and malignant mesothelioma in the Italian National Registry of Mesotheliomas," Occupational and Environmental Medicine, vol. 67, no. 11, pp. 792–794, 2010.
38. J. A. C. Sterne and G. D. Smith, "Sifting the evidence: whats wrong with significance tests?" British Medical Journal, vol. 322, no. 7280, pp. 226–231, 2001.
39. OMS, "Classificazioni delle malattie, traumatismi e cause di morte. Nona revisione," 1975, ISTAT, vol. 1-2, Metodi e Norme, serie C no. 10, 1997.
40. OMS, "Classificazione statistica internazionale delle malattie e dei problemi sanitari correlati," in Decima Revisione, vol. 1–3, Ministero della Sanità, 2001.
41. R. N. Anderson, A. M. Miniño, D. L. Hoyert, and H. M. Rosenberg,

"Comparability of cause of death between ICD-9 and ICD-10: preliminary estimates," National Vital Statistics Reports, vol. 49, no. 2, pp. 1–32, 2001.

42. R. N. Anderson and H. M. Rosenberg, "Disease classification: measuring the effect of the Tenth Revision of the International Classification of Diseases on cause-of-death data in the United States," Statistics in Medicine, vol. 22, no. 9, pp. 1551–1570, 2003.
43. Eurostat, "Guidelines for bridge coding studies. European Statistics on Causes of Death—COD Methodological information. Document prepared by the Task Force on 'ICD-10 up-dates and ACME implementation in Europe'," 2007.
44. "Analisi del Bridge Coding ICD-9 ICD-10 per le statistiche di mortalità per causa in Italia," ISTAT, Metodi e norme, n. 50, 2011.
45. R. Capocaccia and G. Caselli, Popolazione Residente Per età e Sesso Nelle Province Italiane: Anni 1972–1981, Università degli Studi di Roma, Dipartimento di Scienze Demografiche, Roma, Italy, 1990.
46. ISTAT-ISS, Ricostruzione Della Popolazione Residente Per età e Sesso Nelle Province Italiane: Anni 1982–1991, ISTAT, Roma, Italy, 1996.
47. M. De Maria, G. Minelli, and S. Conti, "L'utilizzo dei dati di mortalità a livello comunale in Italia: progetto SENTIERI," Epidemiologia and Prevenzione, vol. 35, supplement 4, no. 5-6, pp. 181–184, 2011.

Chapter 4

THE MEXICAN ENVIRONMENTAL FLOW STANDARD: SCOPE, APPLICATION AND IMPLEMENTATION

María Antonieta Gómez-Balandra, María del Pilar Saldaña-Fabela, Maricela Martínez-Jiménez

Hydrobiology and Environmental Assessment Department, Mexican Institute of Water Technology, Jiutepec, México

ABSTRACT

With the implementation of the Official Mexican Standard NOM-011-CONAGUA-2000 [1], the water balance of 730 basins has been calculated and its water availability agreement is published. This rule points out to allocate water for the environment only as an annual volume since methods for estimating environmental flows were not standardized in the country. For this reason, The Water Agency (CONAGUA) issued the standard NMX-AA- 159-SCFI-2012 [2], to assess environmental flows needed both, at the strategic level in Integrated Water Resources Management (IWRM), or as part of the Environmental Impact Assessment (EIA) of large hydraulic projects. For over ten years, this standard was developed and finally published in September 2012 [3]. It explains different methods from hydrological to holistic approaches, with examples for the country. Its application will cover the urgent need to preserve water for ecosystems in watersheds with high ecological importance and low stress for water use. In this paper, an analysis of the environmental flow standard and examples of the suggested hydrological methods are presented. For its implementation, some steps are taking place, mainly establishing environmental water reserves and building capacities. In addition, environmental allocations are becoming a common practice for all water projects, as well as setting limits to hydrological alterations by hydroelectric dams. The standard promotes the use of technical integration tools to analyze the responses of ecosystems to changes in the flow regime and adaptive management under different scenarios of water use. Although the main steps have been taken, its implementation as mandatory rule will take time.

INTRODUCTION

Environmental, ecological or in stream flow (EF) is the amount of water that is kept flowing down in order to maintain the river in a desired environmental condition in terms of its spatial and timing distribution required to maintain the components, functions and aquatic ecosystems processes from which people obtain benefits ([4-10]. The concept has evolved to recognize the natural variability of flow regimes which consider: magnitude, duration, timing, frequency and rate of change associated to environmental services and ecosystems resilience [11].

Environmental flow assessment (EFA) is based on a scientific process to deal with basin features, probabilistic variability of natural flows and cross section hydromorphology models. These issues determine habitat connectivity and availability to be used by species at several stages of their life cycle. EFA is also a social process since water regulators and users define the level of stress that can be supported by a river in terms of their quantity, quality and ecosystem services, although not many of these decisions have been properly taken. Then, generalized flow standards can be used to allocate water for the environment [12] but, every river can have an environmental flow regime according to what people want from a river in terms of conservation and use [13].

EFA as an escalated set of methodologies is becoming an important toolbox to assess sequentially from rivers' natural flow regime to its ecological integrity (structure, function, processes and alterations). Methods have been classified as: hydrological, hydraulics, for habitat simulation and holistic [9,14-17]). Hydrological approaches are the basis to identify the natural flow regime and its alterations caused by water subtractions or hydraulic infrastructure [11,18]. Results are complemented with hydraulic measurements and associated to habitat availability and connectivity among aquatic, riparian and coastal communities. In addition, holistic approaches consider environmental services for people and comparison of water allocation strategies or altered scenarios by infrastructure [19]. Due to the condition of many rivers which are over-allocated, fragmented, polluted and losing biodiversity, it is recognized that EFA is needed as a part of approaches both at strategic level in the integrated water resource management IWRM and as a part of EIA for large hydraulic projects [20-22]

The EF implementation is on progress and overcoming several drags, since hydrological records are sometimes the only available data for many regions, whereas hydraulic studies are really scarce and its development is time and cost consuming [9]. Besides, these methods need to broaden their scope towards habitat availability for species, not only as hydraulics research. This approach is under development since several disciplines are involved and need to formulate flow alteration—habitat availability—ecological responses hypothesis to be tested through different indicators. The final goal is to deal with an adaptive approach both to operate hydraulic infrastructure and produce conditions for preserving species, maybe in the year by year basis. Thus the final output is to reduce the effects of anthropogenic activities on aquatic ecosystems. Changes can be followed by hydrological parameters, habitat availability, species presence—absence, condition factor, weight—length relationship and food webs, among others [23]

EFA MEXICAN APPROACH

The "environmental water use or use for ecological conservation" is defined by the National Water Law as "the minimum flow or volume required in receiving water bodies, including streams or reservoirs, or the minimum flow of natural discharge from one aquifer that must be maintained to protect environmental conditions and the ecological balance.

The specific definition of Ecological Flow in the Mexican Standard [2], which establishes the procedure to determine the environmental flow is: the quantity, quality, and flow variations or water levels required to preserve environmental services, components, functions, processes and the resilience of aquatic and terrestrial ecosystems. They depend on the hydrological, geomorphological, ecological and social processes. This implies that in addition to provide water for domestic, urban public, livestock and agriculture uses, it is possible to maintain flows from both runoff and aquifer outcropto preserverivers (perennial, intermittent and ephemeral), lenticwater bodies (lakes, ponds, and wetlands), and riparian ecosystems.

The need for ecological flows was a pendant issue since the Water Act publication in 1992 and re -emerged with the application of an Official National Standard [1], to set basins' water availability. This availability comprises a committed natural discharge as a fraction of the hydrogeological unit of natural discharge, committed as superficial water to diverse uses or to be preserved to prevent negative impacts on ecosystems or the migration of bad quality water into the aquifer. The estimated water availability is then published in each hydrometric gauge of the basin and as a total volume.

The Mexican Standard to estimate the EF is a technical guideline rather than a compulsory limit [2]. Its main scientific principles are: the recognition of natural hydrological regime and the gradient of the biological condition.

Therefore, any methodology is valid for the standard as long as it focuses on understanding the ecological significance of each component of the natural flow regime, and it generates proposals for its conservation or restoration in whole or partially, from the functional point of view.

Following a top-down initiative and based on the gradients of ecological importance and water scarcity, an environmental objective (EO) was assigned such as water availability is issued for each basin or sub-basin in the country (Figure 1). The Rio Verde sub-basins were numbered for further analysis.

The EF Mexican standard calls for the application of hydrological methods as the basis to reserve or allocate water for the environment. Therefore, according to the EO's, reference values (% of medium annual volume or flow) as well as the components of a seasonal flow regime are stated as follows:

1) Under a modified Tennant approach (García et al. [5]).
   a) Percentage of mean annual flow (MAF) with different seasonal percentages (Reference values).
   b) Percentage of monthly mean flows MMF.
   c) A base flow during dry season not less than the historical minimum monthly flow (MinF).
2) Under WWF Mexico approach.
   a) A total volume associated to an ordinary regime (TVOR), considering wet, medium and dry years.
   b) A total volume related to the frequency of floods' regime (TVFR) with a return period of 1, 1.5 and 5 years to be reproduced within 10 years return period according to EO's.
3) Under TNC approach [14].
   a) Intra and inter-annual natural variability.
   b) Thresholds for hydrological alterations on hydrological parameters with ecological importance (monthly flows, 1, 7, 30 and 90 days minimum and maximum flows).

These approaches can be applied at basins, sub-basins or river reaches as it is pointed out in Figure 2. At least 20 years of hydrological data are recommended for all these analysis.

The Mexican flow standard also describes some initial experiences in using habitat simulation with BBM building blocks methodology [24]. as holistic approach and recommends the use of habitat simulation methods like PHabSim or others.

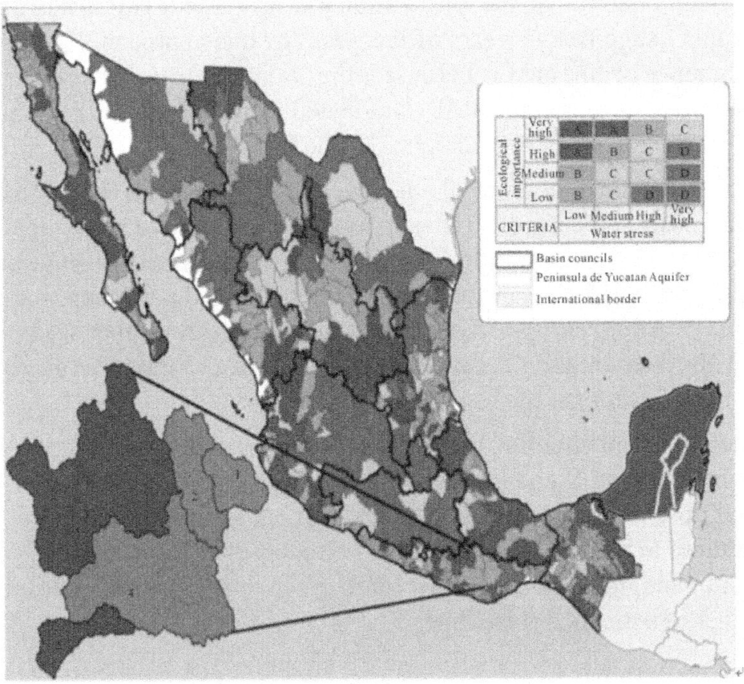

Figure 1: Environmental objectives for the Mexican basins and the Rio Verde Basin (Modified from WWF, [24]).

EFA Hydrological Methods Application

The Environmental objectives for the Rio Verde Basin and sub-basins located in the hydrological region 20 named Costa Chica de Guerrero were stated as is pointed out in Table 1.

The hydrological gauge Paso de la Reina was chosen to show result from the standard application (Figure 3) in this site a proposed hydro project is under planning studies.

Modified Tennant Approach ([5])

General steps:

1) Selection of the site of study.
2) Analysis of the monthly dataset.
3) Determination of monthly and annual flow regime.
4) Formulation of proposals for monthly and annual environmental flow regime.

As mentioned above, the selected site was the Paso de la Reina hydrometric gauge, this gauge has 45 years of records. For this approach, the start of the monthly series begins in May because is the month when the rain season begins. The monthly flow regime (MMF) was stated as well as the mean annual flow (MAF) as the reference started points (Figure 4).

According to environmental objectives (EO), Table 2 shows recommended seasonal flow percentages. Tennant modified by García et al. [5,26], and proposed by [25]. As Paso de la Reina site was classified as Environmental Objective "A" [2], then annual and monthly allocated percentages are shown in Figure 5. After setting percentages it must be reviewed that environmental flow in dry season should never be greater than the monthly average flow (MMF) nor lesser than the base flow.

The environmental flow proposal is based on monthly percentages (light green area in Figure 6 and values of Table 3). For dry years the proposal can be adjusted up to seasonal thresholds (30% and 60% of MAF blueline). For rain season these limits can be set between percentages of mean annual flow MAF (blue line) and monthly mean flow (dark green area). For each stream, a base flow must be estimated (blackline).

WWF Approach

According to the World Wildlife Fund Mexico approach a reference volume as percentage of the medium annual flow should be equal or greater than 40% (Table 4).

The next step is to identify alterations in the hydrological regime trough dividing time series data in two periods. For the first set of 20 years (natural hydrological regimen RHN) obtain percentiles 10 and 90 and check if the number of the monthly means of the second period (present hydrological regime RHA) that are within these limits. If the fulfillment of the actual regime (RHA) <50% is in relation to percentiles (RHN) the river is considered altered. The Figures 7 and 8 pointed out that the Rio Verde at Paso de la Reina hydrometric gauge is not altered.

For seasonal environmental flow estimation, the Total Ordinary Volume Regime (TOVR) and the Total Volume for Flood Regime (TVFR) were obtained using the procedure proposed by World Wildlife Fund Mexico at the study basin level. TOVR was obtained multiplying the frequency by volume for each year condition to be reproduced with regards to the environmental objective (Table 5).

The two steps to calculate TVFR that try to resemble natural pattern, according to the conservation objectives, were:

1) Classification and characterization of peaks: To separate flood types initially used the criterion of magnitude, maximum flow and probability of occurrence over time (frequency). Analysis of statistical distributions are needed to determine magnitude of floods associated return periods as follows:

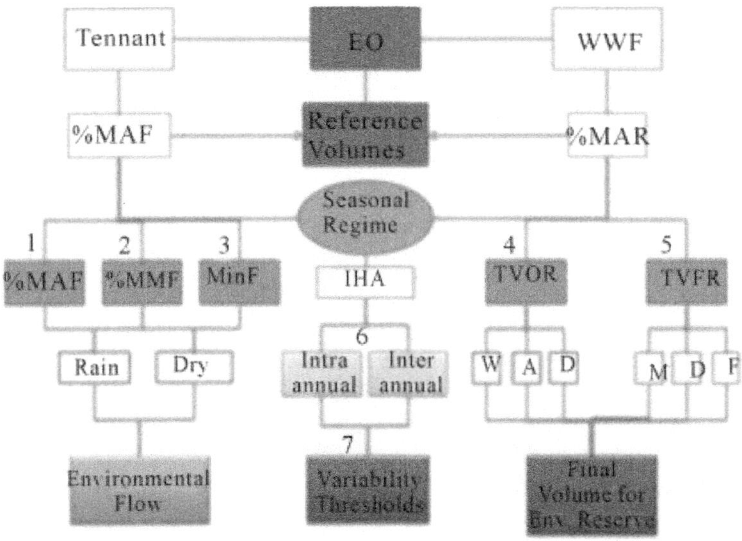

Figure 2: EFA hydrological approaches in the Mexican standard.

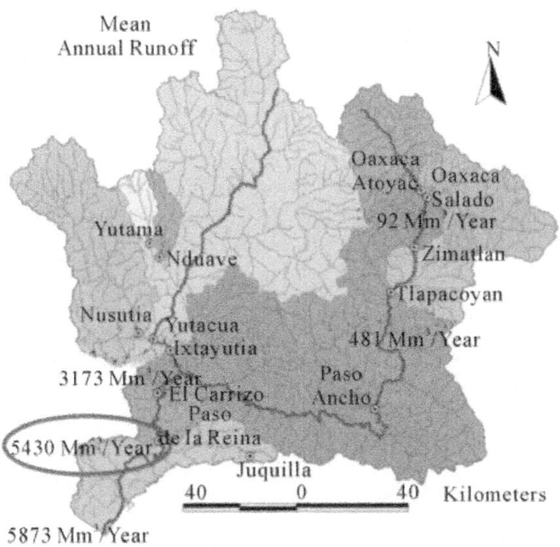

Figure 3: Water availability issued for the basin. CFR: 05/ 07/2013 [25].

Table 1: Environmental objectives by sub-basin

Rio Verde Sub-basins	1	2	3	4	5
Annual Availability (Mm³/year)	92.4	481.1	3173.8	5430.3	5873.3
Ecological Importance	High	High	High	Medium	High
Water use stress	Medium	Medium	Low	Low	Low
Present/Desired Condition	Good	Good	Very Good	Good	Very Good
Environmental Objective	B	B	A	B	A

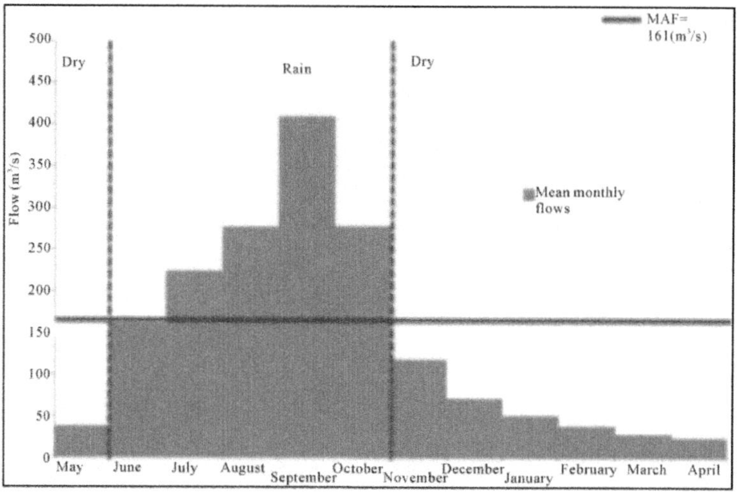

Figure 4: Monthly flow regime and mean annual flow that determines the dry and rainy seasons.

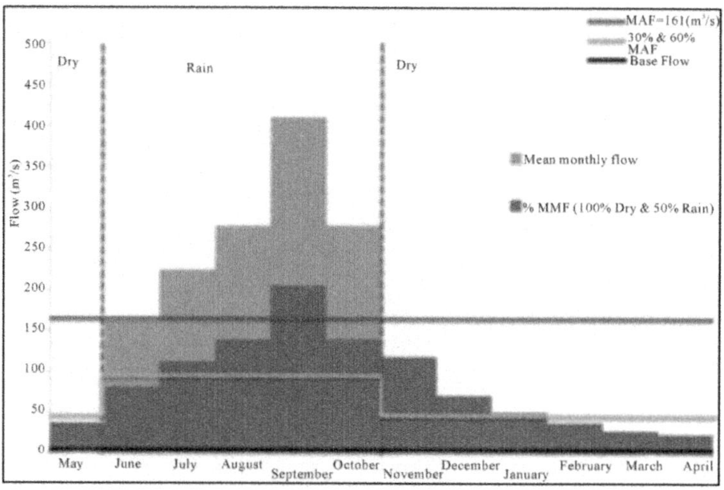

Figure 5: Annual and monthly allocated percentages.

a) Runoff with return period of 1 year (Category I).
b) Runoff with return period of 1.5 years (Category II).
c) Runoff with return period of 5 years (Category III).

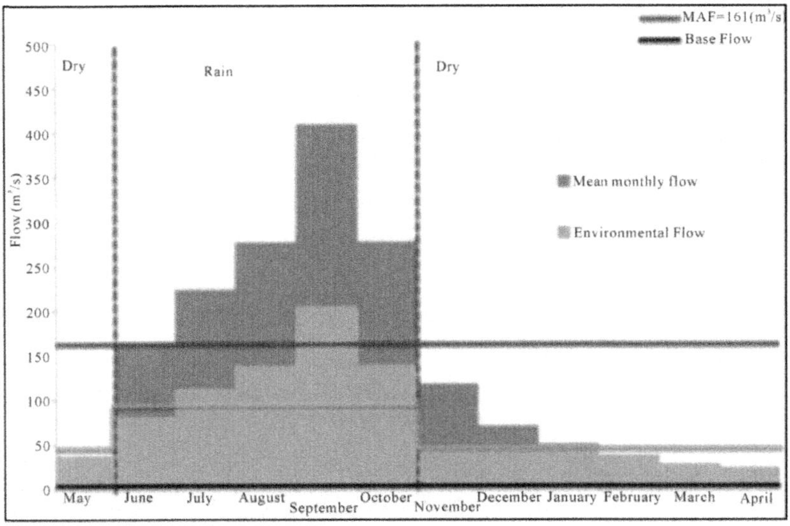

Figure 6: Environmental flow proposal.

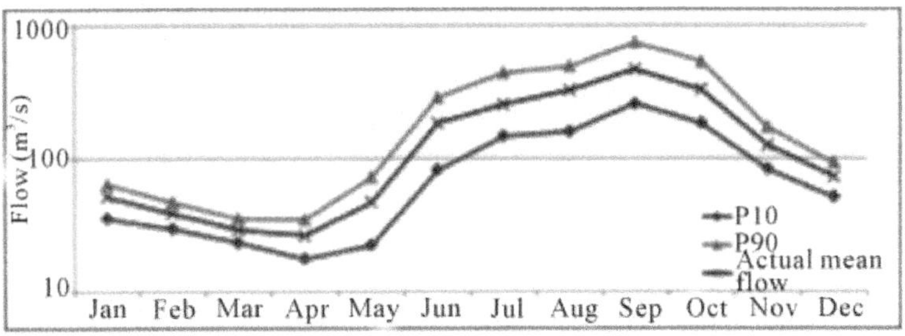

Figure 7: Monthly present mean and RHN limits.

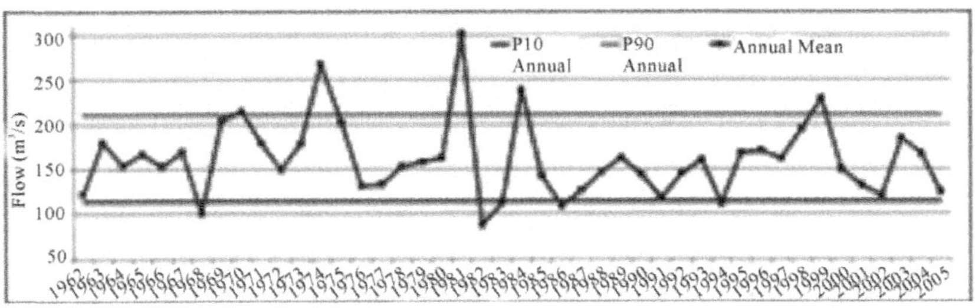

Figure 8: Annual present mean and RHN limits.

2) Duration in days of each runoff period is then obtained and the product of this duration, frequency and volume resulted in the Total Volume for Flood Regime for three different runoffs within 10 years. Table 6 shows the results for these calculations.

The integration of the final reserved volume (FRV) or environmental flow to be considered in the annual water balance of the basin and its water availability is the sum of total ordinary volume regime (TOVR) with the total volume of flood regime (TVFR), as follows:

Table 2: Reference values for each EO

Environmental Objective	Period			
	Dry		*Rain*	
	% MAF	*% MMF*	*% MAF*	*% MMF*
A	30	100	60	50
B	20	80	40	40
C	15	60	30	30
D	5	40	10	20

Table 3: Monthly environmental flow regime in cubic meters per second (m^3/s)

Month	Environmental Flow
January	48.3
February	37.8
March	28.3
April	24.2
May	37
June	80.4
July	111.9
August	138.7
September	204.8
October	139.2
November	48.3
December	48.3

Table 4: Reference values for specific environmental objectives

Environmental Objective	*Conservation Gradient*	*Environmental Flow (% MAF)* *Perennial streams*
A	Very good	>40 > 64.6 m^3/s or >2172 Mm^3/year

Table 5: Total Ordinary Volume Regime

Type of Year	Very dry	Dry	Average	Wet
Percentile	*P0*	*P10*	*P25*	*P75*
Ordinary volume regime (OVR Mm^3/year)	2052	2862	3611	6066
% Mean Annual Runoff	40	56	71	119
Frequency	0.2	0.3	0.4	0.1
Total Ordinary Volume Regime		3319 Mm^3/year		

FRV = TOVR + TVFR

FRV = 3319 + 198 = 3517 Mm^3/year

Table 6: Total Volume of Flood Regime

Attribute of hydrological regime		Category I	Category II	Category III
		1 year *return*	1.5 years *return*	5 years *return*
Magnitude	m^3/s	303	843	1874
	Mm^3/day	26	73	162
Frequency		10	6	2
Duration		3	2	1
Timing			Jul-Oct	
Rate of change	Rise		75	
	Fall		40	
Total Volume for Flood Regime in 10 Years			1980 Mm^3	
TVFR/Year			198 Mm^3	

MAR= 5430 Mm^3/year

FRV = 3517 Mm^3/Year = 64% of MAR

CONCLUSIONS

Although the Mexican Water Act states the environmental use and the Availability Standard points out an annual volume to the environment in each basin, for more than fifteen years, the minimum flow as much as 10% of monthly flows have prevailed to deal with environmental allocation of water for projects with consumptive uses (water supply and irrigation). Availability agreements have been issued and are going to be reviewed every two years. Therefore, implementing EFA is the next challenge to be included in these agreements and then gives some priority of reserving water for the environment [27].

The Mexican environmental flow standard as a technical procedure gives the support to move ahead from this 10% minimum approach and consider the natural flow variability. Through the recognition of this natural variability and applying at least hydrological methods, twofold protection goals can be achieved to: 1) Reproduce seasonal and even monthly variability and 2) Consider thresholds for alteration either seasonal, monthly or inter-annually when ecological process can be vulnerable.

The hydrological methods are also the basis for the regional water resources management under frameworks like the integrated watershed resource management IWRM and the ecological limits of hydrological alteration (ELOHA) [28]

The application of habitat simulation and holistic methods is under development in Mexico. Therefore, experiences at regional and basin level as well as for large infrastructure projects are an urgent need to protect the ecological integrity of ecosystems. These approaches should be integrative and under an interdisciplinary approach.

As a result of the application of habitat simulation or holistic methods, a set of ecological and social indicators can be obtained to follow during the EF implementation. For Mexico, the regional ecosystem changes and social responses will be valuable when the adaptive approach has to consider tradeoffs between ecological conservation and water use tightening further hydraulic development.

REFERENCES

1. NOM-011-CNA-2000, "Norma Oficial Mexicana, Conservación del Recurso Agua. Que Establece las Especificaciones y el Método para Disponibilidad Media Anual de las Aguas Nacionales," DOF 17-Abril-2002, México, p. 20.
2. NMX-AA-159-SCFI-2012, "Norma Mexicana. Que Establece el

Procedimiento para la Determinación del Caudal Ecológico en Cuencas Hidrológicas," DOF 20 de Septiembre de 2012, México, p. 123.

3. P. A. E. Lis, M. A. G. Balandra and P. S. Fabela, "Requerimientos para Implementar el Caudal Ambiental en México," Instituto Mexicano de Tecnología del Agua, Alianza World Wildlife Fund/Fundación Gonzalo Río Arronte. Programa Hidrológico Internacional UNESCO, SEMARNAT, 2007, p. 176.
4. T. Annear, I. Chisholm, H. Beecher and A. Locke, "Instream flows for Riverine Resource Steweardship," Revised Edition, Instream Flow Council, Cheyenne, 2004, p. 268.
5. E. García, R. González, P. Martínez, J. Athala and G. Paz-Soldán, "Guía de Aplicación de los Métodos de cálculo de Caudales de Reserva Ecológicos en México, Colección," CNA-IMTA-SEMARNAP, México D.F., 1999, p. 190.
6. S. Postel and B. Richter, "Rivers for Life: Managing Water for People and Nature," Island Press, Washington DC, 2003.
7. N. L. Poff, "Managing for Variability to Sustain Freshwater Ecosystems," Journal of Water Resources Planning and Management, Vol. 135, No. 1, 2009, pp. 1-4.
8. A. H. Arthington, S. E. Bunn, N. L. Poff and R. J. Naiman, "The Challenge of Providing Environmental Flow Rules to Sustain River Ecosystems," Ecological Applications: A Publication of the Ecological Society of America, Vol. 16, No. 4, 2006, pp. 1311-1138.
9. R. E. Tharme, "A Global Perspective on Environmental Flow Assessment: Emerging Trends in the Development and Application of Environmental Flow Methodologies for Rivers," River Research and Applications, Vol. 19, No. 5-6, 2003, pp. 397-441. http://dx.doi.org/10.1002/rra.736
10. M. Acreman and M. J. Dunbar, "Defining Environmental River Flow Requirements a Review," Hydrology and Earth System Sciences, Vol. 8, No. 5, 2004, pp. 861-876. http://dx.doi.org/10.5194/hess-8-861-2004
11. B. D. Richter, J. V. Baumgartner, R. Wigington and D. P. Braun, "How Much Water Does A River Need?" Freshwater Biology, Vol. 37, No. 1, 1997, pp. 231-249. http://dx.doi.org/10.1046/j.1365-2427.1997.00153.x
12. B. D. Richter, "Short Communication Re-Thinking Environmental Flows: From Allocations and Reserves to Sustainability Boundaries," River Research and Applications, Vol. 26, No. 8, 2010, pp. 1052-1063.
13. J. O'Keeffe and T. Le Quesne, "Keeping Rivers Alive— A Primer on Environmental Flows and Their Assessment," World Wildlife Found—

Water Security Series 2, Surrey, 2009.

14. TNC (The Nature Conservancy), "Indicators of Hydrologic Alteration," Version 7.1 User's Manual, 2009, p. 81.
15. WWF and Fundación Gonzalo Río Arronte, "Guía Para la Determinación de Caudal Ecológico en México Sistematización de Experiencias de la Alianza WWF," Fundación Gonzalo Río Arronte I.A.P., Mexico D.F., 2011.
16. M. J. DUNBAR and M. C. Acreman, "Applied Hydro-Ecological Science for the Twenty-First Century," In: M. C. Acreman, Ed., Hydro-Ecology: Linking Hydrology and Aquatic Ecology (Proceedings of an International Workshop (HW2). Centre for Ecology and Hydrology (Formerly Institute of Hydrology), Wallingford, Oxfordshire, 2001, pp. 1-18.
17. R. Hirji and R. Davis, "Environmental Flows in Water Resources Policies, Plans and Projects. Case Studies," Water Resources, No. 117. Washington DC, 2009, p. 181.
18. N. L. Poff, J. D. Allan, M. A. Palmer, D. D. Hart, B. D. Richter, A. H. Arthington, K. H. Rogers, J. L. Meyer and J. A. Stanford, "River Flows and Water Wars: Emerging Science for Environmental Decision Making," Frontiers in Ecology and the Environment, Vol. 1, No. 6, 2003, pp. 298-306. http://dx.doi.org/10.1890/1540-9295(2003)0010298:RFAWWE.2.0.CO;2
19. A. H. Arthington, R. J. Naiman, M. E. McClain and C. Nilsson, "Preserving the Biodiversity and Ecological Services of Rivers: New Challenges and Research Opportunities," Freshwater Biology, Vol. 55, No. 1, 2009, pp. 1-16. http://dx.doi.org/10.1111/j.1365-2427.2009.02340.x
20. C. Nilsson and B. M. Renöfält, "Linking Flow Regime and Water Quality in Rivers: A Challenge to Adaptive Catchment Management," Ecology and Society, Vol. 13, No. 2, 2008, 20 p.
21. C. J. Vörösmarty, P. B. McIntyre, M. O. Gessner, D. Dudgeon, A. Prusevich, P. Green, S. Glidden, S. E. Bunn, C. A. Sullivan, C. R. Liermann and P. M. Davies, "Global Threats to Human Water Security and River Biodiversity," Nature, Vol. 467, 2010, pp. 555-561. http://dx.doi.org/10.1038/nature09440
22. The Brisbane Declaration, The Brisbane Declaration at 10th International Riversymposium & Environmental Flows Conference, Brisbane, 2007.
23. H. Chen, L. Ma, W. Guo, Y. Yang, T. Guo and C. Feng, "Linking Water Quality and Quantity in Environmental Flow Assessment in Deteriorated Ecosystems: A Food Web View," Plos One, Vol. 8, 2013, No. 7, p. 13.

24. WWF, Fundación Río Arronte, and CONAGUA, "Norma Mexicana de Caudal Ecológico—Una PolíTica Pública Para la Gestión del Agua a Través de la Conservación del Régimen Hidrológico," Mexico D.F., 2012.

25. CONAGUA, "Identificación de Reservas Potenciales de agua Para el Medio Ambiente en México," Comisión Nacional del Agua, Alianza WWF-Fundación Gonzalo Río Arronte I.A.P, Primera Ed. Mexico, 2011, p. 87.

26. D. L. Tennant, "Instream Flow Regimens for Fish, Wildlife, Recreation and Related Environmental Resources," Fisheries, Vol. 1, No. 4, 1976, pp. 6-10. http://dx.doi.org/10.1577/1548-8446(1976)0012.0.CO;2

27. J. King, R. Tharme and M. de Villier, "Environmental Flow Assessments for Rivers: Manual for the Building Block Methodology," Water Research Commission, SAF, Cape Town, 2008.

28. N. L. Poff, B. D. Richter, A. H. Arthington, S. E. Bunn, R. J. Naiman, E. Kendy, M. Acreman, C. Apse, B. P. Bledsoe, M. C. Freeman, J. Henriksen, R. B. Jacobson, J. G. Kennen, D. M. Merritt, J. H. O'Keeffe, J. D. Olden, K. Rogers, R. E. Tharme and A. Warner, "The Ecological Limits of Hydrologic Alteration (ELOHA): A New Framework for Developing Regional Environmental Flow Standards," Freshwater Biology, Vol. 55, No. 1, 2010, pp. 147-170. http://dx.doi.org/10.1111/j.1365-2427.2009.02204.x

Chapter 5

EVALUATION OF SOIL HYDRAULIC PARAMETERS IN SOILS AND LAND USE CHANGE

Fereshte Haghighi[1], Mirmasoud Kheirkhah[2] and Bahram Saghafian[2]

[1]Soil Conservation and Watershed Management Institute, Tehran

[2]Soil Conservation and Watershed Management Research Institute, Tehran, Iran

INTRODUCTION

The knowledge of soil water properties and land-use effects on these properties are important for efficient soil and water management. Furthermore, the use of the pedo transfer functions (PTFs) to estimate soil water content (θ_h) is important to assess. The loosening effect of dryland farming on soil water retention is known. In this chapter we review soil water content, pedotransfer functions and some infiltration models applicability for two land-use types. The land-use effect on soil water retention may be significant at water potentials of -33 kPa and 0 kPa in the soil. At the -1500 kPa pressure head, water content may not be affected by cultivation of rangeland at different soil depths. In addition, pedotransfer functions can be used as a physically based model for soil water retention characterization in the various areas. Moreover, it is essential to evaluate the infiltration models applicability for different soils and various land-uses.

Definition

Soil hydraulic properties

Soil hydraulic properties govern transport processes and water balance in soils. Water retention capacity, infiltration rate, and saturated hydraulic conductivity are important soil hydraulic properties. Soil water retention and saturated hydraulic conductivity (K_s) are necessary input data for the simulations of water flow in soil and water engineering. Characterizing hydrological behavior of catchments requires knowledge of hydraulic parameters.

Soil Water Retention

Soil water retention at field capacity (FC) and permanent wilting point (PWP) are used to estimate the water depth applied by irrigation (Hansen et al., 1980), and to calculate water availability, as a crucial factor to assess the land area suitability for crop producing (Sys et al., 1991).

SOIL WATER RETENTION CAPACITY AND LAND USE

One important soil hydraulic property is water retention capacity, which affects soil productivity and management. Soil water content (θ_h) governs the transport characteristics of water and solutes in soils. The knowledge of water retention capacity and land use effects on this property is important for efficient soil and water management. Upon conversion of natural lands to cultivated fields, water retention capacity is strongly influenced (Schwartz et al., 2000; Bormann and Klaassen, 2008; Zhou et al., 2008). Soil water retention at field capacity (FC) and permanent wilting point (PWP) are important to estimate the irrigation water depth which may be affected by land use change. Soil water retention characteristic, is affected by soil organic matter (SOM) content and porosity, which are significantly influenced by land use type (Zhou et al., 2008).

We conducted a study to evaluate, document, and quantify the effect of cultivation of rangeland on soil water retention in field capacity (FC), permanent wilting point (PWP), and to test the use of the van Genuchten equation to estimate θ_h in cultivated and natural lands in the same soils of the Taleghan watershed in Iran.

Significant differences in the OM and bulk density (BD) were observed between dryland farming and rangeland at both depths of 0 cm - 15 cm and 15 cm - 30 cm. Soil sample water contents at different pressure heads under both land use types are presented in Figures 1. The overall measured and fitted soil water retention curves did not show significant difference within the selected water potentials for both land use types in this study. However, measured θ_s (0 kPa) values were found to be significantly lower for dryland farming when compared with rangeland at depths of 0 cm - 15 cm and 15 cm - 30 cm, respectively. Moreover, the land use effect on soil water retention was significant at a water potential of -33 kPa (FC) based on laboratory measurements only at the top (15 cm depth). The results indicated that the conversion of rangeland to dryland farming led to a significant decrease (16.56% on average) in the FC at a depth of 0 cm - 15 cm. The mean -1500 kPa (PWP) water content was not affected by the land-use type. Figure 1 indicates that the mean total field capacity (FC) was significantly greater in rangeland when compared with dryland farming at a depth of 0 cm - 15 cm. In this study, there were not statistically significant differences in water content at other potentials (-50 kPa, -100 kPa, -500 kPa, and -1000 kPa pressures) between the two types of land use presented in Figure 1. At those pressure heads and at a -1500 kPa water content, the amount of micropores were not affected by cultivation of rangelands (Fig. 1). Overall, the results showed that the soil pore system and reduced total porosity under dryland farming can decrease water storage capacity at water potentials of -33 kPa and 0 kPa. Ndiaye et al., (2007) has shown that improper soil management decreases the soil macroporosity in the long-term affecting the θ_s. The data obtained in our study demonstrated the loosening effect of dryland farming on soil water retention. Previous studies on the effect of land use have demonstrated clear changes in soil physical properties, such as soil porosity, SOM, and BD, in relation to hydraulic properties (Bormann and Klassen 2008; Haghighi et al., 2010b).

PEDOTRANSFER FUNCTIONS (PTFS)

Determination of soil water properties required as input data for simulation models is time consuming and relatively costly (Wösten et al., 1995). Thus, indirect estimation of these characteristics has been proposed as one alternative to direct estimation of the soil hydraulic parameters based on the measured water retention data. Pedotransfer functions (PTFs) are emerged as the relationship between soil hydraulic and other more available measured properties (Bouma, 1989) which can be used to estimate hydraulic parameters. PTFs are useful tools for modeling applications.

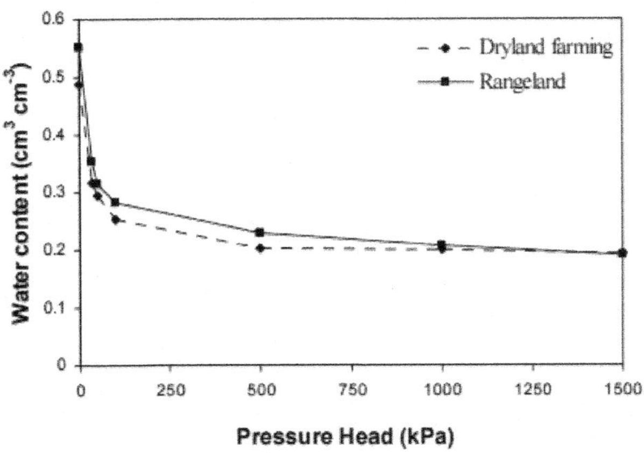

(a)

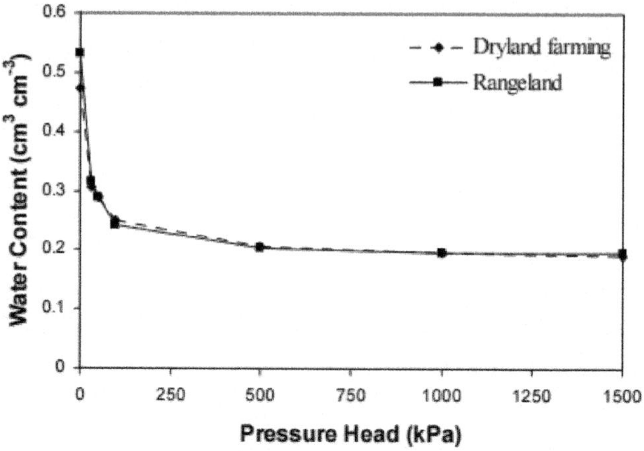

(b)

Figure 1: Soil water content as a function of the pressure head for two landuse types at depths of (a) 0 cm -15 cm and (b) 15 cm - 30 cm.

PEDOTRANSFER FUNCTIONS (PTFS) AND DIFFERENT LAND USES

To estimate the land use effects on soil water retention, van Genuchten model (Van Genuchten, 1980) may be applied. Some researches have correlated van Genuchten parameters with soil organic matter, bulk density (BD), and soil particle size distribution and many researchers have estimated the water retention curve using soil texture, bulk density, and porosity.

Many statistical equations (pedotransfer functions) characterizing the water retention curve have been presented (Kutilek and Nielsen., 1994). PTFs are useful tools for modeling applications. Such analytical functions are derived involving various soil data. Such data are measured in the field and laboratory analysis. Soil hydraulic parameters derived through PTFs can be used to express soil hydraulic properties and water retention (Brooks and Corey, 1964). Consequently, physically based models such as van Genuchten representing a pedotransfer function may be considered as a valuable tool to simulate the soil water properties in different land uses.

The $\theta_{(h)}$ data may be fitted to van Genuchten equation to derive retention curves and parameters (a, n, and θ_r), using the RETC (RETention curve) optimization computer code (Van Genuchten et al., 1991). The van Genuchten model (van Genuchten, 1980) is defined as:

$$\theta(h) = \theta_r + \frac{\theta_s - \theta_r}{(1+|\alpha h|^n)^m} \qquad \theta_{(h)} = \theta_s \qquad h \geq 0 \tag{1}$$

Where $\theta_{(h)}$ ($cm^3\ cm^{-3}$) is the volumetric water content (for h<0), θ_r ($cm^3\ cm^{-3}$) is the residual water content, and θ_s ($cm^3\ cm^{-3}$) is the saturated water content. Here, m is $1-(1/n)$ with $n>1$. a (cm^{-1}) and n are empirical parameters determining the shape of the curve which were obtained for each core. Parameter n is related to steepness of the water retention curve.

$$K(S_e) = K_s S_e [1-(1-S_e^{(1/m)})^m] \tag{2}$$

Where K_s (mm/h) is saturated hydraulic conductivity and S_e is the effective saturation expressed as:

$$S_e = \frac{\theta - \theta_r}{\theta_s - \theta_r} \tag{3}$$

The effect of landuse type on soil water retention and PTF applications have not been documented for different land-uses to the best of our knowledge. In developing countries, there is a lack of large databases that are needed to develop PTFs. Thus, in many developing countries, the use of available PTFs can cause errors for estimating soil hydraulic properties. This encourages further investigations of the model applications and development of suitable point and parametric PTFs for estimating soil hydraulic properties in the studied area. The selection of more suitable PTFs for application where there are not developed PTFs caused by a lack of large databases is difficult. Consequently, it is essential to evaluate the model applicability and to develop point and parametric PTFs for estimating soil hydraulic properties for the soils in various sites. Thus, the estimates may be improved by comprehensive local studies.

Table 1: Mualem-van Genuchten parameters calculated for old grassland (site A), recently reseeded grassland (site B) and previous maize cultivated land (site C) (Sonneveld et al, 2003)

Location (site)	Depth: 10–20 cm					Depth: 25–50 cm				
	K_{sat} (cm/day)	θ_{sat} (–)	α (1/cm)	l (–)	n (–)	K_{sat} (cm/day)	θ_{sat} (–)	α (1/cm)	l (–)	n (–)
A	16.0	0.43	0.007	0.944	1.37	23.1	0.43	0.014	−0.350	1.41
B	16.0	0.42	0.007	0.331	1.36	15.8	0.40	0.009	−0.448	1.33
C	17.4	0.40	0.008	0.093	1.35	14.7	0.38	0.010	−0.482	1.33

EVALUATION OF COMMON INFILTRATION MODELS FOR DIFFERENT LAND-USES

The evaluation of infiltration characteristics as a hydrologic process in soils is necessary in agricultural studies. The knowledge of final steady infiltration rate is important for irrigation water efficiency, designing desirable irrigation systems, and loss of water. Thus, infiltration rate is important factor in sustainable agriculture, effective watershed management, surface runoff, and retaining water and soil resources. Since measuring the final infiltration rate is time consuming, several physical and empirical models have proposed to determine it. The empirical models such as Kostiakov (1932) and Horton (1940), and physical model such as Philip (1957) are the most common models to estimate infiltration rate of the soils.

Kostiakov-Lewis model

The model of Kostiakov modified for long times as follows:

$$f = at^{-b} + f_c \quad (4)$$

Where a and b are the equation's parameters ($a>0$ and $0<b<1$) i_c is the steady infiltration rate (LT^{-1}).

Horton's Model

The Horton's infiltration model (Horton, 1940) is expressed as follows:

$$f = (f_0 - f_c)e^{-kt} + f_c \quad (5)$$

Where i_c is the presumed final infiltration rate (LT^{-1}), i_0 is the initial infiltration rate (LT^{-1}) and t is time (T). k is the infiltration decay factor.

Philip Two-parameter Model

The Philip two-term model is expressed as (Philip, 1957):

$$f = \frac{1}{2}St^{-0.5} + A \quad (6)$$

Where f is the infiltration rate (LT^{-1}) as a function of time.

A= Transmissivity factor (LT^{-1}) as a function of soil properties and water contents,

S = Sorptivity that is function of soil matric suction ($LT^{-0.5}$).

t= time (T)

Singh (1992) expressed that the various models can estimate different values of the final infiltration rate in a soil which seems to be uncorrect, because of the final infiltration rate is a soil-dependent factor. Compared to the previous investigations on soil infiltration properties and models, studies on soil infiltration modelling depending on land use are scarce. Nevertheless, it can be assumed that landuse type have a significant impact on soil infiltration and infiltration models performance. Machiwal et al (2006) observed the infiltration process was well described by the Philip's model in a wasteland of Kharagpur, India. However, different soil management that influences the final infiltration rate is a major reason for different applicability of these models. Long-term effects of land use changes on soil infiltration and infiltration models (e.g., Horton, Kostiakov, and Philip models) can be observed (Navar and Synnott, 2000; Shukla et al, 2003).

Thus, the variability of soil infiltration characteristics and goodness of fit of the infiltration models for different land-uses should be considered during infiltration modelling studies helping on correct predictions of final infiltration for different land uses. Ability of these models for estimating the infiltration rate in different land-uses and soil management has been examined by some researchers. Gifford (1976) observed among the Horton, Kostiakov and Philip's models, the Horton's model was the best model to fit the infiltration data in mostly semi-arid rangelands from the Australia, but only under specific conditions. Shukla et al (2003) evaluated some of the infiltration models at different soil management and landuse systems in Ohio and observed among infiltration models, the Swartzendruber model was the best ones and fitted the observed infiltration data with lower sum of squares and higher model efficiency. Davidoff and Selim (1986) examined the goodness of fit for eight infiltration models on a Norwood soil with four winter cover crop treatments and results of their study showed that the Philip, Kostiakov and Horton's models had best predictions than the other models. Haghighi et al (2010b) evaluated the effects of rangeland and dryland farming land uses on performance of some infiltration models to estimate the final infiltration rate of soils. The study was conducted on some soils of Taleghan watershed, Iran. According to reports (Taleghan watershed study report, 1993), investigated soils are calcareous and classified as Typic Xerorthents. Mean annual rainfall alters from 464 to 796 mm and lands slope is by 15 %. The soil texture varied from clay-loam to silty clay loam.

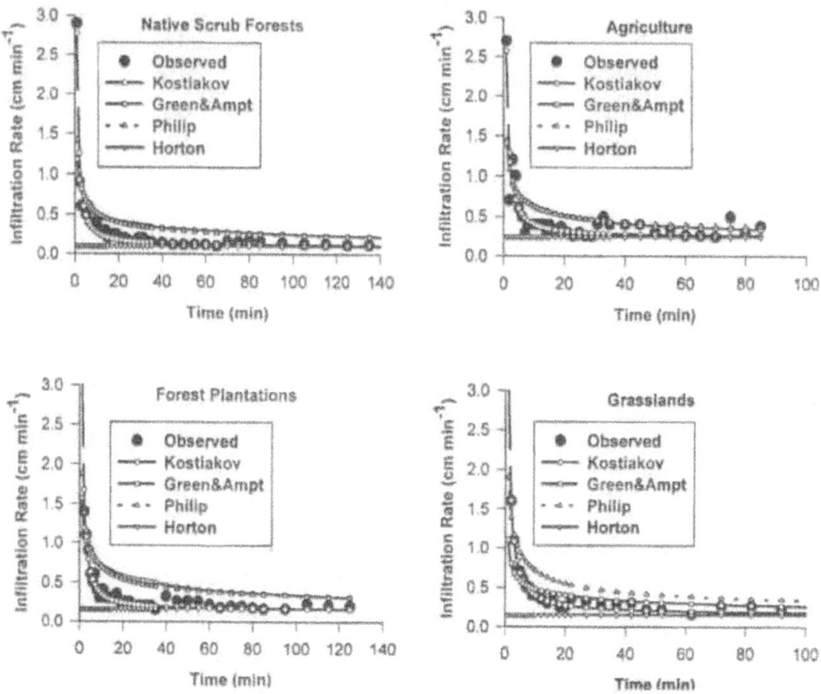

Figure 2: Effects of land use changes on soil infiltration and infiltration models (Navar and Synnott, 2000).

In our study, the goodness of the fit of selected models and ability of them for estimating the final infiltration rate of rangeland and dryland farming soils was evaluated using the root mean squared errors (RMSE). The values of R^2 were determined high (0.99) and equal for all sites and land-uses, but the values of RMSE and the final steady infiltration showed that the estimated infiltration rates by the infiltration model of Horton, approached more closely to the measured ones at the selected area

Table 1.. The Horton's model was the best model selected for both of land-uses. It can be expressed that various models can suppose different final infiltration (f_c) values for a soil, which seems to be not practical, because fc is a soildependent parameter, in general. Common changes in land-use negatively affect soil physical properties and decrease soil infiltration rate and could change modelling performance. Effect of land-use should be well documented aiming on good predictions in the studied areas and elsewhere.

The infiltration models can be used for estimating the infiltration rate in soils, well. But only, one or some of these models are better and appropriate for a specific site. Thus, the infiltration models should be analyzed for their ability to estimate the infiltration rate of each location. The investigation of Haghighi et al (2010a) showed that the Horton's model is the best ones selected for rangeland and dryland farming and land-use type is not an important factor to affect infiltration models efficiency. Due to a few number of investigation in this field of research, there is a need for further investigation on land-use effect on infiltration modelling and for the impact of land-use on soil infiltration characteristics, as well.

Table 2: Parameters of the selected infiltration models in both of land use types (Haghighi et al, 2010a)

Land use	Kostiakov-Lewis model			Philip two-term model		Horton's model			Observed final infiltration rate (cm min^{-1})
	c	b	a	S	A	β	f_0	f_c	
Rangeland 1	0.2052	0.543	1.136	1.217	0.223	0.130	0.894	0.299	0.2803
Rangeland 2	0	0.770	1.058	1.322	0.235	0.021	0.596	0.190	0.2441
Rangeland 3	0.0870	0.606	0.839	0.970	0.127	0.090	0.556	0.183	0.1613
Rangeland 4	5.13×10^{-14}	0.855	0.5076	0.543	0.208	0.045	0.387	0.225	0.2285
Dryland farming 1	0.0585	0.636	1.475	1.741	0.162	0.085	0.909	0.258	0.2347
Dryland farming 2	0	0.522	2.863	2.984	0.018	0.120	1.601	0.200	0.1863
Dryland farming 3	1.37×10^{-12}	0.781	0.4857	0.592	0.119	0.029	0.290	0.120	0.1098
Dryland farming 4	0.0452	0.681	0.2555	0.315	0.073	0.071	0.198	0.089	0.0869

CONCLUSIONS

Soil management and land use change may affect soil water retention at a -33 kPa (FC) potential in the soil based on laboratory measurements and model simulations. Lower water content at the -33 kPa potential would be expected upon conversion of natural lands to cultivated lands. In addition, the saturated soil water content (θ_s) may be affected by cultivation of rangeland. Moreover, because cultivation of natural lands affects soil macroporosity, we suggest measuring soil water retention at higher suction heads to document the land use effect on soil water retention properties in relation to soil macropores. Appropriate technology for dryland farming and suitable measures are necessary to improve soil water retention where cropping is required.

The findings show that the van Genuchten model is useful in describing soil water retention. Thus, use of this model may be considered as a valuable tool to gain more knowledge of hydraulic properties for various soil types. The effect of land use type on soil water retention and PTF applications have not been documented for dryland farming to the best of our knowledge. In many developing countries, such as Iran, the use of available PTFs can cause errors for estimating soil hydraulic properties. This review encourages further investigations of the model applications and development of suitable point and parametric PTFs for estimating soil hydraulic properties. The selection of more suitable PTFs for application where there are not developed PTFs caused by a lack of large databases is difficult. Consequently, it is essential to evaluate the model applicability and to develop point and parametric PTFs for estimating soil hydraulic properties for different land uses.

REFERENCES

1. Bormann, H., Klaassen, K., 2008. "Seasonal and land use dependent variability of soil hydraulic and soil hydrological properties of two Northern German soils." Geoderma. 145, 295-302.
2. Brooks, R.H., Corey, A.T., 1964. "Hydraulic properties of porous media." Civil Engineering Dept., Colorado State University, Fort Collins, CO.
3. Davidoff, B. and Selim, H.M., 1986. Goodness of fit for eight water infiltration models, Soil Sci. Soc. Am. J. 50, 759-764
4. Gifford, GF. 1976. Applicability of some infiltration formulae to rangeland infiltrometer data. Journal of Hydrology. 28, 1-11.
5. Haghighi F, Gorji M, Shorafa M, Sarmadian F, Mohammadi MH. 2010. "Evaluation of some infiltration models and hydraulic parameters." Spanish Journal of Agricultural Research (INIA), 8 (1), 210-217.
6. Haghighi F, Gorji M, Shorafa M. 2010. "A study of the effects of land use change on soil physical properties and organic matter." Land Degradation & Development Journal. In press 10.1002/ldr.999.
7. Horton, R.E. 1940. An approach toward a physical interpretation of infiltration-capacity. Soil Sci. Soc. Am. Proc.5, 339-417.
8. Kostiakov, A. N. 1932. On the dynamics of the coefficient of water percolation in soils and on the necessity of studying it from a dynamic point of view for the purposes of amelioration.
9. Kutílek M. 2004. Soil hydraulic properties as related to soil structure, Soil Till. Re 79, 175–184.
10. Machiwal, D., K.J.., and B.C. Mal. 2006. Modelling infiltration and quantifying spatial soil variability in a watershed of Kharagpur, India. Biosystems Engineering, 95(4), 569-582.
11. Mishra, s.k, J.V.Tyagi, and V.P.Singh. 2003. Comparison of infiltration models. Hydrological processes. 17, 2629-2652 .
12. Navar.J, and T.J. Synnote, 2000. Soil infiltration and land use in Linares, N.L., Mexico, Terra Latinoamericana. 18 (3), 255-262.
13. Ndiaye, B., Molenat, J., Hallaire, V., Hamon, C.G.Y., 2007. "Effects of agricultural practices on hydraulic properties and water movement in

soils in Brittany (France)". J. Soil & Tillage Res. 93, 251-263.

14. Philip, J.R. 1957. The theory of infiltration: 4. Sorptivity and algebric infiltration equations. Soil Sci. 84, 257-264.
15. Schwartz, R.C., Unger, P.W., Evett S.R., 2000. "Land use effects on soil hydraulic properties." ISTRO.
16. Shukla, M.K., R.Lal., and P.Unkefer. 2003. Experimental evaluation of infiltration models for different land use and soil management systems. Soil Sci. 168 (3), 178-191.
17. Singh V.P. 1992. Elementary Hydrology. Prentice Hall: Englewood Cliffs, NJ.
18. Sonneveld MPW., Bachx MAHM., Bouma.J. 2003. Simulation of soil water regimes including pedotransfer functions and land use related preferential flow. Geoderma, 112, p: 97-110.
19. University of Tehran. "Taleghan Watershed study report"., 1993. Irrigation Engineering Department, Iran.
20. Van Genuchten, M.Th., 1980. "A closed-form equation for predicting the hydraulic conductivity of unsaturated soils." Soil Sci. Soc. Am. J. 44, 892–898.
21. Van Genuchten, M.Th., Leij, F. J., Yates, S.R., 1991. "The RETC code for quantifying the hydraulic functions of unsaturated soils." EPA/600/2-91/065. R. S. Kerr, Environmental Research Laboratory. U. S. Environmental Protection Agency, Ada, OK. p: 83.
22. Wösten, J.H.M., Finke, P.A., Jansen, M.J.W., 1995. "Comparison of class and continuous pedotransfer functions to generate soil hydraulic characteristics." Geoderma. 66, 227–237.
23. Zhou, X., Lin, H.S., White, E.A., 2008. "Surface soil hydraulic properties in four soil series under different land uses and their temporal changes." Catena. 73, 180-188.

Chapter 6

ENERGY-RELATED CARBON EMISSIONS OF CHINA'S MODEL ENVIRONMENTAL CITIES

Kevin Lo
Department of Resource Management and Geography, University of Melbourne, Melbourne, VIC 3010, Australia

ABSTRACT

This paper identifies three types of model environmental cities in China and examines their levels of energy-related carbon emissions using a bottom-up accounting system. Model environmental cities are identified as those that have been recently awarded official recognition from the central government for their efforts in environmental protection. The findings show that, on average, the Low-Carbon Cities have lower annual carbon emissions, carbon intensities, and per capita emissions than the Eco-Garden Cities and the Environmental Protection Cities. Compared internationally, the Eco-Garden Cities and the Environmental Protection Cities have per capita emissions that are similar to those of American cities whereas per capita emissions from the Low-Carbon Cities are similar to those of European cities. The result indicates that addressing climate change is not a priority for some model environmental cities. Policy changes are needed to prioritize climate mitigation in these cities, considering that climate change is a cross-cutting environmental issue with wide-ranging impact.

INTRODUCTION

Climate change is one of the most significant challenges facing the world today and cities are increasingly seen as key contributors to the issue. Urban activities, such as motorized transport, industrial production, electricity generation, domestic fuel use, and waste disposal generate significant amounts of greenhouse gases [1–3]. The International Energy Agency (IEA) forecasted that, by 2030, urban areas will account for 76% of global carbon emissions [4].

Thus, to prevent dangerous climate change, cities must bear some responsibility for reducing their impact on climate change and formulating effective responses [1, 5]. Inventorying urban carbon emissions is an important first step to address climate change effectively and fairly. Creating emissions inventories at the city level allows policymakers to identify the sources, establish baselines, monitor changes over time, make cross-comparisons with other localities, set appropriate emissions reduction targets, and formulate appropriate solutions [1]. The importance of inventorying urban carbon emissions is underscored by the large number of studies dedicated to such task [6–12].

China is critically important to the global effort of addressing climate change because of its enormous size of emissions [13]. Since 2006, the government has stepped up measures to promote energy conservation and renewable energy [14, 15]. Yet virtually no cities in China openly publish their greenhouse gases inventory on a regular basis. A number of studies have attempted to fill this gap. The calculation by Li et al. [16] showed that Shanghai's energy-related carbon emissions increased from 110 million tCO_2e in 1995 to 180 million tCO_2e in 2006. Wang et al. [17], also studying Shanghai, found that the city's total carbon emissions during 2000–2008 increased from 136 teragrams of CO_2e to 200 teragrams of CO_2e. In 2008, per capita emissions in Shanghai reached 14.03 tCO_2e, which were higher than both the world average and the average in China. Bi et al. [18] calculated that, in 2009, the total carbon emissions in Nanjing reached 75.43 million tCO_2e and per capital emissions reached 9.78 tCO_2e. Inventorying 12 Chinese cities, Wang et al. [19] found that carbon emissions in all cities rose from 2004 to 2008, but per capita emissions varied widely in different cities. Although China had lower-than-global-average per capita emissions (5.5 tCO_2e), the per capita emissions of many Chinese cities were greater than 8.0 tCO_2e, which were comparable to or even higher than those of many other cities in developed countries. Dhakal [20] approximated the energy-related emissions from 35 major cities using provincial carbon intensity data. The results showed that there were large differences in terms of per capita emissions within these cities, with the carbon-intensive cities largely located in the central and western parts of China, an area that attracted energy-intensive industries with low energy prices and favourable policies [21].

Building on these previous findings, this study aims to examine the levels of energy-related carbon emissions from China's model environmental cities, identified as those that have been recently awarded official recognition from the central government for their efforts in environmental protection. These cities are regarded as the leaders of environmental protection and sustainable development. However, concrete evidence on the environmental qualities of these cities in the present era of climate change has been lacking. As such, one may wonder whether and to what extent these cities are really environmentally friendly from the perspective of low-carbon urbanism. Following this introduction, this paper identifies three types of modelenvironmental cities. It then estimates the carbon emissions from these cities using annual emissions, annual per capita emissions, and carbon intensity. The paper concludes with a discussion of the policy implications of the results.

THREE TYPES OF MODEL ENVIRONMENTAL CITIES

This paper defines model environmental cities as cities that have received official recognition from the central government for their efforts in environmental protection. Although many cities in China declare their environmental credentials, cities that received official recognition have to go through quality assurance system and meet certain specified environmental standards. Therefore, they are considered as the leaders in China with respect to environmental issues. There are three types of model environmental cities: Eco-Garden Cities, Environmental Protection Cities, and Low-Carbon Cities. The remaining portion of this section will examine these different types of model environmental cities more closely.

Eco-Garden Cities

The Eco-Garden City Program was established in 2004 by the Ministry of Construction (now the Ministry of Housing and Urban-Rural Development [MOHURD]). The aim of the program is to award recognition to cities that protect and improve the health of urban ecosystems. Eco-Garden Cities are required to work toward a number of objectives, including construction and maintenance of parkland, improving access for pedestrian and bicyclist, urban greening, improving public transport, increasing the density of developed areas, protection of ecosystems, increasing the proportion of energy-efficient buildings, and increasing the deployment of renewable energy. The MOHURD developed a complicated system of 90 quantitative targets to guide local governments toward the achievement of these objectives. For example, one of the energy and climate targets is to achieve at least 60% coverage by public transportation. Because not every target is compulsory, an Eco-Garden City may fail to meet one or more of the criteria.

A city aspiring to be an Eco-Garden City must first establish an action plan to achieve the specified targets and then must implement the plan for at least three years. The city is then evaluated, first by the Department of Housing and Urban-Rural Development at the provincial-level and then by the MOHURD. The assessment process involves independent surveys, remote-sensing analysis, field inspections, and expert evaluations. A city that passes the assessment earns the right to call itself an Eco-Garden City. However, an Eco-Garden City that is later found to be noncompliant with the requirements would be either cautioned or disqualified, depending on the degree, scale, and nature of the infringement.

Environmental Protection Cities

In 1997, the Environmental Protection Bureau (now the Ministry of Environmental Protection [MEP]) established the Environmental Protection City Program to encourage cities to improve environmental quality through sustainable development. Environmental Protection Cities are required to work toward a number of environmental objectives, including air pollution prevention, water pollution prevention, improving waste management, reducing industrial energy intensity, and increasing the deployment of renewable energy use. The MEP developed 22 quantitative targets for Environmental Protection Cities. All of the targets are compulsory, unlike those of the Eco-Garden City Program. There are two targets relevant to climate change. First, industrial energy intensity must be less than the national average. Second, more than 50% of the energy used must come from sources other than coal.

A city aspiring to be an Environmental Protection City must first prepare a proposal demonstrating concrete plans to meet the specified targets and then submit the proposal to the provincial Environmental Protection Bureau for approval. The city must then implement the proposal and later submit a detailed report on how the criteria have been met. The MEP then assembles a team of experts to evaluate the application. The application is immediately rejected if more than two targets have not been met. However, if only one or two targets are not met, the city is given three months to meet the targets. A program revision in 2011 introduced a five-year lifecycle for the program, under which the Environmental Protection City status automatically expires after five years unless the city reapplies successfully.

Low-Carbon Cities

The National Development and Reform Commission (NDRC) established the Low-Carbon City Pilot Program in 2010 as a key climate change response. The NDRC is a powerful ministry in the central government. It is responsible for planning the economic development of the country and has been designated as the ministry with primary responsibility for energy and climate change. Participating local governments are encouraged to go beyond mere compliance with existing low-carbon policies and programs. The specific obligations and commitments include developing a long-term low-carbon development plan, implementing institutional reform and effective policy instruments to lower carbon emissions, developing low-carbon industries, buildings and transportation, developing carbon emissions accounting and management systems, and promoting low-carbon lifestyles and consumption. The NDRC has yet to release quantified targets for these obligations. The NDRC has promised to provide venues for information exchange and policy learning to help cities achieve their objectives.

Studied Cities

At the time of writing, there were 12 Eco-Garden Cities, among them, 9 were prefecture-level cities (Guilin, Yangzhou, Nanjing, Suzhou, Qingdao, Weihai, Jincheng, Hangzhou, and Shaoxing), and 3 were county-level cities. There were 22 Environmental Protection Model cities; among them 14 were prefecture-level cities (Dongguan, Foshan, Zhongshan, Langfang, Daqing, Yichang, Zhenjiang, Xuzhou, Huaian, Yinchuan, Liaocheng, Linyi, Weihai, and Shaoxing). There were 8 cities in the Low-Carbon Cities Pilot Program and 6 of them were prefecture-level cities (Xiamen, Shenzhen, Guiyang, Baoding, Nanchang, and Hangzhou) and 2 were provincial-level cities (Chongqing and Tianjin). Because of overlaps and the lack of accessibility to statistical yearbooks in some cities, in total, there are 15 cities that serve as the dataset for this study. Figure 1 displays the locations of the studied cities.

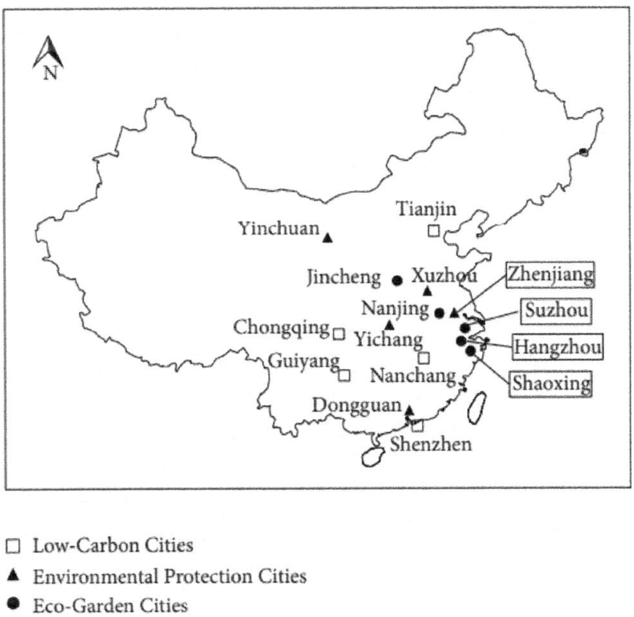

Figure 1: The locations of the studied cities.

METHODS

The present study uses a bottom-up accounting system to measure carbon emissions from stationary and mobile sources at the municipal level. It first calculates stationary emissions from the combustion of primary energy sources (e.g., coal and gasoline). The inventory is then modified because a city may import or export electricity. Emissions from a net export of electricity are deducted from the inventory, and emissions associated with a net import of electricity are added to the inventory. The formula for stationary emissions is

$$\mathrm{GHG} = \left(\sum_{i,j} E_{i,j} C_j\right) + (I - X)\, C_e, \tag{1}$$

where GHG is the total stationary emissions in tons; i represents subsectors (e.g., textile or steel making); represents the energy type (e.g., coal or crude oil); $E_{i,j}$ is the energy consumption per subsector and energy type; C_j is the CO_2 emission factor for specific energy types; I is the quantity of electricity imported into the city; X is the quantity of exported electricity; and C_e is the CO_2 emission factor for electricity.

With regard to mobile emissions, this study only calculates emissions from road transportation because of a lack of data from air, water, and rail transportation. Air, water, and rail transportation is also usually cross-boundary, and it is therefore difficult to attribute their emissions to a particular city. This study distinguishes five modes of road transport: taxis, public buses, other passenger vehicles, trucks, and motorcycles. Each mode of transport has its own unique annual mileage and emission factors. Emissions from road transport are calculated using the following equation:

$$\mathrm{GHG} = \sum_{i} P_i \times D_i \times C_g, \tag{2}$$

where GHG is carbon emissions from road transportation; i is the mode of transport (e.g., taxis or passenger cars); P_i is the number of vehicles in each mode of transportation; D_i is the annual mileage of a type- I vehicle measured in km; and C_g is the CO_2 emission factor measured in ton/km.

Data describing industrial energy consumption; import and export of electricity; numbers of taxis, buses, passenger vehicles, trucks, and motorcycles; GDP; and population are collected from the 2011 statistical yearbook for each municipality [22–36]. Some of the statistical yearbooks do not contain information about export and import of electricity, in which case the import/export data are estimated from the provincial-level electricity balance table that is in the China Energy Statistical Yearbook [37]. Carbon emission factors are obtain from the 2006 IPCC Guidelines for National Greenhouse

Gas Inventories (Table 1) [38]. Carbon emission factors for electricity are acquired from Wang et al. [39], who calculated the emission factors in China's six large national power grids using electricity generation fuel mixes and power exchange data (Table2). Average annual mileage and average gasoline consumption for different modes of transportation are collected from Loo and Li [3] (Table 3).

Table 1: Emission factors for combustion of fuel

Fuel	CO_2 emission factor (kg/TJ)	Fuel	CO_2 emission factor (kg/TJ)
Crude oil	73,300	Natural gas	56,100
Gasoline	74,100	Other petroleum products	73,300
Kerosene	71,500	Raw coal	97,500
Fuel oil	77,400	Washed coal	94,600
Diesel oil	74,100	Other types of washed coal	94,600
Coke oven gas	44,400	Coke	107,000
LPG	63,100		

Table 2: Emission factors of six major power grids

Grid	CO_2 emission factor (ton/million kWh)	Grid	CO_2 emission factor (ton/million kWh)
Northeast China	871	East China	688
North China	874	Northwest China	701
Central China	555	South China	552

Table 3: Average annual mileage and average gasoline consumption different modes of transportation

	Taxis	City buses	Private cars and institutional vehicles	Motorcycles
Average annual mileage (km)	71,175	34,000	18,000	10,000
Average gasoline consumption (km/kg)	15.07	3.91	15.07	50.23

This study focuses on energy-related emissions because the data are mostly complete and reliable. Three types of emissions are not calculated in this study, including emissions from agriculture, forest management, and other land uses (AFOLU), such as CO_2 from deforestation, CH_4 from enteric fermentation and manure, and N_2O from fertilizer use, landfill gases, mainly CH_4, and emissions from industrial processes, such as CO_2 and perfluorocarbon emissions from aluminum smelting and N_2O emissions from the production of nitric acid. These omissions result in an underestimation of the emissions inventories, although not by a significant amount, because energy consumption is the dominant source of carbon emissions in Chinese cities, typically responsible for over 90% of the total emissions [40].

RESULTS

Annual carbon emissions, annual per capita carbon emissions, and carbon intensity of the studied cities are calculated using the methods and data described above. Per capita emissions and carbon intensity are used to facilitate comparisons among the cities, which vary widely in population size and state of economic development. Per capita emissions are calculated by dividing annual emissions by the total population of the city in question. Carbon intensity expresses the emissions generated in the production of 10,000 RMB worth of GDP and is calculated by dividing annual emissions by annual GDP. The results are shown in Figures 2, 3, and4.

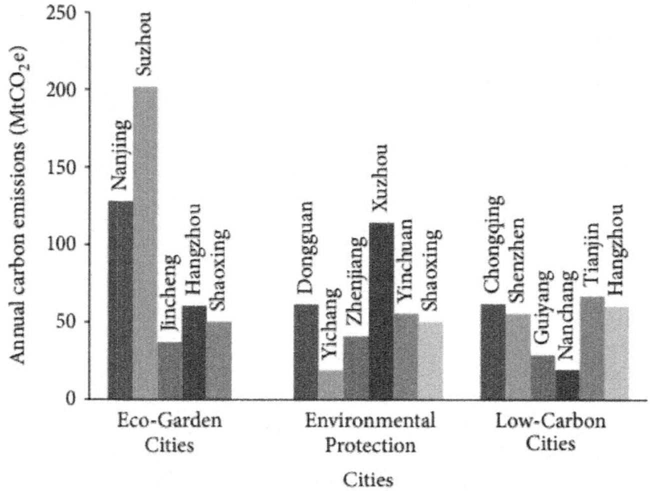

Figure 2: Annual carbon emissions from the model environmental cities.

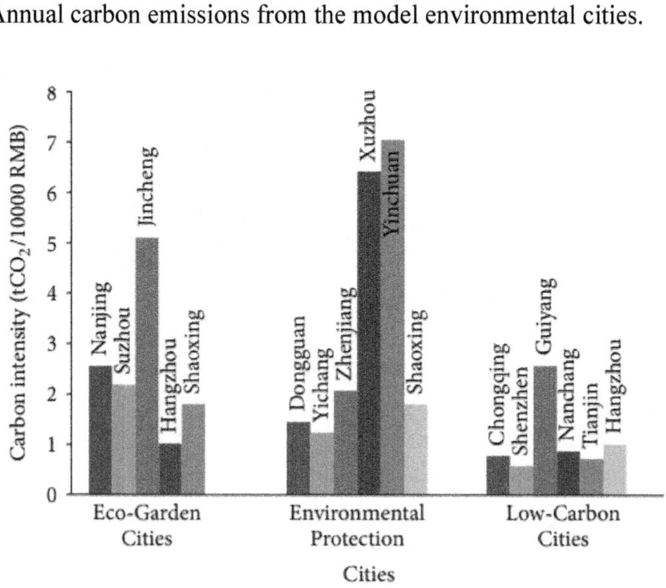

Figure 3: Carbon intensity of the model environmental cities.

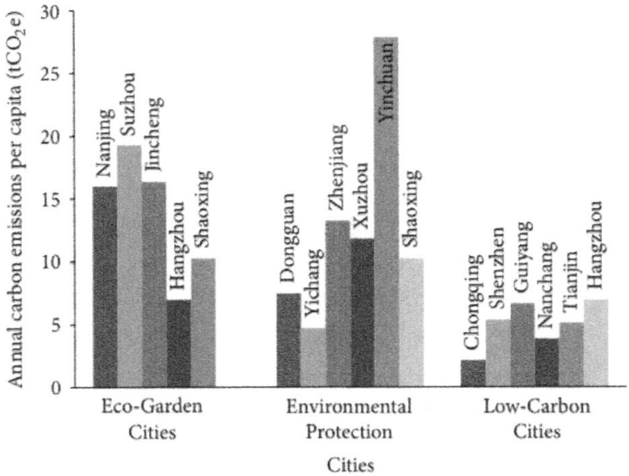

Figure 4: Annual carbon emissions per capita from the model environmental cities.

On average, the Eco-Garden Cities emit far more CO_2 (mean = 95,771,862 tCO_2e) than both the Environmental Protection Cities (mean = 57,102,423 tCO_2e) and the Low-Carbon Cities (mean = 48,991,821 tCO_2e). However, there are significant differences within each group according to the size and economic activity of the city. Nanjing (an Eco-Garden City), Suzhou (an Eco-Garden City), and Xuzhou (an Environmental Protection City), which produce more emissions than any other city in this study, are all located in Jiangsu, one of China's most heavily industrialized area. Yichang, an Environmental Protection City, and Nancheng, a Low-Carbon City, have the lowest levels of emissions. Both of these cities are small-sized cities in inland China.

The Environmental Protection Cities have the highest average carbon intensity (mean = 3.34 tCO_2/10000 RMB), followed by the Eco-Garden Cities (mean = 2.54 tCO_2/10000 RMB) and the Low-Carbon Cities (mean = 1.09 tCO_2/10000 RMB). The most carbon-intensive city is Yinchuan, a coal-dominated Environmental Protection City in the northwest. The least carbon-intensive city is Shenzhen, a wealthy and postindustrial Low-Carbon City in the Pearl River Delta area.

The Low-Carbon Cities have on average substantially lower per capita emissions (mean = 5.03 tCO_2e) than both the Eco-Garden Cities (mean = 13.79 tCO_2e) and the Environmental Protection Cities (mean = 12.58 tCO_2e). Yinchuan is the city with the highest per capita emissions, and, Chongqing, a Low-Carbon City in inland China, has the lowest. Compared internationally, the Eco-Garden Cities and the Environmental Protection Cities have per capita emissions that are similar to American cities, such as Boston (12.23 tCO_2e, 2009 data) and Seattle (11.47 tCO_2e, 2008 data) [41] whereas the levels of per capita emissions from the Low-Carbon Cities are similar to those of European cities, such as Berlin (5.86 tCO_2e, 2007 data) and London (5.98 tCO_2e, 2008 data) [41].

DISCUSSION AND CONCLUSION

The analysis has shown that, on average, the Low-Carbon Cities have lower annual carbon emissions, carbon intensities, and per capita emissions than the Eco-Garden Cities and the Environmental Protection Cities. This reflects the different goals and priorities of the model environmental cities, which are influenced by the programs that award official designations. The Low-Carbon City Program focuses on climate change, and all participating cities are required to implement a number of climate mitigation measures. All Low-Carbon Cities have therefore formulated climate change action plans and committed themselves to carbon intensity targets. Some cities, such as Hangzhou and Nanchang, have announced targets that are more ambitious than the national commitment of reducing carbon intensity by 40–45% of 2005 levels by 2020. Low-Carbon City climate change action plans are usually comprehensive, covering emissions from key sectors, including industry, energy, housing, and transportation. For example, Shenzhen will develop new industries in alternative energy, information technology, biotechnology, new materials, cultural industry, and energy conservation services, promote the deployment of natural gas, nuclear power, and solar energy, and introduce compulsory solar water heating for buildings less than 12 stories in height [42]. Shenzhen is currently the most successful city in the promotion of electric vehicles. The city launched the world's largest green taxi fleet of 300 electric-only vehicles in 2010 and plans to build 250 charging stations and install 12,500 charging posts in the city.

In contrast, climate change is not a focus of the Eco-Garden City Program and the Environmental Protection Program. The Eco-Garden City Program focuses primarily on protecting the health of urban ecosystems. The Environmental Protection City Program concentrates on air and water pollution and waste management. Both programs incorporate the concept of sustainable development and therefore emphasize a broad range of issues, such as economic and social development and environmental quality. While these two programs have energy and climate objectives, such objectives are too narrow and not sufficiently demanding. Furthermore, the objectives can be ignored completely in the Eco-Garden City Program because they are not compulsory. Therefore, the Eco-Garden Cities and the Environmental

Protection Cities have, in general, failed to pay sufficient attention to climate mitigation.

Climate change is a cross-cutting issue that impacts a wide range of economic, social, and environmental aspects of cities. Climate change exacerbates urban heat island effects [43, 44], increases the frequency of flooding due to rising sea levels, leads to more frequent and severe extreme weather events [45], and exacerbates water security problems by changing the amount, variability, timing, form, and intensity of precipitation [46]. For these reasons, climate change mitigation should be made a priority for all types of model environmental cities in China, not just the Low-Carbon Cities. This can be achieved by reforming the Eco-Garden City Program and the Environmental Protection City Program to incorporate more climate-related objectives. These climate objectives should be comprehensive, covering emissions from the industry, building, and transportation sectors, and be sufficiently challenging to motivate cities to experiment with different climate mitigation policies.

REFERENCES

1. H. Bulkeley, Cities and Climate Change, Routledge, Oxon, UK, 2013.
2. K. Lo, "Energy conservation in China's higher education institutions," Energy Policy, vol. 56, pp. 703–710, 2013.
3. B. P. Y. Loo and L. Li, "Carbon dioxide emissions from passenger transport in China since 1949: implications for developing sustainable transport," Energy Policy, vol. 50, pp. 464–476, 2012.
4. International Energy Agency, World Energy Outlook 2008, International Energy Agency, Paris, France, 2008.
5. H. Bulkeley and V. C. Broto, "Government by experiment? Global cities and the governing of climate change," Transactions of the Institute of British Geographers, vol. 38, no. 3, pp. 361–375, 2012.
6. L. Parshall, K. Gurney, S. A. Hammer, D. Mendoza, Y. Zhou, and S. Geethakumar, "Modeling energy consumption and CO_2 emissions at the urban scale: methodological challenges and insights from the United States," Energy Policy, vol. 38, no. 9, pp. 4765–4782, 2010.E. L. Glaeser and M. E. Kahn, "The greenness of cities: carbon dioxide emissions and urban development," Journal of Urban Economics, vol. 67, no. 3, pp. 404–418, 2010.C. Kennedy, J. Steinberger, B. Gasson et al., "Methodology for inventorying greenhouse gas emissions from global cities," Energy Policy, vol. 38, no. 9, pp. 4828–4837, 2010.T. Hillman and A. Ramaswami, "Greenhouse gas emission footprints and energy use benchmarks for eight US cities," Environmental Science & Technology,

vol. 44, no. 6, pp. 1902–1910, 2010.J. Minx, G. Baiocchi, T. Wiedmann et al., "Carbon footprints of cities and other human settlements in the UK," Environmental Research Letters, vol. 8, no. 3, Article ID 035039, 2013.

7. A. Chavez and A. Ramaswami, "Progress toward low carbon cities: approaches for transboundary GHG emissions› footprinting," Carbon Management, vol. 2, no. 4, pp. 471–482, 2011.C. Kennedy, J. Steinberger, B. Gasson et al., "Greenhouse gas emissions from global cities," Environmental Science & Technology, vol. 43, no. 19, pp. 7297–7302, 2009. ·
8. J. S. Gregg, R. J. Andres, and G. Marland, "China: emissions pattern of the world leader in CO_2 emissions from fossil fuel consumption and cement production," Geophysical Research Letters, vol. 35, no. 8, Article ID L08806, 2008.K. Lo and M. Y. Wang, "Energy conservation in China›s Twelfth Five-Year Plan period: continuation or paradigm shift?" Renewable and Sustainable Energy Reviews, vol. 18, pp. 499–507, 2013.
9. K. Lo, "A critical review of China›s rapidly developing renewable energy and energy efficiency policies,"Renewable and Sustainable Energy Reviews, vol. 29, pp. 508–516, 2014.
10. L. Li, C. Chen, S. Xie et al., "Energy demand and carbon emissions under different development scenarios for Shanghai, China," Energy Policy, vol. 38, no. 9, pp. 4797–4807, 2010.
11. Y. Wang, W. Ma, W. Tu, Q. Zhao, and Q. Yu, "A study on carbon emissions in Shanghai 2000–2008, China," Environmental Science & Policy, vol. 27, pp. 151–161, 2013.
12. J. Bi, R. Zhang, H. Wang, M. Liu, and Y. Wu, "The benchmarks of carbon emissions and policy implications for China›s cities: case of Nanjing," Energy Policy, vol. 39, no. 9, pp. 4785–4794, 2011.H. Wang, R. Zhang, M. Liu, and J. Bi, "The carbon emissions of Chinese cities," Atmospheric Chemistry and Physics, vol. 12, pp. 6197–6206, 2012.
13. S. Dhakal, "Urban energy use and carbon emissions from cities in China and policy implications," Energy Policy, vol. 37, no. 11, pp. 4208–4219, 2009.K. Lo, "Deliberating on the energy cap in China: the key to a low-carbon future?" Carbon Management, vol. 4, no. 4, pp. 365–367, 2013.
14. Nanjing Municipal Statistics Bureau, Nanjing Statistical Yearbook 2011, China Statistics Press, Beijing, China, 2011.
15. Suzhou Municipal Statistics Bureau, Suzhou Statistical Yearbook 2011,

China Statistics Press, Beijing, China, 2011.

16. Jincheng Municipal Statistics Bureau, Jincheng Statistical Yearbook 2011, China Statistics Press, Beijing, China, 2011.
17. Hangzhou Municipal Statistics Bureau, Hangzhou Statistical Yearbook 2011, China Statistics Press, Beijing, China, 2011.
18. Shaoxing Municipal Statistics Bureau, Shaoxing Statistical Yearbook 2011, China Statistics Press, Beijing, China, 2011.
19. Dongguan Municipal Statistics Bureau, Dongguan Statistical Yearbook 2011, China Statistics Press, Beijing, China, 2011.
20. Yichang Municipal Statistics Bureau, Yichang Statistical Yearbook 2011, China Statistics Press, Beijing, China, 2011.
21. Zhenjiang Municipal Statistics Bureau, Zhenjiang Statistical Yearbook 2011, China Statistics Press, Beijing, China, 2011.
22. Xuzhou Municipal Statistics Bureau, Xuzhou Statistical Yearbook 2011, China Statistics Press, Beijing, China, 2011.
23. Yinchuan Municipal Statistics Bureau, Yinchuan Statistical Yearbook 2011, China Statistics Press, Beijing, China, 2011.
24. Chongqing Municipal Statistics Bureau, Chongqing Statistical Yearbook 2011, China Statistics Press, Beijing, China, 2011.
25. Shenzhen Municipal Statistics Bureau, Shenzhen Statistical Yearbook 2011, China Statistics Press, Beijing, China, 2011.
26. Guiyang Municipal Statistics Bureau, Guiyang Statistical Yearbook 2011, China Statistics Press, Beijing, China, 2011.
27. Nanchang Municipal Statistics Bureau, Nanchang Statistical Yearbook 2011, China Statistics Press, Beijing, China, 2011.
28. Tianjin Municipal Statistics Bureau, Tianjin Statistical Yearbook 2011, China Statistics Press, Beijing, China, 2011.
29. National Bureau of Statistics, China Energy Statistical Yearbook 2011, China Statistics Press, Beijing, China, 2011.
30. IPCC, "IPCC Guidelines for National Greenhouse Gas Inventories," 2006.
31. K. Wang, X. Zhang, Y.-M. Wei, and S. Yu, "Regional allocation of CO_2 emissions allowance over provinces in China by 2020," Energy Policy, vol. 54, pp. 214–229, 2013.
32. F. Xi, Y. Geng, X. Chen et al., "Contributing to local policy making on GHG emission reduction through inventorying and attribution: a case study of Shenyang, China," Energy Policy, vol. 39, no. 10, pp. 5999–

6010, 2011.C. Kennedy, S. Demoullin, and E. Mohareb, "Cities reducing their greenhouse gas emissions," Energy Policy, vol. 49, pp. 774–777, 2012.

33. Shenzhen Municipal People›s Government, Shenzhen Low-Carbon Development Mid-to-Long Term Plan, Shenzhen Municipal People›s Government, Shenzhen, China, 2012.
34. S. E. Lee and G. J. Levermore, "Simulating urban heat island effects with climate change on a Manchester house," Building Services Engineering Research & Technology, vol. 34, no. 2, pp. 203–221, 2012.
35. J.-M. Feng, Y.-L. Wang, Z.-G. Ma, and Y.-H. Liu, "Simulating the regional impacts of urbanization and anthropogenic heat release on climate across China," Journal of Climate, vol. 25, no. 20, pp. 7197–7203, 2012.
36. V. Czako, "Drowning the suburb: settlement planning and climate change adaptation in a Hungarian metropolitan area," Urban Research & Practice, vol. 6, no. 1, pp. 95–109, 2013.
37. J. S. Risbey, "Dangerous climate change and water resources in Australia," Regional Environmental Change, vol. 11, no. 1, supplement, pp. 197–203, 2011.

Chapter 7

CULTIVATING A VALUE FOR NON-HUMAN INTERESTS THROUGH THE CONVERGENCE OF ANIMAL WELFARE, ANIMAL RIGHTS, AND DEEP ECOLOGY IN ENVIRONMENTAL EDUCATION

Helen Kopnina[1], and Brett Cherniak[2]

[1]Institute Cultural Anthropology and Development Sociology, Social and Behavioural Sciences, Leiden University, Wassenaarseweg 52, Leiden 2300 RB, The Netherlands

[2]47 Jones Street, Hamilton, Ontario, ON L8R 1X9, Canada

ABSTRACT

While the original objective of environmental education (EE) and education for sustainable development (ESD) acquired an awareness of the natural world and its current plight, animal welfare (AW), animal rights (AR), and deep ecology (DE) have often been absent within EE and ESD. AW and AR focus their attention on individual animals, while the DE perspective recognizes the intrinsic value of the environment. In this article, we shall discuss how the integration of these three approaches within EE/ESD can and should be improved, with particular reference to the ethical underpinnings of educational scholarship and practice. This article will argue that these three positions are well placed to enhance the democratic practices of EE/ESD through the adoption of an inclusive pluralism that embraces representation of non-human species and recognizes their interests.

INTRODUCTION

The objectives of environmental education (EE) are defined as helping students acquire: an understanding of basic ecology, an awareness of the natural world and its current plight, a sensitivity to the need for protecting nature, and the skills to help address environmental challenges [1]. Education for environmental sustainability engages both the macro units of study, such as "nature" or "ecosystems," and the micro, such as species and individual animals, plants, fungi, or bacteria. The scope of this article does not allow for a more detailed discussion of what qualifies as "animals" present in animal rights literature. For the purpose of this article, we shall only refer here to animals as sentient beings, thus a category that can also sometimes apply to plants, fungi, or bacteria. Simultaneously growing alongside EE over the decades, education addressing animal welfare and animal rights has emerged as part of a larger environmental concern [2].

The ethical underpinnings of animal rights (AR), animal welfare (AW) and deep ecology (DE) are based on a number of ethical positions relating to EE's original objectives. These ethical positions range from concern for animal welfare and animal rights on one hand to environmental ethics on the other. According to Tom Regan [3], animal rights are defined as a commitment to the abolition of the use of animals in science; dissolution of commercial animal agriculture and elimination of commercial and sport hunting and trapping. Environmental ethics, in turn, range from anthropocentrism (human-centeredness) to ecocentrism or biocentrism (placing ecosystems or biosphere at the heart of moral consideration) as well as the in-between gradations, including strong and weak anthropocentrism. Norton [4] distinguished between "felt" and "considered" preferences. A felt preference is one that may be temporarily satisfied by some specific experience. A considered preference is arrived at after "careful deliberations." According to Norton, an ethic is strongly anthropocentric if it focuses on felt preferences without reference to socio-cultural context (thus, non-consequentialism) and weak anthropocentrism if it focuses on felt and considered preferences, dependent on one's worldview (thus, consequentialism. Naess [5] drew a distinction between deep and shallow ecology, with the former associated with recognition of intrinsic value of environment beyond its unity, and the latter including human interests. While anthropocentrism assigns moral value to humans only, it can also include consideration of the environment when human welfare is affected, or of animals used for food, research, entertainment or companionship. Generally, non-anthropocentric ethical perspectives are concerned with the preservation of biodiversity and conservation—not so much for its utility to us humans, but for the benefit of entire habitats or species themselves. These general ethical orientations consider the moral position, welfare, or legal rights of either environment or individual animals, or both.

AW refers to the desire to prevent unnecessary animal suffering, and while not categorically opposed to the use of animals, wanting to ensure a good quality of life. AW includes not only the state of the animal's body, but also its feelings or sentience [6]. Animal welfare education (AWE) promotes knowledge, understanding, skills, attitudes and values related to human involvement in the lives of animals [7]. A number of educational programs and specialized courses have been established that address animal welfare from ethical, political, and practical perspectives. The Animal Welfare Institute (AWI), founded in 1951, and the International Fund for Animal Welfare (IFAW), founded in 1969, are both involved in education for animal rights and welfare [8,9]. However, their educational programs are not connected to EE nor education for sustainable development (ESD).

AR refers to the philosophical belief that animals should have the right to live their lives free of human intervention, opposing the use of some animals by humans [3]. In education, AR has not been taught independently from AWE, nor has "humane education", defined by the Institute for Humane Education as a form of education that "instills the desire and capacity to live with compassion, integrity, and wisdom, but also provides the knowledge and tools to put our values into action in meaningful, far-reaching ways."

Education for DE has been taught as part of outdoor (experience) education when the students are involved in outdoor activities [10] or philosophy courses [11]. AR, AW and DE have routinely been taught in existing courses such as biology, ethics, or sustainability, with independent focus being scarce (e.g., [12]). Within ESD, environmental sustainability concerns are normally displayed along with social and economic objectives, and AR, AW and DE are presented as an *ad hoc* practice, primarily associated with the conservationist and the naturalist subjects (e.g., [13,14,15,16]).

In this article, we shall address the ethical underpinning of the EE/ESD scholarship and reflect upon the position of AR, AW and DE within the established practice. We shall focus on the questions: How is care for ecosystems, biosphere, species or individual animals positioned within EE/ESD? What ethical positions underlie considerations of environment and animals in EE/ESD?

FROM ETHICS TO SUSTAINABILITY

The complexity of the natural world has necessitated the exploration and development of several nuanced and distinct non-anthropocentric ethics. Kronlid and Öhman [17] (p. 23) have distinguished between five different typologies of non-anthropocentric ethics in the context of EE research: sentientism (non-human animals with the ability to experience pain and suffering have intrinsic value); animal rights (non-human animals with a sense of self have intrinsic value); social animal ethics (non-human animals with the ability to engage in relationships have intrinsic value); biocentrism (all organisms have intrinsic value by virtue of having a good of their own related to their flourishing), and ecocentrism (ecosystems and species have intrinsic value because it is possible to relate to them as separate entities and because they have a capacity to sustain life and well-being in terms of their integrity, stability and beauty). These distinctions allow us to consider a wider range of perspectives, values and situations when discussing the position of non-humans within education practice while bringing to light the many concerns of the natural world.

Many theorists have attempted to demonstrate the precise ways in which deep ecology and animal rights/ethics/liberation are entangled and interdependent, as well as how they differ, for a review of some perspectives see [18]. AR, AW and DE are divided on a number of points but are united in their support of care for the environment and individual animals. The greatest commonality between DE and much of animal ethics is that they remove humans from the moral pedestal and seek recognition of the plight of non-humans. As Jamieson [19] has pointed out, the urgency of environmental predicament and the plight of non-humans does not allow for more internal battles, thus calling different "camps" to combine their knowledge, not to "overwhelm" the status quo with a flurry of "niche" groups.

This plight of non-humans is characterized primarily by the destruction of habitats upon which wildlife is dependent and the abuse of animals used for human consumption through the industrial food production system, e.g., concentrated animal feeding operations (CAFOs). Ecocentric scholars have argued that unless the intrinsic value of non-humans is recognized, there will be no institutional guarantees that the interests of other species will be protected and not continually neglected [20,21,22,23]. The market economy has been represented as the solution to issues of sustainability, embedding economic reasoning deeply into environmental policy, planning, and practice and ignoring the plight of the planet [24]. Such reasoning led to the subordination of non-humans and their habitats to social and economic objectives [1,13,17,25,26].

This trend of presenting nature as capital and commodity has made its way into educational practice. The literature is replete with references to natural *resources*, natural *capital*, and ecosystem *services*, conceptualizing nature through an anthropocentric, utilitarian lens while the recognition of the *intrinsic* value of biodiversity seldom appears in the same space. [1,13,17,27]. Environmental "management" of "natural resources" became interlinked with the market mechanisms of "species banking" and "carbon trading." In line with the Brundtland definition of "development that meets the needs of the present without compromising the ability of future generations to meet their own needs," ESD has shifted its focus towards environmental justice, primarily concerned with the distribution of environmental benefits and burdens among human beings [28]. ESD chiefly promotes sustainability, placing its focus on a "sense of justice, responsibility, exploration and dialogue" [29]. In this context, education concerned with animals is not commonly discussed as an integral part of education for sustainability. We shall argue that it is crucial to see AR, AW, and DE connected to the overall goal of enhanced empathy, which is stressed in social equity, as in Norton's convergence theory [4]. Convergence can be illustrated by the consumers' insistence on transparency in our food system, which is linked to the treatment of farm laborers as well as livestock; or address pollution that affects both human and natural systems. For example, as the World Society for the Protection of Animals (WSPA) has outlined, AWE entered ESD through the avenues of: environmental and agricultural sustainability (responsible animal management); human health (good animal care reduces the risk and spread of diseases that can be transmitted to humans and of food poisoning); poverty and hunger reduction (looking after animals properly improves their productivity and helps farmers to provide a secure food supply and income for themselves, helping to alleviate poverty); disaster preparedness and risk reduction (animals are important for people's lives and livelihoods and must be given due consideration in plans for disaster preparedness and response). It is notable that a view of "useful animals" is interlinked with conceptions of stewardship, management, and other optimistic "innovations" [30].

However, we also warn that when developing a framework for connecting AR, AW, and DE to sustainability education—as, for example, through ESD that emphasizes normative competencies, which include values and issues of morality—we need to be careful not to over-emphasize compromise and token accomplishments when harder choices about economic development *versus* preservation of biodiversity need to be made. "Improving animal productivity" has little relation to ethical concerns about non-humans, giving way to the broad aims embodied by ESD, such as social equality and economic equity (e.g., [31,32]). While there has long been a supposed connection between the conscientious consumption of green products and conservation, empirical evidence from wealthier countries denies that consumer responsibility alone can overcome unsustainability. Nor have multinational corporations shown much transformative power in their ability to alter global unsustainability practices or address the mere scale of animal use in industries ranging from food to pharmaceuticals [2,3,5]. Societal concerns do not naturally and automatically seep into educational practice, and neither can we assume that corporate or political leaders can be "taught" to care about non-humans. What can be achieved, however, is the critical awareness of how normative concepts of "sustainable development" or "animal productivity" are used. Yet, this critical reflection and skepticism about the indoctrinating tendencies of normative concepts should not turn into a fear of all types of instrumentality, as we shall explain in more detail below. Some EE/ESD scholars have expressed their angst of indoctrination leading to behavioral change that "environmentalists" could implant into their pupils' brains, even if this change means a more inclusive recognition of non-humans [33,34,35]. Illustrating the fear of environmentalism, they have warned against "eco-totalitarianism" [33] (p. 225). Breiting [34] has expressed concern with educators and pupils being used as marionettes by the generalized camp of "environmentalists". A telling example is provided by Bob Jickling. In reflecting upon the controversy about shooting wolves in the Yukon area, Jickling felt that advocating the pro-wolf position as a schoolteacher in a local community would be "neither practically viable nor educationally justifiable" [35] (p. 92). Jickling justifies this position by the need to stay neutral in order to teach students democratic and open values and avoid indoctrination.

Other EE/ESD scholars have tried to bridge the gap between the generalized camps of ecocentric environmentalists and "emancipatory" educators by arguing that educational practitioners and theorists should not continually reproduce a simplified notion of anthropocentrism and non-anthropocentrism [36]. Kronlid and Öhman [17] (p. 34) note that when environmental ethics is used in EE research, the "cross-disciplinary work *should* take the complexity and pluralism of environmental ethical issues and the variety of sub-positions produced above into consideration" (italics added). This may be seen as another example of the "latent authoritarian tendency" of environmentalists [37] (p. 126): the presumption that open discussions are guaranteed to solve all of our problems and produce the desired results. Rather than dwelling on the anthropocentrism/ecocentrism dichotomy, they suggest that educators should embrace different ethical perspectives and foster democratic learning.

The onus for drawing attention to AW and AR appears to have been left by educationists to outside pressure groups and the media [2], with DE being relegated to the marginal experiential education [10]. With a few exceptions, a couple of educational programs focus on intrinsic values or the rights of animals. This raises a number of ethical questions in both EE/ESD and within the general enterprise of sustainable development: Should future generations of humans have the right to experience wilderness and biological diversity? Does sustainable development concern other species? Do animals and plants have the right to a secure future? [36] (p. 67). Why is teaching for democracy, gender and racial equality and economic equity an expressed aim in most EE/ESD while the ongoing extinction or treatment of animals in the industrial food production system is a marginal concern?

AR, AW and DE have been present *ad hoc* for a few decades within or outside of EE or ESD, with issues concerning speciesism (discrimination against other species) seen as a footnote in mainstream environmental education journals (for review see [28]).

ANIMAL RIGHTS, ANIMAL WELFARE, AND DEEP ECOLOGY IN ENVIRONMENTAL ETHICS

Critics have pointed out that much of the anthropocentric ethics prevalent in EE/ESD is primarily concerned with questions of social and economic equity [28]. “Social equity,” for example, usually means “universal participation in global markets that will require continued ecological sacrifices” (e.g., [25]). In fact, the moral imperatives of social and economic equality conveniently serve the global market through the global perpetuation of consumer culture [26]. Considerations of non-humans are marginal in the pluralistic perspective dominated by social and economic agendas [27,28].

Will we even notice when species go extinct, and if we do, what will we notice? Indeed, there are plenty of empirical examples which demonstrate that the extinction of species have not affected food supply chains or human welfare. As witnessed during recent extinctions, few direct negative side effects are experienced by people, aside from some marginal benefits for the pharmaceutical industry or zoos. Thompson [38] and Kareiva *et al.* [39] have found that we do not need to save every species as humanity is not dependent on them, but should instead embrace human-managed nature and no longer waste time teaching students to care about what is left of the wilderness. This transition of academic discourse toward human-managed nature is part of a more general trend that embraces the domestication of the planet. Yet critics have argued that this instrumental attitude to nature is precisely what has brought us to our current environmental problems [27].

While many different schools of environmental ethicists agree that humans and environment are intimately interrelated, interconnected, and the relationship between the human and non-human worlds is that of mutual reciprocity, more skeptical critics emphasize that this reconciliatory view tends to ignore how economic development destroys entire habitats and distracts public attention from the plight of farm animals or those used for medical experimentation [25]. According to this critique, it is not a question of whether humans and nature are connected or dependent on nature (in a similar way, we could inquire whether the slave owners were really dependent on slaves), but whether these elements of nature—living creatures—have any significant moral status. Is there not a moral need to recognize that we are not the only species on Earth that needs to be considered? To return to the question posed in the introduction, what ethical positions underlie considerations of the environment and animals in EE/ESD?

THE QUESTION OF MORALITY

While there is a shift in public awareness towards recognition of animal welfare concerns, there is no consistent discussion about the *scale* of instrumental use of other species. The scale of this exploitation, including CAFOs and intensive farming, which relegate billions of animals to short and miserable existences, has increased exponentially alongside human population growth and consumption [25]. While the fate of a single abused dog may capture public attention through the media, there is no consistent discussion about the millions of species "harvested" for consumption, or used for medical experiments.

The philosophical and political difficulty of establishing a non-anthropocentric ethic has led pragmatic theorists to support an "enlightened anthropocentrism" which, whilst recognizing the human-centered reality, claims to be consistent with a high degree of protection for non-human nature (e.g., [4,40,41]). Such a case is based primarily on an empirical claim that the protection of the natural world—including animals—is in the interest of humans. However, the pragmatist thinkers reflect that relying on the intrinsic value of non-human animals and the environment is not sustainable and argue that moral anthropocentrism is unavoidable [41,42]. Enlightened anthropocentrism is largely consistent with "the convergence thesis," stating that preservationist and conservationist policies will tend to converge in the long run [43].

For example, biological and organic meat can be good both for farm animals and for human health. Another familiar example is addressing water, soil or air pollution that is likely to harm humans with the positive side-effects it brings for habitats or species. These lead to the type of convergence thinking that assumes a healthy environment caters to human needs and, conversely, that social or economic needs will serve nature. This reliance on perceived interdependence is often seen in EE/ESD. Famously, the Brundtland report speaks of convergence between the three pillars of economic development, social equity, and environmental protection. A similar effort is made to "balance" opposing perspectives, presenting conflicts not in moral (good *vs.* bad) or rational (right *vs.* wrong) terms but as a broader exercise in social learning [44] or "learning from sustainable development" [45].

However, others have noted that occasionally the interests of humans and animals (and humans and environment more broadly) coincide, anthropocentrism can make a positive contribution to the evaluation and justification of environmental policy in situations dealing with artificial (human-created or controlled) and not natural systems [46]. Examples of policies addressing artificial systems might include urban air pollution, regulation of greenhouse gases, issues of environmental justice, and the environmental impacts of agriculture. These are largely consistent with conventional sustainability policy, where the environment and individual species or animals are kept in orbit with economics at the center; our destructive systems become more sustainable, leading us to believe that we are helping the environment as a whole. The environment in this policy will be considered if and only if a degrading environment might undermine ongoing development [47]. However, in cases of largely natural systems, wilderness preservation and the protection of endangered species, a moral monism [48] embracing ecocentric, biocentric or animal-centric positions offers a wider protection for environment than an anthropocentric-dominant pluralism. According to Katz "there is a clear difference between anthropocentric and nonanthropocentric justifications for environmental policy, and in these situations I remain committed (mercilessly) to nonanthropocentrism" [46] (p. 390).

Naess argued that sustainability hinges on a consistent education curriculum which is tied back to a well-informed understanding of the state of the planet [49]. The non-anthropocentric environmental ethics calls for a new ethic, not beholden to any previously dominant stream of thought and not dependent on the shifting cultural morality of today [19]. An imperative to protect and preserve (or, to use a more popular term, to accept stewardship of) non-humans could stem from human reason, or love, or sense of duty and responsibility [47].

Yet, this ethic is now in short supply if we notice how mercilessly the processes endangering habitats, species and individual animals have expanded their hold within public imagination, and that of EE/ESD scholars. The critics have noted that the brainwashed, anthropocentric ideology that has taken hold of the supposedly free "choices" being made in neoliberal education systems is not likely to offer a way forward as the choices given still find themselves under the influence of this anthropocentric ideology [24,50,51].

REFLECTION

In answering the question, "How is care for ecosystems, biosphere, species or individual animals positioned within EE/ESD?" we note that this possibility is contingent upon a socio-political and cultural context permitting an ecocentric ethic to be recognized as a necessary contribution to environmental integrity. An EE/ESD that is already respective and inclusive of AW, AR and DE could help to create this context. As with other socio-political changes, education plays an essential role in change to society as a whole and the political institutions within. Still, we need to be aware that if inroads are only made in education, this progress might be lost once students leave the classroom if the greater context is not in place, thus a wider societal transformation that offers an ethical support base for AW, AR and DE would be close to a requirement. An education is pointless if students enter a society that does not allow for the inclusive support of the values and perspectives they have learned.

Encouragingly, Andrew Dobson has suggested that non-human animals could be democratically represented through proxy representatives elected explicitly to promote their interests, occurring through real elections in which candidates are elected in the present system [22]. In his later work, Dobson argued that since animals cannot speak themselves, the ability to recognize their "voice" through their representatives, is of crucial importance if they are to be considered as democratic (with "demos" in this case extending to non-humans) participants [23]. In this case we need to note that the voice of animals (as well as fungi, algae, *etc.*) can be extended to habitats and ecosystems by combining common threads from AR, AW and DE. Generally, these common threads are based on the philosophy of inclusion of ecosystems, species or non-human individuals within the sphere of moral consideration, based on biophilia (love of nature), and, particularly in the case of DE, realization of intimate interconnectedness between humans and natural elements.

While DE's philosophy of inclusion might be too weak in the face of dominant anthropocentric neo-liberal ideology, AR might be too "strong." Presently, AR is often perceived as subversive or threatening. For example, Animal Liberation Front (ALF) and Earth Liberation Front (ELF) activists are seen as dangerous and radical [52,53,54]. While such organizations exhibit a concern that is desirable if non-human interests are to be upheld, they do not address the social, economic, or cultural forces that shape the general public's attitudes. However, it is still precisely this type of "radical" attitude that is needed to reverse the impetus of the presently taken-for-granted anthropocentric ethics that governs much of EE/ESD educational practice. This is not to imply that the acts of property destruction that these activists have committed are recommended for educators to teach to students, but rather understanding the type and development of commitment, rage, as well as selfless love of non-human beings, be they clear-cut forests or animals used for laboratory experiments that deserve educators' attention. Richard Kahn discussed the need of radical reorientation of education toward ecological concerns and other species, examining how the essence of ALF and ELF's ideas (although certainly not the methods) can be used to inspire educational practice beyond the current status quo [53]. Integrating the perspectives from groups such as ELF in the curriculum was discussed by Kopnina [55,56]. It was suggested that we can draw inspiration for educational practice from eco-pedagogy, which places ecological integrity as well as animal welfare at the center of sustainability challenges. An example of this is given by Kopnina in her discussion of exposing business students to radical environmental activism through the in-class viewing of the documentary film, *If a Tree Falls* (2011), directed by Marshall Curry and Sam Cullman about the rise and fall of ELF. As one of the reviewers of the film wrote:

What is interesting is that even while we don't agree with what the ELF is doing, the film gives us images that allow us to understand their point of view. We see footage of trees that have stood for thousands of years, blindly cut down. We see horse mills, with hundreds of dead horses hung from the ceiling. We see the heartbreaking sight of a group of legendary trees sawed down to make a parking lot… We see the protesters themselves, camped out in the trees that are to be cut down, beaten unmercifully by the local police… The irony is that the members of the group who are clearly guilty of vandalism haven't done any physical harm to other human beings but are being beaten down by law enforcement as if they were murderers [57]

After viewing the film, the business students exhibited a shift in their ethical thinking and an increase in awareness of ecocentric concerns. The film made students question the activists' reasons behind their level of commitment, contemplating what makes some people join ALF and ELF and others refrain. In written assignments following the viewing of the film, students were able to separate the beliefs of ALF/ELF from the stigma attached to their actions. The students realized that the "radical" or "terrorist" label of the activism featured in the film could indeed be disputed—to quote a law enforcement officer in the film, "One man's terrorist is another man's freedom fighter."

More generally, returning to educational practice and questions which researchers should explore further: Do we need their "strong" views throughout the entire process, or just to begin "the shift"? What creates a positive feeling about non-humans, and how can we educators help to spread that? Would aligning EE/ESD with some of the aims of radical organizations, or only their attitudes, provide an easy target for opposition?

This shift in awareness is also crucial in relation to AR and AW. Regarding animal ethics and welfare in education, Kelly reflected that "those who know it is not necessary to kill animals for food or fight them as entertainment have an ethical responsibility to teach other humans how to eat without killing, how to be entertained without exploiting" [58]. According to Kelly, this type of education triggering a shift in awareness and ethical consideration of non-humans would "allow researchers to devote research to making animals' lives better without worrying about the accusation that they should be devoting their time to less trivial matters; and would function to make education a far less anthropocentric endeavor" [58] (p. 45). This research could form a base, noted Kelly, to teaching about legal protection of animals that (1) emphasize respect for animals, even when others understand this as moral proselytizing; (2) teach that it is wrong to continue practices and habits that are exploitive of animals when alternative practices are possible; (3) design life-skills curricula that help those who cannot envision the feasibility of these alternative practices to do so (such as vegetarian cooking); (4) de-anthropocentrize the mission of educational institutions; and (5) discount any unnecessary piece of knowledge gained at the expense of animals' welfare as unworthy of academic attention [58] (p. 45). These steps will require continuous representation of non-human interests by eco-advocate educators.

Thus, the role of representation of non-human interests through eco-advocates [59,60] is essential if educational researchers and practitioners are to move beyond the status quo of conventional plurality. In discussing environmental justice (equal distribution of environmental risks and benefits, including to the non-human species) and democracy, Dobson emphasized that "if harm is being done, then more justice rather than more talking is the first requirement" [22] (p. 26).

Because other species cannot engage in "pluralistic" discussion due to their inability to speak our language, the participation of educators who represent non-human "voices" is essential. Following this, the insistence that the distinction between anthropocentric and non-anthropocentric ethics is not relevant for EE/ESD (e.g., [17,36]) is counterproductive to the effort to expand the pedestal of moral significance to the non-humans. Questioning anthropocentrism is far more than an academic exercise of debating the dominant cultural motif of placing humans at the center of material and ethical concerns. It is a fertile way of shifting the focus of attention away from the problem-symptoms of our time to the investigation of root causes. And certainly the dominant beliefs, values, and attitudes guiding human action constitute a significant driver of the pressing problems of our day [61].

The misplaced fear of eco-totalitarianism in education [33] fails to account for the one-species' totalitarianism over non-human beings, denying eco-advocates the right of providing the "voice" to the millions in need of urgent protection. The dominant form of pluralism is often reflective of the neoliberal model in which individual choices are intertwined with free market thinking at the expense of ecological concerns [50,51]. Anthropocentric pluralism places humanity outside and above the natural world, and its constructed distinction and artificial perch offer the illusion of superiority and a transient experience of "wealth" at the expense of others, while what is sacrificed is the very source from which true power flows: abundance, creativity, unexpectedness, reciprocity, and mutual flourishing [61]. One-species-only pluralism leads educators to think that they give students choices to choose plural perspectives, while the choices are between different shades of anthropocentrism.

Inclusive pluralism embraces non-human representation, while human eco-advocates "speak for nature" [62] and billions of Earth's citizens. As a starting point, we could discuss what these non-human voices are exactly saying, and whether some should count more than others, and why some educators should claim to speak for them. Without presuming to know the answer to either of these questions, we support one simple claim: that sentient beings, from orangutans to plants, want to survive, and it is our shared responsibility as members of one single species that both endangers and can act on behalf of their survival, to consider them in our 'pluralist' ethics and education. An inclusive pluralism recognizes plurality of choices of *all* Earth's citizens, not just those of one species. Andrew Dobson's work in relation to democratic inclusion of non-humans [22,23] as well as Arne Naess's work [5,49] on how different forms of respect and inclusion of non-humans might be grounded provides hopeful directions for consequent research on inclusive pluralism in educational contexts.

However, we realize that incorporating environmental ethics, animal welfare and animal rights into a seamless synthesis easily taught to students can be difficult. There are great conceptual difficulties involved in extending ethics beyond its current application. The strength of animal ethics lies not in incorporating it as part of environmental ethics, which itself seems difficult to subsume under current societal ethics. How does one extend our social ethic to non-sentient and abstract entities such as wilderness areas or even species? The extension of ethics to animals was accomplished in the 70's by various philosophers including Singer and Rollin when they demonstrated the absence of morally relevant differences between humans and animals, and the presence of morally relevant similarities. It is in fact a far more difficult task to extend moral status to entities not continuous in traits with humans. Yet, if we despair, no change is likely to come.

HOPEFUL DIRECTIONS

Examples of educational practice that integrate inclusive pluralism in which non-human agents are recognized as potential contributors to diversity perspectives include conservation education [63,64,65], outdoor education [66,67], education for deep ecology (e.g., [10,11,68]), AWE [2] and post-humanist education [69,70]. All these different directions offer a hope of restoration—not just that of nature and animal welfare, but also of ourselves as caring educators and students. As Richard Louv [71] (p. 226), the author of a popular book *Last Child in the Woods* has stated, "An environment-based education movement—at all levels of education—will help students realize that school isn't supposed to be a polite form of incarceration, but a portal to the wider world" and, as this article makes clear, that wider world includes the inner world. This wider world includes the possibility that lecturers could transmit to their students the ability to transcend the top-down sustainable development that is usually based on control rather than compassion. Just as many conservationists are more interested in increasing their capacity for empathy than for control, they are also frequently interested in communicating to larger audiences how and why their knowledge of nature leads them to love nature [72]. In the words of Vucetich and Nelson [72] (p. 13), knowing conservation›s role in society requires knowing how and why populations and ecosystems are valuable. In particular, we need to know how they are valuable beyond their utility to humans—and this is the task that EE and ESD are well suited to undertaking. However, EE/ESD need to be reoriented toward a clearer distinction between what is socially and economically defined as sustainability and what is ecologically defined sustainability—in other words, the difference between sustainable development as a slogan that implies sustaining economic growth, and environmental sustainability that underlies long-term survival of social, economic and natural systems [73].

A hopeful direction for non-anthropocentric education includes initiatives by a number of charities and foundations. The Foundation for Deep Ecology (FDE) supports education and advocacy on behalf of wild Nature, educating citizens that stopping the global extinction crisis and achieving true ecological sustainability will require rethinking our values as a society. FDE argues that "present assumptions about economics, development, and the place of human beings in the natural order must be reevaluated." FDE's educational programs support restoration of "traditional knowledge, values, and ethics of behavior that celebrate the intrinsic value and sacredness of the natural world and give the preservation of Nature prime importance," thereby introducing students to broader topics of sustainability with nature and animals as focal points [74]. In a similar effort, the Rewilding Foundation [75] participates in ecological education and participatory citizen science projects.

Another hopeful direction is integrating AR, AW, and DE within national platforms of EE and ESD, as, for example, the inclusion of Animals (with subsections specified e.g., "Animal welfare", and "Endangered species") by the National Association for Environmental Education UK [76]. At present, these topic sections are empty, implying that it is only a gesture of good will. Also in England, charities like the Royal Society for the Prevention of Cruelty to Animals (RSPCA) offer free training and support for professionals who work with children and young people, stating that their educational mission is to help students to develop informed, responsible and active citizens [77].

Tellingly, RSPCA combines the elements of AR, AW and DE by offering both domestic animal-oriented as well as wilderness-oriented curricula through their cooperation with animal shelters, an outdoor education center and a wildlife hospital. People for the Ethical Treatment of Animals (PETA) is the largest animal welfare organization in the world and also allocates education a special role, linking the values encompassing respect for animals, nature, and other human beings, as well as active citizenship:

Teaching kids to have compassion and empathy for their furry, feathered, and finned friends is vital for preventing cruelty to animals as well as in raising them to respect and treat those who are different from them with kindness [78]

These are some among many promising paths that integrate true pluralism by which we mean inclusion of all species into moral consideration and responsible citizenship within and beyond EE and ESD.

CONCLUSIONS

In this article, it was argued that education for sustainability should consider integrating animal rights, animal welfare, and deep ecology (DE) into the core of environment education (EE) and education for sustainable development (ESD) and that these three directions should guide education for sustainability. The instrumental attitude to nature appears to be insufficient for protection of the most vulnerable elements of the environment, or for addressing animal welfare or animal rights, and raising ethical objectives to the anthropocentric view of nature in education. Conservation education, education for deep ecology and education for animal rights can be seen as education that embraces a *truly plural* position that recognizes diversity of *all* species. This position would entail a critical education exposing the deficiencies of the mainstream anthropocentric ethics. This might necessitate a continuous affirmative action program as, unlike disenfranchised humans, non-humans will never be able to speak, even when threatened with extinction.

Mike Appleby, an animal welfare scientist was once asked: “What should we do about animal welfare?” He replied: “The answer, ladies and gentlemen, is ‘More’.” [79]. The same could be said about the question of what should be done about ethical considerations of environment and animals in EE and ESD [60]. What precise shape such recognition will take in both societal and educational discourse still needs to be discussed (e.g., should we oppose all research on animals, or only that which is particularly cruel and not for any meaningful purpose, *i.e.*, cosmetics?). The need for this urgent discussion should be a starting point. While we as researchers and academics would certainly benefit from extended discussion of how the non-human voices can be best represented and by whom, the move toward recognition that the very survival of these non-humans is at stake deserves to be recognized in education for sustainability. This type of change will require affirmative action programs on behalf of what is left of wild Nature and of a wonderful human capacity for empathy and compassion.

AUTHOR CONTRIBUTIONS

Helen Kopnina conceived and set out the initial purpose of the work. Helen Kopnina and Brett Cherniak analyzed the relevant material. Helen Kopnina provided extensive experience in environmental education and animal rights education research. Brett Cherniak contributed critical discourse and experience in the democratic element of environmental education. Helen Kopnina and Brett Cherniak thoroughly edited the article throughout its development and review process. Helen Kopnina and Brett Cherniak agreed and reviewed upon all changes before any round of submissions. Helen Kopnina and Brett Cherniak wrote the paper.

REFERENCES

1. UNESCO. *The Belgrade Charter*; UNESCO: Paris, France, 1976.
2. Marsden, W.E. Animal welfare issues in geography and environmental education: Some British historical perspectives. *Int. Res. Geogr. Environ. Educ.* 2001, *10*, 394–410.
3. Regan, T. A Case for Animal Rights. In *Advances in Animal Welfare Science*; Fox, M.W., Mickley, L.D., Eds.; The Humane Society of the United States: Washington, DC, USA, 1986; pp. 179–189.
4. Norton, B.G. Environmental ethics and weak anthropocentrism. *Environ. Eth.* 1984, *6*, 131–148.
5. Naess, A. The shallow and the deep: Long-range ecology movement. *Inquiry* 1973, *16*, 95–99.
6. Hewson, C. What is animal welfare? Common definitions and their practical consequences. *Can. Vet. J.* 2003, *44*, 496–499.
7. World Society for the Protection of Animals. Education for Sustainable development. Available online: http://www.animalmosaic.org/Images/Education%20for%20Sustainable%20Development_tcm46–29460.pdf (accessed on 21 November 2015).
8. Animal Welfare Institute. Humane education. Available online: https://awionline.org/content/humane-education(accessed on 21 November 2015).
9. International Fund for Animal Welfare. Available online: http://www.

ifaw.org/united-states/our-work/education(accessed on 21 November 2015).

10. LaChapelle, D. Educating for deep ecology. *J. Exp. Educ.* 1991, *14*, 18–22.
11. Drengson, A.R. Introduction: Environmental crisis, education, and deep ecology. *Trumpeter* 1991, *8*, 97–98.
12. Root-Bernstein, M.; Root-Bernstein, M.; Root-Bernstein, R. Tools for thinking applied to nature provide an inclusive pedagogical framework for environmental education. *Fauna Flora Int. Oryx* 2014, *3*, 1–9.
13. Sauvé, L. Currents in environmental education: Mapping a complex and evolving pedagogical field. *Can. J. Environ. Educ.* 2005, *10*, 11–37.
14. Glasser, H. Learning Our Way to a Sustainable and Desirable World: Ideas Inspired by Arne Naess and Deep Ecology. In *Higher Education and the Challenge of Sustainability: Problematics, Promise, and Practice*; Corcoran, P.B., Wals, A.E.J., Eds.; Springer Netherlands: New York, NY, USA, 2004; pp. 131–148.
15. Gorski, P.C. Critical ties: The animal rights awakening of a social justice educator. 2009. Available online: http://www.edchange.org/publications/animal-rights-social-justice.pdf (accessed on 21 November 2015).
16. Hickman, J.A. Personal care products council: An animal cruelty analysis. 2010. Available online: http://www.edchange.org/multicultural/humane/pcpc.pdf (accessed on 21 November 2015).
17. Kronlid, D.O.; Öhman, J. An environmental ethical conceptual framework for research on sustainability and environmental education. *Environ. Educ. Res.* 2013, *19*, 21–44.
18. Garner, R. Environmental Politics, Animal Rights and Ecological Justice. In *Sustainability: Key Issues*; Kopnina, H., Shoreman-Ouimet, E., Eds.; Routledge Earthscan: New York, NY, USA, 2015.
19. Jamieson, D. Animal liberation is an environmental ethic. *Environ. Values* 1998, *7*, 41–57.
20. Eckersley, R. Liberal democracy and the rights of nature: The struggle for inclusion. *Environ. Polit.* 1995, *4*, 169–198.
21. Eckersley, R. *The Green State: Rethinking Democracy and Sovereignty*; MIT Press: London, UK, 2004.
22. Dobson, A. *Citizenship and the Environment*; Oxford University Press: Oxford, UK, 2003.
23. Dobson, A. *Listening for Democracy*; Oxford University Press: Oxford, UK, 2014.

24. Washington, H.G. *Human Dependence on Nature*; Routledge: New York, NY, USA, 2013.
25. Crist, E. Ecocide and the extinction of animal minds. In *Ignoring Nature no More: The Case for Compassionate Conservation*; Bekoff, M., Ed.; Chicago University Press: Chicago, IL, USA, 2013; pp. 45–53.
26. Crist, E. Abundant Earth and Population. In *Life on the Brink: Environmentalists Confront Overpopulation*; Cafaro, P., Crist, E., Eds.; University of Georgia Press: Athens, GA, USA, 2012; pp. 141–153.
27. Bonnett, M. Sustainable development, environmental education, and the significance of being in place. *Curr. J.* 2013,*24*, 250–271.
28. Kopnina, H. Education for sustainable development (ESD): The turn away from "environment" in environmental education? *Environ. Educ. Res.* 2012, *18*, 699–717.
29. Nevin, E. Education and sustainable development. *Policy Pract. A Dev. Educ. Rev.* 2008, *6*, 49–62.
30. Jickling, B.; Wals, A.E.J. Globalization and environmental education: Looking beyond sustainable development. *J. Curric. Stud.* 2008, *40*, 1–21.
31. UNESCO. United Nations Decade of Education for Sustainable Development (2005–2014). In *Framework for the International Implementation Scheme*; UNESCO: Paris, France, 2005; Volume 32C.
32. Wals, A.E.J. *Shaping the Education of Tomorrow: 2012 Full-Length Report on the UN Decade of Education for Sustainable Development*; UNESCO: Paris, France, 2012.
33. Wals, A.E.J.; Jickling, B. "Sustainability" in higher education: From doublethink and newspeak to critical thinking and meaningful learning. *Int. J. Sustain. High. Educ.* 2002, *3*, 221–232.
34. Breiting, S. Issues for environmental education and ESD research development: Looking ahead from WEEC 2007 in Durban. *Environ. Educ. Res.* 2009, *15*, 199–207.
35. Jickling, B. Wolves, Ethics and Education: Looking at Ethics through the Yukon Wolf Conservation and Management Plan. In *A Colloquium on Environment, Ethics and Education*; Jickling, B., Ed.; Yukon College: Whitehorse, YT, Canada, 1996; pp. 158–163.
36. Öhman, J.; Östman, L. Clarifying the ethical tendency in education for sustainable development practice: A wittgenstein-inspired approach. *Can. J. Environ. Educ.* 2008, *13*, 57–72.
37. Torgerson, D. *The Promise of Green Politics: Environmentalism and the*

Public Sphere; Duke University Press: Durham, NC, USA, 1999.

38. Thompson, K. *Do We Need Pandas? The Uncomfortable Truth about Biodiversity*; UIT Cambridge Ltd.: Cambridge, UK, 2010.
39. Kareiva, P.; Lalasz, R.; Marvier, M. Conservation in the anthropocene: Beyond solitude and fragility. *Breakthr. J.* 2011, 27–29. Available online: http://thebreakthrough.org/index.php/journal/past-issues/issue-2/conservation-in-the-anthropocene (accessed on 21 November 2015).
40. Barry, J. *Rethinking Green Politics: Nature, Virtue and Progress*; Sage: London, UK, 1999.
41. Hui, K. Moral anthropocentrism is unavoidable. *Am. J. Bioeth.* 2014, *14*, 25.
42. Light, A. Compatibilism in Political Ecology. In *Environmental Pragmatism*; Light, A., Katz, E., Eds.; Routledge: New York, NY, USA, 1996.
43. Norton, B.G. Conservation and preservation: A conceptual rehabilitation. *Environ. Eth.* 1986, *8*, 195–220.
44. Wals, A.E.J. Between knowing what is right and knowing that it is wrong to tell others what is right: On relativism, uncertainty and democracy in environmental and sustainability education. *Environ. Educ. Res.* 2010, *16*, 143–151.
45. Van Poeck, K.; Vandenabeele, J. Learning from sustainable development: Education in the light of public issues.*Environ. Educ. Res.* 2012, *18*, 541–552.
46. Katz, E. A pragmatic reconsideration of anthropocentrism. *Environ. Eth.* 1999, *21*, 377–390.
47. Rolston, H., III. Environmental Ethics for Tomorrow: Sustaining the Biosphere. In *Sustainability: Key Issues*; Kopnina, H., Shoreman-Ouimet, E., Eds.; Routledge Earthscan: New York, NY, USA, 2015.
48. Callicott, J.B. The pragmatic power and promise of theoretical environmental ethics. *Environ. Values* 2002, *11*, 3–25.
49. Naess, A. Beautiful action. Its function in the ecological crisis. *Environ. Values* 1993, *2*, 67–71.
50. Bansel, P. Subjects of choice and lifelong learning. *Int. J. Qual. Stud. Educ.* 2007, *20*, 283–300.
51. Davies, B.; Bansel, P. Neoliberalism and education. *Int. J. Qual. Stud. Educ.* 2014, *20*, 247–259.
52. Liddick, D.R. *Eco-terrorism: Radical Environmental and Animal*

Liberation Movements; Praeger Publishers: Santa Barbara, CA, USA, 2006.

53. Kahn, R. *Critical Pedagogy, Ecoliteracy and Planetary Crisis: The Ecopedagogy Movement*; Peter Lang: New York, NY, USA, 2010.
54. Kopnina, H. The Lorax complex: Deep ecology, ecocentrism and exclusion. *J. Integr. Environ. Sci.* 2012, *9*, 235–254.
55. Kopnina, H. If a tree falls: Business students' reflections on environmentalism. *Int. J. Environ. Sustain. Dev.* 2014, *8*, 311–329.
56. Kopnina, H. If a tree falls and everybody hears the sound: Teaching deep ecology to business students. *J. Educ. Sustain. Dev.* 2015, *9*, 1–16.
57. IMDB Film Reviews. If a Tree Falls: A Story of the Earth Liberation Front. Available online: http://www.imdb.com/title/tt1787725/reviews (accessed on 21 November 2015).
58. Kelly, S. Geschlecht, speciesism, and animal rights in Leopold von Sacher-Masoch. *J. Crit. Anim. Stud.* 2014, *12*, 28–50.
59. Lidskog, R.; Elander, I. Addressing climate change democratically. Multi-level governance, transnational networks and governmental structures. *Sustain. Dev.* 2009, *18*, 32–41.
60. Kopnina, H.; Gjerris, M. Are some animals more equal than others? Animal rights and deep ecology in environmental education. *Can. J. Environ. Educ.* 2015, *20*, 109–123.
61. Crist, E.; Kopnina, H. Unsettling anthropocentrism. *Dialect. Anthropol.* 2014, *38*, 387–396.
62. O'Neill, J. Who Speaks for Nature? In *How Nature Speaks: The Dynamics of the Human Ecological Condition*; Haila, Y., Dyke, C., Eds.; Duke University Press Books: Durham, NC, USA, 2006.
63. Norris, K.S.; Jacobson, S.K. Content analysis of tropical conservation education programs: Elements of success. *J. Environ. Educ.* 1998, *30*, 38–44.
64. Goodall, J. Caring for People and Valuing Forests in Africa. In *Protecting the Wild: Parks and Wilderness, The Foundation for Conservation*; Wuerthner, G., Crist, E., Butler, T., Eds.; The Island Press: Washington, DC, USA; London, UK, 2015; pp. 21–26.
65. Noss, R.F.; Dobson, A.P.; Baldwin, R.; Beier, P.; Davis, C.R.; DellaSala, D.A.; Francis, J.; Locke, H.; Nowak, K.; Lopez, R.; *et al.* Bolder Thinking for Conservation. In *Protecting the Wild: Parks and Wilderness, The Foundation for Conservation*; Wuerthner, G., Crist, E., Butler, T., Eds.; The Island Press: Washington, DC, USA; London, UK, 2015; pp. 17–21.

66. Sandell, K.; Öhman, J. Educational potentials of encounters with nature: Reflections from a Swedish outdoor perspective. *Environ. Educ. Res.* 2010, *16*, 113–132.
67. Galen D'Amato, L.; Krasny, M.E. Outdoor Adventure Education: Applying Transformative Learning Theory to Understanding Instrumental Learning and Personal Growth in Environmental Education. *J. Environ. Educ.* 2011, *42*, 237–254.
68. Glasser, H. Naess's deep ecology: Implications for the human prospect and challenges for the future. *Inq. Interdiscip. J. Philos.* 2011, *54*, 52–77.
69. Bonnett, M. Retrieving nature: Education for a post-humanist age. *J. Philos. Educ.* 2003, *37*, 551–730.
70. Bonnett, M. Sustainability, the Metaphysics of Mastery and Transcendent Nature. In *Sustainability: Key Issues*; Kopnina, H., Shoreman-Ouimet, E., Eds.; Routledge Earthscan: New York, NY, USA, 2015.
71. Louv, R. *Last Child in the Woods: Saving Our Children from Nature-Deficit Disorder*; Workman Publishing Company: New York, NY, USA, 2008.
72. Vucetich, J.A.; Nelson, M.P. The Infirm Ethical Foundations of Conservation. In *Ignoring Nature No More: The Case for Compassionate Conservation*; Bekoff, M., Ed.; University of Chicago Press: Chicago, IL, USA, 2013; pp. 9–25.
73. Washington, H. *Demystifying Sustainability: Towards Real Solutions*; Routledge: London, UK, 2015.
74. Foundation for Deep Ecology. Available online: http://deepecology.org/mission.htm (accessed on 23 November 2015).
75. Rewilding Foundation. Available online: www.rewildingfoundation.org (accessed on 23 November 2015).
76. National Association for Environmental Education (UK). Available online: http://naee.org.uk/ (accessed on 23 November 2015).
77. Royal Society for the Prevention of Cruelty to Animals. Available online: http://education.rspca.org.uk/education/home (accessed on 23 November 2015).
78. PETA Kids. Available online: http://www.petakids.com/parents/teaching-compassion/#ixzz3LXBoruUv (accessed on 21 November 2015).
79. Hewson, C. Focus on animal welfare. *Can. Vet. J.* 2003, *44*, 335–336.

Chapter 8

AN ECOLOGY FOR CITIES: A TRANSFORMATIONAL NEXUS OF DESIGN AND ECOLOGY TO ADVANCE CLIMATE CHANGE RESILIENCE AND URBAN SUSTAINABILITY

Daniel L. Childers[1], Mary L. Cadenasso[2], J. Morgan Grove[3], Victoria Marshall[4], Brian McGrath[4] and Steward T. A. Pickett[5]

[1]School of Sustainability, Arizona State University, Tempe, AZ 85287, USA

[2]Department of Plant Sciences, University of California, Davis, CA 95616, USA

[3]U.S. Forest Service, Baltimore, MD 21201, USA

[4]Parsons The New School for Design, New York, NY 10011, USA

[5]Cary Institute of Ecosystem Studies, Millbrook, NY 12545, USA

ABSTRACT

Cities around the world are facing an ever-increasing variety of challenges that seem to make more sustainable urban futures elusive. Many of these challenges are being driven by, and exacerbated by, increases in urban populations and climate change. Novel solutions are needed today if our cities are to have any hope of more sustainable and resilient futures. Because most of the environmental impacts of any project are manifest at the point of design, we posit that this is where a real difference in urban development can be made. To this end, we present a transformative model that merges urban design and ecology into an inclusive, creative, knowledge-to-action process. This design-ecology nexus—an ecology *for* cities—will redefine both the process and its products. In this paper we: (1) summarize the relationships among design, infrastructure, and urban development, emphasizing the importance of joining the three to achieve urban climate resilience and enhance sustainability; (2) discuss how urban ecology can move from an ecology *of* cities to an ecology *for* cities based on a knowledge-to-action agenda; (3) detail our model for a transformational urban design-ecology nexus, and; (4) demonstrate the efficacy of our model with several case studies.

INTRODUCTION

Sustainability is a broad ideal. It links social, environmental, and economic integrity through a process of democratic, inclusive goal setting [1,2]. Further, sustainability is a process, not an outcome or endpoint [3] and the transitions that cities undergo as they move towards more sustainable futures can take many forms [4]. In this paper we propose that an important way to enhance both urban sustainability and climate change resilience is for mutual learning between urban design practice and ecological science, with particular attention to adaptive processes and pluralistic governance. We define "design" as far more than a plan or drawing of an object or space before it is built or made. We conceive design to be the purpose, planning, or intention that exists, or is thought to exist, behind any action or object. It is a conscious effort to create order in all human activity and it includes our perceptions, actions, and review of the impacts of those actions (*sensu* [5]). Design can play a substantial role in environmental outcomes. For instance, Thackara [6] estimated that 80% of the environmental impact of any object, including the built environment, is determined at the design stage. Understanding and mitigating that impact demands a process of public deliberation that better integrates design and ecology in order to achieve sustainable urban development.

Throughout the paper, we focus our discussion and examples on challenges that cities face, and will face in the future, associated with climate change—with special emphasis on heat extremes, droughts, and floods. In the case of heat extremes, the urban heat island effect is one of the most well-established features of contemporary urbanization, but its social significance and implications for human health, environmental justice, and urban infrastructure remain important frontiers [7]. In the case of drought and flooding, climate change projections include an increased frequency of extreme precipitation events interspersed by longer, more extreme drought events [8]. Drought is not only a concern of arid-climate cities [9,10]. In fact, the effects of drought as a limit to development and as a periodic threat to the functioning of mesic urban systems are becoming more prevalent [10]. However, drought in its various manifestations is often seen as a temporary pulse to be waited out, rather than as a stimulus for sustainable design, adaptation, and planning [11]. On the other hand, stormwater management is a strong stimulus for sustainable design, adaptation, and planning in all cities [12,13]. Flood risk is a similar stimulus for river cities, as is the inevitable risk of sea level rise and ocean flooding in coastal cities. All are examples of extreme conditions that are expected to intensify in the future. They often have disproportionate impacts on the most vulnerable populations, such as youth and the elderly, low-income people, people of color, and immigrants [14].

Urban design has been defined by and associated with architects, landscape architects, and planning professionals since the mid-20th century. However, as cities worldwide must adapt to economic restructuring, mass migrations, and climate change, urban design is quickly evolving to include civil society, community actors, engineers, and city managers [15,16]. The urban design process, as it is more broadly conceived, has the capacity to incorporate and give physical form to our growing scientific understanding of urban regions as social-ecological systems [17] facing new climate change risks and vulnerabilities. Pre-industrial cities were largely designed to provide human comfort in sometimes-hostile conditions. Later, modern infrastructure put distance between urban design and direct climate response, but recent innovations in climate-appropriate architecture and design are beginning to address this dichotomy. We argue that an integrated design-ecology nexus has the potential to stimulate a new era of sustainable urban development based in novel systems of deliberative decision-making and governance. The rationale for this integration was well articulated by Tanner *et al.* [16], and here we both expand on that rationale and present a model to make it happen. This merging of design and ecology will advance urban sustainability most quickly when a nexus of feedbacks between them is purposefully created to promote adaptive change and risk mitigation in both urban neighborhoods and regions. As our species continues to become more urban [18], this transformational design-ecology nexus is increasingly critical. It will help existing and new cities meet the challenges of growing in sustainable ways while becoming more resilient to severe, climate-induced events. In essence, society cannot adequately meet these challenges if urban design and ecology continue to be practiced as separate fields and in isolation from residents, city governance, and decision-making.

In this paper we: (1) summarize the relationships among design, infrastructure, and urban development with particular emphasis on the importance of joining the three to achieve urban climate resilience and enhance sustainability; (2) discuss the need for urban ecology to move from an ecology *of* cities [19,20] to an ecology *for* cities based on a knowledge-to-action agenda; (3) present a model for a transformational urban design-ecology nexus, and; (4) demonstrate the efficacy of our model with several case studies.

DESIGN, INFRASTRUCTURE, AND SUSTAINABLE URBAN DEVELOPMENT

Urban design became a professional practice in the middle of the 20th century as cities around the world faced the enormous challenges of reconstruction after World War II [21]. Through its development, the large-scale *tabula rasa* or "blank slate" approach to modern urban planning was replaced by code-based district or patch-scale urban design. This shift emphasized reconstructing or reconnecting to traditional cities and their fine-scale streets, blocks, and open space rather than the large-scale infrastructural emphasis of modern city planning. The discipline of landscape ecology helped shape regional planning approaches to, for example, greenfield development [22,23] and more recent movements labeled landscape or ecological urbanism have focused on post-industrial brownfield redevelopment [24,25]. However a more comprehensive and integrative design-ecology nexus is needed to consider greenfield, brownfield, and greyfield areas of cities while operating at scales of existing and new neighborhood development [26].

It is well known that urbanization is increasing globally, and converting ever-greater areas of land worldwide. In addition to direct land conversion, urban processes, lifestyles, and investments are transforming existing cities while also influencing lands far distant from traditionally recognized cities [27,28]. Thus, an urban perspective on and concern for sustainability is increasingly relevant to a large proportion of the Earth's land covers, ecosystems, and populations. Many cities are already addressing the need to advance urban sustainability with formal sustainability plans. However, these plans often focus on particular sectors, infrastructure types, or processes, such as transportation, water, or energy [29]. In many cases these sustainability plans are not adequate to prepare cities for the crises and tipping points that so many cities and new urban areas face currently or will experience in the future [3,4]. These plans often suffer from a lack of integration and holistic vision that connects across sectors and allows a networked approach to planning for the future. Many sustainability plans also tend to have a narrow focus on existing "hard" infrastructures (e.g., transportation, water supply, and sanitary and stormwater treatment systems) and on "low hanging fruit" green infrastructures (e.g., parks and public trees [30,31,32]). Furthermore, these plans have rarely been integrated into the smaller-scale neighborhood approach of contemporary urban design practice [33,34].

Urban infrastructure takes many forms. Engineered infrastructure, or "gray" infrastructure, is the capital investments that move or house people, goods, water, waste, and energy and the associated guidelines for their construction, operation, maintenance, and rehabilitation. In most contemporary cities, what is in place today was largely deployed over the past century during the era of the "Sanitary City" [34], when less was known about the environmental, social, and climate change impacts of the physical design of systems. Gray infrastructure was built to protect against environmental hazards by, in general, overdesigning or designing to control the environmental factors [35]. This reduced the likelihood that infrastructure would fail—a "fail safe" goal. This kind of engineered resilience is rigid and unyielding; it is not adaptive and imparts large inertias on urban systems [3]. New approaches that couple infrastructure and urban ecological services are needed, as evidence accumulates that these twentieth century infrastructure design principles often make urban systems more vulnerable to extreme climate events [16,36,37].

Urban infrastructure that incorporates natural elements is often called "green" infrastructure. This includes traditional examples, such a parks and street trees, but also includes more novel urban components such as community gardens and multi-purpose and multi-function stormwater management facilities. Virtually all cities are also characterized by a variety of water features that provide a range of ecosystem services, including rivers and streams, lakes, and fountains. Collectively these are known as "blue" infrastructure. Because green and blue infrastructure features take advantage of natural structures and ecological processes, they are surprisingly adaptable to a changing future, and thus impart resilience to urban systems far more than do inertia-bound gray infrastructures. In fact, there are numerous options for using green and/or blue infrastructure designs instead of gray infrastructure [38]. We argue, as did Tanner *et al.* [16], that the best strategy for moving in this direction is for the urban design process to expand beyond the realm of engineers and planners in order to include urban ecologists, landscape architects, and deliberative decision-making and governance. Representatives of all disciplines and perspectives should be a part of the design process from beginning to end [39].

Urban design has social impacts across a range of social-ecological scales and can contribute to integrating the ecological functioning of green and blue infrastructure with the finer scales of individuals, urban neighborhoods, and communities. At the level of individuals, environmental stewardship can promote relaxation, mitigate stress, create self-confidence, and strengthen self-sufficiency. At the collective level of neighborhoods and communities, environmental stewardship can help establish trust, strengthen social cohesion, and enhance knowledge sharing [40,41,42]. Investments in urban design, in concert with green, blue, and turquoise infrastructure (Text Box 1), have economic impacts in the real estate sector and the workforce. For example, numerous studies have demonstrated the positive impacts of proximity to parks, community gardens, and tree-lined streets on residential real estate value [43,44] and on commercial activity in business districts [45]. As cities initiate sustainability plans and move towards more green, blue, and turquoise infrastructure designs, society will need a larger urban natural infrastructure management workforce. Many cities have embarked on widespread tree planting and green infrastructure initiatives, with the need for ongoing maintenance by professionals [46,47,48,49]. With a growing interest in new green, blue, and turquoise infrastructure, urban environmental restoration and maintenance jobs offer an opportunity for those with limited formal education or specialized job training to establish what are broadly referred to as "green-collar" careers. Furthermore, engaging in environmental restoration work may provide benefits to workers beyond simple employment, including exposure to and interactions with nature [50], which is a limited "commodity" in many urban locations. In short, cities are habitat for people, so the urban design process should include city residents and integrate a social component into design objectives and actions.

Green and blue infrastructure are familiar concepts and features of cities. These two terms, and color designations, delineate services provided by terrestrial versus aquatic urban ecological features and ecosystems. But these two infrastructure designations may not cover all urban ecological features and associated services. For example, urban wetlands are an important component of "natural" infrastructure in virtually all cities. While some urban wetlands are found in, or have been rehabilitated to be in, somewhat natural states, many of these wetland systems were designed and constructed to provide specific services. These services include wastewater treatment, stormwater management, recreation and aesthetics, and habitat. Wetlands are found where terrestrial and aquatic ecosystems meet, and are considered ecotone ecosystems. They have ecological functions and structures that are both terrestrial and aquatic in character [51]. As such, urban wetlands are both green and blue infrastructure. When one combines green and blue, the result is the color turquoise. Thus, we posit that urban wetlands are unique enough in the urban matrix to warrant the term "turquoise infrastructure".

Box 1: The Colors of Urban Design: Turquoise Infrastructure.

MOVING FROM AN ECOLOGY *OF* CITIES TO AN ECOLOGY *FOR* CITIES

Calls for improved integration of ecological knowledge with the theory and practice of urban design are decades old [52,53], including those from the UNESCO Man and the Biosphere Program (e.g., [54,55]). Yet successful integration is rare and is concentrated in a few institutions and opportunities [56,57,58]. We argue that there is a pressing need to institutionalize these interdisciplinary connections and to extend their scope. Sustaining both human well-being and the integrity of the biosphere continues to be a test of public decision-making and institutions, and a major challenge to the academy and its linkages to civic processes. The scope of this challenge is daunting because of the complexity of sustainability, whose three pillars—environment, society, and economy—encompass many dimensions, including social values, ethics, equity, justice, connectedness and heritage, ecological integrity, and economic viability. Our argument for and approach to a transformational design-ecology nexus is also a call for us to move from an ecology *of* cities, as described by Pickett *et al.* [19] and Grimm *et al.* [20], to an ecology *for* cities where urban ecologists, designers, planners, engineers, residents, and others are actively pursuing a more sustainable future (per [3]).

If sustainability is a socially-generated process, then social-ecological resilience is its mechanism [3]. Resilience as a theory highlights the need for understanding how complex adaptive systems develop and evolve, the role of disturbance and disruption, the drivers of collapse of complex social-ecological systems, the nature of reorganization after collapse, and the various modes of adaptation that permit reorganization [59,60,61]. However, standard resilience theory has largely neglected an explicit consideration of design, and both its intended and unintended consequences, and instead has focused on the more comfortable realm of policy and management. Thus, we need an improved understanding of urban sustainability that extends beyond the founding concerns of resilience theory with natural resources management to include the design process. The design realm often focuses on resilience as a socially- or politically-driven desire to build back after some disaster [62]. This is an important step, and additional analyses of resilience, such as those focused on transportation and associated urban form [63] exemplify additional steps. This transformative integration is required to achieve urban systems resilience, particularly in the face of extreme conditions, such as heat, drought, and floods. In the next section we describe a model for this transformation.

A MODEL FOR TRANSFORMING THE URBAN DESIGN-ECOLOGY NEXUS

Merely positing that a transformational change needs to occur is insufficient if positive outcomes are the goal. Thus, we present a model for how to advance an urban design-ecology nexus to meet the complex needs of cities facing a future of change and uncertainty. We propose that any model needs to have certain performance features. It needs to be flexible so that ecological boundaries—as spaces of action—can change and diversify, mirroring everyday environmental realities. It needs to be adaptive, inclusive, and responsive to our increasingly inequitable societies and human migrations. And it needs to be future thinking in ways that value diverse forms of imagination and communication—for example, drawing has a powerful role in capturing a shared sense of future possibilities [64]. These features of an urban design-ecology approach will substantially improve the capacity of urbanized jurisdictions and urban regions to improve their sustainability plans and actions. Our focus on extreme events, exemplified primarily by heat waves, drought, and floods, turns the attention of the design and policy-making communities to conditions that are not always present, but that nonetheless are important for making sustainable urban designs flexible and adaptable. Because design addresses many spatial scales, myriad clients, and is situated in a diversity of land covers and neighborhoods, engaging ecological science will vastly enhance the effectiveness of both the design process and of urban ecology [65].

First and foremost, our model is inclusive and involves all requisite parties in the urban design process (Figure 1B). As we noted in the section above, this transdisciplinary design process should include urban social and ecological scientists, landscape architects, city residents, and students of all related disciplines in addition to professional architects, engineers, and planners. A key contrast between our model and the status quo (Figure 1A) is that our model of the design process is non-linear and includes critical feedbacks, throughout the process, among key stakeholders, urban ecological research, and knowledge derived from previous projects [39]. In terms of sustainability science, this is the co-production or co-generation of urban design [66] that is manifest by providing a collaborative environment for urban designs to be envisioned and implemented [67,68].

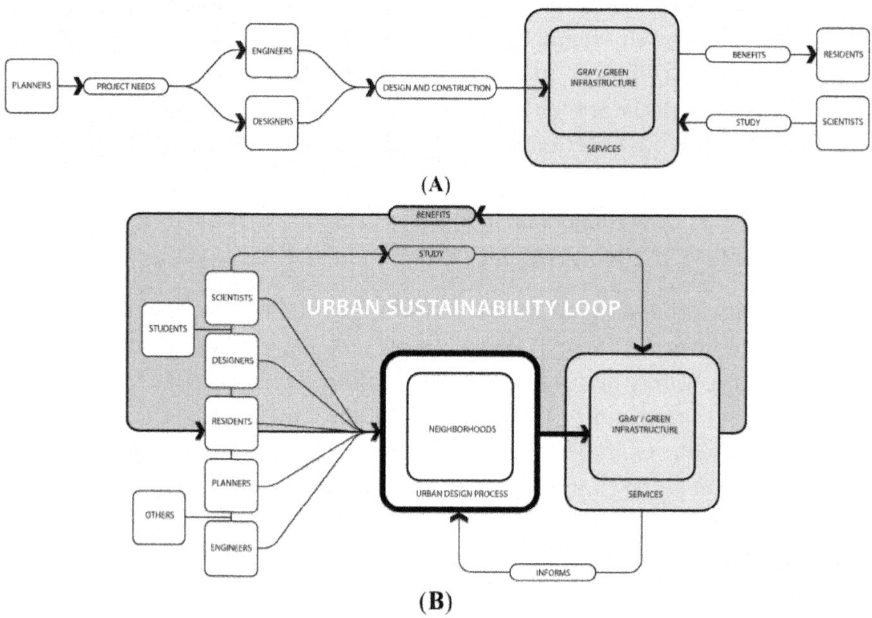

Figure 1: (A) (Top): The typical way in which urban infrastructure is conceived, de-

signed, constructed, and evaluated. Focus is often on only one or a few services and the process is linear; (B) (Bottom): The integrated approach, proposed here, for implementing urban infrastructure in which the design is co-produced in a transdisciplinary way, and the infrastructure is designed to provide multiple services. This approach is not linear; rather, it has several key feedback loops that we show as contained in an "urban sustainability loop"; this loop is highlighted in yellow for emphasis. Note that "Scientists" includes urban social and ecological scientists and "Others" may include elected officials, business leaders, famous individuals, powerful interest groups, *etc.*

Cities are designed to be human habitat, and as such *Homo sapiens* are the "ecosystem engineers" [69] of urban systems. No attempt to create a new, transformative, integrated approach to urban design can be successful without fully considering the social structures and dynamics of the city [57], including those that are value-laden and normative [70]. Paying attention to the well-being of populations that have been historically overburdened by environmental hazards and excluded from policy and planning decisions is also crucial in the design of sustainable and equitable cities and regions [54]. For this reason equity is a key tenet of sustainability [1,71,72]. A transformative urban design process must incorporate justice and equity [73], and strategies for doing so should be informed by history, including socio-spatial analyses of patterns of settlement, displacement, and segregation of vulnerable populations, and the policy and planning procedures that shaped those patterns [74].

Urban governance is defined as a mode of public decision-making that involves public, private, and non-profit actors at all levels of city management [75]. Governance theory can be used as an organizing framework to examine and understand macro-scale shifts from top-down state-centered approaches towards more interactive policy-making processes and hybrid arrangements that involve diverse interactions among multiple social actors [76,77,78]. In the context of urban environmental governance, a shift towards multi-sector activities can be a response to climate change uncertainty and a management strategy for enabling robust urban ecosystem services as well as systemic adaptation and resilience [79,80,81].

While urban planners and policy-makers have long addressed environmental issues, the constellation of groups associated with the governance of urban ecological processes has shifted in recent decades from hierarchical to polycentric structures [82,83,84,85,86]. In many cities, urban environmental stewardship groups have become an essential component of governance that regulates ecosystem services [87]. These groups include organizations that work to conserve, manage, monitor, advocate for, or educate about a range of quality-of-life issues regarding urban resources. They include community development groups focused on environmental issues and environmental groups focused on community-based quality-of-life issues [88]. Taken together, these constellations of civic groups can form a large component of novel urban governance systems and can be important stakeholders in our transformational model for the urban design process [89].

Designers can choose to locate their work within democratic politics in a range of modes. The consensus mode, as an ideal, is sometimes not realizable. In addition, it has been criticized for being hegemonic when there is no possibility of final reconciliation. In these situations designers can work in an agonistic mode of democratic politics by visualizing that which is repressed and destroyed by the consensus of post-political democracy [90], and expanding capacities for collective production and self-governance [89].

As we noted earlier, our model for a transformative urban design-ecology nexus is future-oriented and anticipatory. For this reason, a critical component of our model is creative visioning. Transformation does not happen instantaneously. Thus, any evolution of the urban design process must involve the education and training of a cadre of: (1) ecologically literate urban designers and engineers; (2) design-literate, engineering-conscious ecologists; (3) broad-thinking and holistically inclined planners, and; (4) place-aware and activist city residents. There are myriad ways to accomplish these educational goals [33,90], using both place-based and virtual approaches. Interdisciplinary design studios, workshops, community charrettes, and teaching are all critical, and may be facilitated by Internet-based interactions. This may be particularly valuable for facilitating interactions with government or non-profit employees who may find it difficult to travel to face-to-face studios or charrettes. Although direct interactions may be quite valuable, virtual interactions can also facilitate the contributions of community members in design processes that bring together experts from distant locations who might not otherwise be able to participate. Furthermore, virtual interactions can be digitally archived and used again, thus allowing activities to be virtual in both space and time. Our model does not intend to be prescriptive about the specifics, as long as transformative goals, inclusiveness, and a focus on the future are central to the approach. One might imagine that place-based "transformative design studios" would be located at academic institutions or at larger design firms, with strong connections to local planners and community members. A more ambitious approach would network and integrate a number of these inclusive, place-based studios across cities, with virtual connections to each other and to planning and to decision-making entities in their respective cities.

Regardless of the approach, it is critical that these "transformative design studios" or "arenas of inclusive design" produce: (1) methodologies for rigorously assessing the effectiveness of ecologically-informed designs both on the "drawing board" and in the field; (2) platforms for the transdisciplinary sharing of sustainable ecological design insights; (3) lasting institutional and organizational collaborations, preferably across a spectrum of climatic conditions for the inter-city networks, to maintain these efforts; and (4) an expansion of general theories of inclusive urbanism [17,26,91,92]. This integrated network of students, community members and leaders, graduates, scientists, and teachers should connect across the nation, and ultimately the globe. But at the same time they must engage their local institutions and government agencies that have a history of acting in isolation despite the growing calls for sustainable urban design, revitalization, and regional integration. In short, our ambition for a transformational urban design-ecology nexus has the ultimate goal of changing the way cities are designed, managed, and inhabited based on a deeper social, ecological, technological, and political understanding of urban sustainability and of the urban experience.

EXAMPLES OF CHALLENGES AND SUCCESSES

Fine-scale experimental urban design provides a unique opportunity to explore transformative new approaches to the design process [93,94]. In our inclusive model for transforming the urban design process (Figure 1), all players involved in design would also be continually involved in these on-the-ground design initiatives or experiments. As such, our model includes both the co-production of design and the co-production of knowledge through design experimentation, all happening in a safe space—real or virtual—for collaboration and communication. Below we briefly describe several case studies in which this co-production approach has been attempted or used successfully.

- Goodyear AZ USA: This project was focused on an experimental streetscape project that was initiated at the behest of city planners from the City of Goodyear, AZ, USA. A team of urban ecologists, landscape architects and students, and social scientists was assembled to design and implement four streetscape options with differing landscaping, water use, and microclimate impacts. The streetscape was to be built on a 25 m × 750 m plot of vacant land along Goodyear's main boulevard, near City Hall (Figure 2). City planners and community members from an adjacent historic neighborhood provided ideas on and local knowledge that informed designs drafted by an upper division landscape architecture studio class at Arizona State University. Prior to construction of the experimental streetscape, data were to be collected on human-scale microclimate to quantify physical characteristics that relate to people's perceptions of their immediate climate (Figure 3).

Educational signage was planned to keep the neighborhood and users of this unique streetscape park updated on new findings and project progress, and residents were to be periodically be surveyed about their project-related perceptions and values. After the project was constructed, biophysical and social monitoring were to continue to quantitatively identify the best, and most desirable, streetscape designs, as well as the success of the park as an education and outreach tool. This project should have been a showcase example of how integrating ecology, design, social science, and policy moves us from an ecology *of* cities to an ecology *for* cities (per [3]). Unfortunately, the entire project was ultimately cancelled after more than a year of negotiations between City officials, the landowner, and lawyers. The park was to be located on the edge of a remediated Superfund site, and the legal challenges associated with this finally made the project untenable. As this paper was published, City officials were looking for another site for the park. This is an example of the challenges that can be faced when trying to implement our transdisciplinary ecology-design nexus model.

Figure 2: Street view of the property in Goodyear, AZ, USA where the experimental demonstration streetscape will be constructed. The park was to be on the left and the historical neighborhood is on the extreme right (photo credit: D. Childers).

Figure 3: Micrometeorological towers used to monitor variables related to people's

perceptions of their immediate climate. Towers include sensors that measure air temperature, relative humidity, wind speed, and light at 2 m height, as well as soil temperature, soil moisture, and irrigation water use (photo credit, B. Ruddell).

- Baltimore MD USA: Over ten years, academic urban design studios have been instrumental in building a dialogue about sustainable urban development between scientific research by the Baltimore Ecosystem Study Long-Term Ecological Research Program (BES LTER) and local non-profit and government organizations. The urban design-ecology nexus developed by the BES LTER Urban Design Working Group (UDWG) connects the neighborhood scale of urban design practice with the ecological theory of patch dynamics. Urban neighborhoods in the Baltimore region are distinct land-cover neighborhood patches within larger watershed and human ecosystem frameworks. Early studios organized by the BES UDWG focused on connecting the emerging social infrastructure of the Gwynns Falls trail system to new urban land cover and hydrological data [95]. This work, based on an open urban stream network, continued in the underground piped Watershed 263 in West Baltimore [96]. It also included proposals for the redevelopment of the Middle Branch waterfront with the office of Baltimore City Planning. More recently design studios with the Baltimore Office of Sustainability and two neighborhood groups—The Harlem Park Neighborhood Association/Parks and People and the Historic East Baltimore Action Coalition—provided examples of how ecosystem science can be integrated with local grassroots efforts to rejuvenate neighborhoods suffering from disinvestment. Working with vacant lots and unused city space, social capital and job generation were a critical part of these green and blue infrastructure proposals. Examples of the resulting designs can be found at [97,98]
- The Urban Complex, NJ USA: Between 2002 and 2010, the urban design and landscape architecture firm Till Design undertook two client-based, professional-practice, brownfield redevelopment projects. The first, Monroe Center for the Arts, includes an art community in two old factory buildings and two new buildings surrounded by a plaza on two blocks located on a former wetland on the western edge of Hoboken, NJ. The second, Celadon, is a new high-rise, high-density, transit-oriented, mixed-use development located on a municipal dump on Newark Bay, in Elizabeth NJ. In both projects, urban ecology informed the design in three ways to increase the economic, cultural, and environmental value of the projects:
- • The inclusion of green infrastructure such as stormwater features, roof

gardens, vegetated plazas, and waterfront promenades.

- The inclusion of a phased approach where micro-scale social-ecological engagement was supported from the very start of the project. This allowed the project to debut publicly, with elements such as a drawing club on a parking lot, a temporary film garden, a low-tech kayak launch, and a sound game. Other elements included fountains, roof gardens accessible by public elevators, the co-production of a "bay music" sound signature, and a ferry to Manhattan.
- The inclusion of innovative financing in short, medium, and long concurrent cycles that were designed and launched to allow both of these environmentally activist real-estate development to happen [99,100].
- Sacramento, CA, USA: The Building Healthy Communities site in Sacramento, CA has embarked upon a process of community planning in order to engage residents and develop comprehensive and inclusive neighborhood action plans that inform advocacy. Building Healthy Communities is a program of the California Endowment, a private, state-wide health organization that aims to improve the health status of all Californians by considering what happens outside of the doctor's office and beyond people's individual choices. In 2010, the California Endowment initiated a 10-year initiative called Building Healthy Communities (BHC). It chose 14 cities for its initial application based on the health status and challenges to health faced by the residents of these cities and on the potential to inspire broader policy changes that could create healthier environments. The Sacramento BHC consists of nearly 20 small neighborhoods spanning both incorporated and unincorporated areas of South Sacramento. More than 65,000 residents call the Sacramento BHC home. While these residents experience high rates of economic, educational, and health challenges, they also are represented by engaged and supportive elective officials. The residents are served by many active and collaborative grassroots advocacy organizations working to address these disparities. In 2014, a collection of organizations, led by the non-profit organization Ubuntu Green, held a series of events that culminated in a week-long charrette, engaging residents to develop community action plans around several issues including access to healthy food, economic development and housing, pedestrian and bike safety, and environmental health. Residents could participate in person or by responding to Internet-based surveys and commenting on social media outlets. All materials were presented in three languages and events ran through the day and into the evening, providing as much time as possible for participation. The resulting

community action plans will be released in early 2015 and will be used to inform advocacy for the organizations and resident groups associated with the BHC.

SUMMARY

Urban development takes a wide variety of forms around the world, from the older, traditional compact city, to the bustling metropolis of the 19th and 20th century, to the low density car-based exurb, to the informal megacity. Cities of all forms would benefit from the transformative nexus of ecology and design that we propose here in order to advance climate change resilience and enhance future sustainability. This design-ecology nexus does not prescribe a narrow range of solutions for green design or urban sustainability; rather, it is a transdisciplinary process of reiteration, feedbacks, deliberation, and debate between social and natural systems. An ecology *for* cities is a call for action-based ecological research and knowledge that is part of a new urban design process working at all scales of urban decision-making, from individual households to neighborhoods to regions. We argue that this inclusive, creative process will produce new and innovative solutions that will allow tomorrow's cities to be better prepared for a climate-uncertain future.

ACKNOWLEDGMENTS

The authors of this paper are members of the Urban Sustainability Research Coordination Network that is supported by the U.S. National Science Foundation (NSF) through Grant No. 1140070. Additional support has been provided by the NSF to DLC through the CAP LTER Program (Grant No. 1026865), to MLC, JMG, BM, and STAP through the BES LTER Program (Grant No. 1027188), and to MLC through Grant No. 0844778.

AUTHOR CONTRIBUTIONS

All authors contributed substantially to the ideas, concepts, and work presented in this paper. All authors were also involved in the preparation of the manuscript and have approved the submitted form.

REFERENCES

1. Gibson, R.B. Beyond the pillars: Sustainability assessment as a framework for effective integration of social, economic, and ecological considerations in significant decision-making. *J. Environ. Assess. Policy Manag.* 2006, *8*, 259–280.
2. Leach, M.; Scoones, I.; Stirling, A. *Dynamic Sustainabilities: Technology,*

Environment, Social Justice; Earthscan: New York, NY, USA, 2010.

3. Childers, D.L.; Pickett, S.T.A.; Grove, J.M.; Ogden, L.; Whitmer, A. Advancing urban sustainability theory and action: Challenges and opportunities. *Landsc. Urban Plan.* 2014, *125*, 320–328.
4. Pickett, S.T.A.; Boone, C.G.; McGrath, B.P.; Cadenasso, M.L.; Childers, D.L.; Ogden, L.A.; McHale, M.; Grove, J.M. Ecological science and transformation to the sustainable city. *Cities* 2013, *32*, S10–S20.
5. Papaneck, V. *Design for the Real World: Human Ecology and Social Change*; Academy Chicago Publishers: Chicago, IL, USA, 2005.
6. Thackara, J. *In the Bubble: Designing for a Complex World*; The MIT Press: Cambridge, UK, 2006.
7. Huang, G.L.; Zhou, W.Q.; Cadenasso, M.L. Is everyone hot in the city? Spatial pattern of land surface temperatures, land cover and neighborhood socioeconomic characteristics in Baltimore, MD. *J. Environ. Manag.* 2011, *92*, 1753–1759.
8. Milly, P.C.D.; Betancourt, J.; Falkenmark, M.; Hirsch, R.M.; Kundzewicz, Z.W.; Letternaier, D.P.; Stouffer, R.J. Stationarity is dead: Whither water management. *Science* 2008, *319*, 573–574.
9. Platt, R.H.; Rowntree, R.A.; Muick, P.C. Sustainability in urban ecosystems: Beyond an object of study. In *The Ecological City: Preserving and Restoring Urban Biodiversity*; University of Massachusetts-Amherst Press: Amherst, MA, USA, 1994; pp. 49–65.
10. Hunt, D.V.L.; Rogers, C.D.F. A Benchmarking System for Domestic Water Use. *Sustainability* 2014, *6*, 2993–3018.
11. Tingstad, A.H.; Groves, D.G.; Lempert, R.J. Paleoclimate Scenarios to Inform Decision Making in Water Resource Management: Example from Southern California's Inland Empire. *J. Water Res. Plan. Manag.* 2014, *140*, 04014025.
12. Desimini, J. Civic space in regional frameworks: Resilient approaches to urban design. In *Resilience in Ecology and Urban Design: Linking Theory and Practice for Sustainable Cities*; Pickett, S.T.A., Cadenasso, M.L., McGrath, B., Eds.; Springer: New York, NY, USA, 2013; pp. 307–318.
13. Shannon, K. Eco-engineering for water: From soft to hard and back. In *Resilience in Ecology and Urban Design: Linking Theory and Practice for Sustainable Cities*; Pickett, S.T.A., Cadenasso, M.L., McGrath, B., Eds.; Springer: New York, NY, USA, 2013; pp. 163–182.
14. Adaman, F. Power inequalities in explaining the link between natural

hazards and unnatural disasters. *Dev. Chang.*2012, *43*, 395–407.

15. McGrath, B. *Urban Design Ecologies*; John Wiley & Sons: London, UK, 2012.
16. Tanner, C.J.; Adler, F.R.; Grimm, N.B.; Groffman, P.M.; Levin, S.A.; Munshi-South, J.; Pataki, D.E.; Pavao-Zucherman, M.; Wilson, W.G. Urban ecology: Advancing science and society. *Front. Ecol. Environ.* 2014, *21*, 574–581.
17. Pickett, S.T.A.; Cadenasso, M.L.; McGrath, B. *Resilience in Ecology and Urban Design: Linking Thoery and Practice for Sustainable Cities*; Springer: New York, NY, USA, 2013; p. 499.
18. United Nations. *World Urbanization Prospects: The 2011 Revision Highlights*; United Nations: New York, NY, USA, 2012; p. 50.
19. Pickett, S.T.A.; Burch, W., Jr.; Dalton, S.; Foresman, T.W.; Rowntree, R. A conceptual framework for the study of human ecosystems in urban areas. *Urban Ecosyst.* 1997, *1*, 185–199.
20. Grimm, N.B.; Grove, J.M.; Pickett, S.T.A.; Redman, C.L. Integrated approaches to long-term studies of urban ecological systems. *BioScience* 2000, *50*, 571–584.
21. Shane, D.G. *Urban Design since 1945: A Global Perspective*; John Wiley & Sons: London, UK, 2011.
22. McHarg, I. *Design with Nature*; The Natural History Press: New York, NY, USA, 1969.
23. Forman, R.T. *Land Mosaics: The Ecology of Landscapes and Cities*; Cambridge University Press: Cambridge, UK, 1995.
24. Waldheim, C. *Landscape Urbanism*; Princeton Architectural Press: New York, NY, USA, 2006.
25. Mostafavi, M.; Doherty, G. *Ecological Urbanism*; Lars Muller: Zürich, Switzerland, 2010.
26. Spirn, A.W. Ecological Urbanism: A Framework for the Design of Resilient Cities. Available online: http://www.annewhistonspirn.com/pdf/Spirn-EcoUrbanism-2012.pdf (accessed on 1 November 2012).
27. Seto, K.C.; Reenberg, A.; Boone, C.G.; Fragkias, M.; Haase, D.; Langanke, T.; Marcotullio, P.; Munroe, D.K.; Olah, B.; Simon, D. Urban land teleconnections and sustainability. *Proc. Natl. Acad. Sci. USA* 2012, *109*, 7687–7692.
28. Boone, C.G.; Redman, C.L.; Blanco, H.; Haase, D.; Koch, J.; Lwasa, S.; Nagendra, H.; Pauleit, S.; Pickett, S.T.A.; Seto, K.C.; *et al.* Reconceptualizing land for sustainable urbanity. In *Rethinking Urban*

Land Use in a Global Era; Seto, K.C., Reenberg, A., Eds.; MIT Press: Cambridge, UK, 2014; pp. 313–330.

29. Newman, P.; Jenkins, I. *Cities As Sustainable Ecosystems: Principles and Practices*; Island Press: Washington, DC, USA, 2008.
30. Larson, E.K.; Earl, S.; Hagen, E.M.; Hale, R.; Hartnett, H.; McCracken, M.K.; McHale, M.; Grimm, N.B. Beyond restoration and into design: Hydrological alterations in aridland cities. In *Resilience in Ecology and Urban Design: Linking Theory and Practice for Sustainable Cities*; Pickett, S.T.A., Cadenasso, M.L., McGrath, B., Eds.; Springer: New York, NY, USA, 2013; pp. 183–210.
31. Novotny, V. Water-energy nexus: Retrofitting urban areas to achieve zero pollution. *Build. Res. Inf.* 2013, *45*, 489–604.
32. Sze, J.; Gambirrazio, G. Eco-cities without ecology: Constructing ideologies, valuing nature. In *Resilience in Ecology and Urban Design: Linking Theory and Practice for Sustainable Cities*; Pickett, S.T.A., Cadenasso, M.L., McGrath, B., Eds.; Springer: New York, NY, USA, 2013; pp. 289–297.
33. Bargmann, J. Just ground: A social infrastructure for urban landscape regeneration. In *Resilience in Ecology and Urban Design: Linking Theory and Practice for Sustainable Cities*; Pickett, S.T.A., Cadenasso, M.L., McGrath, B., Eds.; Springer: New York, NY, USA, 2013; pp. 347–354.
34. Melosi, M.V. *The sanitary city: Urban Infrastructure in America from Colonial Times to the Present*; Johns Hopkins University Press: Baltimore, MD, USA, 2000; p. 600.
35. Ahern, J. From fail-safe to safe-to-fail: Sustainability and resilience in the new urban world. *Landsc. Urban Plan.* 2011,*100*, 341–343.
36. Fiksel, J. Sustainability and resilience: Toward a systems approach. *Sustain. Sci. Pract. Policy* 2006, *2*, 14–21.
37. Anderies, J.M.; Janssen, M.A. Robustness of social-ecological systems: Implications for public policy. *Policy Stud. J.*2013, *41*, 513–536.
38. Beatley, T. *Green urbanism: Learning from European cities*; Island Press: Washington, DC, USA, 2000; p. 491.
39. Felson, A.J. The design process as a framework for collaboration between ecologists and designers. In *Resilience in Ecology and Urban Design: Linking Theory and Practice for Sustainable Cities*; Pickett, S.T.A., Cadenasso, M.L., McGrath, B., Eds.; Springer: New York, NY, USA, 2013; pp. 365–382.
40. Glover, T.D.; Shinew, K.J.; Parry, D.C. Association, sociability, and

civic culture: The democratic effect of community gardening. *Leis. Sci.* 2005, *27*, 75–92.

41. Svendsen, E. Cultivating resilience: Urban stewardship as a means to improving health and well-being. *Restorative Commons: Creating Health and Well-Being through Urban Landscapes*; Campbell, L., Wiesen, A., Eds.; General Technical Report. U.S. Department of Agriculture, Forest Service, Northern Research Station: Newtown Square, PA, USA, 2009; pp. 58–87.
42. Peters, K.; Elands, B.; Buijs, A. Social interactions in urban parks: Stimulating social cohesion? *Urban For. Urban Green.*2010, *9*, 93–100.
43. Morales, D.J. The contribution of trees to residential property value. *J. Arbor.* 1980, *6*, 305–308.
44. Anderson, L.M.; Cordell, H.K. Influence of trees on residential property values in Athens, Georgia (USA): A survey based on actual sales prices. *Land. Urban Plan.* 1988, *15*, 153–164.
45. Wolf, K. Business district streetscapes, Trees and consumer response. *J. For.* 2005, *103*, 396–400.
46. Baltimore Center for Green Careers. Available online: http://www.baltimoregreencareers.org (accessed on 12 March 2015).
47. City of Los Angeles. Million Trees LA. Available online: http://www.milliontreesla.org/mtabout1.htm (accessed on 12 March 2015).
48. City of Seattle. City of Seattle Comprehensive Plan. Available online: http://www.seattle.gov/dpd/Planning/Seattle_s_Comprehensive_Plan/ComprehensivePlan/ (accessed on 12 March 2012).
49. City of Phoenix. Tree and Shade Master Plan. Available online: https://www.phoenix.gov/parkssite/Documents/T%20and%20A%202010.pdf (accessed on 12 March 2015).
50. Falxa-Raymond, N.; Svendsen, E.; Campbell, L.K. From job training to green jobs: A case study for a young adult employment program centered on environmental restoration in New York City, USA. *Urban For. Urban Green.* 2013,*12*, 287–295.
51. Mitsch, W.J.; Gosselink, J.G. *Wetlands: Third Edition*; John Wiley and Sons: New York, NY, USA, 2000.
52. Spirn, A.W. *The Granite Garden: Urban Nature and Human Design*; Basic Books: New York, NY, USA, 1984.
53. Steiner, F. Urban landscape perspectives. *Landsc. Urban Plan.* 2014, *3*, 342–350.
54. Di Castri, F.; Baker, F.; Hadley, M. The Man and the Biosphere Program

as an Evolving System. *Ambio* 1981, *10*, 52–57.

55. Boyden, S.; Millar, S.; Newcombe, K.; O'Neill, B. *The Ecology of a City and Its People: The Case of Hong Kong*; Australian National University Press: Canberra, Australia, 1981.
56. Pickett, S.T.A.; McGrath, B.; Cadenasso, M.L. The ecology of the metacity: Shaping the dynamic; patchy, networked, and adaptive cities of the future. In *Resilience in Ecology and Urban Design: Linking Theory and Practice for Sustainable Cities*; Pickett, S.T.A., Cadenasso, M.L., McGrath, B., Eds.; Springer: New York, NY, USA, 2013; pp. 463–489.
57. Boone, C.G. Social dynamics and sustainable urban design. In *Resilience in Ecology and Urban Design: Linking Theory and Practice for Sustainable Cities*; Pickett, S.T.A., Cadenasso, M.L., McGrath, B., Eds.; Springer: New York, NY, USA, 2013; pp. 47–61.
58. Grove, J.M. Ecological and social linkages in urban design projects: A synthesis. In *Resilience in Ecology and Urban Design: Linking Theory and Practice for Sustainable Cities*; Pickett, S.T.A., Cadenasso, M.L., McGrath, B., Eds.; Springer: New York, NY, USA, 2013; pp. 355–360.
59. Holling, C.S. Cross-scale morphology, geometry, and dynamics of ecosystems. *Ecol. Monogr.* 1992, *62*, 447–502.
60. Wu, J.; Wu, T. Ecological resilience as a foundation for urban design and sustainability. In *Resilience in Ecology and Urban Design: Linking Theory and Practice for Sustainable Cities*; Pickett, S.T.A., Cadenasso, M.L., McGrath, B., Eds.; Springer: New York, NY, USA, 2013; pp. 211–229.
61. Pickett, S.T.A.; McGrath, B.; Cadenasso, M.L.; Felson, A.J. Ecological resilience and resilient cities. *Build. Res. Inf.* 2014,*42*, 143–157.
62. Vale, L.J.; Campanella, T.J. *The Resilient City: How Modern Cities Recover from Disaster*; Oxford University Press: New York, NY, USA, 2005; p. 376.
63. Newman, P.; Beatley, T.; Boyer, H. *Resilient Cities: Responding to Peak Oil and Climate Change*; Island Press: Washington, DC, USA, 2009.
64. Marshall, V. Ecological Urban Design Visuality and Landscape. *Nakhara J. Environ. Des. Plan.* 2014, *10*, 113–126.
65. Felson, A.J.; Bradford, M.A.; Terway, T.M. Promoting Earth Stewardship through urban design experiments. *Front. Ecol. Environ.* 2013, *11*, 362–367.
66. Schellnhuber, H.J.; Crutzen, P.J.; Clark, W.C.; Hunt, J. Earth system analysis for sustainability. *Sci. Policy Sustain. Dev.*2005, *47*, 11–25.

67. Lemos, M.C.; Morehouse, B.J. The co-production of science and policy in integrated assessments. *Glob. Environ. Chang.*2005, *15*, 57–68.
68. Robinson, J.; Tansey, J. Co-production, emergent properties and strong interactive social research: The Georgia Basin Futures Project. *Sci. Public Policy* 2006, *33*, 151–160.
69. Lawton, J.H.; Jones, C.G. Linking species and ecosystems: Organisms as ecosystem engineers. In *Linking Species and Ecosystems*; Jones, C.G., Lawton, J.H., Eds.; Chapman and Hall: New York, NY, USA, 1995; pp. 141–150.
70. Rozzi, R.; Pickett, S.T.A.; Palmer, C.; Callicott, J.B. *Linking Ecology and Ethics for a Changing World: Values, Philosophy, and Action*; Springer: New York, NY, USA, 2014.
71. Agyeman, J.; Evans, T. Toward just sustainability in urban communities: Building equity rights with sustainable solutions. *Ann. Am. Acad. Polit. Soc. Sci.* 2003, *590*, 35–53.
72. Agyeman, J. *Sustainable Communities and the Challenge of Environmental Justice*; NYU Press: New York, NY, USA, 2005.
73. Pickett, S.T.A.; Buckley, G.L.; Kaushal, S.S.; Williams, Y. Social-ecological science in the humane metropolis. *Urban Ecosyst.* 2011, *14*, 319–339.
74. Lord, C.H.; Norquist, K. Cities as Emergent Systems: Race as a Rule in Organized Complexity. *Environ. Law* 2010, *40*, 551–597.
75. Stoker, G. Governance as theory: Five propositions. *Int. Soc. Sci. J.* 1998, *50*, 17–28.
76. Rhodes, R.A.W. The new governance: Governing without government. *Polit. Stud.* 1996, *44*, 652–667.
77. Jordan, A. The governance of sustainable development: Taking stock and looking forwards. *Environ. Plan. C Govern. Policy* 2008, *26*, 17–33.
78. Fisher, D.; Svendsen, E. Hybrid arrangements within the environmental state. In *Routledge International Handbook of Social and Environmental Change*; Lockie, S., Sonnenfeld, D., Fisher, D., Eds.; Routledge: London, UK, 2014; pp. 179–189.
79. Bulkeley, H.; Betsill, M. *Cities and Climate Change: Urban Sustainability and Global Environmental Governance*; Routledge: New York, NY, USA, 2003.
80. Adger, W.N.; Arnell, N.W.; Tompkins, E.L. Successful adaptation to climate change across scales. *Glob. Environ. Chang.*2005, *15*, 77–86.
81. Daily, G.C.; Polasky, S.; Goldstein, J.; Kareiva, P.M.; Mooney, H.A.;

Pejchar, L.; Ricketts, T.H.; Salzman, J.; Shallenberger, R. Ecosystem services in decision making: Time to deliver. *Front. Ecol. Environ.* 2009, *7*, 21–28.

82. Weber, E.P. A new vanguard for the environment: Grass-roots ecosystem management as a new environmental movement. *Soc. Natl. Resour.* 2000, *13*, 237–259.
83. Kempton, W.; Holland, D.C.; Bunting-Howarth, K.; Hannan, E.C. Local environmental groups: A systematic enumeration in two geographical areas. *Rural Sociol.* 2001, *66*, 557–578.
84. Sirianni, C.; Friedland, L. *Civic Innovation in America: Community Empowerment, Public Policy, and the Movement for Civic Renewal*; University of California Press: Berkeley, CA, USA, 2001.
85. Andrews, K.T.; Edwards, B. The organizational structure of local environmentalism. *Mobiliz. Int. Quart.* 2005, *10*, 213–234.
86. Svendsen, E.S.; Campbell, L.K. *The Urban Ecology Collaborative Assessment: Understanding the Structure, Function, and Network of Local Environmental Stewardship*; U.S.D.A. Forest Service Northern Research Station: New York, NY, USA, 2005.
87. Ernstson, H.; Sorlin, S.; Elmqvist, T. Social movements and ecosystem services—The role of social network structure in protecting and managing urban green areas in Stockholm. *Ecol. Soc.* 2008, *13*, Article 39.
88. Fisher, D.R.; Campbell, L.K.; Svendsen, E.S. The organizational structure of urban environmental stewardship.*Environ. Polit.* 2012, *12*, 26–48.
89. Svendsen, E. Storyline and design: How civic stewardship shapes urban design in New York City. In *Resilience in Ecology and Urban Design: Linking Theory and Practice for Sustainable Cities*; Pickett, S.T.A., Cadenasso, M.L., McGrath, B., Eds.; Springer: New York, NY, USA, 2013; pp. 269–287.
90. Mouffe, C. Art and Democracy: Art as an Agnostic Intervention in Public Space. *Open Art Public Issue* 2008, *14*, 6–15.
91. Hardt, M.; Negri, A. *Commonwealth*; Harvard University Press: Cambridge, UK, 2009.
92. Miss, M. Remixing messages: A call for collaboration between artists and scientists. In *Resilience in Ecology and Urban Design: Linking Theory and Practice for Sustainable Cities*; Pickett, S.T.A., Cadenasso, M.L., McGrath, B., Eds.; Springer: New York, NY, USA, 2013.
93. Ellin, N. Integral urbanism: A context for urban design. In *Resilience in Ecology and Urban Design: Linking Theory and Practice for Sustainable*

Cities; Pickett, S.T.A., Cadenasso, M.L., McGrath, B., Eds.; Springer: New York, NY, USA, 2013; pp. 63–78.

94. North, A.; Waldheim, C. Landscape urbanism: A North American perspective. In *Resilience in Ecology and Urban Design: Linking Theory and Practice for Sustainable Cities*; Pickett, S.T.A., Cadenasso, M.L., McGrath, B., Eds.; Springer: New York, NY, USA, 2013.

95. Felson, A.J.; Pickett, S.T.A. Designed experiments: New approaches to studying urban ecosystems. *Front. Ecol. Environ.*2005, *3*, 549–556.

96. Hager, G.W.; Belt, K.T.; Stack, W.; Burgess, K.; Grove, J.M.; Caplan, B.; Hardcastle, M.; Shelley, D.; Pickett, S.T.A.; Groffman, P.M. Socioecological revitalization of an urban watershed. *Front. Ecol. Environ.* 2013, *11*, 28–36.

97. Harlem Park, Baltimore, example. Available online: http://sce.parsons.edu/studiocourses/BaltimoreStudio2010/(accessed on 12 March 2015).

98. Baltimore Ecosystem Study Urban Design Working Group. Available online: http://beslter.org/frame4-page_3i_07.html (accessed on 12 March 2015).

99. Felson, A.J.; Oldfield, E.E.; Bradford, M.A. Involving ecologists in shaping large-scale green infrastructure projects.*BioScience* 2013, *63*, 882–890.

100. McGrath, B.; Marshall, V.; Cadenasso, M.L.; Grove, J.M.; Pickett, S.T.A.; Plunz, R.; Towers, J. *Designing Patch Dynamics*; Columbia University: New York, NY, USA, 2007.

Chapter 9

INDUSTRIAL ECOLOGY AND ENVIRONMENTAL LEAN MANAGEMENT: LIGHTS AND SHADOWS

Giuseppe Ioppolo[1], Stefano Cucurachi[2], Roberta Salomone[1], Giuseppe Saija[1] and Luigi Ciraolo[1]

[1]Department of Economics, Business, Environment and Quantitative Methods (SEAM), University of Study of Messina, P.zza Puglatti 1, Messina 98122, Italy

[2]Department CML-Industrial Ecology, Leiden University, 2300 RA Leiden, The Netherlands

ABSTRACT

Current industrial production is driven by increasing globalization, which has led to a steady increase in production volumes and complexity of products aimed at the pursuit of meeting the needs of customers. In this context, one of the main tools in the management of customer value is Lean Manufacturing or Production, though it is considered primarily as a set of tools to reduce the total cost of the resources needed to achieve such needs. This philosophy has recently been enriched in the literature with case studies that link Lean Management (LM) with the improvement of environmental sustainability. The consequence is an expansion of the Computer Integrated Manufacturing (CIM); indeed, CIM, currently, combining and integrating the key business functions (e.g., business, engineering, manufacturing, and information management) with a view of the life cycle, does not highlight the strategic role of the environmental aspects. In order to deal with the increasingly rapid environmental degradation that is reflected in society, in terms of both economy and quality of life, Industrial Ecology (IE) introduced a new paradigm of principles and instruments of analysis and decision support (e.g., Life Cycle Assessment—LCA, Social Life Cycle Assessment -SLCA, Material Flow Account—MFA, *etc.*) that can be considered as the main basis for integrating the environmental aspects in each strategy, design, production, final product, and end of life management, through the re-engineering of processes and activities towards the development of an eco-industrial system. This paper presents the preliminary observations based on a analysis of both theories (LM-IE) and provides a possible assessment of the key factors relevant to their integration in a "lean environmental management", highlighting both positives (lights) and possible barriers (shadows).

INTRODUCTION

In March 2010, the European Commission presented its strategy "Europe 2020", and took note that the Western model of economic development has triggered a process of environmental degradation of difficult resolution. Such problem overlaps, both with the problem of limited natural resources, and also with the inequality of resource availability across geographies. Europe 2020 is a decennary strategy aiming at a structural transformation of the European economy in order to overcome the economic crisis and the challenges of the next decade through competitive and sustainable forms [1]. To achieve this, the Europe 2020 Strategy addresses the issue of growth, declining in three main areas [1]:

(1) Smart growth (promotion of knowledge, innovation, education, and digital society);

(2) Sustainable growth (production more efficient in terms of resources and greater competitiveness);

(3) Inclusive growth (more jobs, skills, and combating poverty).

In this way, the Europe 2020 Strategy identifies the eco-industries (also defined environmental industries or companies environmental technology oriented) as the enabled actors in order to do grow wealth and employment without causing serious damage to the environment [1]. For example, the strategy promotes those industries that produce so-called "*enabling technologies*". These kinds of industries allow to greatly increase the performance and ability of the user without increasing the consumption of resources (as in the case of information technology); moreover, this industrial model allows to reach the famous "decoupling" between economic growth and resource use, including the creation of wealth and environmental impacts.

The priority initiatives indicated in Europe 2020, were further underlined by the European Commission in its Communication "Rio+20: Towards the green economy and better governance" [2]. Indeed, the EU highlights the combination of competitiveness and the green economy, outlining the strategic directions (e.g., shifting the aim on R&D, introducing clean technologies, pointing to an industrial policy for green growth as a means to improve competitiveness, creating new jobs working to achieve a low-carbon and efficient use of resources). The increasing attention to environmental aspects, measured along the entire life cycle of a product/process, either by the company or by all possible stakeholders, needs to align the production model to that of environmental management, in order to create a new business model that is green-oriented. The authors, in order to deal the increasingly rapid environmental degradation, in terms of both economy and quality of life, answering to the Europe 2020 strategy, carry out several key factors useful to the re-engineering of processes and activities towards the development of an eco-industrial system.

A possible way is furnished by the Industrial Ecology (IE) theory, that introduced a new paradigm of principles and instruments of analysis and decision support (e.g., LCA, SLCA, MFA, *etc.*); these tools can be considered as the main basis for assessing and integrating the environmental aspects in each strategy, design, production, final product, and end of life management.

This paper presents the preliminary considerations based on an examination of two accredited theories (also defined strategies and philosophies):

(1) The Lean Management (LM) as a highly competitive production model [3,4];

(2) The Industrial Ecology (IE) as a framework of principles and tools of environmental analysis.

The aim of this paper is to highlight the key factors relevant to their integration in a environmental lean management system, both of positive terms (lights) and possible barrier (shadows).

THEORY AND METHODS-LEAN MANAGEMENT AND INDUSTRIAL ECOLOGY

The Lean Production (LP) is based on the principles and processes introduced in the Toyota Production System (TPS), and was defined as "doing more with less" [5,6,7]. Womack, formalizing the principles of this theory, characterizes it as a system of measures and methods that, in a holistic approach, have the potential to reduce the production factors. It follows that the lean production model is reflected in the degree of competitiveness of the entire business system that adopts it. The LP is therefore a strategy or philosophy that promotes the use of practices, such as the *kanban*, a type of scheduling system, the total quality management (TQM) and just-in-time (JIT), to minimize scrap/waste and improve the performance of a company [8,9]. The LP initially broken down into four areas/strategic phases of production: Product development, supply chain, also called Kanban supermarket [10,11,12], the management of the workshop, and after-sales service. Thanks to a continued and applied research the LP has become a vital model for the entire business system, the Lean Management (LM) [13], emphasizing the expansion of the management production management of the organization in all its activities. In this way, the LM is often seen as a set of tools that compete for the reduction of the overall cost, and at the same time, are intended to improve the quality of manufactured products.

From an operational perspective, the LM is achieved through the adoption of a business model of integrated production or CIM (Computer Integrated Manufacturing), which represents the most complete form of integration between the different areas of a production system in an automated factory (*i.e.*, design, engineering, production, quality control, production planning, and marketing).

The CIM can be considered as a standard for the industry and may be automated through the use of Information Communication Technology (ICT), the LM is performed throughout the business life cycle (from design Computer Aided Design-Computer Aided Engineering-Finite Element Analysis (CAD-CAE-FEA, to the stages of production Computer Aided Manufacturing (CAM), to sales and after-sales service, a unique business model for optimal management of resources, Enterprise Resource Planning (ERP)).

Only recently the scientific literature proposes studies linking the philosophy of the LM with the improvement of environmental sustainability [14,15,16,17,18]. These studies suggest that lean production is more than a set of lean tools to optimize production efficiencies; it is a *modus operandi* and a mindset that must be integrated into production systems, in a systematic way, in order to achieve sustainability [16,17]. Obviously, the goal is to guide the LM towards a green perspective, and involves the crossing of conceptual limits of the same LM. In fact, although certainly the LM ensures better operating results, such as lower inventory through a lean warehouse, higher quality in all business processes, and overall shorter timescale due to a complete synchronization between events, however, it does not internalize the environmental perspective in their principles [6,10].

The above considerations are reflected in the "new inventory paradigm" introduced by Chikan [19], from which one can clearly understand what the relationships to other processes and functions within enterprises geared to profitability are. The paradigm does not focus on environmental issues, but on a number of mediating factors between lean production and financial performance, which is the real driving force in the spread of this theory. The paradigm can be detailed as follows:

(1) Establish a strategic vision: The *Value* must be defined jointly for each

product family, along with a cost-based target on the value perceived by the customer;

(2) Identify and establish teams (identify the flow): *Value Stream*, namely the monitoring and identification of responsibility in all activities that are necessary specifications from the design, management of orders and deliveries, launch, production and final delivery to the customer;

(3) Identify the products (slide the flow): *Flow*, rethink specific work practices and tools to eliminate returns, scrap and arrests (of any kind), so that the design, order and production of a specific product may proceed in a continuous manner;

(4) Identify processes: *Pull*, flow only active when pulled to the next step;

(5) Review the layout of the factory: *Perfection*, the complete elimination of muda (waste) so that all the activities cascade contribute to the creation of value;

(6) Select an appropriate strategy *Kanban* (pull-push adaptive approach);

(7) Improve continuously, while maintaining the excellent results.

In combination with the above principles, companies can apply different types of environmental practices to improve their productivity in the use of natural resources, such as energy and materials, and to reduce the related environmental impacts of their activities [4].

The report "Lean and Green", presented by Zokaei *et al.* [20], and the studies carried out by Glavic and Lukman [21] and Lozano [22] provide an overview of some of these practices (e.g., the use of cleaner forms of production, the introduction of models production-oriented eco-efficiency). For Cagno *et al.* [23], cleaner production is an initiative of environmental protection that the company puts in place with a view to prevention. This initiative is designed to minimize the amount of waste and emissions (negative output), while maximizing production (positive output).

Using the definition given by González del Río [24], clean technology can be assimilated to changes in production processes that reduce the amount of waste and pollution generated in the production process or during the entire life cycle of the product.

This approach directs the company towards a greater focus on the analysis and measurement of the flows of materials and energy produced by the relationship between the company itself and the reference area or crossing. In addition, the management, in a pro-active way, adopts strategies to reduce these flows in their industrial processes, using both the improvements in the management (for example, the integration of the quality management system with the environment management system-TQEM). Another strategy is the introduction of technological advances applied to production (introduction of the Environmental Technology Innovations (TEIs)), in order to develop an eco-friendly company policy (e.g., the sustainable issue are the waste reduction, the improvement in the integrated management of the water cycle, the reduction of greenhouse gas emissions and losses of warmth, the attention to other impacts, such as acoustic, *etc.*).

Korhone [25] rationalizes the above principles in seven critical success factors, which are necessary to achieve eco-efficiency in production:

(1) The reduction in the intensity of the material (de-materialization);
(2) The reduction in energy intensity (de-energized);
(3) The reduction of dispersion of toxic substances;
(4) The improvement the recyclability of materials introduced into the production process;
(5) The substitution of input materials with other resources more environmentally sustainable;
(6) The reduction in the degree of persistence of the pollutants;
(7) The value increasing of the intangible component of each product.

These success factors are combined with the need to incorporate eco-efficiency already in the vision and business strategy, along with a growing availability of cleaner technology, which transforms the green market in the new field on which to compete. In contrast, the process of industrialization, compared with an ever-increasing demand following the enlargement of markets, is linked to the follow main factors of environmental impact:

(1) The high level of air pollution caused by combustion processes;
(2) The pollution of water bodies, determined by discharges of process waste to which are added the discharges of household;
(3) The production and accumulation of waste, increasingly complex, and difficult to recycle;
(4) The production of new materials, chemical products, such as plastics and synthetic products are not biodegradable.

In this way, another important role is played by the community of all stakeholders, which have to be involved in the decision process in order to build a shared knowledge and a collective consciousness changing its behaviors [26].

As a result of the foregoing, the world of scientific research and technological innovation, has set the goal of finding the appropriate solutions, creating new technologies applied to production cycles, making it possible to prevent and/ or reduce pollution and to reduce, to minimal amount, the substances emitted (output) and the natural resources used (input).

At the same time, the increased level of awareness on the environmental risks has spread an ever-increasing attention on environmental aspects along the whole life cycle of a product/process; in this sense, it has reinforced the need to assess and share information, such as [27]:

(1) The environmental damage caused by the technological processes implemented and processed products;

(2) The actions necessary for modernization or modification of the technological process;

(3) The results of the comparative analysis for the definition of alternative technologies that lead to the production of the same product or a product change, reducing environmental impacts.

In response to these pressures from the legislature, the marketplace and the community, and the local, national, and international levels that, among the various theories have assumed a significant role in the Industrial Ecology (IE). The central concept and characterizing for the IE is an industrial system that should not be considered isolated from its surrounding systems, but in a position of continuous exchange. The IE deals with the systematic study of patterns of industry in relation to the natural and social systems involved, and it is designed to optimize such trade in terms of sustainability [28,29].

Tibbs [30] and Ayres and Ayres [15] summarize the concept of industrial ecology and translate it into an "industrial ecosystem", stressing the importance of thinking about how the characteristics of the natural ecosystem can be translated into an industrial ecosystem. The introduction of the concept of time-scale completes this concept, which in the anthropic system (techno-sphere) is relevant and not unlimited, allowing the analysis of the interactions that take place between all parties (competitive interactions and/or cooperation). Focusing on the principles of industrial ecology, they could be considered as the main support for the introduction of environmental considerations in all activities, from strategy to design, to production, to the realization of the product throughout the life cycle, through the king-engineering of processes and activities in an eco-industrial system.

To achieve these goals, it is necessary to understand the factors influenced a sustainable society; these are the scarcity of resources, the need for materials and the growth of energy consumption, the approach to the de-materialization and substitutability for sustainable development. In this sense, there are three major transitions [31,32]:

(1) The transition from fossil fuels to renewable energy sources;
(2) The transition from linear flows of an economy closed-loop material that gives economic value to the new secondary raw materials, with a view to re-use and recycling;
(3) The transition from the exploitation of nature and biodiversity to its protection.

The IE is a paradigm for environmental management principles and makes use of tools for environmental analysis and the definitions of compatible choices. It can then be finalized by the interpretation of the transformation of the industrial system connected to the load capacity of the territorial system in which it is rooted. In this sense, the IE has been regarded as a broad holistic framework to guide the transformation of the industrial system towards a model of sustainable production.

The profound change that is involved in the management of production from a linear model into a closed-loop model closely resembles the cyclical flows of ecosystems, drawing on the biological concept of ecology, which is the branch of biology that deals with the study of relationships that occur between organisms and the physical environment that hosts them [15].

Therefore, the IE seeks to structure the industrial models in a substantially closed loop, to benefit both economically and environmentally, not focusing on individual industrial processes, but proposing a new sustainable economy, based on a clear understanding of the interaction between the world of production and environmental system.

For sighting this result you need to avoid that the analyses are partial and simplistic, ignoring important variables and, above all, leading to unintended consequences.

The IE, in fact, addresses the entire lifecycle of a product-process, focusing on the use of resources and materials in relation to the analysis of energy flows, and through modeling systems, investigates the impact-relations environment, using a multidisciplinary and interdisciplinary approach, which aims to suggest options and more sustainable choices.

The IE uses a set of tools oriented to the product, e.g., the analysis of the life cycle (LCA), which seeks to ensure that, in the examination of an industrial process or product, all its interactions and impacts on the environment are fully accounted, from the extraction of raw materials, the processes of production, use and disposal of the product; this tool has been further specialized with the assessment of costs throughout the life cycle (LCC) and the evaluation of social aspects (SLCA), which are also particularly important in the industrial process in terms of human resource management [33].

Together with the LCA, there are a number of other tools oriented to the study of relationships between environmental systems (e.g., enterprise/ manufacturing district and geographical areas of production and sales). Among these services, one may include the analysis of the flow of materials (MFA), which adopts a macroscopic approach, the analysis of the flow of a substance (SFA) with a microscopic approach, tools based on input-output tables that use statistical data and are most suitable for studying environmental reports at national/international regulations (e.g., physical input-output table (PIOT), and the ecological network analysis (ENA)).

For the analysis of the flows of materials and energy the first law of thermodynamics on the conservation of matter is applied, allowing a consistent and comprehensive collection of input and output flows, and stocks within the study area, *i.e.*, the system under observation [34].

The analysis of the flows of materials and energy can be used at both the global and local scales. Globally, this analysis of flows can help determine the extent to which human activities are influencing/impacting the natural systems of the Earth (e.g., hydrological cycle, the carbon cycle, and the nitrogen cycle). At the local level or enterprise level, these instruments introduce research methods that calculate the mass balances of industrial processes, the results of which can be used to ensure that all resources are fully valued in a sustainable manner.

The IE also introduces tools that are particularly suited to the realization of environmental information that is clear and easily understandable, and especially consumer-oriented. For example, the concept of ecological footprint has been developed by Ayres [35] in response to the debate on the concept of "carrying capacity". The carrying capacity is defined as "the population of a given species that can be sustained indefinitely in a given habitat without permanently damaging the ecosystem on which it depends" [36].

Thus, the ecological footprint expresses the theoretical area (in a single indicator that is "global hectare") used by man to produce biological resources it consumes and to absorb the waste it generates (including CO_2 resulting from its use of energy) [37].

Finally, again, based on the LCA, it is appropriate to highlight the Design for Environment (also called eco-design), which is a widely used approach for the improvement of environmental performance. The Design for Environment integrates environmental considerations throughout the life cycle of the product from the earliest stages of product design [38].

What has been said up to now shows the intrinsic connections between the LM and IE approaches. Based on the previous discussion of both theories, it is possible to identify a set of common evaluation criteria, which use the same data to measure the relationship between the company (techno-sphere) and the environment (eco-sphere).

The full integration of the two theories could represent the real success factor by single operating unit up to complex industrial system (e.g., District), observing such phenomena from the local to global/international level (Figure 1).

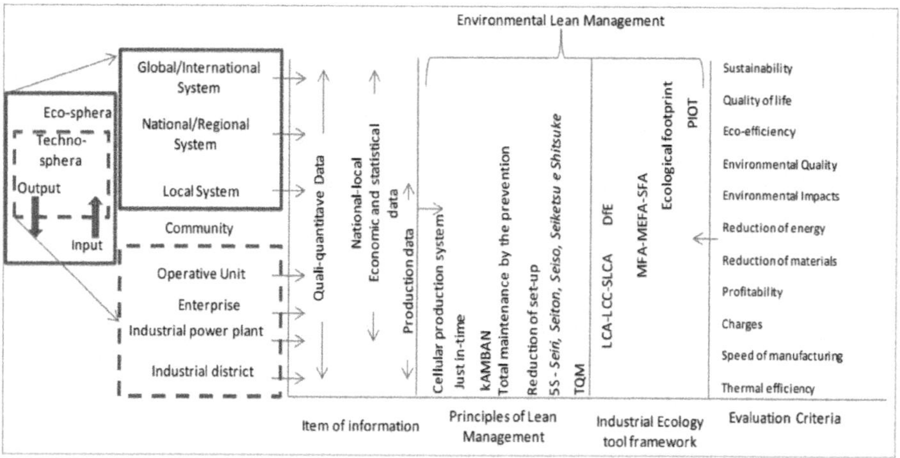

Figure 1: Selection of LM principles and IE tool, on the basis of common evaluation criteria and the same data types in order to support an environmental lean management model.

In this context, a lean company, although not necessarily green-oriented (rather generally, companies must make trade-offs among multiple objectives that are not fully compatible), has time for conversion and response to new market trends, including environmental, quicker and, therefore, is particularly ready to adopt a model of environmental lean management.

RESULTS

Environmental lean management fully integrates with the green-oriented LM model, combining the basic principles of LM (*i.e.*, the five basic principles [39]) with those green principles (e.g., better use of natural resources and the reduction environmental impact [4]) in order to create a unique integrated model.

The environmental model of LM introduces the environmental variable along all processes, imposing eco-efficiency in production and use of resources, materials, and energy, and also introducing the goals of reducing environmental impacts, and of sharing the environmental information for environmental awareness spread along the flow/value chain.

From integration of LM and CT strengths and opportunities, is possible to carry out the common targets' need to push the launch of the Environmental lean management system (Figure 2).

The environmental model of lean management is focused on overcoming the traditional forms of savings, which include reduction of overproduction, waiting time, transport optimization, adequacy in the answers to the problems, sizing inventory flow management, and reduction of defects [40].

This idea is reflected, for example, in the study by Moreira *et al.* [4], which explores other forms of loss in value (environmental impacts, energy consumption, material consumption, and emissions) to demonstrate that environmental liabilities are hidden inside of classic lean manufacturing.

From what has been stated in the previous paragraph, it is possible declare that a lean environment acts as a catalyst to facilitate environmental sustainability.

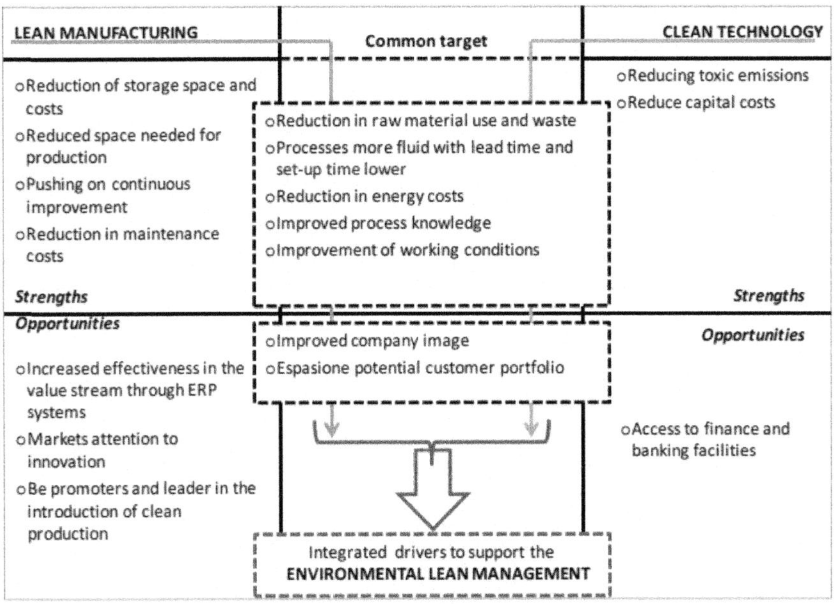

Figure 2: Synthesis of the main strengths and opportunities of LM and CT, in order to support the environmental lean management model.

Combining the LM aims [40] with the needs of sustainability that society asks [41], is possible to identify the vision of the environmental model of LM, that is to realize an efficient production system, effective and sustainable, within the social and economic aspects connected to the environmental and territorial conditions, from a long value chain perspective (from local community to global market).

In this perspective, the IE provides, through its principles, and its analysis and decision support tools, an effective response to transit from one model to that of environmental lean management.

In fact, the IE, in terms of support for the environmental assessment, applied to the entire corporate system allows:

(1) An overall assessment of the environmental risks arising from technological process and finished products;

(2) The quantitative analysis of environmental burdens related to the flows of materials and energy used in the process, as well as the flow of output products (In-put = output + stock);

(3) The evaluation of the recovery of secondary raw materials derived from waste and recycling processes that should be taken into account;

(4) The measurement of the consumption of all the raw materials and the use of recovered energy in the process;

(5) An identification of any other industrial waste streams secondary (solid, liquid, and gaseous fuels) to be taken into account (hidden flows);

(6) An identification of any flows of co-products harmful to the environment.

Then, IE could be considered as a paradigm of tools that environmental lean management adopts to maximize resource productivity and close the cycle of movement of resources within the techno-sphere, preserving the limited natural resources and minimizing the amount of production waste and related emissions.

This model can be used as a basis for determining the prospects and the direction of change necessary for a sustainable society, increasing the productivity of resources through the recirculation of what, before, was "lost" or "released" to the ecosphere.

A company that intends to reconvert according to a model of environmental lean management must start from the identification of the main sources of waste, and proceed with the elimination of the arising inefficiencies, and, then, develop more sustainable products, addressing green markets.

With the introduction of IE in the *modus operandi* of the LM, it is possible to define a new set of objectives that integrate environmental aspects within the financial-productive concerns:

(1) Reduce material and energy use for a product, including services in the course of his life;

(2) Reduce emissions, dispersion and the creation of toxic substances during his lifetime;

(3) Increase the amount of recyclable materials;

(4) Maximize sustainable use of renewable resources;

(5) Minimize the intensity of service for products and services;

(6) Extending the useful life of a product;

(7) Assess and minimize the environmental impact during the life of the product;

(8) Have a "functional economy" is a way to replace the products with services;

(9) Use "reverse logistics", which means that all efforts are used in order to reuse products and materials;

(10) Increase the efficiency of a product during use.

The final result is that both theories-philosophies have a strategic function and totally present, if one (IE) is applied in the other (LM).

DISCUSSIONS OF POSITIVE ASPECTS AND POSSIBLE BARRIERS

The correlation between LM and IE, for the study of new business models, leans green-oriented, points to a number of involved factors. The most important pressures for companies of a certain size, are the standards and rules set by customers (particularly international customers) that require environmental safeguards, in the perspectives of both cost savings and increased profits for shareholders; moreover, the increasing of communication and sharing of policies and performance, in terms of action and responsible environmental protection, with all the stakeholders-community, represents another relevant matter.

Instead, for SMEs, the most important drivers are represented by the target of cost savings. In addition, for such companies, environmental sustainability remains a consequence dictated by the legal obligations that continue to play a decisive role (mandatory system under "command and control").

Availability and accessibility to technology are two other key factors. However, an ever-increasing level of technological development is not proportional and uniform access to it. Limitations on physical-geographical, political-regulated, financial, and historical barriers are also slowing down in the field of technological innovation, the introduction of cleaner technology.

The three dimensions of sustainability (3P-society (people), environment (planet), economics (profit)), should, therefore, be extended by a fourth: The technological dimension [42,43,44]. The European Commission, in this sense, plays an important role as a catalyst for the development of European Technology Platforms (ETP) that promote the TEIs. In doing so, they gather key stakeholders operating in sectors with a high content of technology, innovation, and research, and sectors for which competitiveness and innovation, economic development and social development, depend on technological advances implemented in the medium and long term.

ETPs represent a model of technological governance, in which the legislature is a partner of the voluntary initiative of a European industrial system, with "bottom up" and "technology driven" approaches.

The ETP also try to overcome some of the main obstacles to the development of a lean green-oriented model from the point of view of industrial operators, namely the high costs of implementation of TEIs and the lack of tradition/skills.

In this sense, the ETP represent a positive boost thanks to the mobilization of significant human and financial resources and can help ensure investment in research and development to improve the competitiveness of European companies and bring benefits to European citizens. Another significant aspect, not related to the company size, is represented by the objectives posed by the highly diffused Total Quality Management models. These are an ideal base for expansion and integration with other management models (e.g., environmental, social and ethical, safety, *etc.*), supporting a culture of environmental protection, through shared responsibility between the high direction (top-down) and all the parts inside and outside the company (bottom-up), and a culture of training extended to the entire lifecycle of the product, from the supplier to the end customer (life cycle management).

From an operational perspective, the literature shows that most of the efforts are focused, for instance, on product design and on the introduction of technology on single production step, but not along the entire supply chain; this strategy acts is more oriented on the management of end-life, rather than on an integrated production system. In this sense, the biggest problem is the mentality of the producers, who believe that the main role of industrial systems is to transform inputs into outputs, cost-effectively, using technologies and efficient processes, but does not consider it essential that sustainability can play a key role in competitiveness. This can also be attributed to the lack of widespread knowledge about the forces and mechanisms that support the implementation of practices for sustainable production are only also connected to a non-adequate notice of the improvements achieved by some green-oriented producers.

With regard to instruments introduced by the IE, there are strong resistances due to, e.g., the need for an investment in terms of money (external advice or preparation of a team within the company), or in terms of time and commitment to the collection of data, which are often not easily accessible and do not have adequate strength. For example, in the case of LCA, Udo de Haes [45] emphasizes the significant role to assess the complex interaction between a product and the environment, from cradle to cradle (as closed circle) and provide in-depth information on environmental impacts. LCA can be useful for manufacturing companies because it can demonstrate that the activities, processes, and materials used contribute to the production of large environmental impacts; these results can then serve as targets for improvement for green-oriented businesses.

However, as with other instruments, for the complexity of LCA studies can be particularly difficult for the entrepreneurs [46]. We have tried to overcome this negative aspect, which often prevents the application of a full LCA, through the application of a simplified version that also integrates into the paradigm the key features of lean management. Several other rigorous methodologies have been introduced to simplify LCA, supporting the adoption and dissemination [47].

Once all the difficulties that are typically associated with the application of IE are overcome, the analysis may prove its full ability to capture all environmental aspects. For example the DfE clearly points out the environmental impacts of a product that can be reduced more effectively, through the integration of environmental aspects in the early stages of the process (design) [48,49]. Environmental lean management can be a way to address the barriers and turn them into positive factors of competitiveness. In fact, this philosophy-strategy, if widely adopted, could represent a road map to guide long-term scientific and technological research, according to the needs of the industry, and to promote the application of technologies with a valid commercial return; environmental lean management can be considered an interdisciplinary model on which the industry may be interfaced with the research community and public authorities and regulators, in order to extend the dialog aimed at developing more effective and consistent standards and regulations, even in terms of promotion and development of adequate funding.

CONCLUSIONS

Improving the environmental performance in the manufacturing sector means decoupling economic performance from the environmental load of human activities. Environmental lean management represents a new frontier in terms of production efficiently, and is effectively able to reduce inputs of natural resources, materials and energy, or waste or pollutant outputs, while maintaining the assets of the techno-sphere separate from those of the ecosphere, in order to avoid environmental degradation.

The proposed integration could represent a best solution to achieve the sustainability. Current research in the areas of technologies for the production and management company systemic apparatus are increasingly linked to the productivity of resources, especially energy efficiency, and environmental assessment. Despite the growth in research on sustainable production, as well as efforts in the industrial sector, it is still difficult to find information on how to improve manufacturing operations and the cash flows of resources from the point of view of the producer [50].

The theme, still strongly present, meets a great deal of resistance among entrepreneurs, but it is a springboard for a future in the short term, in which the environmental variable has a weight comparable to or greater than the financial condition.

ACKNOWLEDGMENTS

We are particularly grateful to two anonymous referees for reading and for their comments.

AUTHOR CONTRIBUTIONS

These authors contributed fully and equally to this work. Giuseppe Ioppolo contributed to research design and coordinated with Luigi Ciraolo on the research project. Stefano Cucurachi analyzed the sources and literature. Roberta Salomone and Giuseppe Saija supervised the research project and carried out a detailed revision. All authors wrote the body of the paper and read and approved the final manuscript.

REFERENCES

1. Europa 2020. Available online: http://ec.europa.eu/europe2020/europe-2020-in-a-nutshell/flagship-initiatives/index_it.htm (accessed on 1 October 2013).
2. COM. Rio+20: Verso Un'economia Verde e una Migliore Governance. Available online: http://eur-lex.europa.eu/LexUriServ/LexUriServ.do?uri=COM:2011:0363:FIN:IT:PDF (accessed on 16 December 2013).

3. Gordon, P.J. *Lean and Greendprofit for Your Workplace and Environment*; Berrett-Koehler Publishers: San Francisco, CA, USA, 2001.
4. Moreira, F.; Alves, A.C.; Sousa, R.M. Towards eco-efficient lean production systems. *IFIP Adv. Inform. Commun. Technol.* 2010, *322*, 100–108.
5. Holweg, M. The genealogy of lean production. *J. Oper. Manag.* 2007, *25*, 420–437.
6. Womack, J.P.; Jones, D.T.; Roos, D. *The Machine that Changed the World*; Harper Perennial: New York, NY, USA, 1990.
7. Hopp, W.; Spearman, M.L. *Factory Physics*; McGraw-Hill/Irwin: New York, NY, USA, 2008.
8. Hofer, C.; Eroglu, C.; Hofer, A.R. The effect of lean production on financial performance: The mediating role of inventory leanness. *Int. J. Prod. Econ.* 2012, *138*, 242–253.
9. Guinipero, L.C.; Pillai, K.G.; Chapman, S.N.; Clark, R.A. A longitudinal examination of JIT purchasing practices. *Int. J. Logist. Manag.* 2005, *16*, 51–70.
10. Monden, Y. Adaptable Kanban system helps Toyota maintain just-in-time production. *Ind. Eng.* 1981, *13*, 29–46.
11. Ohno, T. *The Toyota Production System: Beyond Large-Scale Production*; Productivity Press: Portland, IN, USA, 1988.
12. Sugimori, Y.K.; Kusunoki, K.; Cho, F.; Uchikawa, S. Toyota Production System and Kanban System; materialization of just-intime and respect-for-human system. *Int. J. Prod. Res.* 1977, *15*, 553–564.
13. Warnecke, H.J.; Hiiser, M. Lean production. *Int. J. Prod. Econ.* 1995, *41*, 37–43.
14. Graedel, T.E. Industrial ecology: Definition and implementation. In *Industrial Ecology and Global Change*; Socolow, R., Andrews, C., Berkhout, F., Thomas, V., Eds.; Cambridge University Press: Cambridge, UK, 1994; pp. 23–41.
15. Ayres, R.U., Ayres, L.W., Eds.; *A Handbook of Industrial Ecology*; Edward Elgar Publishing: Northampton, MA, USA, 2002.
16. Robèrt, K.-H.; Schmidt-Bleek, B.; Aloisi de Larderel, J.; Basile, G.; Jansen, J.L.; Kuehr, R.; Thomas, P.P.; Suzuki, M.; Hawken, P.; Wackernagel, M. Strategic sustainable development: Selection, design and synergies of applied tools. *J. Clean. Prod.* 2002, *10*, 197–214.
17. Allwood, J.M. What is Sustainable Manufacturing? In *Sustainable Manufacturing Seminar Series*; Cambridge University: Cambridge, UK,

2005; pp. 1–32.

18. Seliger, G. *Sustainability in Manufacturing: Recovery of Resources in Product and Material Cycles*; Springer: Berlin, Germany, 2007.
19. Chikán, A. Managers' view of a new inventory paradigm. *Int. J. Prod. Econ.* 2011, *133*, 54–59.
20. Zokaei, K.; Martinez, F.; Vazquez, D.; Evans, B.; Sarmadi, K.; Peattie, L.; Pampanelli, A.; Hines, P.; Trivedi, N. *Best Practice Tools and Techniques for Carbon Reduction and Climate Change: Lean and Green Report*; Lean Enterprise Research Centre, Cardiff University: Cardiff, UK, 2010.
21. Glavic, P.; Lukman, R. Review of sustainability terms and their definitions. *J. Clean. Prod.* 2007, *15*, 1875–1885.
22. Lozano, R. Towards a better understanding of sustainability into companies' systems: An analysis of voluntary corporate initiatives. *J. Clean. Prod.* 2012, *25*, 14–26.
23. Cagno, E.; Trucco, P.; Tardini, L. Cleaner production and profitability: Analysis of 134 industrial pollution prevention (P2) project reports. *J. Clean. Prod.* 2005, *13*, 593–605.
24. Del Río González, P. Analysing the factors influencing clean technology adoption: A study of the Spanish pulp and paper industry. *Bus. Strateg. Environ.* 2005, *14*, 20–37.
25. Korhone, J. From material flow analysis to material flow management: Strategic sustainability management on a principle level. *J. Clean. Prod.* 2007, *15*, 1585–1595.
26. Ioppolo, G.; Saija, G.; Salomone, R. From coastal management to environmental management: The sustainable eco-tourism program for the mid-western coast of Sardinia (Italy). *Land Use Policy* 2013, *31*, 460–471.
27. Fijał, T. An environmental assessment method for cleaner production technologies. *J. Clean. Prod.* 2007, *15*, 914–919.
28. Ehrenfeld, J.R.; Gertler, N. Industrial ecology in practice: The evolution of interdependence at Kalundborg. *J. Ind. Ecol.* 1997, *1*, 67–79.
29. Graedel, T.E.; Howard-Grenville, J.A. *Greening the Industrial Facility: Perspectives, Approaches, and Tools*; Springer: New York, NY, USA, 2005.
30. Tibbs, H. Industrial Ecology. An Environmental Agenda for Industry. Global Business Network, 1993. Available online: http://www.hardintibbs.com/wp-content/uploads/2009/05/tibbs_indecology.pdf (accessed on 10 November 2013).

31. Kleijn, E.G.M. *Materials and Energy: A Story of Linkages*; Muller Visual: Leiden, The Netherlands, 2012.
32. Ioppolo, G.; Heijungs, R.; Cucurachi, S.; Salomone, R.; Kleijn, R. Urban Metabolism: Many open questions for future answers. In *Pathways to Environmental Sustainability: Methodologies and Experiences*; Salomone, R., Saija, G., Eds.; Springer International Publishing AG: Dordrecht, The Netherlands, 2014; pp. 23–32.
33. Zamagni, A. Life cycle sustainability assessment. *Int. J. Life Cycle Assess.* 2012, *17*, 373–376.
34. Hammer, M.; Giljum, S.; Bargigli, S.; Hinterberger, F. Material Flow Analysis on the Regional Level: Questions, Problems, Solutions. NEDS Working Paper, 2003. Available online: http://seri.at/wp-content/uploads/2009/09/Material-flow-analysis-on-the-regional-level1.pdf (accessed on 18 June 2014).
35. Ayres, R.U. Commentary in the utility of the ecological footprint concept. *Ecol. Econ.* 2000, *32*, 347–349.
36. Rees, W.E. Ecological footprints and appropriated carrying capacity: What urban economics leaves out. *Environ. Urban.* 1992, *4*, 121–130.
37. Wackernagel, M.; Lewan, L.; Hansson, C.B. Evaluating the use of natural capital with the ecological footprint: Applications in Sweden and Subregions. *Ambio* 1999, *28*, 604–612.
38. Allenby, B.R., Richards, D.J., Eds.; *The Greening of Industrial Ecosystems*; National Academy Press: Washington, DC, USA, 1994.
39. Womack, J.P.; Jones, D.T. *Lean Thinking*; Free Press: New York, NY, USA, 1998.
40. Bicheno, J.; Holweg, M. *The Lean Toolbox*; PICSIE Books: Buckingham, UK, 2000.
41. Lozano, R. Envisioning sustainability three-dimensionally. *J. Clean. Prod.* 2008, *16*, 1838–1846.
42. Elkington, J. *Cannibals with Forks: The Triple Bottom Line of 21st Century Business*; Capstone Publishing: Oxford, UK, 1997.
43. Lovins, A.B.; Lovins, L.H.; Hawken, P. A road map for natural capitalism. *Harv. Bus. Rev.* 1999, *77*, 145–158.
44. Jovane, F.; Yoshikawa, H.; Alting, L.; Boë, C.R.; Westkämper, E.; Williams, D.; Tseng, M.; Seliger, G.; Paci, A.M. The incoming global technological and industrial revolution towards competitive sustainable manufacturing. *CIRP Ann.-Manuf. Technol.* 2008, *57*, 641–659.
45. Udo de Haes, H.A. Applications of life cycle assessment: Expectations,

drawbacks and perspectives. *J. Clean. Prod.*1993, *1*, 131–137.

46. Ahlroth, S.; Nilsson, M.; Finnveden, G.; Hjelm, O.; Hochschorner, E. Weighting and valuation in selected environmental systems analysis tools e suggestions for further developments. *J. Clean. Prod.* 2011, *19*, 145–156.
47. Mori, Y.; Huppes, G.; Udo de Haes, H.A.; Otoma, S. Component manufacturing analysis: A simplified and rigorous LCI method for the manufacturing stage. *Int. J. Life Cycle Assess.* 2000, *5*, 327–334.
48. Keoleian, G.A.; Menerey, D. Sustainable development by design: Review of life cycle design and related approaches. *J. Air Waste Manag. Assoc.* 1994, *44*, 645–668.
49. Müller, K.G.; Court, A.W.; Besant, C.B. Energy life cycle design: A method. *J. Eng. Manuf.* 1999, *213*, 415–419.
50. Despeisse, M.; Mbaye, F.; Ball, P.D.; Evans, S.; Levers, A. Emergence of sustainable manufacturing practices. *Int. J. Prod. Plan. Control* 2012, *23*, 354–376.

Chapter 10

ECOLOGY AND ESCIENCE

Christian Mulder

National Institute for Public Health and the Environment (RIVM)

COMPREHENDING THE GIFTS OF ECOLOGY

Stress ecology, climate change, human well-being, and global sustainability are popular items (Naeem et al. 2009). Given all the challenges in a developing world where the global population is supposed to reach 9.15 billion in 2050 (Pimentel et al. 1999; United Nations Population Division 2010), policy makers are, for the first time, keen on concrete assessments of our world, looking with interest and fear to ecological models. Although the discussion between scientists and politicians is known to be difficult, too many recent catastrophes during a single year - from the British Petroleum oil spill in the Gulf of Mexico (De Gouw et al. 2011) up to the ongoing radioactive Fukushima wreckage (Schiermeier 2011) - rapidly forced a better and constructive interaction between applied ecologists and policy makers at different organization levels. Such an interaction is also reflected by the arising use of internet metrics, blogs, tweets, and social networking - all digital tools that are already more or less linked to the thought process that society and policy are currently going through. Scientists are used to the Web of Science for selecting the appropriate papers, and policy makers are using methodologies for weighing opinions (Bollen et al. 2009). The latter authors even defined modern science as a 'gift economy', and they are absolutely right. What else should happen to improve the interactions between policy and research?

eScience: Computation Impacted Science

As can be seen in the plethora of data and the huge degree of multidisciplinarity (Figure 1), *a full and immediate open access across all disciplines* is supposed to be the dream of any eScientist and stakeholder involved in this thought process. At a glance, ecologists should be eScientists *par excellence*. Most ecological disciplines, in fact, overlap and typically (re)use data from other sciences, which leads to a huge increase in science productivity. Ecology often benefits from methods originally developed for mathematics, physics, and chemistry (Cohen 2004; Elser 2006) and might even benefit from models used in the worlds of informatics and finance (Allesina and Pascual 2009; Haldane and May 2011, respectively). Opposite flows occur as well, from ecological knowledge to archaeology and urban systems, toxicology, and medicine (Sjögren and Lamentowicz 2008; Grimm et al. 2000; Van Straalen 2003; Zu Dohna et al. 2009, respectively). Ecology is a perfect eScience, and I believe that no scientific discipline has already so many examples of mutual benefits as all ecological (sub)disciplines: molecular/evolutionary ecology and genetics, ecology and economics, ecology and geology, ecology and paleontology, ecology and atmospheric deposition, ecology and climate regulation, ecology and environmental planning, and ecology and environmental quality (e.g.,Dicke et al. 2011; Costanza et al. 1997; Dommain et al. 2011; Dunne et al. 2008; Elser et al. 2009; West et al. 2011; Taylor Lovell and Johnston 2009; Mulder et al. 2011, respectively). Although most attention is provided to pristine terrestrial biota and extreme environments (e.g., Bai et al. 2004; Yu et al. 2010), ecology provides the necessary knowledge to assess consequences of disturbance and stress, such as the effects of pollution, mining, and engineering on local communities at the landscape level (De Zwart et al. 2006; Dugan et al. 2010; Wang et al. 2010). Moreover, using several case studies from the Great Barrier Reef, Stoeckl et al. (2011) pointed out how important it is to differentiate between information about existing ecosystem services (Adhikari and Nadella 2011) and information about the extent to which existing ecosystem services might change (or have already changed) in response to external events (West et al. 2011).

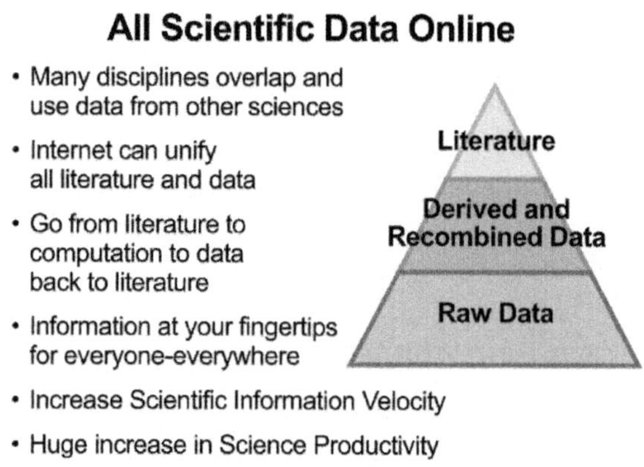

Figure 1: eScience is about global collaboration: a world where data and literature interoperate with each other. (Adapted from Hey and Trefethen 2003; Henzinger and

Lawrence 2004, and references therein; photo credit by Hey et al. 2009; free eBook at http://fourthparadigm.org, courtesy of Microsoft Research.).

The World of Science Changed Radically

Science is at a tipping point at which research will rapidly become more applied. Land conversion for agriculture is projected to rise sharply (Laurance 2001), and strong empirical evidence is available. Thanks to the huge increase in peer-reviewed publications, ecology helps in understanding what is going on, in sustaining current ecosystems and in managing novel ecosystems, and contributes in the societal thought process. Indeed, nobody is questioning that the entire world of science has changed (Laurance 2001; Hey et al. 2009). From this perspective, open-access journals (ranging from cosmology to environment and from medicine to biology) are for sure the best way to share empirical data, models, and thoughts with others (Bourne et al. 2008; Figure 2). Successful linking of (e)Science with decision making will depend on accuracy and communication (Clark et al. 2001).

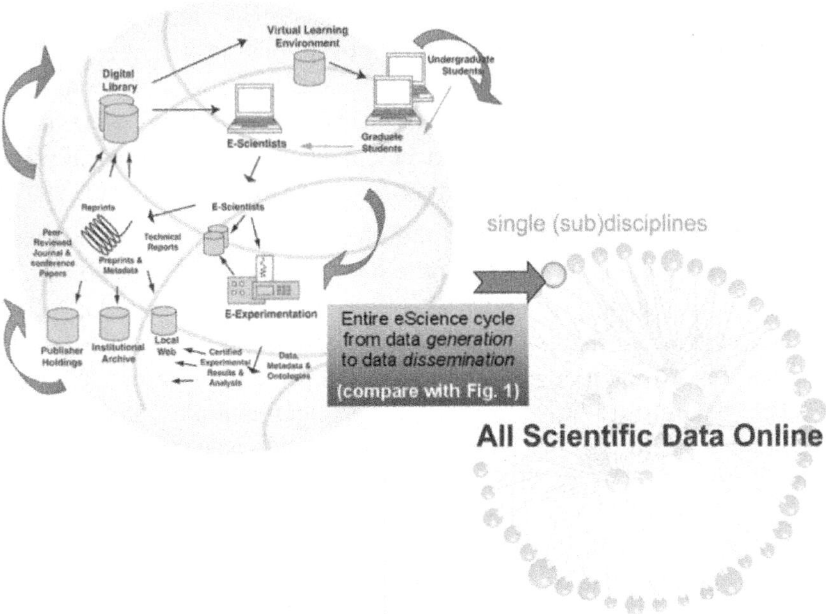

Figure 2: The entire eScience cycle from data generation to data dissemination. eScience remains linked to the requirements for dataBank procedures, i.e., the quality, integrity, identity, provenance, description, and ontology of original scientific data. (Grid photo courtesy of Liz Lyon, e-Bank, UK; modified from Hey and Trefethen 2005; network credit by Alice Boit, University of Potsdam.).

Balancing theory and empirical data at different levels allowed a shift in the scientific inquiries, resulting in an improved performance of current investigation. (Previously, a myriad of topics made navigation more difficult.) Internet, networking, international programs, and the astonishingly high amount of data confluence into this new conceptual framework (Hey et al. 2009), where any traditional (sub)discipline has been forced to change (see Box 1 and references therein; Mulder et al. 2011). Internet can unify such (sub) disciplines in a fractal-like world with increasing complexity and accessible detail (Figure 2), demonstrating the need to unify philosophy of science with the science and technology studies.

Albeit some ecologists seem not to be fully aware of the enhancements of online scholarly communication, open access is really beginning to spread. Still, only one-fourth of the contributions published between 2003 and 2010 by the Public Library of Science [PLoS] (2011) belong to the categories Development and Evolutionary Biology, Ecology, Marine and Aquatic Sciences, or Plant Biology, and less than 0.15% of the seven hundreds of thousand contributions deposited during the same period in the arXiv (2011) belong to the 'Quantitative Biology' category. Seeing how many ecologists appreciate PLoS (and open access in general), these relatively low contributions surprise me. Something comparable occurs with data sharing: empirical data sets are of the highest value (*cf.* Bourne 2005), but ecologists seem to become 'shy' when asked to make their data accessible. This makes me question why in 2010, only 7% of all the participants to one EU-funded project responded to their open data call?

This short overview shows, on the one hand, how many ecologists are already dealing with data sharing and eScience, and on the other hand, how different complementary frameworks seem to coexist. The overlap between schools and frameworks can be rather confounding. For instance, although many features are common to different ways of performing research, metadata focusing on compiled information for local communities are widespread as ecological networks, whereas scientific data derived from simulated or empirical information are commonly defined as food webs. This implies that webs can, in most cases, be seen as networks although not all ecological networks may be seen as traditional food webs. However, in contrast to many nested networks, eScience is mutual and, as such, is expected to grow with vigor because, in contrast to (antagonistic) living organisms, eScience has no discrete boundaries but has digitally cross-connected domains with high modularity.

Thinking Seriously About the Scenario

After all the efforts of producing a research paper (from laboratory and/or field work up to forecasting, networking, feedbacks, and final writing), how are ecologists currently thinking about their outreach? Bourne (2010) provocatively wrote that in contrast to the rather static PDF interface, publishing workflows - and preprint repositories and post-publication commentaries can be seen as a kind of creative workflows - are more powerful but harder to manage and represent, therefore, *a major change for most scientists.*

Apparently, ecologists belong to those scientists, but if they are not able to keep online the attention of their own colleagues, then how can they expect any concrete feedbacks from environmentally interested laypersons and policy decision makers? How can stakeholders expect a concrete use of ecological indicators if most scientists dealing with such an integrative discipline seem to (be willing to) ignore a large part of data and recent literature on environmental impact and ecosystem services (*cf.* Bjorndal et al. 2011)?

Having seen these disputes between institutions, it is not a wonder that Al Gore has succeeded in areas where so many scientists, stakeholders, and NGOs have failed. It is time for a radical departure; to improve the synergy between authors, publishers, and readers, two possible incentives should be considered as novel opportunities:

- Be explicit and do not try to be exhaustive

There is an increasing amount of review papers cited to support empirical data, and most results can be interesting for a wide variety of scientists if the original results could be put in a much wider context. Most manuscripts just aim to bridge knowledge gaps, but overwhelming references are always inappropriate to catch real attention, and chains of citations have to be avoided as they move from journal to journal (Hirsch 2005). We should strive to generate less academic papers and attempt to sample the web uniformly.

- Be wide and do not claim something else

Bibliographic couplings aim to identify structures and communities (Menczer 2004). To avoid strong discipline bias, take substantial precaution. Do not claim shocking lack of knowledge (such terms are not objective) and do not cite only the papers that seem to confirm your results (Jennions and Møller 2002). Your literature shows your horizons. In the framework of eScience, papers always benefit from testing the theory or hypothesis regarding how relationships change, so be wide in the chosen references.

Let us put matters straight: ecological forecasts are imperative (Clark et al. 2001). These authors defined ecological forecasting as 'the process of predicting the state of ecosystems, ecosystem services, and natural capital' (Clark et al. 2001). Although some scientists still prefer to claim that no data are available to specify uncertainties or find that novel ideas cannot be tested properly for lack of appropriate technology (*cf.* Collins 2010), others state that larger data sets (inclusive historic data sets) are meanwhile discoverable (Hunt et al. 2009), making ecological forecasting feasible. Data are interwoven with peer-reviewed scientific papers, are authored in digital form, and are benefits of today's structural use of the entire digital environment (Lynch 2009).

A New Scientific Endeavor Online

As a matter a fact, 'science aims to produce far more than a simple mechanical prediction of correlations' (Ginsparg 2009), and most open-access journals clearly aim to support a better *comprehension* of the exabytes of already available information and to improve strongly the*dissemination* of data, causalities, and implications. Therefore, *Ecological Processes* is governed by three principles: research quality, multidisciplinarity and integration, and open access. (Overarching principles are reflected by all those who have agreed to provide part of their precious time as committed editors.) As editors in chief, we all have plenty of ideas for what *Ecological Processes* means, but we really wish to get your input too, as you like it.

REFERENCES

1. Adhikari B, Nadella K: Ecological economics of soil erosion: a review of the current state of knowledge. *Ann New York Acad Sci*2011, 1219: 134–152. 10.1111/j.1749-6632.2010.05910.x
2. Allesina S, Pascual M: Googling food webs: can an Eigenvector measure species› importance for coextinctions? *PLoS Comput Biol*2009,5(9):e1000494. 10.1371/journal.pcbi.1000494
3. arXiv Cornell University Library, Ithaca, NY; 2011.http://arxiv.org . Accessed 20 Apr 2011
4. Bai Y, Han X, Wu J, Chen Z, Li L: Ecosystem stability and compensatory effects in the Inner Mongolia grassland. *Nature* 2004, 431:181–184. 10.1038/nature02850
5. Bjorndal KA, Bowen BW, Chaloupka M, Crowder LB, Heppell SS, Jones CM, Lutcavage ME, Policansky D, Solow AR, Witherington BE:Better science needed for restoration in the Gulf of Mexico. *Science* 2011, 331: 537–538. 10.1126/science.1199935
6. Bollen J, Van de Sompel H, Hagberg A, Chute R: A principal component analysis of 39 scientific impact measures. *PLoS ONE*2009,4(6):e6022. 10.1371/journal.pone.0006022
7. Bourne P: Will a biological database be different from a biological journal? *PLoS Comput Biol* 2005,1(3):e34. 10.1371/journal.pcbi.0010034

8. Bourne PE: What do I want from the publisher of the future? *PLoS Comput Biol* 2010,6(5):e1000787. 10.1371/journal.pcbi.1000787

9. Bourne PE, Fink JL, Gerstein M: Open access: taking full advantage of the content. *PLoS Comput Biol* 2008,4(3):e1000037. 10.1371/journal.pcbi.1000037

10. Clark JS, Carpenter SR, Barber M, Collins S, Dobson A, Foley JA, Lodge DM, Pascual M, Pielke R Jr, Pizer W, Pringle C, Reid WV, Rose KA, Sala O, Schlesinger WH, Wall DH, Wear D: Ecological forecasts: an emerging imperative. *Science* 2001, 293: 657–660. 10.1126/science.293.5530.657

11. Cohen JE: Mathematics is biology›s next microscope, only better; biology is mathematics› next physics, only better. *PLoS Biol*2004,2(4):e439.

12. Collins JP: Sailing on an ocean of 0s and 1s. *Science* 2010, 327: 1455–1456. 10.1126/science.1186123

13. Costanza R, D›Arge R, De Groot R, Farber S, Grasso M, Hannon B, Limburg K, Naeem S, O›Neill RV, Paruelo J, Raskin RG, Sutton P, Van den Belt M: The value of the world›s ecosystem services and natural capital. *Nature* 1997, 387: 253–260. 10.1038/387253a0

14. De Gouw JA, Middlebrook AM, Warneke C, Ahmadov R, Atlas EL, Bahreini R, Blake DR, Brock CA, Brioude J, Fahey DW, Fehsenfeld FC, Holloway JS, Le Henaff M, Lueb RA, McKeen SA, Meagher JF, Murphy DM, Paris C, Parrish DD, Perring AE, Pollack IB, Ravishankara AR, Robinson AL, Ryerson TB, Schwarz JP, Spackman JR, Srinivasan A, Watts LA: Organic aerosol formation downwind from the Deepwater Horizon oil spill. *Science* 2011, 331: 1295–1299. 10.1126/science.1200320

15. De Zwart D, Dyer SD, Posthuma L, Hawkins CP: Predictive models attribute effects on fish assemblages to toxicity and habitat alteration. *Ecol Appl* 2006, 16: 1295–1310. 10.1890/1051-0761(2006)016[1295:PMAEOF]2.0.CO;2

16. Dicke M, Van Loon JJA, De Jong PW: Ecogenomics benefits community ecology. *Science* 2011, 305: 618–619.

17. Dommain R, Couwenberg J, Joosten H: Development and carbon sequestration of tropical peat domes in south-east Asia: links to post-glacial sea-level changes and Holocene climate variability. *Quater Sci Rev* 2011, 30: 999–1010. 10.1016/j.quascirev.2011.01.018

18. Dugan PJ, Barlow C, Agostinho AA, Baran E, Cada GF, Chen D, Cowx IG, Ferguson JW, Jutagate T, Mallen-Cooper M, Marmulla G, Nestler J, Petrere M, Welcomme RL, Winemiller KO: Fish migration, dams, and loss of ecosystem services in the Mekong basin.*Ambio* 2010, 39: 344–348. 10.1007/s13280-010-0036-1

19. Dunne JA, Williams RJ, Martinez ND, Wood RA, Erwin DH: Compilation and network analyses of Cambrian food webs. *PLoS Biol*2008,6(4):e102. 10.1371/journal.pbio.0060102

20. Elser JJ: Biological stoichiometry: a chemical bridge between ecosystem ecology and evolutionary biology. *Am Nat* 2006, 168:S25-S35. 10.1086/509048

21. Elser JJ, Andersen T, Baron JS, Bergström A-K, Jansson M, Kyle M, Nydick KR, Steger L, Hessen DO: Shifts in lake N:P stoichiometry and nutrient limitation driven by atmospheric nitrogen deposition. *Science* 2009, 326: 835–837. 10.1126/science.1176199

22. Ginsparg P: Scholarly communication. In *The fourth paradigm: data-intensive scientific discovery. Redmond, Microsoft Research* Edited by: Hey T, Tansley S, Tolle K. 2009, 185–191.

23. Grimm NB, Grove JM, Pickett STA, Redman CL: Integrated approaches to long-term studies of urban ecological systems.*BioScience* 2000, 50: 571–584. 10.1641/0006-3568(2000)050[0571:IATLTO]2.0.CO;2

24. Haldane AG, May RM: Systemic risk in banking ecosystems. *Nature* 2011, 469: 351–355. 10.1038/nature09659

25. Henzinger M, Lawrence S: Extracting knowledge from the World Wide Web. *Proc Natl Acad Sci USA* 2004, 101: 5186–5191. 10.1073/pnas.0307528100

26. Hey T, Trefethen A: e-Science and its implications. *Phil Trans R Soc Lond A* 2003, 361: 1809–1825. 10.1098/rsta.2003.1224

27. Hey T, Trefethen A: Cyberinfrastructure for e-Science. *Science* 2005, 308: 817–821. 10.1126/science.1110410

28. Hey T, Tansley S, Tolle K (Eds): The fourth paradigm: data-intensive scientific discovery In *Microsoft Research, Redmond* 2009.

29. Hirsch JE: An index to quantify an individual›s scientific research output. *Proc Natl Acad Sci USA* 2005, 102: 16569–16572. 10.1073/pnas.0507655102

30. Hunt JR, Baldocchi DD, Van Ingen C: Redefining ecological science using data. In *The fourth paradigm: data-intensive scientific discovery. Microsoft Research, Redmond* Edited by: Hey T, Tansley S, Tolle K. 2009, 21–26.

31. Jennions MD, Møller AP: Publication bias in ecology and evolution: an empirical assessment using the ‹trim and fill› method. *Biol Rev* 2002, 77: 211–222.

32. Laurance WF: Future shock: forecasting a grim fate for the Earth. *Trends*

Ecol Evol 2001, 16: 531–533. 10.1016/S0169-5347(01)02268-6

33. Lynch C: Jim Gray›s fourth paradigm and the construction of the scientific record. In *The fourth paradigm: data-intensive scientific discovery. Microsoft Research, Redmond* Edited by: Hey T, Tansley S, Tolle K. 2009, 177–183.
34. Menczer F: Evolution of document networks. *Proc Natl Acad Sci USA* 2004, 101: 5261–5265. 10.1073/pnas.0307554100
35. Mulder C, Boit A, Bonkowski M, De Ruiter PC, Mancinelli G, Van der Heijden MGA, Van Wijnen HJ, Vonk JA, Rutgers M: A belowground perspective on Dutch agroecosystems: how soil organisms interact to support ecosystem services. *Adv Ecol Res* 2011, 44: 277–357.
36. Naeem S, Bunker DE, Hector A, Loreau M, Perring CP (Eds): *Biodiversity, ecosystem functioning, and human wellbeing* Oxford University Press, Oxford; 2009.
37. Pimentel D, Bailey O, Kim P, Mullaney E, Calabrese J, Walman L, Nelson F, Yao X: Will limits of the Earth›s resources control human numbers? *Environ Dev Sustain* 1999, 1: 19–39. 10.1023/A:1010008112119
38. PLoS Public Library of Science, San Francisco, CA; 2011.http://www.plos.org . Accessed 20 Apr 2011
39. Schiermeier Q: Radiation release will hit marine life. *Nature* 2011, 472: 145–146. 10.1038/472145a
40. Sjögren P, Lamentowicz M: Human and climatic impact on mires: a case study of Les Amburnex mire, Swiss Jura Mountains. *Veg Hist Archaeobot* 2008, 17: 185–197. 10.1007/s00334-007-0095-9
41. Stoeckl N, Hicks CC, Mills M, Fabricius K, Esparon M, Kroon F, Kaur K, Costanza R: The economic value of ecosystem services in the Great Barrier Reef: our state of knowledge. *Ann New York Acad Sci* 2011, 1219: 113–133. 10.1111/j.1749-6632.2010.05892.x
42. Taylor Lovell S, Johnston DM: Creating multifunctional landscapes: how can the field of ecology inform the design of the landscape? *Front Ecol Environ* 2009, 7: 212–220. 10.1890/070178
43. United Nations Population Division: World population prospects. 2010. http://esa.un.org/unpd/wpp/index.htm. Accessed 18 Jan 2012
44. Van Straalen NM: Ecotoxicology becomes stress ecology. *Environ Sci Technol* 2003, 37: 324A-330A. 10.1021/es0325720
45. Wang J, Huang J, Wu J, Han X, Lin G: Ecological consequences of the Three Gorges Dam: insularization affects foraging behavior and dynamics of rodent populations. *Front Ecol Environ* 2010, 8: 13–19.

10.1890/070188

46. West PC, Narisma GT, Barford CC, Kucharik CJ, Foley JA: An alternative approach for quantifying climate regulation by ecosystems. *Front Ecol Environ* 2011, 9: 126–133. 10.1890/090015
47. Yu Q, Chen Q, Elser JJ, He N, Wu H, Zhang G, Wu J, Bai Y, Han X: Linking stoichiometric homeostasis with ecosystem structure, functioning, and stability. *Ecol Lett* 2010, 13: 1390–1399. 10.1111/j.1461-0248.2010.01532.x
48. Zu Dohna H, Cecere MC, Gürtler RE, Kitron U, Cohen JE: Spatial re-establishment dynamics of local populations of vectors of Chagas disease. *PLoS Negl Trop Dis* 2009,3(7):e490. 10.1371/journal.pntd.0000490

Chapter 11

A BIBLIOMETRIC ANALYSIS OF GLOBAL FOREST ECOLOGY RESEARCH DURING 2002–2011

Yajun Song[1, 2] and Tianzhong Zhao[1]

[1]School of Information Science & Technology, Beijing Forestry University

[2]Library of Beijing International Studies University

ABSTRACT

Bibliometric is increasingly used for the analysis of discipline dynamics and management related decision-making. This study analyzes 937,923 keywords from 78,986 articles concerning forest ecology and conducts a serial analysis of these articles' characteristics. The articles' records, published between 2002 and 2011, were downloaded from the Web of Science, and their keywords were exported by Java processing programs. The result shows that forest ecology studies focused on forest diversity, conservation, dynamics and vegetation in the last decade. Developed countries, such as the USA, Canada, and Germany, were the most productive countries in the field of forest ecology research. From 2002 to 2011, the number of articles published annually related to forest ecology grew at a stable rate, as indicated by the fit produced by a high determination coefficient (R^2=0.9955). The findings of this study may be applicable for planning and managing forest ecology research and partners involved in such research may use this study as a reference.

INTRODUCTION

Bibliometric analysis is an important part of reference and research services. Forest ecology is closely related to forest management and many studies have been performed from various perspectives, including studies of ecosystems at multiple forest spatial scales (Rodrigues et al. 2011;Sitzia et al. 2010), long term ecosystem change (Diaz et al. 2007;van Oudenhoven et al. 2012), climate change (Cheaib et al. 2012; Şekercioğlu et al.2012), soils (McLachlan and Bazely2003; Wang et al.2011), physiography (Morrissey et al.2009; Rubio and Escudero2005), carbon balance (Mitchell et al.2009; Sillett et al.2010), nutrient cycling (Berger et al.2009; XU and Chen2006), landscape ecology (Loucks et al.2001; Wintle et al.2005) and biodiversity (Hanberry et al.2012; Lamb et al.2005). In addition to these studies, a bibliometric analysis of global forest ecology could provide a fresh look at the current status of global forest ecology research and help identify hot spots.

In recent years, along with its continuously expanding range of application, bibliometric analysis plays an increasingly important role in management and decision-making in science and technology. It has been used to document the development of some research fields (Grandjean et al.2011; Hendrix2008; Narotsky et al.2012; van Eck et al.2010; van Raan2006), including forestry (Dobbertin and Nobis2010; Perez et al.2004).

In this study, we perform a bibliometric analysis of forest ecology research over the last 10 years (2002–2011) aimed at (1) examining the temporal hot topics of forest ecology research by keyword frequency analysis, (2) revealing the distribution of articles by country/region, organization, funding agency, research area, author, year and publication name for articles covering forest ecology research and revealing advancements in forest ecological research, and (3) providing a new keywords frequency analysis method, which may benefit future research.

MATERIALS AND METHODOLOGY

Data Collection

Literature records, our analytical objects, were derived from the Web of Science, an online academic citation index database provided by Thomson Reuters. To define search terms, we used the "thesaurus" tool of Commonwealth Agricultural Bureaux (CAB) Abstracts.

We conducted a search on the word "ecology" in CAB Abstracts and the search produced 41 terms, including 19 narrower terms and 22 other related terms (Figure 1). We selected terms with more than 200 hits and used Microsoft Excel to rank them in descending order. We then removed the words "ecology" and "forest" from the Excel sheet and added the terms "climate," "soils," "physiography," "carbon balance" and "nutrient cycling," based on the concepts related to forest ecology defined by Barnes et al. (1997). Then, we defined the remaining 43 search terms and constructed a new search query. The search was limited to "article" type publications published between 1 January 2002 and 31 December 2011 in English.

ecology	Hits 39976
[Used For]	
ecologia	
[Narrower Terms]	
animal ecology	6052
autecology	1150
chemical ecology	733
community ecology	3503
dendroecology	213
fire ecology	1529
forest ecology	9340
freshwater ecology	2471
human ecology	602
landscape ecology	4586
marine ecology	949
microbial ecology	7890
palaeoecology	3505
phenology	13073
plant ecology	7454
population ecology	3293
restoration ecology	1668
riparian ecology	56
synecology	4362
[Related Terms]	
biocoenosis	344
biodiversity	36118
bioenergetics	536
biogeography	3304
ecological balance	465
ecological disturbance	5000
ecologists	12
ecosystems	25391
ecotypes	3275
environmental degradation	8427
environmental factors	31508
food chains	1293
food webs	2264
habitats	43231
landscape	15144
lowland areas	1719
microenvironments	228
plant communities	21514
populations	6551
predator prey relationships	2929
species diversity	37244
species richness	31934

Figure 1: Narrower terms and 22 related terms of ecology.

The search query included 43 terms (see Appendix A). This query was run in Web of Science, which is a citation database of the Web of Knowledge, and a total of 78,986 forest ecology-related articles were identified.

Using the Web of Science's analysis tools, we exported the 78,986 articles by country/region, organization, funding agency, research area, author, year, and publication. The statistical methods used by the Web of Science for the above statistical indicators of multi-author articles do not distinguish between the order of author's locations, which may result the sum of these statistical result was greater than 78,986. The article records, including title, author, keywords, abstract, and organization, were exported in full record mode from the Web of Science to text files. A total of 158 text files were created, because the Web of Science limits each export to 500 records. In every text file, "author keywords" were marked by "DE," and "keywords plus" were provided by the Web of Science and marked by "ID". Both these two kinds of keywords were considered in this study.

Keywords Analysis

First, the frequency of each keyword was counted in each text file. We developed a java program named count.java (Additional file 1: Appendix B) using Eclipse software, a famous cross-platform integrated development environment. This java program can find and select keywords in the output text file by identifying parameters, and connect each keyword to a long string, while deleting the carriage returns. After detection, the keywords in the string were split by semicolons, and counted using HashMap traversal algorithm. The HashMap traversal result was saved to an array and sorted by the counters; then, the sorted result was exported to an intermediate file.

Second, the 158 intermediate files were merged, and the frequency of each keyword was counted. We developed a java program named merge.java (Additional file 1: Appendix C) using Eclipse software. When this program was run, the intermediate files defined in the input parameters were opened, and the keywords and their counters were saved to a HashMap. Then the keywords were counted again with HashMap traversal algorithm: the counters of the same keywords were added. Then, the HashMap traversal result was saved to an array, sorted by the counters, and exported into a result file.

Third, we developed a program (Additional file 1: Appendix D) to create a java package named frequency.jar to store the compiled java class files which were produced by compiling count.java and merge.java.

Fourth, we developed a batch program named count.bat (Additional file 1: Appendix E) to call the count.class with the input parameters "DE" and "ID". All 158 text files were processed one by one. As a result, 158 intermediate files were created.

Fifth, we developed another batch program named merge.bat (Additional file 1: Appendix F) to call the merge.class with the input parameters, that is, the 158 intermediate files, to merge them. As a result, a final file was created, in which all keywords in 78,986 articles were counted and sorted.

After data processing, 937,923 keywords from those 78,986 articles were merged into 150,974 keywords. All of the keywords were sorted in reverse order based on their frequencies. The 100 most frequently used keywords became the focus of our study.

RESULTS

Keywords Analysis Results

To narrow the research scope, the 100, 200, 300 most frequently used keywords were selected and analyzed. As a result, the 100 most frequently used keywords, 0.07% of the 150,974 unique keywords analyzed here represented 18.54% of the total (937,923) of all keywords harvested (Table 1). We focused on the top 100 keywords to examine the hot topics of forest ecology research (Table 2).

Table 1: The top 100, 200, 300 keyword ratio and their frequencies

Keywords number	Keywords ratio	Keywords frequencies	Frequencies Ratio (%)
100	0.07%(100/150974)	173925	18.54%(173925/937923)
200	0.13%(200/150974)	233042	24.85%(233042/937923)
300	0.20%(300/150974)	271233	28.92%(271233/937923)

Table 2: The top 100 keywords in forest ecology articles indexed using the Web of Science during 2002–2011

	Keywords	Frequencies
1	forest	9302
2	diversity	5424
3	conservation	5135
4	dynamics	4886
5	vegetation	4720
6	biodiversity	4613
7	patterns	4166
8	growth	4069
9	rain-forest	3253
10	management	3236
11	nitrogen	3136
12	forests	3069
13	soil	2793
14	ecology	2677
15	communities	2596
16	carbon	2568
17	climate-change	2412

18	ecosystems	2407
19	disturbance	2389
20	species richness	2381
21	boreal forest	2334
22	landscape	2180
23	biomass	2130
24	model	2100
25	climate	2095
26	fire	2043
27	abundance	1855
28	united-states	1849
29	habitat	1846
30	temperature	1824
31	plants	1782
32	organic-matter	1755
33	populations	1733
34	decomposition	1603
35	climate change	1599
36	dispersal	1590
37	responses	1576
38	regeneration	1531
39	tropical forest	1513
40	land-use	1509
41	habitat fragmentation	1495
42	trees	1486
43	fragmentation	1473
44	forest soils	1441
45	evolution	1408
46	succession	1384
47	deforestation	1375
48	ecosystem	1362
49	birds	1333
50	population	1276
51	competition	1273
52	water	1235

53	variability	1210
54	deciduous forest	1190
55	forest manage-ment	1189
56	community structure	1178
57	behavior	1140
58	community	1131
59	restoration	1127
60	tropical forests	1107
61	photosynthesis	1093
62	seed dispersal	1081
63	usa	1067
64	productivity	1054
65	microbial bio-mass	1040
66	density	1034
67	impact	1019
68	brazil	1018
69	models	988
70	carbon-dioxide	978
71	phosphorus	971
72	size	971
73	predation	947
74	classification	943
75	respiration	932
76	scale	927
77	drought	920
78	national-park	918
79	plant	910
80	selection	909
81	tree	902
82	deposition	889
83	history	888
84	recruitment	875
85	norway spruce	874

86	soil respiration	870
87	australia	868
88	consequences	864
89	tropical rain-forest	839
90	survival	834
91	quality	830
92	mexico	819
93	costa-rica	813
94	impacts	812
95	new-zealand	796
96	forest soil	794
97	mortality	788
98	soils	787
99	grassland	786
100	assemblages	785

Articles Analysis Result

By Country/Region

The 78,986 articles were analyzed by countries or regions and sorted in reverse order by their total numbers and Table 3 lists the results for the top 20 countries. We supplemented a column in the original table and classified these 20 countries/regions by their respective continents, which showed that North America and 12 European countries had about 44.71% and 42.35% of all the articles, respectively, indicating published articles related to forest ecology in North America and Europe predominate.

Table 3: Top 20 countries/regions publishing articles on forest ecology indexed using the web of science during 2002–2011

	Countries/Regions	Records	Ratio (%)	Continents
1	USA	28060	35.53	North America
2	Canada	7255	9.19	North America
3	Germany	6311	7.99	Europe
4	Brazil	4561	5.77	Africa
5	Australia	4375	5.54	Australia

6	England	4229	5.35	Europe
7	Peoples R China	4122	5.22	Asia
8	France	3930	4.98	Europe
9	Japan	3504	4.44	Asia
10	Spain	3402	4.31	Europe
11	Sweden	2708	3.43	Europe
12	Finland	2417	3.06	Europe
13	Italy	2230	2.82	Europe
14	Netherlands	1921	2.43	Europe
15	Switzerland	1871	2.37	Europe
16	India	1798	2.28	Asia
17	Mexico	1572	1.99	South America
18	Russia	1554	1.97	Europe
19	Scotland	1455	1.84	Europe
20	New Zealand	1421	1.80	Europe

The combined frequency of keywords related to tropical forest, represented by "rain-forest" (3,253), "tropical forest" (1,513), "tropical forests" (1,107), and "tropical rain-forest" (839), totaled 6,712 keyword entries, which was exceeded only by the keyword "forest" with 9,302 entries (Table 2). This indicates that tropical forest was the main focus of research in forest ecology studies. Tropical forest is mainly distributed in Southeast Asia, Central America, South America, Australia, Africa. However, the main countries with strong research capabilities related to tropical forest research were not located in those areas, but were found in North America and Europe.

By Organization

Forest ecology studies were conducted by 7,598 organizations, and Table 4 lists the top 20 organizations and their related countries. The University of California System, the Chinese Academy of science, and US Forest Service produced the most articles. Eight organizations were from the USA, two each from Canada, Brazil, and Germany, and the remaining six were from China, Sweden, Finland, Russia, Spain, and France.

Table 4: Top 20 organizations publishing articles on forest ecology indexed using the web of science during 2002–2011

	Organizations	Records	Ratio (%)	Counties
1	Univ Calif System	2749	3.48	USA

2	Chinese Acad SCI	2359	2.99	China
3	US Forest Serv	2203	2.79	USA
4	Swedish Univ Agr SCI	1342	1.70	Sweden
5	Oregon State Univ	1200	1.52	USA
6	Univ Helsinki	1055	1.34	Finland
7	Univ British Columbia	1008	1.28	Canada
8	Univ Wisconsin System	978	1.24	USA
9	Univ Alberta	973	1.23	Canada
10	Russian Acad SCI	925	1.17	Russia
11	Univ Florida	905	1.15	USA
12	USDA	905	1.15	USA
13	Univ Sao Paulo	896	1.13	Brazil
14	US Geol Survey	883	1.12	USA
15	Univ Fed Santa Maria	868	1.10	Brazil
16	Smithsonian Inst	867	1.10	USA
17	Max Planck Society	808	1.02	Germany
18	Univ Gottingen	785	0.99	Germany
19	INRA	771	0.98	France
20	CSIC	766	0.97	Spain

USDA United States department of agriculture, *INRA* Institut National de la recherche agronomique, *CSIC* consejo superior de investigaciones científicas.

By Funding Agency

6,356 funding agencies subsidized forest ecology studies, and the top 20 were exported for closer analysis. Because many articles used abbreviations for the funding agencies the top 20 were merged into 15 (Table 5). Examples include the National Science Foundation (NSF), the Conselho Nacional de Desenvolvimento Científico e Tecnológico (CNPq), the European Union (EU), and the Natural Sciences and Engineering Research Council of Canada (NSERC).

Table 5: The 15 most productive agencies funding forest ecology research indexed by the web of science during 2002–2011

	Funding agencies	**Articles number**	**Ratio (%)**	**Countries**
1	National Science Foundation	2240	2.84	USA

2	National Natural Science Foundation of China	831	1.05	China
3	Natural Sciences and Engineering Research Council of Canada	807	1.02	Canada
4	CNPq	744	0.94	Brazil
5	European Union	601	0.76	EU
6	Chinese Academy of Sciences	372	0.47	China
7	NASA	357	0.45	USA
8	European Commission	337	0.43	EC
9	Academy of Finland	311	0.39	Finland
10	Australian Research Council	265	0.34	Australia
11	CAPES	221	0.28	Brazil
12	National Basic Research Program of China	196	0.25	China
13	FAPESP	192	0.24	Brazil
14	Russian Foundation for Basic Research	185	0.23	Russia
15	USDA Forest Service	172	0.22	USA

CNPQ: Conselho Nacional de Desenvolvimento Científico e Tecnológico, Brazil; NASA: National Aeronautics and Space Administration, USA; CAPES: Coordenação de Aperfeiçoamento de Pessoal de Nivel Superior, Brazil; FAPESP: Fundação de Amparo à Pesquisa do Estado de São Paulo, Brazil; EC: European Commission.

The National Science Foundation (USA), National Natural Science Foundation of China (China), Natural Sciences and Engineering Research Council of Canada (Canada), Conselho Nacional de Desenvolvimento Científico e Tecnológico (Brazil), and European Commission were more prolific in forest ecology than other funding agencies. Combining the number of articles in Table 5 by country/region demonstrates that the USA (2,769), China (1,399), Brazil (1,157), Canada (807), and EU (601) were also the top five countries/regions and provided more financial aid to forest ecology research than other countries.

By Research Area

In the analysis, forest ecology was related to 72 research areas identified by the Web of Science data. Table 6 lists the top 20 research areas and clearly shows that forest ecology studies were related to a wide range of disciplines. Environmental sciences ecology (31,172 or 39.47% of all articles), forestry

(13,164, 16.67%), agriculture (8,354, 10.58%), and plant sciences (8,027, 10.16%) were the top four major related research areas.

Table 6: The top 20 research areas related to forest ecology indexed using the web of science during 2002–2011

	Research areas	Articles number	Ratio (%)
1	Environmental Sciences Ecology	31172	39.47
2	Forestry	13164	16.67
3	Agriculture	8354	10.58
4	Plant Sciences	8027	10.16
5	Zoology	6470	8.19
6	Biodiversity Conservation	6005	7.60
7	Geology	5660	7.17
8	Meteorology Atmospheric Sciences	3654	4.63
9	Physical Geography	3453	4.37
10	Water Resources	2521	3.19
11	Marine Freshwater Biology	2271	2.88
12	Entomology	2176	2.76
13	Engineering	1981	2.51
14	Life Sciences Biomedicine Other Topics	1650	2.09
15	Evolutionary Biology	1631	2.07
16	Remote Sensing	1611	2.04
17	Science Technology Other Topics	1319	1.67
18	Biochemistry Molecular Biology	1269	1.61
19	Imaging Science Photographic Technology	1205	1.53
20	Genetics Heredity	1079	1.37

By Author

A total of 48,373 authors participated in forest ecology related studies. Among the 20 authors publishing the most articles, five were from the USA, four were from Canada, and two each were from Belgium, Finland, and England (Table 7).

Table 7: The 20 most productive authors of research papers related to forest ecology indexed using the Web of Science during 2002–2011

	Authors	Authors' countries	Articles number	Ratio (%)
1	Bergeron Y	Canada	146	0.19
2	Kulmala M	Finland	123	0.16
3	Hermy M	Belgium	114	0.14
4	Lindenmayer DB	Australia	110	0.14
5	Black TA	Canada	103	0.13
6	Coops NC	Canada	95	0.12
7	Asner GP	USA	91	0.12
8	Verheyen K	Belgium	91	0.12
9	Reich PB	USA	87	0.11
10	Penuelas J	Spain	85	0.11
11	Vesala T	Finland	85	0.11
12	Leuschner C	Germany	81	0.10
13	Peres CA	England	81	0.10
14	Chen JM	Canada	80	0.10
15	Ciais P	France	80	0.10
16	Groffman PM	USA	79	0.10
17	Law BE	USA	78	0.10
18	Malhi Y	England	78	0.10
19	Fahey TJ	USA	77	0.10
20	Yu GR	China	77	0.10

By Year

From 2002 to 2011, the annual number of published articles about forest ecology was growing at a stable rate (Table 8), as the fit produced a high determination coefficient from the collected data ($R^2=0.9955$). The best fit for forest ecology was found to be: $y=629.75x - 1.2557exp+06$, where y is the article number and x is the number of years since 2002. Extrapolating from the model, the number of articles about forest ecology in the following years could be forecasted (Figure 2).

Table 8: Annual number of articles on forest ecology indexed using the Web of Science during 2002–2011

	Years	Articles number	Ratio (%)
1	2002	5245	6.64
2	2003	5729	7.25

3	2004	6250	7.91
4	2005	6816	8.63
5	2006	7555	9.57
6	2007	8098	10.25
7	2008	8970	11.36
8	2009	9311	11.79
9	2010	10096	12.78
10	2011	10915	13.82

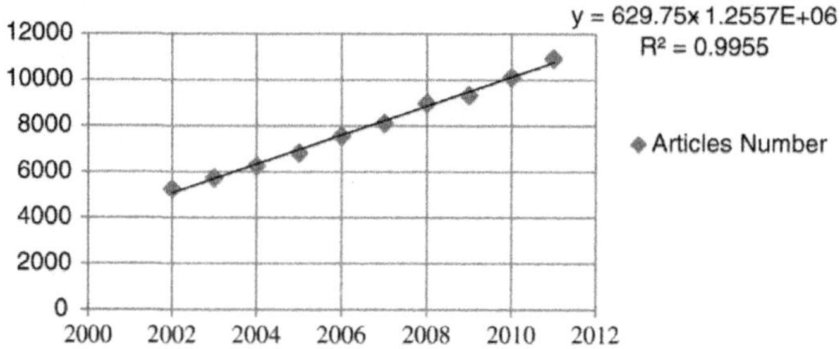

Figure 2: A linear relationship between articles number and years during 2002-2011.

By Publication

The number of journals publishing forest ecology related articles each year increased from 430 in 2002 to 856 in 2011. Table 9 shows the top 20 major journals indicating that *Forest Ecology and Management* (3,876, 4.91%) was the top journal on forest ecology by article count, followed by *Canadian Journal of Forest Research* (1,399, 1.77%) and *Biological Conservation* (1,399, 1.77%).

Table 9: The top 20 journals related to forest ecology analyzed using the Web of Science during 2002–2011

	Publications	**Articles number**	**Ratio (%)**
1	Forest Ecology and Management	3876	4.91
2	Canadian Journal of Forest Research	1399	1.77
3	Biological Conservation	933	1.18
4	Soil Biology Biochemistry	929	1.18
5	Biodiversity and Conservation	928	1.18
6	Global Change Biology	824	1.04

7	Ecology	750	0.95
8	Oecologia	741	0.94
9	Biotropica	666	0.84
10	Plant and Soil	653	0.83
11	Ecological Applications	636	0.81
12	Plant Ecology	614	0.78
13	Ecological Modeling	599	0.76
14	Remote Sensing of Environment	598	0.76
15	Argicultural and Forest Meteorology	589	0.75
16	Journal of Tropical Ecology	543	0.69
17	Journal of Geophysical Research Atmospheres	523	0.66
18	Conservation Biology	516	0.65
19	Journal of Biogeography	510	0.65
20	Tree Physiology	508	0.64

DISCUSSION

The results of this study pointed to several significant hotspots in global research related to forest ecology based on an analysis of article keywords for articles published during 2002–2011, and revealed the distribution of the articles from seven aspects listed above. The keyword analysis method and the java analysis program could be extended to other related research fields.

In the keywords analysis, we presumed that a keyword appeared only once in the keywords list of an article (Campbell1963). Therefore the frequency of a keyword could show the number of articles that had used this keyword. For example, the frequency of "forest" was 9,302, meaning that 9,302 articles had used "forest" as a keyword in 73,740 articles.

It was undisputed that "forest" was the most frequently used keyword (9,302 articles). Most writers used this word to express the concept of "forest" instead of its plural "forests"; therefore, "forest" appeared in articles three times more than "forests" (3,069). The next four most frequently used words were "diversity" (5,424), "conservation" (5,135), "dynamics" (4,886), and "vegetation" (4,720) indicating forest diversity, forest conservation, forest dynamics and forest vegetation were the focus of forest ecological studies.

The frequency of "patterns" (4,166), "model" (2,100), and "models" (988) demonstrated that these words were widely used in forest developmental pattern and model studies. The keywords "management" (3,236), "ecology" (2,677), "ecosystems" (2,407), and "ecosystem" (1,362) were also frequently used in macro research (9,682 times), accounting for 1.03% in all keywords indicating large numbers of studies had been carried out in these aspects of forest research in last ten years.

USA" (2,916), "Brazil" (1,018), "Australia" (868), "Mexico" (819), "Costa Rica" (813) and "New Zealand" (796) appeared more frequently than the names of other countries showing that many studies focused on those countries. During the early twenty-first century, the warm droughts in the United States, Europe and Australia have been recognized as a considerable change from the climatological conditions and variability of the late twentieth century (Dai2011), and the focus of forest ecology studies in those regions were impacted accordingly. From a regional point of view, we can see that the total frequencies of "rain-forest" (3,253), "tropical forests" (1,107), and "tropical forest" (1,513) were 5,873, 2.5 times more frequent than "boreal forest" (2,334), indicating that forest ecology studies concerning tropical forests were produced more frequently than those related to boreal forests.

In 2005, large-scale, warm droughts occurred in North America, Africa, Europe, Amazonia and Australia, resulting in major effects on terrestrial ecosystems, carbon balance and food security (Breshears2005). The words "nitrogen" (3,136), "carbon" (2,568), and "phosphorus" (971) were used frequently in the studies concerning elemental nutrients. There were numerous studies related to how the climate is affecting forest ecology, as indicated by the frequencies of "climate-change", "climate", and "climate change," which were 2,412, 2,095 and 1,599, respectively.

This study did reveal some problem areas. Some keywords were not being used consistently, such as soil, soils, forest soil and forest soils, which all pointed to the same thing: forest soil. Another example was that tropical forest and tropical forests also expressed similar meanings. The use of multiple keywords for a single concept might be related to the writing styles and habits of different authors, but this creates difficulty in statistical analysis.

The USA, Canada, and Germany were the top three most productive countries of forest ecology related research. The most three productive organizations were the University of California System, Chinese Academy of Sciences, and the US Forest Service. The three most productive funding agencies were the National Science Foundation, the National Natural Science Foundation of China, and the Natural Sciences and Engineering Research Council of Canada. Environmental science / ecology, forestry, and agriculture were the top three most popular categories. The spatial clusters of authors were mainly in the USA and Canada. *Forest Ecology and Management*, *Canadian Journal of Forest Research*, and *Biological Conservation* were the top three journals with the most publications related to forest ecology research. In the article analysis, the results by country/region, organization, funding agency, author distribution, and sources titles, was clustered in developed countries, apparently because these countries have economic strength required to invest in science and technology.

In this study, the limitations of search term expressions and the English language made it impossible to include all related keywords in the field of forest ecology research, especially in other languages. This study did not analyze the effects of cooperation between authors and joint papers by authors from multiple nations. In the journal sort, the impact factor of the journal was not considered.

CONCLUSIONS

A serial java program was developed and applied to conduct keyword frequency analysis. That improved the efficiency of data processing and provided an analysis method. Keyword analysis offered insight into forest ecology research areas of interest, while the abundance of less frequent keywords suggested a lack of continuity in research and a wide disparity in the focus of forest ecology research. The top 100 keywords in the keyword analysis were almost all included in the top 20 research areas in the article analysis, so one could conclude that keyword frequency analysis is consistent with article research area analysis. Their difference is the former is concrete and the latter is abstract.

APPENDIX A

(TS = (habitats) or TS = (species diversity) or TS = (biodiversity) or TS = (species richness) or TS = (environmental factors) or TS = (ecosystems) or TS = (plant communities) or TS = (landscape) or TS = (phenology) or TS = (environmental degradation) or TS = (plant) or TS = (populations) or TS = (animal) or TS = (ecological disturbance) or TS = (landscape) or TS = (synecology) or TS = (palaeo ecology) or TS = (community) or TS = (biogeography) or TS = (population) or TS = (ecotypes) or TS = (predator prey relationships) or TS = (microbial) or TS = (freshwater) or TS = (food webs) or TS = (lowland areas) or TS = (restoration) or TS = (fire) or TS = (food chains) or TS = (autecology) or TS = (marine) or TS = (chemical) or TS = (human) or TS = (bioenergetics) or TS = (ecological balance) or TS = (bio coenosis) or TS = (microenvironments) or TS = (dendro ecology) or TS = (climate) or TS = (soils) or TS = (physiography) or TS = (carbon balance) or TS = (nutrient cycling) and (TS = (forest).

ACKNOWLEDGEMENTS

This study was supported by the Forestry Commonweal Programs (No. 200904003) from State Forestry Administration, P.R.China. The authors greatly appreciate the technical support of Peng Shi, Yingni Zou, and Ying Pan. The authors are grateful to Yungang Liao for his helpful suggestions. The authors would also like to thank the chief editor of *SpringerPlus* and anonymous reviewers for their valuable comments.

AUTHORS' CONTRIBUTIONS

YS carried out the bibliometric analysis, and drafted the manuscript. TZ gave financial assistance and some advice to this manuscript. Both authors read and approved the final manuscript.

REFERENCES

1. Barnes BV, Zak DR, Denton SR, Spurr SH: *Forest ecology*. 1997. http://www.cabdirect.org/abstracts/19980607188.html
2. Berger TW, Inselsbacher E, Mutsch F, Pfeffer M: Nutrient cycling and soil leaching in eighteen pure and mixed stands of beech (Fagus sylvatica) and spruce (Picea abies). *Forest Ecol Manag* 2009, 258(11):2578-2592. 10.1016/j.foreco.2009.09.014
3. Breshears DD: Regional vegetation die-off in response to global-change-type drought. *Proc Natl Acad Sci* 2005, 102: 15144-15148. 10.1073/pnas.0505734102
4. Campbell DJ: Making your own indexing system in science and technology (classification and keyword systems). *ASLIB Proc* 1963., 15(10):
5. Cheaib A, Badeau V, Boe J, Chuine I, Delire C, Dufrêne E: Climate change impacts on tree ranges: model intercomparison facilitates understanding and quantification of uncertainty. *Ecol Lett* 2012, 15(6):533-544. 10.1111/j.1461-0248.2012.01764.x
6. Dai A: Drought under global warming: a review. *WIRES Climate Change* 2011, (2):45-65.
7. Diaz S, Lavorel S, de Bello F, Quetier F, Grigulis K, Robson M: Incorporating plant functional diversity effects in ecosystem service assessments. *Proc Natl Acad Sci USA* 2007, 104(52):20684-20689. 10.1073/pnas.0704716104
8. Dobbertin M, Nobis M: Exploring research issues in selected forest journals 1979–2008. *Ann Forest Sci* 2010, 67(8):800. 10.1051/

forest/2010052

9. Grandjean P, Eriksen M, Ellegaard O, Wallin J: The Matthew effect in environmental science publication: A bibliometric analysis of chemical substances in journal articles. *Environ Health* 2011, 10(1):1. 10.1186/1476-069X-10-1
10. Hanberry BB, Kabrick JM, He HS, Palik BJ: Historical trajectories and restoration strategies for the Mississippi River Alluvial Valley. *Forest Ecol Manag* 2012, 280: 103-111. 10.1016
11. Hendrix D: An analysis of bibliometric indicators, National Institutes of Health funding, and faculty size at Association of American Medical Colleges medical schools, 1997–2007. *J Med Libr Assoc* 2008, 96(4):324-334. 10.3163/1536-5050.96.4.007
12. Lamb D, Erskine P, Parrotta J: Restoration of degraded tropical forest landscapes. *Science* 2005, 310(5754):1628-1632. 10.1126/science.1111773
13. Loucks CJ, Lu Z, Dinerstein E, Wang H, Olson DM, Zhu C: Ecology. Giant pandas in a changing landscape. *Science* 2001, 294(5546):1465. 10.1126/science.1064710
14. McLachlan SM, Bazely DR: Outcomes of longterm deciduous forest restoration in southwestern Ontario, Canada. *Biol Conserv* 2003, 113(2):159-169. 10.1016/S0006-3207(02)00248-3
15. Mitchell SR, Harmon ME, O'Connell KEB: Forest fuel reduction alters fire severity and long-term carbon storage in three Pacific Northwest ecosystems. *Ecol Appl* 2009, 19(3):643-655. 10.1890/08-0501.1
16. Morrissey RC, Gauthier M, Kershaw JA Jr, Jacobs DF, Seifert JR, Fischer BC: Grapevine (Vitis spp.) dynamics in association with manual tending, physiography, and host tree associations in temperate deciduous forests. *Forest Ecol Manag* 2009, 257(8):1839-1846. 10.1016/j.foreco.2009.02.004
17. Narotsky D, Green PH, Lebwohl B: Temporal and geographic trends in celiac disease publications: a bibliometric analysis. *Eur J Gastroenterol Hepatol* 2012, 24(9):1071-1077. 10.1097/MEG.0b013e328355a4ab
18. Perez M, Fu M, Xie J, Yang X, Belcher B: The relationship between forest research and forest management in China: an analysis of four leading Chinese forestry journals. *Int For Rev* 2004, 6(3–4):341-345.
19. Rodrigues RR, Gandolfi S, Nave AG, Aronson J, Barreto TE, Vidal CY: Large-scale ecological restoration of high-diversity tropical forests in SE Brazil. *Forest Ecology Manag* 2011, 261(10):1605-1613. 10.1016/j.

foreco.2010.07.005

20. Rubio A, Escudero A: Effect of climate and physiography on occurrence and intensity of decarbonation in Mediterranean forest soils of Spain. *Geoderma* 2005, 125(3–4):309-319.
21. Şekercioğlu ÇH, Primack RB, Wormworth J: The effects of climate change on tropical birds. *Biol Conserv* 2012, 148(1):1-18. 10.1016/j.biocon.2011.10.019
22. Sillett SC, Van Pelt R, Koch GW, Ambrose AR, Carroll AL, Antoine ME: Increasing wood production through old age in tall trees. *Forest Ecol Manag* 2010, 259(5):976-994. 10.1016/j.foreco.2009.12.003
23. Sitzia T, Semenzato P, Trentanovi G: Natural reforestation is changing spatial patterns of rural mountain and hill landscapes: A global overview. *Forest Ecol Manag* 2010, 259(8):1354-1362. 10.1016/j.foreco.2010.01.048
24. van Eck NJ, Waltman L, Dekker R, van den Berg J: A comparison of two techniques for bibliometric mapping: Multidimensional scaling and VOS. *J Am Soc Inf Sci Technol* 2010, 61(12):2405-2416. 10.1002/asi.21421
25. van Oudenhoven APE, Petz K, Alkemade R, Hein L, de Groot RS: Framework for systematic indicator selection to assess effects of land management on ecosystem services. *Ecol Indic* 2012, 21: 110-122.
26. van Raan AFJ: Comparison of the Hirsch-index with standard bibliometric indicators and with peer judgment for 147 chemistry research groups. *Scientometrics* 2006, 67(3):491.
27. Wang Q, Wang S, Yu X: Decline of soil fertility during forest conversion of secondary forest to Chinese fir plantations in substropical China.*Land Degradation Dev* 2011, 22(4):444-452. 10.1002/ldr.1030
28. Wintle B, Bekessy S, Venier L, Pearce J, Chisholm R: Utility of dynamic-landscape metapopulation models for sustainable forest management. *Conserv Biol* 2005, 19(6):1930-1943. 10.1111/j.1523-1739.2005.00276.x
29. Xu Z, Chen C: Fingerprinting global climate change and forest management within rhizosphere carbon and nutrient cycling processes. *Environ Sci Pollut Res* 2006, 13(5):293. 10.1065/espr2006.08.340

Chapter 12

SIZE-ENERGY RELATIONSHIPS IN ECOLOGICAL COMMUNITIES

Brent J. Sewall[1,2], Amy L. Freestone[1], Joseph E. Hawes[3], Ernest Andriamanarina[4]

[1] Department of Biology, Temple University, Philadelphia, Pennsylvania, United States of America

[2] Department of Wildlife, Fish, and Conservation Biology, University of California Davis, Davis, California, United States of America

[3] School of Environmental Sciences, University of East Anglia, Norwich Research Park, Norwich, United Kingdom

[4] De´partement des Sciences de la Nature et de l'Environnement, Universite´ d'Antsiranana, Antsiranana, Madagascar

ABSTRACT

Hypotheses that relate body size to energy use are of particular interest in community ecology and macroecology because of their potential to facilitate quantitative predictions about species interactions and to clarify complex ecological patterns. One prominent size-energy hypothesis, the energetic equivalence hypothesis, proposes that energy use from shared, limiting resources by populations or size classes of foragers will be independent of body size. Alternative hypotheses propose that energy use will increase with body size, decrease with body size, or peak at an intermediate body size. Despite extensive study, however, size-energy hypotheses remain controversial, due to a lack of directly-measured data on energy use, a tendency to confound distinct scaling relationships, and insufficient attention to the ecological contexts in which predicted relationships are likely to occur. Our goal, therefore, was to directly evaluate size-energy hypotheses while clarifying how results would differ with alternate methods and assumptions. We comprehensively tested size-energy hypotheses in a vertebrate frugivore guild in a tropical forest in Madagascar. Our test of size-energy hypotheses, which is the first to examine energy intake directly, was consistent with the energetic equivalence hypothesis. This finding corresponds with predictions of metabolic theory and models of energy distribution in ecological communities, which imply that body size does not confer an advantage in competition for energy among populations or size classes of foragers. This result was robust to different assumptions about energy regulation. Our results from direct energy measurement, however, contrasted with those obtained with conventional methods of indirect inference from size-density relationships, suggesting that size-density relationships do not provide an appropriate proxy for size-energy relationships as has commonly been assumed. Our research also provides insights into mechanisms underlying local size-energy relationships and has important implications for predicting species interactions and for understanding the structure and dynamics of ecological communities.

INTRODUCTION

Body size has strong, widespread, and predictable relationships to organisms' physiological traits and life history characteristics [1], [2], and has therefore been proposed as a primary driver of diverse patterns in community ecology and macroecology [2], [3]. Relationships between body size and energy use in particular have long intrigued ecologists [4], [5], because they promise to reveal an underlying order in highly complex ecological systems, and to predict quantitative patterns of species interactions from local to global spatial scales.

One prominent hypothesis relating body size to energy use was proposed by Damuth, who focused on three variables of interest – the density (*D*) of a population, the mean metabolic (respiration) rate (*R*) of individuals of that population, and the total amount of energy used by all individuals of that population, or population energy use (*PEU*) – as well as the relationships of each of these three variables to body size, as indicated by body mass (*M*). In a widely-cited paper [6], he suggested that population energy use could be estimated as the product of population density and mean metabolic rate (i.e., $PEU = D * R$). To relate these variables to body size, he compiled a global dataset of population densities of mammalian herbivores from published accounts, and found that population density (*D*) scaled globally with body mass (*M*) to the power of −0.75 [6]. Noting that metabolic rate (*R*) was already understood to scale with body mass (*M*) to the power of +0.75 [1], Damuth then hypothesized that the total energy used by a population should generally be independent of body size [6], because $PEU = D * R \propto M^{-0.75} * M^{0.75} = M^0$. This finding, later called the 'energetic equivalence rule' [7], has been highly influential, because it suggests that underlying the complexities of ecological communities, populations of animals of all body sizes are on equal footing in the competition for energy (i.e., there is no optimal body size where energy use is maximized) [6]. The energetic equivalence rule has also garnered widespread attention since it provides a quantifiable prediction for a point of great interest in ecology – the distribution of energy among populations in a community – on the basis of a relatively easily-measured variable, body size. This energetic equivalence rule, and its emphasis on three-quarter-power scaling with body mass, has provided the theoretical foundation for explaining a range of ecological patterns, from population growth [8]to community structure [9] to global biodiversity patterns [10].

Later authors [5], [11], [12] expanded on this idea, noting that population energy use would also be independent of body size wherever both scaling exponents were exactly inversely correlated. Thus, if $D \propto M^a$, $R \propto M^b$, and $PEU \propto M^c$, then $M^a * M^b = M^c$ and population energy use will be independent of body size whenever $c = 0$. This will occur whenever $-a = b$, even if $-a$ and b are not equal to 0.75 in the system being studied. Thus, the 'energetic equivalence rule' is a special case (where $a = -0.75$, $b = +0.75$) of a broader hypothesis known as 'energetic equivalence' [5], where $-a = b$. This energetic equivalence hypothesis predicts that inverse scaling relationships of population density and metabolic rate are widespread, and that population energy use by different species is broadly independent of body size in ecological communities [13].

While the energetic equivalence hypothesis has been supported in some studies [14], [15], this hypothesis is only one possible expectation for energy use within ecological communities [4]. Other studies have proposed and provided evidence in support of alternative hypotheses: that energy use increases with body size (e.g., [16]), decreases with body size (e.g., [12]), or peaks at an intermediate body size (e.g., [17]). Thus, despite intense interest in size-energy hypotheses, and despite their potential importance for understanding community structure and dynamics, past studies have produced conflicting results and the influence of body size on energy use in ecological communities remains unclear.

In addition, size-energy hypotheses have been controversial because they are often disconnected from the proposed mechanism underlying these hypotheses – namely, that size-energy relationships result from energetic tradeoffs due to resource competition among organisms of different body sizes in a community [2], [5], [6] but see [18]. This mechanism implies that size-energy hypotheses apply only in certain ecological contexts – namely those where all species can directly compete for shared, limiting resources. Previous studies of size-energy hypotheses, however, have tended to examine large spatial scales by compiling data from many different communities around the globe (e.g., [6]), though species selected for analysis from separate communities cannot plausibly compete [19]. Further, those studies that have examined the local scale have not controlled for resource availability (e.g., [20]), though the use of different resources implies a lack of energetic tradeoffs due to competition [2], [20]. Inferences about size-energy hypotheses from large spatial scale or multi-resource studies should therefore be interpreted carefully.

Size-energy hypotheses, and past tests of these hypotheses, have also been controversial on methodological grounds, in part due to a tendency to confound distinct scaling relationships in two primary ways. First, all previous studies have indirectly used size-density relationships, rather than size-energy relationships, to test size-energy hypotheses (e.g., [11], [21], [22]). This method (hereafter, the 'indirect method') has spread because in practice it is easier to measure the density scaling relationship (i.e., the exponent a) than the energy scaling relationship (c) in a community. In such studies, the exponent of the metabolic scaling relationship (b) is assumed to be a universal constant (+0.75) (e.g., [6], [14], [15], [23]), or is taken from published metabolic scaling relationships when available (e.g., [24]), such that c can be inferred directly from a via the equation $M^a * M^b = M^c$, or $a+b = c$. However, recent study suggests that metabolic scaling exponents are highly variable [25], and so a specific exponent cannot be assumed for all systems (i.e., b is not constant across all systems, and thus a is insufficient to determine c). This approach also suffers from a lack of data on species-specific metabolic scaling exponents and is subject to the hidden non-linearity, propagation of error variances, and inherent imprecision of multiplying allometric relationships [21], [26]. This practice is also circular, since in the analysis, body size serves both as the predictor variable (M) and as a means to calculate the response variable (PEU); results therefore will tend to be biased to suggest a positive correlation. As a result, although size-density relationships are interesting in their own right [5], and can provide insight into mechanisms underlying energy use, their utility for estimating the form of the size-energy relationships that are the focus of size-energy hypotheses or for providing a robust test of these hypotheses remains unclear [19]. Direct measurement of energy use (hereafter, the 'direct method') could avoid these problems, but measurement of energy use by all species in a community is technically challenging [4], and no test of size-energy hypotheses using direct measurements of energy use has previously been attempted [27].

Second, two distinct approaches to evaluating size-energy hypotheses have emerged on the basis of the underlying assumptions for how energy use is regulated in ecological communities. Competition for energy implies zero-sum dynamics [13], [28], where limited resources are allocated to individual organisms in the community [29] and thus increases in energy consumed from a shared, limiting resource by one individual are necessarily offset by equal and opposite decreases in energy consumed by other individuals. Authors of studies on terrestrial animals have generally assumed that such competition for energy is regulated by competition among species populations in a community (e.g., [6]), while authors of studies on trees and marine systems have generally assumed that energy use is regulated by competition among size classes in a community (e.g., [30], [31]). For the former assumption, population energy use (*PEU*) is used to evaluate the size-energy relationship, whereas for the latter assumption, a related but distinct variable – size class energy use (*SCEU*) – is used. Unlike *PEU*, which is calculated as the sum of the energy used by all individuals in a species population, *SCEU* is calculated as the sum of the energy used by all individuals in a given size class, regardless of species identity. Because size classes may vary in the number of species they contain, and individuals of a single species may be distributed among multiple size classes, tests of size-energy hypotheses using each approach may provide differing results for the same community[32]. Despite these differences, reviews of size-energy hypotheses have typically not distinguished between the population and size-class approaches, leading to confusion in the literature [5]. Comparison of the two approaches in the same system could clarify underlying assumptions of how energy use is regulated in communities. Such comparisons have been completed for size-density relationships in multi-trophic communities [32] (see also [5], [33] fordiscussion), but not for size-energy relationships.

Our goal in this project was to directly evaluate size-energy hypotheses while clarifying how results would differ with alternate methods and assumptions. We had three objectives in support of this goal. First, we sought to complete the first analyses of size-energy hypotheses to directly measure how energy use scales with body size. Specifically, we evaluated whether energy use is independent of body size, increases with body size, decreases with body size, or peaks at an intermediate body size. Second, we sought to evaluate the influence of indirect versus direct methods on conclusions about size-energy hypotheses. Specifically, we compared results of our novel direct method with those from the indirect method conventionally used to test size-energy hypotheses. Third, we sought to determine the influence of assumptions about energy regulation on conclusions about size-energy hypotheses. Specifically, we compared results obtained using the population and size-class approaches. We addressed these objectives by examining scaling relationships among a guild of vertebrate frugivores as they foraged on a common set of shared, limiting resources in a tropical forest in Madagascar.

MATERIALS AND METHODS

Ethics Statement

This study was conducted at all times in strict accordance with the animal welfare protocols of the University of California, Davis and with the laws of the participating countries. Animal research was non-manipulative and completely observational, and solely examined wild animals in their natural habitat. The study was conducted on protected land owned by the Madagascar government and managed by Madagascar National Parks, and was approved by the Madagascar Ministère de l'Environnement, Direction Générale des Eaux et Forêts, under permits N° 143/MINEV.EF/SG/DGEF/DPB/SCBLF/RECH, N° 225/06/MINEV.EF/SG/DGEF/DPB/SCBLF/RECH, N° 024N-EV01/MG06, and N° 308N-EV11/MG06.

Study System

We focused on frugivores, or fruit eating animals, due to their central ecological importance for the maintenance and regeneration of tropical forests [34], [35]. Specifically, we studied the vertebrate frugivore guild of Ankarana National Park, near Mahamasina in northern Madagascar, within 175 ha of semi-evergreen primary forest. This region undergoes two distinct seasons: a short wet season from January–April when almost all of the ~1890 mm of annual rainfall occurs, and a longer dry season during the remaining months ([36]; pers. obs.). The study period was the end of the dry season (October–December 2005 and October 2006-early January 2007), at a time when available fruit is in limited supply and competition for shared resources is expected to be greatest. Frugivores are abundant and easily observable within the park, where they directly compete for clearly identifiable resources – primarily fruit from the canopies of fruiting trees. The frugivore guild comprised five bird, five primate (lemur), and two fruit bat species, a taxonomically diverse fauna with distinct foraging behaviors and body sizes spanning two orders of magnitude (Table 1).

Table 1. Vertebrate frugivore guild of Ankarana National Park, Madagascar, with taxa, activity, mass, and size class

Species	English name	Taxon	Activity	Mass (g)[a]	Size class
Saroglossa aurata	Madagascar Starling	Bird	Diurnal	40	A
Hypsipetes madagascariensis	Madagascar Bulbul	Bird	Diurnal	43.5	A
Microcebus tavaratra[b,c]	northern rufous mouse lemur	Lemur	Nocturnal	64.5	B
Rousettus madagascariensis	Madagascar rousette	Bat	Nocturnal	65	B
Cheirogaleus medius[c]	fat-tailed dwarf lemur	Lemur	Nocturnal	198	C
Treron australis	Madagascar Green Pigeon	Bird	Diurnal	236	C
Coracopsis nigra	Lesser Vasa Parrot	Bird	Diurnal	254	D
Eidolon dupreanum	Madagascar straw-colored fruit bat	Bat	Nocturnal	295	D
Phaner electromontis	Amber Mountain fork-marked lemur	Lemur	Nocturnal	425	E
Coracopsis vasa	Greater Vasa Parrot	Bird	Diurnal	530	E
Eulemur coronatus	crowned lemur	Lemur	Cathemeral	1450	F
Eulemur sanfordi	Sanford's brown lemur	Lemur	Cathemeral	2150	G

[a]Based on the midpoint of the body mass range of adults of each species [72,73,74].
[b]We considered all *Microcebus* observations at Ankarana to be of *M. tavaratra* on the basis of recent taxonomic analyses [75].
[c]These or related species may enter torpor seasonally at other sites [73], but we observed them foraging throughout our study period.
doi:10.1371/journal.pone.0068657.t001

Indirect method for estimating size-energy relationships from size-density relationships

We first used the conventional indirect method for testing size-energy hypotheses, by measuring size-density relationships using frugivore population densities. For this portion of the study, we focused on frugivorous lemurs and bats. To determine lemur population densities, we established 12 standardized line transects of 500 m in length in primary forest, ≥100 m from each other and the forest edge. We walked each transect three times, during peak periods of frugivore foraging just after sunrise (05:00–08:00), and just before (15:00–18:00) and just after sunset (18:30–21:30). To control for animal behavior and detectability, we did not conduct observations during rain or wind, or during moonlit nights. We walked transects at 10 m per minute, searching both sides of the transect line for visual or auditory signs of lemurs. During nocturnal observations, we used two observers to ensure adequate coverage [37], and headlamps to facilitate lemur detections via direct sighting and eyeshine [38]. Upon detection, we identified each lemur to species and estimated perpendicular distance from the line to lemur individuals. To calculate densities of lemur populations from transect data, we determined a standard suite of detection functions separately for each species. We controlled for different effort during diurnal and nocturnal sampling with a multiplier [39], and we calculated densities of the two cathemeral species, *Eulemur coronatus* and *E. sanfordi*, solely from diurnal periods, when they were most active. We selected the best detection models following well-established methodology for distance sampling analyses [39], [40], and using Distance 6.0.2 software [41].

Since fruit bats were not easily detectable with distance sampling methods, we used roost count data to estimate their population densities. Because both species, *Rousettus madagascariensis* and *Eidolon dupreanum*, are capable of flying long distances, we used published estimates of their abundance at all roosts near Ankarana National Park [42], divided by the area of all primary and secondary forest in the same region [43].

We then converted densities to energy use with the conventional indirect approach, multiplying population density by individual field metabolic rate to determine energy use per unit area (kJ/(hr * km^2)). As in past studies, we determined metabolic rates from allometric scaling relationships of similar species in the literature (equation 2 for eutherian mammals in [24]).

Direct Method for Measuring Size-Energy Relationships

We went beyond this conventional indirect method by developing a direct method for measuring the size-energy relationship. Size-energy hypotheses are typically formulated in relation to 'energy use' (e.g., population energy use, *PEU*), the amount of energy processed by all organisms of a population during metabolism, but this metric cannot easily be measured directly in the field. However, 'energy intake', the amount of metabolizable energy consumed by an organism, is both directly observable and closely correlated with energy use [44]. Energy intake is also directly linked to the proposed mechanism (of resource competition) that underlies size-energy hypotheses [5], [6], since energy consumed by one organism from a shared, limited resource in a habitat is not available to another organism, regardless of whether that consumed energy is actually processed. We therefore used energy intake as our metric by measuring fruit consumption during both diurnal and nocturnal foraging periods by the entire frugivore guild – including frugivorous birds, lemurs, and bats.

To determine the size-energy relationship, we measured energy intake by frugivore species while foraging at fig (*Ficus*, Moraceae) trees as a proxy for total energy intake. This approach assumes that figs were accessible to all frugivores and that the relative portion of energy intake at fig trees by frugivores was proportional to their total energy intake during the study period. The general importance of the frugivore-fig interaction, and several characteristics of our site – including strong preferences by frugivores for fig fruit, the abundance of fig fruit, scarcity of other resources, and evidence of rapid fig consumption – together suggest this assumption is justified in our study system. Specifically, the frugivore-fig interaction is critical to tropical forest communities as figs are widely distributed [45] and common in tropical forests [46]. Figs also play a keystone role in maintaining frugivore populations [47], [48], as they provide non-defended, easy-to-digest syconia (hereafter, fruit) year-round that is critical to a wide variety of frugivores during periods of fruit scarcity [48]. At our site, observations of frugivore foraging in fig trees and in the eight other most common non-fig species of fruiting trees during the dry season indicated strong preferences by all frugivore species for fig fruit, and no body-size trend among frugivores in relative preferences for fig fruit (KE Reuter, AA Gudiel, S Nieves, C Stanley, BJ Sewall, unpublished). During our study period in the late dry season, non-fig fruits were scarce: on the basis of fruit crop measurements at fig trees, and fruit counts at 10 randomly-placed transects of 50 m×2 m observed at two-week intervals, the three most common fig species provided a mean of 62.6% of all ripe fruits (range: 45–84%) in the habitat. During this period, fig fruit was highly prized: frugivores completely depleted even the largest fig fruit crops (up to ~80,000 fruits) within one month after the onset of ripening, and often within 1–2 weeks. Thus fig fruits represent a shared, limited resource for the entire frugivore guild. Of the six fig tree species in the study area, we focused on the three most common: *Ficus grevei*,*F. polita* (formerly *F. megapoda*), and *F. reflexa*, which together comprised 81% of all fig tree individuals, and 95% of all fig trees to reach peak fruiting during the study period.

We determined energy intake from fruit consumption via direct, unobtrusive observations on free-ranging frugivores in the field at focal fig trees. We observed foraging by frugivores during peak fruiting stages at each of 34 trees (18 *F. grevei*, seven *F. polita*, and nine *F. reflexa*). At each tree, we conducted observations during peak periods of both diurnal and nocturnal foraging. We used focal observations of individual frugivores at each tree to determine the fruit consumption rate (the rate at which an individual of a frugivore species consumed fig fruit) for each frugivore species at each fig species. At each tree we also used repeated scan censuses of the tree canopy to determine the residency rate (the number of minutes spent in the tree's canopy by all individuals of a species per hour) for each frugivore species at each tree. For each fig species, we used laboratory analysis of collected fruit samples to determine the metabolizable energy available per fruit. We then multiplied the fruit consumption rate (fruits/individual-min), the residency rate (individual-min/hr), and the metabolizable energy per fruit (kJ/fruit) to determine the energy intake rate by each frugivore species at each tree (kJ/hr). We then summed across all fruiting figs in the habitat and converted to energy intake per unit area (kJ/(hr * km^2)). Further details of these data collection protocols are in Appendix S1.

Analyses

To test size-energy hypotheses, we determined size-energy relationships with both the indirect and direct methods, and with both the population and size-class approaches, rendering four size-energy scaling relationships. All variables were natural log transformed prior to analysis. For the population approach, population density or population energy intake was regressed onto the body mass of each species. For the size-class approach, we assigned each frugivore species to a size class on the basis of adult body mass (Table 1), and we then summed density or energy intake across all species in the size class. We used normalized-logarithmic binning to assess the form of the size-class relationships. We chose normalized-logarithmic binning because linear binning is known to provide inaccurate estimates of parameters [49]. We did not use maximum likelihood estimation because we expected exponents >-1 and methods for handling these exponents are not commonly available and require estimating a maximum size, which is unknown. Size classes were 0.5 natural log unit bins, normalized for width of the bins (following [49]). Size-class density or size-class energy intake was then regressed onto the midpoint of each size-class bin.

For each of the four size-energy scaling relationships, we first compared the hypothesized linear relationships (energetic equivalence, energy intake increases with body size, energy intake decreases with body size) with a quadratic model for the hypothesized non-linear relationship (energy intake peaks at an intermediate body size). The relative suitability of the linear and quadratic models given the data was evaluated with a model estimation and selection process using information theory (following [50]). While such a model selection process is suitable for the population approach, the process is an ad hoc method for model selection in the size-class approach (see [51] for discussion of issues related to model selection in size-class models). Conclusions would not have differed had we not used a model selection procedure and instead simply used the full model, due to low support for the quadratic term. Since linear models were always selected via the selection procedure (see Results), we then determined the slope (scaling exponent) of the relationship for the selected models from the log-log plot. Because the sign of the slope of these relationships was not known *a priori* and under some hypotheses was expected to be zero, it was not appropriate to use standardized major axis regression [52]. Therefore, we determined slopes in the conventional manner, with ordinary least squares regression (following [6], [14], [16]). To evaluate correlations between the indirect and direct methods, we compared energy use values for all species or size classes common to each pair of size-energy relationships with Pearson's correlation coefficient. Finally, to understand factors that may affect correlations among the size-energy relationships, we used the mathematical links between scaling exponents ($a+b = c$; see Introduction) in the population approach to calculate the size-metabolism exponent b from the measured size-density exponent a and the measured size-energy exponent c. We completed all statistical analyses with JMP 8.0 [53] and R 2.11.1 [54] statistical software.

RESULTS

Indirect Method for Estimating Size-Energy Relationships from Size-Density Relationships

We detected 389 frugivorous lemurs in transects. Overall, frugivorous mammal densities ranged from 65.2 individuals per km^2 in *E. dupreanum* to 189.7 individuals per km^2 in *E. coronatus* (Appendix S2). Based on these data, the linear models for the size-energy relationship with the indirect method received more support than the quadratic models (Table 2). The slope of the size-density relationship using the population approach ($a = -0.11$, $F_{1,5} = 23.23$, $p = 0.005$; Fig. 1A) was shallower (less negative) than expected under the energetic equivalence rule (where a is hypothesized to equal -0.75). Evidence for a difference from the energetic equivalence rule was weaker in the size-density relationship, however, when this relationship was examined with the size-class approach ($a = -0.40$, $F_{1,4} = 2.60$, $p = 0.18$; Fig. 1B). The slope of the size-energy relationship calculated with the indirect method was greater than expected under energetic equivalence (where c is hypothesized to equal 0) with the population approach ($c = 0.66$, $F_{1,5} = 24.85$, $p = 0.004$; Fig. 1C), suggesting that results from the indirect method support the hypothesis that population energy use increases linearly with body size. Evidence for such a difference from energetic equivalence in the size-energy relationship with the indirect method, however, was weaker when this relationship was examined with the size-class approach ($c = 0.37$, $F_{1,4} = 2.91$, $p = 0.16$; Fig. 1D).

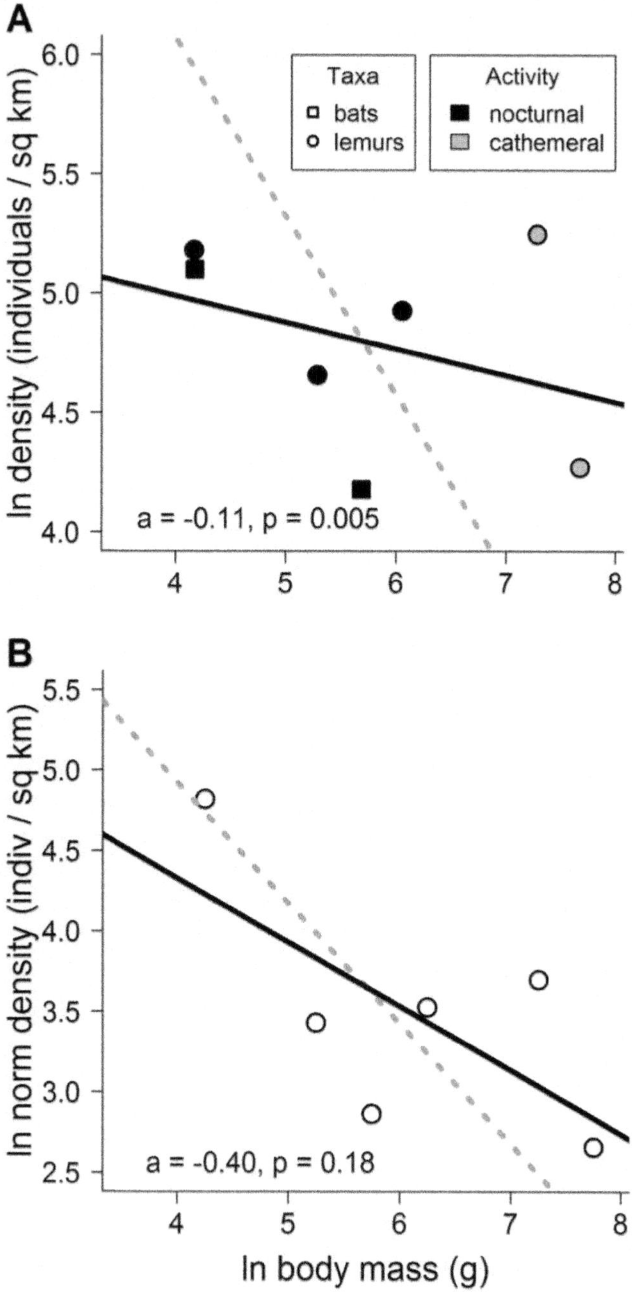
A
ln density (individuals / sq km)
6.0
5.5
5.0
4.5
4.0
Taxa
bats
lemurs
Activity
nocturnal
cathemeral
a = -0.11, p = 0.005
4
5
6
7
8
B
ln norm density (indiv / sq km)
5.5
5.0
4.5
4.0
3.5
3.0
2.5
a = -0.40, p = 0.18
4
5
6
7
8
ln body mass (g)

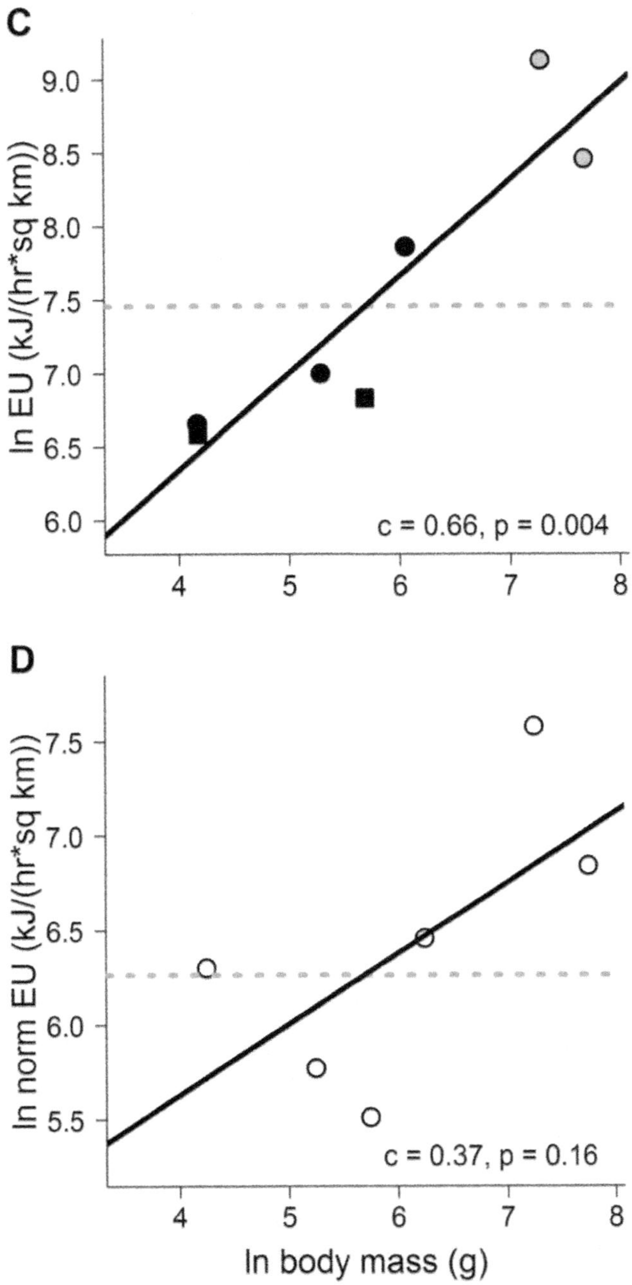

Figure 1. Size-energy relationships, as determined from the indirect method of inference from size-density relationships. (A) The regression slope (solid line) of the size-density relationship with the population approach was significantly shallower (less negative) than the expected slope under the energetic equivalence rule ($a = -0.75$, dot-

ted line). (B) Evidence for a difference from the energetic equivalence rule was weaker in the size-density relationship with the size-class approach. (C) When calculated with the indirect method, the size-energy relationship (where the size-energy scaling exponent, c, was equal to 0.66) with the population approach was significantly more positive than expected under energetic equivalence ($c = 0$, dotted line). (D) Evidence was weaker for a difference from energetic equivalence for the size-energy relationship when examined with the size-class approach ($c = 0.37$). Note that EU = energy use, and the y-axes in the size-class analyses (B and D) are normalized for bin width.

Table 2. Comparison of linear and quadratic models for log-log regressions of size-energy relationships, examined with both the indirect and direct methods and both the population and size-class approaches

Method	Approach	Model Type	AICc	Δ_i	w_i
Indirect	*Population*	Linear	20.21	0	0.999
		Quadratic	33.46	13.26	0.001
	Size class	Linear	27.12	0	1.000
		Quadratic	55.18	28.06	0.000
Direct	*Population*	Linear	63.26	0	0.861
		Quadratic	66.92	3.66	0.139
	Size class	Linear	37.19	0	0.999
		Quadratic	50.65	13.46	0.001

Best models were selected on the basis of the small sample size corrected Akaike's Information Criterion (AICc); Δ_i is the difference in AICc between the best and the alternate model, and w_i is the Akaike weight, the weight of evidence for each model given the data (where 1.000 represents the highest likelihood of the model relative to the alternate model).

doi:10.1371/journal.pone.0068657.t002

Direct method for measuring size-energy relationships

At our 34 focal trees, we conducted 272 person-hours of observations, with 4080 scan sample observations and 241 focal observations of frugivore foraging of at least 30 seconds in duration. We also collected 1159 fruits from nine fig trees for fruit content analysis (Appendix S1). Population energy intake ranged from 0.25 kJ/(hr * km^2) by *Phaner electromontis* to 937 kJ/(hr * km^2) by *Treron australis* (Appendix S2). Based on these data, the linear models for the size-energy relationship with the direct method received more support than the quadratic models (Table 2). Using the direct method, the slopes of the size-energy relationship using both the population approach ($c = 0.31$, $F_{1,10} = 0.28$, p

= 0.61; Fig. 2A) and the size-class approach ($c = -0.08$, $F_{1,5} = 0.036$, $p = 0.86$; Fig. 2B) were not different from zero, suggesting that results from the direct method were consistent with the energetic equivalence hypothesis.

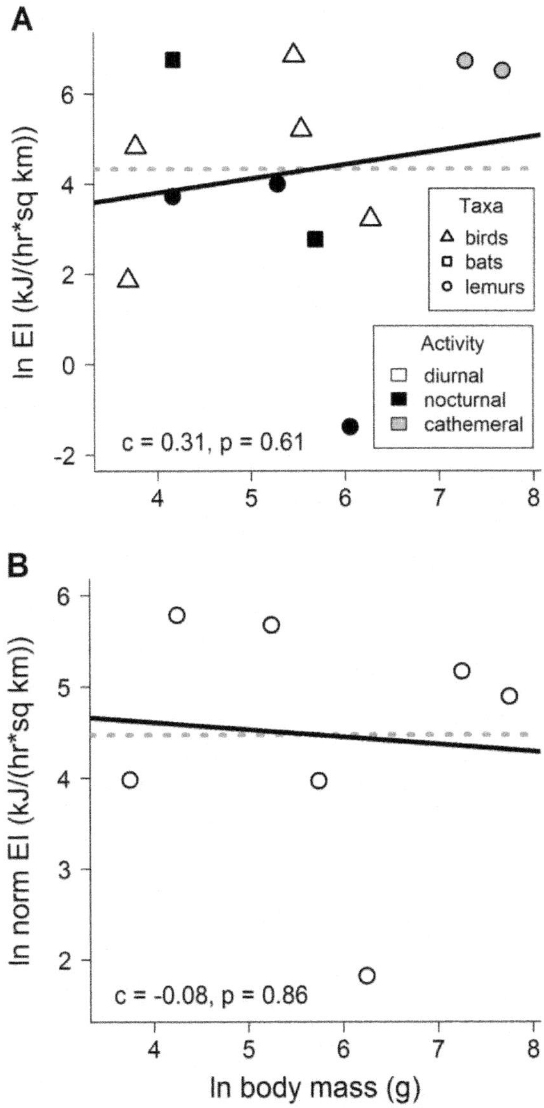

Figure 2. Size-energy relationships, as determined from the direct method of measuring energy intake. Regression slopes (solid lines) of the size-energy relationship with the (A) population approach (where the size-energy scaling exponent, *c*, was equal to 0.31), and (B) size-class appr($c = -0.08$) were both not significantly different from zero, and therefore were both consistent with the expected slope under energetic

equivalence ($c = 0$, dotted lines). Slopes were not qualitatively different if potential outliers (i.e., low values in A and B) were removed. Note that EI = energy intake, and the y-axis in the size class analysis (B) is normalized for bin width.

Correlations among Size-Energy Relationships and Implications for Size-Metabolism Relationships

The indirect and direct methods of estimating size-energy relationships were uncorrelated (population approach, $r = 0.19$, $p = 0.68$; size-class approach, $r = 0.07$, $p = 0.89$). Finally, in the population approach, the slopes of our measured size-density relationship (Fig. 1A) and our measured size-energy relationship (Fig. 2A) imply the slope of the size-metabolism relationship$b$ is 0.42.

DISCUSSION

Size-Energy Hypotheses—Direct Method

Size-energy hypotheses promise to clarify complex patterns of resource use in ecological communities and to provide quantitative predictions of species interactions, but such hypotheses have remained controversial. Rigorous empirical evaluation of size-energy hypotheses requires direct measurement of energy use by foragers during competition for shared, limiting resources [2], [19], but until now tests using direct measurements of energy use had not been completed. We conducted the first such direct test of size-energy hypotheses, and our results (Table 2, Fig. 2) were consistent with the central prediction of the energetic equivalence hypothesis, namely that energy use does not vary systematically with body size. These findings correspond with predictions of metabolic theory [2] and models of energy distribution in ecological communities [55], which suggest a general principle for community ecology: that body size does not confer an advantage to populations or size classes in the competition for energy from shared, limiting resources in communities.

Size-Energy Hypotheses—Comparison of Direct and Indirect Methods

For comparison with previous studies, we also estimated energy use with the conventional indirect method of inference from size-density relationships and assumptions of constant allometric relationships with metabolism. Previous studies of size-energy hypotheses, which have all used this method, have produced variable results, with most community-scale studies indicating that energy use peaks at intermediate body sizes [13], [17], [20], [26], [30] or increases with body size [11], [16], [21], [22], [56]. These results have been invoked to explain a number of broad ecological patterns, including the island rule (the common pattern of dwarfism and gigantism on isolated islands; [17]), Cope's Rule (the common evolutionary cycle in which large-bodied organisms emerge and become extinct more rapidly than small-bodied species;[16]), and the distribution of colony sizes in eusocial species [22].

Our results from the indirect method and population approach indicated that the slope of the size-energy relationship was linear (Table 2) and positive ($c>0$; Fig. 1C), suggesting that populations of animals of large body size dominate others in the competition for energy. These results were similar to many previous community-scale studies [11], [16], [21], [22], [56], yet these results contrasted with our directly-measured results from the same community (Fig. 2). The indirectly-calculated slope with the size-class approach was also linear (Table 2) but did not differ significantly from the expectation of energetic equivalence (Fig. 1D). Finally, in all cases, results from the indirect method were uncorrelated with results from the direct method (all $r \leq 0.19$, all $p \geq 0.68$, see Results).

The discrepancy in results from the direct and indirect methods – at least in the population approach – may occur due to the inherent biases and assumptions of the indirect method. Notably, the direction of the difference (relatively greater energy use at larger body sizes with the indirect method) represents just the kind of positive correlation between body size and energy use that is expected to derive from the use of body size as a component of both the predictor and response variable in the conventional method of determining and analyzing size-energy relationships (see Introduction).

Alternatively, this discrepancy could also reflect the slightly different focus of both methods. Our study focused on long-lived species, and thus the indirect method, due to its reliance on population densities to estimate energy use, may integrate energy use over all resources and over multiple years. In contrast, our direct method targeted key food resources during a critical period of the year for frugivore survival. Mismatches between the two methods could arise if organisms respond differently to temporal variation in food abundance on the basis of body size, such as for example, if small and large organisms differ in ability (e.g., due to different home range sizes or search areas) to access patchy and unpredictable fruit resources during periods of fruit scarcity [57]. Thus, differences in size-energy relationships resulting from the indirect and direct methods may represent size-dependent temporal variation in energy use.

The contrasting results from the two methods suggest size-density relationships are not appropriate proxies for size-energy relationships. This is not to argue against the examination of size-density relationships, which have provided important insights into key patterns of broad significance for ecology and macroecology [5]. A variety of other promising approaches are also emerging to address energy use in communities [28], [29]. However, our results highlight the limitations and challenges of applying density measurements to understand size-energy relationships, and call into question theories for size-energy relationships that have been derived from measured densities rather than energy use. Without further efforts to empirically validate or refute hypothesized relationships between the two variables with independent lines of evidence, the utility of the density-based methods for estimating local energy scaling relationships will remain unclear.

Processes of Energy Regulation—Comparison of Population and Size-Class Assumptions

Different versions of size-energy hypotheses imply alternative assumptions about processes of energy regulation in communities, and analyses of size-density relationships in multi-trophic communities indicate that energy regulation assumptions may strongly influence observed scaling patterns [2], [23], [32]. In size-energy relationships, population and size-class approaches assume that energy use is driven by metabolic processes related to an organism's taxonomic identity or its body size, respectively [2]. We provided the first empirical comparison of these alternative assumptions for size-energy relationships in the same system, through the use of both population and size-class approaches to quantifying scaling relationships. Although size structure was skewed in our guild toward species with smaller body sizes (Table 1), as is typical of many guilds [58], directly-measured results from the population approach (Fig. 2A) and the size-class approach (Fig. 2B) nonetheless were not different, suggesting that our conclusions about size-energy hypotheses are robust to the energy regulation assumption used. Future comparisons of energy-regulation assumptions could benefit from new techniques for the size-class approach that account for intraspecific variation in body size during species sorting into size class bins [59].

Mechanisms Underlying Size-Energy Relationships

Theory suggests that the energy use of populations or size classes is dependent on their density and individual metabolic rate [2], [6]. While results from our measured size-energy relationship are consistent with energetic equivalence (Fig. 2), results from our measured size-density relationship (Fig. 1A, 1B) add to accumulating evidence [5], [21] that the slope of the size-density relationship is shallower (less negative) at local scales than the −0.75 slope expected from global-scale studies [60], metabolic theory [2], and the energetic equivalence rule [6]. The combination of these two sets of results is unexpected because the conventional indirect method assumes the slope of the size-metabolism relationship, *b*, is +0.75, a value which corresponds with global patterns in size-metabolism relationships [1], [6], [24]. However, our combined results can be explained if the slope of the local size-metabolism relationship in our system was shallower (less positive) than expected. Specifically, our results for the population approach imply a slope of the local size-metabolism relationship of +0.42 (see Results). Local-scale patterns of size-metabolism relationships for interacting species within a community have not previously been explored, but such differences in local-scale variation from global-scale patterns could be common, since metabolic rate is widely variable among taxa [25]and with activity level [61]. In our system the shallower-than-expected slope of the size-metabolism relationship may result from the relatively slow metabolism of frugivores and arboreal species [62], [63] or the energetic constraints imposed by food-scarcity at the end of the dry season [48], [62].

Challenges in Testing Energetic Equivalence

In addition to the methodological and analytical challenges discussed above, several other factors may complicate tests of scaling relationships in general and the energetic equivalence hypothesis in particular. First, statistical methods for scaling relationships have been controversial, with particular debate over means to reduce bias in estimation of scaling exponents (e.g, [12], [64]). Statistical testing of the energetic equivalence hypothesis poses an additional challenge, as the hypothesis posits a slope of the relationship between body size and energy use of zero, yet regression, the statistical test used to evaluate such relationships, is not designed to enable detection of a zero slope but rather to distinguish deviations from zero [65]. Energetic equivalence is treated statistically as a null hypothesis, and although non-significant results such as those we obtained for the direct measurement of energy use (Fig. 2) have often been taken as positive evidence of the invariance of energy use with body size (e.g., [14], [15]), they cannot be distinguished statistically from a failure to detect a difference from zero [27]. Further application of statistical methods to address these issues is needed [27], [64].

Second, tests of energetic equivalence often exhibit substantial variation around mean trends (e.g., [15]), and both simulation models [66] and empirical evidence [67] predict higher variation at smaller body size ranges. Such variation may decrease power and make it harder to detect scaling relationships empirically in community-scale studies [27]. Greater power could be obtained by examining a greater number of species or a greater body mass range, but this will not be possible to accomplish while controlling for resource use in many communities such as ours, where species numbers and body mass ranges in a guild are limited (Table 1; [66]).

In our study, data were highly variable for all relationships (Figs. 1, 2), but variance around the mean scaling trend was much greater for the direct method (standard deviation >2 natural log units; Fig. 2A, 2B) than for the indirect method (st. dev. <0.5 natural log units; Figs. 1C, 1D). This discrepancy in variances in the two methods may be due to the use of an assumed constant metabolic scaling exponent, *b*, of +0.75 for all species to calculate energy use in the conventional indirect method (see Introduction), despite the highly variable nature of this exponent by taxon [25] or community (as noted above, *b* differed from this assumption for our system). By ignoring such community-scale variation in size-metabolism relationships, the conventional indirect approach may obscure an important source of error and greatly overestimate the precision of reported size-energy relationships. Any future analyses using the indirect method should explicitly account for this expected variance in the metabolic scaling exponent.

While size-energy relationships offer an explanation for the mean scaling trend among species in a community, they do not provide an explanation for the high levels of residual variation in the data. Such variation may be due to differences in a number of evolutionary and ecological factors, including relatedness, niche differences, dispersal patterns, predation risk, or trade-offs among growth, mortality, fecundity and other life history variables [28], [29], [66], [68]. Understanding the relative influence of body size and each of these other factors on energy use remains an important question in community ecology [29]. Further insights can be gained through studies comparing the relative influence of each variable on energy use, and through additional data collected across communities and over time [27], [28].

CONCLUSIONS

Investigations of local scaling relationships enable prediction of community patterns of population density and resource use; provide important links among individual-, population-, and community-level processes in ecology; and clarify mechanisms underlying resource partition in ecological communities [5]. By providing the first direct measurement of the size-energy relationship, and by controlling for factors that have often confounded previous studies, our study enabled new insights into the form of the local size-energy relationship. Specifically, our study is consistent with the idea that energy use is independent of body size. Our results were robust to the energy regulation assumption used, and suggest that shallower-than-expected relationships of body size to density and metabolism may combine to produce energetic equivalence at local scales. Our study focused on foragers on shared, limiting resources in communities because competition for resources is the proposed mechanism underlying energetic equivalence, but such results can be extended to multi-trophic communities [2], [23], [32], [64], [69], [70].

In addition to directly evaluating the size-energy relationship, our study compared alternative methods to understand this relationship at the community scale. Our direct method of measurement of the size-energy relationship contrasts sharply with interpretations that would have been derived from the conventional indirect method, suggesting that the size-density relationship in a community cannot be used to infer the size-energy relationship. Further, the differences we observed in size-density relationships between our local study and past global studies highlight that body size relationships may describe distinct phenomena, be generated by differing mechanisms, and exhibit contrasting patterns at different spatial scales.

More broadly, our focus in this study was on evaluating size-energy hypotheses, and our finding that energy intake was independent of body size (Fig. 2) was consistent with the energetic equivalence hypothesis. This finding is of particular interest for ecological theory, since energetic equivalence implies that small and large animals will produce similar amounts of biomass over time [6]; population growth and carrying capacities will be related predictably to body size [8]; resources will be used at similar rates by groups of organisms of different sizes[2]; and community assembly will not be driven by or oriented around an optimal body size [71]. Our findings therefore have important implications for predicting species interactions and for understanding many aspects of community structure and dynamics.

ACKNOWLEDGMENTS

We thank Madagascar National Parks, the Universities of Antsiranana and Antananarivo, Madagascar Institut pour la Conservation des Ecosystèmes Tropicaux, Madagascar CAFF/CORE, Direction des Eaux et Forêts, Missouri Botanical Garden, and the villagers of Mahamasina for cooperation; J. Jamildine, M. Jaoravo, J. Robert, P. Randrianantenaina, R. Andrianjaka, S. Rakotondriabe, G. Papan'i Dada, and K. Arison for field assistance; A. Abdel Cader, L. Miadana, and the Ankarana staff and guides for logistical support; D. Van Vuren for advice and support; and D. Van Vuren, D. Kelt, A. Sih, R. Greenberg, N. Isaac, E. White, and an anonymous reviewer for comments on previous versions of the manuscript.

AUTHOR CONTRIBUTIONS

Analyzed the data: BJS. Wrote the paper: BJS. Conceived of project: BJS. Developed conceptual framework: BJS ALF. Designed observations: BJS ALF JEH EA. Managed field teams: BJS JEH. Collected data: BJS ALF JEH EA. Edited the manuscript: BJS ALF JEH.

REFERENCES

1. Kleiber M (1975) The Fire of Life, 2nd edition. New York: Krieger.
2. Brown JH, Gillooly JF, Allen AP, Savage VM, West GB (2004) Toward a metabolic theory of ecology. Ecology 85: 1771–1789. doi: 10.1890/03-9000
3. Brown JH, Maurer BA (1989) Macroecology: the division of food and space among species on continents. Science 243: 1145–1150. doi: 10.1126/science.243.4895.1145
4. Blackburn TM, Gaston KJ (1996) On being the right size: different definitions of 'right'. Oikos 75: 551–557. doi: 10.2307/3545899
5. White EP, Ernest SKM, Kerkhoff AJ, Enquist BJ (2007) Relationships between body size and abundance in ecology. Trends in Ecology & Evolution 22: 323–330. doi: 10.1016/j.tree.2007.03.007
6. Damuth J (1981) Population-density and body size in mammals. Nature 290: 699–700. doi: 10.1038/290699a0
7. Nee S, Read AF, Greenwood JJD, Harvey PH (1991) The relationship between abundance and body size in British birds. Nature 351: 312–313. doi: 10.1038/351312a0
8. Savage VM, Gillooly JF, Brown JH, West GB, Charnov EL (2004) Effects of body size and temperature on population growth. American Naturalist 163: 429–441. doi: 10.1086/381872
9. Enquist BJ, West GB, Brown JH (2009) Extensions and evaluations of a general quantitative theory of forest structure and dynamics. Proceedings of the National Academy of Sciences of the United States of America 106: 7046–7051. doi: 10.1073/pnas.0812303106
10. Allen AP, Brown JH, Gillooly JF (2002) Global biodiversity, biochemical kinetics, and the energetic-equivalence rule. Science 297: 1545–1548. doi: 10.1126/science.1072380
11. Russo SE, Robinson SK, Terborgh J (2003) Size-abundance relationships in an Amazonian bird community: implications for the energetic equivalence rule. American Naturalist 161: 267–283. doi: 10.1086/345938

12. Griffiths D (1992) Size, abundance, and energy use in communities. Journal of Animal Ecology 61: 307–315. doi: 10.2307/5323
13. Marquet PA, Quinones RA, Abades S, Labra F, Tognelli M, et al. (2005) Scaling and power-laws in ecological systems. Journal of Experimental Biology 208: 1749–1769. doi: 10.1242/jeb.01588
14. Enquist BJ, Brown JH, West GB (1998) Allometric scaling of plant energetics and population density. Nature 395: 163–165.
15. Ernest SKM, Enquist BJ, Brown JH, Charnov EL, Gillooly JF, et al. (2003) Thermodynamic and metabolic effects on the scaling of production and population energy use. Ecology Letters 6: 990–995. doi: 10.1046/j.1461-0248.2003.00526.x
16. Brown JH, Maurer BA (1986) Body size, ecological dominance and Cope's rule. Nature 324: 248–250. doi: 10.1038/324248a0
17. Damuth J (1993) Cope's rule, the island rule and the scaling of mammalian population density. Nature 365: 748–750. doi: 10.1038/365748a0
18. DeLong JP, Vasseur DA (2012) A dynamic explanation of size-density scaling in carnivores. Ecology 93: 470–476. doi: 10.1890/11-1138.1
19. Loeuille N, Loreau M (2006) Evolution of body size in food webs: does the energetic equivalence rule hold? Ecology Letters 9: 171–178. doi: 10.1111/j.1461-0248.2005.00861.x
20. Ernest SKM (2005) Body size, energy use, and community structure of small mammals. Ecology 86: 1407–1413. doi: 10.1890/03-3179
21. Hayward A, Khalid M, Kolasa J (2009) Population energy use scales positively with body size in natural aquatic microcosms. Global Ecology and Biogeography 18: 553–562. doi: 10.1111/j.1466-8238.2009.00459.x
22. King JR (2010) Size-abundance relationships in Florida ant communities reveal how ants break the energetic equivalence rule. Ecological Entomology 35: 287–298. doi: 10.1111/j.1365-2311.2010.01177.x
23. Brown JH, Gillooly JF (2003) Ecological food webs: high-quality data facilitate theoretical unification. Proceedings of the National Academy of Sciences of the United States of America 100: 1467–1468. doi: 10.1073/pnas.0630310100
24. Nagy KA, Girard IA, Brown TK (1999) Energetics of free-ranging mammals, reptiles, and birds. Annual Review of Nutrition 19: 247–277. doi: 10.1146/annurev.nutr.19.1.247
25. Isaac NJB, Carbone C (2010) Why are metabolic scaling exponents so controversial? Quantifying variance and testing hypotheses. Ecology Letters 13: 728–735. doi: 10.1111/j.1461-0248.2010.01461.x

26. Marquet PA, Navarrete SA, Castilla JC (1995) Body size, population density, and the Energetic Equivalence Rule. Journal of Animal Ecology 64: 325–332. doi: 10.2307/5894

27. Isaac NJB, Storch D, Carbone C (2013) The paradox of energy equivalence. Global Ecology and Biogeography 22: 1–5. doi: 10.1111/j.1466-8238.2012.00782.x

28. Ernest SKM, Brown JH, Thibault KM, White EP, Goheen JR (2008) Zero sum, the niche, and metacommunities: long-term dynamics of community assembly. American Naturalist 172: E257–E269. doi: 10.1086/592402

29. Isaac NJB, Carbone C, McGill BJ (2012) Population and community ecology. In: Sibly RM, Brown JH, Kodric-Brown A, editors. Metabolic Ecology: A Scaling Approach. West Sussex, U.K.: John Wiley & Sons. pp. 77–85.

30. Ackerman JL, Bellwood DR, Brown JH (2004) The contribution of small individuals to density-body size relationships: examination of energetic equivalence in reef fishes. Oecologia 139: 568–571. doi: 10.1007/s00442-004-1536-0

31. West GB, Enquist BJ, Brown JH (2009) A general quantitative theory of forest structure and dynamics. Proceedings of the National Academy of Sciences of the United States of America 106: 7040–7045. doi: 10.1073/pnas.0812294106

32. Reuman DC, Mulder C, Raffaelli D, Cohen JE (2008) Three allometric relations of population density to body mass: theoretical integration and empirical tests in 149 food webs. Ecology Letters 11: 1216–1228. doi: 10.1111/j.1461-0248.2008.01236.x

33. Yvon-Durocher G, Reiss J, Blanchard J, Ebenman B, Perkins DM, et al. (2011) Across ecosystem comparisons of size structure: methods, approaches and prospects. Oikos 120: 550–563. doi: 10.1111/j.1600-0706.2010.18863.x

34. Cordeiro NJ, Ndangalasi HJ, McEntee JP, Howe HF (2009) Disperser limitation and recruitment of an endemic African tree in a fragmented landscape. Ecology 90: 1030–1041. doi: 10.1890/07-1208.1

35. Nunez-Iturri G, Howe HF (2007) Bushmeat and the fate of trees with seeds dispersed by large primates in a lowland rain forest in western Amazonia. Biotropica 39: 348–354. doi: 10.1111/j.1744-7429.2007.00276.x

36. Hawkins AFA, Chapman P, Ganzhorn JU, Bloxam QMC, Barlow SC, et al. (1990) Vertebrate conservation in Ankarana Special Reserve, northern Madagascar. Biological Conservation 54: 83–110. doi: 10.1016/0006-3207(90)90136-d

37. Nekaris KAI, Blackham GV, Nijman V (2008) Conservation implications of low encounter rates of five nocturnal primate species (*Nycticebus* spp.) in Asia. Biodiversity and Conservation 17: 733–747. doi: 10.1007/s10531-007-9308-x

38. Radhakrishna S, Goswami AB, Sinha A (2006) Distribution and conservation of*Nycticebus bengalensis* in northeastern India. International Journal of Primatology 27: 971–982. doi: 10.1007/s10764-006-9057-9

39. Buckland ST, Anderson DR, Burnham KP, Laake JL, Borchers DL, et al.. (2001) Introduction to Distance Sampling: Estimating Abundance of Biological Populations. New York: Oxford University Press.

40. Thomas L, Buckland ST, Rexstad EA, Laake JL, Strindberg S, et al. (2010) Distance software: design and analysis of distance sampling surveys for estimating population size. Journal of Applied Ecology 47: 5–14. doi: 10.1111/j.1365-2664.2009.01737.x

41. Thomas L, Laake JL, Rexstad E, Strindberg S, Marques FFC, et al.. (2010) Distance 6.0. Release 2: Research Unit for Wildlife Population Assessment, University of St. Andrews, UK. Program Distance website. Available: http://www.ruwpa.st-and.ac.uk/distance/. Accessed 2010 Aug 13.

42. Cardiff SG, Ratrimomanarivo FH, Rembert G, Goodman SM (2009) Hunting, disturbance and roost persistence of bats in caves at Ankarana, northern Madagascar. African Journal of Ecology 47: 640–649. doi: 10.1111/j.1365-2028.2008.01015.x

43. Fowler SV, Chapman P, Checkley D, Hurd S, McHale M, et al. (1989) Survey and management proposals for a tropical deciduous forest reserve at Ankarana in northern Madagascar. Biological Conservation 47: 297–313. doi: 10.1016/0006-3207(89)90072-4

44. Clauss M, Schwarm A, Ortmann S, Streich WJ, Hummel J (2007) A case of non-scaling in mammalian physiology? Body size, digestive capacity, food intake, and ingesta passage in mammalian herbivores. Comparative Biochemistry and Physiology A-Molecular & Integrative Physiology 148: 249–265. doi: 10.1016/j.cbpa.2007.05.024

45. Berg CC (1989) Classification and distribution of *Ficus*. Experientia 45: 605–611. doi: 10.1007/bf01975677

46. Janzen DH (1979) How to be a fig. Annual Review of Ecology and Systematics 10: 13–51. doi: 10.1146/annurev.es.10.110179.000305

47. Shanahan M, So S, Compton SG, Corlett R (2001) Fig-eating by vertebrate frugivores: a global review. Biological Reviews 76: 529–572. doi: 10.1017/s1464793101005760

48. Terborgh J (1986) Keystone plant resources in the tropical forest. In: Soule ME, editor. Conservation Biology: The Science of Scarcity and Diversity. Sutherland, Massachusetts: Sinauer Associates. pp. 330–344.
49. White EP, Enquist BJ, Green JL (2008) On estimating the exponent of power-law frequency distributions. Ecology 89: 905–912. doi: 10.1890/07-1288.1
50. Burnham KP, Anderson DR (2002) Model Selection and Multimodel Inference: A Practical Information-Theoretic Approach. 2nd edition. New York: Springer. 488 p.
51. Edwards AM, Freeman MP, Breed GA, Jonsen ID (2012) Incorrect likelihood methods were used to infer scaling laws of marine predator search behaviour. PLOS One 7. doi: 10.1371/journal.pone.0045174
52. Warton DI, Wright IJ, Falster DS, Westoby M (2006) Bivariate line-fitting methods for allometry. Biological Reviews 81: 259–291. doi: 10.1017/s1464793106007007
53. SAS Institute (2008) JMP. Version 8.0. Statistical software 1989–2008. Cary, NC.
54. R Development Core Team (2010) R Version 2.11.1. R Foundation for Statistical Computing. Vienna, Austria. R Project website. Available: http://www.R-project.org. Accessed 2010 July 20.
55. Damuth J (2007) A macroevolutionary explanation for energy equivalence in the scaling of body size and population density. American Naturalist 169: 621–631. doi: 10.1086/513495
56. Arneberg P, Skorping A, Read AF (1998) Parasite abundance, body size, life histories, and the energetic equivalence rule. American Naturalist 151: 497–513. doi: 10.1086/286136
57. Saracco JF, Collazo JA, Groom MJ (2004) How do frugivores track resources? Insights from spatial analyses of bird foraging in a tropical forest. Oecologia 139: 235–245. doi: 10.1007/s00442-004-1493-7
58. Blackburn TM, Gaston KJ (1994) Animal body size distributions: patterns, mechanisms, and implications. Trends in Ecology & Evolution 9: 471–474. doi: 10.1016/0169-5347(94)90311-5
59. Thibault KM, White EP, Hurlbert AH, Ernest SKM (2011) Multimodality in the individual size distributions of bird communities. Global Ecology and Biogeography 20: 145–153. doi: 10.1111/j.1466-8238.2010.00576.x
60. Damuth J (1987) Interspecific allometry of population density in mammals and other animals: the independence of body mass and population energy use. Biological Journal of the Linnean Society 31: 193–246. doi: 10.1111/j.1095-8312.1987.tb01990.x

61. Nagy KA (2005) Field metabolic rate and body size. Journal of Experimental Biology 208: 1621–1625. doi: 10.1242/jeb.01553

62. McNab BK (1986) The influence of food habits on the energetics of eutherian mammals. Ecological Monographs 56: 1–19. doi: 10.2307/2937268

63. McNab BK (2003) Metabolism - Ecology shapes bird bioenergetics. Nature 426: 620–621. doi: 10.1038/426620b

64. Arim M, Berazategui M, Barreneche JM, Ziegler L, Zarucki M, et al.. (2011) Determinants of Density-Body Size Scaling Within Food Webs and Tools for Their Detection. In: Belgrano A, Reiss J, editors. Advances in Ecological Research, Vol 45: The Role of Body Size in Multispecies Systems. pp. 1–39.

65. Fisher RA (1925) Statistical Methods for Research Workers, 14th edition. Reprinted in 1990 in Statistical Methods, Experimental Design and Scientific Inference. Oxford, U.K.: Oxford University Press.

66. Tilman D, Hillerislambers J, Harpole S, Dybzinski R, Fargione J, et al. (2004) Does metabolic theory apply to community ecology? It's a matter of scale. Ecology 85: 1797–1799. doi: 10.1890/03-0725

67. Hayward A, Kolasa J, Stone JR (2010) The scale-dependence of population density-body mass allometry: Statistical artefact or biological mechanism? Ecological Complexity 7: 115–124. doi: 10.1016/j.ecocom.2009.08.005

68. Cotgreave P (1993) The relationship between body size and population abundance in animals. Trends in Ecology & Evolution 8: 244–248. doi: 10.1016/0169-5347(93)90199-y

69. Hechinger RF, Lafferty KD, Dobson AP, Brown JH, Kuris AM (2011) A common scaling rule for abundance, energetics, and production of parasitic and free-living species. Science 333: 445–448. doi: 10.1126/science.1204337

70. Reuman DC, Mulder C, Banasek-Richter C, Blandenier MFC, Breure AM, et al.. (2009) Allometry of Body Size and Abundance in 166 Food Webs. In: Caswell H, editor. Advances in Ecological Research, Vol 41. pp. 1–44.

71. Kelt DA (1997) Assembly of local communities: consequences of an optimal body size for the organization of competitively structured communities. Biological Journal of the Linnean Society 62: 15–37. doi: 10.1111/j.1095-8312.1997.tb01615.x

72. Ravokatra M, Wilmé L, Goodman SM (2003) Bird weights. In: Goodman SM, Benstead JP, editors. The Natural History of Madagascar. Chicago: The University of Chicago Press. pp. 1059–1063.

73. Mittermeier RA, Konstant WR, Hawkins AFA, Louis EE, Langrand O, et al.. (2006) Lemurs of Madagascar. Washington, D.C.: Conservation International. 520 p.

74. Garbutt N (2007) Mammals of Madagascar: A Complete Guide. New Haven, Connecticut: Yale University Press.

75. Rasoloarison RM, Goodman SM, Ganzhorn JU (2000) Taxonomic revision of mouse lemurs (*Microcebus*) in the western portions of Madagascar. International Journal of Primatology 21: 963–1019.

Chapter 13

SEISMIC MICROZONATION OF BREGINJSKI KOT (NW SLOVENIA) BASED ON DETAILED ENGINEERING GEOLOGICAL MAPPING

Jure Kokošin[1] and Andrej Gosar[2,3]

[1]Sweco Norge AS, Dronningensgate 52/54, 8509 Narvik, Norway

[2]Faculty of Natural Sciences and Engineering, University of Ljubljana, Aškerčeva 12, SI-1000 Ljubljana, Slovenia

[3]Slovenian Environment Agency, Seismology and Geology Office, Dunajska 47, SI-1000 Ljubljana, Slovenia

ABSTRACT

Breginjski kot is among the most endangered seismic zones in Slovenia with the seismic hazard assessed to intensity IX MSK and the design ground acceleration of 0.250 g, both for 500-year return period. The most destructive was the 1976 Friuli Mw = 6.4 earthquake which had maximum intensity VIII-IX. Since the previous microzonation of the area was based solely on the basic geological map and did not include supplementary field research, we have performed a new soil classification of the area. First, a detailed engineering geological mapping in scale 1: 5.000 was conducted. Mapped units were described in detail and some of them interpreted anew. Stiff sites are composed of hard to medium-hard rocks which were subjected to erosion mainly evoked by glacial and postglacial age. At that time a prominent topography was formed and different types of sediments were deposited in valleys by mass flows. A distinction between sediments and weathered rocks, their exact position, and thickness are of significant importance for microzonation. On the basis of geological mapping, a soil classification was carried out according to the Medvedev method (intensity increments) and the Eurocode 8 standard (soil factors) and two microzonation maps were prepared. The bulk of the studied area is covered by soft sediments and nine out of ten settlements are situated on them. The microzonation clearly points out the dependence of damage distribution in the case of 1976 Friuli earthquake to local site effects.

INTRODUCTION

The Breginjski kot (NW Slovenia, Figure 1) is located close to a seismically very active area of Friuli in NE Italy, but also the Julian Alps in Slovenia have recently experienced an increase of seismic activity. In this area, in fact, the seismic hazard in Slovenia is the highest. Most of numerous settlements in the area are built on soft sediments which can significantly enhance the ground shaking. The whole area was highly damaged during the Friuli 1976 Mw = 6.4 earthquake and its aftershock sequence. For a realistic assessment of side effects and earthquake hazard, a thorough seismic microzonation is therefore essential. On the other hand the existing microzonation of Breginjski kot [1] was based solely on the basic geological map of Slovenia in scale 1: 100.000, without supplementary field investigations or subsurface information, thus making it fairly inaccurate. Therefore we have decided to perform a new microzonation study based on a detailed engineering geological mapping in scale 1: 5.000. Such a map would enable soil classification according to two different approaches without using expensive geotechnical drilling or geophysical investigations.

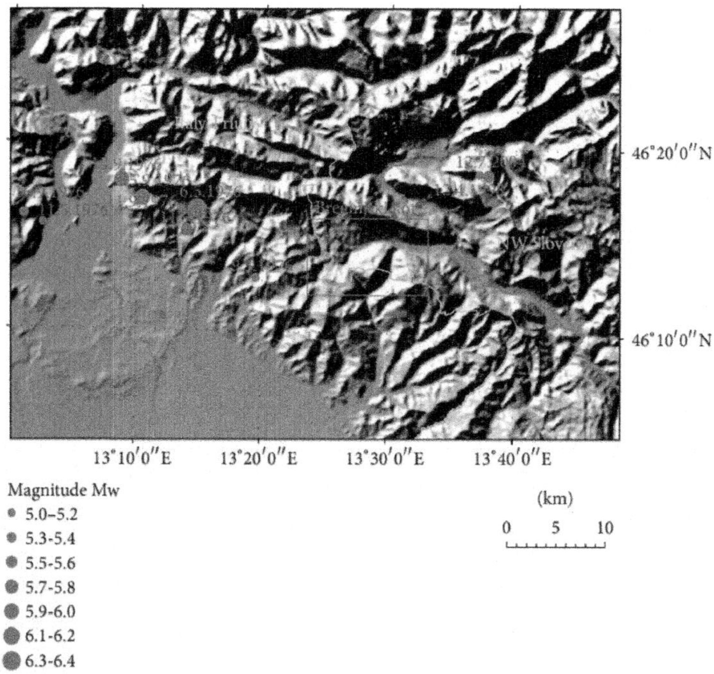

Figure 1: Seismicity (Mw ≥ 5.0) in Friuli and NW Slovenia with location of Breginjski kot study area.

The seismic microzonation according to Eurocode 8 was based on a new European seismic standard [2] which has been applied in Slovenia from 2008. This standard defines the seismic hazard by the peak ground acceleration, whereas the site effects are expressed by the soil factor. The product of both values is used in the design of earthquake resistant buildings. The seismic microzonation according to the Medvedev method [3] denotes seismic hazard by increments of seismic intensity. Although the seismic intensity in Slovenia is not used for building design anymore, it is still important for seismological analyses and for the civil protection purposes.

SEISMOLOGICAL CHARACTERISTICS

The Breginjski kot is located in one of the three areas with the highest seismic hazard in Slovenia. This is mainly due to the proximity of the seismically very active area of Friuli located 10–30 km to the west in NE Italy. In this area the Mw = 6.4 Friuli earthquake occurred in May 1976. The seismic sequence that followed (Table 1, Figure 1) consisted of six events of Mw between 5.0 and 6.0 during the same year and one event of Mw = 5.3 that occurred in September 1977 [4, 5]. The intensity of the main shock in Breginj was VIII EMS-98, but the cumulative intensity which includes also damage from aftershocks reached IX EMS-98. The highest intensity in Slovenia from the main shock was VIII-IX in Robidišče [6]. A complete review of 1976 earthquake intensities in Breginjski kot settlements is summarized in Table 2.

Table 1: Data for earthquakes with Mw ≥ 5.0 in the vicinity of Breginski kot (after [4, 7,8])

Date	Time (UTC)	Lat (°N)	Lon (°N)	Depth (km)	Mw	Region
06/05/1976	20:00	46.275	13.246	6.5	6.4	Friuli
09/05/1976	00:53	46.214	13.326	8.5	5.1	Friuli
11/05/1976	22:44	46.260	13.041	6.0	5.0	Friuli
11/09/1976	16:31	46.275	13.224	6.5	5.2	Friuli
11/09/1976	16:35	46.256	13.233	4.3	5.6	Friuli
15/09/1976	03:15	46.284	13.173	5.0	5.9	Friuli
15/09/1976	09:21	46.300	13.145	7.0	6.0	Friuli
16/09/1977	23:48	46.268	13.016	8.0	5.3	Friuli
12/04/1998	10:55	46.309	13.632	7.6	5.6	Krn Mts.
12/07/2004	13:04	46.310	13.620	11.0	5.2	Krn Mts.

Table 2: Intensities of 1976 earthquake in different settlements of Breginjski kot

Settlement	Intensity EMS-98
Borjana	VI
Kred	VI
Robič	VI
Potoki	VI+ (MSK-78)
Logje	VI-VII
Staro Selo	VII
Sedlo	VII-VIII
Breginj	VIII
Podbela	VIII
Robidišče	VIII-IX

For the NW Slovenia relatively weak rates of seismicity were observed before April 1998 when a Mw = 5.6 earthquake occurred in Krn Mountains, only 7 km NW from Kobarid [9, 10]. It was followed by a Mw = 5.2 event in July 2004 in the same epicentral area (Figure 1). Both earthquakes occurred on the NW-SE trending near-vertical dextral strike-slip Ravne fault [11] in the depth range 7.6–11 km. The intensity of the 1998 earthquake in Kobarid (located at eastern margin of the Breginjski kot) was VI-VII EMS-98 [7] and of the 2004 event VI EMS-98 [8].

NW Slovenia and Friuli region are located at the kinematic transition between E-W striking thrust faults of the Alpine system (Friuli earthquakes) and NW-SE striking strike-slip faults of the Dinarides system (Krn Mountains earthquakes) (Figure 1). The strongest earthquake ever recorded in the Alps-Dinarides junction area was the 1511 western Slovenia earthquake (Mw = 6.8); the exact location and mechanism of this event are still debated [12, 13] due to early occurrence and thus very little or no historical documents.

According to the old seismic hazard map of Slovenia for a 500-year return period showing expected intensities on the MSK scale [14], Breginjski kot is estimated to have IX MSK intensity. According to the new seismic hazard map for a 475-year return period [15], a design ground acceleration value for a rock site in Breginjski kot is 0.250 g. The important methodological difference between both maps is that intensities are assessed for a "medium" soil type, whereas design ground acceleration is assessed for a "rock" site according to Eurocode 8.

Accelerographs were installed in Kobarid after the main Friuli shock in 1976 and after the Krn Mountains earthquake in 1998. The strongest ground motion in the Friuli seismic sequence was recorded for the September 15, 1976 aftershock (Mw = 6.0, distance = 37 km) as 0.138 g peak ground acceleration [16]. For the July 12, 2004 Krn Mountain earthquake (Mw = 5.2, distance = 7 km) a peak ground acceleration of 0.152 g was recorded [17].

The previously existing microzonation of Breginjski kot [1] was based solely on the basic geological map in scale 1: 100.000, without supplementary field research or subsurface geological and geophysical information, thus making it fairly inaccurate. It was prepared to be used together with the old seismic hazard map showing intensities on the MSK scale. This microzonation showed that the maximum expected intensity due to the effects of soft sediments can be increased by up to one intensity degree.

ENGINEERING GEOLOGICAL MAPPING

A detailed engineering geological map of Breginjski kot in scale 1: 5.000 was prepared by outcrop observing and geological boundaries tracking method. Since seismic microzonation was the final goal, the main focus of the field survey was to characterise local geological conditions in vicinity of settlements and other substantial infrastructure like roads, power lines, water collectors, and so forth. Consequently some less relevant lithological boundaries were summarised from the manuscript geological map in scale 1 : 25.000, which was the basis for published basic geological map in scale 1 : 100.000 [18]. Based on the mapping accomplished, rocks and sediments were classified by their characteristics determined in the field. Properties such as structure, dip, strength and weathered material thickness were defined for rocks, whereas granulation, roundness, and thickness were described for unconsolidated sediments. Mapped lithological units were compared with the Bovec basin, located 15 km to the NE in the Soča valley for which intensity increment microzonation was performed after the 1998 earthquake [19]. Lithologies described below (Figures 2 and 3) are arranged by age order and are presented on a detailed engineering geological map (Figure 4) of Breginjski kot. Their names and age classification mainly coincide with lithologies described in the explanatory text of the basic geological map of Slovenia [20].

Figure 2: Typical lithologies according to the engineering-geological mapping: (a) Megalodontidae shell in Dachstein limestone in Robič, (b) limestone breccia interbedded with shalestone near Robidišče, (c) folds in flyschoid formation near Borjana, (d) heterogeneous composition of diamicton in Sedlo, (e) lacustrine chalk SW of Podbela, and (f) proluvium in Podbela.

(a)

(b)

(c)

(d)

Figure 3: (a) Karst sinkhole in coral limestone represents a depositional place for significant amount of weathered material; Robidišče in the background. (b) Limestone breccia in Nadiža valley south of Logje is typically bedded. (c) Diamicton near Breginj, in the background Mt. Stol built of Dachstein limestone. (d) Alluvial fan sediments form gentle slopes above Kred polje, which is composed of fine grained alluvial sediments which promote to alluvial sediments near Nadiža River channel.

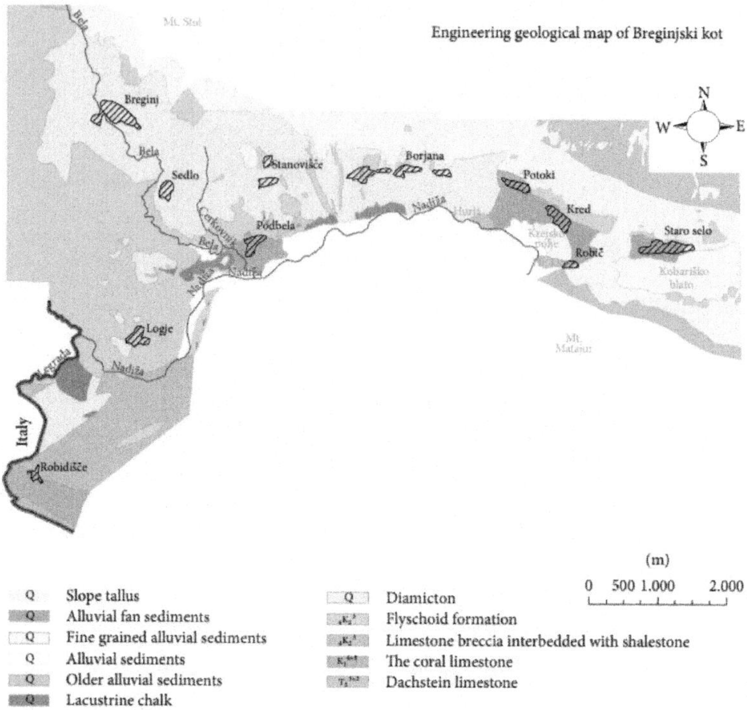

Figure 4: Engineering geological map of Breginjski kot.

Dachstein limestone (Figure 2(a)) from Late Triassic is the oldest formation in the studied area. It forms Mt. Stol, Mt. Matajur, and Der Hill west of Robič. Rock has a light grey colour and a crystalline texture. Dachstein limestone is characterized by Lofer development which consists of three parts. Cyclothem of B member contains stromatolites whereas cyclothem of C member contains well-preserved Megalodontidae shells. Cyclothem of A member consists of basal breccias but was not observed in the field. Weathered material thickness was estimated to be less than 0.5 m.

The Coral limestone of Early Cretaceous is situated in a small part of the examined area near Robidišče. Coral limestone is fine grained, massive and grey in colour. Outcrops are rare because they are overlain by significant thickness of weathered material. Hence it was distinguished from limestone breccias by numerous karst sinkholes which determine its morphology (Figure 3(a)). Moreover the weathered material thickness can be altered promptly as a consequence of a karst relief. On the slopes less than 1 m of weathered material is expected, whereas more than 5 m of weathered material is estimated for the flatland.

Limestone breccia interbedded with shalestone (Figure 2(b)) from Upper Cretaceous represents mainly the area from Napoleon Bridge west of Podbela to Robidišče (Figure 3(b)). Limestone breccias are of light grey or grey colour. Breccia is clast supported and composed of angular to subangular clasts which range from 2 mm to 1 m in size. Breccia formation is mostly bedded, rarely massive. If bedded, its thickness ranges from 10 cm to several meters in size, moreover brownish green shalestones are often interbedded. Formation generally dips to the north. On the slopes less than 1 m of weathered material is expected, whereas from 2 m to 3 m of weathered material is estimated for the flatland.

Flyschoid formation (Figure 2(c)) is the next formation of Upper Cretaceous and is situated mainly west of Breginj-Sedlo-Podbela line, while plenty outcrops are exposed in erosion windows of Mt. Stol and Mt. Matajur, especially in ravines with high-energy creeks. The Flyschoid formation is divided into series of rocks: shalestone, sandstone, and breccia, which are stratified and in sequence one after the other. Shalestone is dark grey and appears with thickness of beds ranging from 1 cm to 5 cm. Sandstone, which mostly consists of carbonate minerals, is grey with thickness of beds ranging from 5 cm to few ten centimetres. Breccia is grey to dark grey and consists of carbonate clasts ranging from 0.5 cm up to few meters. Coarse grained breccia is clast supported. Thickness of breccia beds ranges from 1 m to few ten meters. Flyschoid formation dips generally towards the north, moreover thin bedded series in particular are often found folded. Thickness of weathered material varies from 0.5 m to 1 m on the slopes and from 2 m to 3 m on the flatland.

Diverse Quaternary sediments in Breginjski kot are the result of postglacial era when water erosion and other transformational processes formed specific relief. Sediment deposits are mostly abundant on the slopes of Mt. Stol where most settlements are located.

Diamicton (Figure 2(d)) is an unlithified sediment which origins from glacial till mixed with other Quaternary sediments and transported by mass flow. It is deposited on the slope of Mt. Stol (Figure 3(c)), confined by Potoki to the east, a large area of flyschoid formation to the west and Nadiža River to the south. In addition some erosion patches can be found between Sedlo and Logje. According to the estimation on the field, diamicton is composed of more or less subrounded grains which are very poorly sorted. 70% of grains represent fine gravel, sand, and silt fractions, which ratio can vary significantly on studied locations. The remaining 30% is mainly gravel which ranges from

1 cm to 5 cm in size and boulders which ranges from 20 cm to several meters in size. Fragments consist of carbonate in general. Besides, there are many clastic fragments and diamictite ones, which are rather scarce. Glacial striations on fragments were not observed. Diamicton thickness varies significantly, for instance: along the Nadiža River only a few meters can be reached, in vicinity of flyschoid formation erosion windows it can vary between 5 m and 10 m, whereas away from ravines it can reach more than 20 m. The thickness depends on paleorelief and selective sedimentation processes, therefore a large amount of material has been deposited there. For instance in Cerkovnik creek flyschoid formation is not exposed, despite more than 20 m deep eroded ravine. According to ravine morphology, the biggest thickness is expected in the vicinity of Breginj.

Diamictit has the same origin as diamicton but it is lithified. It is not presented on the engineering-geological map, because it is scattered around the Diamicton area in small patches of about 10 m^2 in size. It consists of around 70% of gravel and boulder sized clasts which are matrix supported. The latter is composed of silt, sand and fine gravel sized grains. Subrounded clasts are poorly sorted, those in the range from 0.5 cm to 5 cm prevail. Glacial striations on clasts are rarely preserved.

Lacustrine chalk (Figure 2(e)) is deposited in patches along broader zone of Nadiža River and in the largest portion besides Legrada creek. This sediment is characterised by laminated silt, grey in colour with a blue tinge which has moisture and can be slightly molded. According to Casagrande's Soil Classification System (ASTM) [21] it was defined as ML, inorganic silt and very fine sands with low plasticity. Besides laminated silt there can be less than 5% of rounded pebbles found, mainly ranging from 1 cm to 5 cm and rarely to 10 cm. Lacustrine chalk is covered by diamicton; main evidence can be found near to Hurja farm.

Older alluvial sediments can generally be found between Kred and Podbela up to 320 m above sea level as erosion patches around 40 m^2 in size and are consequently not thicker than 5 m. They include conglomerate, but sands and gravel are more rare. First is clast supported conglomerate which is good sorted. The clasts, which range from 0.2 to 5 cm in size, are well rounded and consist mainly of carbonate and flyschoid fragments.

Alluvial sediments are located along former and actual steam channels of Nadiža River and Bela creek and in alluvial terraces of Kred polje (Figure 3(d)) and Podbela area. The volume ratio of sand and pebbles is 60: 40. Fragments, which consist of carbonate and flyschoid formations, are well-rounded and moderately sorted. Thickness of alluvium deposits is estimated to be more than 10 m.

Fine grained alluvial sediments are deposited in Kred polje (Figure 3(d)) and Kobariško blato which represent a sedimentary basin between Mt. Stol and Mt. Matajur. The flat relief induces rare outcrops, whereas some exposures were seen in old melioration trenches. Some 4 m pits, which were excavated in the frame of geological investigations for planned wastewater treatment plant, were also examined. A particular sediment found in trenches and pits is characterised by silt of bluish grey colour. According to ASTM it is defined as ML, inorganic silt with low plasticity. On the foothills of Mt. Stol the formation is overlain by alluvial fan sediments.

Alluvial fan sediments (proluvium) (Figure 2(f)) are deposited by torrential creeks on the foothills of Mt. Stol from west of Staro Selo to Potoki (Figure 3(d)). In Podbela alluvial fan sediments are found as well. The formation consists of flyschoid and carbonate fragments, whereas the ratio depends on geological setting of hinterlands. Proluvium is poorly sorted since it consists of fragments generally ranging up to 10 cm, of which up to 40% is fine sand and silt. Boulders ranging up to 60 cm can also be observed. In addition, fragments are medium rounded. Thickness of formation is estimated on less than 20 m.

Slope talus (gravel) is deposited below steep slopes of Mt. Stol and Mt. Matajur. The lower boundary of the slope talus on the slope of Mt. Stol is increasing in elevation from east to west, whereas on the slope of Mt. Matajur it is decreasing in the same direction. Gravel consists of limestone and dolomite fragments of Late Triassic. They were distinguished by the reaction with dilute HCl. Besides gravel sized fragments, also significant amount of cobbles and some boulders can be found. Additionally patches of lithified slope tallus (gravel breccia), which have the same origin as unlithified, can be observed, but they are not presented in the map due to their negligible size and dispersion.

The described lithologies are presented on the detailed engineering geological map (Figure 4), which is the basis for the seismic microzonation of the studied area.

INTENSITY INCREMENTS SEISMIC MICROZONATION

According to the old seismic hazard map of Slovenia for a 500-year return period [14] Breginjski kot is estimated to have IX MSK intensity. It is well-known that variations in local site characteristics have influence on observed earthquake intensity. Medvedev [3] proposed a classification that improves estimation on intensity considering engineering geological conditions. According to this classification, the soil is divided into three categories. The 2nd category corresponds to the medium stiff ground, which agrees with the given intensity on the seismological map [14], whereas for the 1st category an intensity decrease, and for the 3rd category an intensity increase is expected.

According to this scheme Breginjski kot is subdivided into three categories (Table 3) and presented in the map (Figure 5). In addition, engineering geological conditions, topography, and soil saturation are given descriptively in the text below, according to Medvedev [3] and Mayer-Rosa and Jimenez [22].

Table 3: Soil categories for seismic microzonation according to Medvedev method (after [3])

Soil category	Soil type	Intensity increment according to seismic area		
		VII	VIII	IX
I	hard rocks	VI	VII	VIII
II	medium-hard rocks, lithified Quaternary sediments, alluvial sediments	VII	VIII	IX
III	unlithified Quaternary sediments	VIII	IX	X

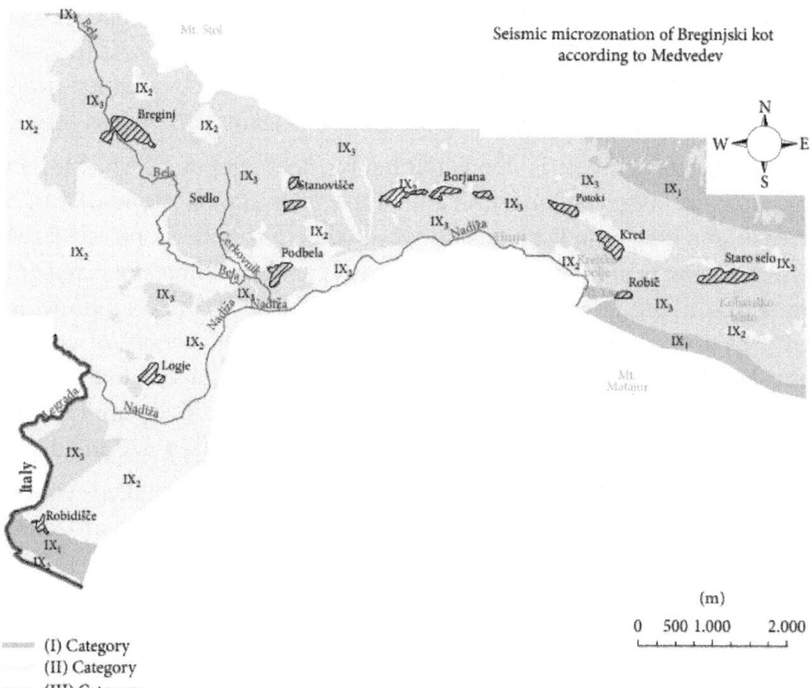

Figure 5: Intensity increments seismic microzonation of Breginjski kot according to the Medvedev method.

Category I represented by Dachstein limestone and coral limestone which are in general seismically insensitive. But, due to limestone karstification some intensity increments may be expected also for these rocks. In addition, terrain subsidence is expected because of cave collapses. Severe earthquakes may also trigger rockfalls and rock slides which can endanger settlements or road facilities.

Category II is characterised by flyschoid formations, limestone breccia interbedded with shalestone, older alluvial sediments, and alluvial sediments. Besides them, diamictit and slope tallus (gravel breccia) are also within this category, but are not presented on the map because of negligible size and dispersion. Intensity increments may be significant due to the high- degree saturation of soil along Nadiža River or Bela creek and limestone breccia karstification. Severe earthquakes may also trigger rockfalls and rock slides which can endanger settlements or road facilities. Areas with increased thickness of weathered material are prone to landside occurrences. Many caves, abysses, and sinkholes are observed in the Robidišče plateau (Figure 3(a)), where the thickness of weathered material in general exceeds 5 m and can even reach 20 m in sinkholes. A local increment of intensity can be therefore expected.

Category III is represented by lacustrine chalk, fine grained alluvial sediments, alluvial fan sediments (proluvium), diamicton, and slope tallus (gravel). Intensity increments might be larger at contacts between different sediments, especially in Podbela, Krejsko polje, and Kobariško blato which are located on the foothills of Mt. Stol. Moreover, prominent topography changes like in Sedlo, which is situated in crest vicinity, can cause an additional intensity increment.

Since significant diversities in geological conditions and topography are observed in the studied area, which might cause an increase of intensity, five main reasons for site effects in different settlements of Breginjski kot are given in Table 4. Soft sediments should be considered as a complex set of depositional events of distinctive origin, hence different sediment layers exist, divided by a certain number of sediment contacts. Generally, the larger the number of contacts is, the larger the intensity increment is. In addition, prominent topography with steep slopes, sharp relief changes and abundant karst features are also reasons for an intensity increase.

Table 4: Main reasons for significant site effects in settlements of Breginjski kot

Settlement	Main reasons for site effects				
	Prevailing lithology	No. of sediment contacts	Thickness to the basement rock	Topography	Karst features
Borjana	Diamicton	1	5–10 or more	Steep-moderate slopes	no
Kred	Proluvium	2	<10 m	Flat	no
Robič	Gravel and limestone	2	<5 m	Flat	no
Potoki	Proluvium and flyschoid f.	1	0–10 m or more	moderate slopes	no
Logje	Flyschoid f.	0*	<2 m	slight slopes-flat	no
Staro Selo	Proluvium	3	>20 m	slight slopes-flat	no
Sedlo	Diamicton	1	10–20 m or more	slight slopes, sharp relief change	no
Breginj	Diamicton	1	>20 m	slight slopes-flat	no
Podbela	Proluvium	2	>20 m	flat	no
Robidišče	Limestone and limestone breccia	0*	>5 m	flat	yes

Since 2008 seismic intensity is not used in Slovenia for building design anymore, but it is still important for seismological analysis and for civil protection. Broader applicability of the method is also in comparison with similar studies. The new seismic microzonation according to Medvedev method improves the older seismic microzonation [1] in many aspects: (a) mapping was performed on a larger scale, (b) lithologies descriptions are more precise and new interpretations are made and (c) map is digitalized. Moreover, the Bovec basin seismic microzonation [19] can also be compared to Breginjski kot due to geographic proximity and similar geological structures, though many different geotechnical and geophysical data were also acquired in the Bovec basin, as well as a study based on the microtremor method [23].

EUROCODE 8 SEISMIC MICROZONATION

The new seismic hazard map of Slovenia defines the seismic hazard with the peak ground acceleration. Furthermore, local ground conditions account for the additional influence on the seismic ground motion. In Europe ground types with different soil factors are usually determined according to the seismic standard Eurocode 8 [2] and National Annex [24].

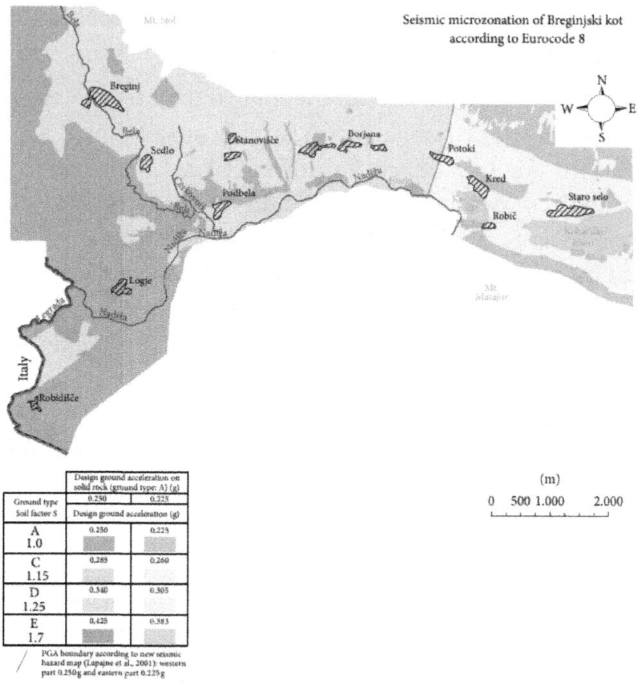

Figure 6: Eurocode 8 seismic microzonation of Breginjski kot.

According to the seismic hazard map for a 475-year return period [15], Breginjski kot is characterized in its larger western part by a design ground acceleration value for a rock site of 0.250 g and in the smaller easternmost part by 0.225 g. Ground types are classified on the basis of their geomechanical properties and weathered material thickness. Thus four different ground types, namely A, C, D, and E were assessed for the studied area (Figure 6).

Class A represents the least vulnerable ground type. It is characterised by limestone, limestone breccia interbedded with shalestone and flyschoid formation, and less than five meters of weathered material. According to the map (Figure 6) the concentration of class A is the largest in the westernmost part, though small patches are present throughout the area. Logje is the only settlement which is located entirely on class A ground. Only small parts of Robič, Potoki, and Borjana are lying on class A ground.

Class C is characterized by alluvial sediments, several tens of meters thick. The area which corresponds to the class C is situated along Nadiža River or Bela creek where no settlement is located.

Class D is characterized by diamicton, alluvial fan sediments, and slope tallus (gravel). This is the largest area in the map, which is mainly located on the slopes of Mt. Stol and Mt. Matajur. In addition it can be found in two patches between Sedlo and Robidišče. Due to its wide extent, most settlements belong to this class, for example, Staro Selo, Kred, Robič, Potoki, Borjana, Podbela, Sedlo, and Breginj.

Class E has the highest soil factor according to Eurocode 8. It comprises soil type C or D with a thickness varying from 5 m to 20 m underlain by a rock of type A. In addition, fine grained alluvial and lacustrine sediments belong to the latter type as a result of their poor geomechanical properties and heterogeneity. The area of class E includes Kobariško blato, Krejsko polje, Robidišče plateau, and some smaller patches along Nadiža valley. Robidišče is the only settlement which corresponds to the class E.

In addition to the upper classification some additional risks persist for each class, such as terrain subsidence in areas covered by limestone due to karst caves collapse and possibility for rockfalls. Final design acceleration values for each ground type are computed in accordance to the following expression:

$$\mathrm{DGA_F} = \mathrm{DGA_A} \times S \times T_{\mathrm{NCR}}, \tag{1}$$

where final design ground acceleration (DGA_F) corresponds to design ground acceleration on type A ground spectral acceleration multiplied by the soil factor (S) and the factor for the reference return period (T_{NCR}). Topographic factor is neglected in this study but it should be taken into account for important structures according to Eurocode 8 [25]. Final design ground acceleration values are shown in Table 5. According to the National Annex [24] the soil factor for ground type E is 1.7, whereas in the original Eurocode 8 standard [2] it is 1.4; for all other ground types the soil factors are the same.

Table 5: Final design ground acceleration values for each soil type in Breginjski kot.

Soil type	S	T_{NCR}	DGA_A [g]	
			0.250	0.225
			DGA_E [g]	
A	1.0	475 years	0.250	0.225
C	1.15		0.285	0.260
D	1.35	1.00	0.340	0.305
E	1.70		0.425	0.383
A	1.0	1000 years	0.315	0.280
C	1.15		0.360	0.325
D	1.35	1.25	0.420	0.380
E	1.70		0.531	0.478

DISCUSSION OF RESULTS

According to the engineering geological map of Breginjski kot, twelve different lithologies are presented and described in detail. Several of them are interpreted anew compared to previous studies (e.g., [18]).

Since typical glacial sediments are missing due to mass flows, alternative term diamicton for pseudoglacial sediments is suggested. Diamictit is also presented but only locally and in small patches. A recent study has shown that similar types of soft sediments or rocks are also identified in nearby Bovec basin [26]. Furthermore, gravel breccia was also observed in small patches similar to diamictit. Neither of them are presented in the engineering geological map but are only presented descriptively, because of their small size and sparse occurrence.

Fine grained alluvial sediments in Krejsko polje and Kobariško blato were deposited after Nadiža River had changed the current direction from the relatively wide Kobariško blato valley into the recent Nadiža valley, which is very narrow, with steep and rocky slopes [20]. Hence relatively large area was covered by swamplands and oxbow lakes where fine grained material was deposited and accumulated throughout Quaternary, although intense melioration in modern ages results in controlled river channel and thus in decreased sedimentation.

Because of detailed mapping, lithologies are defined more accurately and restricted in comparison with previous studies [20]. Therefore, their position and geometry can be better understood, thus contributing to the adequate estimation of the number of sediment contacts, their thicknesses, and abundance of karst phenomena, which may result in additional site effects

(Table 4). Moreover, thicknesses of weathered material were estimated for each rock formation, because this also has influence on seismic effects.

A significantly large thickness of weathered material is located in the Robidišče plateau due to extreme weathering of shale stone which outcrops at the edge of the plateau. The Robidišče plateau is characterized by limestone breccia and coral limestone, which includes karst features like sinkholes where the thickness of weathered material generally exceeds 5 m. Moreover, the thickness of weathered material varies significantly, since in some areas bedrock outcrops are observed. Karst phenomena like caves and abysses can also contribute to an increase of seismic ground motion. According to the geological mapping, limestone breccia interbedded with shale stone is less subjected to karst feature occurrences or such features are scarce. In the settlements of Breginjski kot small thickness of weathered material was observed especially in Logje, where outcrops of rock formation are abundant. On the other hand, the outcrops in Borjana, Potoki, and Robič were limited, thus soft sediments are more abundant. Their cumulative thickness of less than 5 m is expected. However, it should be noted that geological characteristics of sediments and their thickness are estimated from the engineering geological mapping, whereas for a more detailed analysis systematic geotechnical drilling and geophysical investigations should be performed.

The damage to buildings in settlements of Breginjski due to the 1976 Friuli earthquake was unevenly distributed, although their distance from the epicentre does not vary significantly. For comparison, the soil classification is determined for each settlement of Breginjski kot according to both seismic microzonations. A comparison between intensities of the 1976 earthquake and soil classifications is given in Table 6.

Table 6: Intensities of 1976 earthquake in different settlements of Breginjski kot and soil classification after Medvedev and Eurocode 8 from both seismic microzonation maps

Settlement	Intensity EMS-98	Soil classification	
		After Medvedev	After Eurocode 8
Borjana	VI	III, II	D, A
Kred	VI	III	D
Robič	VI	III, I	D, A
Potoki	VI+ (MSK-78)	III, II	D, A
Logje	VI-VII	II	A
Staro Selo	VII	III	D
Sedlo	VII-VIII	III	D
Breginj	VIII	III	D
Podbela	VIII	III	D
Robidišče	VIII-IX	II	E

Soil classification according to Medvedev method does not show a clear correlation of ground type to observed intensities. The main reason for this is that Medvedev soil classification does not take into account the thickness of soft sediments or weathered material. In addition it does not consider sufficiently engineering geological conditions which are annotated descriptively only. On the contrary, Eurocode 8 soil classification clearly shows a correlation between the ground type and observed intensities. This standard classifies ground types more precisely, also since it considers the thickness of soft sediments and weathered material. A better evaluation of local soil effect is therefore achieved. Based on the 1976 earthquake intensities, the settlements of Breginjski kot can be divided into two categories. The first group consists of settlements which show relatively low site effects like Borjana, Kred, Robič, Potoki, and Logje. They are located either on ground type A, or on the weathered material less than 5 m thick. The indicators of small weathered material thickness are frequent erosion windows. The second group consists of settlements which show relatively high site effects like Staro Selo, Sedlo, Breginj, Podbela, and Robidišče. They are located on soft sediments with considerable thickness. Additional causes of significant site effects in each particular settlement are described in Table 4.

CONCLUSIONS

This study represents a contribution to the evaluation of the seismic hazard in Breginjski kot, which is among the most endangered seismic areas in Slovenia. Two new seismic microzonations were prepared, because the previous seismic microzonation was based solely on the basic geological map and did not include supplementary field research, thus making it fairly inaccurate. In this study detailed engineering geological mapping of Breginjski kot in scale 1 : 5.000 was conducted. The results of mapping showed a diverse composition of sediments and rocks. Mapped units were described in detail and some of them are interpreted anew. Descriptions of mapped units also include the estimated thickness of soft sediments and weathered material. The results can be enhanced and validated by drilling and geophysical investigations, which is recommended especially in the settlements that were badly damaged during the 1976 Friuli earthquake. In addition, a microtremor method study would be appropriate to apply in the flatlands of Krejsko polje and Kobariško blato to determine soil resonance frequencies.

On the basis of engineering geological mapping, seismic microzonation was prepared according to the Medvedev method and the Eurocode 8 standard. According to the first classification the soil in Breginjski kot was divided into three categories, and according to the latter classified into categories A, C, D, and E. Two maps of seismic microzonation, based on both soil classifications, were made in the GIS platform and are therefore ready to use in urban planning. Seismic microzonation according to Eurocode 8 is more appropriate for the application, because it is in accordance with the modern seismic hazard standards and because it better correlates thickness of soft sediments or weathered material with site amplification. Moreover, ground types are classified more precisely, and thus the correlation with the geological map is better. The new seismic microzonation reveals the correlation of site effects and observed intensities. Sites located on softer ground experienced a higher degree of damage during the Friuli 1976 earthquake. Although the microzonation based on seismic intensities is not used in earthquake engineering anymore, it is still important for civil protection purposes and in seismological analysis. The new map improves the old microzonation and enables a comparison of geologically similar areas of Breginjski kot and the Bovec Basin.

ACKNOWLEDGMENT

The study was realized with the support of the research program P1-0011 financed by Slovenian Research Agency.

REFERENCES

1. R. Vidrih, M. Godec, and J. Lapajne, Earthquake Risk in Slovenia, Seismological Survey of Slovenia, Ljubljana, Slovenia, 1991.
2. CEN, Eurocode 8—Design of Structures for Earthquake Resistance, Part 1: General Rules, Seismic Actions and Rules for Buildings, European Standard, EN, 1998-1: 2004 (E), Stage 64, European Committee for Standardization, Brussels, Belgium, 2004.
3. S. V. Medvedev, Engineering Seismology, Građevinske knjige, Belgrade, Serbia, 1965.
4. B. Perniola, G. Bressan, and S. Pondrelli, "Changes in failure stress and stress transfer during the 1976–77 Friuli earthquake sequence," Geophysical Journal International, vol. 156, no. 2, pp. 297–306, 2004. G. B. Carulli and D. Slejko, "The 1976 Friuli (NE Italy) earthquake," Giornale di Geologia Applicata, vol. 1, pp. 147–156, 2005.
5. V. Ribarič, "Earthquakes in Friuli in 1976 and the short history of seismicity at the margin of Eastern Alps," Potresni Zbornik, pp. 17–80, 1980.
6. P. Zupančič, I. Cecić, A. Gosar, L. Placer, M. Poljak, and M. Živčić, "The earthquake of 12 April 1998 in the Krn Mountains (Upper Soča valley, Slovenia) and its seismotectonic characteristics," Geologija, vol. 44, pp. 169–192, 2001.

7. I. Cecić, M. Živčić, T. Jesenko, and J. Kolar, "Earthquakes in Slovenia in 2004," Potresi V Letu 2004, pp. 16–40, 2006.
8. J. Bajc, A. Aoudia, A. Saraò, and P. Suhadolc, "The 1998 Bovec-Krn mountain (Slovenia) earthquake sequence," Geophysical Research Letters, vol. 28, no. 9, pp. 1839–1842, 2001.
9. A. Gosar, "Site effects and soil-structure resonance study in the Kobarid basin (NW Slovenia) using microtremors," Natural Hazards and Earth System Sciences, vol. 10, no. 4, pp. 761–772, 2010.
10. V. Kastelic, M. Vrabec, D. Cunningham, and A. Gosar, "Neo-Alpine structural evolution and present-day tectonic activity of the eastern Southern Alps: the case of the Ravne Fault, NW Slovenia," Journal of Structural Geology, vol. 30, no. 8, pp. 963–975, 2008.
11. F. Fitzko, P. Suhadolc, A. Aoudia, and G. F. Panza, "Constraints on the location and mechanism of the 1511 Western-Slovenia earthquake from active tectonics and modeling of macroseismic data,"Tectonophysics, vol. 404, no. 1-2, pp. 77–90, 2005.
12. R. Camassi, C. H. Caracciolo, V. Castelli, and D. Slejko, "The 1511 Eastern Alps earthquakes: a critical update and comparison of existing macroseismic datasets," Journal of Seismology, vol. 15, no. 2, pp. 191–213, 2011.
13. V. Ribarič, Seismological Map of Slovenia for 500 Years Return Period, Seismological survey of Slovenia, Ljubljana, Slovenia, 1987.
14. J. Lapajne, B. Šket-Motnikar, and P. Zupančič, "Design ground acceleration map of Slovenia," Potresi V Letu 1999, pp. 40–49, 2001.
15. N. Ambraseys, P. Smit, R. Sigbjornsson, P. Suhadolc, and B. Margaris, "Internet-Site for European Strong-Motion Data," 2002,
16. B. Šket-Motnikar and T. Prosen, "Accelerations in Posočje on July 12 2004," Potresi V Letu 2004, pp. 105–113, 2006
17. S. Buser, Basic Geological Map of Yugoslavia 1: 100. 000—Sheet Tolmin and Videm, Geological Survey of Slovenia, Ljubljana, Slovenia, 1986.
18. M. Ribičič, R. Vidrih, and M. Godec, "Seismogeological and geotechnical conditions of buildings in Upper Soča Territory, Slovenia," Geologija, vol. 43, no. 1, pp. 115–143, 2000.
19. S. Buser, Basic Geological Map of Yugoslavia 1: 100. 000—Sheet Tolmin and Videm Explanatory Text, Geological Survey of Slovenia, Ljubljana, Slovenia, 1986.
20. A. Casagrande, "Classification and identification of soils," Transactions of the American Society of Civil Engineers, vol. 113, pp. 901–930, 1948.

21. D. Mayer-Rosa and M. J. Jimenez, Seismic Zoning: State-of-the-Art and Recommendations for Switzerland, Federal Office for Water and Geology, Swiss National Hydrogeological and Geological Survey, Berne, Switzerland, 2000.
22. A. Gosar, "Microtremor HVSR study for assessing site effects in the Bovec basin (NW Slovenia) related to 1998 Mw5.6 and 2004 Mw5.2 earthquakes," Engineering Geology, vol. 91, no. 2–4, pp. 178–193, 2007
23. SIST, Slovenian Standard SIST EN, 1998-1 Eurocode 8: Design of Structures for Earthquake Resistance—Part 1: General Rules, Seismic Actions and Rules for Buildings—National Annex, Slovenian Institute for Standardization, Ljubljana, Slovenian, 2004.
24. CEN, Eurocode 8—Design of Structures For Earthquake Resistance, Part 5: Foundations, Retaining Structures and Geotechnical Aspects, European Standard, EN, 1998-5: 2004 (E), Stage 64, European Committee for Standardization, Brussels, Belgium, 2004.
25. M. Bavec, "New temporal, and genetic determinations of some late Quaternary sediments in the Bovec basin and its surroundings (NW Slovenia)," Geologija, vol. 45, no. 2, pp. 291–298, 2002.

CITATION

CHAPTER 1

Mohammad Ali Zare Chahouki, Multivariate Analysis Techniques in Environmental Science, ISBN: 978-953-307-468-9.

CHAPTER 2

Khageshwar Singh Patel, Shobhana Ramteke, Yogita Naik, Bharat Lal Sahu, Saroj Sharma, Jutta Lintelmann, Matuschek Georg, Contamination of Environment with Polycyclic Aromatic Hydrocarbons in India, http://dx.doi.org/10.4236/jep.2015.611111.

CHAPTER 3

Roberta Pirastu, Pietro Comba, Ivano Iavarone, et al., "Environment and Health in Contaminated Sites: The Case of Taranto, Italy," Journal of Environmental and Public Health, vol. 2013, Article ID 753719, 20 pages, 2013. doi:10.1155/2013/753719.

CHAPTER 4

María Antonieta Gómez-Balandra, María del Pilar Saldaña-Fabela, Maricela Martínez-Jiménez, The Mexican Environmental Flow Standard: Scope, Application and Implementation, http://dx.doi.org/10.4236/jep.2014.51010.

CHAPTER 5

Fereshte Haghighi, Mirmasoud Kheirkhah and Bahram Saghafian, Evaluation of Soil Hydraulic Parameters in Soils and Land Use Change, ISBN: 978-953-307-468-9.

CHAPTER 6

Kevin Lo, "Energy-Related Carbon Emissions of China's Model Environmental Cities," Geography Journal, vol. 2014, Article ID 204745, 7 pages, 2014. doi:10.1155/2014/204745.

CHAPTER 7

Helen Kopnina, and Brett Cherniak, Cultivating a Value for Non-Human Interests through the Convergence of Animal Welfare, Animal Rights, and Deep Ecology in Environmental Education, doi:10.3390/educsci5040363.

CHAPTER 8

Daniel L. Childers, Mary L. Cadenasso, J. Morgan Grove, Victoria Marshall, Brian McGrath and Steward T. A. Pickett, An Ecology for Cities: A Transformational Nexus of Design and Ecology to Advance Climate Change Resilience and Urban Sustainability, doi:10.3390/su7043774.

CHAPTER 9

Giuseppe Ioppolo, Stefano Cucurachi, Roberta Salomone, Giuseppe Saija and Luigi Ciraolo, Industrial Ecology and Environmental Lean Management: Lights and Shadows, doi:10.3390/su6096362.

CHAPTER 10

Christian Mulder, Ecology and eScience, DOI: 10.1186/2192-1709-1-1.

CHAPTER 11

Yajun Song and Tianzhong Zhao, A bibliometric analysis of global forest ecology research during 2002–2011, DOI: 10.1186/2193-1801-2-204.

CHAPTER 12

Sewall BJ, Freestone AL, Hawes JE, Andriamanarina E (2013) Size-Energy Relationships in Ecological Communities. PLoS ONE 8(8): e68657. doi:10.1371/journal.pone.0068657.

CHAPTER 13

Jure Kokošin and Andrej Gosar, "Seismic Microzonation of Breginjski Kot (NW Slovenia) Based on Detailed Engineering Geological Mapping," The Scientific World Journal, vol. 2013, Article ID 626854, 12 pages, 2013. doi:10.1155/2013/626854

INDEX

Encyclopaedia of
APPLIED GEO ENVIRONMENTAL STUDIES

Encyclopaedia of
APPLIED GEO ENVIRONMENTAL STUDIES

Volume III: Environmental, Groundwater, and Engineering Geology

Contributors

Mohammad Ali Zare Chahouki, Paulo Cesar Fernandes da Silva, and Bizhan Shirdel et al.

AURIS
Reference

www.aurisreference.com

Encyclopaedia of Applied Geo Environmental Studies

Volume III: Environmental, Groundwater, and Engineering Geology

Contributors: Mohammad Ali Zare Chahouki, Paulo Cesar Fernandes da Silva, and Bizhan Shirdel et al.

Published by Auris Reference Limited

www.aurisreference.com

United Kingdom

Printed in 2017 for Sale in the Indian Subcontinent

Environmental, Groundwater, and Engineering Geology

ISBN: 978-1-78154-980-3 (Set of 3 Volumes)

British Library Cataloguing in Publication Data
A CIP record for this book is available from the British Library

Printed in the United Kingdom

Exclusively distributed by CBS Publishers & Distributors Pvt. Ltd.

Sales & Distribution Rights only for India, Pakistan, Bangladesh, Sri Lanka, Nepal and Bhutan.This book is not to be sold outside these territories.

Contents

List of Abbreviations

AMD	Acid mine drainage
DWR	Department of Water Resources'
DEM	Digital Elevation Model
IEC	Electrotech- nical Commission
FCT	Foundation for the Science and Technology
GIS	Geographic Information System
GPS	Global Positioning System
GMS	Groundwater Modeling System
HRU	Hydrologic response units
HRU	Hydrologic response units
IMS	Integrated management system
IGWMC	International Ground Water Modeling Center
IDW	Inverse Distance Weighting
KPK	Khyber Pakhtunkhwa
MSSs	Man- agement systems standards
MSL	Mean Sea Level
MOSTI	Ministry of Science, Technology and Innovation
NSF	National Science Fellowship
OHS	Occupational health and safety
QA	Quality assurance
QC	Quality control
QMS	Quality Management System
RQD	Rock quality designation
RMSE	Root Mean Square Error
SAFCA	Sacramento Area Flood Control Agency
SRCSD	Sacramento Regional County Sanitation District
SRM	Sacramento Regional Model
SRF	Stress reduction factor
TIN	Triangulated Irregular Network
USDA	United States Department of Agriculture
UTM	Universal Transverse Mercator
WGS	World Geodesic System

List of Contributors

Bizhan Shirdel
Department of Geology, Karaj Branch, Islamic Azad University, Karaj, Iran

Sayed A. Selim
Geology Department, Faculty of Science, Aswan University, Aswan, Egypt

Ali M. Hamdan
Geology Department, Faculty of Science, Aswan University, Aswan, Egypt

Ahmed Abdel Rady
Aswan Water and Sanitation Company, Aswan, Egypt

Saad Merayyan
California State University, Sacramento, CA, USA

Samsor Safi
Sacramento Area Sewer District, Sacramento, CA, USA

Sarva Mangala PRAVEENA
School of Science and Technology, Universiti Malaysia Sabah, Kota Kinabalu, Sabah, Malaysia.

Mohd Harun ABDULLAH
School of Science and Technology, Universiti Malaysia Sabah, Kota Kinabalu, Sabah, Malaysia.

Ahmad Zaharin ARIS
Department of Environmental Sciences, Universiti Putra Malaysia, Selangor, Malaysia

Kawi BIDIN
School of Science and Technology, Universiti Malaysia Sabah, Kota Kinabalu, Sabah, Malaysia.

Jacques Édoukou Djémin
Department of Sciences and Water Technology and Environment Engineering (Laboratory of Remote Sensing and Spatial Analysis Applied to Hydrogeology), Félix Houphouët-Boigny University, Abidjan, Côte d'Ivoire

Jean Kan Kouamé
Department of Sciences and Water Technology and Environment Engineering (Laboratory of Remote Sensing and Spatial Analysis Applied to Hydrogeology), Félix Houphouët-Boigny University, Abidjan, Côte d'Ivoire

Kouakou Serge Deh
Department of Sciences and Water Technology and Environment Engineering (Laboratory of Remote Sensing and Spatial Analysis Applied to Hydrogeology), Félix Houphouët-Boigny University, Abidjan, Côte d'Ivoire

Amani Tawa Abinan
Department of Sciences and Water Technology and Environment Engineering (Laboratory of Remote Sensing and Spatial Analysis Applied to Hydrogeology), Félix Houphouët-Boigny University, Abidjan, Côte d'Ivoire

Jean Patrice Jourda
Department of Sciences and Water Technology and Environment Engineering (Laboratory of Remote Sensing and Spatial Analysis Applied to Hydrogeology), Félix Houphouët-Boigny University, Abidjan, Côte d'Ivoire

Manuel Ferreira Rebelo
CLEGI, Lusíada University, Vila Nova de Famalicão, Portugal

Gilberto Santos
CLEGI, Lusíada University, Vila Nova de Famalicão, Portugal
College of Technology, Polytechnic Institute of Cávado and Ave, Barcelos, Portugal

Rui Silva
CLEGI, Lusíada University, Vila Nova de Famalicão, Portugal
College of Technology, Polytechnic Institute of Cávado and Ave, Barcelos, Portugal

Abdollah Yazdi
Department of Geology, Kahnooj Branch, Islamic Azad University, Kerman, Iran

Mohammad Ali Arian
Department of Geology, North Tehran Branch, Islamic Azad University, Tehran, Iran

Mahmoud M. Rezapour Tabari
Department of Engineering, Shahrekord University, Shahrekord, Iran

Leila Khodapanah
Department of Environmental Sciences, Faculty of Environmental Studies, Universiti Putra Malaysia, Serdang, Malaysia

Wan Nor Azmin Sulaiman
Department of Environmental Sciences, Faculty of Environmental Studies, Universiti Putra Malaysia, Serdang, Malaysia

Hamid Reza Nassery
Faculty of Earth Science, Shahid Beheshti University, Tehran, Iran

Arshad Ashraf
Water Resources Research Institute, National Agricultural Research Center, Islamabad, Pakistan

Luc Descroix
IRD / UJF, Grenoble, France

Ibrahim Bouzou Moussa
UAM University, Niamey, Niger

Pierre Genthon
IRD-HSM, Montpellier, France

Daniel Sighomnou
Niger Basin Authority, Niamey, Niger

Gil Mahé
IRD-HSM, Montpellier, France

Ibrahim Mamadou
University of Zinder, Niger

Jean-Pierre Vandervaere
UJF-LTHE, Grenoble, France

Emmanuèle Gautier
Université Paris 8, France

Oumarou Faran Maiga
UAM University, Niamey, Niger

Jean-Louis Rajot
IRD-BIOEMCO, Créteil, France

Moussa Malam Abdou
IRD / UJF, Grenoble, France

Nadine Dessay
IRD-ESPACE-DEV, Montpellier, France

Aghali Ingatan
UAM University, Niamey, Niger

Ibrahim Noma
UAM University, Niamey, Niger

Kadidiatou Souley Yéro
IRD / UJF, Grenoble, France

Harouna Karambiri
2iE International high School, Ouagadougou, Burkina Faso

Rasmus Fensholt
University of Copenhague, Denmark

Jean Alberge
IRD-LISAH, Montpellier, France

Jean-Claude Olivry
IRD, France

Wei-Zu Gu
Institute for Hydrology and Water Resources, Nanjing Hydraulic Research Institutes, Nanjing, China

Jiu-Fu Liu
Institute for Hydrology and Water Resources, Nanjing Hydraulic Research Institutes, Nanjing, China

Jia-Ju Lu
Institute for Hydrology and Water Resources, Nanjing Hydraulic Research Institutes, Nanjing, China

Jay Frentress
Oregon State University, Corvallis, USA

Jian Liu
Faculty of Geosciences and Environmental Engineering, Southwest Jiaotong University, Chengdu, Sichuan 610031, China

Dan Liu
Faculty of Geosciences and Environmental Engineering, Southwest Jiaotong University, Chengdu, Sichuan 610031, China

Kai Song
Faculty of Geosciences and Environmental Engineering, Southwest Jiaotong University, Chengdu, Sichuan 610031, China

Muhammad Aqeel Ashraf,
Department of Chemistry, University of Malaya, Kuala Lumpur 50603, Malaysia

Mohd. Jamil Maah,
Department of Chemistry, University of Malaya, Kuala Lumpur 50603, Malaysia

Ismail Yusoff
Department of Geology, University of Malaya, Kuala Lumpur 50603, Malaysia

Preface

Encyclopaedia of Applied Geo Environmental Studies summarizes the development of groundwater and its utilization, and explains the theory and applied aspects of engineering geology. Environmental geology is an applied science concerned with the practical application of the principles of geology in the solving of environmental problems. Engineering geology is the application of the geological sciences to engineering study for the purpose of assuring that the geological factors regarding the location, design, construction, operation and maintenance of engineering works are recognized and accounted for. The text *Environmental, Groundwater, and Engineering Geology* focuses on new concepts and recent advances in environmental, groundwater, and engineering geology. Analysis of engineering geology indices in three units of Atamir Formation in Kope Dagh Zone has been focused in first chapter. Second chapter examines the rise in the level of the groundwater in the Quaternary aquifer at Aswan city, Upper Egypt. Third chapter evaluates the feasibility of groundwater banking in the Central Basin. Fourth chapter reviews the types of groundwater solution techniques in terms of advantages and limitations. Fifth chapter presents the assessment of the groundwater intrinsic vulnerability in the Dabou region from the DRASTIC method which requires seven hydrogeological parameters in its application. The aim of sixth chapter is to present and justify a designed methodology to be used by organizations to support the integration of various management systems (MSs). Seventh chapter studies Hormoz Island in terms of geological features and geotourism potentials. The objective of eighth chapter is to develop the hydrogeological framework of the groundwater system in Shariar, Iran and to estimate groundwater balance as a scientific database for future water resources development programs. In ninth chapter, SWAT model developed by United States Department of Agriculture (USDA) has been used to evaluate surface runoff generation, soil erosion and quantify the water balance of a Himalayan watershed in the Northern Pakistan. The purpose of tenth chapter is to provide an overview of hydrological behavior throughout West Africa based on point, local, meso and regional scales observations. Current challenges in experimental watershed hydrology have been revealed in eleventh chapter. Last chapter focuses on evaluating the influence resulting from tunnel excavation on groundwater environment, by means of employing an indicator system.

Chapter 1

ANALYSIS OF ENGINEERING GEOLOGY INDICES IN THREE UNITS OF ATAMIR FORMATION IN KOPE DAGH ZONE

Bizhan Shirdel

Department of Geology, Karaj Branch, Islamic Azad University, Karaj, Iran

ABSTRACT

This is a field report on a comprehensive study of the Atamir Formation from the engineering geology perspective using the related indices. The Atamir Formation of the Cretaceous Period, which has outcropped in the form of thick frequencies of grey-knotted sandstone and black shales, is situated in the Kope Dagh zone. A survey of discontinuities together with bedding was carried out to study slope stability. The layers have a general east-west trend with a gentle slope towards the south. Because of the tectonic and stratigraphic differences, and with the purpose of facilitating surveys related to joint study of the outcrop, the formation in the study region was divided into three units. The lower unit is made of shale, the middle of sandstone, and the upper of marlstone. All three units were studied from the perspective of geomechanical classification, rock mass indices, geological strength, geomechanical indices, and wedge instability analysis under dry and wet conditions, and the results were investigated in the form of various images and figures. The Dips software was used to display the rose diagram and stereographic projection of each unit, the Swedge software to analyze instability of the wedges, and the Roctab software to analyze the geomechanical parameters and present the outputs along with the description of each unit.

INTRODUCTION

Based on field surveys and geological surveys such as outcrop studies, the Atamir Formation was divided into three separate units to study the characteristics of the discontinuities and of the rock mass in each one. This study intended to review results of research carried out on the Atamir Formation in detail and present a field report. Therefore, the units were studied and the results were compared.

RESULTS

The Atamir Formation Unit One K (at) 1

This lithology included black shales. In the study region, there were three dominant discontinuity sets (two joint sets together with bedding) and random discontinuities. Table 1 lists the characteristics of these discontinuities. Table 2 lists geotechnical characteristics of the various discontinuities in the rock masses. Figures 1-3 show the Stereographic characteristics of Figures 4-7 show statistical analysis.

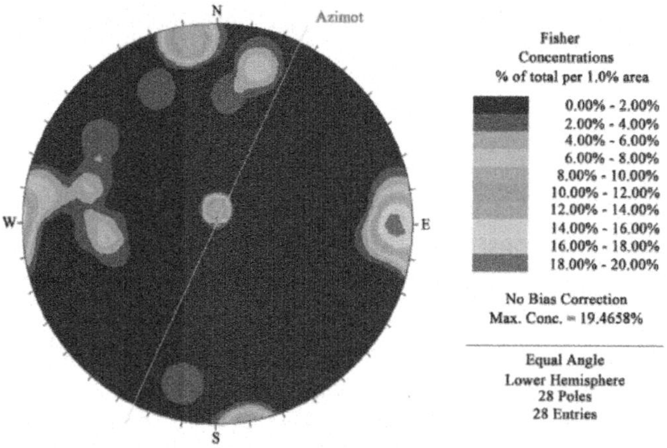

Figure 1: Stereographic image of the discontinuities in unit one of the Atamir Formation K (at) 1.

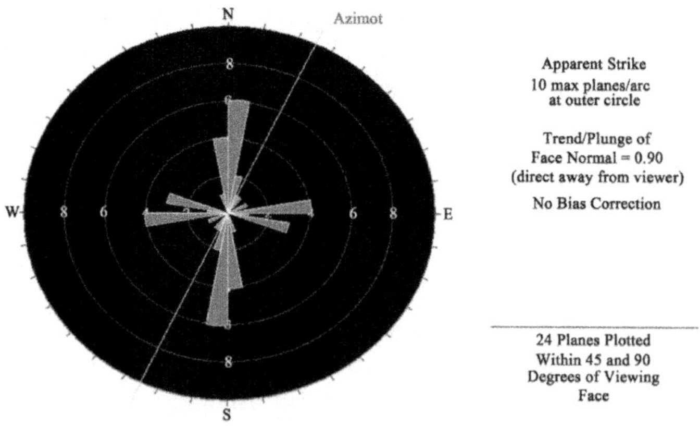

Figure 2: Rose diagram of the discontinuities in unit one of the Atamir Formation K (at) 1.

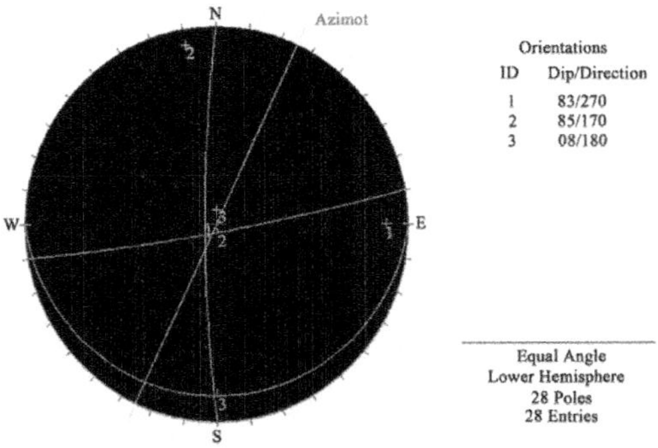

Figure 3: Stereography of discontinuities in unit one of the Atamir Formation K (at) 1.

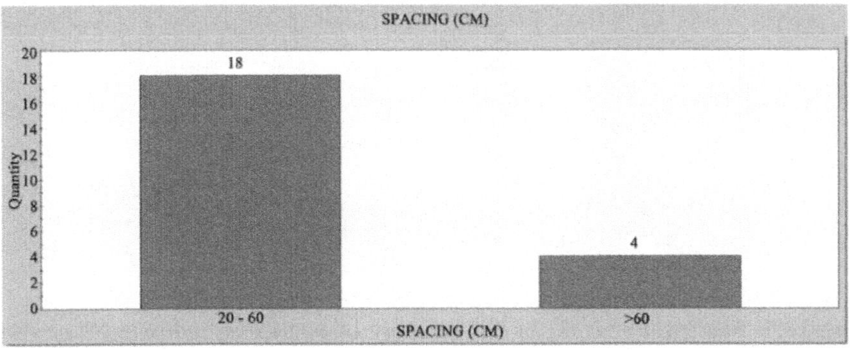

Figure 4: Statistical analysis of discontinuity spacings in unit one of the Atamir Formation K (at) 1.

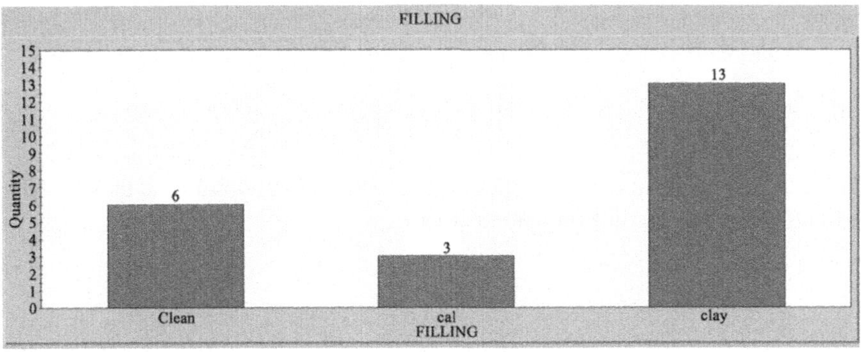

Figure 5: Statistical analysis of discontinuity fillings in unit one of the Atamir Formation K (at) 1.

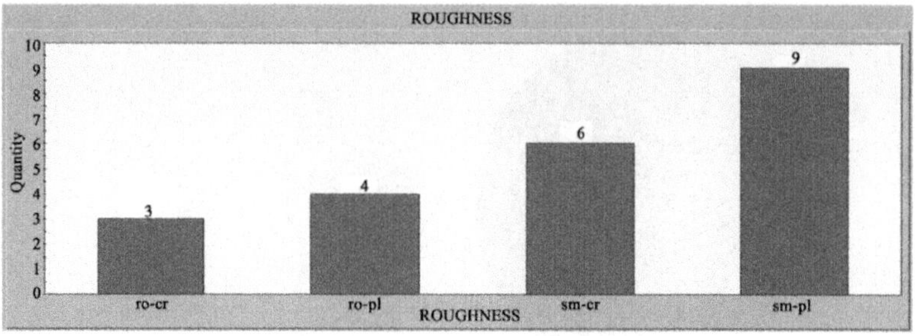

Figure 6: Statistical analysis of discontinuity roughness in unit one of the Atamir Formation K (at) 1.

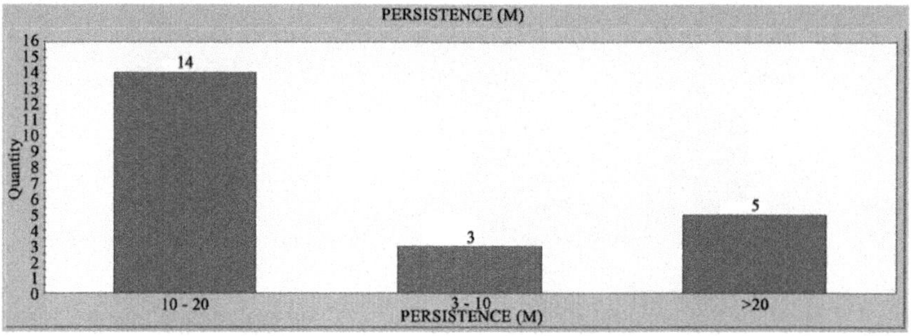

Figure 7: Statistical analysis of discontinuity orientation in unit one of the Atamir Formation K (at) 1.

Table 1: Characteristics of the discontinuities in unit one of the Atamir Formation K (at) 1.

Discontinuity	ID	Dip	Dip Direction
Js1	1	83	270
Js2	2	85	170
Bedding	3	8	180

Table 2; Geotechnical characteristics of the various discontinuities in the rock masses of unit one in the Atamir Formation K (at) 1

Joint Sets	Dip Direction	Joint Frequency	Persistence (m)					Joint Roughness (%)				
			>20%	10% - 20%	3% - 10%	1% - 3%	<1%	Very Rough	Rough	Rough-Smooth	Smooth	Slickenside
Js1	270\83	50	-	80	20	-	-	-	-	20	80	-
Js2	170\85	30	-	70	30	-	-	-	30	70	-	-
Bedding	180\8	20	100	-	-	-	-	-	-	100	-	-

Joint Alteration	Joint Water	Aperture (mm)					Joint filling					Joint spacing (m-mm)				
		>5 mm %	1 - 5 mm %	0.1 - 1 mm %	<0.1 mm %	Close	Soft filling mm 5 < %	Soft filling mm 5 > %	Hard filling mm 5 > %	Hard filling mm 5 < %	Clean %	<60 mm %	60 - 200 mm %	200 - 600 Mm %	0.6 - 2 m %	>2 m %
		-	100	-	-	-	-	100	-	-	-	-	-	100	-	-
W2	Dry	40	60	-	-	-	-	-	40	-	60	-	-	50	50	-
		-	100	-	60	-	100	-	-	-	-	-	-	-	-	100

Rock masses were classified based on the methods proposed by Bieniawski (RMR), Barton (Q), and Hoek (GSI), all of which are discussed below.

Geomechanical Classification (RMR)

- Compressive strength of intact rock
    - The value of this parameter was determined based on the strength characteristic of outcropped masses and on performing tests.
- Rock quality designation
    - The rating of 8 may be given depending on the state of the discontinuity system.
- Joint spacing
    - The rating for this parameter was the least value of spacing. In the study region, there were three sets of discontinuities (two sets of dominant discontinuities together with bedding), and the minimum discontinuity spacing was less than 10 centimeters. Considering the geomechanical classification parameters, the rating of 8 - 10 was given to this parameter.
- Condition of discontinuities
    - The joints in the rock masses in unit one of the Atamir Formation, K (at) 1, were 10 - 20 meters long, clay, calcite, and few traces of iron oxide were the main filling, the joints were smooth to a little rough and, with respect to euhidrality, the rock masses were planar and moderately weathered. Considering the existing parameters, the rating given to this parameter was 10.
- Groundwater
    - Presence of groundwater in a region is undoubtedly very important, especially if permeability is high. Considering the conditions of the rock masses in the region, the seepage state was determined and given the rating of 4.
- Discontinuity orientations

This parameter is determined by the mode of transgression of the structure towards the jointed rock mass, by the dips of the layers (and of the discontinuities), and by the orientation of the structure (which has been used by Feenstra & Wickham in 1975 [1] and suggested by Bieniawski 1973) [2] .

The presented parameters 2 - 1 to 2 - 5 are used for determining the value of RMR that, for the rock masses in unit one of the Atamir Formation, K (at) 1, was 34 - 36. Based on this, these rock masses were placed in the poor class; pertaining data are shown in Table 3.

Qualitative Index of the Rock Mass (Q)

The index Q for the rock masses in unit one of the Atamir Formation, K9 (at) 1, was estimated based on the following 6 parameters.

- Rock quality designation (RQD)

  Considering the condition of the rock masses, the value of 43 - 48 was given to the RQD parameter of unit one in the Atamir Formation, K (at) 1.

Table 3. Geomechanical classification of the rock masses in unit one of the Atamir Formation K (at) 1

Parameter	Value	Rating
UCS (Mpa)	25 - 50	4
RQD	25 - 50	8
Spacing (mm)	60 - 200	8 - 10
Condition of discontinuities		10
Groundwater	Flow	4
RMR	34 - 35	
Class no.	Poor rock	
Cohesion of the rock mass (KPa)	115	
Friction angle of the rock mass (deg)	15 - 25	

- Number of joint sets (J_n)

  Based on the data obtained from the geological field surveys, this parameter received the value of 12 for unit one of the Atamir Formation, K (at) 1.

- Discontinuity surface roughness (J_r)

  This parameter shows the degree of roughness of the discontinuities and, based on Barton's theory, the weakest set of critical discontinuities or discontinuities filled with clay materials in the related zone must be determined. Considering the conditions, the value of one was given to this parameter for unit one of the Atamir Formation, K (at) 1.

- Joint alteration number (J_a)

  Considering the joint filling and the apparent conditions of discontinuity surfaces, this parameter received the value of 4 for unit one of the Atamir Formation, K (at) 1.

- Joint water reduction factor (J_w)

This parameter measures water pressure, which inversely affects shear strength of the discontinuity. Based on data collected in the study region, the value of 0.66 was considered for this parameter.

- Stress reduction factor (SRF)

  This parameter is determined based on the characteristics of the study region including the existing stresses, the tectonic conditions, the shear zones, etc. Since the necessary tests were not performed, the estimated value of one was used for this factor.

Considering the ratings of the six parameters, and using the presented relationship, the expected value determined for the Q index was 0.592 - 0.66 for the rock masses in unit one of the Atamir Formation, K (at) 1, which places them in the class of very poor rocks. Data are shown in Table 4.

Geological Strength Index (GSI) of the Rock Masses

Using the following three methods, the value of the geological strength index for the rock masses in unit one was determined.

- Geological strength index of the rock mass (GSI) (Figure 8)

  The GSI system is the only classification system that is directly related to engineering parameters such as Mohr-Coulomb and Hoek-Brown parameters. Conventional methods for determining strength parameters and rock mass modulus in Iran and other countries include plate-loading tests and block shear in situ. However, these tests can be performed only if exploration drilling has been carried out and, in addition to that, they are very costly. Therefore, many studies have been conducted and numerous attempts made for developing and introducing methods to indirectly determine engineering parameters of rock masses, one of which is the GIS system [3] . In the following sections, the GSI of rock masses in unit one will be discussed.

- Determining GSI based on field observations

  Based on field studies and observations, the value of geological strength index of rock masses in unit one of the Atamir Formation, K (at) 1 was 35 - 45.

Table 4: The Q index of the rock masses in unit one of the Atamir Formation, K (at) 1

Parameter	Description	Value
RQD	43 - 48	43 - 48
J_n	Three joint sets with random	12
J_r	Smooth or rough to **Craw** or Planar	1
J_a	Softening or low friction clay mineral coatings	4
J_w	Inflow	0.66
SRF	Medium stress, favorable stress condition	1
Q	0.592 - 0.66	
Class No.	Very poor rock	

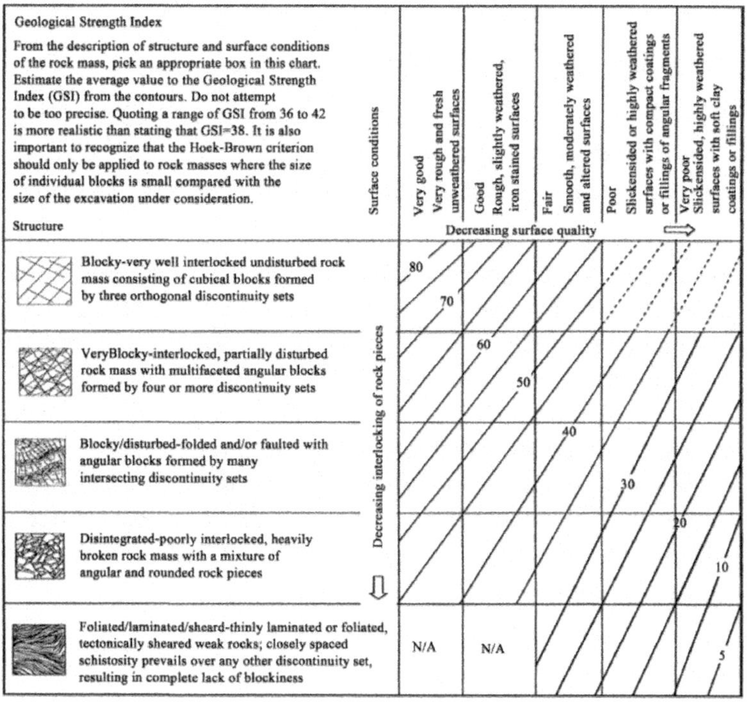

Figure 8: The GSI range of the rock masses in unit one of the Atamir Formation K (at) 1 in the GSI classification diagram.

Determination of Geomechanical Parameters of the Rock Masses in Unit One of the Atamir Formation K (at) 1

To determine these parameters, the GSI was considered 35 - 45. Since cohesive strength and angle of internal friction are among the input parameters in most methods of analysis, the Mohr-Coulomb and Hoek-Brown criteria had to be fitted to determine these two parameters. Results are listed inTable 5 and shown in Figures 9-11.

Analysis of Wedge Instability in Unit One of the Atamir Formation, K (at) 1, under Dry and Wet Conditions

Analysis of wedge instability was performed using SWEDGE. This software, which was developed by the ROCSCIENCE Company, analyzes wedge stability in rock slopes, degree of cohesion, geomechanical characteristics, etc., studies stability from various aspects, analyzes areas with rock fall potential using the StereoNet network, follows discontinuities and determines and analyzes intersection points of wedges, and plots the StereoNet using input data. Figure 12 shows stereographic images of the joint sets. Applying the primary parameters, the characteristics of the largest wedges that can be formed in the walls are determined. According to the performed analysis, wedge slide will happen along the interface between two discontinuities, and the factor of safety (FS) will be 0.07743. Figures 13-18 show other views.

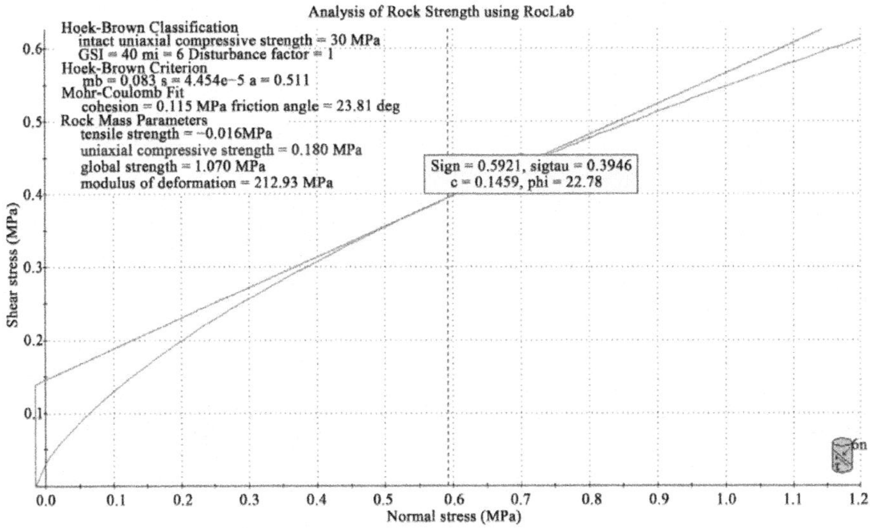

Figure 9: Fitting the data on the rock masses in unit one of the Atamir Formation, K (at) 2, to the Hoek-Brown and Mohr- Coulomb criteria based on GSI = 40 and normal shear stress.

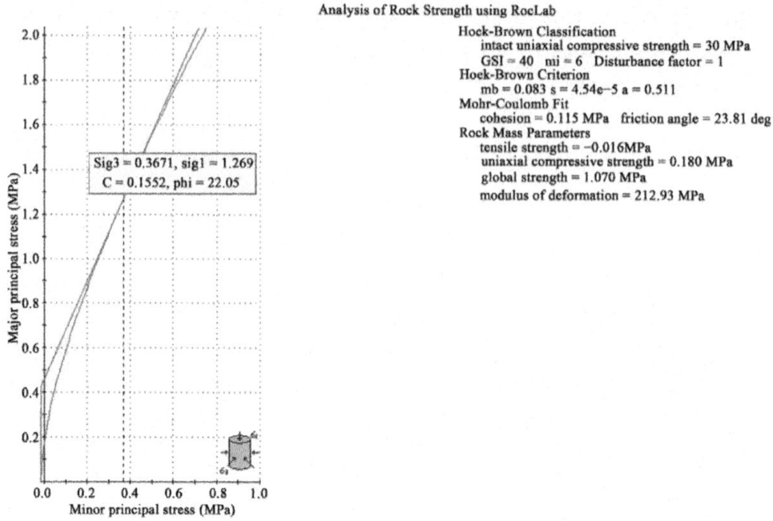

Figure 10: Fitting the data on the rock masses in unit one of the Atamir Formation, K (at) 2, to the Hoek-Brown and Mohr- Coulomb criteria based on GSI = 40 and major principal stress.

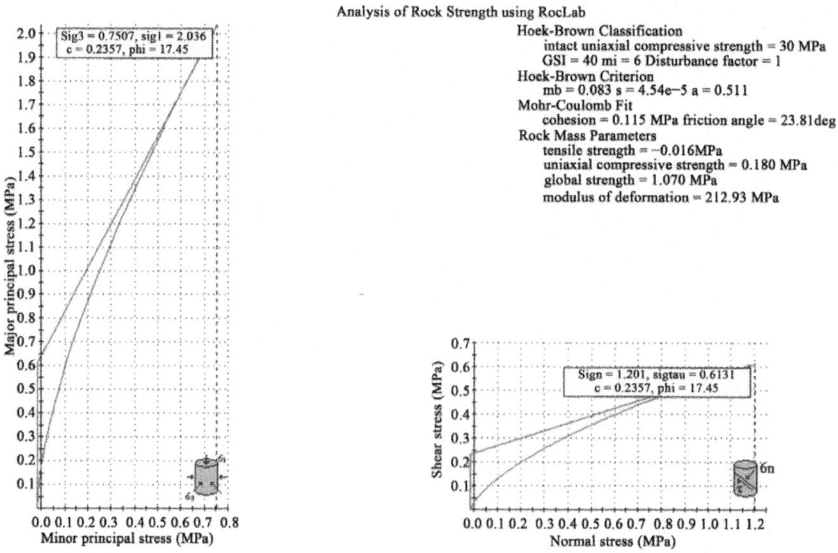

Figure 11: Analysis of output data for determining geomechanical parameters of the rock masses in unit one based on GSI = 40.

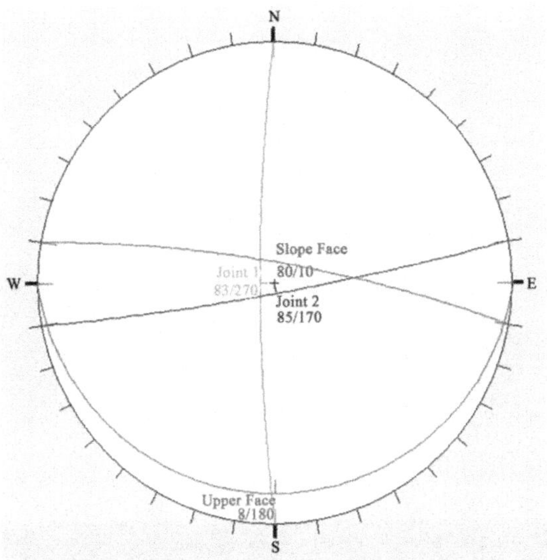

Figure 12: Stereographic images of joint sets.

Figure 13: Top view of wedge slide analysis.

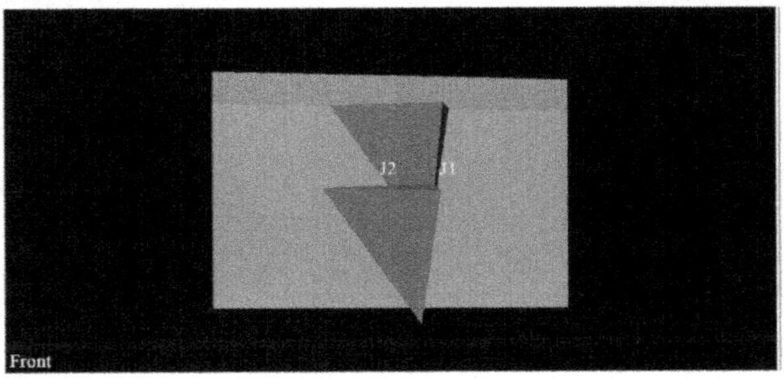

Figure 14: Front view of wedge slide analysis.

Figure 15: Wedge slide: front view.

Figure 16: Side view of wedge slide analysis.

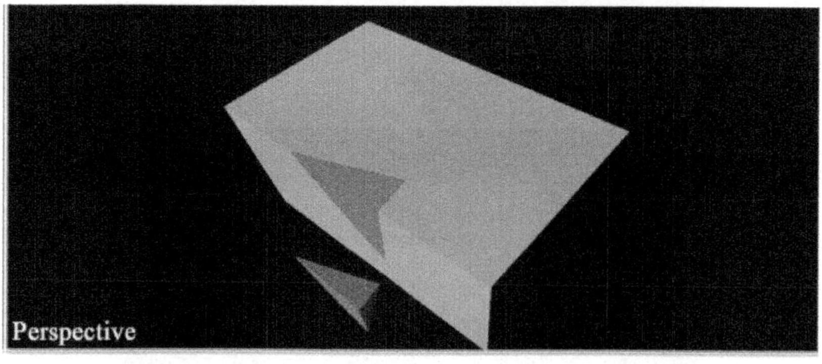

Figure 17: Perspective view of wedge slide analysis.

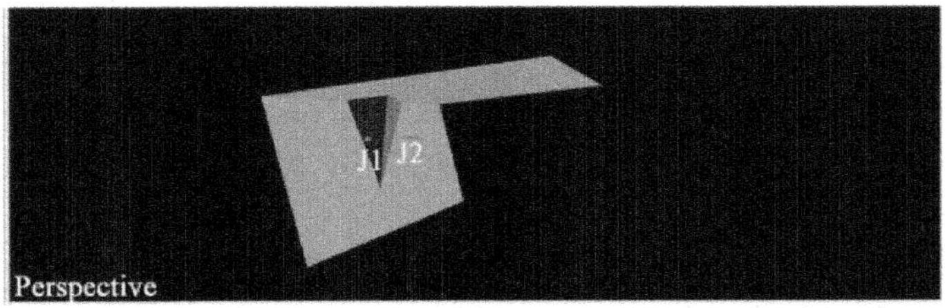

Figure 18: A view of wedge slide analysis obtained by applying directions of water penetration.

Table 5: Geotechnical parameters of the rock masses in unit one of the Atamir Formation K (at) 1

Parameter		Min	Max	Ave
GSI		35	45	40
Intact uniaxial compressive strength (Mpa)		10	50	30
Mi		3	9	6
Material constants	m_b	0.0235	0.923	0.083
	s	2.65	6.73	4.54
	A	0.515	0.502	0.511
Cohesion (Mpa)		0.089	0.431	0.115
Friction angle (Deg)		12.76	62.76	23.81
Rock mass parameter (MPa)	Tensile strength	−0.010	−0.05	−0.016
	Uniaxial compressive strength	0.09	0.98	0.180
	Global strength	0.9	2.1	1.070
	Modulus of deformation	180.97	560.98	212.93

Analysis of Wedge Instability under Wet Conditions in Unit One of the Atamir Formation, K (at) 1

Although water is not directly involved in mass (slope) movement, it is an important factor because:

- Increased water content of slopes caused by rain and snowmelt makes them heavy. Water can penetrate into pores and fractures and replace air and, since water is heavier than air, weights of rocks increase. Because weight is the same as force, and force applies stress on surfaces, increased stress can cause instability.
- Water can alter slope angle (angle of the slope at which the slope is stable).
- Water can be absorbed or repelled by soil mineral substances. Water absorption causes the electrical poles of water molecules to stick to

the surfaces of mineral materials and penetrate into them. Therefore, increased water content makes rocks heavier, leading to reduced rock strength. In general, wet clay soils have less strength compared to dry clay soils. Therefore, water absorption reduces strength and, since clay is the material that fills the spaces between the joints in this unit, increased water content gains more importance.

- Groundwater is present almost everywhere under the surface of the ground. This water fills the empty spaces between rock particles and even the cracks within rocks. Groundwater level changes due to rainfall and rises in wet seasons leading to greater water penetration. In dry seasons, groundwater level falls causing less water penetration. These changes in groundwater levels can be an important factor in slope stability.

If water penetrates through joint surfaces, the factor of safety will become zero (FS = 0) (Figure 19).

Unit Two of the Atamir Formation, K (at) 2 (from 300 + 3 to 500 + 5 Meters)

This lithology includes dark grey sandstones, there are three dominant discontinuity sets in the study region (two joint sets together with bedding), and the discontinuities are random. Table 6and Table 7 shows the characteristics of the discontinuities. It is shown in Figures 20-22.

Figure 19: Water penetration into joint sets in unit one of the Atamir Formation.

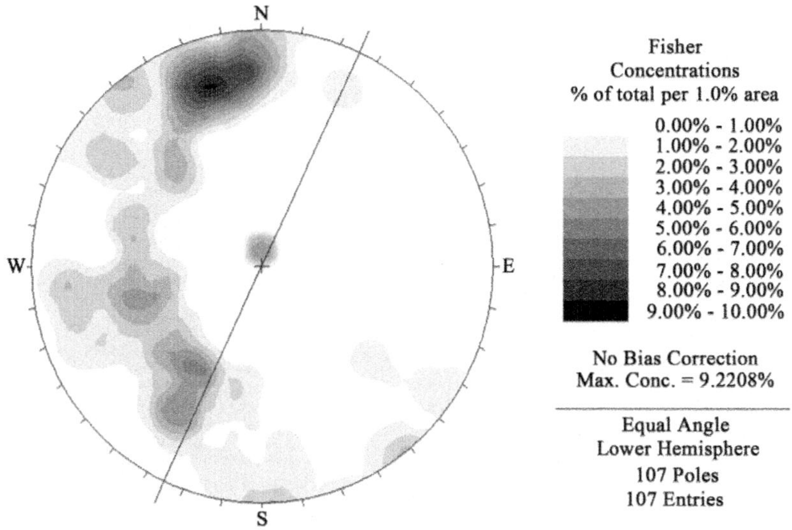

Figure 20: Stereographic image of the discontinuities in unit two of the Atamir Formation K (at) 2.

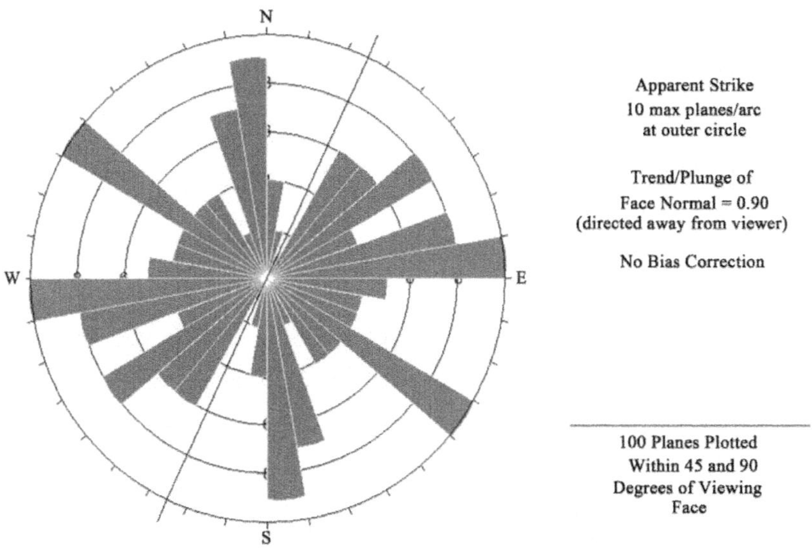

Figure 21: Rose diagram of the discontinuities in unit two of the Atamir Formation K (at) 2.

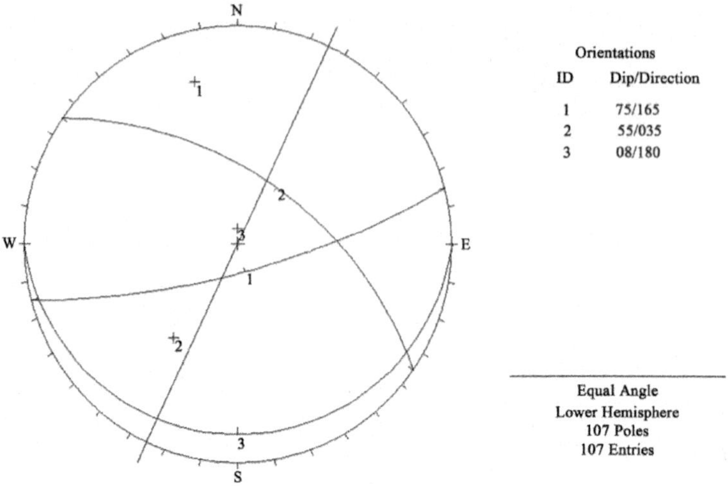

Figure 22: Stereographic image of the discontinuities in unit two of the Atamir Formation K (at) 2.

Table 6: Characteristics of the discontinuities in unit two of the Atamir Formation K (at) 2

Discontinuity	ID	Dip	Dip direction
Js1	1	75	165
Js2	2	55	035
Bedding	3	8	180

Table 7: Geotechnical characteristics of the various discontinuities in the rock masses of unit two in the Atamir Formation K (at) 2

Joint sets	Dip Dip direction		Persistence (m) >20%	10% - 20%	3% - 10%	1% - 3%	<1%	Very rough	Joint roughness (%) Rough	Rough-Smooth	Smooth	Slickenside
Js 1	165\75	60	80	20	-	-	-	-	80	20	-	-
Js 2	035\55	20	30	70	-	-	-	-	30	70	-	-
Bedding	180\8	20	100	-	-	-	-	-	100	100	-	-

Joint alteration	Joint water	Aperture (mm) >5 mm %	1 - 5 mm %	0.1 - 1 mm %	<0.1 mm %	Close	Joint filling Soft Filling mm 5 < %	Soft Filling mm 5 > %	Hard Filling mm 5 > %	Hard Filling mm5< %	Clean %	Joint spacing (m-mm) <60 mm %	60 - 200 mm %	200 - 600 mm %	0.6 - 2 m %	>2 m %
		90	10	-	-	-	-	50	-	50	-	-	-	100	-	-
W1	Dry	60	40	-	-	-	-	-	70	-	30	-	-	70	30	-
		-	100	-	-	-	70	30	-	-	-	-	-	-	-	100

Figures 23-26 show statistical analysis for unit 2.

Tables 8-17 show information of data for the Atamir Formation K (at) 2 like the order of previous zone.

The GSI range of the rock masses in unit two of the Atamir Formation is shown in Figure 27.

Fitting data related to the rock masses in unit two of the Atamir Formation is shown in Figures 28-30.

Stereographic images of joint sets are shown in Figure 31. Wedge slide analysis is also shown in Figures 32-36.

Unit Three of the Atamir Formation, K (at) 3 (from 500 + 3 to 540 + 3)

This lithology consists of dark grey sandstones, and there are four dominant discontinuity classes (three joint classes together with bedding) and random discontinuities.

Stereography of the discontinuities is shown in Figure 37-39.

Statistical analysis of discontinuity from Figures 40-43.

Range of GSI in the rock masses of unit three in the Atamir Formation is shown in Figure 44.

Fitting the data on rock masses in unit three is shown in Figures 45-47.

Stereographic images of the joint sets for unit 3 is shown in Figure 48.

Views of wedge slide analysis are shown in Figures 49-53.

CONCLUSIONS

All three units had very poor to poor rocks that required rapid stabilization at most 10 hours after blasting operations. The following geomechanical and geotechnical characteristics were observed:

- Layered sedimentary rocks sloping towards the outer domain with slopes close or equal to dip slope.
- Joints and foliations that form weak extended surfaces and intersect the domain surface.
- Intersecting joints that cause wedge slide failure.
- Presence of soft, layered, strongly jointed shales having weak layers.
- Saturation or wetting of joint surfaces having clay filling by water from rainfall, snowmelt, changes in groundwater levels, and reduced frictional force (considering the failure mechanism, and ground surveys,

the effects of last spring's rain on the occurrence of this phenomenon cannot be denied).

- Wet and semi-humid climates and successive snowfalls and, in addition, drought and wet periods.
- Formations such as marl and shale ones, or intermittent marl and shale formations, are susceptible to mass movements, and the more unsaturated clay fills the joints, the more susceptible the formation will be to mass movement.
- The tectonic condition (topography) of the dip slope (slopes exceeding 30%), and slope and orientation of the faults, layers, and fractures.
- Failure to start stabilization in time according to the RMR table.

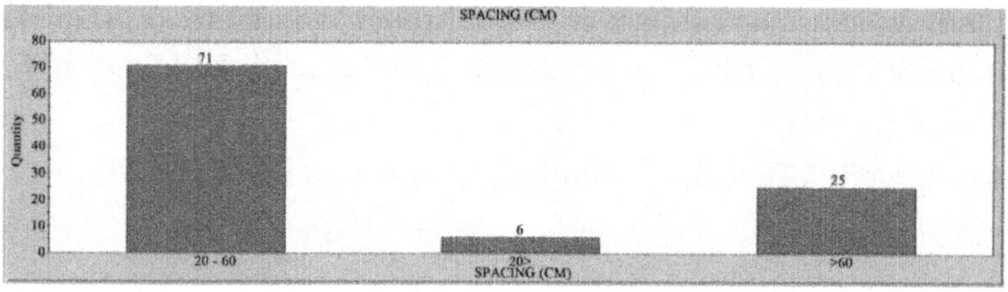

Figure 23. Statistical analysis of discontinuity spacing in unit two of the Atamir Formation K (at) 2.

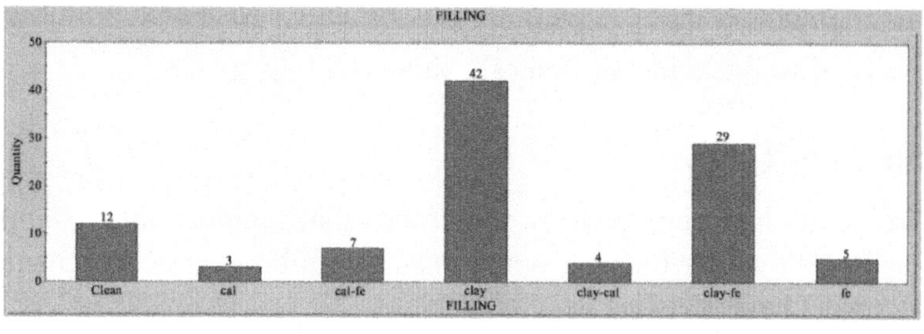

Figure 24. Statistical analysis of discontinuity filling in unit two of the Atamir Formation K (at) 2.

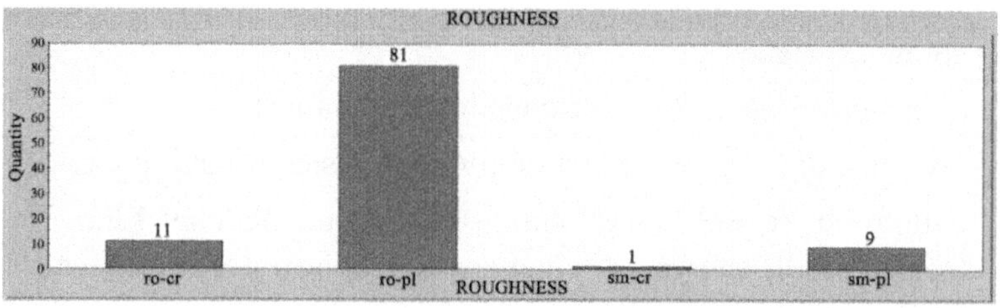

Figure 25. Statistical analysis of discontinuity roughness in unit two of the Atamir Formation K (at) 2.

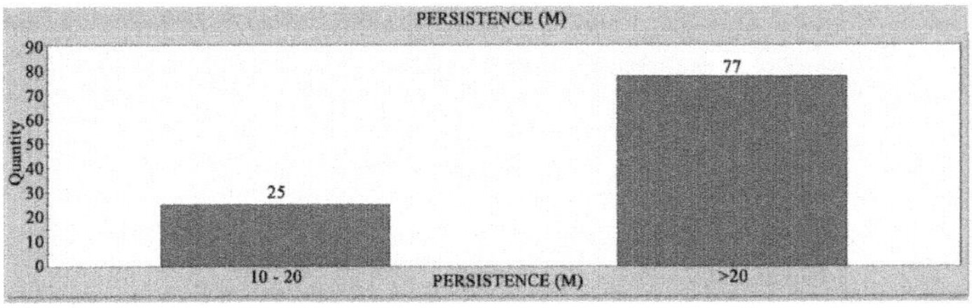

Figure 26. Statistical analysis of discontinuity orientations in unit two of the Atamir Formation K (at) 2.

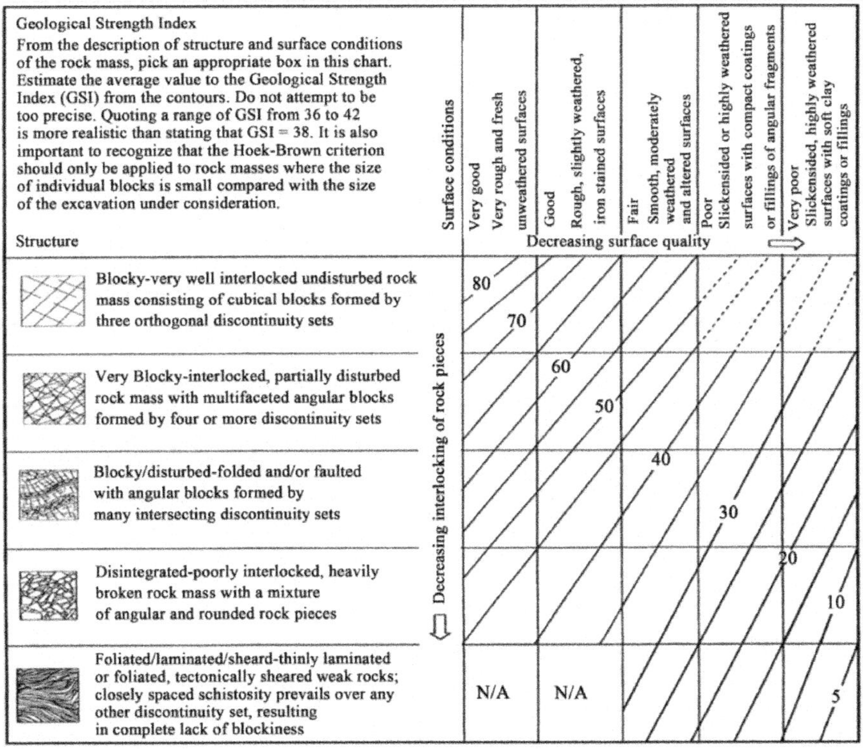

Figure 27. The GSI range of the rock masses in unit two of the Atamir Formation K (at) 2 in the GSI classification diagram.

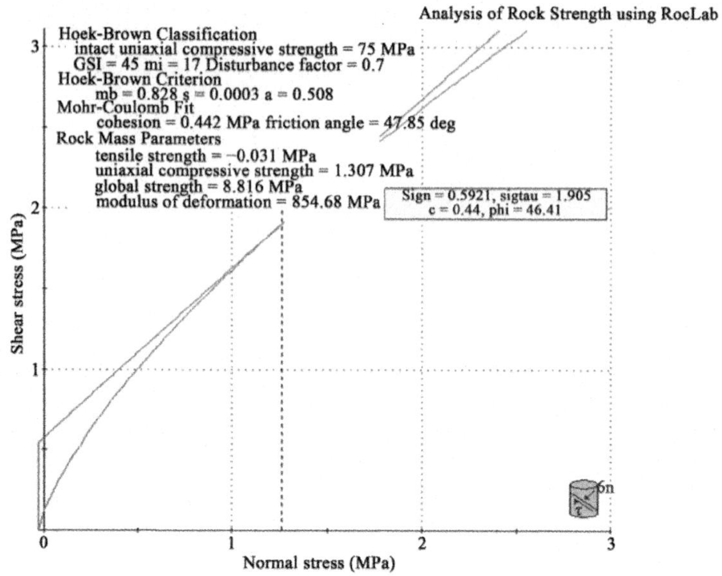

Figure 28. Fitting data related to the rock masses in unit two of the Atamir Formation, K (at) 2, to Hoek-Brown and Mohr- Coulomb criteria.

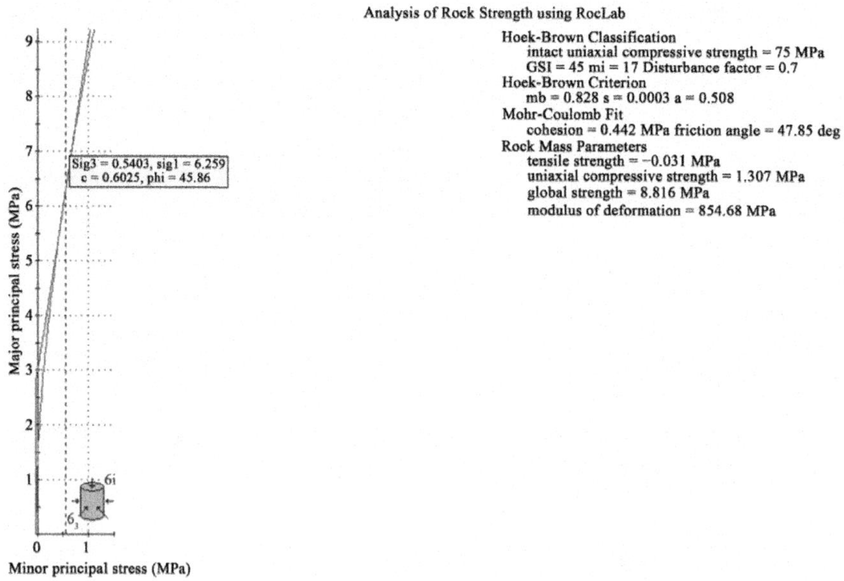

Figure 29. Fitting the data related to the rock masses of unit two to the Hoek-Brown and Mohr-Coulomb criteria based on GSI = 45 and on major principal stress.

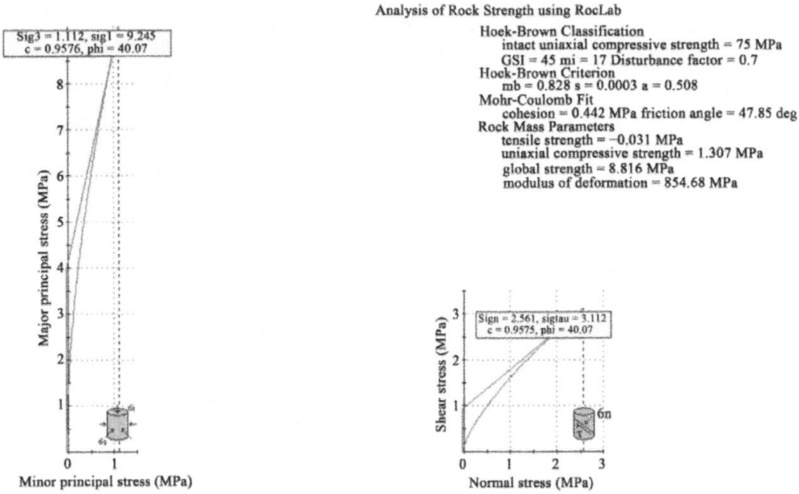

Figure 30. Analysis of output data determining geomechanical parameters of the rock masses in unit two based on GSI = 45.

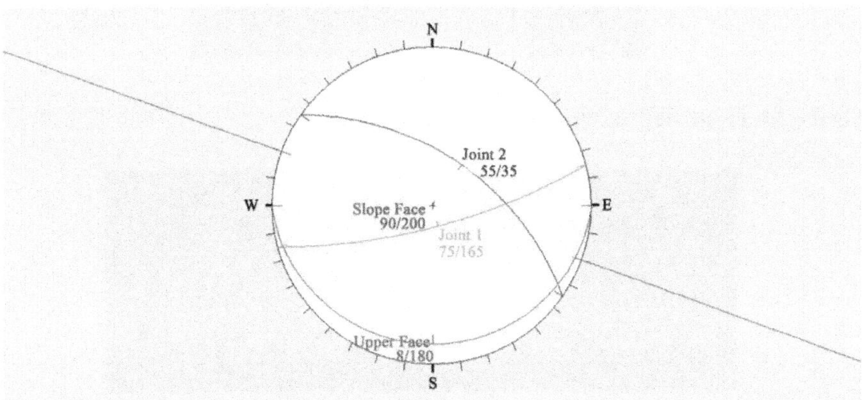

Figure 31. Stereographic images of joint sets.

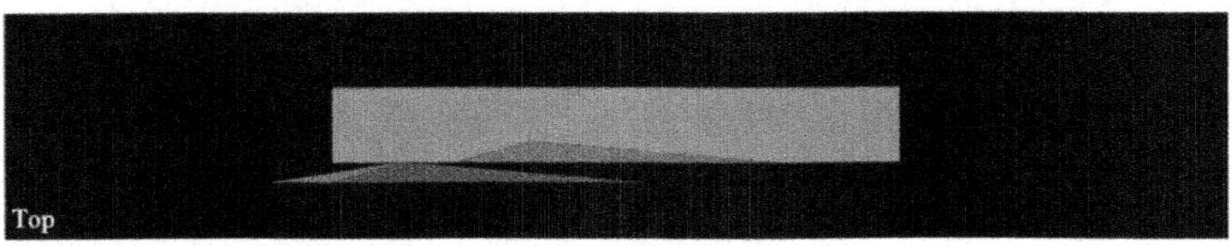

Figure 32. Top view of wedge slide analysis.

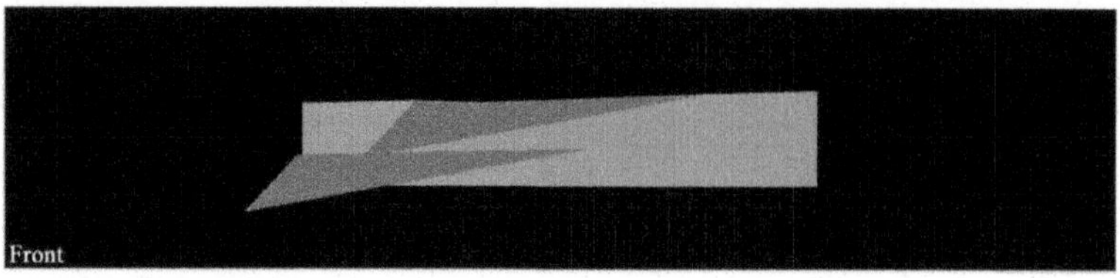

Figure 33. Front view of wedge slide analysis.

Figure 34. Front view of wedge slide.

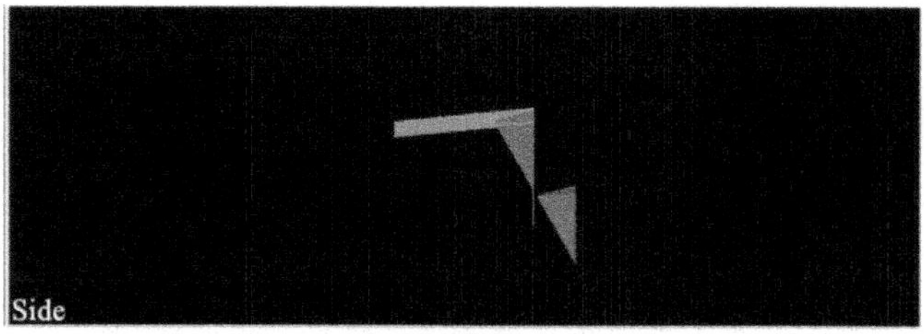

Figure 35. Side view of wedge slide analysis.

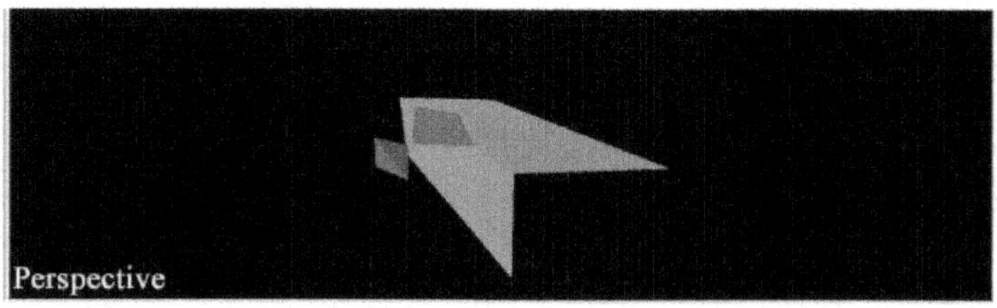

Figure 36. Perspective view of wedge slide analysis.

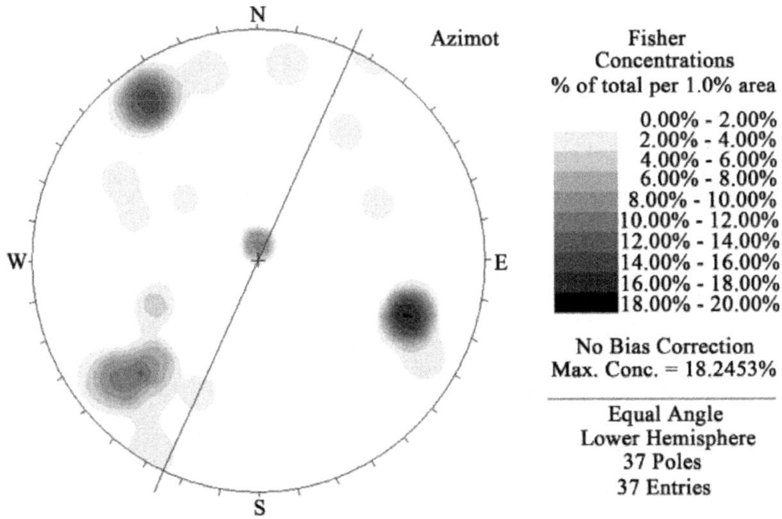

Figure 37. Stereographic image of the discontinuities in unit three of the Atamir Formation K (at) 3.

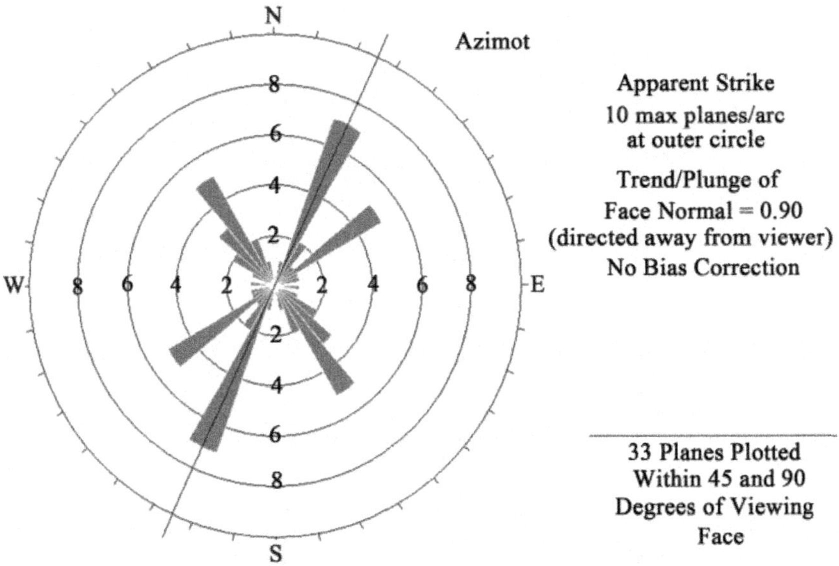

Figure 38. Rose diagram of the discontinuities in unit three of the Atamir Formation K (at) 3.

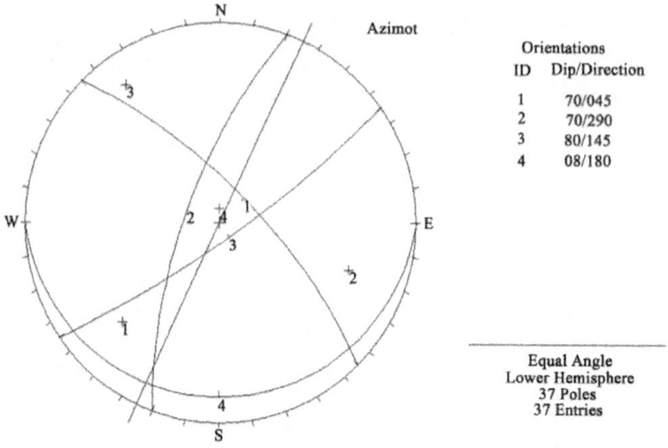

Figure 39. Stereography of the discontinuities in unit three of the Atamir Formation K (at) 3.

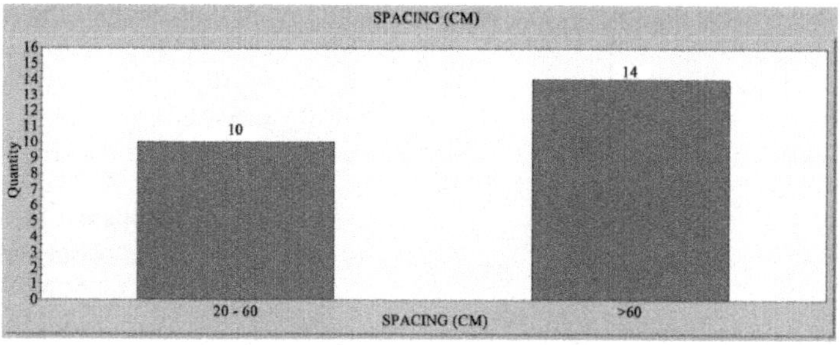

Figure 40. Statistical analysis of discontinuity spacings in unit three of the Atamir Formation K (at) 3.

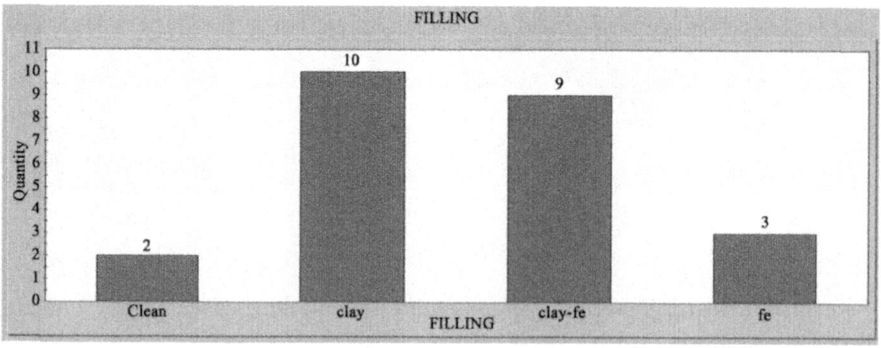

Figure 41. Statistical analysis of discontinuity filling in unit three of the Atamir Formation K (at) 3.

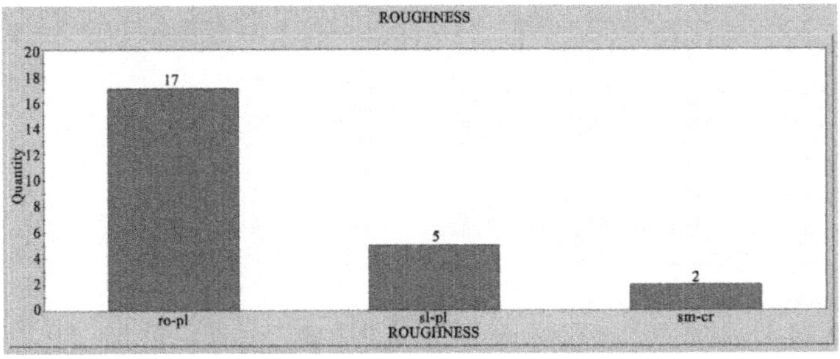

Figure 42. Statistical analysis of discontinuity roughness in unit three of the Atamir Formation K (at) 3.

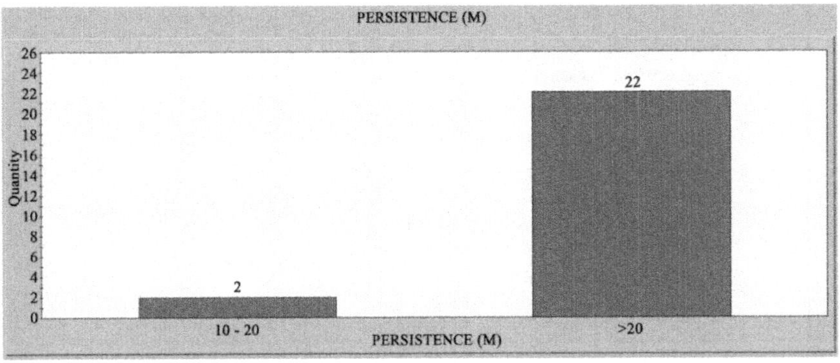

Figure 43. Statistical analysis of discontinuity orientation in unit three of the Atamir Formation K (at) 3.

Table 8. Geomechanical classification of the rock masses in unit two of the Atamir Formation K (at) 2

Parameter	Value	Rating
UCS (MPa)	40 - 60	4 - 6
RQD	25 - 50	8
Spacing (MM)	60 - 200	8 - 10
Condition of discontinuities	-	10
Groundwater	Dam	7
RMR	37 - 41	
Class No.	Dam	
Cohesion of the rock mass (KPa)	442	
Friction angle of the rock mass (deg)	15 - 25	

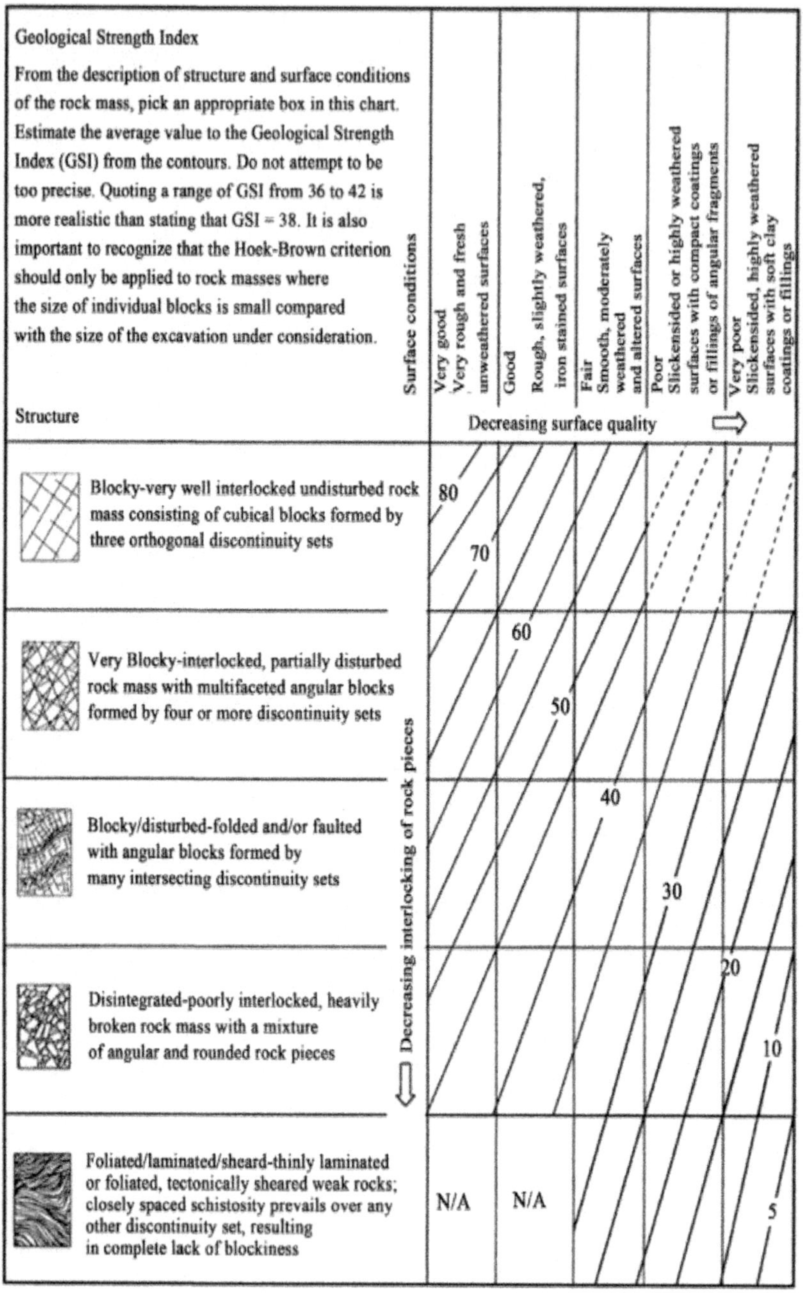

Figure 44. Range of GSI in the rock masses of unit three in the Atamir Formation K (at) 3.

Table 9. The Q index of the rock masses in unit two of the Atamir Formation K (at) 2

Parameter	Description	Value
RQD	40 - 50	40 - 45
J_n	Three joint sets with random to four joint sets	12
J_r	Smooth to planar	1
J_a	Softening or low friction clay mineral coatings	4
J_w	Dry to medium inflow of pressure	1 - 0.66
SRF	Medium stress, favorable stress condition	1
Q		0.62 - 0.83
Class No.		Very poor rock

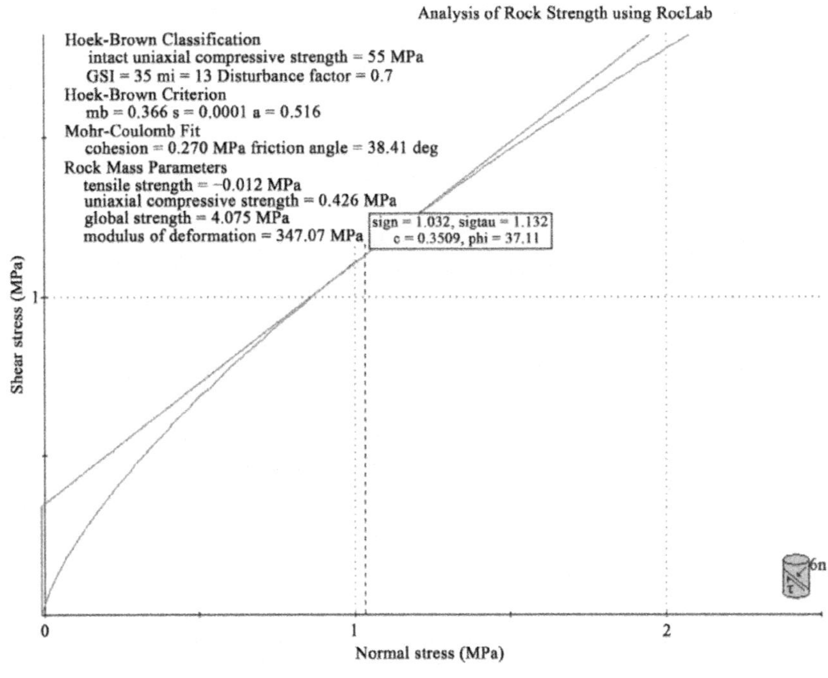

Figure 45. Fitting the data on rock masses in unit three to the Hoek-Brown and Mohr-Coulomb criteria, based on GSI = 35 and on normal shear stress.

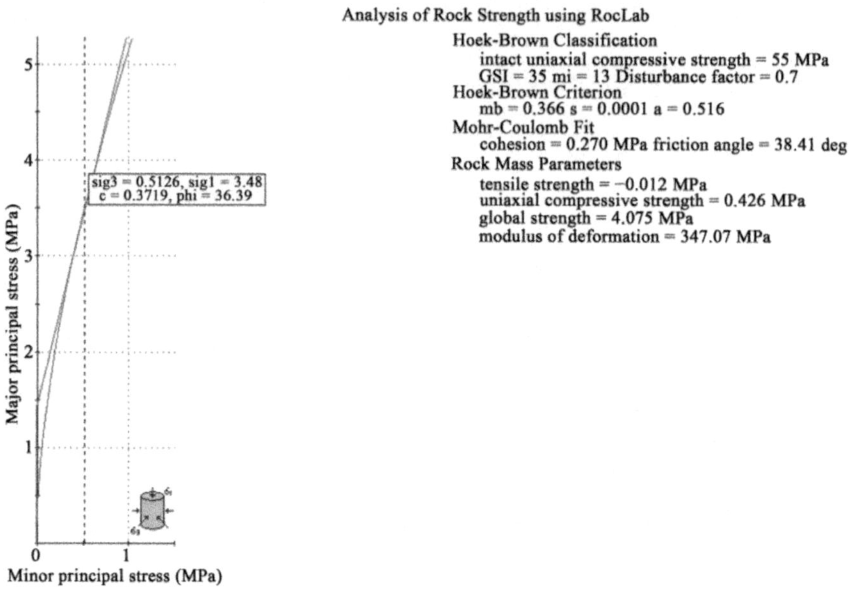

Figure 46. Fitting the data on rock masses in unit three to the Hoek-Brown and Mohr-Coulomb criteria, based on GSI = 35 and on major principal stress.

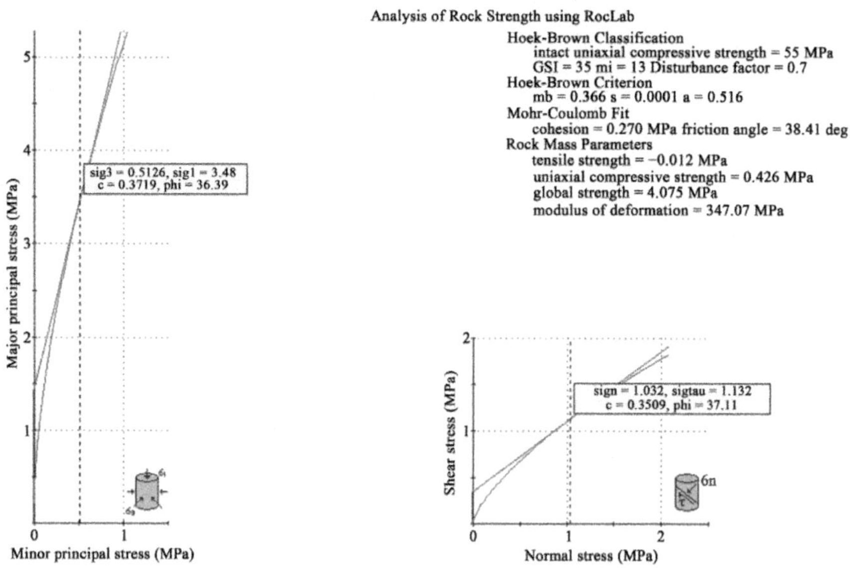

Figure 47. Analysis of output data determining the geomechanical parameters of rock masses in unit three, based on GSI = 35.

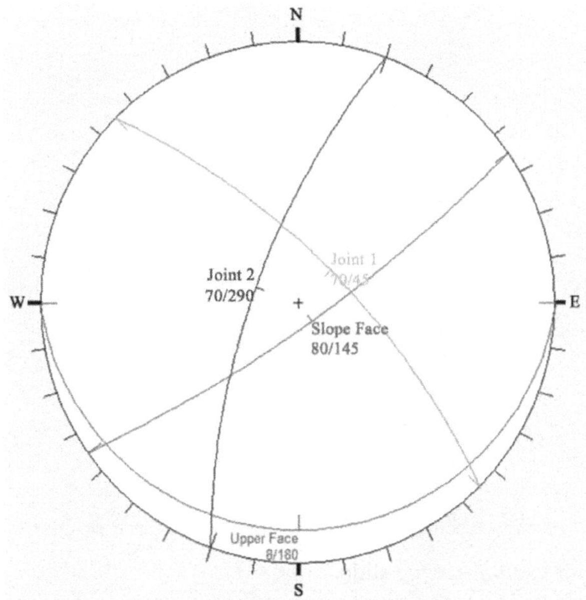

Figure 48. Stereographic images of the joint sets.

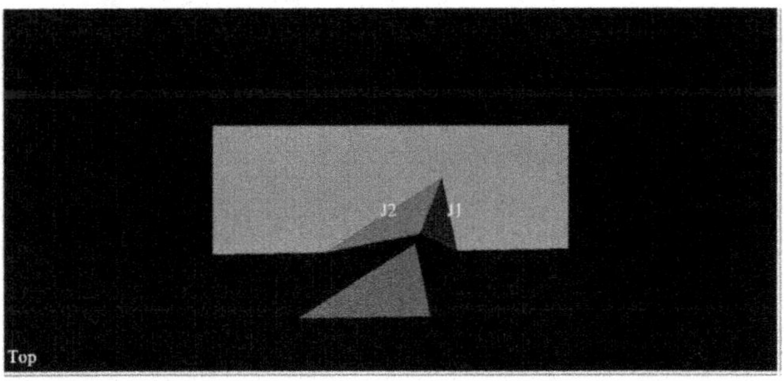

Figure 49. Top view of wedge slide analysis.

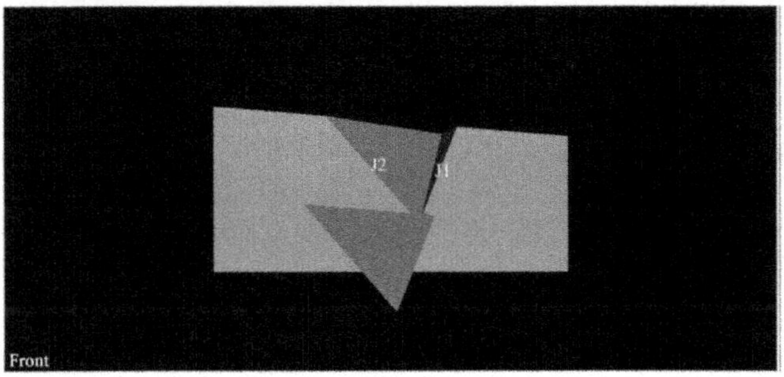

Figure 50. Front view of wedge slide analysis.

Figure 51. Front view of wedge slide.

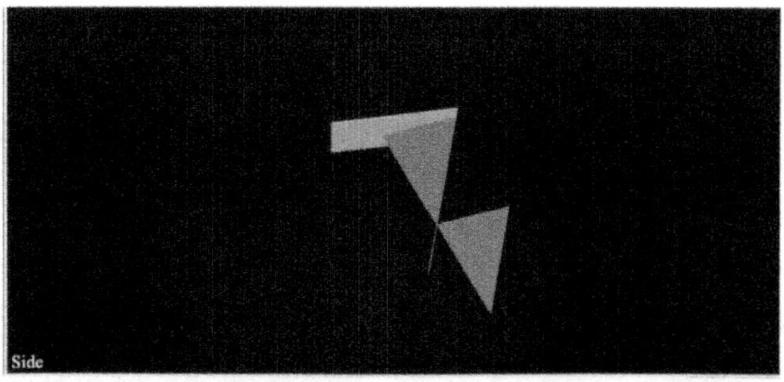

Figure 52. Side view of wedge slide analysis.

Figure 53. Perspective view of wedge slide analysis.

Table 10. Geomechanical parameters of the rock masses in unit two of the Atamir Formation K (at) 2

Parameter		Min	Max	Ave
GSI		40	50	45
Intact uniaxial compressive strength (Mpa)		50	100	75
Mi		11	23	17
Material constants	m_b	0.154	1.754	0.828
	S	0.0001	0.0014	0.0003
	A	0.478	0.612	0.508
Cohesion (Mpa)		0.210	0.501	0.442
Friction angle (deg)		62.87	90.63	47.85
Rock mass parameter (Mpa)	Tensile strength	−0.019	−0.08	−0.031
	Uniaxial compressive strength	0.096	2.68	1.307
	Global strength	6.98	10.67	8.816
	Modulus of deformation	600.75	1000.5	854.68

Table 11. Characteristics of the discontinuities in unit three of the Atamir Formation K (at) 3

Discontinuity	ID	Dip	Dip direction
Js1	1	70	45
Js2	2	70	290
Js3	3	80	145
bedding	4	8	180

Table 12. Geotechnical characteristics of various discontinuities in rock masses of unit three in the Atamir Formation K (at) 3

Joint sets	Dip Dip direction		Persistence (m)					Joint roughness (%)				
			>20%	10% - 20%	3% - 10%	1% - 3%	<1%	Very rough	Rough	Rough-Smooth	Smooth	Slickenside
Js1	45\70	50	100	-	-	-	-		100	-	-	-
Hs2	290\70	25	70	30	-	-	-	-	20	80	-	-
Js3	145\80	10	-	100	-	-	-	-	-	100	-	-
Bedding	180\8	15	100	-	-	-	-	-	100	-	-	-

Joint alteration	Joint water	Aperture (mm)					Joint filling					Joint spacing (m-mm)				
		>5 mm %	1 - 5 mm %	0.1 - 1 mm %	<0.1 mm %	Close	Soft filling mm 5 < %	Soft filling mm 5 > %	Hard filling mm 5 > %	Hard filling mm 5 < %	Clean %	<60 mm %	60 - 200 mm %	200 - 600 mm %	0.6 - 2 m %	>2 m %
		-	100	-	-	-	-	50	-	50	-	-	-	-	100	-
W1	Dry	100	-				100	-	-	-	-	-	-	100	-	-
		40	60	-	-	-	-	-	70	-	30	-	-	30	70	-
		-	100	-	-	-	70	30	-	-	-	-	-	-	-	100

Table 13. Geomechanical classification of rock masses in unit three of the Atamir Formation K (at) 2

Parameter	Value	Rating
UCS (Mpa)	40 - 60	4 - 6
RQD	25 - 50	8
Spacing (mm)	>60 - 200	8
Condition of discontinuities		10
Groundwater	Dam	7
RMR	35 - 39	
Class No.	Dam	
Cohesion of the rock mass (KPa)	190	
Friction angle of the rock mass (Deg)	15 - 25	

Table 14. The Q index of rock masses in unit three of the Atamir Formation K (at) 3

Parameter	Description	Value
RQD	37 - 43	37 - 43
J_n	Three joint sets with random to four joint sets	12
J_r	Smooth to planar	1
J_a	Softening or low friction clay mineral coatings	4
J_w	Dry to medium inflow or pressure	1 - 0.66
SRF	Medium stress, favorable stress condition	1
Q	0.51 - 0.89	
Class No.	Very poor rock	

Table 15. Geomechanical parameters of the rock masses in unit three of the Atamir formation K (at) 3

Parameter		Min	Max	Ave
GSI		30	40	35
Intact uniaxial compressive strength (MPa)		30	80	55
Mi		9	17	13
Material constants	m_b	0.457	0.754	0.366
	S	0.0001	0.0009	0.0001
	A	0.489	0.654	0.516
Cohesion (Mpa)		0.190	0.451	0.270
Friction angle (Deg)		19.05	50.63	38.41
Rock mass parameter (MPa)	Tensile strength	−0.021	−0.09	−0.012
	Uniaxial compressive strength	0.023	1.98	0.426
	Global strength	1.65	9.76	4.075
	Modulus of deformation	600.75	1000.5	347.07

Table 16. Geomechanical parameters of the rock masses in the three units of the Atamir Formation

Unit	RMR	Q	GSI	Factor of safety (FS) of the wedge
Unit one of Atamir Formation K (at) 1	34 - 36 Poor	0.592 - 0.66 Very poor	35 - 45 Poor	0.037
Unit two of the Atamir Formation K(at) 2	37 - 41 Poor	0.62 - 0.83 Very poor	40 - 50 Poor	0
Unit three of the Atamir Formation K (at) 3	35 - 39 poor	0.51 - 0.89 Very poor	30 - 40 Poor	0

Table 17. Geomechanical parameters of the rock masses in unit three of the Atamir Formation K (at) 3

Set number	Characteristics and parameters of rock mass	Rating rock masses				
		81 - 100 (I)	61 - 80 (II)	41 - 60 (III)	21 - 40 (IV)	<20 (V)
1	Rock classification	Very good	Good	Relatively good	Poor	Very poor
2	Average time of self supporting capacity	10 years for active opening of 15 meters	6 months for active opening of 8 meters	1 week for active opening of 5 meters	10 hours for active opening of 2.5 meters	30 minutes for active opening of 1 meter
3	Cohesion of rock mass (Mpa)	>0.4	0.3 - 0.4	0.2 - 0.3	0.1 - 0.2	<0.1
4	Angle of internal friction	>45 degrees	35 - 45 degrees	25 - 35 degrees	15 - 25 degrees	<15 degrees
5	Load bearing capacity (T/M^2)	440 - 600	280 - 440	135 - 280	45 - 135	30 - 45

Therefore, rockslide has been unavoidable in rock masses of all three units. The point of great importance is the mass movement of the rocks towards the road and channel that has blocked them and will cause floods during future rainfalls. Based on observations made, the process of landslide under discussion has not ended, and it seems that a long time has to pass (from several months to several years) for stability to be established.

Condition of the rock masses (Table 16).

Considering the RMR stability table (Table 17).

REFERENCES

1. Feenstra, R. and Wickham, J. (1975) Evolution of Folds around Broken Bow Uplift, Ouachita Mountains, Southeastern Oklahoma. AAPG Bulletin, 59, 974-985.
2. Bieniawski, Z.T. (1973) Engineering Classification of Jointed Rock Masses. Civil Engineer in South Africa, 15, 353.
3. Cai, M., Kaiser, P.K., Uno, H., Tasaka, Y. and Minami, M. (2004) Estimation of Rock Mass Strength and Deformation Modulus of Jointed Hard Rock Masses Using the GSI System. International Journal of Rock Mechanics and Mining Sciences, 41, 3-19.http://dx.doi.org/10.1016/S1365-1609(03)00025-X

Chapter 2

GROUNDWATER RISING AS ENVIRONMENTAL PROBLEM, CAUSES AND SOLUTIONS: CASE STUDY FROM ASWAN CITY, UPPER EGYPT

Sayed A. Selim[1], Ali M. Hamdan[1], Ahmed Abdel Rady[2]

[1]Geology Department, Faculty of Science, Aswan University, Aswan, Egypt

[2]Aswan Water and Sanitation Company, Aswan, Egypt

ABSTRACT

This paper examines the rise in the level of the groundwater in the Quaternary aquifer at Aswan city, Upper Egypt. Since the 1960's, the areal extent of Aswan City and the urban populations are growing at a high pace which introduces new sources of water that increase groundwater recharge. As a result of leakages or infiltrations from different sources, the natural groundwater balance is overturned into an unbalance where the input to water table is comparatively much more than the natural groundwater flow towards the Nile River. The present study shows a variation in the groundwater level, from 1971 up to 2014, where the water table rising ranges between 12.55 and 13.69 m. Also, it shows an abrupt increase in the water levels in 2010 continuing up till now. The groundwater rising phenomena that happened in 2010 can be directly refereed to the cessation of groundwater pumping from El-Shallal wells, and to the reduction of pumping from KIMA factory wells. Generally, the rate of water rising is much higher in the western side of the city and in Kima factory area, where they are characterized by low relief and dense population. The most troublesome groundwater mounds under urban areas are likely to develop in low-lying areas of relatively high permeability aquifer, which is not exploited for water supply. These damages will become more widespread if the rising groundwater table remains uncontrolled. The environmental impact of the water rising includes: forming ponds in low lying areas (Kima and El Shallal ponds), flooding building's basements, and inundating underground infrastructure. A general deterioration in groundwater quality was identified.

INTRODUCTION

The area under investigation (Aswan City) is situated along the Nile Valley in the southern part of Egypt. It is located adjacent to the east bank of the Nile River, bounded by latitudes 24°01'30" and 24°06'30"N and longitudes 32°52' and 32°56'E (Figure 1).

Over last few years, a rise in groundwater levels has been observed in several parts. Raising groundwater levels have the potential to inundate underground infrastructure, flooding basements and submerging sewer pipes and utility lines that deliver water and electricity. The magnitude of anthropic impacts upon their environment makes humans the major geologic agent on the surface of the planet [1] .

This work aims to review the causes of changes in groundwater levels in Aswan region which is under the subtropical arid region conditions. The geo-environmental impact of the rising water-level and their consequences on the built environment will be considered.

Over the last few years, a steady rise in groundwater levels has been observed in several parts of Aswan City. This reflect environmental problems existing in many areas of the city, where it creates swamps and ponds and affects the basement of many buildings as it is shown in El-Seil, KhorAwada, Phatemic graves, El-Aqad buildings, Blood Bank, Military building, El Shallal and KIMA factory area (Figure 1 and Figure 2). Rising groundwater levels are expected to be a chronic problem and will likely be a major issue for residential areas of Aswan city.

STUDY AREA

Geomorphologically, Aswan City area can be divided into three main units: low elevated lands (Aswan plain); high elevated lands; and the Nile River channel (Figure 1).

1) Low lands (Aswan plains): are underlain by the urban areas of Aswan city and can be divided into two main parts. The western plain that extends along El Sadat road up to Atlas area northward, while the eastern plain extends along El Samad road (Figure 1). These two plains were the location of the old channel of the Nile River and their width determined the original width of the old Nile gorge. The maximum width of these plains is encountered near the center of the town (about 3.5 Km), and the minimum one is along El Sadat road (less than 1 kilometer). Generally, these plains are characterized by great variation in ground elevation and show a general slope from south to north (Figure 3).

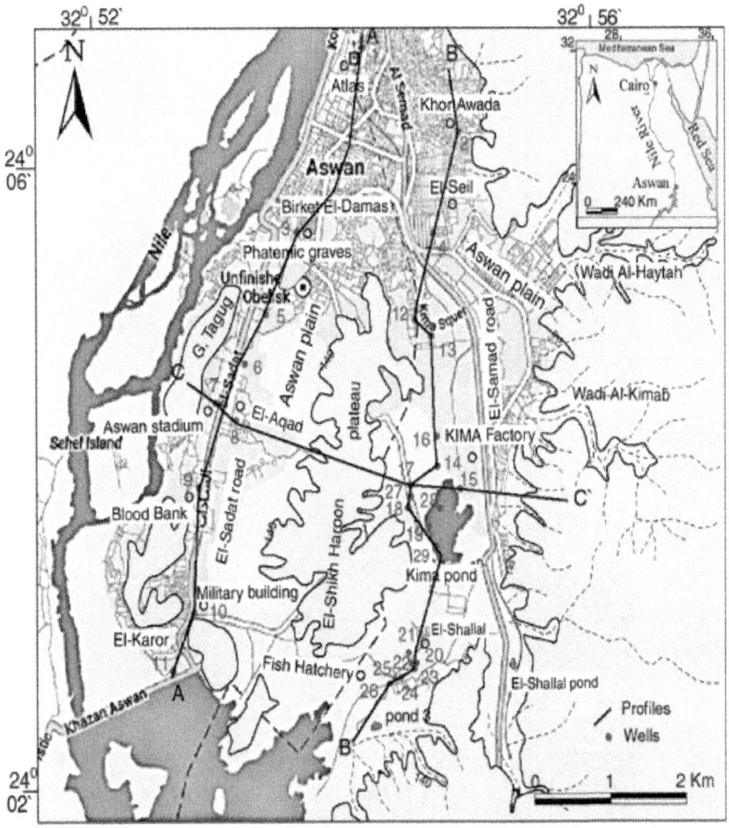

Figure 1: Location map of drilled wells and profiles in the study area.

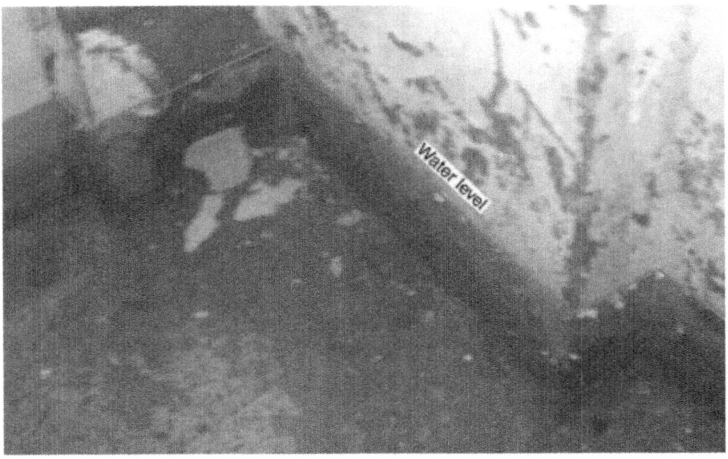

(a)

(b)

(c)

(d)

Figure 2: Rising groundwater level affecting areas: (a) KhorAwada, (b) El-Seil, (c) Phatemic graves area, and (d) Aswan stadium.

2) The highlands: the more extensive one bounded the study area from its eastern side (elevations vary from 164 to 188 m). Few wadis incise this highland and directed from east to west, W. El-Keimab and W. Al-Haytah (Figure 1). The second main highland occupies the central parts of the study area and represented by El-Shikh Haroon plateau (elevations vary from 142 to 167 m). Other small highlands are located to west of the study area close to the Nile River as Gebel Tagug (150 m) and El Karor (135 m). The high land areas are underlain either by basement rocks and or by Nubian sandstones.

3) The Nile River Course forms the western limit of Aswan City area (the approximately water level is +85 m). Aswan Dam (Khazan Aswan) with its water reservoir is situated to the south of Aswan city where the water level ranges between 106 and 118 m.

The general geology of Aswan area is relatively well known and has been studied by several investigators [2] - [10] .

The following lithological units were exposed in the study area (Figure 4).

- The basement rocks represent the oldest rocks (Precambrian age) and mainly exposed in the many parts of the study area (Figure 4). They are characterize by the presence of many fissures and joints, and can play an important role in accumulation of water in many parts of the study area.
- The Nubian Sandstone rocks of Cambrian to cretaceous age [11] are exposed in the central and eastern parts of the study area.

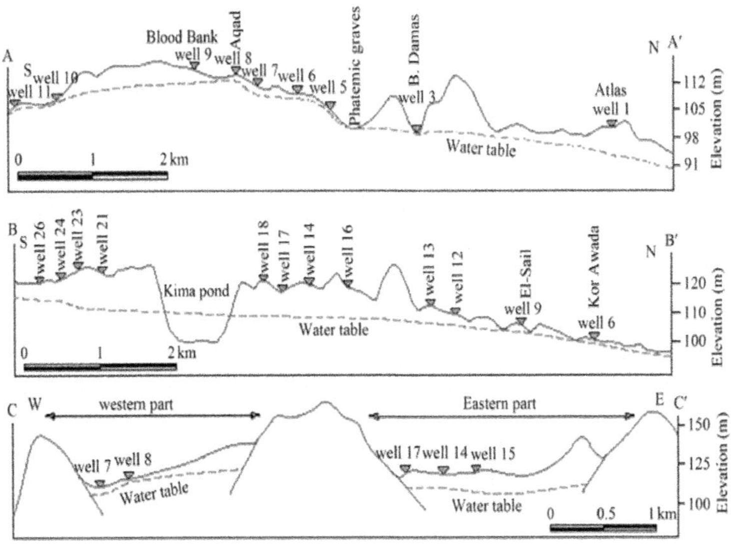

Figure 3: Hydrogeological profiles along the study area (2014).

It exposed at high lands in the central part (El Shikh Haroon plateau) and the eastern part and unconformably lies over the basement complex (Figure 4). It is made of ferruginous sandstone, sandstone and clays and composed into three Formations from base to top, Abu Agag Formation; Timsah Fomation; and Um Barmile Formation and has a total thickness ranges between 20 and 85 meters [12] .

- The Quaternary sediments are represented by sands, gravels and clays of the Pleistocene time, and by mud and Aeolian sediments of the recent. They are underlain by a thick bed of Pliocene clays [13] . These sediments are well exposed in the area through El-Samad and El Sadat roads with a thickness reached up to 248 m. The Quaternary gravels are composed of coarse well-rounded pebbles, ranging from 15 to 20 centimeters in diameter. The recent deposits comprise a small portion of the surface of the study area and are represented by alluvial and Aeolian deposits. The alluvial deposits, sand and mud form the cultivated land in the northern part of the study area. The Aeolian deposits are represented by blown-sand.

Hydrogeologicaly, few has been published on the hydrogeology of Aswan city area due to the limited amount that groundwater used for public water supply. The Quaternary sediments represent the main aquifer in the studied area and outcropped in two geomorphologic low lands (Aswan plains) which will be outlined in the following.

At the western plain (along El Sadat road): The quaternary aquifer is consists of unconsolidated sediments of sands, gravels, and clays enriched of smectite minerals group of the Pleistocene time. These sediments are unconformably overlies the basement rocks and fill the floor of the low lands with a thickness vary from few meters to less than 20 m. basement rocks outcrops in many parts along the western part of the study area as in El Karor and G. Tagug.

The eastern border of the aquifer is El-Shikh Haroon plateau (basement rocks caped by The Nubian Sandstone rocks), but westward is bounded by G. Tagog and El Karorplateaux. The width of the aquifer is relatively small and varies from 1.5 km (at Blood Bank) to 1.67 km (at El-Aqad). It reached 4.52 km at the central part of the city and 1.13 km at the northern extreme (Atlas).

At the eastern plain (along El-Samad road): The quaternary aquifer in the eastern part of the study area studied by [14] [15] . They concluded that the Quaternary aquifer in the eastern part is mainly composed of unconsolidated material of sands, gravels, and clays intercalation. It is not covered by impermeable layers in the major part of the area; therefore, it is under unconfined condition. They stated that the transmissivity (T) values of the Quaternary aquifer range between 1996.4 and 3029 m^2/day which means that the aquifer is of high potential class (more than 500 m^2/day) according to the classification of [16] .

The thickness of the Quaternary aquifer vary from 100 m in the southern part to 137 m in the central part and it underlined by the Pliocene clay, as detected from the lithological log of deep well No. 27 (Figure 5). The east- ern border of the quaternary water-bearing sediments represented by the basement rocks which caped with Nubian sandstone, while westward is bounded by high land of El-Shikh Haroon plateau.

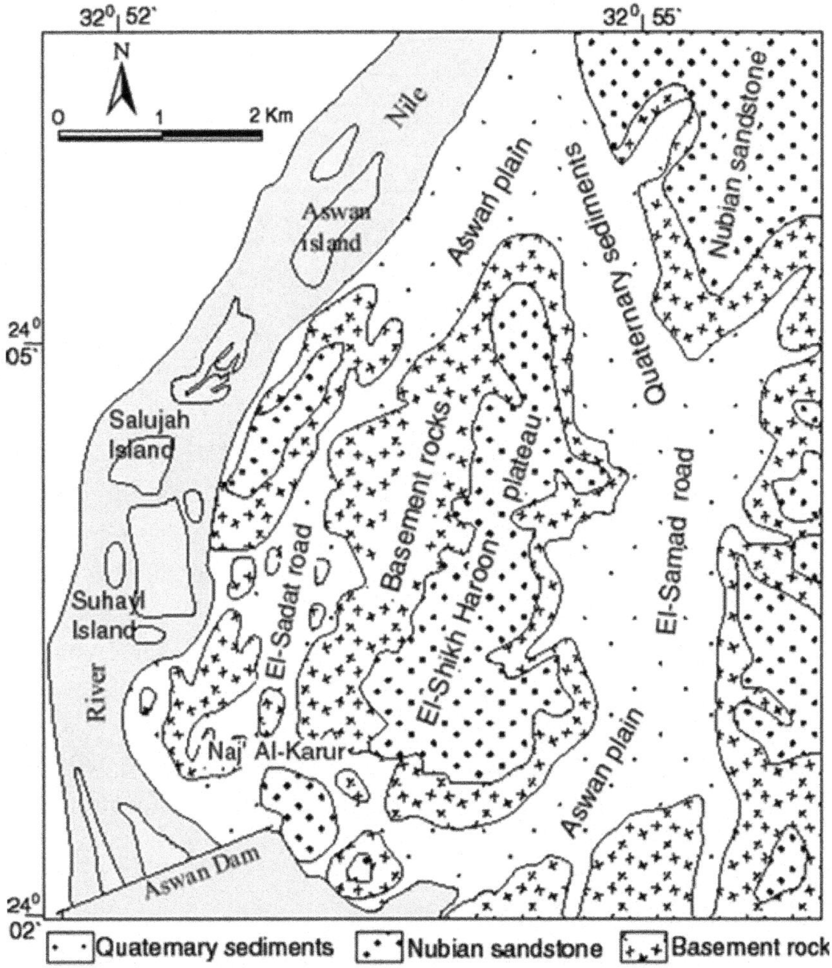

Figure 4: Geological map of the study area.

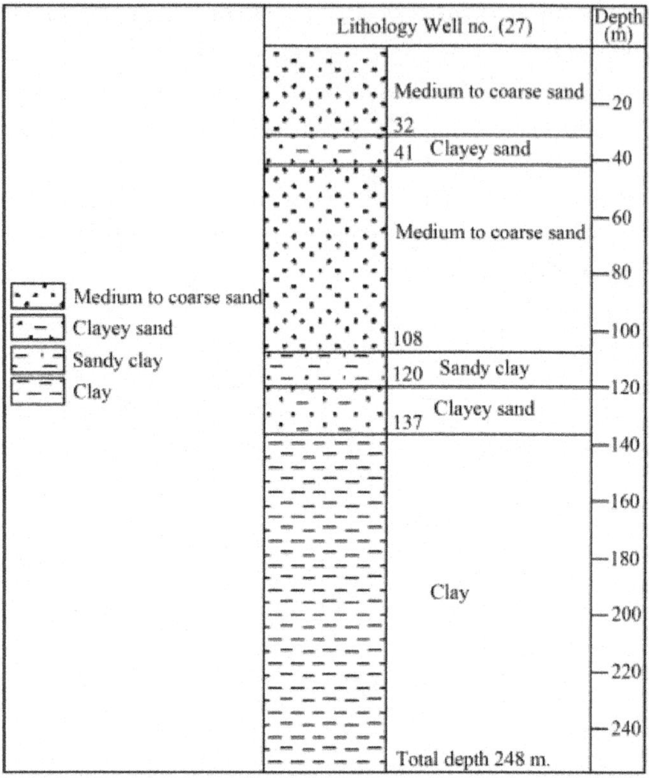

Figure 5: Lithological log of well No. 27.

The width of the aquifer is small and ranges between 2.05 km northward and 1.37 km southward.

The natural recharge to the Quaternary aquifer is from the Aswan Dam Lake, which has water level ranges between 108 and 118 m asl, towards the urban areas of the city.

The aquifer discharged through the natural groundwater outflow to the Nile River, in the northern part of the city in addition to groundwater abstraction by pumping at Kima and El Shallal areas.

MATERIALS AND METHODS

During this study, the data recorded from 1971 to 2014, from six wells at Kima area are evaluated to determine the general trend in groundwater levels and the change in the water-table over a 43-year period (Table 1 and Figure 6). These data also helped to identify general and local groundwater flow directions, where the groundwater level contour map is prepared (Figure 7).

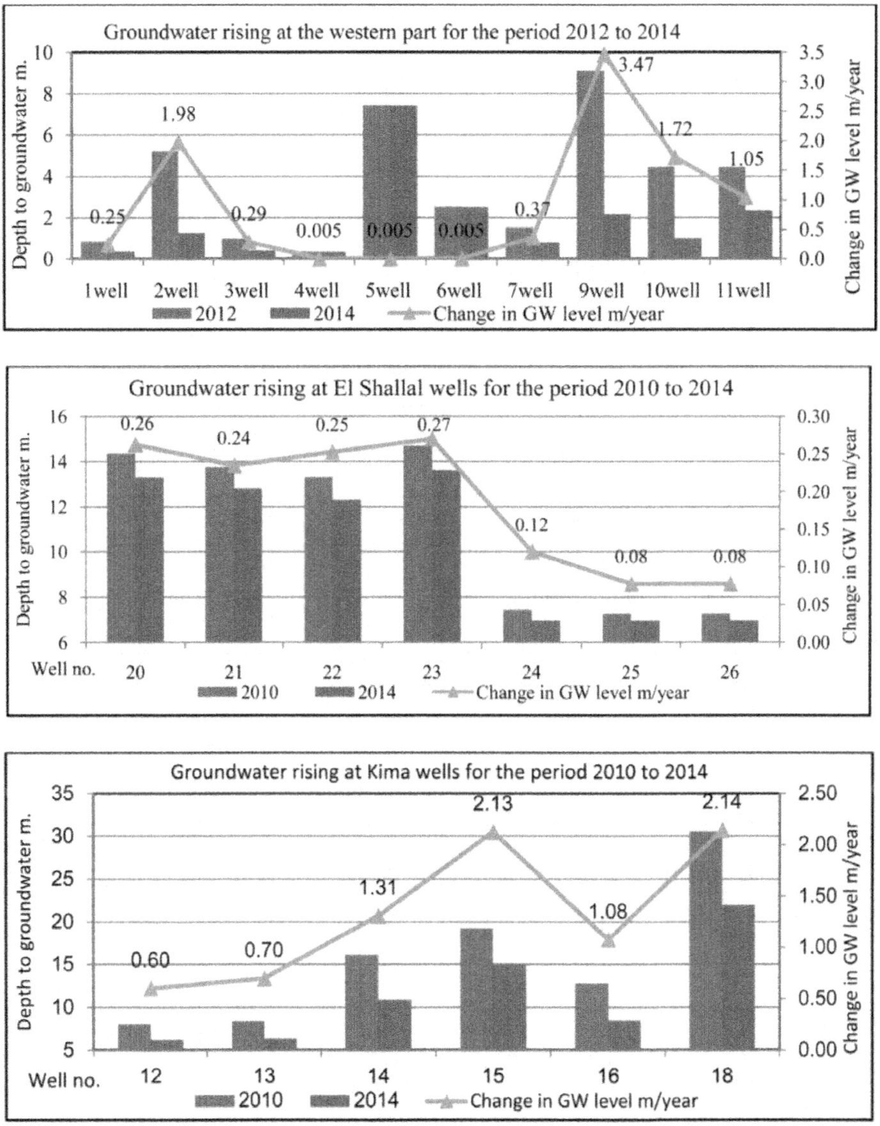

Figure 6: Groundwater rising and changes in GW level per year in the study area.

Table 1: The hydrogeological data of drilled wells of the study area

Date	Water depth	Water Level	Date	Water depth	Water Level	Date	Water depth	Water Level
Well No. 1			Well No. 26			Well No. 13 (continue)		
1/6/2012	0.85	101.86	18/7/2010	7.26	113.74	20/10/2012	6.49	114.51
22/2/2014	0.36	102.35	15/1/2014	6.95	114.05	3/11/2012	6.42	114.58
Well No. 2			Well No. 12			1/12/2012	6.37	114.63
1/6/2012	5.21	104.74	17/10/1971	19.0	91.0	21/1/2013	6.42	114.58
22/2/2014	1.25	108.70	26/6/1975	18.75	91.25	16/2/2013	6.4	114.6
Well No. 3			16/1/1978	15.5	94.5	16/3/2013	6.36	114.64
1/6/2012	0.98	99.01	14/8/1979	15.9	94.1	13/4/2013	6.33	114.67
22/2/2014	0.4	99.59	5/1/1980	15.4	94.6	25/5/2013	6.31	114.69
Well No. 4			18/11/1981	16.1	93.9	Well No. 14		
1/6/2012	0.35	91.82	2/1/1982	14.6	95.4	24/11/1981	26.0	93.0
22/2/2014	0.44	91.73	28/7/1984	16.3	93.7	3/1/1982	25.0	94.0
Well No. 5			2/4/1985	15.6	94.4	25/10/1983	26.0	93.0
1/6/2012	7.42	84.86	31/5/1986	16.3	93.7	2/7/1984	27.0	92.0
22/2/2014	8.23	84.05	1/10/1987	16.05	93.95	1/4/1985	24.57	94.43
Well No. 6			25/1/1988	15.47	94.53	29/9/1987	22.6	96.4
1/6/2012	2.53	98.44	18/1/1989	14.9	95.1	24/1/1988	22.0	97.0
22/2/2014	2.52	98.45	14/3/1990	14.7	95.3	8/2/1989	22.8	96.2
Well No. 7			21/9/1991	13.33	96.67	9/12/1989	23.2	95.8
1/6/2012	1.53	97.05	17/12/1992	11.94	98.06	12/2/1991	22.23	96.77
22/2/2014	0.8	97.78	16/5/1993	11.77	98.23	5/2/1992	21.35	97.65
Well No. 8			15/2/1999	11.8	98.2	18/12/1993	20.67	98.33
1/6/2012	8.76	95.56	6/2/2001	12.0	98.0	5/8/1996	21.65	97.35
22/2/2014	closed	104.32	8/3/2004	11.8	98.2	11/7/2001	22.0	97.0
Well No. 9			15/2/2010	8.0	102.0	19/5/2010	16.1	102.9
1/6/2012	9.1	90.03	19/5/2010	7.5	102.5	28/6/2010	15.98	103.02
22/2/2014	2.17	96.96	28/6/2010	7.4	102.6	5/7/2010	15.87	103.13
Well No. 10			8/1/2011	6.28	103.72	28/4/2011	13.68	105.32

1/6/2012	4.44	94.97	21/6/2011	6.45	103.55	15/9/2011	13.0	106.0
22/2/2014	1.0	98.41	Well No. 13			3/1/2012	12.4	106.6
Well No. 11			6/6/1971	20.0	101.0	7/2/2012	12.9	106.1
1/6/2012	4.43	109.91	9/2/1985	16.27	104.73	25/8/2012	14.55	104.45
22/2/2014	closed	114.34	17/11/1986	17.02	103.98	20/10/2012	14.37	104.63
Well No. 20			24/9/1992	12.48	108.52	16/2/2013	13.93	105.07

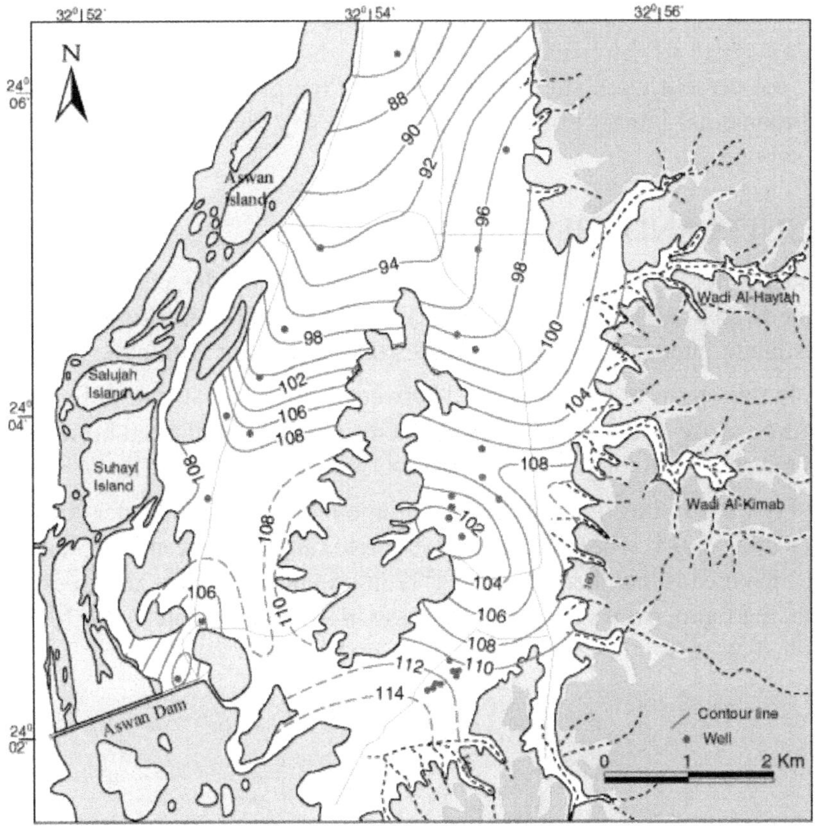

Figure 7: Groundwater level map in the study area (2014).

In order to assess the rate of groundwater rise, the present authors measured the hydrogeological data (depth to water, groundwater level, ground elevation, and total depth) from the drilled wells through many trips from 2010 up till now (Table 1). The collected data covered all affected areas through monitoring the cases of groundwater rising and their environmental impact and the illustrations were prepared (Table 1 and Figure 8, Figure 2, and Figures

9-14).

To achieve the causes of this problem in Aswan city, hydrogeological data were collected from 11 wells drilled along the study area for the period from 1971 to 2009 (Figure 15).

A total of 13 water samples were collected from some of affected areas and bore wells from various localities of the city and chemical and microbiological analyses carried out to assess the groundwater quality. They analyzed for different physio-chemical parameters (TDS, pH, E.C, TH), major cations (Ca^{2+}, Mg^{2+}, Na^{+}, K^{+}), major anions $(HCO_3^-, SO_4^{2-}, Cl^-)$, and some trace ions as (PO_4), (NO_2), (NO_3), Fe, and Mn (Table 2).

The present work studied the causes of the groundwater level rising, the environmental impact of the problem on the city and gave recommendations for solving it.

RESULTS AND DISCUSSION

Groundwater level rise results provide a basis for the characterization of groundwater variation within the Quaternary aquifer in the studied area. It is possible to interpret different processes that may occur within the aquifer.

In this study, the data recorded between 1971 and 2014, from five wells at Kima area, are evaluated in order to determine the general trend in water levels and the change in the water-table over 43-year period (Table 1).

Figure 15 shows a noticeable variation in the groundwater level, from 1971 up to 2014, where the water table rise ranges between 12.55 and 13.69 m which means that there is about 1.37 m groundwater table rise in each year. Also, the figure shows an abrupt increase in the water levels in the 2000's and reached its peak in 2010 and continuing up till now.

(a)

(b)

Figure 8: Rising groundwater level creating ponds: (a) Kima pond, (b) El Shallal pond.

Figure 9: Increase population and Urbanization activity at the study area.

Figure 10: Rash of sewage network system on streets.

From the groundwater level contour map for the year 2014 (Figure 7) the following points can be identified.

- The general groundwater flow direction is generally from the south towards the north, i.e. toward the Nile River, Other direction of water flow comes from the highlands, which is characterized by high urbanization.
- The highest water level is recorded at El Shallal area, close to the area of recharge (Aswan Dam Lake), and the lowest water level is noticed at the northern part of the city, close to the area of discharge, the Nile river.
- The contour lines are more or less regular in the northern part of the city, while they are condensed in El Shallal, Kima and El Aqad areas where these areas are characterized by dense population and industrial activities.

(a)

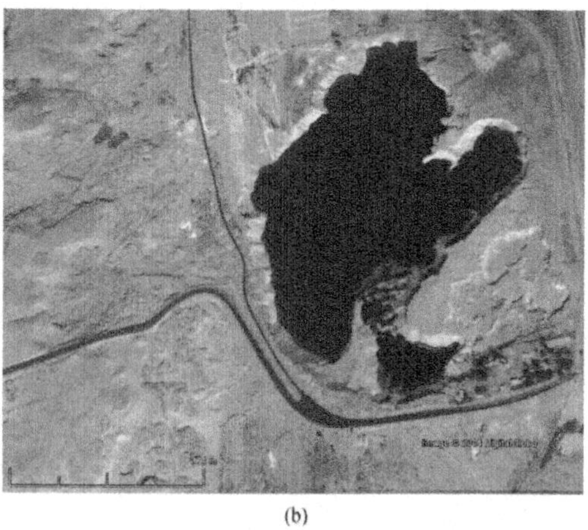

(b)

Figure 11: Satellite image for KIMA factory pond before flooding (a) 2009 and after five years of flooding (b) 2014.

Figure 12: Salt crystals cause the building materials to split, flake and crack (Khor Awada and El Sail areas).

- Two water depressions are noticed in Kima and El Karor areas, these two areas are characterized by low ground elevations.

The authors measured the depth to groundwater in 26 piezometers, all over Aswan city, for the period from 2010 up to 2014 to assess the rate of the water rising in the city (Table 1 and Figure 6).

(a)

(b)

Figure 13: Examples from different locations where algae growth in the rising water at the affected areas. (a) Khor Awada; (b) El Sail.

(a)

(b)

Figure 14: Groundwater rising and their environmental impact: (a) Displacement occurs between two buildings, (b) Example of flooding buried basements of many buildings. (a) El Aqad; (b) Aswan stadium.

The table and figure show that the rise in groundwater level ranges between 8 cm/year (wells 25 & 26) and 27 cm/year (well 23) at El Shallal area, and it ranges between 0.60 cm/year (well 12) and 2.14 m/year (well 18) at Kima area, while it reached 3.46 m/year (well 4) in the western plain and 0.01 cm/year (well 1) at the northern side of the city (Figure 6). We can concluded that the rate of the water rising is much higher in the western side of the city, along El Sadat road, and in Kima factory areas where they are characterized by low relief and by dense population. The minimum rising of the groundwater is noticed in the northern part of Aswan city because this area is of low population and represents the discharge area for the Quaternary aquifer; to the Nile River. During the last five years of monitoring, some wells at KIMA factory area are submerged under the rising water (wells No. 19, 28 and 29).

In 2009, 40 productive wells at El Shallal area were turned off, as a result of deteriorated water quality, they were pumped an amount of 12.44×10^6 m^3/year. Moreover, groundwater withdrawals from KIMA factory wells were reduced from 13.23×10^6 m^3/year to 9.12×10^6 m^3/year. The former discussion proves that the phenomena of groundwater rising that happened in 2009 at Aswan city, can be directly refereed to the cessation of groundwater pumping from El-Shallal wells, and to reduction of pumping from KIMA factory wells.

Causes of the Water Level Rising Problem

The problem of increasing the groundwater level and flooding over the ground surface is observed in many lowlands along the study area as El-Seil, Khor Awada, Birket El-Damas, Phatemic graves, El-Sadat road, El-Aqad, Aswan stadium, Blood Bank, Military building, Kima, El Shallal (Figure 1 and Figure 2). The main causes of the water rising at the Aswan city can be related to the following.

Table 2: Chemical analysis of the surface and groundwater samples in the study area

Sample No.	Physio-chemical parameters				Major Cations (ppm)				Major Anions (ppm)			Trace ions (ppm)					e%
	pH	EC µmho/cm	TDS (ppm)	TH (ppm)	Ca	Mg	Na	K	HCO_3	SO_4	Cl	(PO_4)	NO_2	NO_3	Fe	Mn	
well 1	7.65	540	454	231.9	50	26	16	7.4	275	22.8	15.2	0.11	0.1	3.52	0.016	0.005	0.93
well 2	7.86	5250	3383	993.8	277	73.4	700	14.5	908	550	860	0.13	0.15	3.74	0.011	0.006	0.07
well 3	7.9	8800	5632	1482	424	103	1320	50	360	1650	1700	0.15	0.14	0.19	0.012	0.004	0.05
well 4	8.4	6510	4166	1198	320	97	925	20	340	1382	1068	0.19	1.12	86.4	0.151	0.053	0.15
well 5	7.45	4760	3046	900.2	250	67	650	14.3	320	1042	690	0.17	0.13	0.16	0.017	0.006	0.21
well 6	7.84	4360	2812	833.3	248	52	620	14.8	321	900	657	0.177	0.16	14.3	0.537	0.036	1.66
well 7	8.36	1170	748	270.1	62	28	120	7.4	200	128	160	0.21	0.13	0.19	0.173	0.044	1.58
well 9	7.6	2530	1619	442.3	98	48	340	19	395	332	375	0.19	0.17	0.18	0.113	1.038	0.29
well 10	7.63	3940	2568	522.6	145	39	640	26	380	685	653	0.18	0.17	0.14	0.02	0.006	0.03
well 11	7.39	1250	845	338.8	78	35	135	7.4	265	155	170	0.2	0.18	8.7	0.02	0.004	1.80
Pha-temic graves	7.85	9150	5856	3146	530	443	603	42.3	1827	1181	1255	3.49	0.08	2.12	0.085	0.09	0.13
Khor Awada	8.09	4420	2828	763	106	121	636.4	37.3	792	538	597.5	0.81	0.37	15.8	0.18	0.21	3.33
Aqad	8.03	3810	2418	659	114	91	530.8	35.8	628	456	562	0.72	0.48	14.1	0.11	0.25	2.09

Variation of Direct Groundwater Recharge and Discharge

The principal cause is the great variation of the difference between the recharge and discharge to and from the main aquifer under Aswan city area, where a big and continues recharge amount of water regardless to a small discharge amount of water by time lead to increasing the water level and appear in low elevated lands in the study area causing a concern problem. The groundwater recharge from potable and different supplies may exceed both the natural rate of recharge from Aswan Dam Lake and the natural rate of groundwater discharge. So that, less discharge of groundwater indirectly leads to increase the water level.

The Hydrogeological Parameters of the Quaternary Aquifer

The hydrogeological parameters of the Quaternary aquifer, in Aswan area, play significant role in the calculations of groundwater rise. The pumping tests analyses reveal that the Quaternary aquifer at Aswan area has high hydraulic conductivity (ranges between 0.0001 and 0.0004 m/s), and high transmissivity values (0.02 to 0.04 m^2/s), [14] . Increasing hydraulic conductivity and transmissivity lead to increasing the rate of groundwater flow which act as an important role in the problem of water rising.

Cessation of Groundwater Pumping

In 2009, 40 productive wells at El Shallal area were turned off, as a result of deteriorated water quality, they were pumped an amount of 12.44×10^6 m^3/year. Moreover, groundwater withdrawals from KIMA factory wells were reduced from 13.23×10^6 m^3/year to 9.12×10^6 m^3/year. This proves that the phenomena of groundwater rising that happened in 2009 at Aswan city, can be directly refereed to the cessation of groundwater pumping from El-Shallal wells, and to reduction of pumping from KIMA factory wells.

Groundwater Recharge from the Urban Activities of High-Land

Urbanization has a great effect on the groundwater regime especially to subsurface components of infiltration and percolation leading to groundwater recharge.

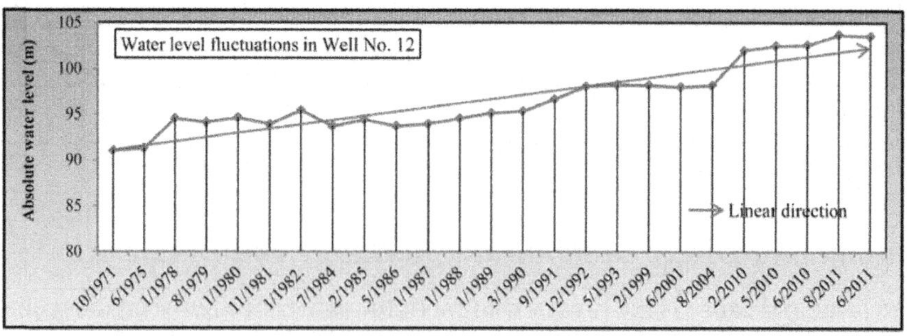

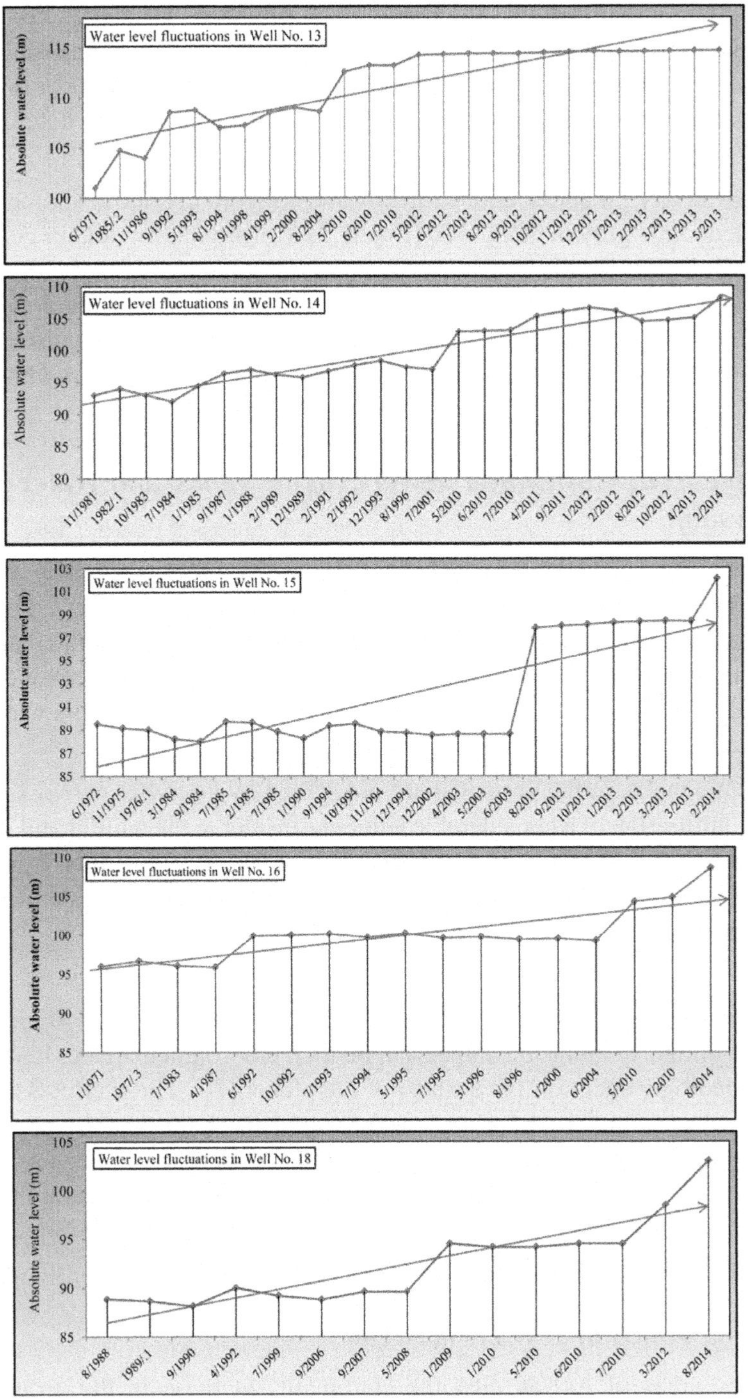

Figure 15: Water level fluctuations in different wells from 1971 till now.

In any urbanization, sanitation and drainage arrangements are of fundamental importance. Quantifying groundwater recharge in urban areas is especially challenging because the urban environment is quite complex as a large variety of land uses coexist and overlap and because of the heterogeneity of the shallow underground.

Aswan city expanded both vertically and horizontally due to unprecedented rapid economic growth after 1970 (Figure 9). Unfortunately, neither geological nor hydrological features are taken into consideration in such developments. Water supply from alternative sources and their distribution through the mains cause to water leakages within urban areas, which enhance groundwater table rise. Moreover, the aquifer beneath the urban area is unexploited which leads to continuous groundwater level rising under these areas.

Leakage from Wastewater Sewers, Septic Tanks, and Water Supply System

When sewer lines are located below the water table, they may infiltrate groundwater, and when located above the water table they may leak. Therefore, the seepage from sewage network system itself conceder another problem. Recently in Aswan city area a phenomena of rash of sewage network system on streets (above ground surface) for many times leads to raise the water level and become as other recharge source (Figure 10).

Septic tanks, in the parts that are not served by a sewage network system, lead to infiltration of a considerable amount of water to the aquifer and increase the water level. In many parts of the study area, seepage water from water supply networks, even if small amount, give another reason of water rising and flow water to accumulate and appear in low lands.

Environmental Impacts of Water Level Rising

The magnitude of anthropic impacts upon their environment makes humans the major geologic agent on the surface of the planet [1] . These effects are most severe where population concentrates, today, half of the world's populations live in urban areas. Water quality is a prime issue in urban settings as shallow aquifers and surface waters in cities are subject to pollution by a multitude of point and non-point sources. At Aswan city, the environmental impact of the water level rising includes the following points.

Forming Ponds in Low Lying Areas

In many areas of Aswan city, the continuous increase in the groundwater rising leads to creation of new ponds e.g. Kma, El Shallal, and pond 3 (Figure

1 and Figure 8). These ponds have bad environmental impact to the area, where theymay threaten the community public health. At KIMA factory area, the continuous rising of groundwater level leads to increasing its area and covering new lands (Figure 11).

Impact on the Groundwater Quality

A total of 13 water samples were collected from some of affected areas and bore wells from various localities of the city and chemical analyses carried out to assess the groundwater quality.

They analyzed for different physio-chemical parameters (TDS, pH, E.C, TH), major cations (Ca^{2+}, Mg^{2+}, Na^{+}, K^{+}), major anions $(HCO_3^-, SO_4^{2-}, Cl^-)$, and some trace ions as (PO_4), (NO_2), (NO_3), Fe, and Mn (Table 2).

In the study area, the average value of TDS is 2798.1 mg/l (Table 2). The relatively high TDS of the analyzed water samples have a bad environmental impact on the infrastructures of the area. The capillary action draws the salty groundwater to the surface and into the porous walls and foundations of the buildings. The dry desert heat causes the evaporation of the water from the walls material leaving behind salt crystals that cause split, flake and crack of these walls (Figure 12).

The salt crystals combinations are calculated for both the surface and groundwater samples along the study area. They revealed the presence of different groups of salt assemblages (Table 2). The hypothetical salt combination revealed the presence of different salts arranged in terms of their predominant as NaCl, $Ca(HCO_3)_2$, Na_2SO_4, $MgSO_4$, $CaSO_4$, $Mg(HCO_3)_2$, and KCL; where the average of equivalent percentage is 42.18%, 19.16%, 13.68%, 10.42%, 9.01%, 4.34%, and 1.37% respectively. In many spots of the study area, high nitrate, sulphate, chloride and sodium concentrations indicate anthropogenic aquifer pollution and it is most severe where population concentrates.

Growth of Micro-Organism and Algae

Some of the collected water samples were subjected to microbiological measurements for the total bacteria and total Coliform in the study area (Table 3). The total Bacteria vary from 20 to 1000 no of colony/100 ml. The total Coliform ranges between 0 and 470 no. of colony/100 ml. In Phatemic graves and KhorAwada, the total bacteria and total Coliform are too numerous to count. The results obtained from the microbiological analysis reflect a presence of bacterial activity in the accumulated water due to rising groundwater level in the study area.

Once a building has been exposed to a large volume of water, may species of micro-organism and algae can growth (Figure 13). They are causing bad environmental impact on building and human health. In addition of directly environmental effect on building and human health of pollutants water of micro-organism, the probability of pollutants water to seeps to the drinking fresh water pipes network increase.

Flooding Building's Basements and Inundate Underground Infrastructure

In many locations of the study area, water level rising leads to inundation of the basement of many buildings (Figure 2 and Figure 14(b)). In any urban area, presence of the groundwater for long time has bad environmental impact on building, where it causes direct damage to building. Figure 14(a) shows clear displacement occurs between the two building as result of water level rising and flooding their basements for long time.

Table 3: The analytical of total bacteria and total Coliform in some water samples.

Sample No.	Total Coliform (colony/100 ml)	Total Bacteria (colony/100 ml)	Sample No.	Total Coliform (colony/100 ml)	Total Bacteria (colony/100 ml)
well 2	0	20	well 10	470	1000
well 4	0	20	well 11	20	40
well 5	10	50	Aqad	273	400
well 6	250	450	Phatemic graves	Too Numerous to count	Too Numerous to count
well 7	0	500	KhorAwada	Too Numerous to count	Too Numerous to count

Underground infrastructures that lie beneath many parts in Aswan City, as Communication networks, wastewater sewers system, pipes of drinking water distribution networks, high voltage electrical cables, and others, are flooded by rising groundwater level.

CONCLUSIONS

Aswan city expanded both vertically and horizontally due to unprecedented rapid economic growth after the year of 1970. Unfortunately, neither geological nor hydrological features are taken into consideration in such developments.

As a result of leakages or infiltrations from different sources, the natural groundwater balance is overturned into an unbalance where the input to water table is comparatively much more than the natural groundwater flow.

In this study, the data recorded between 1971 and 2014, are examined in order to determine the general trend in water levels and the change in the water-table over 43-year period. It shows a noticeable variation in the groundwater level during this period, and the water table rising ranges between 12.55 and 13.69 m, which means that there is about 1.37 m groundwater table rise in each year. Recently, an abrupt increase in the water levels is noticed and reached its peak in 2010 continuing up till now. This study proves that the phenomena of groundwater rising can be directly refereed to the cessation of groundwater pumping from El-Shallal wells, and to the reduction of pumping from KIMA factory wells. During these last five years, all the low-lying areas of Aswan city are suffering from groundwater rising. The continuous increase of groundwater level can lead to appearance of new affected areas in other parts at Aswan city.

The environmental impact of the water rising includes: forming ponds in low lying areas (Kima and El Shallal ponds), flooding building's basements, and inundating underground infrastructure. A general deterioration in groundwater quality was identified. Much higher sulphate and alkali and alkaline earth metal concentrations were found in many spots of the study area and it is most severe where population concentrates. Furthermore, the bacteriological investigations show that the total Bacteria vary from 20 to 1000 no. of colony/100 ml and the total Coliform range between 0 and 470 no. of colony/100 ml indicating that local sources are strongly influencing the observed bacteriological data. In some locations (Phatemic graves and KhorAwada), the total bacteria and total Coliform are too numerous to count.

RECOMMENDATION

The necessary recommendations to decrease and/or prevent water rising problem at Aswan City include re-exploitation of groundwater from El Shallal and Kimawells. Leakages from the networks must be controlled and necessary maintenances are obtained. In the future continuous monitoring of the water level in all the available drilled wells should be taken into consideration where, the continuous increase of groundwater level can lead to appearance of new affected areas in other parts at Aswan city. Moreover, all urbanization activities in the high elevated lands should be under control to decrease water seepage and/or flow to the low elevated lands.

REFERENCES

1. Heiken, G., Fakundiny, R. and Sutter, J. (2003) Earth Science in the City. American Geophysical Union Special Publication Series, 56, 440 p. http://dx.doi.org/10.1029/056SP
2. Attia, M.L. (1954) Deposits in the Nile Valley and the Delta. Geological Survey, Egypt, Cairo, 356 p.
3. Said, R. (1962) The Geology of Egypt. Elsevier Publishing Company, Amesterdam, 337 p.
4. Said, R. (1981) The Geological Evolution of the River Nile. Springer-Verlag, New York, 151 p.
5. Butzer, K.W. and Hanson, C.L. (1968) Desert and River in Nubia. University of Wisconsin Press, Madison, 562.
6. El Ramly, I.M. (1973) Geomorphology, Hydrogeology, Planning for Groundwater Resources and Land Reclamation in Lake Nasser Region and Its Environs. Report from Desert Research Institute to Lake Nasser Development Center Aswan, Cairo.
7. El Shazly, E.M., Adel Hady, M.A., El Ghaay, M.A. and El Kassas, I.A. (1974) Geological Interpretation of ERTS-1 Satallite Images for West Aswan Area, Egypt. Proceeding of the Ninth International Symposium on Remote Sensing of Environment, Michigan, 15-19 April 1974, 119-131.
8. Van Houten, F.B. and Bhattacharyya, D.P. (1979) Late Cretaceous Nubia Formation at Aswan South Eastern Desert, Egypt. Annals Geological Survey, IX, 408-431.
9. Issawi, B. (1981) Geology of the Southwestern Desert of Egypt. Geological Survey, Egypt, Cairo, XI, 57-66.
10. Barber, W. and Carr, D. P. (1981) Water Management Capabilities of the Alluvial Aquifer System of the Nile Valley, Upper Egypt. Technical Report No. 11, Water Master Plane, Ministry of Irrigation, Cairo, Egypt.
11. Issawi, B. and Jux, U. (1982) Contribution to the Stratigraphy of the Paleozoic Rocks in Egypt. Geological Survey, Egypt, Cairo, Vol. 64, 1-28.
12. Klitzsch, E.H. and Wycisk, P. (1987) Geology of the Sedimentary Basins of Northern Sudan and Bordering Areas. Berliner Geowissenschaftliche Abhandlungen, Reihe A, 75, 97-136.
13. RIGW (1988) Hydrogeological Map of Egypt. First Edition, Scale 1:2000,000, Research Institute for Groundwater, Ministry of Public Works and Water Resources, Cairo, Egypt.

14. Selim S. (1995) Geological and Hydrogeological Studies of the Quaternary Aquifer in Aswan Town Area, Egypt. Egyptian Journal of Geology, 39, 631-645.
15. Hamdan, A. and Abdel Rady A. (2013) Vulnerability of the Groundwater in the Quaternary Aquifer at El Shalal-Kema Area, Aswan, Egypt. Arab Journal of Geoscience, 6, 337-358.http://dx.doi.org/10.1007/s12517-011-0363-y
16. Gheorhge, A. (1979) Processing and Synthesis of Hydrogeological Data. Abacus Press, Preston, 390 P.

Chapter 3

FEASIBILITY OF GROUNDWATER BANKING UNDER VARIOUS HYDROLOGIC CONDITIONS IN CALIFORNIA, USA

Saad Merayyan[1], Samsor Safi[2]

[1]California State University, Sacramento, CA, USA

[2]Sacramento Area Sewer District, Sacramento, CA, USA

ABSTRACT

This study evaluates the feasibility of groundwater banking in the Central Basin. The Central Basin is located in Sacramento County in northern California, USA. The study basin is bounded by three rivers (the Sacramento, the American, the Consumes and Mokelumne rivers), and by the Sierra-Nevada mountain range. This study focuses on the potential for groundwater recharge in the Central Basin for three water years (critical, wet, above normal). For that purpose, a 3-D Groundwater Modeling System (GMS) with MODFLOW was created. Three recharge wells were added to the calibrated groundwater model to recharge the water table with 10,000 Acre-Feet (AF) of water to the Central Basin. The banking of 10,000 AF during the critical and wet years was effective in raising the water table elevation in the cone of depression area without causing any negative impact elsewhere in the basin. According to the findings of the Central Basin model, banking up to 10,000 AF of groundwater during any year type is feasible. More than 10,000 AF of groundwater banking might cause more negative impacts than positive benefits.

INTRODUCTION

This project investigates the feasibility of groundwater banking in the Central Sacramento County Basin (Central Basin). The area of the Central Basin overlies the California State Department of Water Resources' (DWR) Groundwater Basin Number 5-21.65, the South American Sub-basin. There are some differences between the Central basin size and boundaries to that of the DWR Groundwater Basin Number 5-21.65. The reason for the differences in the boundaries is due to the fact that the Central Basin was developed based on the Sacramento County IGSM Grid. [1] . For the purpose of this study, the area of DWR's South Sub-Basin American sub-basin was chosen due to its natural boundaries of three perennial rivers. The Central Basin is defined as the area bounded on the west by the Sacramento River, on the north by the American River, on the south by the Consumes and Mokelumne rivers, and on the east by the Sierra-Nevada mountain range. Figure 1presents the general location of the area Central Basin. The surface area of the Central Basin is 388 square miles (248,000 acre). The average annual precipitation in the basin ranges from about 14 inches along the western boundary to greater than 20 inches along the eastern boundary [1] . The eastern basin boundary is defined by the uprising foothills of the Sierra Nevada mountain range. This represents the approximate edge of the alluvial basin, where little groundwater flows into or out of the groundwater basin from the Sierra Nevada foothills [2] . The western portion of the Central Basin consists of nearly flat floodplain deposits from the Sacramento, American, and Consumes rivers. The DWR publication Bulletin 118-3, 1974, indicated that the groundwater movement in the Central Basin occurs in a shallow aquifer underlying by a deeper aquifer [3] .

The shallow aquifer extends approximately 200 to 300 feet below the ground surface. While the base of deep aquifer averages approximately 1400 feet below the ground surface. The deep aquifer is separated from the shallow aquifer by a discontinuous clay layer, which serves as a semi-confining layer for the deep aquifer. The Central Basin aquifer generally receives its recharge from the rivers around the basin and precipitation infiltration. There might be some interaction with other basins but at very deep elevations [4] . Groundwater elevation in the Central Basin has been declining since the 1950s, with some recovery during the mid-1990s [1] . The recovery in 1990s was attributed to the increase use of surface water. Since then, the southern and central portions of the basin have an increased dependence on groundwater. Due to this reason, there has been a continuous trend of declining groundwater levels and a formation of a cone of depression. The groundwater levels generally vary between 10 and 90 feet below mean sea level in this south-central portion of the Central Basin, whereas conditions in the west and north have been stable [1] .

The Central Basin's potential average groundwater extraction rate is 273,000 AF/year with the extractions rates currently close to upper extraction limits. With the continuous increasing demand on water resources in the region, the Central Basin could face future depleted supply and major challenges of the sustainability of this supply. Therefore, it is important to explore other venues to maintain the basin's long-term sustainability and meet future demand. Groundwater banking has become an integral component of any integrated water resources management plan to address both local and statewide water supply issues. Stakeholders from privately owned properties (or well owners) to local and state agencies are interested in the feasibility of groundwater banking in their respective basins.

Relative to the construction of surface water reservoirs, groundwater banking in subsurface aquifers is a less controversial, lower cost, and more environmentally benign approach. Groundwater banking has numerous economic and environmental advantages compared to surface water storage. It reduces losses to evaporation. It allows for greater regulation of natural inflows, without the need to construct of large surface reservoirs [5] . As with any water storage systems, the main purpose of groundwater banking is to store surface water from precipitation and rivers, when water is abundant, and utilize those groundwater sources when the surface water supply stocks is scarce. The major obstacles to groundwater banking are the water conveyance from the source to the area of injection, cost of construction and maintenance of injection wells and regulatory and permitting requirements. The current study assumes that groundwater recharge aspect is an acceptable water management technique.

METHODS

This study was prepared to evaluate the feasibility of groundwater banking in the Central Basin. A three dimensional (3-D) model of the Central Basin was developed using the GMS with MODFLOW (from here on referred as GMS) software package [6] . The model was simulated with three actual historical data scenarios: above normal, wet, and critical years. Once the model was developed and calibrated, it was utilized to examine the impact of groundwater banking. Additional groundwater was banked through injection wells for each scenario to study its feasibility in the Central Basin. In order to carry out groundwater banking analysis, the Central Basin data were collected and compiled into a model, which is described as follows.

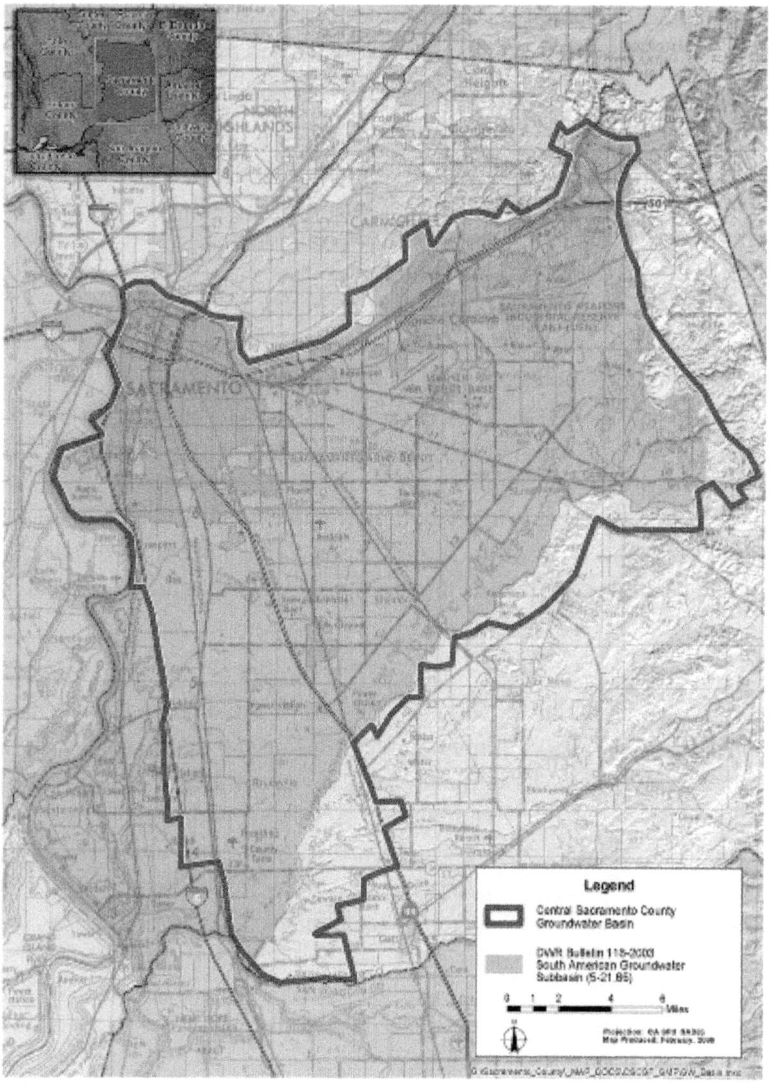

Figure 1: Study area.

Water Years Selection

The Sacramento Valley Water Year Type is determined by DWR's California Data Exchange Center's website (CDEC) based on Sacramento River and tributary runoff necessary to meet the Delta outflow criteria. Table 1 summarizes all possible water year types for the past seven years and the selected years of 2005, 2006, and 2008 [7].

Geology

Review of DWR publications was conducted to re-construct the complex geological formations in the Central Basin region. DWR publications provide detail information on layers of geologic formations in the region, but exact thicknesses of layers is not easily accessible from these publications. Therefore, the elevations of the different layers were estimated from a 2010 City of Roseville groundwater report called Sacramento Regional Model (SRM) [8] . In 2010, Aquaveo performed this groundwater study for City of Roseville to assist in the planning efforts for an aquifer storage and recovery. The SRM study included an area of approximately 1360 square miles in the Greater Sacramento Metropolitan Region which encompasses the area of this study (Central Basin). SRM used the information on the thickness of the post-Eocene continental deposits defined in the DWR Bulletin 118-6. Furthermore, SRM turned this geological information into a 3-D stratigraphic model of the region with estimated thicknesses for the layers of the geological formations. This study extracted those estimated thicknesses for the Central Basin region from the SRM. Table 2 lists ten layers of these geologic formations that were reproduced for the Central Basins model. The estimated geological formation thicknesses were entered into the GMS model as boreholes, which were then converted into a solid stratigraphy of the Central Basin.

Surface Elevations

The ground surface elevations are needed as an input parameter to the GMS model. The elevations were incorporated into the Central Basin region as the top of the surface soil for all the borehole data. The surface elevations were obtained from numerous sources including USGS and a number Geotechnical reports within the Central Basin region [9] -[13].

Precipitation

The precipitation data for the selected three water years 2005, 2006 and 2008 were retrieved from DWR's CDEC website. Table 3 shows the average annual precipitation for the selected years [7] . Average annual precipitation data for each of the selected water years were multiplied by the area (248,000 acres) of the Central Basin to obtain a volume of precipitation in acre-feet. Monthly variation of precipitation is not accounted for in this study. Table 4 summarizes the precipitation data and recharge rates [7].

River Stage

River stage elevations (annual average) along with the bottom elevations and river conductance are the key elements needed as river data into GMS. DWR's CDEC website provides stage elevation in feet for river stations, using the three letters station IDs. The river stage elevations used to develop the river system in Central Basin model are provided in Table 5 [7] . The locations of the river stations are shown in Figure 2 [7].

River Conductance

River conductance is related to the rate at which a unit of riverbed material can transmit fluids, and is used mainly in hydrology in relation to river and lake beds. It is an application of intrinsic permeability to a unit of material with a defined area and thickness. In hydrology, the magnitude of conductance affects the rate of groundwater recharge or interaction with groundwater. This parameter is used in computer modeling codes as GMS. Table 6 shows the initial River Conductance values used as starting point in Central Basin model.

Groundwater Usage

Groundwater use data is one of the key elements in developing the Central Basin GMS model. The SCGA's2007-2008 Basin Management Report provided a comprehensive groundwater use data, including the agricultural use of groundwater in the Central Basin (Table 7) [14] . The total groundwater and agricultural uses are annual averages; hence, monthly variation is not taken into consideration as part of this study.

Table 1: DWR water year type

Water year	Index	Year Type
	Million AF	
2004	7.51	Below Normal
2005	8.49	Above Normal
2006	13.2	Wet
2007	6.19	Dry
2008	5.16	Critical
2009	5.78	Dry
2010	7.05	Below Normal

Table 2: Ten geologic formation layers

Year	Groundwater Levels Predicted Contour Elevations from MSL (ft)		Groundwater Levels Predicted with 10,000 AF Contour Elevation from MSL (ft)	
	Max	Min	Max	Min
2006	160	−99	160	-92
2008	142	−123	142	-112

Table 3: Rain data for the selected year type

Water Year	Annual Average	Year Type
	inches	
2005	23.29	Above Normal
2006	25.36	Wet
2008	13.8	Critical

Table 4: Precipitation data.

Year	Average Annual Precipitation	Central Basin Area	Rain Volume	15% of Volume	15% of Volume	Recharge Rate
	(Inches)	(acre)	(AF/year)	(AF/year)	(ft^3/day)	(ft/day)
2005	23.29	248,000	481,327	72,199	8,616,951	0.0007977
2006	25.36	248,000	524,107	78,616	9,382,820	0.0008685
2008	13.8	248,000	285,200	42,780	5,105,793	0.0004726

Table 5: River stations & stage elevation

Station ID	Station Description	Lat.	Long.	Gage Elev.	Operating Agency	Stage Elev. 2005	Stage Elev. 2006	Stage Elev. 2008
AFO	American River At Fair Oaks	38.6	−121.272.0		USGS	99	105	91
MHB	Consumnes River at Michigan Bar	38.5	−121.0168.0		USGS & DWR	172	178	160
BEN	Mokelumne River near Thornton	38.3	−121.40.0		CA DWR	8	12	1
IST	Sacramento River at I Street	38.6	−121.527.0		CA DWR	12	16	2
SDC	Sacramento River above Delta Cross Channel	38.3	−121.510.0		USGS	5	9	−5

Table 6: Initial river conductance

River Name	River Cross-Sectional Area	b	K	C_b
	ft^2	ft	ft/day	ft^2/day
Sacramento	3,000	20	0.1	15
American	2,000	20	0.1	10
Mokelumne	500	10	0.1	5
Consumes	200	5	0.01	0.4

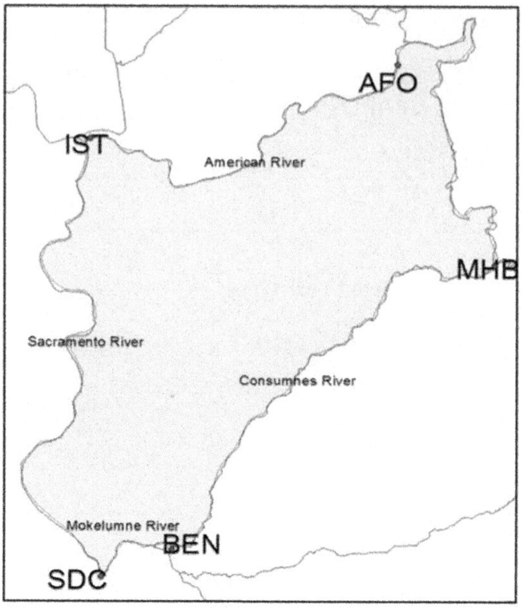

Figure 2: DWR's river station IDs & locations.

Model Development and Setup

The GMS model setup could be achieved using two approaches: the grid approach or the conceptual model approach [15] . The grid approach involves working directly with the 3-D grid, where conceptual model approach involves using GIS tools to construct the model and then it is converted into a 3-D grid. In the current study, a conceptual model approach was selected due to its applicability and flexibility to setup. The datum for the data needed for the GMS model was based on the State Plane Coordinate System with North American Datum 1983 (NAD83). A GIS shape-file containing rivers, streams and lakes in the Sacramento region was imported into GMS to provide information related to these water bodies. A background map was imported into the GMS model in order to provide ground surface elevations, visual guidance and spatial reference. The delineated project boundaries are shown in Figure 3.

Table 7: Groundwater usage

Year	Total Groundwater AF/year	Agriculture
2005	244,026	167,062
2006	245,382	166,148
2008	247,067	164,320

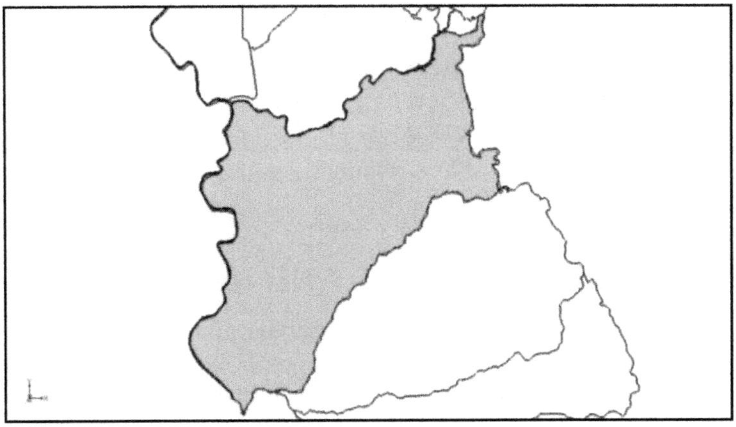

Figure 3: Central basin area & boundaries.

Boreholes for Sub-Surface Materials

Table 8: Well extraction flow rate

Year	Total Groundwater Use	Agricultural Original Use	Agricultural Use Revised	Final Total Agriculture Use		Well Extraction Rate
	(Acre-feet)			(Acre-feet)	(ft^3/day)	(ft^3/day)
2005	244,026	167,062	116,943	193,907	23,142,848	46,193
2006	245,382	166,148	116,304	195,538	23,337,413	46,582
2008	247,067	164,320	115,024	197,771	23,603,969	47,114

A total of twenty-two (22) boreholes were created and spread equally around Central Basin area to produce sufficient representative stratigraphy of the region. The borehole profiles are made of the geologic formation thicknesses estimated from City of Roseville study called SRM. Figure 4 shows the locations of all the boreholes across the Central Basin. Figure 5 shows the profile view of these boreholes.

Triangulated Irregular Network and Solids

Triangulated Irregular Network (TIN) method was used to calculate an intermediate ground surface elevations between points of known elevations by linear interpolation. Figure 6 shows the top soils with the interpolated ground surface elevations using TINs. Each of the 22 boreholes has specific elevations according to the estimated thickness of the geologic formation and ground surface elevation. TIN method uses those borehole elevations and connects them to create solids for different layers stratigraphy as shown in the Figure 7.

Pumping and Observation Wells

Five hundred and one (501) pumping wells were created in the conceptual model. The wells were classified based on the ownership entity: the DWR Wells (359), and the Non-DWR Wells (142).Figure 8 shows the locations of the DWR wells [7] , and Table 8 summarizes the wells extraction flow rate data [14].

THREE-DIMENSIONAL GRID AND MODFLOW MODEL

After entering all the center blocks (materials and their respective layers, rivers, wells, precipitation recharge, and boundary and initial conditions) of the Central Basin model, then model now has all the components for various simulations. The conceptual approach was then converted into 3-D Grid.Figure 9 shows the 200 × 200 × 14 grid containing the 3-D region of the Central Basin. Figure 10shows the model converted into the 3-D grids.

Model Calibration

DWR's spring of 2004 contour map was used as baseline for the model calibration, see Figure 11. The model elevations was run through manual process by deleting the 3-D grid, under which all MODFLOW simulations are stored and regenerating it for the next calibration.

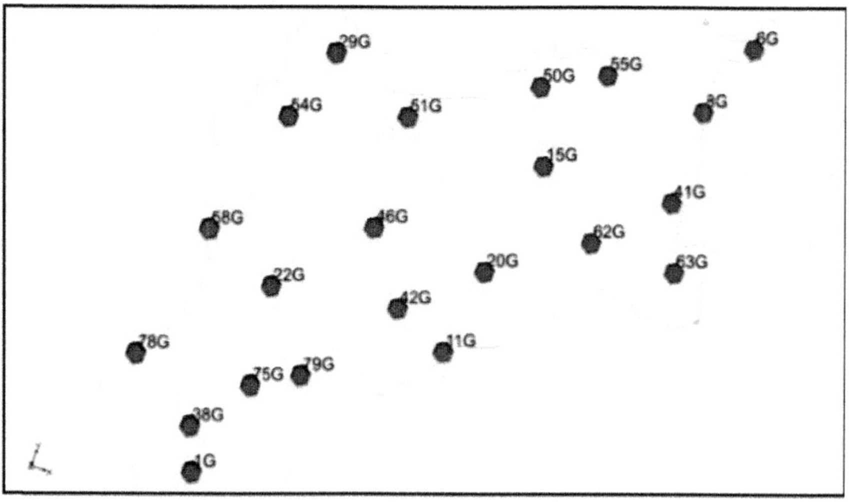

Figure 4: Borehole locations and labels.

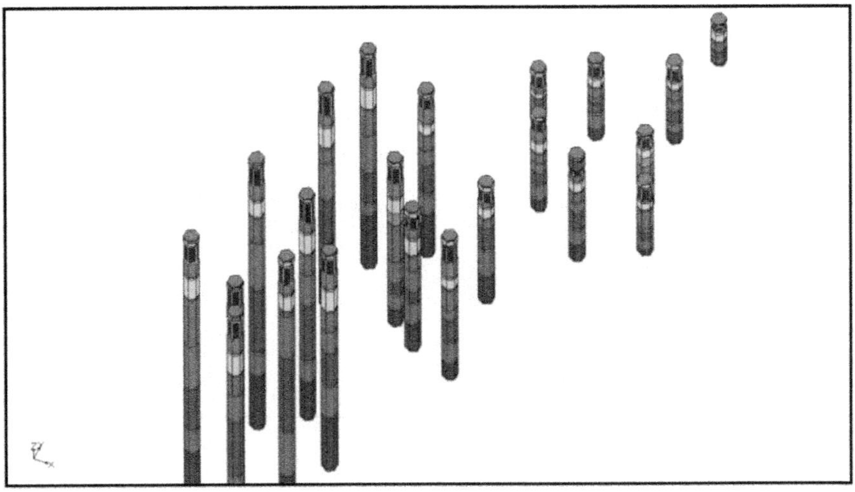

Figure 5: Profile of borehole.

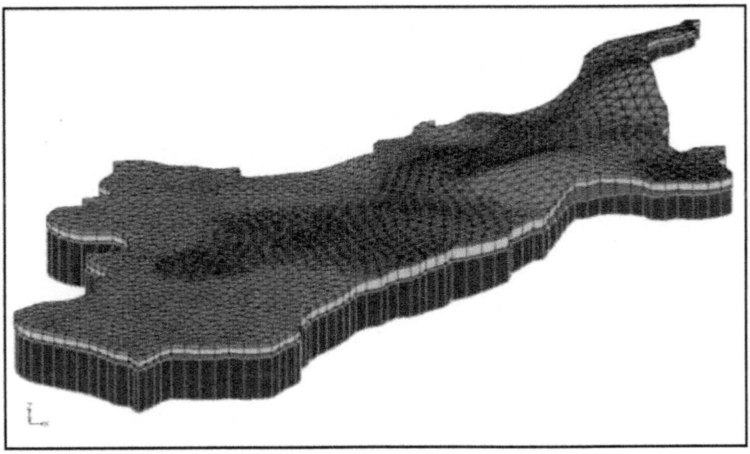

Figure 6: Top soil surface.

This process was repeated about two hundred (200) times in order to produce solutions that closely resemble the observed targeted contour map values as shown in Figure 11.

Parameter sensitivity analyses (approximately 200 runs) were performed and found that the model is more sensitive to the hydraulic conductivity (K value, feet/day) of the soil and geologic formations layers. The model was less sensitive to other elements such as river conductance, river stage elevations, porosity, specific yield and storage.

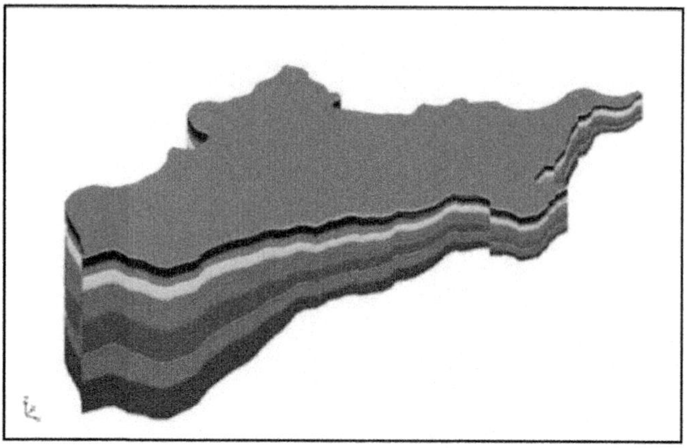

Figure 7: Solids stratigraphy.

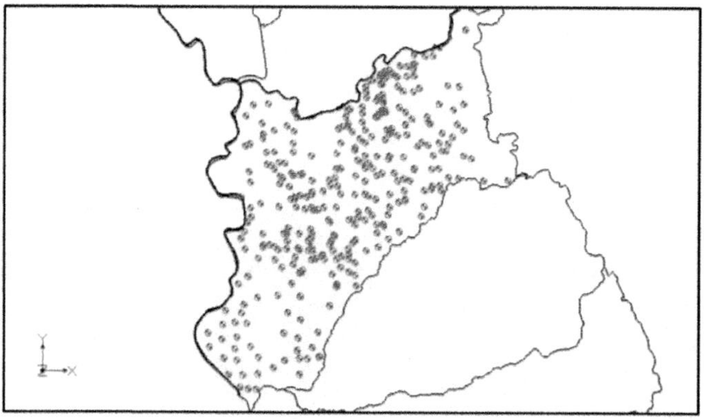

Figure 8: Locations of DWR wells.

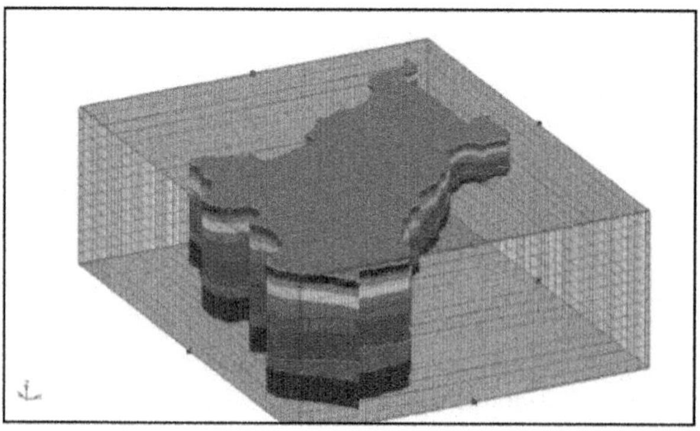

Figure 9: 3-D grid box.

Initial estimates of the hydraulic conductivities started at 0.1 - 10 ft/day and were adjusted until reasonable values 8 - 59 ft/day range were determined to match the existing conditions. The findings were in agree ment with a comprehensive research of the region's soil properties prepared for the Sacramento Area Flood Control Agency (SAFCA).

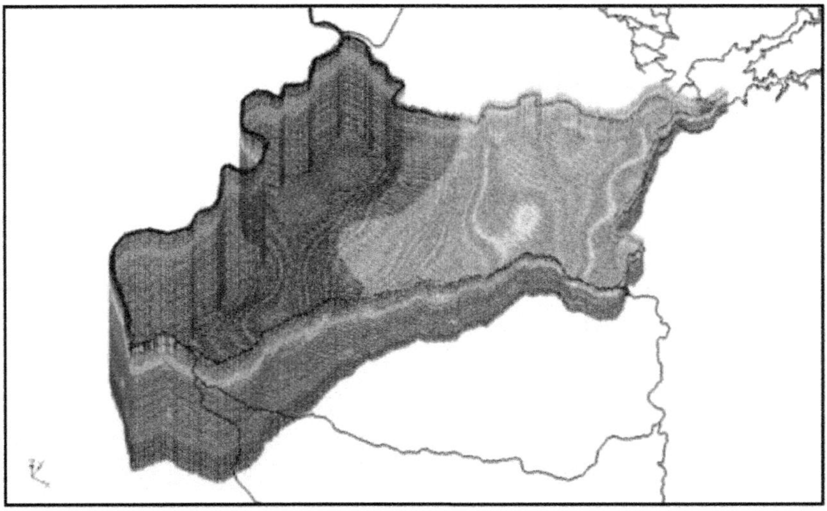

Figure 10: 3-D grid model of central basin.

The SAFCA study evaluated a broad range of reports and they concluded that the K values in the Sacramento area ranges from 5 - 139 ft/day [16].

Above Normal and Wet Years Calibration

The results of the 2005 and 2006 water years were very similar to the observed target (with a maximum difference of 10%), with the only exception of the location of the cone of depression.Figure 12 shows the model result for 2005 water year. In general, the formation of the cone of depression in all of the models was in agreement with historical records of SCGA and DWR. The results showed that the cone of depression of the historical record is shifted to the north when compare to the observed target cone of depression.

Critical Year Calibration

Results of the critical year type (2008) were also in agreement when compared to 2005 and 2006 water years' results. The cone of depressions was deeper, and the inflow from rivers and precipitation to the aquifer was smaller due to decrease in the amount precipitation and the river stage elevations.

Banking Wells

Banking wells were added to the Central Basin model at the location of the cone of depression. Sacramento Regional County Sanitation District (SRCSD), a regional wastewater treatment plant located in the city of Elk Grove, is currently delivering 15,000 AF of recycled water for agricultural use to local cities [17] . Therefore, the 10,000 acre-feet of water was selected as a starting goal of this study to bank. The groundwater banking wells for this study, were constructed to allow the use of any amount more than 10,000 AF if needed.

RESULTS

Critical Year with Banking

The process of injecting additional flow into the model as groundwater banking was successful and seems to produce conclusive results.

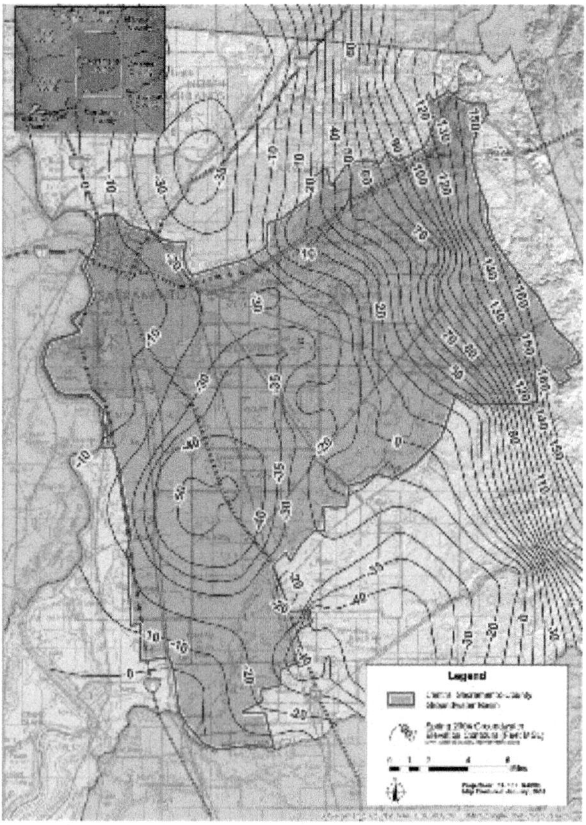

Figure 11: DWR contour calibration map.

The locations of the banking wells were chosen to be in the center of the cone of depression, where it was found to be the most effective locations for groundwater banking. The cone of depression for all three selected years (above normal, wet and critical) was in the same vicinity. The injection of 10,000 AF of groundwater in the effective area seems to have marginally altered the shape of the depression cone. There were no significant changes in the water table elevation elsewhere in the basin. Figure 13 shows the contour map of the basin with the injection of 10,000 AF for banking.

Wet Year with Banking

The results of the groundwater banking are summarized in Table9 It is clear that groundwater banking was more effective during the critical year since the water table elevation increased by 11 ft from 123 ft below Mean Sea Level (MSL) to 112 ft below MSL. While in the wet year (Figure 14), the change in groundwater table elevation was less dramatic. The change in groundwater table was only 7 ft from 99 ft below MSL to 92 ft below MSL.

CONCLUSIONS

This study presents the development, calibration, simulation and examination of the feasibility of groundwater banking for the Central Basin. The hydrologic analysis of the Central Basin was comprehensive and utilized gauge data that were available for rivers, precipitation and existing groundwater use as well as their interaction with the aquifer.

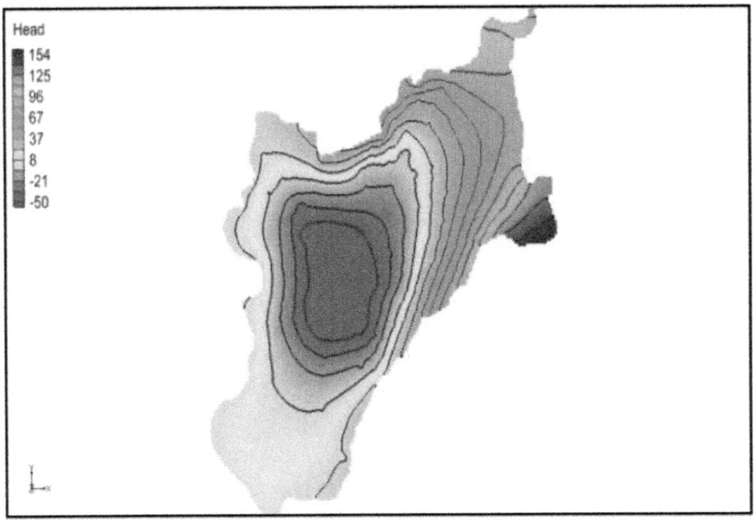

Figure 12: 2005 background contour map.

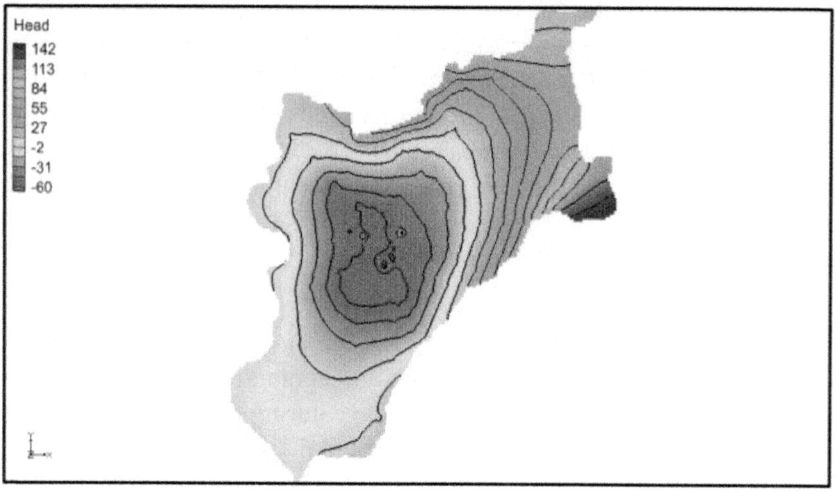

Figure 13: 2008 contour map with 10 K AF.

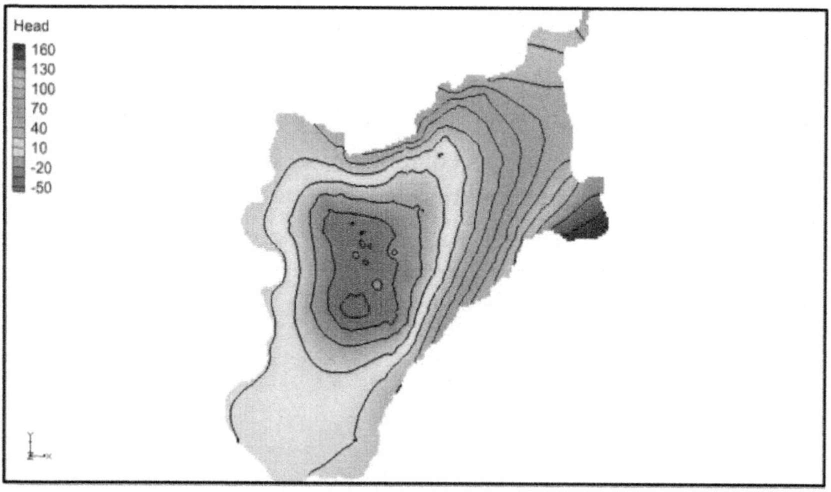

Figure 14: 2006 contour map with 10 K AF.

Table 9: Results of groundwater banking

Year	Groundwater Levels Predicted Contour Elevations from MSL (ft)		Groundwater Levels Predicted with 10,000 AF Contour Elevation from MSL (ft)	
	Max	Min	Max	Min
2006	160	−99	160	-92
2008	142	−123	142	-112

In addition, the GMS model was constructed with complex geologic formation, soil conditions of the region and all of the associated material properties. The model was constructed in a unique way that allows for future modification depending on data availability.

The results of this study showed that the cone of depression formulated during the different simulations is consistent with historical records. Historical analysis suggests that the formulation of the cone of depression is due to the excess of groundwater drafting in certain areas rather than the availability of resources for recharge. The size of the cone of depression was deeper during the critical year when compared to the above normal and wet years. It is further concluded that the banking up to 10,000 AF of groundwater during all three water year types (wet, above normal, and critical) is feasible in the Central Basin. The groundwater table elevation increased due to water banking in the all the scenarios studied. The highest increase in the predicted water table level was during the wet year scenario. The results also show that banking of 10,000 AF at the cone of depression will not cause negative impact elsewhere in Central Basin aquifer.

RECOMMENDATIONS

There were some limitations to this study, which could have influenced the result to a certain degree. First, the precipitation data used in the Central Basin model were annual averages; hence, monthly variations were not taken into consideration. Second, groundwater extraction rate was equally applied to all wells; hence, heavy users (i.e., Elk Grove Water District) were not taken into consideration. However, there are a number of recommendations that could be implemented in order to further improve this study. These recommendations are listed below:

- The extraction rate for each well is available and can be applied to individual wells. It will further increase the accuracy of the model.
- The computer that was used in this study to develop the Central Basin model that was not sufficient to refine the grid around the wells. It is recommended to use a computer with significantly larger storage and processing power to allow for more refinement of the GMS grid to improve accuracy.
- GMS is a strictly groundwater software and has limitation in applying precipitation and precipitation data. Therefore, if better surface runoff and percolation software is used and integrated with the GMS, it may improve the model further.

- Building an accurate stratigraphy of the region is one of the most complex tasks of groundwater model construction. This is subject to change according to the availability of new information from all sources including but not limited to federal, state, local governments and private property owners. Therefore, it is recommended to have significantly more man-hours to improve the stratigraphy of the Central Basin.

ACKNOWLEDGMENTS

The authors of this study would like to acknowledge the staff for help of the City of Roseville, California and for providing the data needed to complete the study. Furthermore, the authors would like to thank Aquaveo LLC for providing valuable information, resource and support towards this study and for donating the GMS software.

REFERENCES

1. Central Sacramento County Groundwater Management Plan (CSCGMP) (2006).
2. California DWR (2004) California's Groundwater: South American Groundwater Sub-Basins Number: 5-21.65. Bulletin 118-6.
3. California Department of Water Resource (DWR) (1974) Evaluation of Ground Water Resources: Sacramento County. Bulletin 118-3.
4. DWR (1978) Evaluation of Ground Water Resources: Sacramento Valley. Bulletin 118-6.
5. Natural Heritage Institute (NHI) (1998) Feasibility Study of a Maximal Program of Groundwater Banking.
6. Groundwater Modeling System (GMS) with MODFLOW software. Version 8.0. (2011) Aquaveo, LLC, Provo, Utah.
7. California DWR's California Data Exchange Center Website. http://cdec.water.ca.gov/
8. City of Roseville (2010) Sacramento Regional Model Groundwater Modeling Report. Aquaveo LLC.
9. Morris, D.A. and Johnson, A.I. (1967) Summary of Hydrologic and Physical Properties of Rock and Soil Materials, as Analyzed by the Hydrologic Laboratory of the U.S. Geological Survey, 1948-1960. USGS Water Supply Paper: 1839-D.
10. Folsom Cordova Unified School District (2008) Geologic Hazard and Geotechnical Engineering Report. Wallace Kuhl & Associate Inc.

11. Youngdahl Consultating Group Inc. (2007) Geotechnical Engineering Study—Update 2007. Seasons at Laguna Ridge.
12. CSA Water Pipeline and Florin Road Sewer (2009) Geotechnical Baseline Report. ENGEO Inc.
13. Elk Grove Promenade Center (2006) Geotechnical Engineering Report. Wallace Kuhl & Associate Inc.
14. Sacramento Central Groundwater Authority (SCGA) (2008) Basin Management Report 2007-2008.
15. Aquaveo LLC (2011) GMS 8.0 Tutorials. Retrieved from Aquaveo Website.http://www.aquaveo.com/gms-learning
16. Sacramento Area Flood Control Agency (SAFCA) (2009) Appendix A: Groundwater Impact Analysis. Luhdorff & Scalmanini Consulting Engineers.
17. Sacramento County Regional Sanitation District (SRCSD) (2011) South Sacramento County Agriculture & Habitat Lands Recycled Water Project Programmatic Feasibility Study.

Chapter 4

GROUNDWATER SOLUTION TECHNIQUES: ENVIRONMENTAL APPLICATIONS

Sarva Mangala PRAVEENA[1], Mohd Harun ABDULLAH[1], Ahmad Zaharin ARIS[2], Kawi BIDIN[1]

[1]School of Science and Technology, Universiti Malaysia Sabah, Kota Kinabalu, Sabah, Malaysia.

[2]Department of Environmental Sciences, Universiti Putra Malaysia, Selangor, Malaysia

ABSTRACT

Groundwater models provide a scientific tool for various groundwater studies which include groundwater flow, solute transport, heat transport and deformation. However, without a good understanding of a model, modeling studies are not well designed or the model does not represent the natural system which being modeled long term effects may results. Thus, this review has focused and reviewed the types of solution techniques in terms of advantages and limitations. The findings are vital to improve the model conceptualization and understanding of the uncertainty in model results. On the same hand, it acts as guide and reference to groundwater modeler, reduces the time spent in understanding the solution technique and complexity of groundwater models, as well as focus ways to address the groundwater problems and deliver modeling output more efficiently.

INTRODUCTION

According to [1], groundwater modeling covers different aspects of the system behavior. Groundwater modeling studies have four potential relevance processes which include groundwater flow, solute transport, heat transport and deformation. According [2,3], groundwater modeling has turn out to be a crucial tool in decision making and planning in environmental management. Decision making and planning processes in environmental management are associated with water resource allocation, complex development and requiring multidisciplinary information for evaluating their effects on a social, economic and environmental level [4]. Generally, most of the groundwater modeling studies are conducted using either deterministic models, based on precise description of cause-and-effect or stochastic models based on the probabilistic nature of a groundwater system [5,6]. The main components of groundwater modeling are selecting the natural system which the model is designed, creating the conceptual representing the natural system, models representing the controlling mechanism, solution of the model, calibration and validation of the model along with simulation [7,8].

There are enormous amount of groundwater models to study the cause and effect or the probabilistic nature of a groundwater system. It is an ad-vantage to classify them in groups based on criterias such as aquifer type, techniques used, type of aquifer simulated and the dimension of the problem [9]. [10] stated that the classification of groundwater models can be done based on model objectives, processed modeled, physical system characteristics modeled and mathematical approaches. According to International Ground Water Modeling Center (IGWMC), there are many various manners in groundwater models classifications (flow, media, transport, temperature, phases, chemical reaction, dispersion, thermodynamics, fractured rock, vapor transport, variable saturated, saturated) that a specific and systematic classification cannot be developed. A detailed explanation of these classifications can be found in [10].

Various solution techniques are a crucial component in groundwater models [6]. Solution techniques in groundwater modeling activities are to follow a multilevel approach. Multi-level approach involves data collection of groundwater flow and mass, contaminant transport and advection-dispersion equations, evaluation of the data and final decision to select the model. An understanding of various solution techniques is vital due to complexity in groundwater modeling and universal importance perspective. Era of numerous groundwater models development has been stimulated by high advance of computer technology and programming techniques. Yet the current numerous model development and groundwater complexity often leave those involve in groundwater studies spend a lot of time in understanding the solution techniques. This increased time resulted in less time spent in understanding the system. Thus, there are many gaps in our understanding of groundwater modeling which limits our capacity. Various groundwater models development have exposed with many reviews on the favors and disfavors of these models [6,8,11]. However, there are limited reviews on the solution techniques of these groundwater models although they are crucial components utilized in groundwater modeling. While a number of these solution techniques are focused on the types of models and applications in real world [11–14], a lack of quantitative information on the advantages and limitations of these tools impedes the use of these tools for real-world applications.

An understanding of various solution techniques is crucial due to complexity in groundwater modeling. This work was intended primarily as a guide and reference for the practitioner who is trying to simulate groundwater in their site of interest. This attempt is a way to lessen the time spent in understanding the solution technique and complexity of groundwater models, as well as focus ways to address the groundwater problems to render modeling output more effectively. The conceptual framework of the review was based on the types of solution techniques available in groundwater studies. An assessment of mutual understanding, advantages and limitations of all the solution techniques is applied to all kind of groundwater modeling studies and not limited to any particular purpose or equations. It is an attempt to reduce the time spent in understanding the solution technique and complexity of groundwater models and represent focus ways to address the groundwater problems and render modeling output more effectively.

VARIOUS SOLUTION TECHNIQUES ASSESSMENT

According to [8], the term model has different meanings. Combinations of all model components are suitable for groundwater model. However, term model is also used in a part of various solution techniques. Thus, the term model will also be used in a part of solution technique in this review. Numerous sophisticated solution techniques or model are currently available to overweigh the accuracy of the groundwater system representation [15]. The groundwater solution techniques comprise from simple to complex [6]. According to [2] until early 1970s, physical and analog models were widely used as mathematical models solving groundwater problems. As groundwater modeling techniques boosted with extensive computer programmings, various solution techniques have been developed to solve the systems of mathematical equations. The simplest classification was done by [12] and [14], where the solution techniques are divided into two broad groups namely physical models and mathematical or numerical models. Solution techniques grouping done by [11] listed that groundwater models are divided into four broad groups which are porous media, analog, electric analog and digital models. Along with the advent of computers, groundwater modeling has focused on the numerical models expressing the groundwater flow and transport studies. However, these models (analytical, physical, analog, porous, empirical and mass balance) are still needed to investigate and validate new models. The requirements are to examine and analyze whether certain assumptions underlie the new models are valid. The conceptual framework of this review was based on the types of solution techniques listed by [8] as showed in Figure 1.

SOLUTION TECHNIQUES EVALUATIONS

It is very important to have strong understanding with a model in order to know the advantages and limitations of each solution techniques. Perspectives of advantages and limitations of the solution techniques were evaluated in this review.

Analytical Models

Analytical models are the rapid way to analyze physical characteristics and conceptual behavior of groundwater system compare to other models. This is because it uses an exact analytical solution for specific field applications. On the other hand, analytical models are only limited to steady and uniform groundwater problems involving small parts of study area and bulky to transport problems. Table 1 presents other points of advantages and limitations of analytical models.

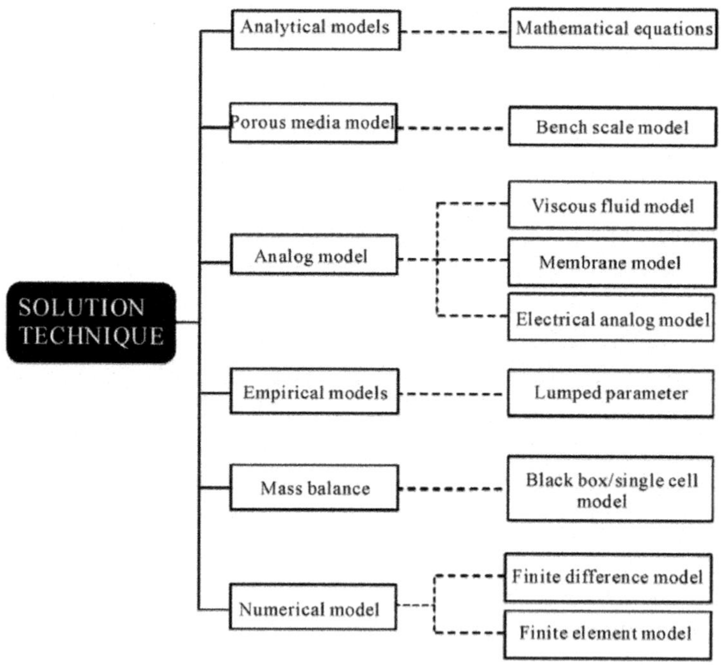

Figure 1: Types of groundwater models.

Porous Media Models

Porous media or bench-scale models belong to the group of hydraulic models which has been widely used in hydraulic engineering. Porous media models are suitable to use at any dimensionality, any type of groundwater flow and transport problems (variable saturated, heterogeneity, anisotropy, phreatic, steady, unsteady, advection, dispersion, sorption, decay and reactions). Information about porous media is presented in Table 2.

Analog Models

In terms of demonstration and education tools, analog models are still widely used for groundwater studies. Analog models (viscous fluid, membrane and lumped models) are not suitable for groundwater transport. The models have limited capability to involve with advection, dispersion, sorption, decay and reactions studies in groundwater. The views on advantages and disadvantages of analog models are detailed in Table 3.

Table 1: Applicability of analytical models

Model type	Analytical model
Advantages	• Simple [6,16] • Economical/ inexpensive [2,3,6] • Rapid way to analyze physical characteristics of groundwater [2,3,20] • More efficient than other models [6,9,16] • Can form useful complements to any numerical models [25,26] • Can used either for verification or being part of numerical models [16,17] • An important and useful tool for estimating fate and transport parameters from field or laboratory data [16,17] • Provide more insight into conceptual behavior of the groundwater system [3] • Does not introduce errors due to the numerical diffusion and approximation by the finite difference model [12]
Limitations	• An exact analytical solution may outweighed by the errors introduced by simplifying assumptions of complex field environment [9,10] • Complex and cumbersome in transport problems [2] • Limited to cases with steady and uniform flow problems [2] • Relatively simple initial and assumptions in boundary conditions. Hydrogeological boundary conditions must be idealized to fit the model [2] • Professional judgment and experience in field application are needed to apply the analytical model [2] • Suitable to solve groundwater problems involving small parts of aquifer systems or small area extent [9,18] • Could not handle spatial/temporal variations in groundwater system [18]

Empirical Models

Empirical models are useful to use when detailed site specific data are lacking or impractical situation to simulate fine-scale processes. Lack of understanding in the processes involve in study area, these models can be misused or misunderstood as the models are easy to employ as well as lumping process together will mask the disadvantages of these models. Table 4 summarizes the information on empirical models.

Table 2: Applicability of porous media models

Model type	Porous media model
Advantages	• Relatively straightforward and simple [19, 20] • Allow the study of special aspects of groundwater flow and transport under almost natural condition [19,20] • Useful to enhance site characterization and features [9] • Good demonstration and education tools for students [4,7,20] • Obeys laws that govern other physical systems including laminar flow of fluids and heat [4,6,7] • Good starting point for groundwater modeling beginners [4]
Limitations	• Capillary rise takes place in such models is far larger than that which actually occurs in a real field situation [13] • Difficult to visual and identify the water table [7,13] • Time consuming and prohibitively costly [5]

Table 3: Applicability of analog models

Model type	Analog model
Advantages	• Illustrative and still widely used for demonstration purposes of groundwater flow [4,21] • Versatility and can readily study a variety of aquifer conditions [8] • True for groundwater flow without natural recharge if the weight of the membrane is small [4] • Inexpensive tools to use to visualize groundwater stress [4] • Useful tool to help the inexperienced earth scientist to understand about groundwater hydraulics [4] • Solves problems concerning the phreatic surface for transient and steady flow conditions[4,7,21]
Limitations	• A good care is required in the model construction because flow rate varies with the cube width [4,7] • Temperature is also another factor need to be focused [4,5] • Limitation on applications involving nonlinear conditions of varying transmissivity in unconfined aquifers and two-fluid flow problems [7,13] • Also limited applications in groundwater lowering in construction field [21] • Electric potential is unaffected by gravity, therefore it requires adjustments [22]

Mass Balance Models

Mass balance model is also known as the black box or single-cell model. It is also a numerical model in its simplest form. In mass balance models, the averaging of an entire area is a crude approximation. Evaluation of field data is only involves in and out fluxes. Table 5 details the information of mass balance models.

Table 4: Applicability of empirical model

Model type	Empirical model
Advantages	• Impact the accuracy of the model predictions [23,24] • Suitable to use when detailed site specific data are lacking and appropriate when it is impractical to simulate fine-scale processes [4] • Representing an entire groundwater problem employs a series of physical laws, empirical laws and conservative assumptions to represent the problem of interest [1,4,23,24] • A good alternative method [23,24] • Provide useful predictions without the costly calibration time [23,24]
Limitations	• Lack of understanding of process involved and only a temporary solution to assist analysis [7, 24] • Can be misused and misunderstood because they are easy to employ [4] • Lumping processes together will mask the limitations of these models [7]

Table 5: Applicability of mass balance model

Model type	Mass balance model
Advantages	• The simplest form of numerical model. The best fitted in numerical modeling [4,14] • Very useful which leads to an examination of the global mass balance [14] • Easy to use [4,14] • Efficiently aid in the analysis of the impact of the management options [14] • Suitable to use when detailed site-specific data are lacking or impractical situation to simulate fine-scale processes [14] • An important part in more complexes of numerical models [8]
Limitations	• Lack of understanding of the processes involved [4] • Acts as a temporary solution to aid analysis [4,14] • Can be misused or misunderstood because they are easy to use [25] • Applicable only in limited circumstances and masked by lumping process together [10,14]

Numerical Models

Among of the solution techniques assessment, numerical models were found to have more advantages over other solution techniques. They are such as it solves both simple and complex groundwater problems, capable to used almost of any type of groundwater system and impose no restrictions on the initial conditions, boundary types as well as characteristics of the groundwater. The most advantage in numerical models is that the models utilize the latest advances in computer technology without writing any computer codes. Numerical models which employ the latest computer technology also have limitations in terms of accuracy, errors and codes. Accuracy of numerical output mainly depends on the availability of soil hydraulic information, errors in numerical dispersion are hard to be identified as well as special codes are need for specific groundwater problem (Table 6).

Table 6: Applicability of numerical model

Model type	Numerical model
Advantages	• Employed with the latest and recent advances in computer technology [4,5,11,13] • Solves both simple and complex groundwater problems [4,7,13,26,27] • Dominated the complex study of groundwater problems as it solves both simple and complex one, two or three dimensional problems [4,7,13, 15] • Capable to simulate almost any type of groundwater situation [5,7,17] • Well suited to exploring hypothetical scenarios [15,27] • Can easily handle spatial or temporal variations of groundwater system [6,11] • Impose no restrictions on the initial conditions, boundary types, characteristics of the groundwater or investigated solute [5,10] • Computer programs for most groundwater problems are available easily and the users can apply relevant computer programs without writing any computer code [4,7,13,26,27]
Limitations	• Time consuming for data collection and input [4,7,11,13] • Require much information to characterize the system [28] • Expensive models [28] • Special codes are required for specific problems, such as density-dependent flow and coupled saturated unsaturated flow [15,27] • Accuracy of the results of numerical models mainly depends on the availability of information about the hydraulic properties of the subsoil [28] • Errors in numerical dispersion [28] • Uncertainty of the model predictions is hard to quantify [28]

CONCLUSIONS

This review has focused and reviewed the types of solution techniques available in groundwater modeling studies. Assessment of six solution techniques namely analytical, porous media, analog, empirical, mass balance and numerical models was done to give a clear understanding of each solution techniques. Advantages and limitations of all the solution techniques were listed and analyzed. Analytical, porous media and mass balance models are simple and appropriate to use in groundwater modeling studies. In terms of demonstration and education tools, porous media and analog models are still widely used for groundwater studies. Empirical and mass balance models are useful to use when detailed site specific data are lacking or impractical situation to simulate fine-scale processes. The most benefit of numerical models is it utilizes the latest advances in computer technology without writing any computer codes as well as solves both simple and complex of any groundwater problems. On the other hand, limitations of analytical models are only limited to steady and uniform groundwater problem involving small parts of study area. Porous media and numerical models face time consuming for data collection and expensive as their constraints in the applications. Empirical and mass balance models face lack of understanding in the processes involve in study area and can be misused or misunderstood. In the view of analog models, they are not suitable for groundwater transport. Moreover, errors in numerical dispersion are hard to be identified as well as special codes are need for specific groundwater problems. As a final note, it is important to point out that a good understanding of various solution techniques act as guide and reference to groundwater modeler. Besides, it reduces the time spent in understanding the solution technique and complexity of groundwater models, as well as focus ways to address the groundwater problems and render modeling output more effectively.

ACKNOWLEDGEMENT

The first author gratefully acknowledges the support by National Science Fellowship (NSF) Scholarship under sponsorship of Ministry of Science, Technology and Innovation (MOSTI), Malaysia for her doctoral study. Sincere appreciation is also extended to the reviewers for their helpful comments and suggestions which have improved the quality of this paper.

REFERENCES

1. L. W. Canter, D. M. Fairchild, and R. C. Knox, "Ground water quality protection," CRC Press, Boca Raton, Florida, 1988.

2. J. Bear, M. S. Beljin, and R. R. Ross, "Fundamentals of groundwater modeling," United States Environmental Protection Agency, 1992.
3. P. K. M. van der Heijde, "Quality assurance in computer simulations of groundwater contamination," Environmental Software, Vol. 2, pp. 19–25, 1987.
4. E. Manoli, P. Katsiardi, G. Arampatzis, and D. Assimacopoulos, "Comprehensive Water Management Scenarios for Strategic Planning," Global NEST Journal, Vol. 7, pp. 369–378, 2005.
5. S. M. Praveena, M. H. Abdullah, A. Z. Aris, and L.C. Yik, "A brush up on seawater intrusion models," in the Proceeding of Third Regional Symposium on Environment and Natural Resources, Kuala Lumpur, pp. 313–324, 2008.
6. C. P. Kumar, "Pitfalls and sensitivities in groundwater modeling," Civil Engineering, Vol. 84, pp. 116–120, 2003.
7. V. S. Singh and C. P. Gupta, "Groundwater in a coral island," Environmental Geology, Vol. 37, pp. 72–77, 1999.
8. K. Spitz and J. Moreno, "A practical guide to groundwater and solute transport modeling," John Wiley and Sons, New York, 1996.
9. M. E. Thangarajan, "Resource evaluation, augmentation, contamination, restoration, modeling and management," Capital Publishing Company, 2007.
10. J. R. Boulding and J. S. Ginn, "Practical handbook of soil, vadose zone, and ground-water contamination: Assessment, prevention and remediation," Lewis Publishers, Boca Raton, Florida, 2004.
11. D. K. Todd, "Groundwater hydrology," Second Edition. John Wiley & Sons, New York, 1980.
12. M. P. Anderson and W. W. Woessner, "Applied groundwater modeling: Simulation of flow and advective transport," Academic Press, Inc., San Diego, 2002.
13. N. Krešić, "Hydrogeology and groundwater modeling," CRC Press, Boca Raton, Florida, 2006.
14. W. C. Walton, "Groundwater resource evaluation," McGraw-Hill Education, 1976.
15. K. McGillicuddy and T. Sovich, "Strategies for operation of orange county water district Talbert seawater intrusion barrier, California," ASCE, New York, 1996.

16. A. M. M. Elfeki, G. J. M. Uffink, and F. B. J. Barends, "Groundwater contaminant transport: Impact of heterogeneous characterization: A new view on dispersion," Taylor & Francis, 1997.
17. N. Emekli, N. Karahanoglu, H. Yazicigil, and V. Doyuran. "Numerical simulation of saltwater intrusion in a groundwater Basin," Water Environment Research, Vol. 68, pp. 855–866, 1996.
18. L. F. Konikow and T. E. Reilly, "Groundwater modeling," In: The handbook of groundwater engineering, CRC Press, Boca Raton, Florida, 1995.
19. L. Konikow and J. Mercer, "Groundwater flow and transport modelling," Journal of Hydrology, Vol. 100, pp. 379–409, 1988.
20. J. J. Fried, "Groundwater pollution: Theory, methodology, modelling, and practical rules," Elsevier Scientific Publishing Company; Amsterdam-Oxford-New York, 1975.
21. R. Bowen, "Groundwater," Springer; London, 1986.
22. J. Wainwright and M. Mulligan, "Environmental modeling: finding simplicity in complexity," John Wiley and Sons, New York, 2004.
23. K. R. Rushton, "Groundwater hydrology: Conceptual and computational models," John Wiley & Sons, New York, 2003.
24. Environmental Protection Agency, "Models and computers in groundwater investigations," 1991, http://www. cepis.ops-oms.org/muwww/fulltext/repind46/models/models.html.
25. V. Batu, "Applied flow and solute transport modeling in aquifers: fundamental principles and analytical and numerical methods," CRC Press, Boca Raton, Florida, 2006.
26. G. B. Maxey, W. Back, and D. A. Stephenson "Contemporary hydrogeology," The George Burke Maxey memorial volume, Elsevier, 1979.
27. M. Kasenow, "Determination of hydraulic conductivity from grain size analysis," Water Resources Publication, 2002.
28. E. Holzbecher, "Modelling density-driven flow in porous media," Springer Publisher, Heidelberg, 1998.

Chapter 5

CONTRIBUTION OF THE SENSITIVITY ANALYSIS IN GROUNDWATER VULNERABILITY ASSESSING USING THE DRASTIC METHOD: APPLICATION TO GROUNDWATER IN DABOU REGION (SOUTHERN OF CÔTE D'IVOIRE)

Jacques Édoukou Djémin, Jean Kan Kouamé, Kouakou Serge Deh, Amani Tawa Abinan, Jean Patrice Jourda

Department of Sciences and Water Technology and Environment Engineering (Laboratory of Remote Sensing and Spatial Analysis Applied to Hydrogeology), Félix Houphouët-Boigny University, Abidjan, Côte d'Ivoire

ABSTRACT

The groundwater constitutes the main source of drinking water for the populations in the Dabou region which is marked by a multiplication of socio-economic activities. The quality of groundwater is increasingly tested by diverse sources of pollution caused by these human activities. In order to preserve their quality against any form of contamination, the present study aims to assess the groundwater vulnerability in this region and to highlight the relative importance of hydrogeolo- gical parameters which will be taken into account in this assessment. The assessment of the intrinsic vulnerability is to identify the most sensitive zones in order to prevent the groundwater pollution risks on the surface of the ground. To do it, the DRASTIC method is applied through a GIS. The GIS has also used to perform sensitivity analysis through the map removal and the single- parameter sensitivity analysis tests. The indexes calculated for the DRASTIC vulnerability map vary from 95 to 187 of the North towards the South. This vulnerability map presents four classes: very high (26.22%) in the South and the East, high (37.71%) in the Center, the North-East and the North-West, moderate (34.73%) to the North and the West and low (1.34%) in the North. The DRASTIC vulnerability map is heavily influenced by the impact of vadose zone and the depth to water table according to the first test. For the second test, it is the impact of vadose zone, the aquifer media and the soil media which have a more significant impact on the vulnerability map. Both sensitivity analysis tests confirm that the impact of vadose zone therefore sediment type is more implied in this assessment of the groundwater vulnerability in the Dabou region.

INTRODUCTION

The groundwater is the main source of drinking water for some populations. The estimates indicate that nearly two billion people in the world depend directly on the groundwater [1] . Groundwater used not only for human consumption but also the development of socio-economic activities. However, they are exposed to many sources of diffuse and punctual pollution resulting from human activities, particularly the agricultural practices [2] . The progress and the generalization of industrialization have also increased the risk of groundwater pollution by using of hazardous substances [3] . The introduction of pollutants into water tables deteriorates the quality of groundwater and reduces their consumable nature [4] . Given this situation, it is necessary to study the aquifers vulnerability, which is defined as the set of natural geological and hydrogeological characteristics that determine susceptibility of groundwater to contamination by human activities [5] . The vulnerability assessment therefore aims to prevent potential risks of pollution by the delimitation of sensitive areas in order to control the activities at the ground surface to protect the groundwater quality. The groundwater of the study area is exploited in both urban and rural areas to satisfy the water requirements of the populations. However, in some cases, nitrate concentrations of exploited water can reach 51.7 $mg{\cdot}L^{-1}$ [6] . This high concentration of nitrates in collected water can be related to several factors such as the intensive use of agricultural inputs and pesticides in agro-industrial plantations for the improvement of the productions. Moreover, the uncontrolled urbanization favoring the installation of inadequate systems of cleansing and the insufficiency of these systems for a better management of the domestic and industrial waste water rejections can be the causes of a real deterioration of these waters quality. Added to this is the storage of household waste promoting the production of leachate (lixiviates) which can cause the groundwater contamination in this region. Indeed, to limit and prevent contamination risks of these waters in order to preserve their quality, it is necessary to establish the vulnerability map which is under the influence of several hydrogeological parameters. This is firstly to assess the groundwater vulnerability from the DRASTIC method that uses seven hydrogeological parameters and secondly to assess the relative importance of each parameter based on sensitivity analysis tests. The sensitivity analysis will make it possible to apprehend the most determining parameters in the assessment of the intrinsic vulnerability in this region. Application of DRASTIC method and sensitivity analysis tests requires the use of a Geographic Information System (GIS) which is the gateway to the database and output decisional thematic maps.

PRESENTATION OF STUDY AREA

The study area is the Dabou region which is a coastal area of southern Côte d'Ivoire (Figure 1). It is between longitudes 4°10' and 5°00' West and latitudes 5°00' et 5°40' North with an estimated area at 2435 km^2. The population is estimated at 238,294 inhabitants according to the last population census [7] . The study area consists of three departments: Dabou which occupies much of the area, Jacqueville and Grand Lahou. The climate is equatorial transition characterized by four seasons with two dry seasons which alternates with two rainy seasons. The average rainfall ranges around 1400 mm. The direct infiltration feeding the water tables is estimated on average at 382 mm a year which represents 26.30% of the rainfall. This coastal area is characterized by a hydrographic network with an average drainage density equal to 1.7 km.km^{-2} dominated by Ira River and Ebrie Lagoon. The Agneby and Bandama rivers respectively to the East and West of the area are hydraulic limits of the basin.

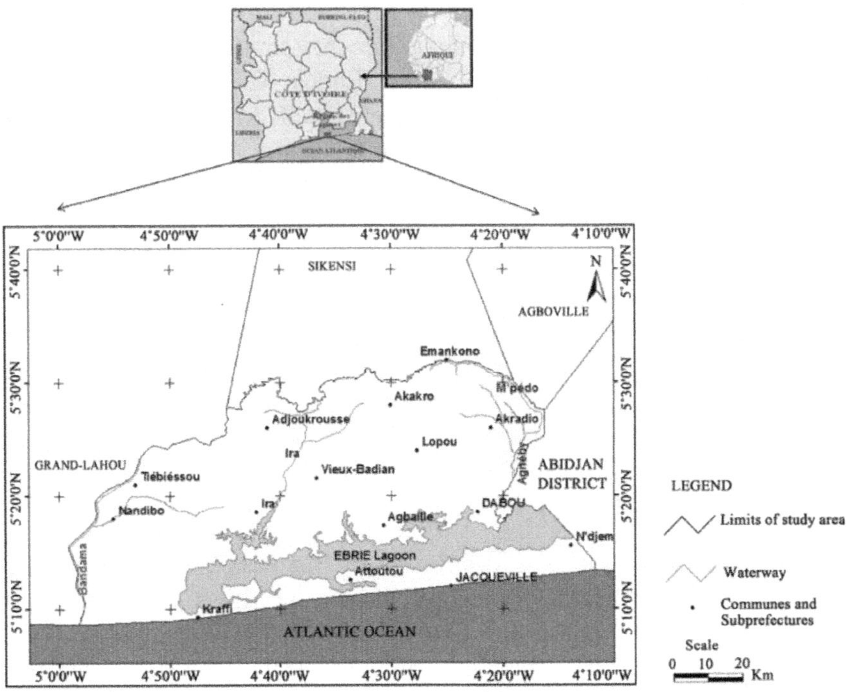

Figure 1: Location map of the study area.

Geological and hydrogeological, the study area belongs to the coastal sedimentary basin composed of post- Eburnean formations characterized by quaternary clayey sands and vases and Continental Terminal clayey sands [8] . This cretaceous to quaternary basin has enormous potential groundwater [9] . These groundwater resources are contained in aquifers three levels of unequal importance [10] . The aquifer system contains the quaternary water table and that of the continental terminal. The aquifer system is exploited by more than 217 boreholes (drillings) with an average speed of 18.79 $m^3 \cdot s^{-1}$.

MATERIALS AND METHOD

Data Acquisition and Materials

The drillings data used in this study were collected from the Water Resources Direction. The drillings data such as transmissivity which were used to estimate the hydraulic conductivity values starting from the thicknesses of calculated water table were collected in previous study. Identification of geological formations that make up the aquifer system in the study area required the drilling logs that exist in previous research of [6] . The geological map (scale: 1/200,000) provided by the Cartography and Remote sending Center has delineated the study area. The Digital Elevation Model (DEM) used in this study is a SRTM (Shuttle Radar Topography Mission) image. It was collected in 2008 by the American Spatial Shuttle "Endeavour" available from (http://strm.csi.cgiar.org) and was used to delineate the DEM of the study area which enabled subsequently estimating slopes of this zone.

The GIS software used in this study is ArcGIS 9.3 offered in three levels of licenses (ArcView, ArcEditor, and ArcInfo). The three modules used in this study are: ArcCatalog that enables the management of the database or the geodatabase; ArcMap used for geoprocessing and ArcScene used for overlaying the information planes. ArcGIS 9.3 has several ex-tensions are those used "Spatial Analyst or GRID and 3D Analyst". Excel software has allowed the entry of alphanumeric data that is imported into the ArcGIS environment through its module ArcCatalog.

DRASTIC Method

The DRASTIC method was developed by the services of the American Agency of Environmental Protection USEPA [11] . It estimates the potential for pollution and to assess the vertical groundwater vulnerability [12] . It takes into account most of the hydrogeological factors that affect and control the flow of groundwater [13] . It is a method to index weighting with seven parameters whose initials form the acronym DRASTIC: Depth to water table

(D), net aquifer Recharge (R), Aquifer media (A), Soil media (S), Topography (T), Impact of vadose zone (I) and Finally the hydraulic Conductivity of the aquifer (C). The development of the intrinsic vulnerability map using the DRASTIC method can be summarized into four phases: (i) data acquisition; (ii) the preparation of the map database; (iii) calculating the vulnerability index and; (iv) the classification index according to DRASTIC classes to determine vulnerability map.

- Definition of the DRASTIC data

The seven DRASTIC parameters are defined as follows:

- The Depth to water table: represents the vertical distance traversed by a contaminant over ground to reach the water table (saturated zone of the aquifer). The larger this depth is, the higher the contaminant puts enough time to reach the water table. It corresponds to the static level in drillings.
- Net recharge is the main vehicle for the transport of contaminant. The larger this recharge is, the more the risk of contamination is high. The net recharge determined in the zone considered is of 382 mm a year.
- The Aquifer media identifies the size of the saturated ground. It is involved in trapping the pollutant that can escape the power of absorption or adsorption of the soil. The larger the particle size is finer, trapping the pollutant is great. This parameter was identified from fifteen drill logs made in previous study.
- The Soil media controls the downward movements of contaminants. Indeed, the presence of fine materials (clays, silts and silt) and organic matter in the soil reduces the intrinsic permeability and retards the migration of contaminants, by physico-chemical processes (adsorption, ion exchange, oxidation, biodegradation). For this parameter, geotechnical map has been used to identify soil types in the study area.
- The Topography or slope (%) favors the runoff of surface waters at the expense of infiltration. The steeper the slope of the land, the more water runoff is important and therefore the groundwater contamination is low. The slope map is obtained by extraction of the contours of the DEM and transforming them into fashion TIN (Triangular Irregular Network) with the extension "3D Analyst" of GIS. The "Slope" of this extension allows converting the TIN in a slope map.
- The Impact of vadose zone is defined as the fraction between the water table and the surface where the pores are partially saturated with water. The permeability of the unsaturated zone controls the flow of pollutants and their arrival at the water table. Its impact is determined from the

texture of the lands that constitute it. Percolation of the pollutant to the water table is so much greater that the texture is favorable. This parameter has been identified from the drill logs produced in the area.

- The hydraulic conductivity is the rate at which water flows through an aquifer. It refers to the ability of the geological formations of the aquifer to transmit water. More this setting, the greater the transfer of the pollutant is fast. The conductivity is determined by the ratio of the transmissivity and the estimated thickness of the water tables.

- Developing of the database maps

The map database is performed in the coordinate system of Universal Transverse Mercator (UTM), World Geodesic System (WGS) 84, Zone 30 and Northern Hemisphere. The geological map was georeferenced and used to scan the survey area using the functions "Georeferencing and Editor" of ArcInfo/ArcGIS 9.3. The method requires spatial interpolation or the DRASTIC parameters. Interpolation is the procedure for estimating the value of a variable in a particular place, from a number of measurement points to calculate each point in the study area [14] . In this study, the chosen interpolation method is the Inverse Distance Weighting (IDW) which is an estimate by linear combination of the values of the known points and is available in "Spatial Analyst" of ArcInfo/ArcGis 9.3".

- Distribution of water points

The study was conducted on 217 drillings which spatial distribution is shown in Figure 2. The values of DRASTIC parameters of each water point made it possible to carry out the interpolations by the chosen method for determining the thematic maps.

- Developing of the thematic maps

The database developed in the framework of this study has two components. A descriptive or alphanumeric component developed under Excel. These data were then imported into ArcInfo/ArcGIS through its ArcCatalog module intended to store and manage this numerical data. The raw values of each DRASTIC parameter were transferred in ArcMap where the interpolations have been performed by setting the spatial resolution (pixel) at 250 m × 250 m.

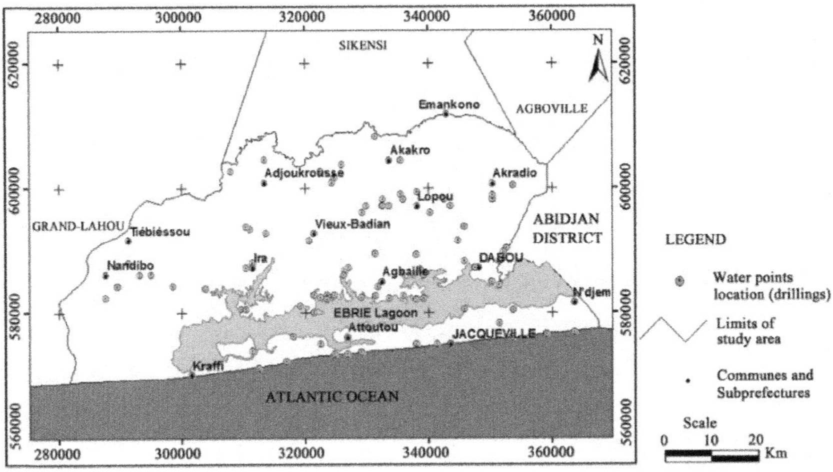

Figure 2: Water points distribution map in the study area.

Thereafter, the classifications according to the method were performed (Table 1) with the function “Reclassify of Spatial Analyst” of the GIS.

Each class is assigned an index named as a symbol (c), typically from 1 to 10 depending on the degree of its impact on the pollution of the aquifer. A weighting factor (p) or fixed multiplication factor of 1 to 5 is assigned to each parameter according to the method in order to relativize their respective importance in terms of vulnerability. This factor reflects the influence degree of each of them in the vulnerability assessment and is imposed by the method. DRASTIC parameter maps are obtained after this manipulations series in the GIS.

Thematic maps of various parameters are established based on DRASTIC ratings. The areas characterized by low scores indicate the favorable hydrogeological conditions to natural groundwater protection. However, for high ratings, these conditions are more and more critical as the vulnerability is high and groundwater is likely to be reached by possible pollution coming from the soil surface.

- Calculation of the DRASTIC vulnerability index

After mapping the parameters according to the rating system of the DRASTIC method, they were weighted. The weighting was to multiply each thematic map (DRASTIC ratings map) by its respective weight. The superposition of different layers (thematic maps) on a result-layer will allow calculating DRASTIC index map. The superimposed layers must obviously have the same cartographic characteristics namely the same projection system, the same units of length, the same geographical extent and also the same resolution as all calculations will be made on the same matrix [15] . The final vulnerability index or noted DRASTIC Vulnerability Index is equal to the weighted sum of seven parameters. The DRASTIC Vulnerability Index (DVI) is calculated by the Equation (1) [11] :

$$\mathrm{DVI} = Dc \times Dp + Rc \times Rp + Ac \times Ap + Sc \times Sp + Tc \times Tp + Ic \times Ip + Cc \times Cp \quad (1)$$

With: D, R, A, S, T, I, C are the parameters mentioned above;

p is weighting factor given to each parameter;

c is rating assigned to each parameter.

- Classification index based on DRASTIC classes

The possible range of values vulnerability index is found between 23 and 226 in the case of the standard version. The values obtained are grouped in eight classes which each one corresponds to a vulnerability degree [11] . The reclassification of the index map within the ranges defined by the method used to develop the groundwater vulnerability to pollution map.

Table 1: Standardization of the parameters using the DRASTIC method [11] (modified)

DEPTH TO WATER TABLE (meter)			TOPOGRAPHY (slope %)		
Range (meter)	**Rating**	**Weight**	**Range (%)**	**Rating**	**Weight**
0 - 1.5	10		0 - 2	10	
1.5 - 4.5	9		3 - 6	9	
4.5 - 9	7		7 - 12	5	1
9 - 15	5	5	13 - 18	3	
15 - 23	3		>18	1	
23 - 30	2		IMPACT OF VADOSE ZONE		
>30	1		**Range**	**Rating**	**Weight**
NET RECHARGE (millimeter·year^{-1})			Sandy clays	3	
Range (millimeter·year^{-1})	**Rating**	**Weight**	Metamorphic rock Sandstone, Clay	4	
0 - 50	1		Armours and Gravel	5	
50 - 100	3	4	Ferruginous sandstone and coarse sands bitumen	6	5
100 - 180	6		Clayey Sand	7	
AQUIFER MEDIA			Sand and Gravel	8	
Range	**Rating**	**Weight**	coarse sands	9	
Sand and Gravel	8	3	HYDRAULIC CONDUCTIVITY (meter·day^{-1})		
Clayey Sands	4		**Range (meter·day^{-1})**	**Rating**	**Weight**
SOIL MEDIA			<4	1	
			4 - 12	2	
Range	**Rating**	**Weight**	12 - 29	4	3
Sands (10% to 45% of Clay)	7		29 - 41	6	
Organic matter (peats)	8	2	41 - 82	8	
Sands (0% to 8% of Clay)	9		>82	10	

Validation Methods of the Vulnerability Map: Sensitivity Analysis Tests

A sensitivity analysis by two tests was conducted to appreciate the effect of each parameter on the intrinsic sensitivity maps. These tests have been applied in recent studies [16] - [21] to analyze the reliability of vulnerability criteria and validate developed vulnerability maps. They are:

- Map removal sensitivity analysis: This test was developed by [22] . It identifies the sensitivity of the DRASTIC vulnerability map for each of the seven parameters. It is therefore to verify the relevance of the parameters in assessing the intrinsic vulnerability and to deduce the most representative in the application of the method. Thus, the degree of influence of a given parameter on the vulnerability index was evaluated by removal the parameter in overlay process. The comparison of the new vulnerability index with the original index provides a direct measure of the parameter influence. This measure corresponding to the sensitivity index (S) of the parameter is determined from the Equation (2):

$$S = \left| \frac{V}{N} - \frac{V_{xi}}{n} \right| \qquad (2)$$

With:

S is the sensitivity index of the parameter;

V is the intrinsic vulnerability index of the method;

N is the total number of parameters used to calculate V;

V_{xi} represents the intrinsic vulnerability index obtained after removal of the parameter X.

- The single parameter sensitivity analysis: This test was developed by [23] . This test consists to assess the impact of each parameter in calculating the vulnerability index by comparing the calculated weight of the input parameter in each polygon with the theoretical weight assigned by the analytical model [2] . This is an analysis of the real influence of weighting compared to the weights assigned to each parameter. The real or effective weight (W) of a parameter is calculated from Equation (3):

$$W = \frac{Xr + Xw}{V} \times 100 \qquad (3)$$

With:

W is the effective weight of the parameter;

Xr and Xw are respectively the rate and the weight assigned to the parameter X and V intrinsic vulnerability index according to the method.

RESULTS AND DISCUSSION

- DRASTIC vulnerability index map

The superposition of the seven punctuated and weighted maps provides DRASTIC index map, which is carried out on the basis of a linear combination between different data values. Each pixel is characterized by a DRASTIC index value. The values of DRASTIC index are defined according to eight classes. These values represent the measurement of the hydrogeological aquifer vulnerability. The vulnerability increases with the index. These values range from 95 to 187 over the entire study area. This value range lies between the extreme values determined by the method that are 23 and 226. To the North and West of the study area, vulnerability index are relatively low and vary from 95 to 131. In the South and East of this zone, the indexes are increasingly important and vary from 131 to 187.

- Groundwater vulnerability map according to DRASTIC method

The application of the DRASTIC method combined to a GIS allowed obtaining the groundwater vulnerability to pollution map of the study area (Figure 3). This map is used to view the main risk areas linked to high index. This map is characterized by four vulnerability classes defined as follows:

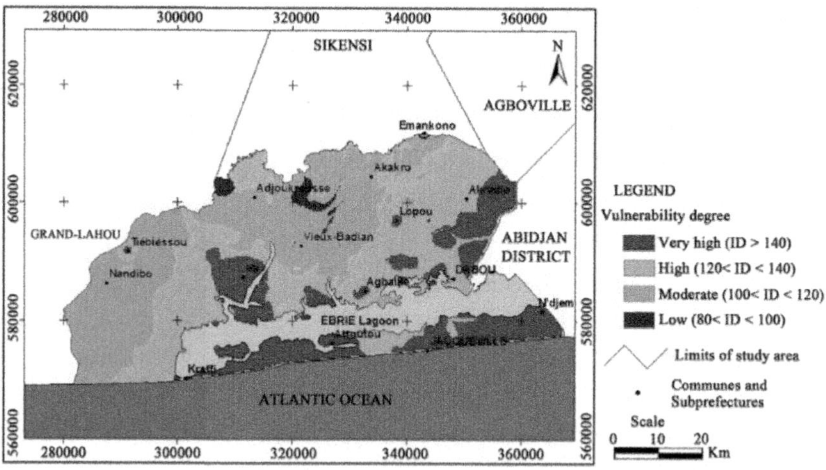

Figure 3: Groundwater vulnerability to pollution map of the study area.

- D > 140: very high vulnerability;
- 20 < ID < 140: high vulnerability;
- 00 < ID < 120: medium or moderate vulnerability;
- 0 < ID < 100: low vulnerability.

The analysis of this map shows a predominance of high and very high vulnerability classes where the index vary from 120 to more than 140 compared to moderate and low vulnerability classes where these index vary from 80 to 120. This map is in raster fashion characterized by a spatial resolution or pixel of 250 m × 250 m allowing knowing the area bounded by each class depending on the number of pixels given by the GIS.

These results show that: very high vulnerability class occupies 26.22% of the study area. This class is repre- sented in the South and East of the region where pollution risks incurred by the groundwater is very significant. The entire area contains almost all the department of Jacqueville where Quaternary formations are only represented. This extreme vulnerability can be explained by the depth to water table relatively very weak in the majority of drillings and where the clay fraction likely to attenuate pollution in the sandy formations is weak (0% - 8%). The high vulnerability class represents the modal class of the spatial distribution of vulnerability. It occupies 37.71% of the study area. This class occupies the central part of the study area. The vulnerability is moving towards the North-East and the South-West. The clay content likely to slow down the migration of the pollutant is slightly higher (10% to 45%) in sandy formations as in the previous case. The moderate vulnerability class is represented generally in North and West of the study area. It occupies 34.73% of the study area. The depth to water table becomes increasingly important with the presence of significant clay content in the formations. Hydraulic conductivity becomes low contrary to the first two classes. All these characteristics confer on the water tables a less significant vulnerability. Low vulnerability class is weakly represented in islands in North with only 1.34% of the study area. The area occupied by this class is almost negligible compared to those of other classes.

- Results of sensitivity analysis tests of DRASTIC vulnerability map

The maps resulting from the sensitivity analysis test by map removal for each of the seven parameters DRASTIC method are presented in Figures 4-10. These maps show a variation of the spatial distribution of the different classes and index calculated based on the removal of these parameters one. Table 2 summarizes the minimum, average and maximum partial index obtained.

The results in Table 2 indicate that the removal of the topography parameter or the hydraulic conductivity parameter of the overlay process were used to obtain more important average partial index values (respectively 121.48 and 123.53) compared to removal of each other parameters. However, removal the impact of vadose zone parameter of this process has to have the lowest average partial index value (97.50). The partial indexes determined were used to calculate subsequently the sensitivity index for each of the seven DRASTIC parameters (Table 3).

The analysis of the results in Table 3 shows that the depth to water table and the impact of the vadose zone have a strong influence on the vulnerability map with respective sensitivity index of 2.57 and 2.42. The topography and the hydraulic conductivity seem to have a moderate influence with respective sensitivity index of 1.58 and 1.26. The net Recharge, the aquifer media and the soil media with respective sensitivity index of 0.84, 0.52 and 0.59 to a lesser extent affect the DRASTIC vulnerability map index.

Table 2: Vulnerability partial index calculated by removing each of the seven DRASTIC parameters

Parameter Removal	Index			
	Minimum	Average	Maximum	Standard Deviation (SD)
D	66	110.36	144	10.87
R	71	106.98	163	14.29
A	71	108.88	163	16.95
S	81	115.53	173	10.07
T	87	121.48	177	14.22
I	60	97.50	147	10.70
C	89	123.53	167	13.37

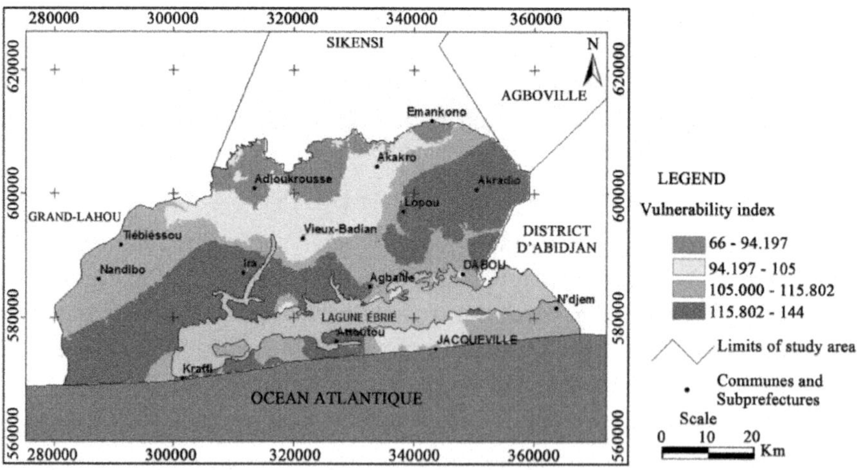

Figure 4: Sensitivity map by removing depth to water.

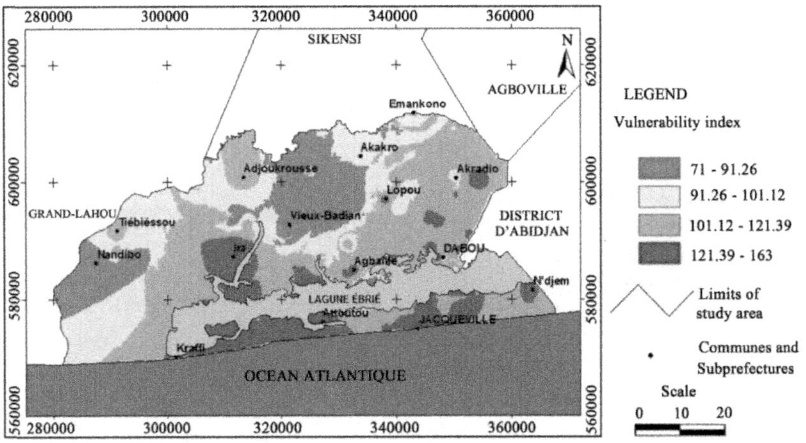

Figure 5: Sensitivity map by removing net recharge.

Table 3: Sensitivity index according to the map removal sensitivity analysis test for DRASTIC vulnerability map

Parameters	Sensitivity Index			
	S Minimum	S Average	S Maximum	Standard Deviation (SD)
D	0.28	2.57	2.71	0.26
R	0.45	0.84	1.74	0.30
A	0.45	0.52	1.74	0.75
S	0.07	0.59	2.12	0.27
T	0.93	1.58	2.79	0.30
I	2.21	2.42	3.57	0.29
C	1.12	1.26	1.92	0.15

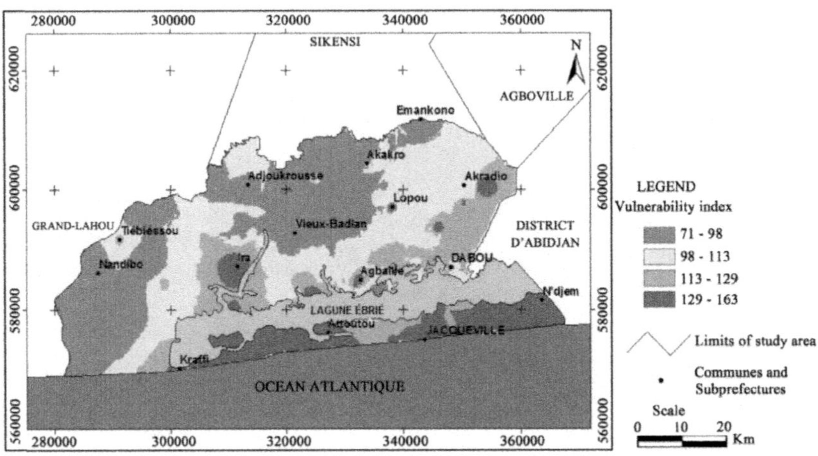

Figure 6: Sensitivity map by removing aquifer media.

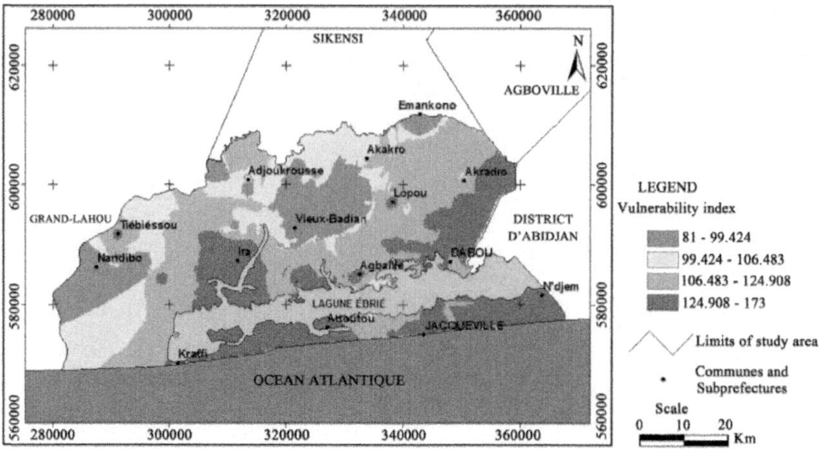

Figure 7: Sensitivity map by removing soil media.

The aquifer media is the parameter that has recorded the highest value of the standard deviation which indicates an important dispersion or a large variation of the results at the regional scale. Table 4 summarizes the results of the single parameter sensitivity analysis test.

The results in Table 4 show that the parameters such as the aquifer media (13.77%), the soil media (12.24%) and the impact of the vadose zone (22.96%) have their most important effective weight that their theoretical weight (13.04%, 8.7% and 21.74% respectively). For other parameters such as the depth of water table (20.20%), the net recharge (15.30%), the topography (4.28%) and the hydraulic conductivity (11.84%), the calculated effective weight are lower than the theoretical weight (21.74%, 17.39%, 4.35% and 13.04% respectively).

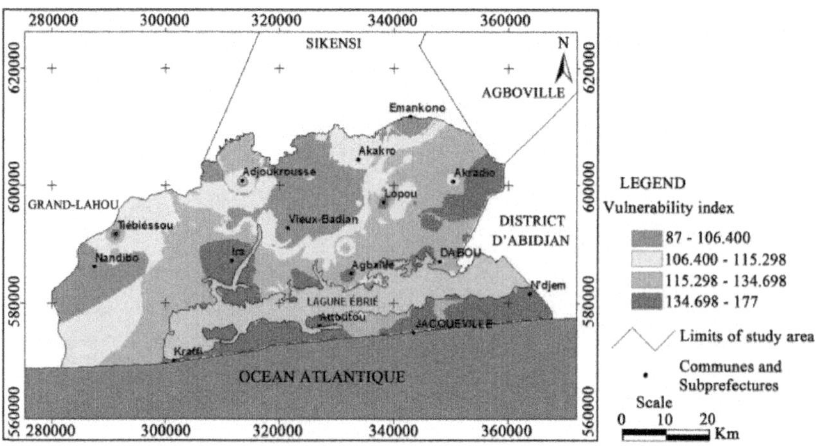

Figure 8: Sensitivity map by removing topography.

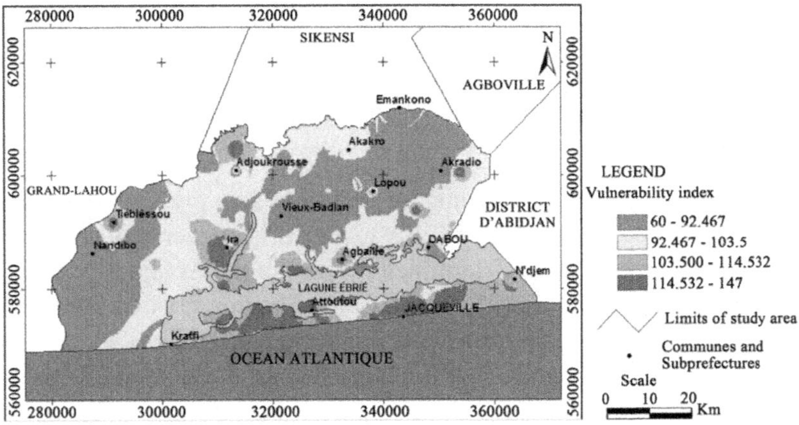

Figure 9: Sensitivity map by removing Impact of vadose zone.

Table 4: Effective weight compared to theoretical weight according to the single parameter sensitivity analysis test for the DRASTIC vulnerability map

Parameters	Weight				
	Theoretical Weight	Theoretical Weight (%)	Effective Weight (%)		
			W_{min}	W_{moy}	W_{max}
D	5	21.74	5.26	20.20	26.74
R	4	17.39	4.21	15.30	12.83
A	3	13.04	12.63	13.77	12.83
S	2	8.7	14.74	12.24	9.63
T	1	4.5	1.05	4.28	5.35
I	5	21.74	15.79	22.96	24.06
C	3	13.04	3.16	11.84	16.04

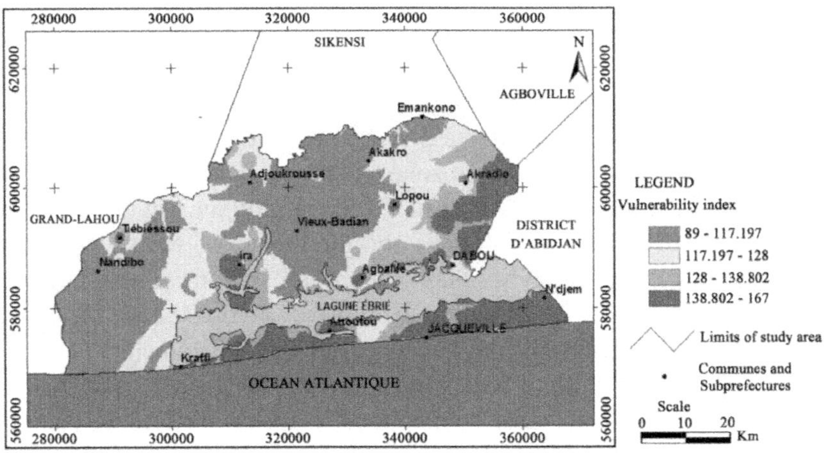

Figure 10: Sensitivity map by removing hydraulic conductivity.

Indeed, these results demonstrate that modeling in the GIS effectively reduces the subjectivity associated with the allocation of the weights to the various parameters using the DRASTIC method. These weights in fact are readjusted during modeling in the GIS to enable better assessment of the vulnerability.

The DRASTIC vulnerability map of the study area presents four classes: very high, high, moderate and low. This region is characterized by relatively low index compared to those determined by [24] in the Abidjan District (113 - 192) and those obtained by [25] in the Bonoua region (101 to 209) in Côte d'Ivoire. There exists all the same a bringing together of the index values. Indeed, it is in neighboring areas of the Ivorian coastal sedimentary basin that almost have the same hydrogeological characteristics justifying the bringing together of the index values in the assessment of the intrinsic vulnerability. The vulnerability index increases to the East or decreases to the West in this basin. These results show that the ground waters of the study area are less vulnerable than those of the Abidjan District and of the Bonoua region where calculated indexes are higher.

Moreover, it is clear from sensitivity analysis that the depth of water table and impact of the vadose zone have a high sensitivity on the maps of groundwater vulnerability in the study area. When each of these two parameters is removed from the overlay process, this leads to a significant decrease in vulnerability index. These results justify the allocation of greater weight to these two parameters in the application of the DRASTIC method and demonstrate the suitability of coupling GIS and DRASTIC method. These results are similar to those obtained by [20] who have shown that in addition to these parameters, the net recharge and the topography have a strong influence on vulnerability index in the Sub-Prefecture of Songon in the Abidjan District. The studies of [19] revealed that the topography and the depth of the water table are the parameters that have a strong sensitivity to the DRASTIC vulnerability index in the Aboisso District (Côte d'Ivoire) located on the rocky plinth. For [18] , it is the impact of the vadose zone and the hydraulic conductivity which significantly influences the groundwater vulnerability map in the Lifen Basin (China). For [16] , the aquifer media and the impact of vadose zone have a high sensitivity on vulnerability map of Russeifa area in Jordan. These same results differ from those obtained by [2] who showed that the net recharge, the soil media and the topography are the most influential parameters in assessing the aquifer vulnerability in Kakamigahara (central Japan), and that other parameters have a low to moderate impact. Indeed, the parameters that have a strong sensitivity to the DRASTIC vulnerability maps vary from one study to another. That is why [2] also point out that the variability of these parameters depends on the hydrogeological characteristics of each study area.

That same report is also valid in the comparison the theoretical and actual (or effective) weights to deduce the parameters actually involved and that have a significant impact in the assessment of vulnerability from the point of view of weighting. In this study, the parameters such as the soil media, the aquifer media and especially the impact of the vadose zone have been identified as having a significant impact in the vulnerability mapping. These results are in conformity with those of [26] and [20] who have shown that these parameters are, from this point of view, the most deter-mining in the assessing the groundwater vulnerability. [19] also showed that the net recharge made from these parameters while [18] is limited to parameters such as the aquifer media and the soil media. These results come nearer those of [27] and [16] that retained the parameters such as the depth of the water table, the aquifer media or the impact of the vadose zone respectively in the Turbio River valley (Mexico) and in the Russeifa region (Jordan). However, they differ from those of [2] which showed the significant impact of net recharge and hydraulic conductivity in assessing the aquifer vulnerability in Kakamigahara. In this study, the results of the both sensitivity analysis tests show that the impact of the vadose zone is the most important parameter in assessing the vulnerability in the study area. The groundwater vulnerability to pollution is linked to the nature of the sediments (hydrogeological formations) which condition the other parameters essentially the aquifer media and the soil media. According to [21] , the impact of the vadose zone reduces the subjectivity in assigning ratings or weight and increases the reliability in assessing of the groundwater vulnerability to pollution. Although, this parameter remains the most important, it requires all other parameters for a more objective assessment of vulnerability. According to [19] the removal of each of the seven parameters involves a sensitivity concerned with this parameter on the vulnerability map. That is why [23] confirmed that the seven DRASTIC parameters have to be considered very important in the assessment of the vulnerability. The DRASTIC vulnerability maps have already been the subject of several criticisms. For [23] these criticisms concern the inevitable subjectivity associated with the selection of seven parameters and assigning ratings and weights used to calculate the vulnerability index. Despite these limitations, the DRASTIC vulnerability map of the study area reflects in our opinion the ground realities. It can therefore be used as tool of decision-making aid for the management or the prevention of groundwater pollution risks. It helps to have an idea on sensitive areas which it will have to take into account during the regional planning [28] .

CONCLUSION

This study presents the assessment of the groundwater intrinsic vulnerability in the Dabou region from the DRASTIC method which requires seven hydrogeological parameters in its application. The established DRASTIC vulnerability map presents four classes: very high (26.22%) in the South and the East, high (37.71%) in the Center, the North-East and North-West, moderate (34.73%) in the North and West and low (1.34%) in the North. The indexes calculated during the establishment of this map vary from 95 to 187 indicating a greater vulnerability. The sensitivity analysis by map removal has shown that the impact of the vadose zone and the depth to water table has a much larger influence on the assessment of the groundwater intrinsic vulnerability in the study area. The vulnerability of groundwater in this region is strongly linked to these two parameters that allow for higher index therefore know the most sensitive areas. The single parameter sensitivity analysis shows that the aquifer media and the soil media have a more significant impact that the net recharge, the topography and the hydraulic conductivity in terms of weighting. The GIS used in this study by allowing the application of the DRASTIC method provide certain objectivity in the assessment of vulnerability. It allows easily per- forming the sensitivity analysis of groundwater vulnerability map of the Dabou region. This map is now a powerful tool of assistance in the prevention of the groundwater pollution risks. The vulnerability map constitutes a tool for monitoring human activities in the regional planning for the sustainable conservation the groundwater quality.

ACKNOWLEDGEMENTS

I would like to thank my supervisor Professor Jean Patrice JOURDA of Félix Houphouët-Boigny University (Côte d'Ivoire) for his support and guidance in completing this project. I would also like to thank Professor Jean Kan KOUAMÉ and Dr. Kouakou Serge DEH for their important contribution and all the members of the Laboratory of Remote Sensing and Spatial Analysis Applied to Hydrogeology for the success of this project.

REFERENCES

1. Morris, B.L., Lawrence, A.R., Chilton, P.J.C., Adams, B., Calow, R.C. and Klinck, B.A. (2003) Groundwater and Its Susceptibility to Degradation: A Global Assessment of the Problem and Options for Management. Early Warning and Assessment Report Series, RS. 03-3, United Nations Environmental Program, Nairobi, Kenya, 126. http://www.unep.org/DEWA/water/groundwater/pdfs/Groundwater_INC_cover.pdf

2. Babiker, I.S., Mohamed, M.A.A., Hiyanna, T. and Kato, A. (2005) A GIS Based DRASTIC Model for Assessing Aquifer Vulnerability in Kakamigahara Heights, Gifu Prefecture, Central Japan. Science of the Total Environment, 345, 127-140. http://dx.doi.org/10.1016/j.scitotenv.2004.11.005

3. Dridi, L. and Schäfer, G.L. (2006) Vapor Flux of the Chlorinated Solvents Quantization from a Source Porous Aquifer to the Atmosphere: Bias Relative to the Non-Uniformity of the Water Content and the Non-Stationary of the Transfer. Mechanical Report, 334, 611-620.

4. Melloul, A. and Collin, M. (1994) Water Quality Factor Identification by the "Principal Components" Statistical Method. Water Science Technology, 34, 41-50.

5. Foster, S.S.D. (1987) Fundamental Concepts in Aquifer Vulnerability, Pollution Risk and Protection Strategy. Hydrological Resources Processes and Informations, 38, 69-86.

6. Tapsoba, S.A. (1995) Contribution to the Geological and Hydrogeological Study of Dabou Region (Southern of Côte d'Ivoire): Hydrochemistry, Isotopic and Cationic Aging Index of Groundwater. PhD Thesis, 3rd Cycle National University of Côte d'Ivoire, 200.

7. Statistic National Institute (SNI) (2010) General Census of Population and Housing. http://fr.wikipedia.org/wiki/Dabou http://fr.wikipedia.org/wiki/Jacqueville_.

8. Tastet, J.P. (1979) Quaternary Sedimentary and Structural Environments of the Coastline of the Guinea Gulf (Côte d'Ivoire, Togo, Benin). PhD Thesis, (State) of Science University of Bordeaux 1, 181.

9. Jourda, J.P. (1987) Contribution to the Geological and Hydrogeological Study of the Area of Greater Abidjan (Côte d'Ivoire). PhD Thesis, 3rd Cycle Scientific, Technical and Medical University of Grenoble, France, 319.

10. Aghui, N. and Biémi, J. (1984) Geology and Hydrogeology of Groundwater in the Region of Abidjan and Contamination Risks. Annals of the National University of Côte d'Ivoire Series C, 20, 331-347.

11. Aller, L., Bennett, T., Lehr, J., Petty, R.J. and Hackett, G. (1987) DRASTIC: A Standardized Method for Evaluating Potential Ground Water Pollution Using Hydrogeological Settings. EPA/600/2-87-035, 622 p.

12. Schnebelen, N., Platel, J.P., Lenindre, Y. and Baudry, D. (2002) Groundwater Management in Aquitaine Year 5. Segment Operation,

Protection of the Water Table of the Oligocene in the Bordeaux Region. Report BRGM/RP-51178-FR, 20 p.

13. Mohamed, R.M. (2001) Evaluation and Mapping of Alluvial Aquifer Vulnerability to Pollution of the Plain of El Madher, Algerian North-East, Using the DRASTIC Method. Science and Planetary Change/Drought, 12, 95-101.

14. European Network for Research on Global Change (ENRICH) (2001) Climate-West Africa; A Network for: Harmonization of Climate Prediction for Mitigation of Global Change Impacts in Sudano-Sahelian. Energy, Environment and Sustainable Development Program Key Action 2. Global Change Climate and Biodiversity, University of East Anglia, Norwich.

15. Smida, H., Abdellaoui, C., Zairi, M. and Dhia, B.H. (2010) Mapping of Agricultural Areas Vulnerable to Pollution by the DRASTIC Method Coupled with a Geographic Information System (GIS): Case the Chaffar Water Table (South of Sfax, Tunisia). Science and Planetary Change/ Drought, 21, 131-146.

16. El-Naqa, A., Hammouri, N. and Kuisi, M. (2006) GIS-Based Evaluation of Groundwater Vulnerability in the Russeifa Area, Jordan. Revisita Mexicana de Ciencias Geológicas, 23, 277-287.

17. Akbari, G.H. and Rahimi-Shahrbabaki, M. (2011) Sensitivity Analysis of Water at Higher Risk Subjected to Soil Contamination. Computer Methods in Civil Engineering, 2, 83-94.

18. Samake, M.H., Tang, Z., Hlaing, W., Ndoh, M.I., Kasereka, K. and Balogun, W. (2011) Groundwater Vulnerability Assessment in Shallow Aquifer in Linfen Basin, Shanxi Province, China Using DRASTIC Model. Journal of Sustainable Development, 4, 53-71. http://dx.doi.org/10.5539/jsd.v4n1p53

19. Doumouya, I., Dibi, B., Kouame, I.K., Saley, B., Jourda, J.P., Savane, I. and Biémi, J. (2012) Modeling of Favorable Zones for the Establishment of Water Points by Geographical Information System (GIS) and Multicriteria Analysis (MCA) in the Aboisso Area (South-East of Côte d'Ivoire). Environmental Earth Science, 67, 1763-1780. http://dx.doi.org/10.1007/s12665-012-1622-2

20. Dibi, B., Kouame, K.I., Konan-Waidhet, A.B., Savane, I., Biémi, J., Nedeff, V. and Lazar, G. (2012) Impact of Agriculture on the Quality of Groundwater Resources in Peri-Urban Zone of Songon (Côte d'Ivoire). Environmental Engineering and Management Journal, 11, 2173-2182. http://www.academia.edu/18674209/Impact_of_agriculture_on_the_

quality_of_groundwater_resources_in_peri-urban_zone_of_Songon_Cote_D_ivoire_

21. Kouamé, K.I., Douagui, G.A., Koffi, K., Dibi, B. and Kouassi, K.K. (2013) Modeling of Quaternary Groundwater Pollution Risk by GIS and Multicriteria Analysis in the Southern District Share of Abidjan (Ivory Coast). Journal of Environmental Protection, 4, 1213-1223. http://dx.doi.org/10.4236/jep.2013.411139
22. Lodwick, W.A., Monson, W. and Svoboda, L. (1990) Attribute Error and Sensitivity Analysis of Map Operations in Geographical Information Systems: Suitability Analysis. International Journal of Geographical Information Systems, 4, 413-428. http://dx.doi.org/10.1080/02693799008941556
23. Napolitano, P. and Fabbri, A.G. (1996) Single Parameter Sensitivity Analysis for Aquifer Vulnerability Assessment Using DRASTIC and SINTACS. In: Kovar, K. and Nachtnebel, H.P., Eds., HydrolGis Application of Geographic Information Systems in Hydrology and Water Resources Management, IAHS Publication, Wallingford, 559-566.
24. Kouamé, K.J. (2007) Contribution to the Integrated Water Resources Management (IWRM) of the Abidjan District (South Côte d'Ivoire): Tools for Decision for the Prevention and Protection of Groundwater against Pollution. PhD Thesis, University of Cocody, Abidjan, 226 p.
25. Aké, G.E. (2009) Impacts of Climate Variability and Anthropogenic Pressures on Water Resources of Bonoua Region (South-East of Côte d'Ivoire). PhD Thesis, University of Cocody, Abidjan, 201 p.
26. Al-Hanbali, A. and Kondoh, A. (2008) Groundwater Vulnerability Assessment and Evaluation of Human Activity Impact (HAI) within the Dead Sea Groundwater Basin, Jordan. Hydrogeology Journal, 16, 499-510. http://dx.doi.org/10.1007/s10040-008-0280-7
27. Ramos, L.J.A. and Castillo, R.R. (2003) Aquifer Vulnerability Mapping in the Turbio River Valley, Mexico: A Validation Study. Geofísica Internacional, 42, 141-156.
28. Jourda, J.P. (2005) Application Methodology Remote Sensing Techniques and Geographical Information Systems to the Study of Fissured Aquifers of West Africa. Hydrotechnic Concept of Space: The Case of Côte d'Ivoire Testing Areas. PhD Thesis, University of Cocody, Abidjan, 430 p.

Chapter 6

A METHODOLOGY TO DEVELOP THE INTEGRATION OF THE ENVIRONMENTAL MANAGEMENT SYSTEM WITH OTHER STANDARDIZED MANAGEMENT SYSTEMS

Manuel Ferreira Rebelo[1], Gilberto Santos[1,2], Rui Silva[1]

[1]CLEGI, Lusíada University, Vila Nova de Famalicão, Portugal

[2]College of Technology, Polytechnic Institute of Cávado and Ave, Barcelos, Portugal

ABSTRACT

Traditionally the global management system of an organization is frequently split into a number of individual management systems that are defined and implemented according to specific man- agement systems standards (MSSs) as well as managed independently. The individual implemen- tation of MSSs is an option that leads to several inefficiencies and sub-optimization of the global management system of an organization. As referred by ISO [1] the interested parties' require- ments increase. A more effective and efficient option for an organization is to integrate, into an integrated management system (IMS), the implementation and management of requirements of multiple MSSs. Certain difficulties are associated to the structuring process, implementation, verification, evaluation, improvement and progressive development of an IMS in the organizations. Several scholars have proposed various theoretical approaches regarding the integration of individual management systems (MSs) leading to the conclusion that there is not a common practice for all organizations as they encompass different characteristics. This paper aims to present and justify a designed methodology to be used by organizations to support the integration of various MSs. Among them are highlighted: the Environmental Management System (EMS) according ISO 14001 [2] , the Quality Management System (QMS) according ISO 9001 [3] , and the Occupational Health and Safety Management System (OH & SMS) according OHSAS 18001 [4] . The methodology was designed in the context of a Portuguese company, on sequence of an organizational diagnosis and a research that was performed through a questionnaire. The strategy and the research methods took into consideration the case study.

INTRODUCTION

Due to the demand of the market itself or by other internal reasons, there are many organizations that implement different standardized MSs [5] . Standards arise through the development of detailed descriptions of particular characteristics of a product or service by experts from companies and scientific institutions [6] . According to ISO [7] , the domain of standardized management systems (MSs) has expanded greatly over the last years and nowadays there exist a relevant number of MSSs for individual MSs, which apply to any type of organization independently of its external and internal context. The objective of the development of standards is to support both individuals and companies when procuring products and services [6] . According to ISO publication [1] a common objective of MSSs is to assist organizations to manage the risks associated with providing products and services to customers and other interested parties.

As the number of MSSs versus standardized MSs increases, their integration becomes a necessity [1] [8] . In the literature and in MSSs are presented several definitions/descriptions of the concept of management system (MS) and integrated management system (IMS). Empirical studies were conducted with the aim of the integra- tion of individualized MSs around the world [9] and several tangible and intangible gains for organizations, as well as to their internal and external interested parties, are achieved with the integration of the individual stan- dardized MSs [10] - [15] , among others. Integration of MSs promotes synergies and cost savings, as well as a re- duction of the time spent when managing the systems [16] . Olaru [17] summarized forty benefits that an or- ganization can gain from the implementation of an IMS. Organizations therefore need a framework to integrate these MSs and facilitate their contribution to the operation of the overall business MS [18] . On the other hand an essential element in the strategy of any organization is the minimization of business risk to a level that ensures the security market [19] . According to Suditu [20] for organizations that want to survive and compete in the ac- tual market, it is necessary to continually improve performance of their business and MSs in a sustainable way, taking in consideration the necessities of the interested parties. In turn, according to Oliveira [21] , integration is justified as a function of the benefits that it provides; certifiable management systems that work separately are more bureaucratic and costly, and generate poorer results than those obtained employing integration. Study conducted by Simon [22] concluded that Organizations prefer integration of MSs to managing them separately and the integration of systems is one of the major strategies for ensuring survival and savings for the organizations of the sample.

Quality is no longer, as formerly, a redundant and restricted concept and must be managed in a global per- spective and of sustainability not only focused on satisfying customers, but on a whole range of interested par- ties [11] . Are example, those identified in the ISO 9004 [7] , several others exist. On the other hand, the increas- ing global competition potentiates, therefore, an increase in the expectations of all the interested parties of or- ganizations. On Table 1 there are listed several interested parties and associated needs and expectations to be satisfied by organizations according to the requirements of related MSSs.

So, more than ever, business sustainability gains increased importance and focus is shifting from their finan- cial results. These results will not verify if that focus does not prioritize also, the satisfaction in a balanced and integrated fashion of customers and others relevant interested parts, that are clearly and objectively the employees for example. In this context of real and new paradigms of management—the Global Quality Management—it is required a constant search for Business Excellence [11] .

Hence, in a not distant past, some organizations in Portugal and other countries, although in a small percent- age, began to integrate their individual standardized management systems like: EMS; QMS; OH & SMS; CSRMS-Corporate Social Responsibility Management System, among others. For this purpose, organizations began to conceive integrated procedures in order to make the integration of two systems (QMS & EMS or EMS & OHSMS) and whenever possible, the three standardized Management Systems: EMS, QMS, and OHSMS [23] . This reveals the growing interest that has been demonstrated by organizations in the adoption of the MSSs ISO 14001, ISO 9001 and OHSAS 18001.

Table 1: Examples of interested parts and their needs and expectations

Interested Parties	Needs and Expectations
Customers	Quality; price; delivery performance of products
Owners/shareholders	Sustained profitability; transparency
People in the organization	Good work environment; job security; recognition and reward
Suppliers and partners	Mutual benefits and continuity
Society	Environmental protection; ethical behaviour; compliance with statutory and regulatory requirements
Competitors	Ethical behaviour; fair competition; zero ethical faults
Government Labor unions Regulators	Attractive employer; business continuity; compliance with statutory and regulatory requirements; energy efficiency; mutual benefits; on time payment of taxes and others fees; risk management; sustained profitability; transparency

Note: Adapted and upgraded from ISO [7].

On the other hand, the integration of MSs, supported by those MSSs in a single system, taking into account the correspondence and the level of compatibility between them and potential tangible and intangible gains resulting from this integration will be an added value that organizations cannot ignore [11] . On the other hand, regulations based on ISO 9000 have been created to guide companies in developing systems for management and prevention of worker risks. Annex A and B of ISO 9001 [3] gives various clauses and subclauses related to the necessary elements of this standard [24] .

Human resources are the main subject of an organizations activity. Quality improvement and the efficiency of the organization activity depend greatly on the quality of human resources [25] . Top management support and commitment are thus essential for the initiation of the integration process, completed and subsequently main- tained. Managers consequently need to recognise that for the IMS to be implemented and maintained, they must continuously push it forward [26] . One interesting finding in research conducted by Alolayan [27] was the fairly strong correlation between top management commitment and the adoption of continuous improvement programs in the work organization. Sustainable management is the combination of management theory and the concept of sustainable development [28] , cited by Tsai and Chou [29] and according to Salomone [30] , a cultural shift is underway and the number of companies with more than one certification is constantly increasing. Many of them are advancing towards integration.

Within the past decade, the application of certification has spread from documenting quality standards to ad- ditional areas, including the management of occupational health and safety (OHS) [31] . To satisfy the require- ments of each standardized MSs like: EMS, QMS, and OHSMS organizations have to assure a lot of docu- mented procedures and other documentation, checking processes and associated records forms, among other several paperwork [32] . As stated by Zeng [33] major problems for enterprises to operate multiple parallel MSs include: it causes complexity of internal management, it lowers management efficiency, it incurs in cultural in- compatibility and it causes employee hostility and increases management costs. In a recent past, in Portugal and other countries, some companies have begun to integrate their individual MSs [34] . Figure 1 [9] presents the evolution on the number of certifications of integrated MSs (Quality, Environment, and OH & S) for 2007, 2010, and 2011 in Portugal.

IMSs and its certification are increasingly used by organizations namely to document and develop optimized conformance and business risk management, in a lean and sustainable way, in a variety of different management areas considering, in general, its internal and external context and the needs and expectations of interested par- ties, in particular. According to Jørgensen [37] the third and most ambitious level, the integration, concerns the creation of a culture of learning, focus on interested parties, continuous improvements, and synergies between the subject areas. This integration creates a sound basis for working towards a more sustainable MS.

APPROACH TO STRUCTURING A METHODOLOGY TO DEVELOP AN IMS

The standards ISO 14001 [2] , ISO 9001 [3] and OHSAS 18001 [4] , and the identification of common areas and requirements versus correspondences between them allowed to structure from the existing individual MSs in organizations, a methodology to develop the integration of the EMS with others standardized MSs like QMS and OH & SMS.

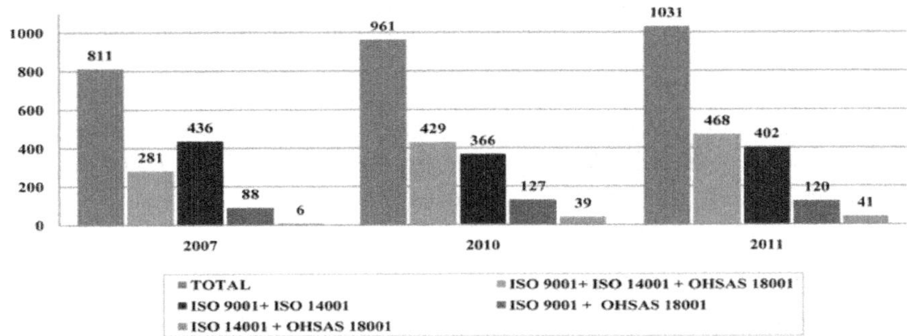

Figure 1: Evolution on the number of certified companies in Portugal related to integrated MSs (Quality, Environment, and OH & S) for 2007, 2010, and 2011. Source: Data from Portuguese Certified Companies Guide [35] [36].

There is a convergence in terms of the management model of the standards: ISO 14001 [2] , ISO 9001 [3] , OHSAS 18001 [4] , specifically in terms of the objectives associated with process efficiency, and to the fact that the three MSSs are supported on the continuous improvement principle of the Deming's Cycle-PDCA (Plan/Do/Check/Act), as described in Table 2.

The ISO 14001 [2] specifies requirements for an EMS to enable an organization to develop and implement a policy and objectives which take into account legal requirements and other requirements to which organizations subscribe, and information about significant environmental aspects. It applies to those environmental aspects that organizations identify as those which can be controlled and those which it can be influenced; the ISO 9001 [3] specifies requirements for a QMS where an organization: a) needs to demonstrate its ability to consistently provide products that meet customer and applicable statutory and regulatory requirements; and b) aims to en- hance customer satisfaction through the effective application of the system, including processes for continual improvement of the system and the assurance of conformity to customer and applicable statutory and regulatory requirements and the OHSAS 18,001 [4] specifies requirements for an OH & SMS to enable an organization to develop and implement a policy and objectives which take into account legal requirements and information about OH & S risks. So, there are several MSSs domains with potential for the effective integration of EMS with QMS and OH&SMS and integration gives a true usefulness and added value to the organization's business, more easier manageable and securely enhances the improvement of conditions in organizations in terms of management, the prevention component of EMS; QMS and OH & SMS [11] . According to Santos [23] , the com- patibility among different MSSs for an effective integration of the individual standardized MSs should be done in moderation and Almeida [38] [39] states that the success of the integration of the MSs is significantly related to the true motivations that leads organizations to integration. To achieve sustained success, top management should establish and maintain a mission, a vision and values for the organization. These should be clearly under- stood, accepted and supported by people in the organization and, appropriate to other interested parties [7].

MATERIALS AND METHOD

A preliminary investigation was conducted in the business environment, in a company, localized in the northern region of Portugal. Over the years the company—a SME, has been progressively adopting, in whole or in part, individualized MSSs and others specifications to implement independent MSs. Relevance to the ISO 14001 [2] for the EMS; ISO 9001 [3] for the QMS; OHSAS 18001 [4] for the OH & SMS, and ISO/International Electrotechnical Commission (IEC) 17025 [40] for Laboratories MS and Accreditation.

While it was imperative to assess the perception of employees of the Company on the methodology, structur- ing, implementation and evaluation of the integration model of IMS and its validation in a real business envi- ronment, it was developed an internal investigation by questionnaire, previously tested and validated, to a repre- sentative sample, of employees. Figure 2 shows the distribution of collaborators surveyed from the different le- vels of the organizational structure.

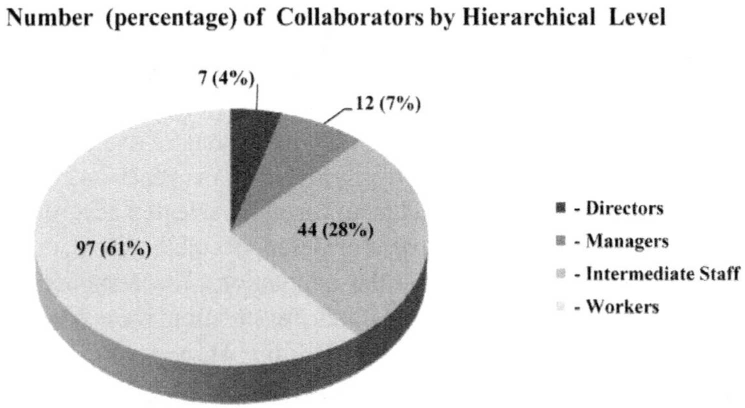

Figure 2: Distribution of collaborators by hierarchical level. Source: [15].

Table 2: PDCA (Plan/Do/Check/Act) convergence at the level of the management model by the correspondent MSSs-ISO 14001 [2] , ISO 9001 [3] , and OHSAS 18001 [4]

PDCAI Cycle	Management Systems Standards		
	ISO 14001 [2]	ISO 9001 [3]	OHSAS 18001 [4]
PLAN To understanding the context of the organizations, to establish the necessary objectives and plans to achieve them, the processes, and their sequences and interconnections, as well as the required criteria and methods to guarantee their control and effectiveness	PLANNING Environmental targets and objectives, identification of environmental aspects and assessment of potential impacts; extent to which material and financial resources are affected; competence, training and awareness of the Human Resources; planning of the activities	RESOURCE MANAGEMENT Personnel competence, training and awareness; material resources; infrastructure and work environment	PLANNING OH & S objectives; hazard identification and risk assessment; extent to which material and financial resources are affected; training, awareness and competence of the Human Resources; communication, participation and consultation; planning of the activities
DO Implement the processes, ensuring the availability of support resources and appropriate documentation and the required documented information for realization and monitoring	IMPLEMENTATION AND OPERATION Environmental management programs; control of documents; operational control; emergency preparedness and response	PRODUCT AND/OR SERVICE REALIZATION Key processes; purchasing; service provision; measurement and monitoring equipment	IMPLEMENTATION AND OPERATION OH & S management programs; control of documents; operational control; emergency plans
CHECK Perform the measurements of the processes and monitoring their performance. Analyze and assess on a regular basis the obtained results	CHECKS AND CORRECTIVE ACTIONS Monitoring and measurement; compliance-nonconformities assessment, corrective and preventive actions; records control; internal audits	MEASUREMENT, ANALYSIS AND IMPROVEMENT Client and employee satisfaction; corrective actions for non-conformities; preventive actions; internal audits	CHECKS AND CORRECTIVE ACTIONS Monitoring and measurement, accidents, nonconformities and corrective and preventive actions; records control; internal audits
ACT Take the needed actions to meet the objectives, achieve the expected results and to encourage improvements	POLICY Environmental aspects; legal and other requirements; environmental policy	MANAGEMENT RESPONSIBILITY Quality policy Decentralized management; communication and information	POLICY Hazards and risks; legal and other requirements; OH & S policy
IMPROVE Continual improvement and innovation of MSs processes	Continual improvement and innovation of the MSs, with the main objectives of its optimization and satisfaction of all interested parties with focus on them for consequent development and sustained success of the organizations.		

Note: Adapted and upgraded from [11] [13].

Note: Adapted and upgraded from [11] [13] .

It was considered a Likert scale on the questionnaire with five levels: 1) irrelevant; 2) not so relevant; 3) rele- vant; 4) ery relevant; 5) determinant. The population was the total of the collaborators of the company from whom the objective was to draw conclusions about the project objectives and the issues that are being re- searched. After the questionnaire had been tested and improved in some of its questions, it was sent by e-mail to each one the company collaborators of the sample that had been carefully selected according to their position in the hierarchy. The sample considered 49 collaborators, representing 30.62% of the total collaborators—the population. The responses rate was 86%. In the data collection, analyses and presentation were considered the guidelines of the Portuguese standard—NP 4463 [41] .

There were considered four main questions: Question 1—importance of the twelve factors identified as moti- vation for the implementation of the IMS; Question 2—influence of nine identified interested parties on the performance and evolution of the IMS; Question 3—main difficulties in a group of seven potentials, in the con- text of the development and implementation of the IMS model. Question 4—potential benefits with the imple- mentation of the IMS-QES.

The main final objective of this preliminary research was to contribute to the integration of EMS, QMS, and OH&SMS in a specific organizational context, supported on a structured methodology of development of the integration, implementation and evaluation of the designed integration model and its validation. The survey re- sults, by them self, justify, validate and prioritize enormously the structure of the designed methodology and model of IMS.

MATRIX OF COMPATIBILITY OF REQUIREMENTS-SUPPORT TO INTEGRATION

One of the activities that forms part of the scope and objectives of this preliminary research to which we have paid particular attention is the compatibility of the requirements of the MSSs, in context and framework of the characterization of the company's situation, backed up by an analysis of these MSSs. This compatibility, as pre- sented in Table 3, represent a starting point for consequents activities of integration, simplification and optimi- zation, to achieve a level of the strictly necessary and consequently the three sub MSs—EMS, QMS, and OH & SMS are integrated to the maximum extent possible.

On the matrix of Table 3 it is shown the requirements of the ISO 14001 [2] , ISO 9001 [3] , and OHSAS 18001 [4] , as well as the established correspondences, made them compatible with each other and associated with the phases of the cycle PDCA-"Plan-Do-Check-Act". This matrix orientate and align the organizational structure of the enterprise in the same direction and in addition creates a structured and useful referential methodology of work to support an effective alignment and correspondences of the sub management systems of Environment, Quality, and Safety with consequent compatibilities between each other, for consequent design and implementa- tion of the IMS. From this matrix it can also be depicted a correspondence with the PDCA cycle, in this circum- stance for the IMS, as well as a set of stages (1.1; 2.1...2.4; 3.1...3.7; 4.1...4.6 and 5.1) associated with each of the phases of the PDCA cycle.

MODEL OF DEVELOPMENT OF THE IMS

The continuous improvement of the global performance of an organization is an objective always present in the development of an IMS. The organization should therefore potentiate for each stage: Plan, Do, Check, Act, to be carefully and methodically analyzed in their differences that effectively can be observed in terms of MSSs re- quirements under clauses equivalent involved and for each phase and each stage of development of the IMS, according to the model of Figure 3 [11] [12] , to ensure its compliance and evidence of it, in full conformity.

Organizational diagnosis is an exercise done to check an organization's current health. A complete diagnosis not only checks the current health, but also suggests corrective measures [42] . So, first of all, the understanding of context of an organization, and definition, approval and communication of the integrated management policy, attentive that is common requirement to the different normative references. The leadership of top management is extremely important for the improvement of management quality. If the quality of top management is bad even if the management system of organization is good, it cannot continue to supply good outcomes such as products or services. As a result, it is thought that consequent profit is not ensured [25] . So, top management should demonstrate a strong commitment leadership and personal involvement through a defined Strategy, Policy, Ob- jectives and Targets for Quality, Environment and Safety, as well as, to make available all the needed resources to achieve the objectives [14] . The policy has to be coherent with the Mission, Vision, and Values of the com- pany, these supported on a strategy and specific objectives which in turn, support the implementation of that policy and its consequent effectiveness and continual improvement. The planning of activities in the aim of the Integrated Management System-Phase I—PLAN, is perhaps the most important of all [11] [12] . In fact, a neglected planning, including the non-identification of critical success factors, will lead to inefficiencies that can be translated into potential deviations to the objectives and consequent unsuccessful implementation of IMS.

Table 3: Matrix of compatibility of MSSs requirements and of support to the integration of the individual standardized MSs: EMS, QMS, and OH & SMS

Phase	MSSs	ISO 14001 (IMS)	ISO 9001 (IMS)	OHSAS 18001 (IMS)
Phase I-PLAN	**1-INTEGRATED MANAGEMENT POLICY**			
	1.1-Identification of organizational context; Management commitment and leadership; Continual improvement.	4.2 4.3.1	4.1 5.1 5.3 8.5.1	4.2 4.3.1
	2-PLANNING			
	2.1-Identification of: Needs and expectations of interested parties; Scope of the IMS aspects, impacts, hazards and risks and their assessment.	4.1 4.3.1	4.1 5.2 7.2.1 7.2.2	4.1 4.3.1
	2.2-Identification, access to and updating of legal requirements and other requirements of interested parties.	4.3.2		4.3.2
	2.3-Definition of objectives, targets and programmes of IMS and improvement.	4.3.3	5.4.1 5.4.2 7.2 7.3 8.5.1	4.3.3
	2.4-Definition of the plans of response to emergency situations	4.4.7	8.3	4.4.7
Phase II - DO	**3-IMPLEMENTATION AND OPERATION**			
	3.1-Resources, organizational structure, roles, responsibilities and authority.	4.4.1	5.1 5.5.1 5.5.2 6.1 6.3	4.4.1
	3.2-Training, awareness, competence and qualifications.	4.4.2	6.2.1 6.2.2	4.4.2
	3.3-Communication, participation and consultation of the interested parties.	4.4.3	5.5.3 7.2.3	4.4.3
	3.4-Documentation of the IMS.	4.4.4	4.2.1	4.4.4
	3.5-Control of documents.	4.4.5	4.2.3	4.4.5
	3.6-Product realization. Operational control.	4.4.6	7.1 to 7.5.5	4.4.6
	3.7-Operationalization of contingency plans.	4.4.7	8.3	4.4.7
Phase III - CHECK	**4-CHECKING AND CORRECTION**			
	4.1-Performance monitoring and measurement of processes and products.	4.5.1	7.6 8.1 8.2.3 8.2.4 8.4	4.5.1
	4.2-Evaluation of compliance .	4.5.2	8.2.3 8.2.4	4.5.2
	4.3-Incident investigation.	---	---	4.5.3.1
	4.4-Non-conformities; corrections; corrective and preventive actions. Control of nonconforming products.	4.5.3	8.3 8.4 8.5.2 8.5.3	4.5.3.2
	4.5-Records control.	4.5.4	4.5.4	4.5.4
	4.6-Combined internal audits.	4.5.5	8.2.2	4.5.5
Phase IV - ACT	**5-MANAGEMENT REVIEW**			
	5.1-Critical analysis and IMS review. Improvement and innovation.	4.6	5.1 5.6.1 5.6.2 5.6.3 8.5.1	4.6
	CONTINUAL IMPROVEMENT OF THE IMS			

Note. Adapted and upgraded from [12] [9].

It is therefore fundamental to invest resources and expertise at this stage, via a thorough and careful work, in order to respond effectively, within an integrated approach, to all interested parties requirements arising from the in- volved standards and others applicable specifications, including legal requirements, in this phase of the planning of the IMS with particular focus on environmental issues, customer satisfaction, and occupational health and safety of the collaborators and their families [11] [12] .

Following is the Implementation and Operation—"Do", the organization should, in this Phase II—DO, pro- mote the "Make/Do" in coherence with what was previously planned, attentive the scope of the IMS. Corresponds mainly to clauses: 4.4—Implementation and operation of ISO 14001 [2] , 7—Product Realization, of ISO 9001 [3] , and 4.4—Implementation and operation of OHSAS 18001 [4].

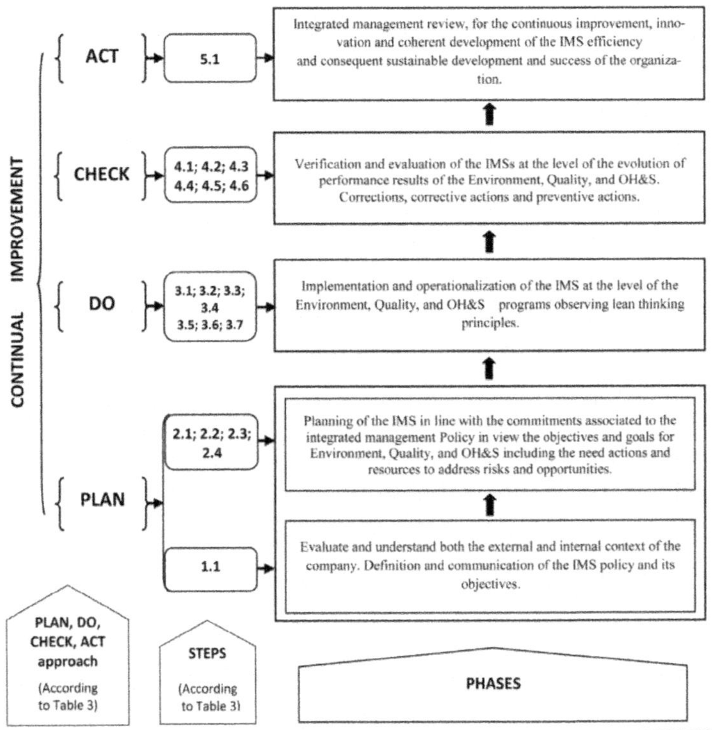

Figure 3: Model of development of the IMS Source: Adapted and upgraded from [11] [12].

In the case of ISO 9001 [3] it should be consid- ered associated with product realization, other complementary clauses, particularly in the context of resource al- location (6.1, 6.2, 6.3, 6.4) and management commitment (5.1, 5.5.1). Different activities of the organization should be set out as processes and put into operation, as per the applicable requirements, in order to strictly comply with the policies and instructions, whether they are documented or not, such as, to ensure that the or- ganization's objectives and targets are achieved and the different Stakeholders are satisfied [14] .

In the Phase III—CHECK, are considered six steps (4.1 to 4.6) designed to meet requirements of clauses: 4.5—Checking of the ISO 14001 [2] , 8—Measurement, analysis and improvement of ISO 9001 [3] , and 4.5—Checking of OHSAS 18001 [4] . With the exception of step 4.3—Investigation of incidents resulting from a specific sub-section, the 4.5.3.1—Incident investigation, the OHSAS 18001 [4] has no correspondence in the ISO 9001 [3] and ISO 14001 [2] . Critical information in the aim of the IMS should be identified, collected and analysed. Consistent with the integrated policy and the commitment to compliance, organization should estab- lish, implement and maintain documented records to demonstrate the improvements and compliance with all the applicable internal and external requirements. The use of KPIs—key process indicators, to monitor the processes, their control and continual improvement should be made systematically and guaranteed by the process owners, through the active involvement and participation of collaborators [14] . Particular relevance for the internal and external combined audits of the different components and areas of the IMS, and suppliers and subcontractors. Audits shall be scheduled at a frequency that takes into account the risk of the business. It must be conducted to assess the level of implementation and compliance of the IMS, its evolution, effectiveness and potentiate the identification of the necessary corrections and opportunities for improvement, which must be listed and priori- tized on its evaluation and implementation [14] [15] .

Finally, in the Phase IV—ACT, it was identified step 5.1—Critical analysis and review of the integrated management system, which refers to requirements of clauses: 4.6—Management review of ISO 14001 [2], 5.6— Management review of ISO 9001 [3] , and 4.6—Management review of OHSAS 18001 [4] . Top management should ensure that processes of assessment, improvement and innovation on the different components of the management system. Management reviews should be conducted at planned intervals, to identify opportunities for improvements based namely in the lean philosophy, to assess the need for modifications to the IMS and to the integrated management policy, to ensure that it continues to be appropriate, suitable, effective and efficient [14] . Top management should review the integrated management system, defining the review inputs according to the requirements of each MSSs. Management review requires, in itself, a very careful preparation phase, par- ticularly, to the level of various information that supports the inputs [11] [12] . Records of the management re- view should be retained [2] - [4].

DISCUSSION AND CONCLUSIONS

Management systems standards have developed in an unprecedented manner in the last few years. There exist at least, one for each interested part and new ones are going to be published. There is no international ISO standard with a specific structural model for Integrated Management Systems for Quality, Environment and Safety, or for other management areas such as: Risk Management; Information Security Management; RDI Management; and Social Responsibility Management, among others. The impact generated by environmental, quality, and safety and other MSSs is demonstrated by the importance of such standards worldwide, ISO 14001 [2] , ISO 9001 [3] , and OSHAS 18001 [4] . Organizations have to understand its context, the needs and expectations of interested parties and associated requirements, and consequently to determine the scope of the integrated management sys- tem, and formalise a policy to be communicated to internal and external interested parties.

The continuous improvement of global performance of organizations must be always a present goal in a per- spective of sustainability [43] . The development methodology to integrate various sub-management systems supported on a model of integration of the EMS, QMS, and OH & SMS at organizations should therefore poten- tiate, for each phase: Plan, Do, Check, Act, a careful and methodical analysis of the differences that effectively are observed at the level of normative requirements under the equivalent clauses and for each step of their de- velopment as the advocated model of integration.

Making MSSs requirements compatible through the analysis of their similarities promotes integration and can be depicted from the compatibility matrix presented—ISO14001 [2] , ISO 9001 [3] , and OHSAS 18001 [4] , es- tablishing correspondences, matching them with each other and associate the following phases of the PDCA cy- cle—Plan, Do, Check, Act: Policy and principles; Planning, Implementation and Operation, Performance Eva- luation, Improvement, Management Review. This is one of the activities that in the aim and objectives of the in- tegration model was given special attention in context of characterization and framework of the situation diag- nosed in the company in which the research was conducted. That compatibilization constitutes, the starting point for subsequent activities of integration, simplification and optimization, to a level of the strictly necessary and consequent integration maximized as desired of the three individual MSs—EMS, QMS, and OH & SMS, in con- text of strong competitiveness and changing of the business environments, which occurs at an accelerated and turbulent manner, enhancing also the sustained development of the business with added value for the relevant interested parts.

The integration of the three individual MSs represent added value both in the present and, fundamentally, for the future, not only for the company, as well as for a whole range of interested parties [43] . Examples are also highlighted by the surveyed respondents: the elimination of conflicts between individual systems with optimiza- tion of Resources; the improvement at the level of the coordinated and integrated management of the risk asso- ciated to the safety of the persons and company assets, environment and quality of the products; the reduction on the number of internal and/or external audits and audits to suppliers, and spent time versus associated costs; the creation of added value for the business through the elimination of several types of waste.

From the statistical analyses, resulting from the responses to the survey, there are shown a set of conclusions that by them self reveal: the importance, presently and for the future, of various "motivating factors" that were evaluated and alone justify and validate the model of implementation of the integration of the EMS, QMS, and OH & SMS in the company, either from internal aspects such as rationalization and optimization of resources, reduction of costs and bureaucracy; otherwise from external aspects, such as increasing competitiveness, to sat- isfy the growing demands of customers and others stakeholders. There were also identified a relevant number of difficulties, as well as a range of potential benefits resulting from the integration of the EMS, QMS, and OH & SMS into an IMS.

One of the major problems that organizations are facing with the integration of several MSSs is regarding the conception of a methodology and the implementation of an adequate structure of IMS to overcome the problems resulting from multiple MSSs. The continuous improvement of the global performance of organizations must be always a present goal in a perspective of sustainability and the route that should be taken by maximizing the in- tegration of the several individual MSs [13] , supported in an model of IMS, flexible, integrator and lean [15].

ACKNOWLEDGEMENTS

This work had the financial support of the Portuguese Foundation for the Science and Technology (FCT) through the Strategic Project-UI 4005-2014, Project Reference PEst-OE/EME/UI4005/2014.

REFERENCES

1. International Organization for Standardization (ISO) (2008) The Integrated Use of Management System Standards. In- ternational Organization for Standardization, Geneva.

2. International Organization for Standardization (ISO) (2004) ISO 14001: Environmental Management Systems. Re- quirements with Guidance for Use. 2nd Edition, ISO Copyright Office, Geneva.
3. International Organization for Standardization (ISO) (2008) ISO 9001: Quality Management Systems—Requirements. 4th Edition, ISO Copyright Office, Geneva.
4. British Standards Institution (BSI) (2007) BS OHSAS 18001: Occupational Health and Safety Management Systems—Requirements. 2nd Edition, BSI Limited, London.
5. Karapetrovic, S. and Casadesus, M. (2009) Implementing Environmental with Other Standardized Management Sys- tems: Scope, Sequence, Time, and Integration. Journal of Cleaner Production, 17, 533-540. http://dx.doi.org/10.1016/j.jclepro.2008.09.006
6. Disterer, G. (2013) ISO/IEC 27000, 27001 and 27002 for Information Security Management. Journal of Information Security, 4, 92-100.http://dx.doi.org/10.4236/jis.2013.42011
7. International Organization for Standardization (ISO) (2009) ISO 9004: Managing for the Sustained Success of an Or- ganization—A Quality Management Approach. 3rd Edition, ISO Copyright Office, Geneva.
8. Bernardo, M., Casadesus, M., Karapetrovic, S. and Heras, I. (2009) How Integrated Are Environmental, Quality, and Other Standardized Management Systems? An Empirical Study. Journal of Cleaner Production, 17, 742-750.http://dx.doi.org/10.1016/j.jclepro.2008.11.003
9. Rebelo, M.F., Santos, G. and Silva, R. (2014) Integration of Individualized Management Systems (MSs) as an Aggre- gating Factor of Sustainable Value for Organizations: An Overview through a Review of the Literature. Journal of Modern Accounting and Auditing, 10, 357-384.
10. Khanna, H.S., Laroiya, S.C. and Sharma, D.D. (2010) Integrated Management Systems in Indian Manufacturing Or- ganizations: Some Key Findings from an Empirical Study. The TQM Journal, 22, 670-686. http://dx.doi.org/10.1108/17542731011085339
11. Rebelo, M.F. (2011) Contribution to the Structuring of a Model of Integrated Management System QES. Master Thesis, Polytechnic Institute of Cávado and Ave, Barcelos.
12. Santos, G., Rebelo, M.F., Barros, S. and Pereira, M. (2012) Certification and Integration of Environment with Quality and Safety—A Path to Sustained Success. In: Curkovic, S., Ed., Sustainable Development—Authoritative and Leading Edge Content for Environmental Management, InTech, Croatia, 193-218. http://dx.doi.org/10.5772/48414

13. Rebelo, M.F. and Silva, R.G. (2012) Integration of Individual Management Systems—An Organizational Pillar for the Competitiveness and the Sustainability of Business. The International Conference on Innovation for Sustainability—IS 2012, Porto, 27-28 September 2012, 1-12.
14. Rebelo, M.F., Santos, G. and Silva, R. (2014) A Generic Model for Integration of Quality, Environment and Safety Management Systems. TQM Journal, 26, 143-159.http://dx.doi.org/10.1108/TQM-08-2012-0055
15. Rebelo, M.F., Santos, G. and Silva, R. (2014) Conception of a Flexible Integrator and Lean Model for Integrated Management Systems. Total Quality Management & Business Excellence, 25, 683-701. http://dx.doi.org/10.1080/14783363.2013.835616
16. Simon, A., Bernardo, M., Karapetrovic, S. and Casadesus, M. (2011) Integration of Standardized Environmental and Quality Management Systems Audits. Journal of Cleaner Production, 19, 2057-2065. http://dx.doi.org/10.1016/j.jclepro.2011.06.028
17. Olaru, M., Maier, D., Nicoara, D. and Maier, A. (2014) Establishing the Basis for Development of an Organization by Adopting the Integrated Management Systems: Comparative Study of Various Models and Concepts of Integration. Procedia-Social and Behavioral Sciences, 109, 693-697. http://dx.doi.org/10.1016/j.sbspro.2013.12.531
18. Asif, M., Bruijn, E.J., Fisscher, O.A.M. and Searcy, C. (2010) Meta-Management of Integration of Management Sys- tems. The TQM Journal, 22, 570-582.http://dx.doi.org/10.1108/17542731011085285
19. Nowicki, P. (2013) Risk Management—An Important Issue in Quality Management Systems. The 7th International Quality Conference, Kragujevac, 24 May 2013, 267-272.
20. Suditu, C. (2007) Positive and Negative Aspects Regarding the Implementation of an Integrated Quality-Environ- mental-Health, and Safety Management System. Annals of the Oradea University, Fascicle of Management and Tech- nological Engineering, 6, 2013-2017.
21. De Oliveira, O.J. (2013) Guidelines for the Integration of Certifiable Management Systems in Industrial Companies. Journal of Cleaner Production, 57, 124-133.http://dx.doi.org/10.1016/j.jclepro.2013.06.037
22. Simon, A., Bernardo, M., Karapetrovic, S. and Casadesus, M. (2013) Implementing Integrated Management Systems in Chemical Firms. Total Quality Management and Business Excellence, 24, 294-309. http://dx.doi.org/10.1080/14783363.2012.669560

23. Santos, G., Mendes, F. and Barbosa, J. (2011) Certification and Integration of Management Systems: The Experience of Portuguese Small and Medium Enterprises. Journal of Cleaner Production, 19, 1965-1974. http://dx.doi.org/10.1016/j.jclepro.2011.06.017

24. Vinodkumar, M.N. and Bhasi, M. (2011) A Study on the Impact of Management System Certification on Safety Management. Safety Science, 49, 498-507.http://dx.doi.org/10.1016/j.ssci.2010.11.009

25. Esaki, K. (2013) General Frame Work of New TQM Based on the ISO/IEC25000 Series of Standard. Intelligent Information Management, 5, 126-135.http://dx.doi.org/10.4236/iim.2013.54013

26. Zeng, S.X., Xie, X.M., Tam, C.M. and Shen, L.Y. (2011) An Empirical Examination of Benefits from Implementing Integrated Management Systems (IMS). Total Quality Management and Business Excellence, 22, 173-186.http://dx.doi.org/10.1080/14783363.2010.530797

27. Alolayan, S., Hashmi, S., Yilbas, B. and Hamdy, H. (2013) An Empirical Evaluation of the ISO 9001 Quality Man- agement Systems for Certified Work Organizations in Kuwait as Benchmarked against Analogous Swedish Organiza- tions. Journal of Service Science and Management, 6, 80-95. http://dx.doi.org/10.4236/jssm.2013.61009

28. Daub, C.H. and Ergenzinger, R. (2005) Enabling Sustainable Management through a New Multi-Disciplinary Concept of Customer Satisfaction. European Journal of Marketing, 39, 998-1012. http://dx.doi.org/10.1108/03090560510610680

29. Tsai, W.H. and Chou, W.C. (2009) Selecting Management Systems for Sustainable Development in SMEs: A Novel Hybrid Model Based on DEMATEL, ANP, and ZOGP. Expert Systems with Applications, 36, 1444-1458.http://dx.doi.org/10.1016/j.eswa.2007.11.058

30. Salomone, R. (2008) Integrated Management Systems: Experiences in Italian Organizations. Journal of Cleaner Production, 16, 1786-1806. http://dx.doi.org/10.1016/j.jclepro.2007.12.003

31. Granerud, L. and Rocha, R.S. (2011) Organisational Learning and Continuous Improvement of Health and Safety in Certified Manufacturers. Safety Science, 49, 1030-1039. http://dx.doi.org/10.1016/j.ssci.2011.01.009

32. Karapetrovic, S. and Jonker, J. (2003) Integration of Standardized Management Systems: Searching for a Recipe and Ingredients. Total Quality Management & Business Excellence, 14, 451-459. http://dx.doi.org/10.1080/1478336032000047264

33. Zeng, S.X., Shi, J.J. and Lou, G.X. (2007) A Synergetic Model for Implementing an Integrated Management System: An Empirical Study in China. Journal of Cleaner Production, 15, 1760-1767. http://dx.doi.org/10.1016/j.jclepro.2006.03.007
34. Santos, G., Barros, S., Mendes, F. and Lopes, N. (2013) The Main Benefits Associated with Health and Safety Man- agement Systems Certification in Portuguese Small and Medium Enterprises Post Quality Management System Certi- fication. Safety Science, 51, 29-36.http://dx.doi.org/10.1016/j.ssci.2012.06.014
35. Cempalavras, Comunicação Empresarial (2012) Portuguese Certified Companies Guide. 7th Edition. http://www.cempalavras.pt/GEC_2012/EN/index.html
36. Cempalavras, Comunicação Empresarial (2013) Portuguese Certified Companies Guide. 8th Edition. http://www.cempalavras.pt/GEC_2013/EN/index.html
37. Jørgensen, T.H. (2008) Towards More Sustainable Management Systems: Through Life Cycle Management and Integration. Journal of Cleaner Production, 16, 1071-1080.http://dx.doi.org/10.1016/j.jclepro.2007.06.006
38. Almeida, J., Sampaio, P. and Santos, G. (2012) Integrated Management Systems—Quality, Environment, and Health and Safety: Motivations, Benefits, Difficulties, and Critical Success Factors. Book of Abstracts, The International Symposium on Occupational Safety and Hygiene—SHO 2012, Guimarães, 9-10 March 2012, 13-15.
39. Almeida, J., Domingues, P. and Sampaio, P. (2014) Different Perspectives on Management Systems Integration. Total Quality Management and Business Excellence, 25, 338-351.http://dx.doi.org/10.1080/14783363.2013.867098
40. International Organization for Standardization/International Electrotechnical Commission [ISO/IEC] (2005) ISO/IEC 17025: General Requirements for the Competence of Testing and Calibration Laboratories. ISO Copyright Office, Geneva.
41. Instituto Português da Qualidade (2009) NP 4463: Linhas de orientação sobre técnicas estatísticas para a ISO 9001: 2000 (ISO/TR 10017:2003). Instituto Português da Qualidade, Caparica.
42. Saeed, B. and Wang, W. (2014) Sustainability Embedded Organizational Diagnostic Model. Modern Economy, 5, 424- 431. http://dx.doi.org/10.4236/me.2014.54041

43. Rebelo, M.F. and Santos, G. (2012) Integration of the Occupational Health and Safety Management System with the Quality Management System and Environmental Management System—From the Theory to the Action. Book of Abstracts, The International Symposium on Occupational Safety and Hygiene—SHO 2012, Guimarães, 9-10 March 2012, 372-374.

Chapter 7

GEOLOGICAL AND GEOTOURISM STUDY OF IRAN GEOLOGY NATURAL MUSEUM, HORMOZ ISLAND

Abdollah Yazdi[1], Mohammad Ali Arian[2], Mahmoud M. Rezapour Tabari[3]

[1]Department of Geology, Kahnooj Branch, Islamic Azad University, Kerman, Iran

[2]Department of Geology, North Tehran Branch, Islamic Azad University, Tehran, Iran

[3]Department of Engineering, Shahrekord University, Shahrekord, Iran

ABSTRACT

Iran is a country that benefits from nice nature, diverse continent, areas full of unique geological phenomena. Thus, it is necessary to study these attractions for better recognition of them. In this regard, Hormoz Island with valuable geoheritage, biodiversity, cultural, historical and political diversity is very important. The accumulation of these attractions and its being located in Persian Gulf strategic area made it of considerable significance in national and international communities. Hormoz Island is a spherical salt dome which is located in Hormoz strait. This Island is composed of evaporites, igneous rocks and sedimentary rocks, and sediments mainly belong to Mishan and Aghajari formations and salt, gypsum and, to a lesser extent, limestone evaporites. In sedimentary formations of Hormoz, Aouthigenic minerals such as Pyrite, Dolomite, quartz, Anhydrite, gypsum and halite are frequently seen. Mineralization of volcanic leads to formation of high temperature minerals, such as oligiste, pyroxene, amphibole and low temperature and hydrothermal minerals such as Pyrite, quartz... in the tracks of igneous rocks. This unique geodiversity in rocks and mineral which made various colors in Hormoz Island made it a mineraogical reservoir. Beside considerable mineralogical attractions, there are other potentials like ochre mine, coral reefs, rock seashore, sea caves, salt caves, plant cover and wild life which have added to various tourism capabilities of this Island and made it a unique place in the world. This paper studies Hormoz Island in terms of geological features and geotourism potentials.

INTRODUCTION

In recent years, tourism industry has become one of the main poles of development from various respects, especially economic source. It is such that most authorities believe that tourism will become dominant industry in near future and will have various socioeconomic effects (Bayati Khatibi, 2010) [1] . Due to job creation and relatively rapid profiting characteristic, tourism is a proper ground for foreign investment and can accelerate tourism development, promote its economic criteria and bring out new ideas, technologies and markets (Papeli Yazdi, 2011) [2] . Geotourism, as a subcategory of tourism, is considered one of the new methods in providing tourism attraction (Servati, 2008) [3] and has allocated a main part of tourism studies to itself. Iran has a nice nature, diverse continent and areas full of unique geological phenomenon so that it seems necessary to investigate these attractions for their better cognition and geotourism development (Yazdi, 2012) [4] .

Iran is one of the few countries of the world with beautiful natural and geological phenomenon due to its unique and excellent geographical condition. In this regard, Hormoz Island has become geological paradise due to certain geological condition and unique diversity of minerals and rocks and attracted many geotourists. This spherical island has been located in Persian Gulf' entrance in 56°25' to 56°30' east longitude and 27°2' to 27°6' north altitude. Its long diameter is maximumly 9 km, its short diameter is almost 5.5 km, and it covers an area about 41 km (Figure 1). This Island is an adult salt dome and its morphological, mineralogical and lithological diversity is significant. Besides geological attractions, this area benefits from historical attractions like Portuguese castle and natural attractions such as alga seashore, wildlife, etc. which will be investigated in this paper.

METHODOLOGY

This research is practical and developmental and the methodology used in it is descriptive-analytical. Based on this, library-documentary studies, interpretation and analysis of satellite images, various field studies and direct observation of phenomenon have been used.

GEOTOURISM

Geotourism is a relatively new concept in tourism industry which has considerable growth in recent decade. Geotourism has a certain definition with geological tourism at its centre (Newsome & Dowling, 2006) [5] and deals with the investigation of related forms and consequences to earth, geomorphologic and geological phenomena. According to Gates (2006) [6] , geotourism means "tourism in geological outlooks". Geotourism, according to Dowling & Newsome (2006) [7] , deals with geology, geomorphology, natural outlooks and the forms of earth surface, layers with fossil, rocks and minerals with emphasis on the creating processes. Furthermore, it can be argued that geotourism is informed and responsible tourism in nature with the aim of looking at recognition of geological phenomena and processes and learning their formation and revolution (Amrikazemi, 2009) [8] . According to the above definitions, geotourism is not only is new part of tourism market, it is a principal guidance to help maintaining nature and sustainable development, which is compatible with the economic equilibrium, social condition and ecology and complements them.

GEOLOGY OF HORMOZ ISLAND

This Island is one of the biggest and most famous salt domes of Persian Gulf whose formation has begun from

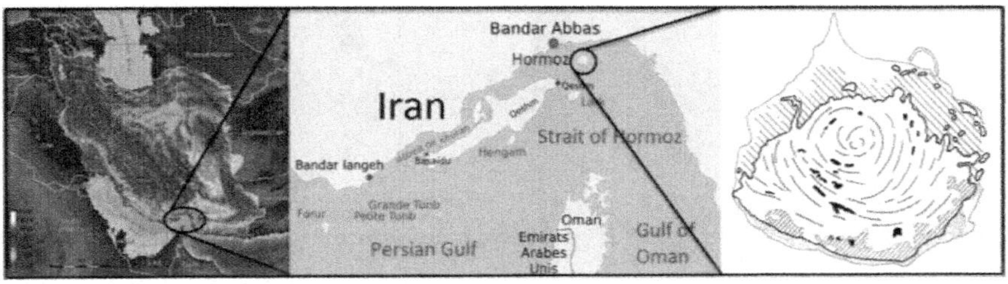

Figure 1: The location of Hormoz Island in Iran.

late Precambrian and come out of water about 50,000 years ago. Hormoz Island is almost structurally concentric with evaporites, igneous rocks and sedimentary rocks layers. Bland Ford in 1872 called the collection of igneous, metamorphic and evaporites rocks of this area as Hormoz Series.

According to Elyasi et al. (1975) [9] , this Island is composed of evaporites rocks, ferroginous formations and Miocene-Pliocene sediments (Figure 2). Beside evaporites, sedimentary rocks and igneous are can be find which has been previosly called Hormoz formation. Stocklin [10] in his study from 1961 to 1968 concluded that expression Hormoz series is true for underlying salt layers and gypsum and sandy rocks and limestone with fossil can be separated and limit the expansion of evaporites in late Precambrian in northern-southern extent of wide area which reaches Oman-Naiband fracture from east and Qatar-Kazeron fracture from west (Stocklin, 1972) [10] . In overall, Hormoz series collection can be attributed to a row of evaporites, igneous and rarely metamorphic rocks which have been deposited after late Precambrian. These rocks began with a complete cycle with dark and usually dolomite carbonate rocks and sediment on chalk, salt and shale. Concerning the similarity of Hormoz series fossils and fossils on lower part of Valdai series in Russia platform and late part of Precambrian in Australia, Hormoz series can be considered as belonging to late Precambrian.

Various magma activities in Magma Island are seen in form of extrusive and intrusive. Igneous events happened in two phase, the first one includes analysed basalt and diabase and the second one includes rhyolite, rhyodacite and trachyte which respectively belong to Permian and Triassic. The mentioned rocks, except rhyolite (Figure 3) which is somehow intact, are mainly decomposed.

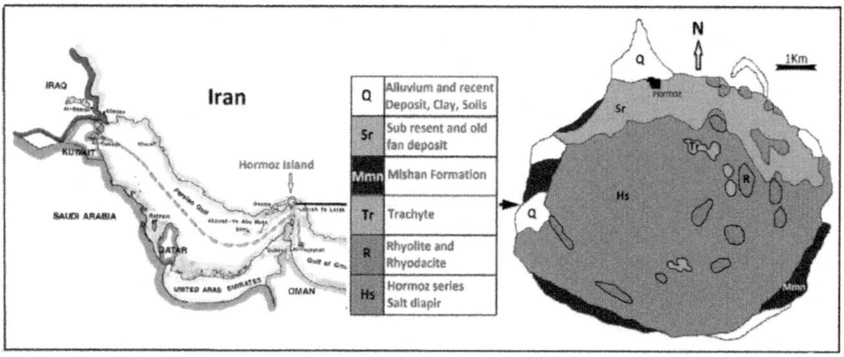

Figure 2: Geology map of Hormoz Island by corrections (Elyasi et al., 1975) [9] .

Figure 3: An overview of Trakyt-Riolite mass of Hormoz Island.

HORMOZ SLAT DOME

In southern Zagros area, there is about 130 salt domes which erosion has created beautiful perspectives in them (Figure 4) such as salt cave in these salt domes. In lower layers of Persian Gulf area, there is a thick bed of salt in 4 - 10 km depth which due to special weight less than lateral rocks and the weight pressure of the upper layers, penetrated in higher layers and reached surface and created various domes. This salt movement might happen in faults through sedimentary layers... The width of most salt domes is about 1 to 10 km and their big dome-like mountains are usually salt lump however, their surface is usually covered by red soils.

The main soil of its covering rock with several meter thickness is in fact the residual of insolvable materials from the main salt beds. The dominant soil of this bed is clay and silt with usually more than 50% plaster-rock. The red colour is up to 15% related to iron oxide that is usually in form of hematite. Some black salt domes are frequently seen. The lump values for salt domes in southern Zagros are usually 2 to 6 mm. Although it seems that in some sample, 15 mm growth has happened (Waltham, 2008) [11] .

Hormoz salt dome has come out of water about 50,000 years ago and due to severity of salt tectonic movements, Neogenes sedimentary layers reached 75-degree slope. Raining has been able to be effective in salt domes in creation of cavities and salt caves as can be seem in seashore salt domes "Namakdan and Hormoz" in Qeshm and Hormoz Islands which have been identified and drawn by geologists from Czech (Figure 5).

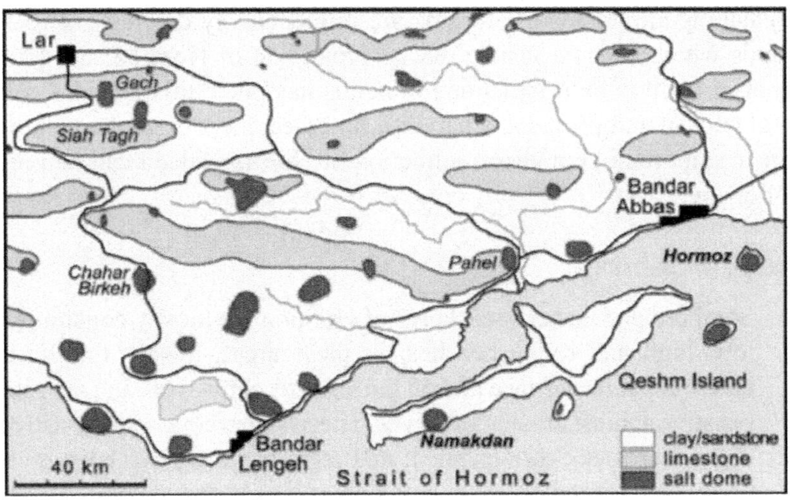

Figure 4: The situation of salt domes of south Zagros and Persian Gulf.

Figure 5: Layering of salt and iron oxide in roof of salt cave of Hormoz salt dome.

GEOLOGICAL AND GEOTOURISM ATTRACTION OF HORMOZ ISLAND

In addition to attractive biological outlooks of Hormoz Island, the geological attractions and perspectives of this Island is very interesting and extraordinary. Eye-catching attractions of Hormoz are influenced by diversity of rocks and minerals which became visible due to protrusion of Hormoz salt done. The diversity of soil colour, rocks and minerals has made this Island a colourful natural tableau and provided a paradise for researchers of geology. In overall, the most important geotourism attraction of Hormoz Island can be presented as following.

Hormoz Seashores

- Southern and eastern seashores of Hormoz are mostly constituted from low, high and rocky beaches. In these areas, mostly fossilized lime layers are at the surface and on the less-strength layers and are relatively resistive against erosion of waves. These layers cover non-resistive units like cover rock. Various small and large cavities have been created in lower parts of rocks due to collision of waves and erosion. These dents have been formed at the bottom of rocks and would become larger over time (Figure 6 & Figure 7).

- Another part of Island seashore near to ocher (red soil) mine is known as red seashore. Due to collision of seawater and this soil which has high iron oxide, the colour of this beach turns red and creates a beautiful scene (Figure 8).
- Other parts of Hormoz Island are known as colourful beaches. Each area of its colourful mountain represents a beautiful and extraordinary beach and has unique diversity (Figure 9).

Figure 6. Precipice cliffs and rocks of eastern south of Hormoz (Aghajari formation has creamy colour and Mishan formation is green in farther areas).

Figure 7. A nice view of southern seashores of Hormoz Island.

Figure 8. Red seashore around Hormoz ochre/red soil mine.

Figure 9: Colourful mountainous seashores of Hormoz Island.

- Other parts of Hormoz beach is constituted from beautiful ripple mark covered with oligiste crystals which make this area shiny especially at the time of sunshine (Figure 10).

Red Soil Mine (Ochre Mine)

Currently, this mineral material has been seen in Persian Gulf area in salt domes and some Islands. Among various mines in this area, Hormoz ochre mine can be referred to which is the most significant one. This mine has a reservoir about 390,000 tons and is unique in terms of quality and application in industries all over the world (Figure 11). The soil of this mine has application in some industries such as cosmetics, paint and rust, colourful glasses, paper, enamel, ceramic, rubber, colouring construction materials, chemical fertilizer and...

(Aqanabati, 2006) [12] . Some years ago the soil of this mine was exported to countries such as Britain, France, Portuguese, India, EAU, and Pakistan and so on, however, now the export of this mineral material is controlled and it is less exported.

This mineral material is seen in form of lens shape masses and has been created due to washing of ferrous rocks by penetrating waters and transfer of feo^{+3} to the earth. Hydration of ferrite rick leads to creation of different kinds of hydrated oxides like limonite and goethite (Moeinvaziri, 1996) [13] which have created yellow and red mountains (especially in Qeshm-Hormoz path).

Native habitants of this Island call the red soil of this area as "Gelak" and use it in different kinds of food as spice. One of these foods is "Suragh" delicious food which is prepared by soaking Sardine fish in Hormoz red soil and orange and is served with bread (Figure 12).

The Diverse Minerals of Hormoz Island

Mineralization processes in igneous rocks lead to creation of various beautiful minerals such as oligiste, pyroxene, amphibole, pyrite, quartz, etc. in tacks of igneous rocks. This unique diversity in rocks and minerals has

Figure 10: Ripple marks of seashore covered with oligiste mineral.

Figure 11: A layout of ochre mine of Hormoz Island.

Figure 12: The area where native habitants of the region use its soil for food.

created various colours in Hormoz Island and made it a treasure of minerals (Yazdi, 2013) [14] .

Gypsum: These crystals can be found in some parts of Islands in form of rhombohedral. These crystals will turn to anhydrite white powder due to dehydration (Figure 13).

Pyroxene: In southern west of Hormoz Island, pyroxene has been formed between tracks of rhyolite masses. These crystals can be seen in 5 cm length along with oligiste and alpha quartz (Moeinvaziri, 1996) [13] , (Figure 14).

Oligiste: High temperature oligiste crystals can be seen in form of bipyramid, hexagonal or in form of simple hexagonal pyramid, low temperature oligiste crystals can be seen in form of thin hexagonal aglet with rhombohedral surfaces in igneous rocks or in sedimentary rocks (shale and dolomite) (Figure 15).

Pyrite: This mineral material has been created in form of didodecahedral and octahedral in Hormoz igneous tuffs by fumerole. In addition to tuffs, pyrite automorph crystals can be seen in sedimentary rocks (Figure 16).

Apatite: Apatite crystals are frequent in ferrous riolite breccia and due to high frequency of these crystals, even they can be thought as a source for rare earth elements (Figure 17).

Dolomite: Dolomite is seen in form of big and automorph crystals in form of grayish and white rhombohedral and pinacoidal in sediments. Small and automorph crystals can be seen in some cavities (Figure 18).

Quartz: It is seen in form of hexagonal pyramids with rhombohedral surface.

Salt Diverse Structure in Hormoz Island

Since Hormoz Island is a salt dome, considerable structures of various salt forms can be seen in central parts of Island (Figure 19).

Figure 13: Gypsum beautiful crystals.

Figure 14: Pyroxene crystals of Hormoz Island.

Figure 15: Wrinkled layers containing oligiste.

Figure 16: Pyrite crystals.

Figure 17: A sample Apatite in Hormoz Island.

Colourful Soils of Hormoz Island

One of the amazing and wonderful geological phenomena of Iran is colourful mountains of Hormoz Island which have created landscape attractions of universe (Figure 20). It is an Island which is natural for geologists and an original scene for artists. What distinguishes Hormoz from other Islands of Persian Gulf is its colourful rainbow soil which has been formed due to diverse rocks and minerals. Thus, most artists design various shapes or carpets with this colourful soil which is unique and outstanding (Figure 21).

Figure 18: A sample Dolomite in Hormoz Island.

Figure 19: Diverse samples of solved salts and gypsum in Hormoz Island.

Figure 20: A view of unique diversity of coloured mountains of Hormoz.

Figure 21: Carpets made of coloured soils of Hormoz.

CONCLUSION

What was mentioned in this paper indicates that considering the global significance of geotourism industry, Hormoz Island which is one of the most beautiful and important salt domes of universe, has high potential for geotourism development. Diversity in geological perspectives and numerous minerals of Hormoz Island, which made it a natural museum for fonder of geology, is a strong potential for making this region be one of the world geotourism poles and a proper area for researchers and scholars. In this way, researchers are able to take proper actions for scientific development of geotourism in this area with stronger foundations obtained from research finding and transfer scientific grounds of phenomenon and geo-sites to visitors. In this condition, geoconservation (conservation of geological phenomena) has been properly performed by tourists. Moreover, geotourism finds its real meaning besides economic development, thus, Hormoz Island can be a good pattern for achieving the above objectives.

REFERENCES

1. Bayati Khatibi, M., Shahabi, H. and Qaderi Zadeh, H. (2010) Geotourism: A New Approach in Utilization of Geomorphologic Attractions. Journal of Geographical Space, 29, 27-50.
2. Papeli Yazdi, M.H. and Saghaei, M. (2011) Tourism, Nature & Concepts. 6th Edition, Samt Publication, Tehran, 6.
3. Servati, M.R. and Qasemi, A. (2008) Geotourism Strategies in Fars. Journal of Geographical Space, 2, 6.
4. Yazdi, A. (2012) A Study of Iran's Lut Desert: Geomorphological and Geotourism Attractions. Proceedings of Annual International Conference on Geological & Earth Sciences (GEOS2012), Singapore, 3-4 December 2012, 35-41.
5. Newsome, D. and Dowling, R.K. (2006) The Scope and Nature of Geotourism. In: Dowling, R.K. and Newsome, D., Eds., Geotourism, Chapter One, Elsevier, Oxford, 3-25.
6. Gates, A.E. (2006) Geotourism: A Perspective from the USA. In: Dowling, R.K. and Newsome, D., Eds., Geotourism, Chapter Nine, Elsevier, Oxford, 157-179.http://dx.doi.org/10.1016/B978-0-7506-6215-4.50017-8
7. Dowling, R.K. and Newsome, D. (2006) Geotourism's Issues and Challenges, Geotourism, Chapter Thirteen, Elsevier, Oxford, 242-254.

8. Amrikazemi, A. (2009) Atlas of Geopark & Geotourism Resources of Iran. Geological Survey of Iran Publication, Tehran, 22-23.

9. Elyasi, J., Aminsobhani, E., Behzad, A., Moeinvaziri, H. and Meysami, A. (1975) Geology of Hormoz Island. Geological Survey of Iran Publication, Tehran, 1, 13.

10. Stocklin, J. (1972) Lexique Stratigraphique International. Vol. 3, Geological Survey of Iran Publication, Tehran, 6, 15.

11. Waltham, T. (2008) Salt Terrains of Iran. Geology Today, 24, 188-194. http://dx.doi.org/10.1111/j.1365-2451.2008.00686.x

12. Aqanabati, S.A. (2006) Geology and Mineral Potential of Hormozgan Province. Journal of Development of Geology Training, 12, 4-11.

13. Moeinvaziri, H. (1996) An Introduction of Magmatism in Iran. University of Tarbiat Moalem Publication, Tehran, 440.

14. Yazdi, A., Emami, M.H. and Jafari, H.R. (2013) IRAN, the Center of Geotourism Potentials. Journal of Basic and Applied Scientific Research, 3, 458-465.

Chapter 8

HYDROGEOLOGICAL FRAMEWORK AND GROUNDWATER BALANCE OF A SEMI-ARID AQUIFER, A CASE STUDY FROM IRAN

Leila Khodapanah[1], Wan Nor Azmin Sulaiman[1], Hamid Reza Nassery[2]

[1]Department of Environmental Sciences, Faculty of Environmental Studies, Universiti Putra Malaysia, Serdang, Malaysia

[2]Faculty of Earth Science, Shahid Beheshti University, Tehran, Iran

ABSTRACT

Climate changing and associated factors combined with considerably increases in water demand have been accompanied by severe depletion of reservoir storage of the most groundwater supplies of Iran. Shahriar aquifer in west of Tehran is a representative aquifer of these kinds. In order to meet water demand of the area and protecting groundwater from quantity and quality deterioration, precision recognition of geology, hydrologic and hydrogeologic characteristics of the aquifer is first step. The basic objective of this study is to develop the hydrogeological framework of the groundwater system in Shariar, Iran and to estimate groundwater balance as a scientific database for future water resources delevopment programs. Based on this research lateral groundwater inflows, direct infiltration of rainfall, stream bed infiltration, irrigation return and surplus drinking and industrial water are the recharging factors of the aquifer. Subsurface outflows, domestic and industrial pumping wells and agricultural abstraction are the main parameters discharge the aquifer system. Water balance in the Shahriar aquifer system is in disequilibrium and a deficit of about 24.7 million cubic meters exists.

INTRODUCTION

Development of hydrogeological investigation with emphasis on recharge rate estimation is one of the basic steps for a suitable water management [1]. This constitutes a major issue in regions with large demands for ground water supplies, such as in semiarid areas, where such resources are the key to agricultural and industrial development [2-4]. There are various methods for aquifer recharge estimations. Physical and chemical methods are the two main categories for estimating aquifer recharge rate [5,6]. Water table fluctuation is one of the most applied methods in recharge estimation. This method calculates the ground water storage change with considering water table fluctuations and the storage parameter. This method is considered to be one of the most promising and attractive due to its accuracy, ease of use and low cost of application in semiarid area [7]. The water table fluctuation method was first used to estimate ground water recharge and has since then has been used in numerous studies for the same purpose [8].

Shahriar plain, situated in west of Tehran and between two rivers Kan and Karaj, was someday considered as one of the main centers of producing sapling and seed in the country. In 1960, Amir Kabir multipurpose dam was constructed on Karaj river for flood storage and regulation, supplying drinking water for Tehran, capital of Iran, agriculture water regulation and supplying hydroelectric power [9]. After several years despite of covering forenamed aims, downstream lands including Shahriar plain was encountered water shortage. Lack of surface water in downstream and feasibility of drilling deep wells has deployed using of Shahriar's groundwater, as these supplies has become main water source of area [10]. Over-exploitation of groundwater resources as a consequent of population growth and rapid development in agriculture and industry has caused water scarcity and drop in groundwater level. Semiarid climate and not recharging aquifer through Karaj River has exacerbated the condition of Shahriar aquifer.

In view of this, there is an urgent need to reevaluate the status of water resources, quantity and quality monitoring of groundwater and better recognition of present aquifer condition for optimum utilizing of water resources in this area. The purpose of this study is to define a conceptual framework of the Shahriar aquifer, to consider the present groundwater resource potential and hydrodynamic characteristics of geological formation of study area. The discharge and recharging parameters of aquifer and water balance of Shahriar groundwater are also assessed.

MATERIALS AND METHODS

Study Area

The Shahriar plain covers an area of 580 km^2 and lies between longitudes 50° and 50°15' east and latitudes 35°30' and 35°45' north (Figure 1). The study area is bounded southern submontane of Central Alborz mountains to the north and conglomerate hills to the south, Karaj River to the west and Kan River to the east. The main towns in the area are Shahriar, Saeid Abad, Eslamshahr, Ghale Hasan Khan and Robat Karim. Topographic elevations range from 1020 to 1200 m above sea level. Mean plain elevation is 1100 m above sea level. The slope of study area in northeastern and eastern regions is toward south and in northwestern and western is toward southeastern. The general slope is roughly 0.7%.

Meteorological and Hydrologic Conditions

Prevailing climate of study area has semi arid characteristics. According to data of three meteorological stations

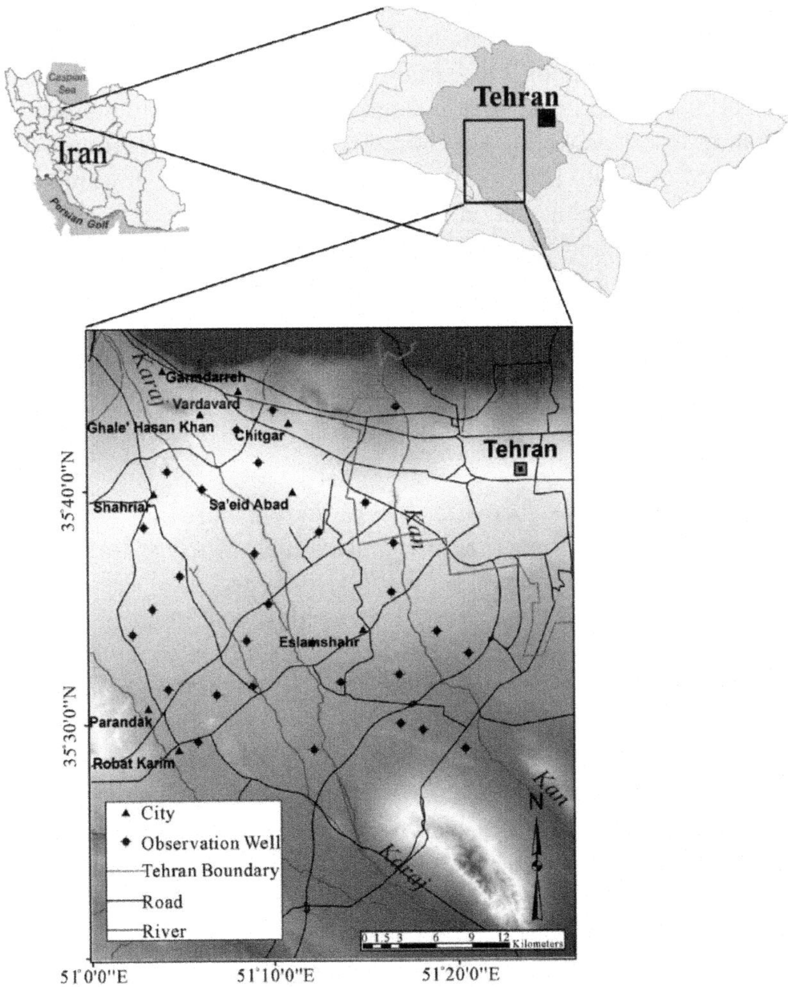

Figure 1: Study area map showing the location of monitoring boreholes.

existing in the area, average annual precipitation recorded for the years 1967-2005 is about 230 mm. The wet season begins from November and continues till May. The maximum rainfall occurs in February (15% - 18%) and winter is the most rainy season of the year. The average minimum temperature is 3.5°C in January and the average maximum is 30°C in August. Mean monthly relative humidity is about 67% in January and minimum relative humidity is 26% during the July-September period [11].

The average annual pan evaporation rate measured at Karim Abad station is about 2270 mm. According to Demarton and Emberger climate classification [12], the study area is dry and cold.

There are three rivers in northern mountains which include the Kan, Chitgar and Karaj from east to west. Kan and Karaj are perennial rivers. During wet season Kan joins to Jajrud river in south of Tehran, meanwhile the others become dry in lower part in south of the region due to percolation and evaporation losses as well as diversion of the water for irrigation. All rivers in the region originate from northern catchments and flow toward south and southeast. Three gauging station on main river are Bileghan station on Karaj river and Kahrizak and Sooleghan stations both on the Kan river. The mean discharges of these stations are 15.87, 0.8 and 2.46 m^3 per second respectively (Figure 2). The mean discharge of Chitgar river is about 0.5 m^3 per second which is calculated based on flow-duration relationships between basins with similar characterisics.

Geological Setting

Shahriar plain is located between Alborz and Central Iran tectonic zones. Present condition and topographic characteristics of study area in south of Central Alborz mountain has been resulted from tectonic processes and the three main river alluvial fans during time, as illustrated in the geological map (Figure 3). The study area is bounded between Alborz mountain and some small conglomerate hills to north and height of different lithology to the south. Central and southern smooth parts of plain form main aquifer and originate from the filling of a tectonic depression with quaternary alluvial. The oldest formation in the area outcrop on the Alborz mountain and southern height and are represented by tuff, andesitic and pyroclastic rocks (Eocene) [13,14]. These rocks which has deposited during the Laramian (laramid) orogenic phase are called Karaj Formation. In the Miocene epoch, there was a marine regression and a change to continental condition, mainly lacustrine, with the deposition of high-colored marls, gypsum, conglomerate and sandstone.

The sediments deposited in this area in the Pliocene epoch caused by Pasadenian orogeny phase are called Hezar Darreh formation. Due to resuming sedimentation till prior quaternary, these sediments are also known as Plio-Quaternar or Late Quaternary Alluvial [15]. Hezar Darreh formation, derived from the uplift of the Alborz Range on the north (Karaj formation tuff), consist of thick coarse light-colored conglomerates and sandstones

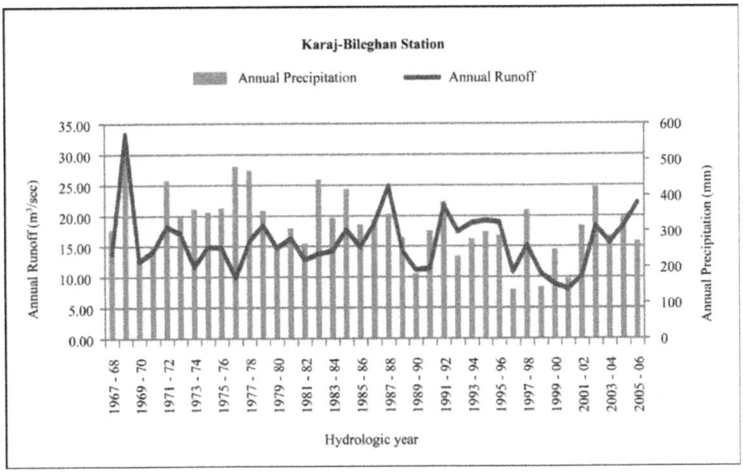

Figure 2: Annual precipitation and runoff in the Karaj-Bileghan station.

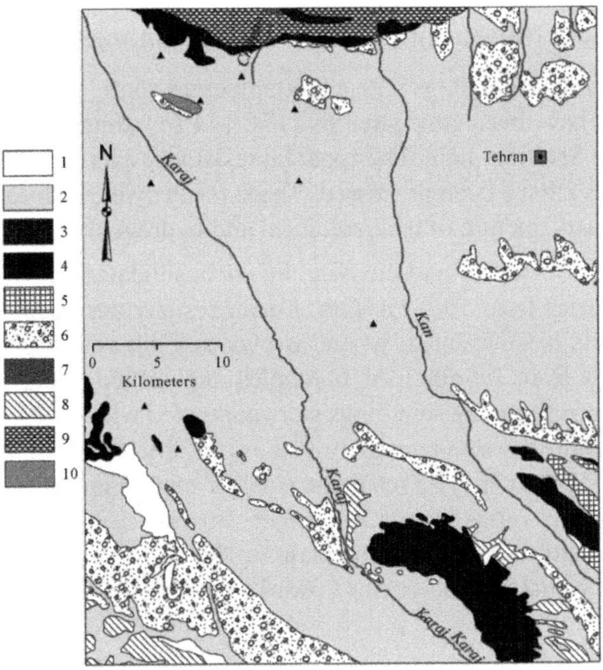

Figure 3: Geological map of the Shariar plain. 1) Alluvial terraces; 2) Alluvium (including TEHRAN ALLUVIUM); 3) Andesite/dacite lava and pyrocalstics marl; 4)

Andesitic lava and pyrocalstics; 5) Conglomerate (including HEZARDARREH FORMATION; Mio-Pliocene); 6) Conglomerate (including KAHRIZAK FORMATION; plio-pleistocene); 7) Mudstone, shale, dacite, pyroclastics; 8) Mudstone, gypsiferous; 9) Undifferentiated mudstone, shale, dacite, pyroclastics marl and andesite lava; 10) Upper tuff member: andesite/dacite lava and pyroclastics, shale, marl, sandstone.

more than 1000 - 2000 meters thick near the mountains but increasing to 5000 meters in the eastern Tehran (type profile). This conglomerate alluvium exposed in north of study area (Chitgar and between Vardavard and Garm Darreh) and in southern part as single conglomerate hills.

Most of the Shahriar plain is built on an alluvial fan complex of Recent and Quaternary age that thickens towards the south and thins towards the Alborz Mountains. The Quaternary sediments can be divided to three groups: Kahrizak alluvial formation, Tehran alluvial formation and Recent alluvials [16]. Kahrizak deposits consist of silt and clay and outcrop in south of area near Kahrizak and Mafi Abad. Tehran alluvial, derived from Hezar Darreh erosion, consist of red coarse gravel, sand and silt with good gradation. These deposits host the main aquifer system of the area. Recent Alluvial, the youngest sediments exposed, occur along the Karaj and Kan streams, consists of gravel and sand in northern plain and silt and clay in southern plain.

Geoelectrical Measurements and Interpretations

Vertical electrical soundings with a Schlumberger configuration in the Tehran-Karaj plains have been conducted by [17]. 144 soundings in 11 profiles were conducted in Shahriar plain. The apparent resistivity data obtained for different values of AB/2 have been processed. These results were subsequently used to obtain a realistic picture of the geological and hydrogeological framework.

Based on different particle size in unconsolidated horizons, electrical resistivity varies from 10 - 250 Ωm. Lower resistivities occur in central and southern parts of study area, where as clay and silt zones are predominant. Data derived from lithology of boreholes and drilled wells located in the vicinity of corresponding soundings were compared with geoelectrical surveys results to obtain the true resistivity values, the thickness and the limitations of the alluvial aquifer. The resultant isopach map (**Figure 5**(a)) shows that aquifer thickness varies from a few meters to about 250 m. It increases from 25 - 100 m in northern edge of the plain to 100 - 250 m in the central parts of plain. Aquifer thickness decreases toward south of area where it reaches to its minimum.

Hydrogeologial Setting

The rapid increase in the population, caused by migration of people from adjacent provinces to the Tehran province, has led to large-scale groundwater developments in the study area over past two decades. As a result, aquifer condition has been changed egregiously during these years.

Based on time-drawdown and recovery data from 8 observation wells, transmissivity is about 750 $m^2 \cdot d^{-1}$ in northern parts and up to 3000 $m^2 \cdot d^{-1}$ in central parts of the plain (Figure 5(b)). It decreases again toward south where it is about 500 $m^2 \cdot d^{-1}$ in outlet of basin. Storage coefficient ranges between 0.25 - 0.35.

Drilled wells logs, water level in observing wells and geophysical investigations shows that main aquifer of Shahriar plain is an unconfined aquifer whose alluvial particle size varies from gravel and sand in north to silt and clay in south. The groundwater occurrence changes according to topographical and subsurface geological conditions. For example the water level is over 100 m below groundwater in the northernmost parts of the area and gradually gets closer to the ground surface toward the south of outlet of plain where it is about 10 m below ground level (Figure 5(c)).

Water levels of 30 observing wells have been used to draw isopotential maps of groundwater level in Shahriar plain. Based on this map, ground water flows from the northern heights and recharge areas of Karaj and Kan rivers toward the central parts and finally discharges to the outlet of the plain. The hydraulic gradient in the study area has its maximum value (13.5% - 18%) in north. While in central zone, due to increasing of saturated thickness of aquifer, its minimum values are found. Towards the study area's southern border and outlet of plain, hydraulic gradient increases again (3% - 8%).

Long-term and seasonal average areal water-level fluctuations recorded in 30 observing wells are plotted during the period 1989-2002. The mean values are calculated by creating Thiessen polygons for observation well stations. According to hydrograph analysis of observing wells it is concluded that there is an average decline of about 13.2 m in the groundwater level. Figure 4shows the water level fluctuation of the Shahriar plain.

According to Tehran Regional Water Authority [18], pumping wells are the main and only way to groundwater exploitation. All qanats and springs existing in the area have been dried in recent years due to intensive water level decline. The total withdrawal from deep and shallow pumping wells are 223, 57 and 38 million m^3/ year for agriculture, drinking and industrial pumping wells respectively.

Groundwater Balance

In this study, a groundwater balance has been prepared for the Shahriar aquifer based on available inflows, outflows and changes in the aquifer groundwater storage. The study was conducted for the period 1999 - 2000. In a specified period changes in volume storage of an aquifer depends on average flows entering and leaving the system through the different sources and sinks. Therefore the groundwater balance has two components: the total

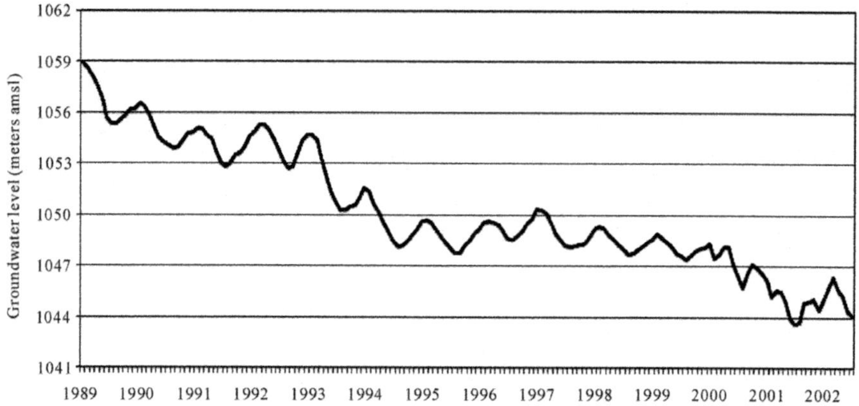

Figure 4: Groundwater level fuctuation of the Shahriar aquifer (Data derived from 30 monitoring boreholes).

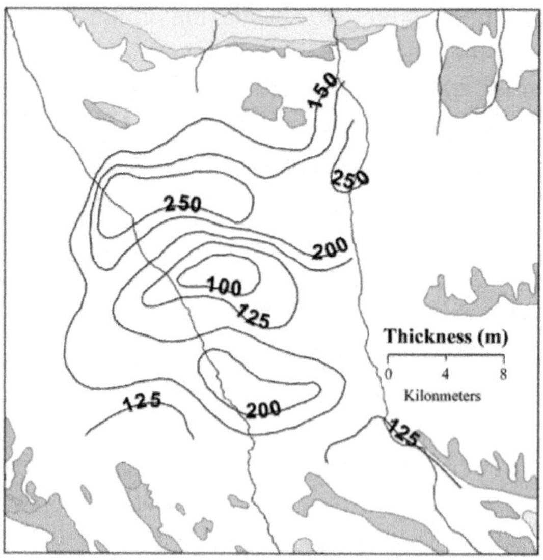

(a)

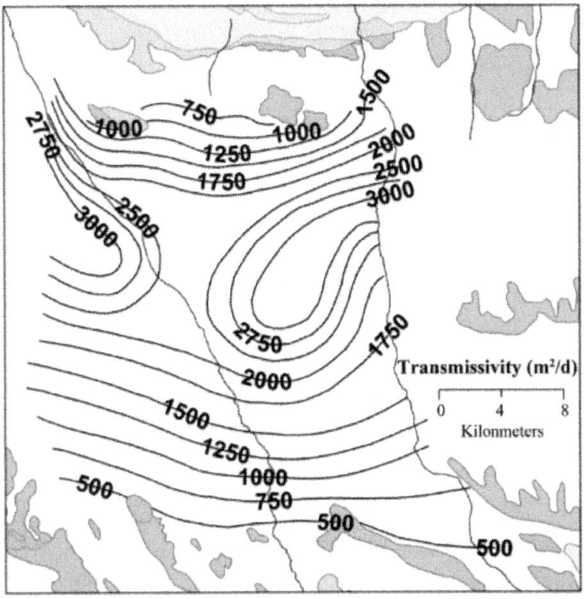
750
1000
1500
1000
2750
1250
2000
1750
2500
3000
2500
3000
2750
1750
Transmissivity (m²/d)
0 4 8
Kilonmeters
2000
1500
1250
1000
500
750
500
500

(b)

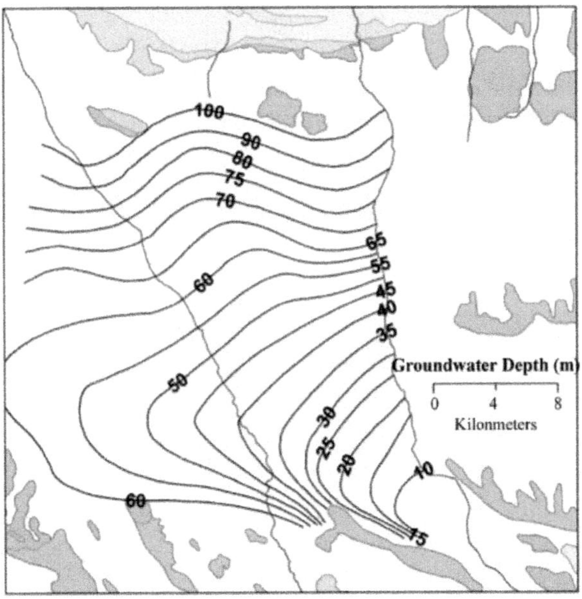
100
90
80
75
70
65
55
60
45
40
35
Groundwater Depth (m)
0 4 8
Kilonmeters
50
30
25
20
10
60
15

(c)

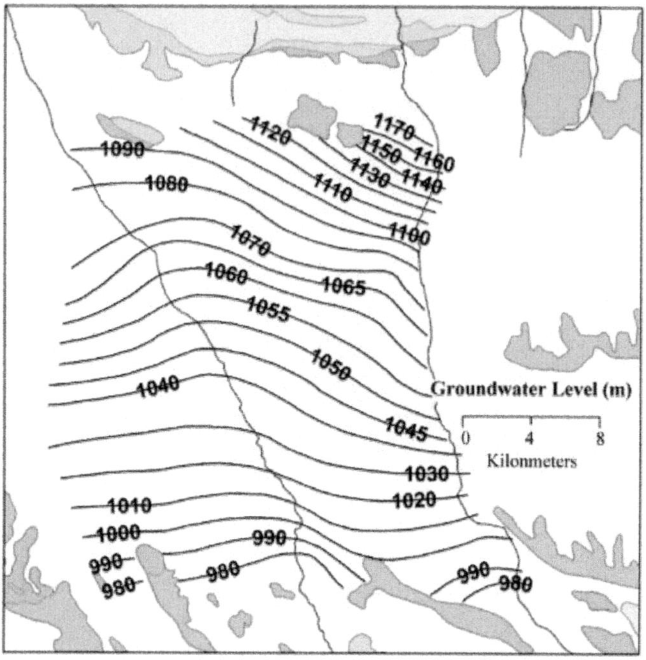

(d)

Figure 5: Hydrogeological maps of the study area: (a) Thichness isopack map; (b) Transmissivity; (c) Groundwater table depth; (d) Groundwater level.

input (ΣQ_{in}) and the total output (ΣQ_{out}). The equation for the groundwater balance during a specified hydrological period can be written as [19]

$$\Sigma Q_{in} - \Sigma Q_{out} = \pm \Delta V \tag{1}$$

where ΔV is the change in groundwater storage.

In Shahriar plain the total groundwater input (ΣQ_{in}) consists of: lateral subsurface inflows (Q_1), rainfall recharge (Q_2), Mountain-front recharge (Q_3), recharge due to irrigation returns (Q_4), recharge due to domestic and industrial wells returns (Q_5) and streambeds and canal infiltrations (Q_6)

$$\Sigma Q_{in} = Q_1 + Q_2 + Q_3 + Q_4 + Q_5 + Q_6 \tag{2}$$

The total groundwater output (ΣQ_{out}) consists of subsurface outflows (Q_7) and pumping groundwater (Q_8)

$$\Sigma Q_{out} = Q_7 + Q_8 \tag{3}$$

The change in groundwater storage (ΔV) can be estimated as

$$\Delta V = A \cdot S_c \cdot \Delta h \quad (4)$$

where:

A = Area of aquifer

Δh = Difference between the average water level at during the hydrological period and S_c = Average storativity of aquifer

RESULTS AND DISCUSSION

Groundwater recharge and discharge in the Shahriar aquifer occurs via seven and two components respectively. These components are contributing in groundwater system as explained below:

Lateral Subsurface Inflows (Q_1)

Inflow and outflow cross sections were determined using equipotential groundwater level map (**Figure 5**(d)) for October 1999 (23 cross sections). Employing Darcy's law and transmissivity map, an estimate of the volume of lateral inflows from the upstream boundary of the aquifer was performed:

Inflow amount (Q_1) = average transmissivity (T) × cross sectional length (L) × gradient (i) × water balance period (t).(5)

For each cross section volume of inflow was calculated using mentioned equation. The total lateral inflows to the aquifer system were calculated to be Q_1 = 167 Million $m^3 \cdot yr^1$ for the hydrological year 1999-2000.

Rainfall Recharge (Q_2)

Various parameters such as annual rainfall depth, slope and soil permeability affect the rainfall recharge amount. Accepting annual rainfall depth about 225 mm for year 1999-2000 and the area about 580 km^2 the annual rainfall volume equals 130 million m^3.

The infiltration ratio considering the lithology and soil texture is estimated 10% of the annual rainfall. Thus in this study the annual recharge of the aquifer system from rainfall was estimated to be Q_2 = 13 million $m^3 \cdot yr^{-1}$.

Mountain-Front Recharge (Q_3)

As mentioned in hydrology section, the annual volume of entering runoff to study area via three sub-basins named Kan-Chitgar, Chitgar, and Chitgar-Karaj are about 1.5, 13.5 and 3 respectively and totally about 18 million m^3. The total recharge through these runoffs is estimated about 15 million m^3.

Recharge Due to Irrigation Returns (Q_4)

According to FAO technical paper about irrigation yield and percolate water losses in farms based on irrigation method and soil texture [20], the irrigation returns are estimated to be as high as 25% of the applied irrigation (223 million $m^3 \cdot yr^{-1}$ from agricultural wells, 29 million $m^3 \cdot yr^{-1}$ through water pumping from streams and canals). Hence, recharge from irrigation returns is 63 million $m^3 \cdot yr^{-1}$.

Recharge Due to Domestic and Industrial Wells Returns (Q_5)

Conciderably percent of water applying for drinking and household uses, turns to wastewater and to some extent percolate to the aquifer. Groundwater extracted for municipal uses from the region aquifer is about 57 million $m^3 \cdot yr^{-1}$. The amount of return flow from these well is estimated about 37 million $m^3 \cdot yr^{-1}$ (65% of the total amount).

The total amount of groundwater abstracted for industrial uses is about 38 million $m^3 \cdot yr^{-1}$ which is estimated 35% of this amount (13 million $m^3 \cdot yr^{-1}$) percolate to the aquifer. Therefore the recharge due to domestic and industrial wells calculated as $Q_5 = 37 + 13 = 50$ million $m^3 \cdot yr^{-1}$.

Streambed and Canals Infiltrations (Q_6)

The total volume of inflow to study area through Navvab canal is 24 million $m^3 \cdot yr^{-1}$. From this amount 5 million $m^3 \cdot yr^{-1}$ is being used for irrigation that 1.3 million $m^3 \cdot yr^{-1}$ (25%) participates in aquifer recharging. From the rest of total volume (19 million $m^3 \cdot yr^{-1}$), 1.9 million $m^3 \cdot yr^{-1}$(tantamount 10%) infiltrates to aquifer system.

Eight Streams are taking supply from Bileghan diversion dam in north west of study area. The recharge to the aquifer system from streambed infiltration was estimated to be 2.1 million $m^3 \cdot yr^{-1}$(10% of total streams flow). From the estimates made above, the annual input to the Shahriar alluvial aquifer is $\Sigma Q_{in} = Q_1 + Q_2 + Q_3 + Q_4 + Q_5 + Q_6 = 328.3$ million $m^3 \cdot yr^{-1}$

The output parameters of the aquifer are groundwater extraction and groundwater discharge across sub-basin boundaries. According to water table depth map of study area which shows minimum depth of water table is more than 5 meter in the region, Evaporation from groundwater has been considered negligible.

Lateral Subsurface Outflows (Q_7)

The total lateral subsurface outflows to the aquifer system were calculated to be $Q_7 = 20$ million $m^3 \cdot yr^{-1}$ according to Darcy's law as it mentioned in lateral inflows.

Pumping Groundwater (Q_8)

Groundwater exploitations are performed via numerous production wells in order to cover the regional needs as below.

Based on the data of pumping wells from the Tehran

Table 1: Groundwater balance

Water input (million $m^3 \cdot yr^{-1}$)		Water output (million $m^3 \cdot yr^{-1}$)	
Lateral groundwater fluxes	167	Subsurface outflows	20
Recharge due to rainfall	13	Agricultural abstractions	223
Mountain-front recharge	15	Domestic use	57
Irrigation return	63	Industrial use	38
Domestic and industrial wells returns	50		
Streambed and canal infiltration	5.3		
	313.3		338

Regional Water Board inventory report, groundwater abstracted from the Shahriar aquifer is 223, 57 and 38 million $m^3 \cdot yr^{-1}$ for agricultural, Domestic and industrial consumptions respectively.

From the estimates made above, the mean annual output from the Shahriar alluvial aquifer is $\Sigma Q_{out} = Q_7 + Q_8 = 20 + 318 = 338$ million $m^3 \cdot yr^{-1}$

As illustrated in Table 1, a negative groundwater balance has been calculated in the studied aquifer system for the year 1999-2000. Groundwater overexploitation by abstractions from over 3500 production wells has resulted in considerable head decline that reached 1.42 m during the hydrological year 1999-2000. It can be concluded that the current abstraction is not sustainable, thus groundwater is withdrawn from storage. The impacts of the overexploitation are the drying up of upper aquifer system wells and qanats, the land subsidence [21,22] and the severe deterioration of groundwater quality.

There are some strategies in order to make the Shahriar groundwater system sustainable. Groundwater exploitation controlling based on plant water demand and restriction the irregular pumping wells' drilling can play an important role in decreasing the depletion of water table. Utilization of the treated wastewater and the application of water-saving techniques such as spray irrigation and drip irrigation can decrease the groundwater quantities for irrigation use [23,24].

Artificial recharge is another alternative as a valuable water management tools that effectively help to offset increased demands for water [25]. Availability, quality and quantity of source water available, resulting water quality (reactions with native water and aquifer materials), clogging potential, underground storage space available, depth to underground storage space, transmission characteristics, and costs are some factors controlling the feasibility of artificial recharge method [26]. The lithological studies in the study area reveals that the northern parts especially in Chitgar and the region between Saeid Abad, Ghale Hasan Khan and Shariar can be considered as suitable recharging zones due to thick and permeable horizons of coarse grain deposits.

Therefore, an appropriate set of management strategies, including water conservation measures, regulation of existing development, improvement of current legislation and public education should be adopted [3]. Future investigations of the groundwater recharge in the alluvial aquifer system would benefit by improvement in hydrological data monitoring, the application of isotopic analysis and computer modeling to simulate the water cycle and groundwater flow.

ACKNOWLEDGEMENTS

The authors wish to thank Tehran Regional Water Authority for supplying the existing relevant data. Also, the constructive comments of anonymous reviewers will be appreciated.

REFERENCES

1. E. Wendland, C. Barreto and L. H. Gomes, "Water Balance in the Guarani Aquifer Outcrop Zone Based on Hydrogeologic Monitoring," Journal of Hydrology, Vol. 342, No. 3-4, 2007, pp. 261-269. doi:10.1016/j.jhydrol.2007.05.033
2. J. C. Marechal, B. Dewandel, S. Ahmed, L. Galeazzi and F. K. Zaidi, "Combined Estimation of Specific Yield and Natural Recharge in a Semi-Arid Groundwater Basin with Irrigated Agriculture," Journal of Hydrology, Vol. 329, No. 1-2, 2006, pp. 281-293.doi:10.1016/j.jhydrol.2006.02.022
3. A. T. Tizro, K. S. Voudouris and M. Eini, "Groundwater Balance, Safe Yield and Recharge Feasibility in a Semi- -Arid Environment: A Case

Study from Western Part of Iran," Journal of Applied Sciences, Vol. 7, No. 20, 2005, pp. 2967-2976.

4. G. R. Lashkaripour, "Contamination of Groundwater Resource in Zahedan City Due to Rapid Development," Journal of Applied Science, Vol. 3, No. 5, 2003, pp. 341-345.doi:10.3923/jas.2003.341.345

5. G. B. Allison, "A Review of Some of the Physical Chemical and Isotopic Techniques Available for Estimating Ground Water Recharge," In: I. Simmers, Ed., Estimation of Natural Ground Water Recharge, Reidel, Dordrecht, 1988, pp. 49-72.

6. S. S. D. Foster, "Quantification of Ground Water Recharge in Arid Regions: A Practical View for Resource Development and Management," In: I. Simmers, Ed., Estimation of Natural Ground Water Recharge, Reidel, Dordrecht, 1988, pp. 323-338.

7. R. W. Healy and P. G. Cook, "Using Groundwater Levels to Estimate Recharge," Hydrogeology Journal, Vol. 10, No. 1, 2002, pp. 91-109. doi:10.1007/s10040-001-0178-0

8. C. Leduc, J. Bromley and P. Schroeter, "Water Table Fluctuation and Recharge in Semi-Arid Climate: Some Results of the HAPEX-Sahel Hydrodynamic Survey (Niger)," Journal of Hydrology, Vol. 188-189, 1997, pp. 123-138. doi:10.1016/S0022-1694(96)03156-3

9. Tehran Regional Water Authority Website, 2009. http://www.tw.org.ir/dams/selectdam_en.asp

10. Tehran Regional Water Authority, "Annual Report of Groundwater Resources Exploitation," Deputy for Water Resources Study and Research, Tehran Regional Water Authority, 2005.

11. Water Resources Studies Division, "Climatology and Hydrologic Report of Tehran, Varamin and Shahriar Basins," Tehran Regional Water Authority, 1997.

12. A. Alizadeh, "The Principle of Applied Hydrology," Astan Ghods Razavi Publications, Mashhad, 2009.

13. A. Darvishzadeh, "Geology of Iran," Amirkabir Publication, Tehran, 2003.

14. S. A. Aghanabati, "Geology of Iran," Geological Survey of Iran Publication, Tehran, 2004.

15. C. Vita-Finzi, "Late Quaternary Alluvial Chronology of Iran," Geologische Rundschau, Vol. 58, No. 2, 1969, pp. 951-973. doi:10.1007/BF01820740

16. E. H. Rieben, "The Geology of the Tehran Plain," American Journal of Science, Vol. 253, 1955, pp. 617-639. doi:10.2475/ajs.253.11.617

17. Water and Soil Engineering Unit of Jihad-e-Sazandegi, "Geoelectrical Studies in Tehran-Karaj Plain," Jihad-eSazandegi Ministry, Tehran, 1983.

18. Tehran Regional Water Authority, "Detailed Data Collection from Discharges of Pumping Wells and Qanats in Tehran Province," Tehran Regional Water Authority, Tehran, 2001.

19. J. J. Carrillo-Rivera, "Application of the Groundwater-Balance Equation to Indicate Interbasin and Vertical Flow in Two Semi-Arid Drainage Basins, Mexico," Hydrogeology Journal, Vol. 8, No. 5, 2000, pp. 503-520. doi:10.1007/s100400000093

20. A. Alizadeh, "Relation between Water, Soil and Plant," Astan Ghods Razavi Publications, Mashhad, 2007.

21. A. Shemshaki and I. E. Soltani, "Land Subsidence Review at Sahariar Plain," Geological Survay of Iran, Tehran, 2004.

22. R. Umar, M. M. A. Khan, I. Ahmed and Sh. Ahmed, "Implications of Kali–Hindon Inter-Stream Aquifer Water Balance for Groundwater Management in Western Uttar Pradesh," Journal of Earth System Science, Vol. 117, No. 1, 2008, pp. 69-78. doi:10.1007/s12040-008-0014-1

23. P. K. Naik, and A. K. Awasthi, "Groundwater Resources Assessment of the Koyna River Basin, India," Hydrogeology Journal, Vol. 11, No. 5, 2003, pp. 582-594.doi:10.1007/s10040-003-0273-5

24. M. N. Bhutta, M. Saeed and M. Rafiq, "Evaluation of Groundwater Balance—A Case Study of Mona Drainage Basin," Pakistan Journal of Water Resources, Vol. 11, No. 2, 2007, pp. 19-26.

25. K. Voudouris, P. Diamantopoulou, G. Giannatos and P. Zannis, "Groundwater Recharge via Deep Boreholes in the Patras Industrial Area Aquifer System (NW Peloponnesus, Greece)," Bulletin of Engineering Geology and the Environment, Vol. 65, No. 3, 2006, pp. 297-308. doi:10.1007/s10064-005-0036-8

26. M. P. O'Hare, D. M. Fairchild, P. A. Hajali and L. W. Canter, "Artificial Recharge of Groundwater: Status and Potential in the Contiguous United States," Lewis Publishers, Chelsea, 1986.

Chapter 9

CHANGING HYDROLOGY OF THE HIMALAYAN WATERSHED

Arshad Ashraf[1]

[1]Water Resources Research Institute, National Agricultural Research Center, Islamabad, Pakistan

INTRODUCTION

The Himalayan region is a source of ten major river systems that together provide irrigation, power and drinking water for 1.3 billion people i.e. over 20% of the world's population. The supply and quality of water in this region is under extreme threat, both from the effects of human activity and from natural processes and variation [1]. Population growth is already putting massive pressure on regional water resources, affecting water resource in terms of demand, water-use patterns and management practices. The change in hydrological cycle may affect river flows, agriculture, forests, biodiversity and health besides creating water related hazards [2]. The need for suitable strategies for climate resilient development has policy and governance implications [3]. Adaptation to climate change is the area that should be strengthened through policy advocacy supported by evidence through rigorous research and verified information.

Re-assessment of true catchments yields under existing and future scenarios of landuse and climate changes is very essential to devise watershed management strategies which can minimize adverse impacts both in terms of quantity and quality. Since trends are still unclear, the extent to which changes can be attributed to variable environmental changes is difficult to determine. It has become imperative to assess ongoing hydrological changes and changes that might occur in future to devise appropriate adaptation measures to foster resilience to future climate change, thereby enhancing water security.

In the present study, SWAT model developed by United States Department of Agriculture (USDA) [4] has been used to evaluate surface runoff generation, soil erosion and quantify the water balance of a Himalayan watershed in the

Northern Pakistan. The response of watershed yield to historical landuse evolution and under variable landuse and climate change scenarios has been studied in order to mitigate the negative impacts of these changes and promote development activities in this region. The study would provide basis to recommend changes in the water management regimes so as to address future adaptation issues.

Modeling Hydrological Processes

Dealing with water management issues requires analyzing of different elements of hydrologic processes taking place in the area of interest. As such processes are taking place in a combine system that exists at a watershed level, thus the analysis must be carried out on a watershed basis. Understanding of relationship between various watershed characteristics such as morphology, landuse and soil, and hydrological components are very essential for water resources development in any area. Since the hydrologic processes are very complex, their proper comprehension is essential and for this watershed models are widely used. Most of the watershed models basically simulate the transformation of precipitation into runoff, sediment outflow and nutrient losses. Changes in landuse including urbanization and de(/re)forestation continue to affect the nature and magnitude of surface and subsurface water interactions and water availability influencing ecosystems and their services. One can formulate water conservation strategies only after understanding the spatial and temporal variations and the interaction of these hydrologic components. The alarming rate of soil erosion in context of changing landuse and climate in the Himalayan region calls for urgent attention for this problem. Assessment of erosion is a very difficult task when executed using conventional methods and requires to be done repetitively. The use of an appropriate watershed model is thus essential to deal with such problems.

Choice of watershed development model depends upon the hydrologic components to be incorporated in the water balance. The most important hydrologic elements from the water management point of view are surface runoff, lateral flow, baseflow and evapotranspiration. In presenting an appropriate view of reality, model must remain simple enough to understand and use. There are a number of integrated physically based distributed models, among which researchers have identified Soil and Water Assessment Tool (SWAT) as the most promising and computationally efficient [4]. The model is an integrated physically based distributed watershed model that has an ability to predict the impact of land management practices on water, sediment yield and agricultural chemical yield [5]. Distributed models also take the spatial variability of watershed properties into account.

Model Description

The SWAT is a process-based continuous daily time-step model that offers distributed parameter, continuous time simulation, and flexible watershed configuration [6]. It has gained international acceptance as a robust interdisciplinary watershed modeling tool. Two methods are used for surface runoff estimation in SWAT i.e. the SCS curve number and Green-Ampt infiltration. This study is based on the use of curve number for surface runoff and hence stream flow simulation. A SWAT model can be built using the Arc-View interface called AVSWAT which provides suitable means to enter data into the SWAT code. Main processes include water balance calculations (i.e. surface runoff, return flow, percolation, evapotranspiration, and transmission losses), estimation of sediment yield, nutrient cycling and pesticide movement.

The spatial heterogeneity is represented by means of observable physical characteristics of the basin such as landuse, soils and topography etc. Model inputs include physical characteristics of the watershed and its sub-basins i.e. precipitation, temperature, soil type, land slope, Manning's n values, USLE K factor, and management inputs like crop rotations, planting and harvesting dates, tillage operations, irrigation, fertilizer use, and pesticide application rates. Model outputs include sub-basin and watershed values for surface flow, ground water and lateral flow, sediment, nutrient and pesticide yields. The main basin is divided into sub-basins which are further divided into hydrologic response units (HRU) composed of homogeneous landuse, soil types, relevant hydrological components and management practices. Sediment yield is estimated by the Modified Universal Soil Loss Equation (MUSLE; [7]. The model has been applied worldwide for the purpose of simulating sediment flow [8], modeling hydrologic balance [9], evaluation of the impact of landuse and landcover changes on the hydrology of catchments [10]. The model provides a flexible capability for creating climate change scenarios evaluating a wide range of "what if" questions about how weather and climate could affect our systems.

Equations of Watershed Hydrology

The hydrologic process in a watershed is simulated by the following water balance equation:

$$SW_t = SW + \sum_{i=1}^{t}\left(R_i - Q_i - ET_i - P_i - QR_i\right) \tag{1}$$

where: SW_t is the final soil water content (mm), SW is the initial soil water content minus the permanent wilting point water content (mm), t is time in days, R is rainfall (mm), Qi is surface runoff (mm), ETi is evapotranspiration (mm), Pi is percolation (mm) and QRi is lateral flow (mm). The surface runoff is predicted by the following equation:

$$Q = \frac{(R - 0.2s)^2}{R + 0.8s} \text{ for } R > 0.2s \tag{2}$$

$$Q = 0.0 \quad \text{for } R < 0.2s$$

$$s = 254\left(\frac{100}{CN} - 1\right) \tag{3}$$

Where, Q = daily surface runoff (mm): R = daily rainfall (mm), S = retention parameter (mm); CN = curve number.

Lateral flow is predicted by:

$$q_{lat} = 0.024 \frac{(2SSC \sin\alpha)}{\theta_d L} \tag{4}$$

Where, qlat = lateral flow (mm/ day); S = drainable volume of soil water per unit area of saturated thickness (mm/day), SC = saturated hydraulic conductivity (mm/h); L = flow length (m); α = slope of the land: ɵd = drainable porosity

The base flow is estimated by:

$$Q_{gwj} = Q_{qwj-1} \cdot e^{(-\alpha_{gw} . \Delta t)} + w_{rchrg} \cdot \left(1 - e^{(-\alpha_{gw} . \Delta t)}\right) \tag{5}$$

Where, Qgwj = groundwater flow into the main channel on day j; αgw = base flow recession constant; Δt= time step. The computed runoff from each element is integrated using a finite difference form of the continuity equation relating moisture supply, storage and outflow.

Description of Study Area

Rawal watershed covers an area of 272 sq km within longitudes 73° 03′ - 73° 24′ E and latitudes 33° 41′ - 33° 54′ N comprising parts of Margalla hills and Murree mountains in the southern Himalayas of Pakistan (Figure 1). About 47% of the watershed area lies in the Islamabad Capital Territory while the rest in Punjab and Khyber Pakhtunkhwa (KPK) provinces, so it is well connected through a metalled road with other parts of the country. Korang is the main river flowing in the watershed that receives runoff from watershed via four major and 43 small streams [11]. Rawal dam is constructed on Korang river, which supplies 22 million gallons per day of water for drinking and other household needs to Rawalpindi city and a limited water for irrigation use to Islamabad area. The elevation ranges between 523 meters and 2145 meters above mean sea level (masl). Physiographically, the watershed comprises of 34% hilly area (Elev. <700 masl), 62% Middle mountains (Elev. within 700-2000 masl) and 4% High mountains (Elev. >2000 masl).

The Himalayas serve as a divide between Central Asia and South Asia. The Indo-Eurasian plates collision resulted in the formation of new relief and topography, which consists of series of mountain ranges located in the north and west of Pakistan, commonly known as the Himalayan Mountain System [12]. The principal uplift occurred during the middle or late Tertiary period, 12 to 65 million years ago. The study area lies in sub-humid to humid sub-tropical continental highlands. The hottest months are May, June and July. The mean maximum temperature ranges between 17.6°C and 40.1°C while mean minimum temperature between 2.1°C and 21.6 °C. The winter months are from October to March. The highest temperature was recorded as 46.6°C in 2005 and the lowest as -3.9°C during 1967 [13]. Mean annual rainfall of 1991-2010 period is about 1232 mm. The occurrence of rainfall is highly erratic both in space and time. Over 60 percent of the annual rainfall occurs during monsoon season i.e. from July to September. Most of the rainfall is drained out rapidly due to steep slopes and dissected nature of the terrain. Springs and streams are the main source of water for drinking and other domestic requirements. A prolonged dry season may cause water shortage in some parts of the area.

Underlying rocks consist of poorly compressed and highly folded and faulted Murree series that are moderately to severely eroded, shallow clayey loams of very low productivity [14]. The soils formed over shale are clayey while those developed on the sandstone are sandy loams to sandy clay loam in texture. The flora is mainly natural with xeric, broad-leaved deciduous, evergreen trees and diverse shrubs on the southern slopes. The dominating plant species are Carissa spinarum (Granda), Dodonaea viscosa (Sanatha) and Olea ferruginea (Wild Olive). Sub-tropical pine zone occupies steep and very steep mountain slopes [15]. Agriculture is practiced in small patches of land as terrace cultivation.

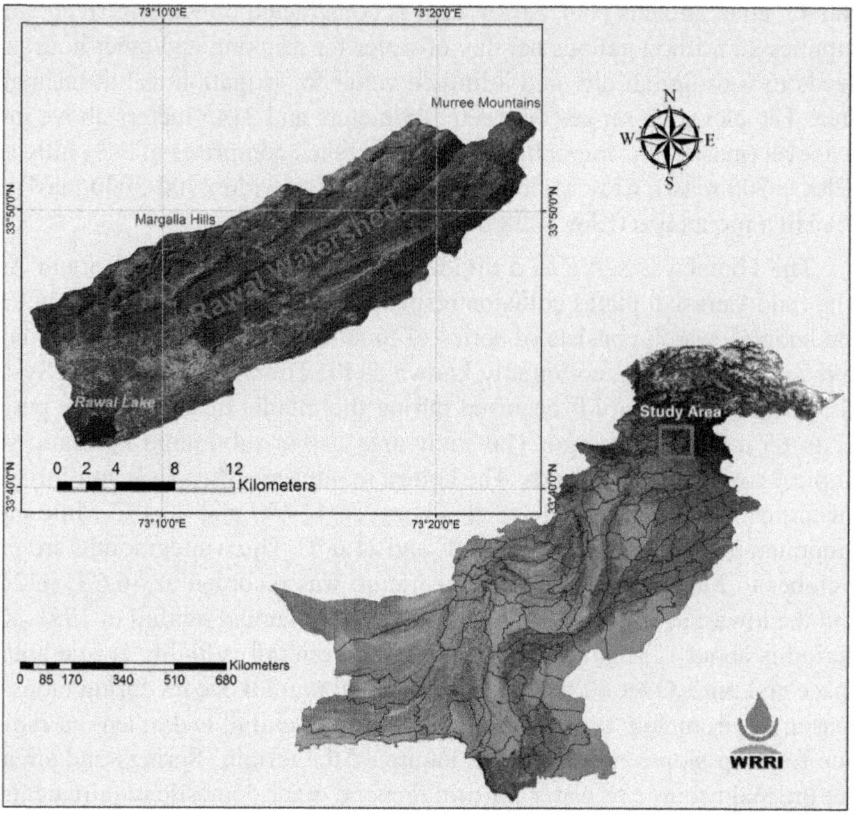

Figure 1: Location map of the study area.

Main Environmental Issues

The watershed is confronting problems of rapid urban development and deforestation due to which its landuse is changing gradually. The population growth and addition of a number of housing colonies in the Rawal Lake catchment area are adversely affecting the regime of water coming into Rawal Lake. The activities like illegal cuttings due to high market value of forest wood and intensive use of forest wood for household needs (cooking, heating, timber etc.), ineffective forest management and forest disease etc. are accelerating the deforestation rate in the watershed area [16-17]. Destruction of aquatic habitat and a reduction of water quality are some of the negative impacts of deforestation. Extensive cattle grazing and fuel wood cutting by the local communities have deformed the plants to bushes [18]. The removal of a forest cover from steep slopes often leads to accelerated surface erosion and dramatically increases the frequency of land sliding and surface runoff.

The storage capacity of the Rawal Lake which was 47,230 acre-ft when it was developed in 1960, has been reduced almost 34 percent due to sedimentation generating from natural and human induced factors in its catchment area [16]. The use of pesticides and herbicides in agriculture is a source of toxic pollution [19]. Many housing schemes, recreational pursuits e.g. Lake view point, Chatter and Valley parks etc. and farmhouses have been developed in the watershed. The construction of roads, pavements and other structures reduce the infiltration area that ultimately affect the recharging of the aquifer of the twin cities. No systematic study has been undertaken yet to document the landuse variability in the watershed.

MATERIALS AND METHODS

Data Used

In the present study, the basic watershed data used to extract spatial input for SWAT model were hydrologic features, soil distribution, landuse information, and topography. The remote sensing technique has potential application in landuse monitoring and assessment at desired scales. RS images of LANDSAT-TM (Thematic Mapper) of period 1992 and LANDSAT-ETM+ (Enhanced Thematic Mapper Plus) of 2000 and 2010 periods (Path-Row: 150-37) were used to delineate landuse/landcover of the watershed area on temporal basis. The LANDSAT ETM+ sensor is a nadir-viewing, 7-band plus multi-spectral scanning radiometer (upgraded ver. of TM sensor) that detects spectrally filtered radiation from several portions of the electromagnetic spectrum. The spatial resolution (pixel sizes) of the image data includes 30 m each for the six visible, near-infrared, and short-wave infrared bands, 60 m for the thermal infrared band, and 15 m for the panchromatic band. The climatic parameters i.e. daily temperature (max& min) and precipitation data recorded at Satrameel observatory (73° 12′ 50” E, 33° 45′ 57” N & Elev: 610 m) maintained by Water Resources Research Institute had been collected for period 1991-2010. The discharge data of Korang river available on monthly basis from Small dams organization was acquired for the same period for model calibration and validation. The soil map developed by Soil Survey of Pakistan was utilized to extract soil data attributes for the study area.

Data Preparation

The base map of the study area was prepared through generating and integrating thematic layers of elevation, physiography and infrastructure using ArcGIS 9.3 software. An integrated hydrological, spatial modeling and field investigations approach was adopted to achieve the study output. The boundaries of the watershed and sub-basins were delineated using Aster 30m DEM of the area in SWAT model 2005 software. Elevation map comprising of four classes i.e. >1600m, 1200-1600m, 800-1200m and <800m range, was prepared from Aster 30m DEM data (Figure 2). The image data was georeferenced using Universal Transverse Mercator (UTM) coordinate system (Zone 43 North). The satellite images were analyzed through visual and digital interpretation techniques to observe spatial variability of landuse. The visual interpretation was performed for qualitative analysis while digital interpretation for quantitative analysis of the image data. The false color composite of 5, 4, 2 (RGB) of LANDSAT image data was selected to extract signatures of representative landcover types from the image. In this bands combination, landcover is visible in true color i.e. vegetation in green, soil in pale to reddish brown and water in shades of blue color. The built-up area is shown in mixed pattern of white, brown, and purple colors due to variable types and density of constructed area, mixing of new and old settlements, presence of land features like lawns, parking sides, water ponds, roads/tracks etc. The signatures were evaluated using error matrix and an overall accuracy of more than 95 percent was achieved.

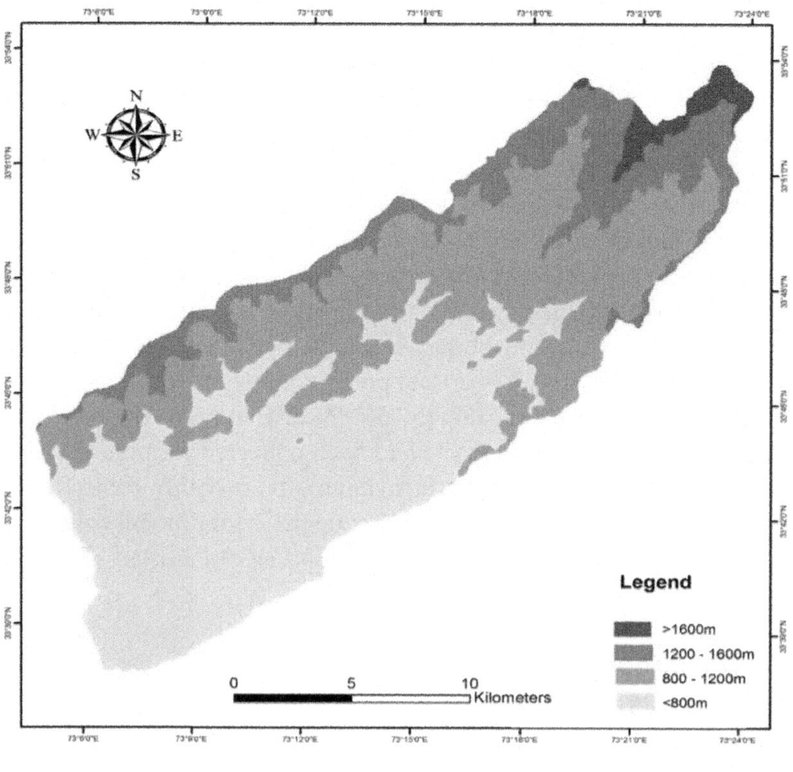

Figure 2: Elevation increases gradually towards northeast in the study area.

The classification of the images was performed using supervised method following maximum likelihood rule mostly used to acquire reliable classification results. The classification output was supported with Normalized difference vegetation index - NDVI data that helps in segregating vegetative areas from non-vegetative [20]. The index which is based on the spectral characteristics of green vegetation cover in the area uses TM3 and TM4 bands of LANDSAT ETM+ image as given in the following equation:

$$NDVI = (TM4 - TM3) / (TM4 + TM3) \tag{6}$$

The classification of the images was performed to obtain seven major landuse/landcover classes which include conifer forest, scrub forest, agriculture, rangeland, soil/rocks, settlement and water. The images were recoded and later filtering technique was applied to remove noisy/misclassified pixels from the recoded image data. The doubtful classes were modified after ground truthing i.e. performing field survey in the target areas. Finally change analysis of landuse/landcover classes was performed using spatial modeling functions of ERDAS Imagine 9.2 software.

Model Baseline Establishment

Main procedures in the model running includes: (a) development of streams and sub-basins databases, (b) landuse and soil data input within sub-basins, (c) Input variable parameters of climate and management options, (d) compilation of input data and running the model for generating output results. The entire watershed had been divided into 15 sub-basins by choosing a threshold area of 500 ha. A total of 73 HRUs were generated in those sub-basins. A threshold of 5% was defined landuse distribution and 15% for soil distribution over sub-basin area. The low percentage for landuse was used to accommodate conifer coverage distributed in patches over northeastern parts of the watershed area. The importance of land uses lies mainly in the computation of surface runoff with the help of SCS curve during the model operation [6]. Three soil classes were identified and mapped i.e. sandy clay loam over northwestern hilly terrain, sandy clay loam over valley area in the northeast and sandy loam over low plains in the southwestern part of the watershed area.

The subcomponents of the water balance identified for use in analyses are total flow (water yield) consisting of surface runoff, lateral and base flow, soil water recharge; and actual evapotranspiration. These components are expressed in terms of average annual depth of water in millimeters over the total watershed area. For estimation of sediment yield, C factor values were used on the basis of soil erosion study [21] carried out previously in Pothwar region. The C value of 0.176 was assigned to soil/rocks while 0.2 was assigned to agricultural land class. Higher C values indicate more risk of soil erosion. The conservation practice factor P was assigned value of 1 on account of no significant conservation practice present in the watershed area [22].

Model Calibration

Calibration and validation of the SWAT model was performed using monthly river flows data of 1991-2010 period. Data pertaining to year 1991 to 2006 had been used for calibration and the rest for validation of the model. The purpose of model validation is to assess whether the model is able to predict field observations for time periods different from the calibration period [23]. Although the model was run for years 1991 to 2006, the first 6 years of the simulated output were disregarded in the calibration process, since these are required by the model as a warm-up period. This period is essential for the stabilization of parameters (e.g groundwater depth), as the results sometimes vary significantly from the observed values. Thus the final calibration period was from January 1997 to December 2006. The calibration accuracy was checked by calculating several indexes which include Nash & Sutcliffe coefficient (NTD), Root Mean Square Error (RMSE) and the correlation coefficient R^2 of the time series. The Nash & Sutcliffe coefficient [24] is an estimate of the variation of a time series from another as given by following equation:

$$NTD = 1 - \frac{\sum_{i=1}^{n}\left(Q_{obs,i} - Q_{sim,i}\right)^2}{\sum_{i=1}^{n}\left(Q_{obs,i} - \bar{Q}_{sim,i}\right)^2} \tag{7}$$

And root mean square error- RMS was computed using following equation:

$$RMSE = \sqrt{\frac{1}{n}\sum_{i=1}^{n}\left[Wi\left(Q_{sim,i} - Q_{obs,i}\right)\right]^2} \tag{8}$$

Where, Q_{sim} = simulated time series, Q_{obs} = observed time series, $\bar{Q}_{sim}$= numerical mean for the simulated time series, W = weight and n = total number of measurements. A Nash & Sutcliffe coefficient approaching unity indicates that the estimated time series is almost identical to the observed one. The results of these tests are summarized in Table 1. The NTD index reached the value of 0.80, signifying a quite precise calibration. Later the model was validated using the same indexes, for the period of January 2007 to December 2010. The results of statistical analysis indicated a Nash Sutcliff efficiency of 0.80. The simulated river flows matched well with the observed values (Figure 3).

TABLE 1: Criteria for examining the accuracy of calibration and validation processes.

Index	Calibrated period	Validated period
NTD	0.80	0.80
RMSE (mm)	17.0	30.4
$^R2$	0.81	0.91

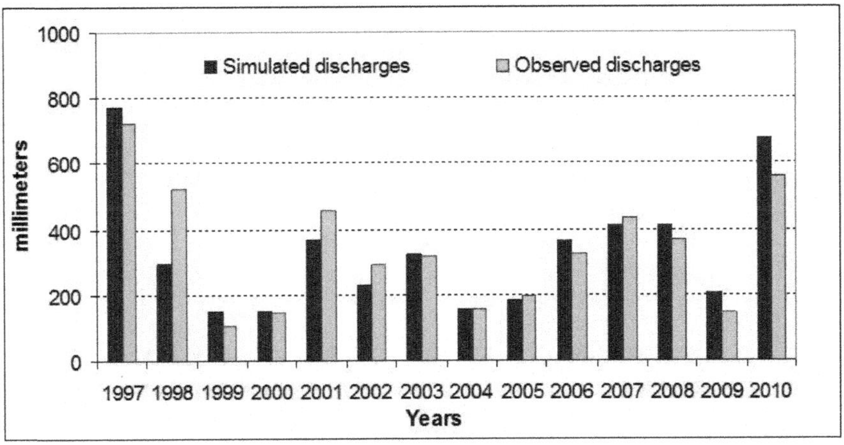

Figure 3: Time series of simulated and observed annual discharges for the Rawal watershed, period 1997-2010.

RESULTS AND DISCUSSION

Assessment of Changing Landuse/Landcover

Comprehensive information on the spatial and temporal distribution of landuse/ landcover is essential for estimating hydrological changes at watershed level. The landuse/landcover condition of the watershed was estimated for three different periods i.e. 1992, 2000 and 2010 (Figure 4). Major landcover change was observed in the scrub class which indicated a reduction of about 4,515 ha during 1992-2010 period (Table 2). The rate of decrease in its coverage was about 1.5% per annum. The scrub wood is mostly used as fuel at local level due to non-availability of other energy sources. Major part of it had been converted into agriculture and built-up land, while in some areas it has changed into rangeland due to extensive wood cutting. These results are verified by the findings of [25] which highlighted maximum decrease in scrub forest during 30-year period i.e. 1977-2006 in Rawalpindi area. The settlement class had shown almost four times increase in coverage i.e. from 2.6% in 1992 to 8.7%

in year 2010. The average rate of increase in this class was about 90 ha y^{-1}. The rate was over 45 ha y^{-1}during 1992-2000 while it was about 125 ha y^{-1} during 2000-2010 period indicating rapid urbanization in the last decade (Figure 5). The conifer forest had shown a decline at a rate of about 2.1% y^{-1} within last two decades. Although FAO [26] had reported deforestation at a rate of about 1.5% annually in the country, but due to high urban development, the rate of forest decline was higher in the watershed area. The agriculture coverage indicated an average increase of about 26 ha annually during 1992-2010 period. The rate of increase was about 3.4% y^{-1} during 1992-2000 while it was 0.3% y^{-1} during 2000-2010 period. The situation indicates intense agriculture activity in the former decade that seems replaced by rapid growth in urban development in the later decade.

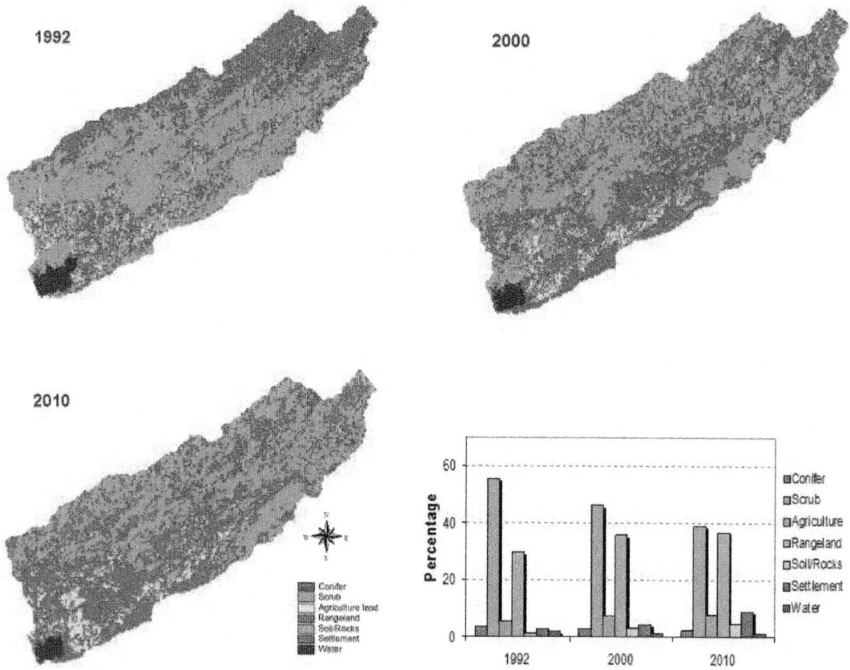

Figure 4: Spatio-temporal variations in landuse/landcover in Rawal watershed area during 1992-2010 period.

Table 2: Detail of landuse/landcover variations during 1992-2010 period

Landuse	1992		2000		2010		1992-2010	
	Area(ha)	%	**Area(ha)**	%	**Area(ha)**	%	**Change (ha)**	**Change** %
Conifer	1006	3.7	762	2.8	626	2.3	-381	-1.4
Scrub	15069	55.4	12485	45.9	10554	38.8	-4515	-16.6
Agriculture	1496	5.5	1958	7.2	2013	7.4	517	1.9
Rangeland	8024	29.5	9629	35.4	9982	36.7	1958	7.2
Soil/Rocks	326	1.2	870	3.2	1306	4.8	979	3.6
Settlement	762	2.8	1170	4.3	2421	8.9	1659	6.1
Water	517	1.9	326	1.2	299	1.1	-218	-0.8
Total	27200	100	27200	100	27200	100	-	-

Figure 5: Growth of urbanization is causing rapid landuse change in the Rawal watershed area.

The changes in landuse/landcover were variable on different elevation ranges during 1992-2010 period. The conifer forest has shown a decrease from 134 ha to 102 ha at greater than 1600m elevation range while this was from 343 ha to 238 ha within 1200-1600m elevation range during 1992-2010. The scrub class indicated a decrease of about 11 percent within 800-1200m range while 65% in less than 800m elevation range. In contrary to this, agriculture class had shown a increase of about 65% within 800-1200m range while 29%

increase in less than 800m elevation range. About 86% settlement class was found below 800 m elevation during year 2010 indicating most of the urban development in the low lying areas of the watershed.

Model Simulation

The model simulated an average water yield of about 378.6 mm/yr using base landuse of 2010 in the watershed area. About 49% of the yield was contributed by surface runoff and the rest by groundwater in the form of sub surface flows and springs etc. More than 70% of the annual yield was contributed during months of July, August and September. The surface runoff was found higher in the month of August i.e. over 83 mm while it was about 51 mm during July and 31 mm in September. The runoff was dominant over lower sub-basins likely due to higher impervious cover here than in the upper sub-basins of the watershed.

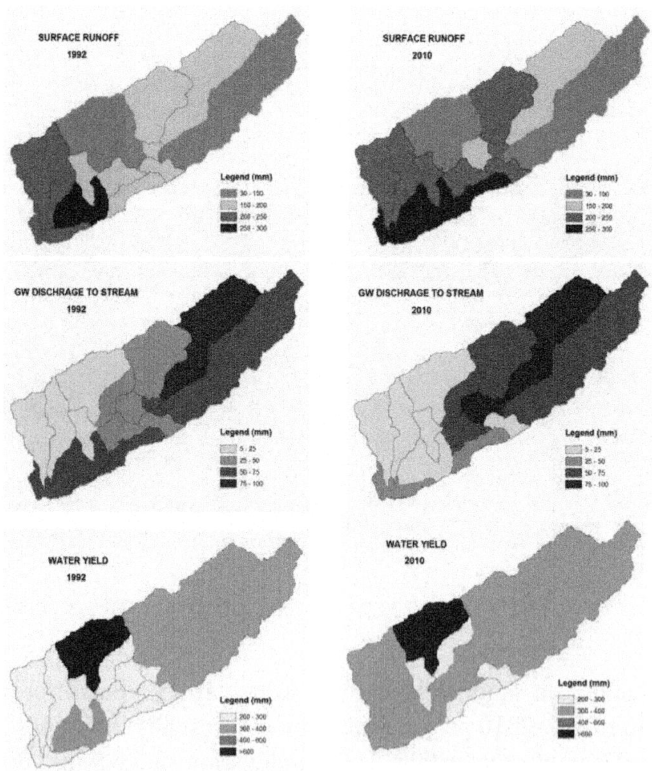

Figure 6: Comparison of the hydrological response of Rawal watershed to landuse conditions of 1992 and 2010 indicates dominant impact of landuse changes (i.e. urban development) in the southern low lying sub-basins on various hydrological parameters.

The groundwater discharge to stream flows was maximum in the month of September and more than 70% of the discharge occurred during period from August to December. The long-term average soil loss in the watershed was estimated over 17 tons ha^{-1} y^{-1} i.e. ranging from 0.4 to 36 tons ha^{-1} y^{-1} in different sub-basins. These estimates of soil loss matched closely with the results of [22] which indicated soil loss ranged from 0.1 to 28 tons ha^{-1} y^{-1} averaging 19.1 tons ha^{-1} y^{-1} at Satrameel study site in this watershed.

The model simulations showed a strong correlation between landuse evolution and the watershed runoff at the outlet. The change in landuse between years 1992 and 2010 indicated an increase of about 6% in the water yield and 14.3% in the surface runoff. The sub-basin wise hydrological response of the watershed during 1992-2010 period is shown in Figure 6. The sub-basins in the southern valley plains of the watershed indicate increase in surface runoff and water yield while decrease in groundwater contribution to the streams. The situation shows higher influence of urban landuse on hydrology of the low lying sub-basins as compared with sub-basins at higher elevations in the northeast of the watershed. Hydrologic changes due to increased impervious area and soil compaction generally lead to increased direct runoff, decreased groundwater recharge, and increased flooding, among other problems [27]. The combined effect of landuse and hydrological variations had exaggerated the problem of soil erosion resulting in an increase of about 17.4% in the sediment yield of watershed during 1992-2010 period.

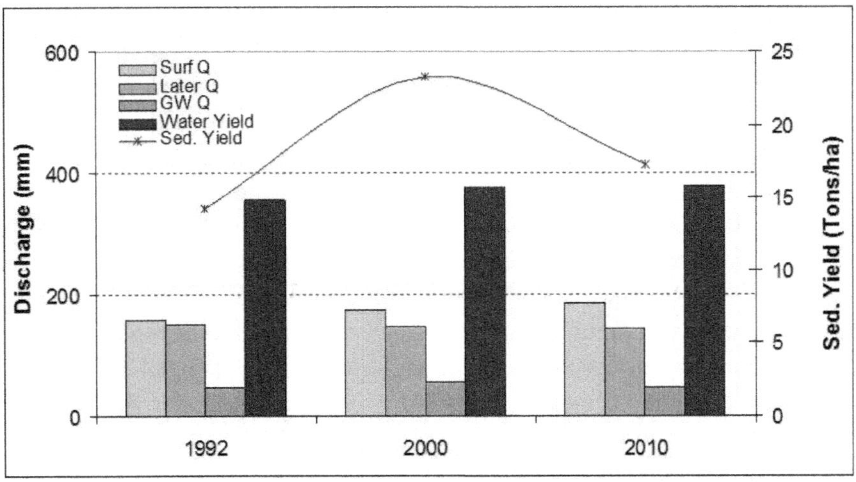

Figure 7: Hydrological parameters like surface runoff, water yield and sediment yield indicate an overall rise in values in reponse to landuse changes occurred within 1992-2010 period.

The increase in sediment yield can be attributed to the increase in surface runoff condition during this period (Figure 7). The zone of low sediment yield i.e. <5 tons $ha^{-1}y^{-1}$ has shown a significant decrease while zones of medium sediment yield i.e. 5-10 tons $ha^{-1}y^{-1}$ and high sediment yield i.e. >15 tons $ha^{-1}y^{-1}$ a relative increase in the southeastern sub-basins of the Rawal watershed during 1992-2010 period (Figure 8).

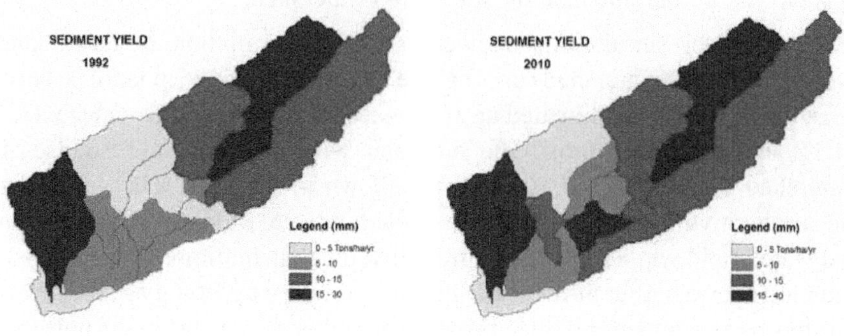

Figure 8: Temporal analysis of avarage annual sediment yield in the watershed during 1992-2010 peiod.

Scenarios of Extreme Conditions

Different scenarios of landuse and climate change were developed to observe the response of water and sediment yields to the expected extreme conditions in future. The first three scenarios are related to probable changes in landuse/ landcover of the watershed in future. As most of the landuse changes are taking place due to growth in urbanization i.e. development of built-up, agriculture land, deforestation/ afforestation in the area, so it formed the basis of these scenarios. The percentage coverage of landuse in the watershed under base line and three scenarios is shown in Table 3 and in map form in Figure 9. The other scenarios are based on the prediction scenarios of climate change for this region i.e. +0.9 °C and +1.8 °C change in temperature during 2020 and 2050 [28] and changes in precipitation. These were formulated in consultation with experts from the Intergovernmental Panel on Climate Change (IPCC) and are consistent with the scenarios generated using the Model for Assessment of Greenhouse gas Induced Climate Change (MAGICC) software. The analysis of different scenarios is given below:

- In the first scenario, all the rangeland below 800m elevation is assumed to be converted into built-up land (About 20% increase in the settlement class). It is based on our study findings that most of the urban development has been occurred in the low valley areas below 800m elevation during the last two decades. The runoff estimates in urban areas are required for comprehensive management analysis. The scenario indicates a decrease of about 0.1% in the water yield while an increase of about 12.1% in the sediment yield from that of the base year 2010 (Table 4). The surface runoff has shown an increase of about 0.2% while lateral discharge a decrease of about 2% due to increase in the impervious area during 2010 in the watershed.

- In the second scenario, all the scrub forest below 1200m elevation is assumed to be converted into agriculture land (About 31% increase in agriculture land) keeping other landuse conditions same as of base year 2010. This scenario is also based on our study findings that major agriculture development has occurred below 1200 m elevation during the last two decades in the watershed area. The scenario indicates an increase of about 3.6% in the water yield and about 73.6% in the sediment yield from the base year 2010. The surface runoff increases by 4.4% while lateral discharge decreases by 5.6% due to decrease in the scrub forest coverage.

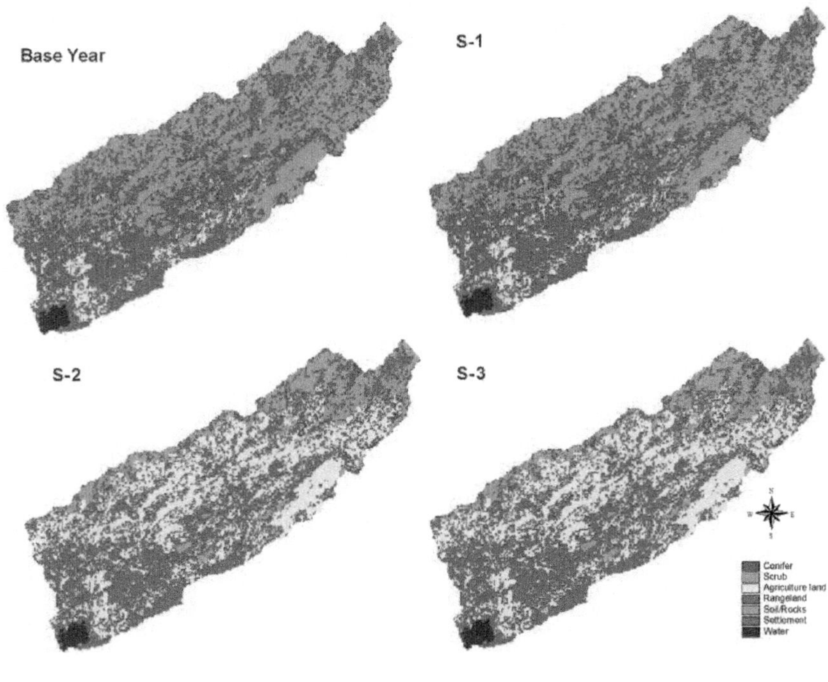

Figure 9: Landuse/landcover status during base year and three landuse change scenarios.

Table 3: Percentage coverage of landuse in base year 2010 and under three landuse change scenarios

Landuse	**Base year 2010**	**Senario-1**	**Senario-2**	**Senario-3**
Conifer	2.3	2.3	2.3	2.3
Scrub	38.8	38.8	7.7	7.7
Agriculture	7.4	7.4	38.5	38.5
Rangeland	36.7	16.3	36.7	16.3
Soil/Rocks	4.8	4.8	4.8	4.8
Settlement	8.9	29.3	8.9	29.3
Water	1.1	1.1	1.1	1.1
Total	100.0	100.0	100.0	100.0

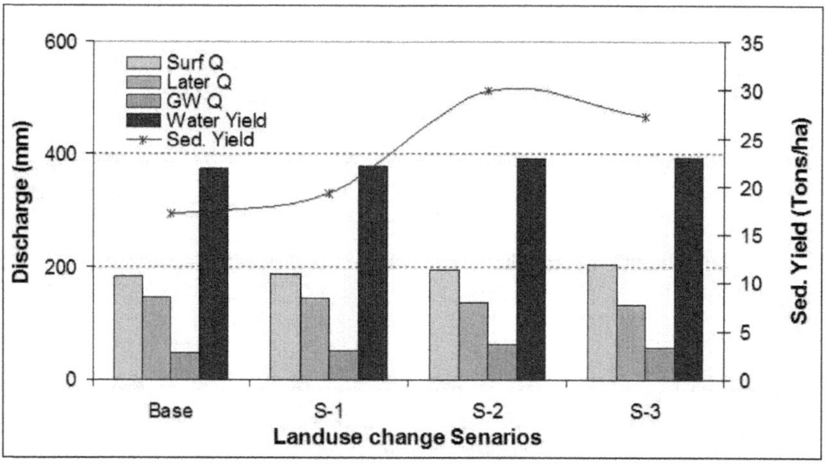

Figure 10: Hydrological response of the watershed under base year and three landuse change senarios.

Table 4: Percentage changes projected for surface runoff, water yield & sediment yield under different landuse and climate change scenarios using base conditions of 2010

No.	**Scenarios**	**Surf. Runoff %**	**Water Yield %**	**Sediment Yield %**
S-1	Urban develop. Below 800m elevation (converting rangeland into built-up land)	0.2	-0.1	12.1
S-2	Agriculture develop. below 1200m elevation (converting scrub forest into agriculture land)	4.4	3.6	73.6
S-3	Combining scenarios 1 & 2	10.5	4.1	58.1
S-4	+0.9°C temperature in 2020	-0.8	-1.3	13.1
S-5	+1.8°C temperature in 2050	-2.1	-3.0	28.3
S-6	Increase of 10% rainfall in 2020 with no change in temperature	24.5	19.1	26.1
S-7	Increase of 0.9°C temperature & 10% rainfall in 2020	23.6	17.8	41.8
S-8	Increase of 1.8°C temperature & 10% rainfall in 2050	22.1	15.8	58.1
S-9	Increase of 0.9°C temperature & decrease of 10% rainfall in 2020	-22.7	-19.0	-13.6
S-10	Increase of 1.8°C temperature & decrease of 10% rainfall in 2050	-23.8	-20.6	-1.5

- The third scenario is based on the combination of the 1st and 2nd scenarios i.e. increase in built-up and agriculture land below 800m and 1200m elevations, respectively. The scenario indicates an increase of about 4.1% in the water yield and 58.1% in the sediment yield of watershed. The surface runoff indicates an increase of about 10.5% while lateral discharge decrease of about 9% due to decline in the scrub forest cover and growth in the urban development. The scenarios 2&3 have also indicated an increase of 1.1% and 0.6% in actual evapotranspiration due to temperature variations. The hydrological response against different landuse change scenarios is shown graphically in Figure 10.
- The scenarios 4 to 10 are based on future climate changes in the watershed area and respective response of the water and sediment yields with the assumption that no change in landuse/landcover of base year 2010 will take place over the time. The rise of about 0.9°C temperature in year 2020 and 1.8°C in year 2050 indicates decrease of about 1.3% and 3.0% in the water yields of the watershed. Increase in temperature may result in higher evaporation rates that would affect the behavior of water yield.

- In scenario-6, increase of 10% in rainfall during 2020 keeping same temperature conditions as of base year 2010 has shown an increase of about 19% in the water yield and 26% in the sediment yield of the watershed. Similar increase in rainfall with same temperature conditions as of scenarios-4 and 5 i.e. +0.9°C in 2020 & +1.8°C in 2050, projects nearly 18% and 16% increase in the water yields and about 24% and 22% increase in the surface runoff as shown under scenarios-7 and 8 in Table 4. The sediment yield has also shown an increase ranging between 41% and 59% in these scenarios. The increase in rainfall usually causes increase in magnitude of floods which ultimately creates soil erosion and land degradation problems.
- On the other hand decrease of 10% in rainfall with same temperature conditions as of scenarios-4 and 5 projects decrease of about 19% and 21% in the water yields which ultimately lowered the sediment yield from that of base condition of 2010.

Risk Mitigation of Sediment Yield

Appropriate strategies have to be defined separately for different landuse conditions for minimizing the risk of soil loss and sediment yield involved in scenarios of extreme conditions. In order to reduce high sediment yield from sub-basins with rapid urban development, the unplanned urbanization needs to be controlled by appropriate laws and means. In the non-urban areas, proper soil and water conservation measures can be adopted to mitigate the risk of soil erosion.

In high risk zone of sediment yield, mainly dissected gullies are more susceptible to soil erosion. The problem of gully erosion can be solved to great extent through restoration of vegetative cover for which proper structures could be placed to provide protection long enough to give vegetation a start. The structures may be of temporary or permanent in nature keeping in view the nature of problem. Conservation structures to reduce velocity of the runoff can also be developed to control the extent of gully erosion. In medium risk zone of soil erosion, modifying the cross section and grade of channel to limit the flow velocities can be performed to stabilize the gullies. The conservation measures like terracing, contour bunding and diversion channels can be adopted in the contributory watershed to control excessive surface runoff causing gully erosion. These practices will also provide additional moisture for growing crops and vegetative cover thus help reducing gully erosion. In low risk zone of soil erosion, contour benches having small bunds crossway the slope of the land on a contour may be established to reduce the erosion risk. High intensity rainfall during monsoon season invariably cause over saturation harmful for plants. The situation can be avoided through provision of water ways and grassy outlets to dispose off the excessive runoff. The risk of erosion can also be minimized through adopting practices like strip farming, terracing, contour farming besides modifying bunds and minor land leveling in the cultivated area of watershed.

CONCLUSIONS

The recent changes in landuse/landcover conditions have brought significant impact on water flows, sediments and threat to eco-hydrology of the Himalayan area. The rapid growth in urbanization has increased the demand for land for development purposes consequently forest and water resources are coming under enormous pressure. The general trends of landuse change are gradual decline in coverage of scrub and coniferous forest, increase in urban development and somewhere in agriculture area. The increase in built-up land in the valleys has reduced the recharge source of groundwater which needs to be protected through controlling unplanned growth of urbanization. The rise in global warming accompanied with high variability in precipitation projects extreme changes in water balance and ultimately deterioration of the land quality. It is essential to regulate the urban development properly, affordable substitute-fuels for household use should be made available and an extensive community reforestation programme is undertaken to improve the fragile eco-system of the region. An integrated adaptation strategy needs to be developed at national and regional levels to cope with future implications of hydrological changes through focusing key policy areas and improving adaptive capacities of the communities at risk. Existing knowledge and data gaps need to be filled by systematic observations and enhanced capacities for research since these will be fundamental for developing climate change adaptation and mitigation programmes for the Himalayan region in future.

REFERENCES

1. Behrman, N. The Waters of The Third Pole: Sources of Threat, Sources of Survival: 2010; 48. www.chinadialogue.net/UserFiles/File/third_pole_full_report.pdf (accessed 17 October, 2012).
2. ICIMOD. Climate Change and the Himalayas: More Vulnerable Livelihoods, Erratic Climate Shifts for the Region and the World. Newsletter, Sustainable Mountain Development 2007; No.53: p55. ISSN 1013-7386 2007.
3. Singh, SP; Bassignana-Khadka, I; Karky, BS and Sharma. Climate change in the Hindu Kush-Himalayas: The state of current knowledge, Kathmandu: ICIMOD; 2011.

4. Neitsch, S.L.; Arnold, J.G.; Kiniry, J.R.; and Williams, J.R. Soil and water assessment tool theoretical documentation. available at: http://swatmodel.tamu.edu/media/1292/ SWAT2005theory.pdf; 2005.
5. Neitsch S.L.; Arnold J.G. and Williams J.R. Soil and Water Assessment Tool, User's Manual; 1999.
6. Arnold, J.G.; J.R. Williams; R. Srinivasan; K.W. King and R.H. Griggs. SWAT - Soil and Water Assessment Tool, USDA, Agricultural Research Service, Grassland, Soil and Water Research Laboratory, 808 East Blackland Road, Temple, TX 76502, revised 10/25/94; 1994.
7. Williams, J.R. Chapter 25. The EPIC Model. p. 909-1000. In Computer Models of Watershed Hydrology. Water Resources Publications. Highlands Ranch, CO.; 1995.
8. Ndomba, P.M. Modelling of Sediment Upstream of Nyumba Ya Mungu Reservoir in Pangani River Basin, Nile Basin Water Science & Engineering Journal 2010; Vol. 3, No. 2: 25-38.
9. Setegen, S.G.; R. Srinivasan and B. Dargahi. Hydro-logical Modelling in Lake Tana Basin, Ethiopia Using SWAT Model, The Open Hydrology Journal 2008; Vol. 2: 49-62.
10. Odira, P.M.A.; M.O. Nyadawa; B. Okello; N.A. Juma and J.P.O. Obiero. Impact of land use/cover dynamics on stream flow: A case study of Nzoia River Catchment, Kenya, Nile Water Science and Engineering Journal 2010; Vol. 3, No. 2: 64-78.
11. Aftab, N. Haphazard colonies polluting Rawal Lake, Daily Times Monday, March 01. http://www.dailytimes.com.pk/default.asp...009_pg11_1. 2010.
12. Roohi, R.; Ashraf, A.; Naz, R.; Hussain, S.A. and Chaudhry, M.H. Inventory of glaciers and glacial lake outburst floods (GLOFs) affected by global warming in the mountains of Himalayan region, Indus Basin, Pakistan Himalaya. Report prepared for ICIMOD, Kathmandu, Nepal; 2005.
13. Ghumman, A.R. Assessment of water quality of Rawal Lake by long term monitoring. Environmental Monitoring Assessment Journal. DOI 10.1007/s10661-010-1776-x.; 2010.
14. Khan, M.I.R. Pakistan Journal of Forestry 1962; 12: 185.
15. Soil Survey Report. Reconnaissance soil survey of Haro Basin 1976, Preliminary Ed. Soil Survey of Pakistan, Lahore; 1978.
16. IUCN. Rapid environmental appraisal of developments in and around Murree Hills, IUCN Pakistan; 2005.

17. Tanvir, A.; B. Shahbaz and A. Suleri (2006), Analysis of myths and realities of deforestation in northwest Pakistan: implications for forestry extension. International Journal of Agriculture & Biology.1560–8530/2006/08–1–107–110. http://www.fspublishers.org.
18. Shafiq, M.; S. Ahmed; A. Nasir; M.Z. Ikram; M. Aslam and M. Khan. Surface runoff from degraded scrub forest watershed under high rainfall zone. Journal of Engineering and Applied Sciences 1997; 16(1): 7-12.
19. EPA (2004), Report on Rawal lake catchment area monitoring operation, Pakistan Environmental Protection Agency, Ministry of Environment, Islamabad: pp. 19.
20. Roohi, R.; A. Ashraf and S. Ahmed. Identification of landuse and vegetation types in Fatehjang area, using LANDSAT-TM data. Quarterly Science Vision 2004; 9(1): 81-88.
21. Oweis, T. and M. Ashraf (eds). Assessment and options for improved productivity and sustainability of natural resources in Dhrabi watershed Pakistan, ICARDA, Aleppo, Syria; 2012.
22. Nasir A.; K. Uchida and M. Ashraf. Estimation of soil erosion by using RUSLE and GIS for small mountainous watersheds. Pakistan Journal of Water Resources 2006; 10(1): 11-21.
23. Donigian Jr., A. S. Watershed model calibration and validation: The HSPF experience. National TMDL Science and Policy, Phoenix, AZ. November 13–16, 2002: 44-73.
24. Nash, J.E. and Sutcliffe V. River flow forecasting through conceptual models, I. A discussion of principles. Journal of Hydrology 1970; 10: 282-290.
25. Arfan, M. Spatio-temporal Modeling of Urbanization and its Effects on Periurban Land Use System. M.Sc Thesis. Department of Plant Sciences. Quaid-e-Azam University, Islamabad; 2008.
26. FAO. State of the world's forests – 2005. Food and Agricultural Organization (FAO), Rome, Italy; 2005.
27. Booth, D. Urbanization and the natural drainage system - impacts, solutions and prognoses. Northwest Environmental Journal 1991; 7: 93-118.
28. INC Report. Pakistan's Initial National Communication on Climate Change to UNFCCC, Ministry of Environment, Islamabad; 2003.

Chapter 10

IMPACT OF DROUGHT AND LAND – USE CHANGES ON SURFACE – WATER QUALITY AND QUANTITY: THE SAHELIAN PARADOX

Luc Descroix[1], Ibrahim Bouzou Moussa[2], Pierre Genthon[3], Daniel Sighomnou[4], Gil Mahé[3], Ibrahim Mamadou[5], Jean-Pierre Vandervaere[6], Emmanuèle Gautier[7], Oumarou Faran Maiga[2], Jean-Louis Rajot[8], Moussa Malam Abdou[1], Nadine Dessay[9], Aghali Ingatan[2], Ibrahim Noma[2], Kadidiatou Souley Yéro[1], Harouna Karambiri[10], Rasmus Fensholt[11], Jean Albergel[12] and Jean-Claude Olivry[13]

[1]IRD / UJF, Grenoble, France

[2]UAM University, Niamey, Niger

[3]IRD-HSM, Montpellier, France

[4]Niger Basin Authority, Niamey, Niger

[5]University of Zinder, Niger

[6]UJF-LTHE, Grenoble, France

[7]Université Paris 8, France

[8] IRD-BIOEMCO, Créteil, France

[9]IRD-ESPACE-DEV, Montpellier, France

[10]2iE International high School, Ouagadougou, Burkina Faso

[11]University of Copenhague, Denmark

[12]IRD-LISAH, Montpellier, France

[13]IRD, France

INTRODUCTION

West Africa has been experiencing drought conditions since the end of the 1960s. This pattern has been particularly evident in the Sahel, but appears to have attenuated in the last decade in the eastern and central parts of this region.

On the other hand, annual rainfall remains very low in the western part of the Sahel [1].

A corresponding decrease has also been observed in the mean annual discharge of the Senegal and Niger rivers, which are the largest in the region and primarily fed by water originating from tropical humid areas. However the percentage decrease in mean annual discharge was almost twice as large as the decrease in rainfall [2] for the period 1970-2010. Similar trends have been observed on smaller river systems.

In contrast, even though the Sahel and most of West Africa also have experienced substantial drought over the past 40 years, runoff coefficients and stream flows have increased in most Sahelian areas. This phenomenon has been named "The Sahelian Paradox" after the increase of the groundwater table in Niger since the 1960s was named the Niamey paradox and attributed to substantial changes in land-use. The HAPEX-Sahel (Hydrological and Atmospheric Pilot Experiment) and the AMMA (African Monsoon Multidisciplinary Analysis) programs have provided, among many comprehensive results, valuable measurements dealing with the spatial and temporal variations in Sahelian soil water content as well as with the infiltration of water through deep soil layers of the vadose zone.

The purpose of this chapter is to provide an overview of hydrological behaviour throughout West Africa based on point, local, meso and regional scales observations.

BACKGROUND

The paradoxical increase in runoff despite drought conditions in sub-Saharan Africa was first noted in a paper by Albergel [3], analysing decadal series of runoff measurements in experimental sites of Burkina Faso. He noticed that this increase was observed in Sahelian areas, but not in the more humid Sudanian regions.

"The decrease in rainfall during the 1969-1983 period seems to be largely offset by the evolution of surface features in the functioning of small catchments. These changes favoured the conditions of runoff in the Sahelian basins; there are due to both the human actions and the climatic conditions. The reduction of vegetation cover and the widespread crops areas cause soil surface settling and the appearance of impervious superficial layers, as well as the extension of eroded areas. Some sahelian basins have nowadays [in 1987] the common characteristics of basins located northward, with great areas of bare soils; perennial graminaceae are replaced with annual ones, and combretaceae with prickly bush species" [3].

Albergel [3] attributed the contrasting behaviour of Sudanian (mean annual rainfall > 750 mm) and Sahelian (mean annual rainfall < 750 mm) areas to increasing bare soils and decreasing vegetation cover in Sahelian basins.

This hypothesis was confirmed in 1999 by Mahé and Olivry [4] and then in 2002 by Olivry [2], who remarked that the discharge of right bank tributaries of Middle Niger River had been increasing since the beginning of the Drought (1968). Similarly, Amani and Nguetora [5] noted that runoff coefficients were increasing significantly in right bank tributaries and showed that the onset of the annual flood was occurring earlier than in previous decades.

Mahé *et al.* [6] analysed the runoff evolution of eight right bank tributaries of the Middle Niger River and noted that the decrease in rainfall did not lead to a decrease in runoff under the Sahelian climate as commonly observed in other basins in the world. Rather, these tributaries exhibited increasing runoff coefficients and in discharges, while "Sudanian" climate tributaries suffered a decrease in discharge and in runoff coefficient [6].

MATERIAL AND METHODS

This study is mainly based on two sources of data:

- field measurements and observations made during the AMMA (African Monsoon Multi-disciplinary Analysis) experiment at the Niger experimental site (Niger River middle stretch and Niamey square degree), and:
- rainfall and discharge data collected on the operational network of Niger basin, provided by the Met Offices of Niger, Mali and Burkina Faso and by NBA (Niger Basin Authority).

The methods included the following:

- Analyse of runoff and river discharge data (in order to characterize the trends in the river discharge records) at several scales:
    - At the point scale: infiltration tests (using disk infiltrometers at multiple suctions) and soil water content monitoring provides data on soil hydraulic conductivity and other physical properties [7] [8];
    - At the local scale: Tondi Kiboro and Wankama catchments, as well as 20 experimental plots of 10 and 100 m^2, located in the same catchments. These data were collected during the AMMA experiment (2004-2010). On the plots, the measurements were made after each event; on the catchment, stream gauges allowed the monitoring of the discharges[9];

- At the meso scale:
  - Some small direct tributaries of the Niger River; we overall use here the regional balance allowed by the stations located in the Niger River upstream (Kandadji) and downstream (Niamey) the studied stretch; however, some discharge data of small direct tributaries were collected for this study [10];
  - The main tributaries of Niger River's middle stretch.
- At the regional scale: the Niger River, the Senegal River, etc, existing data, allowed these analyses [10] [11].

- Analyses of land cover data (including agricultural data, NDVI etc.) and a map of land cover in the square degree of Niamey were realised during AMMA experiment;
- Analyses of precipitation trends across the Sahel ;
- Analysis of endorheism breaks was carried out in the region of Niamey [11].

DECREASE IN RAINFALL, INCREASE IN RUNOFF

Increasing Streamflows

The Great Drought of the Sahel is considered one of the most significant climatic events worldwide [12]. For at least 25 years, more than 3 millions km^2 of semi-arid Sub-Saharan area has suffered a rainfall deficit ranging from 10 to 30%, depending on the location. Figure 1 shows a partial offset of the deficit since the mid-1990s. However, the overall deficit remains and the interannual variability has increased during this period. Not shown in this figure, the intensity of the drought has been largely attenuated in the eastern part of West Africa since the 2000s, but it persists in the western part of the region. In spite of the severe drought, a significant increase in the runoff coefficient and a general increase in stream discharge, have been observed (see background) since the 1960s, in the Sahelian basins (Figure 2).

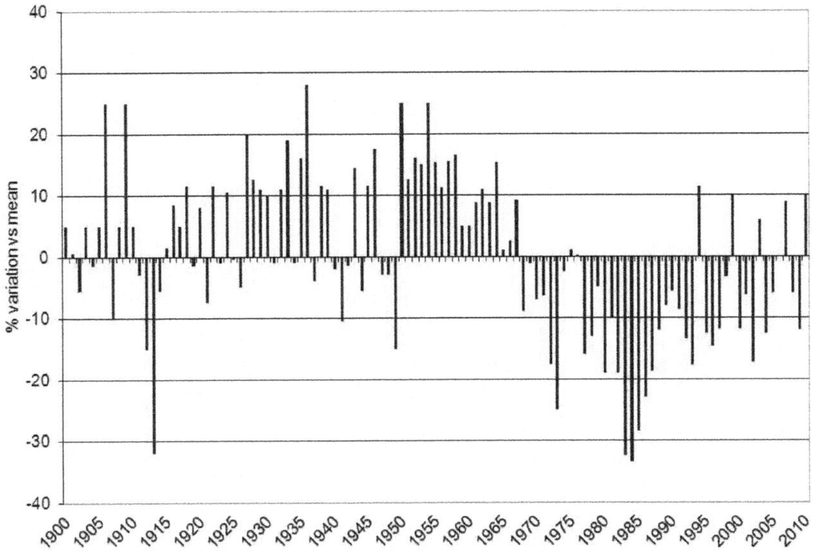

Figure 1: Evolution (1900-2010) of the Standard Precipitation Index for the whole Niger River basin.

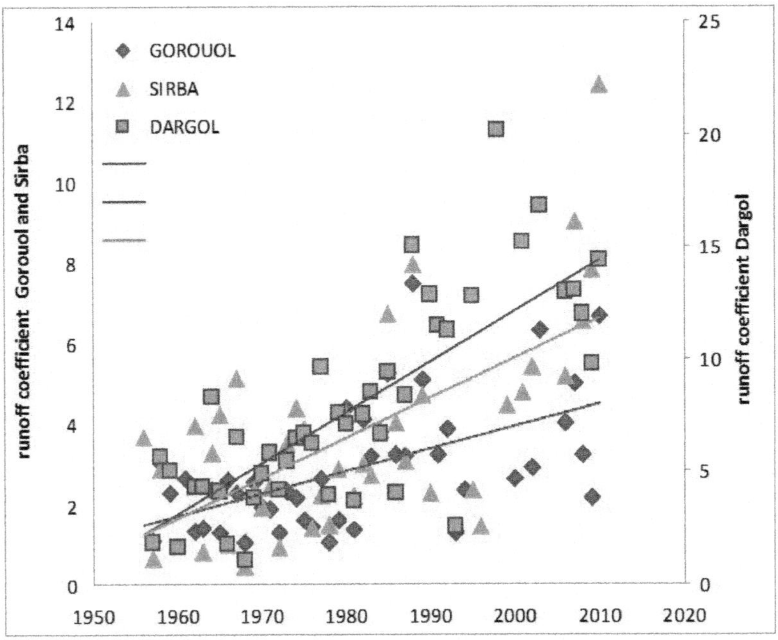

Figure 2: Evolution (1955-2010) of the runoff coefficient of three main right bank tributaries of Niger River.

Figure 3 shows the location of the middle Niger River basin in West Africa. Tributary rivers accounted here are located in the eastern blue circle representing the middle basin. Due to the great loop created by the Niger river northward to the margin of the Saharan desert, the annual flood downstream from the Niger Inner Delta -a large humid area located in the northern reach of the river-, in the Middle Niger River, has two flood peaks. The first one, termed the red or local flood, arises from local rainfall draining through a series of tributaries (Fig. 3, enlargement) and occurs between August and September. The second, termed the Guinean flood, originates from precipitation in the Fouta Djallon (Guinea) area during the rainy season (June–September). Delayed by the crossing of the Niger Inner Delta (see Fig. 3) in Malian territory, the Guinean flood takes place around January. The clear separation between these two flood events makes it possible to distinguish the local Sahelian effect from the more remote trend.

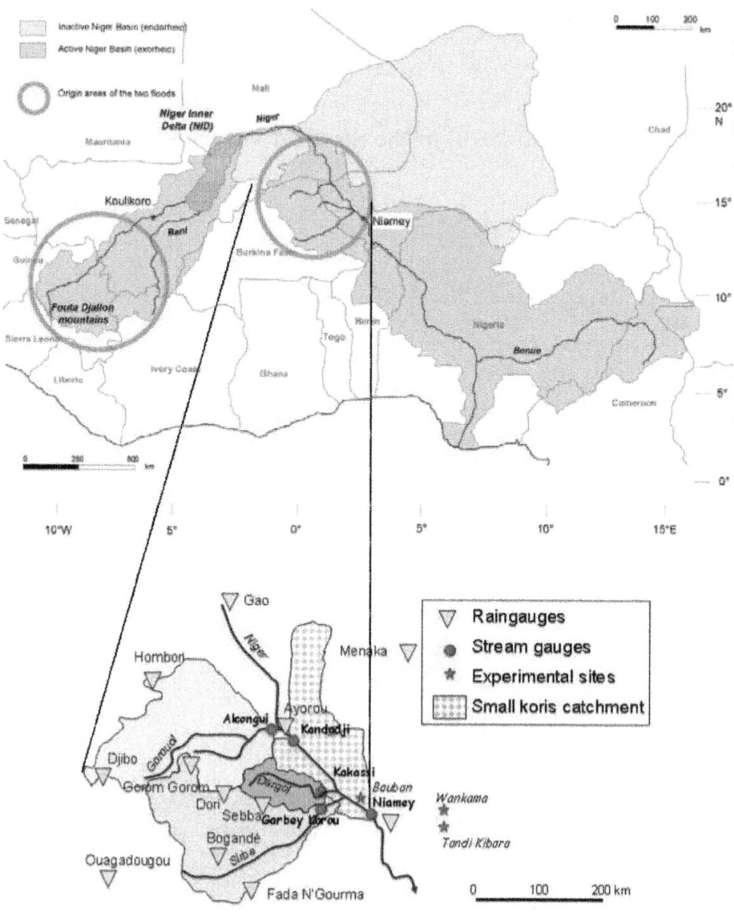

Figure 3: The Niger River basin and (enlargement) its middle basin.

Earlier Flooding

Another change observed during the drought is the earlier onset of yearly flooding, compared with previous periods. Figure 4 shows that the first flood is now arriving approximately forty days earlier than it did forty years ago. This observation is consistent with a decreased soil water holding capacity in the river basin.

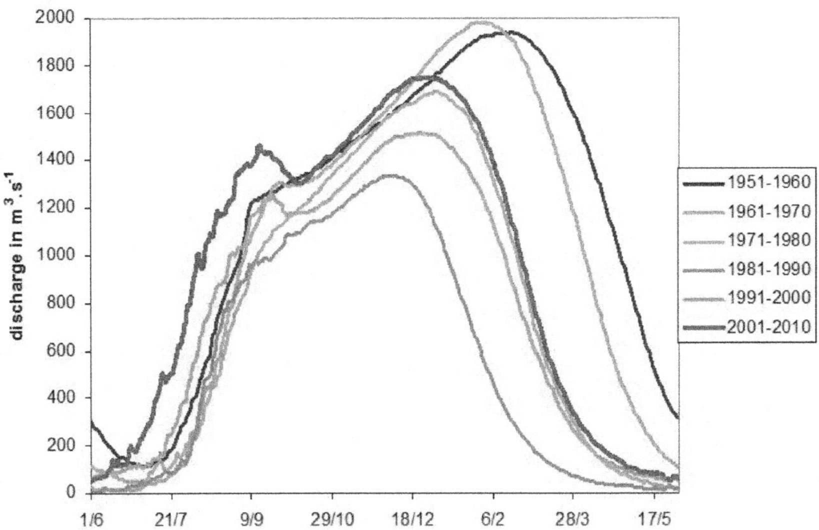

Figure 4: The Niger River decadal hydrographs at Niamey station (Republic of Niger).

Sahelian Paradox is Not Due To Rainfall

Because rainfall has decreased significantly since 1968, rainfall amount does not explain the increase in runoff and stream flow or the earlier flood occurrence. For example, the three main tributaries affected by the semi-arid Sahelian climate (The Gorouol, the Dargol and the Sirba rivers: see enlargement of figure 3) experienced a significant increase in runoff coefficient during the drought, despite a 20% reduction in rainfall (Table 1).

To identify the drivers of increased stream discharge and early flood onset, we analysed the trends and evolution of rainfall for twelve stations with daily rainfall data from 1950 onward. These stations are indicated in figure 3 [11].

A forward shift in the timing of the monsoon rains does also not appear to explain the Sahelian paradox. Nicholson [13] as well as Ali and Lebel [1] observed in recent decades a reduction of rainfall in August and a relative increase in rainfall amount in June and July. A similar forward shift in monsoon timing was observed in the Middle Niger River basin (figure 5). However the total amount of rainfall in June and July remains lower during the last two decades than during the 1950s and 1960s (figure 6).

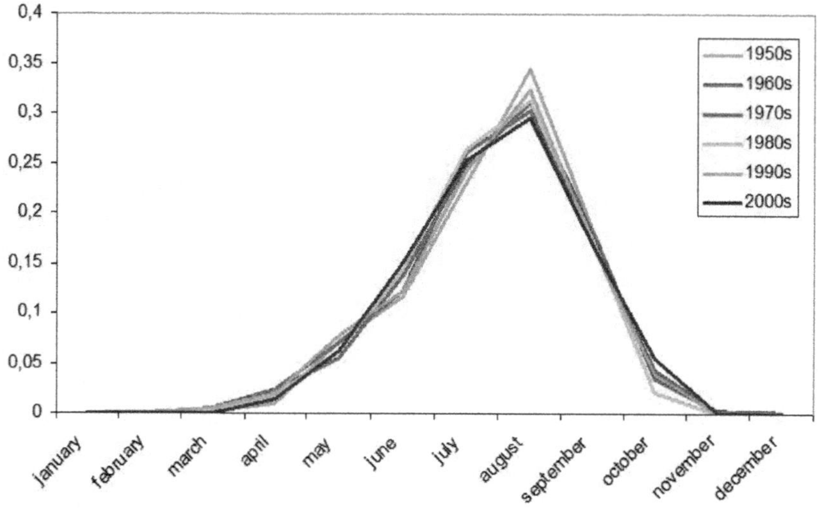

Figure 5: Evolution per decades of the rainfall monthly index (middle Niger River basin).

- Because the runoff coefficient increases with increasing rainfall intensity and amount, a rise in the number of extreme events might explain the higher volume of the floods. However a general decrease in the number of extreme rainfall events has been observed since the 1950s, for each class of events (figure 7), ranged by total amount of the event (classes 20-30 mm; 30-40 mm; 40-60 mm and more than 60 mm). However at the whole Sahel scale, a current study shows an increase in the rainfall amount for events upper than 30 mm during the 2000s.
- An increase in extreme events at the beginning of the rainy season also could explain the early flood occurrence. An increase in the total amount of rainfall fallen in events > 40 mm has been observed in the last decade, but only in June. However, the runoff coefficient is nowadays two or three times higher in the Middle Niger River basin than during the 1950s. Thus, the modest increase in rainfall amount (event > 40 mm) observed in June during the 2000s cannot alone explain the timing and magnitude of the recent floods.

Table 1: Evolution of runoff coefficients of the three main Niger River right bank tributaries from 1957

Basin	Gorouol	Sirba	Dargol
Period 1957-1979	1.9	2.6	5.0
Periods 1980-1994	3.6	4.2	8.8
Periods 1995-2010	4.3	6.0	14.2

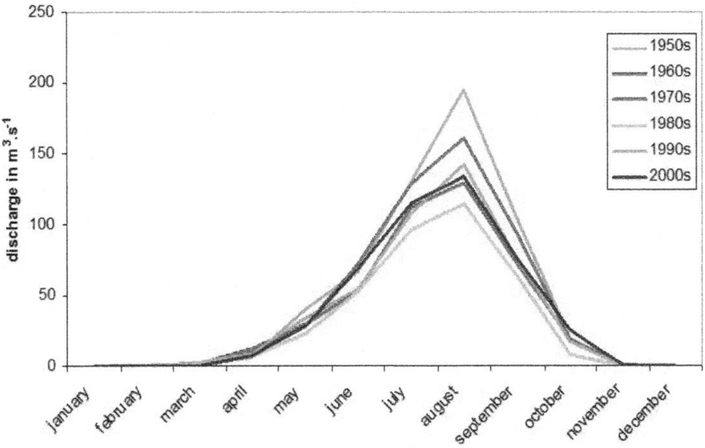

Figure 6: Evolution per decades of the monthly mean rainfall amount (middle Niger River basin).

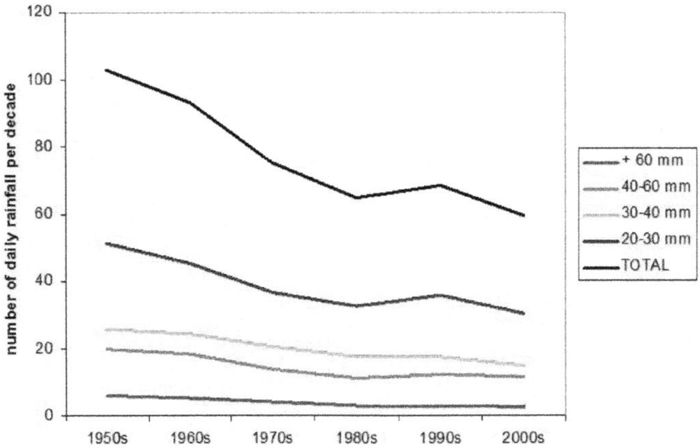

Figure 7: Evolution per decades of the number of extreme events (ranged by total rainfall amount of the rainy event), middle Niger River basin.

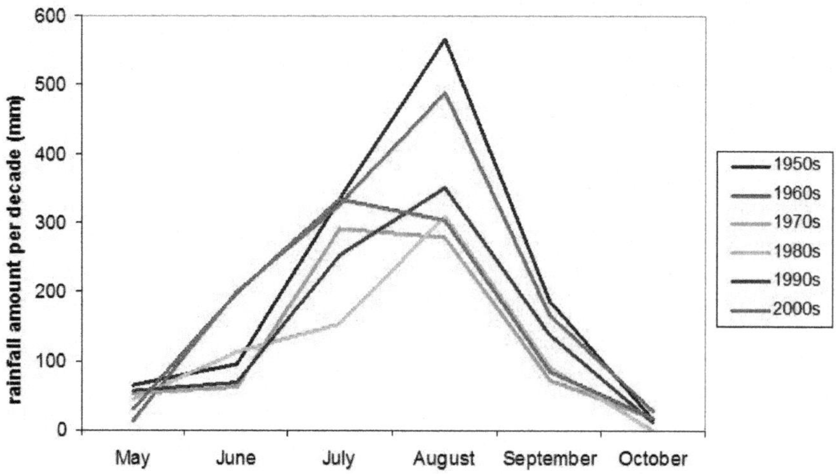

Figure 8: Evolution per decades of monthly rainfall amount of events > 40 mm Middle Niger River basin.

CONTRIBUTION OF CHANGING SOIL CHARACTERISTICS TO SAHELIAN PARADOX?

In Sahelian areas at all scales, runoff coefficient generally increased along with river discharges (Table 2). In most cases, these changes correlate with a decrease in vegetation cover, due to land use changes including increasing crop area, overgrazing and wood harvesting. At the regional scale, the change in vegetation cover is not obvious. The runoff coefficient of degraded soils in the Sahel (close to 60% for the ERO type crusted soils –as defined by Casenave and Valentin [14])- is much higher than that observed for millet crops (4%) and for bush and fallows (10%; see Table 1 and figure 9). Small crusted soil areas can alone explain increased stream flows. Therefore increasing runoff coefficients may be consistent with the re-greening documented in remote sensing studies (higher values of NDVI), the remaining soils being covered by higher millet crops density, graminaceae, herbaceae and annual plants than during previous periods.

Runoff coefficients are measured in plots of 10 and 100 m^2; saturated hydraulic conductivity is measured in at least 20 points for each surfaces class [11].

Figure 9: Opposition between a "fallow on common sandy soil" (left) plot and a "fallow erosion (ERO) crusted soil" plot (right).

Table 2: Comparison of the hydrodynamic properties of non-crusted and crusted soils

Soil surface feature	**Runoff coefficient** %	**saturated hydraulic conductivity mm.h-1**
Millet on common sandy soil	4.0 +/- 1.4	172 +/- 79 (20)*
Fallow on common sandy soil	10 +/- 4	79 +/- 41 (20)
Old fallow with bioderm	25 +/- 7	18 +/- 12 (30)
Millet and fallow erosion (ERO) crusted soils	60 +/- 8	10 +/- 5 (30)

[i] - number of repetitions

Table 3: Runoff characteristics for the periods 1 (1991-1994) and 2 (2004-2009) on the three small Tondi Kiboro basins

1991-1994	**Eain-fall**	**Runoff depth**	**Kr***	**Rain-fall/ runoff**	**r² R = a P + b**	**Yearly runoff total duration in hours**
TK amont	513	180	0,36	R = 0,56 P - 2,61	0,82	34,9
TK aval	513	133	0,26	R = 0,43 P - 2,3	0,79	28,1
TK bodo	485	185	0,38	R = 0,53 P - 2,14	0,68	62,7

2004-2009	rainfall	Runoff depth	Kr*	Rain-fall/ runoff	r² R = a P + b	Yearly runoff total duration in hours
TK amont	495	231	0,47	R = 0,77 P – 4,9	0,85	34,2
TK aval	491	132	0,27	R = 0,49 P – 3,5	0,74	18,2
TK bodo (2007-2009)	520	242	0,47	R = 0,87 P - 7	0,81	25,9

[i] - Kr = runoff coefficient

Hydrodynamic characteristics differ substantially between different types of soil surface features. Given the strong difference in runoff coefficients, land use /land cover evolution could probably explain the runoff increase and "re-greening" is evident in the Sahelian area (see section 9 below). However, what was rather observed within the AMMA experimental sites in Niger is a degradation of soils and vegetation during the last decades, without a noticeable recovery since the mid 1990s.

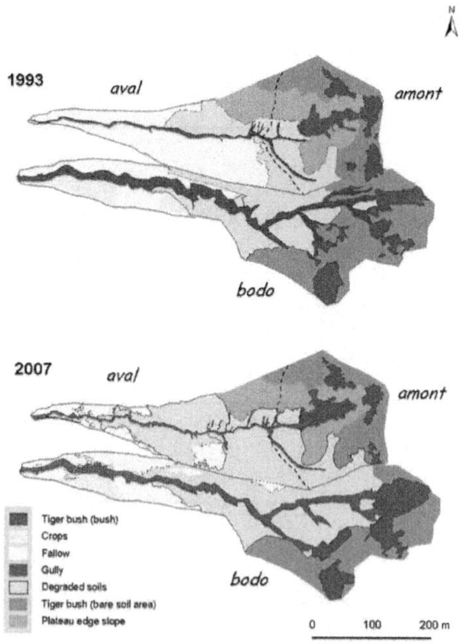

Figure 10: Land use over the Tondi Kiboro experimental catchments (Western Niger) in 1993 (up) and 2007 (down).

In the Tondi Kiboro catchments, the area of degraded soils (mostly ERO crusted soils) in 2007 was twice that observed in 1993 (figure 10). As a matter of fact, the runoff coefficients were significantly increasing from the 1991-1994 to the 2004-2009 period (Table 3), as a consequence of the reduction of soil water holding capacity and the rise in crusted surfaces.

In the Wankama experimental catchment, the same evolution of land cover is observed; but there is no historical hydrological data available for comparison. However, figure 11 allows comparing the vegetation cover in 1950 (aerial photos) and in 2007 (pictures taken from a PIXY® drone; [15]).

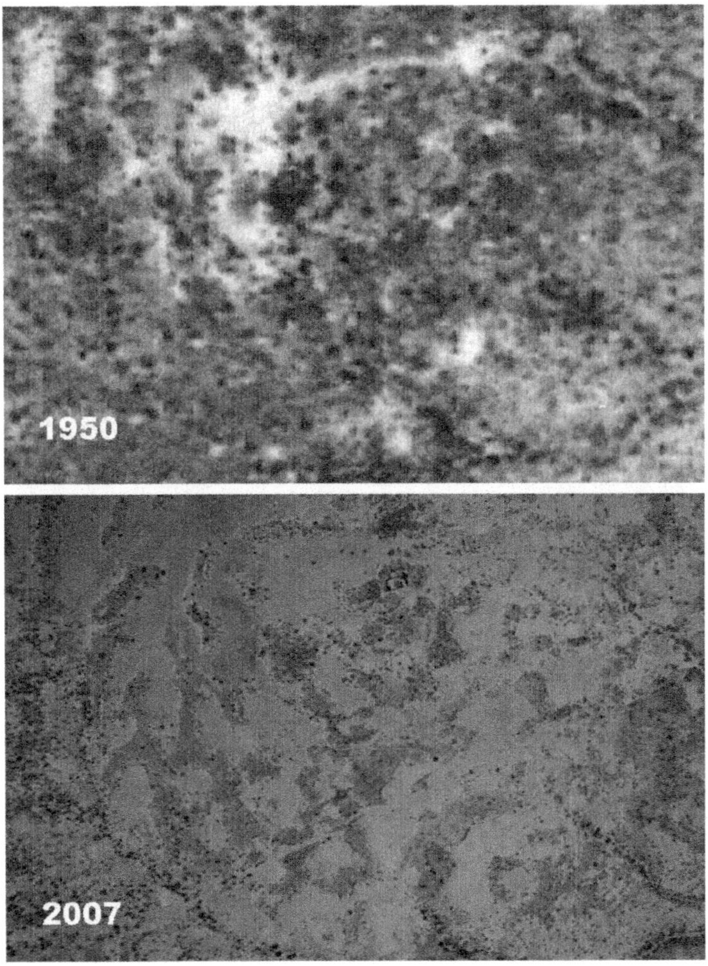

Figure 11: The lower part of Wankama experimental catchment in 1950 (up, aerial photo by IGN France) and in 2007 (photo taken from a PIXY drone); the generalisation of ERO crusted soils in a way of "hydro-aeolian depressions" is noticeable.

In the lower part of this basin, the "ERO" crust areas are widespread and constitute a very active contributing area; 70% of the surface is covered by "ERO" crust (see Fig. 9 at right hand side). The ERO crust runoff coefficient is approximately 60% (Table 2) while it is only 4% in pearl millet crops and 10% in the fallow. An increase in "ERO" crusted area must then have strong hydrological consequences.

The "small koris" catchments (see Fig.3) are not gauged. However, taking the difference between the discharge at Niamey station, on the other hand, the sum of discharges of Niger at Kandadji and those of the Sirba and the Dargol (the two tributaries feeding Niger river between Kandadji and Niamey, Fig.3)[11] showed that a clear change in the behaviour of these "small koris" behaviour occurred after 1997. In general, between 1975 and 1996 the input volume of the Niger River at the Kandadji station was greater than the output volume at Niamey, presumably due to infiltration and evaporation losses. Since 1998, the opposite is observed, presumably due to new input from small tributaries, where crusted soils areas have increased in recent years. Some of these basins have shifted from endorheic prior to and during the 1990s to exorheic in recent years (see 6. section below), increasing the contribution to the Niger River from degraded areas with high runoff. These small tributaries have recently provided several billions cubic meters per year to the Niger River discharge [11] (Fig.12).

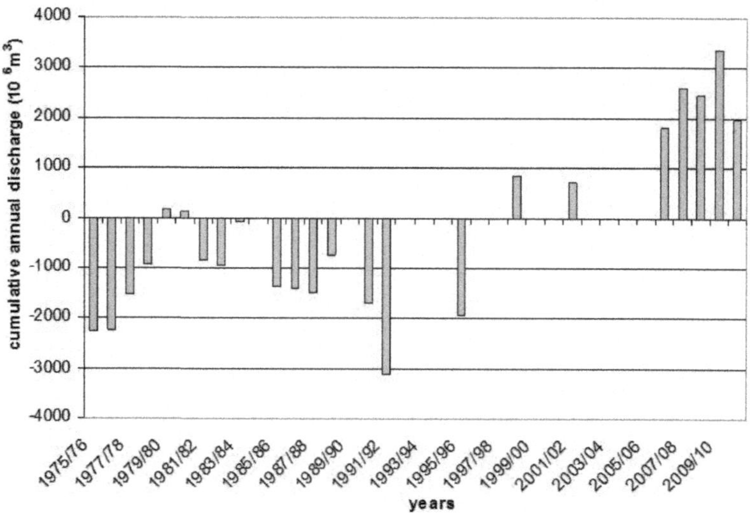

Figure 12: Evolution of the remaining water balance between Kandadji and Niamey from 1975 to 2010, after subtracting the discharge of the two main right bank tributar-

ies (Dargol and Sirba); discharge of the Dargol and Sirba rivers were measured at the Kakassi and Garbey Korou stations (see Fig. 3).

This cited study [11] was dedicated to the severe flooding of August and September 2010 in Niamey, were 300 ha in the river right bank were inundated. Twice, the level of Niger River reached a maximum during the rainy season (2030 $m^3.s^{-1}$ in early August, 2130 $m^3.s^{-1}$ in early September, the maximum previously registered value being 2000 $m^3.s^{-1}$). However, during August 2012, this maximum was widely exceeded. The discharge value reached 2473 $m^3.s^{-1}$ on 18th August, causing severe damage in the city of Niamey and extensive flooding downstream in Niger, Benin and Nigeria (Figure 13) [16].

Thus, previous studies show that from the measurement point scale to the meso-scale basin, the runoff increase is observable in the Sahelian area of the Middle Niger basin:

- at the point scale, it is shown that the new surface features created by land use change (crusted soils areas particularly) have a very low hydraulic conductivity and consequently high runoff coefficients (Table 2), the latter being measured in plots of 10 and 100 m^2;
- at the small basin scale, the experimental basins of Tondi Kiboro (12 ha) exhibited an increase in runoff and discharge probably due to the extension of crusted soil areas (Figure 10 and Table 3);
- the direct middle Niger river tributaries experienced a strong runoff increase after 1997 (Figure 12); the corresponding scale ranges from 10 to 2000 km^2;
- the meso-scale Niger right bank tributaries basins range from 7000 km^2 (Dargol river basin) to 38,500 km^2 (Sirba) and 45,000 km^2 (Gorouol); they have shown a strong discharge increase since the beginning of the drought (figure 2); Amani and Nguetora [5] demonstrated that the flood was occurring during the 1980s almost one month earlier than during the 1960s, as a consequence of both decreasing vegetation cover and reduction in soil water holding capacity;
- the previous statements explain the significant rise in the first, red flood of the Niger river in its Middle basin (figure 13) and the earlier occurrence of this first flood (40 days earlier than during the 1970s as seen in figure 4).

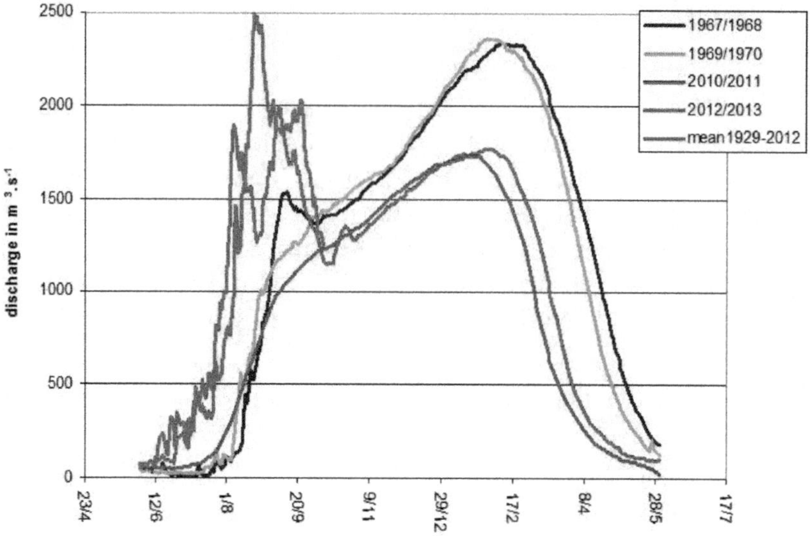

Figure 13: The 2012 red flood compared with previous remarkable years, included the two severe floods during the monsoon of 2010, and the pattern of the Guinean or black flood [16]

AN INCREASE IN FLOOD HAZARD

"As it has been supposed from the end of the 1980's, the change in the hydrological functioning and the newly twin peak hydrograph" (of the Middle Niger river) "are linked to human factors, mostly to land use changes, particularly the land clearing and the extension of crops due to demographic pressure; this led to a soil baring and a fallow shortening which caused a soil crusting resulting in a severe decrease of soil infiltrability. This results in an increase in flood hazard" [11].

The record high 2012 flood must remind policy makers that this hazard is becoming a big social and environmental concern to be accounted in land planning policy. Overall, environmental engineering must be performed and improved in order to offset and mitigate the effects of this trend: urban areas are firstly affected by flooding, but rural areas are those which have to be land managed in order to increase the soil water holding capacity. This should allow increasing land productivity and reducing the flood hazard downstream.

The immediate causes of the 2012 actual flood are not yet determined; however, rainfall amount was high, probably the highest measured since 1968 in the Middle Niger Basin. We focus here on the 2010 monsoon flood.

The first, small, mid July 2010 peak was due to all the sub-basins. It is worth noticing that the first of the two higher rainy season peaks (early August) was firstly due to discharges coming from Malian territory, upstream from Kandadji station and, then, to stream flows coming from the Gorouol basin. The second peak (early September) was mostly produced by flows coming from upstream from the Gorouol confluence, thus from the arid area of eastern Mali, and secondarily by the Sirba and Dargol rivers discharges. The small koris contribution is only estimated, as shown, by the difference between the discharge at Niamey and the sum of Niger at Kandadji, the Dargol and the Sirba. However it is obvious that these small koris had a large contribution to the first peak (figure 14).

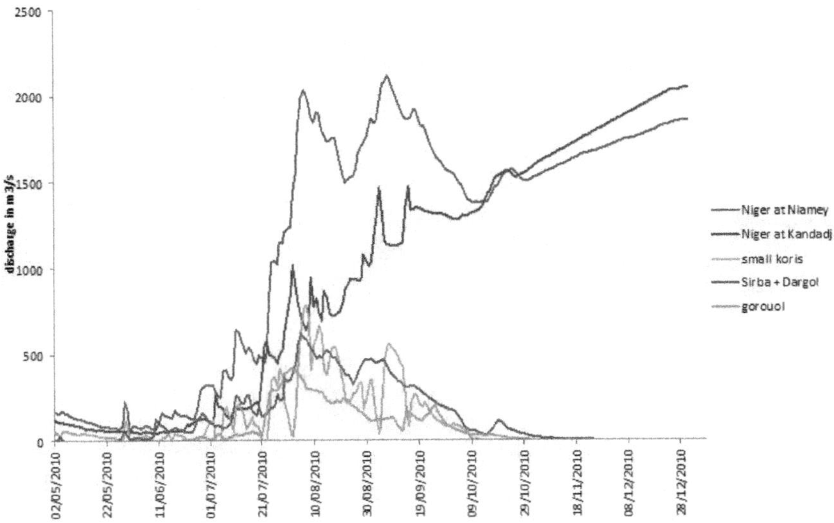

Figure 14: flood decomposition between different contributing areas.

The long term causes of these floods are the land use change and the increase in crusted soil areas. But recently, flood hazard was accentuated by three factors:

- the urbanisation
- the silting up
- the endorheic bursting

Urban population in Sub-Saharan Africa remains the lowest in the world. It is expected to increase strongly in the next years and decades. The percentage of urban population in the Sahel is only 30% in average (but 17% in Niger and 20% in Burkina Faso), compared with 50% in the southern West African countries on the shore of the Gulf of Guinea. The urban population in the Sahel is expected to reach 40% in 2030 and 50% in 2050. As the total population is increasing by 2.5-3% per year, it duplicates every 20-25 years. The urban population doubles every 14-15 years. Most of new resident come from rural areas and settle familial or informal housing. The latter is commonly settled on non-drained and non-buildable areas, which constitute the first areas being inundated in case of flooding. This was the case in 2003 at Saint Louis (Senegal), the 1st September 2009 at Ouagadougou (Burkina Faso) [17] or Agadès (Niger), in August 2010 and August 2012 at Niamey (Niger) [16], and at Dakar (Sénégal). Most of informal housing is built in adobe, straw or iron sheet, which is destroyed or severely damaged by flooding (Fig.15).

Figure 15: Flooded areas in Niamey (shore of Niger River) the 18th August 2012 (left) and in Agadès (right) the September 1st 2009 (*photo Ibrahim Noma and Baptiste Nay*).

In southern Sudanian areas [18] [19], the ongoing increase in the occurrence of inundations was highlighted during the last years. Flood related fatalities in Africa, as well as associated economic losses, have increased dramatically over the past half century [20]. This trend associated with urbanisation is expected to have dramatic consequences.

Figure 16: Evolution of the alluvial fan of the Kourtéré kori into the Niger river bed, just upstream from Niamey (Niger) [10].

The second increasing flood hazard is the current silting up of river beds in the Sahelian region, linked to the observed erosion stage [21]. As runoff is rising, erosion is exacerbated, increasing sediment transport, leading to sandy deposits on the river beds, and contributing to flooding. This was the case for the last Niger River floods in Niamey, due to the invasion of its bed by alluvial fans coming from tributaries upstream from Niamey (Figure 16).

A third increasing flood hazard in the Sahel is the increased exorheism. Some endorheic valleys became exorheic in recent years and decades, increasing the Niger river catchment area and the number of tributaries. Due to soil crusting, these new contributing areas have high runoff coefficients causing a rise in the available discharge for the Niger River (figure 17).

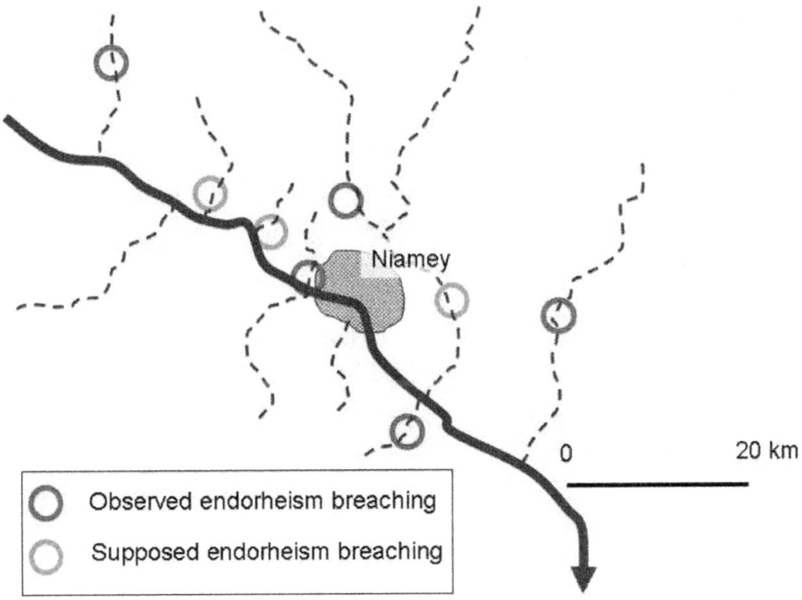

Figure 17: Observed and supposed endorheism bursting in the Niamey area of Niger River valley.

REGIONALISATION

The runoff increase is not observed only in the Middle Niger river basin. The important contributions of Gil Mahé [4] [6] [22] [23] [24], before a first attempt of synthesis of Amogu *et al.,* [10], show the large extension of the increase in rivers discharge, justifying the termed "Sahelian Paradox".

We propose here a regionalisation of such mechanisms and an analysis of the respective role of natural (climatic) and Human (land use changes) factors in the appearance, evolution and geographical distribution of these hydrological behaviours.

Mahé and his colleagues highlighted the rise in the Nakambé River (Burkina Faso) discharge ([22];figure 18), after documenting the same process on the Sahelian Niger river right bank tributaries [6]. More recently, they observed a similar evolution in the western part of the Sahel, in southern Mauritanian rivers ([24]; figure 19). In a newly published paper [23], the discharges in the Sokoto river in Nigeria showed an increasing trend, similar to those of the right bank tributaries.

The stream discharges of Sudanian rivers (Sudanian climate is characterized by annual rainfall amount exceeding 750 mm) were found to be decreasing [10], which was proposed by Mahé [25] to result from a drop in the water level in the aquifers sustaining the rivers during the low flow period. These areas have a "Hewlettian" hydrological behaviour instead of the "Hortonian" functioning typical for the Sahelian semiarid regions. Runoff only onset when the soil is saturated; in these areas, the reduction in rainfall affected firstly the part of rainy water previously dedicated to runoff, explaining the significant runoff and discharges decrease in these areas.

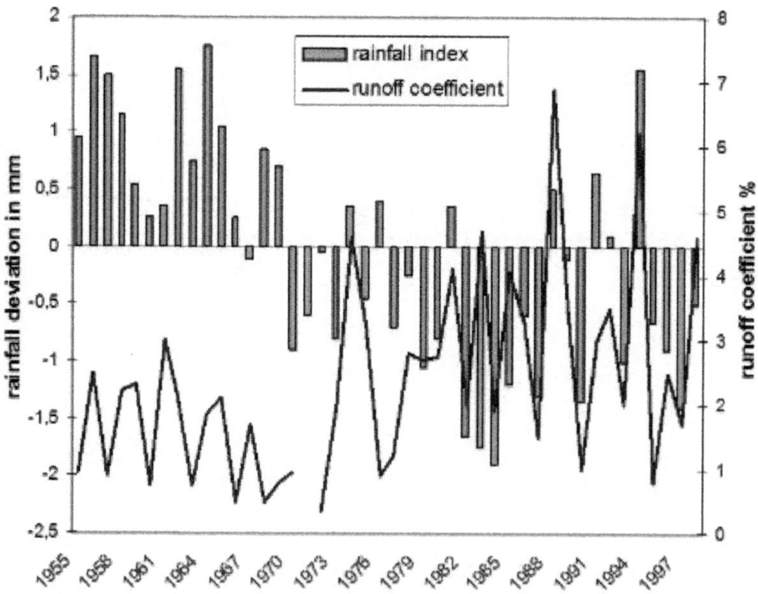

Figure 18: Runoff coefficients from observed measurements and rainfall indexes over the Nakambé river basin (Burkina Faso) [22].

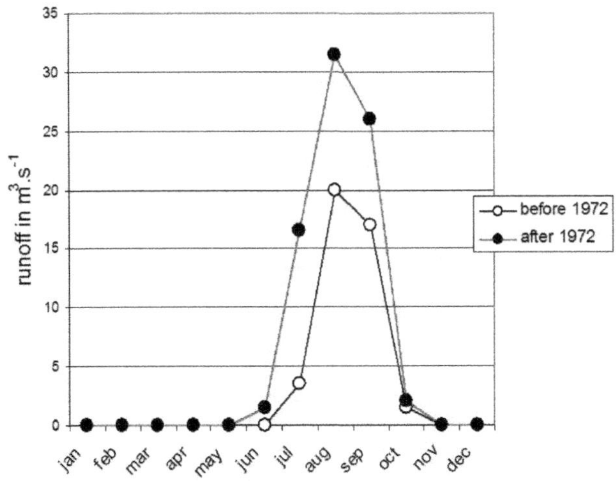

Figure 19: Monthly averaged discharges of seven Mauritanian tributaries of Senegal River, before and after 1972 [23].

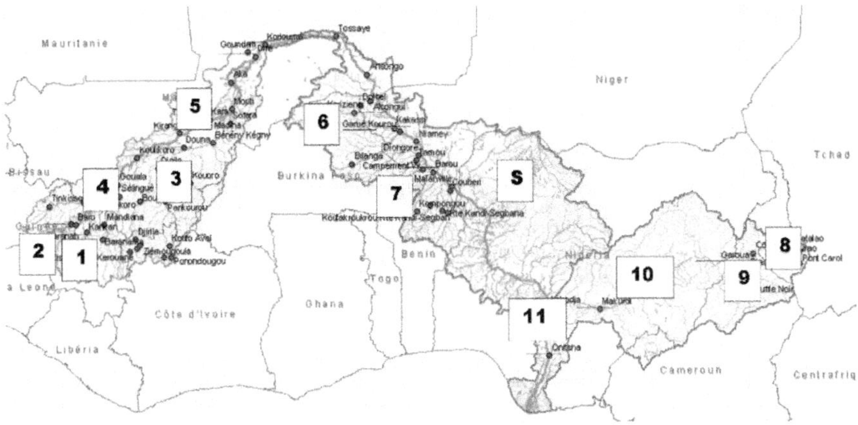

Figure 20: The Niger River basin [26]. Stations cited in Table 4 are underlined and the position of cited basins is indicated with its number in the table.

The synthesis presented in fig.20 and Table 4 shows that, except in Sahelian areas, the discharge has been decreasing as expected, since the beginning of the drought in 1968. Figure 21 shows that there is a clear regional distribution of hydrological behaviour, with the Sahelian region showing a strong increase in runoff, the Sudanian ones exhibiting a significant decrease in river discharge, while Guinean areas (rainfall > 1300 mm) generally show a slight decrease in discharges.

Table 4: mean discharges for two periods : before (until 1969) and after (since 1970) the Drought for 11 basins of Niger River [23]

Basin	**Niger**	**Niger**	**Niger**	**Niger**	**Niger**	**Niger**	**Niger**	**Benue**	**Benue**	**Benue**	**Niger**
River	Milo	Nian-dan	Bani	Niger	Niger	Sahe-lian basin	Mek-rou	Mayo Kébi	Benue	Benue	Niger
Station	Kank-an	Baro	Douna	Kou-likoro	Diré	Rive droite	Barou	Cossi	Garoua	Makur-di	Onitsha
Country	Guin-ea	Guinea	Mali	Mali	Mali	Burki-na-Niger	Niger	Cam-eroon	Camer-oon	Nigeria	Nigeria
Area km²	9900	12600	101600	120000	366500	90500	10500	25100	64000	303600	1388300
Numéro fig1	1	2	3	4	5	6	7	8	9	10	11
years obs.	1950-2000	1950-2000	1922-2000	1907-2000	1924-2000	1956-1995	1961-1999	1955-2000	1946-1991	1955-1995	1950-1987
mean(-69)	211	271	639	1552	2244	35,4	41	104	388	3549	6651
mean(70-)	155	186	235	1039	1349	38,0	23	74	244	2816	5016
(70-)/(-69) %	-27	-31	-63	-33	-40	+7	-42	-30	-37	-21	-25

AND IN ENDORHEIC BASINS?

As the "Sahelian Paradox" seems to apply in the whole Sahel, it must apply also in endorheic areas. In certain endorheic areas in the Sahel, the water table level has been found to be rising over the last several decades despite the strong reduction in rainfall observed after 1968. This phenomenon has been previously defined as the "Niamey Paradox" [27]. The excess in runoff has significantly increased the number of ponds. While ponds are the main zones of deep infiltration, their increase explains the rise of the water table level (figure 22).

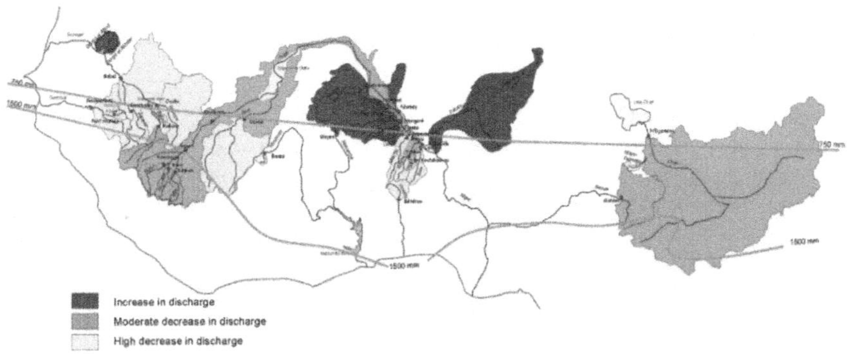

Figure 21: The regional distribution of hydrological behaviour in West Africa; increasing discharges indicate a "hortonian" functioning as well expected decrease in discharge ([10] updated).

Indeed, the current active erosion processes are leading to the appearance of many new gullies and new spreading areas where sediments extracted by aeolian and hydric erosion are then transported in the gullies, and deposited. The gully beds are currently characterized by sand deposits ranging from 2 to 4 metres wide, several tens of centimetres deep (up to 1 to 2 m) and hundreds of metres long. Spreading areas are formed by these newly created streams when they reach gentler slopes, because their transport capacity becomes suddenly insufficient to carry such significant volumes of sand. They form sandy deposits of a magnitude of hundreds of square metres, up to several hectares, and, in some cases, tens of centimetres deep. These areas constitute new deep infiltration areas, accelerating the rise in water table [28].

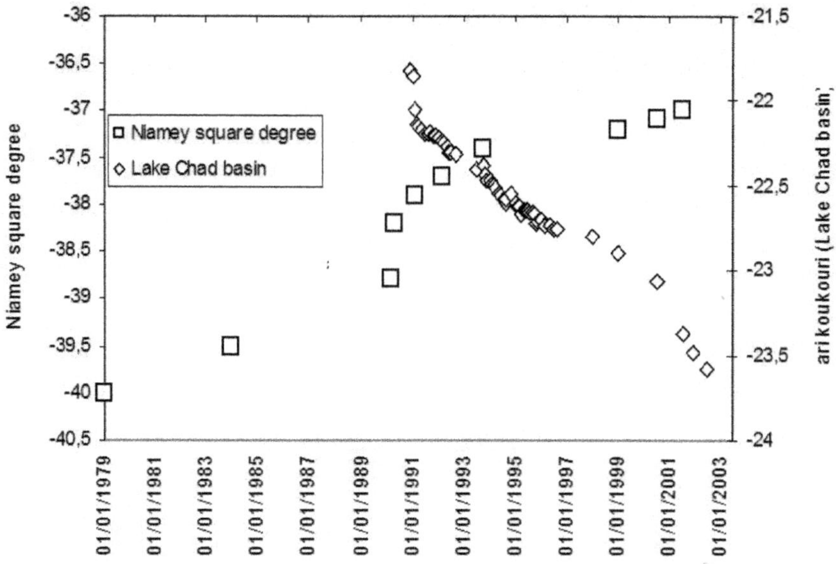

Figure 22: Evolution of water table level under the Niamey square degree area and under the Lake Chad [7] [28] [29], pers. Comm. of G. Favreau, 2007.

In contrast, the water table level is falling in the Lake Chad area (figure 22). This lake and the Niamey square degree are located at the same latitude, but the behaviours of their respective groundwater systems are completely different. The area of the Niamey square degree is composed of small endorheic basins with only local water contribution. While Lake Chad is mostly water fed by tropical humid areas of the Upper Logone and Chari basins (95%). Recently it was shown that discharges are decreasing in Sudanian areas [10] and, thus, water supply to Lake Chad is decreasing. The Ari Koukouri well is located at some tens kilometres of the northern part of the Lake, dried repetitively since the 1980s. The difference in the water feeding patterns explains the opposite behaviour of groundwater under Lake Chad and under the Niamey square degree.

A PERSISTENT DESERTIFICATION?

The evolution of land use/land cover in the Sahel remains the subject of an ongoing debate.

At the Sahel regional scale, cultivated areas have been increasing significantly for more than 50 years due to population growth and very low crop yields (see Fig. 23 for the Niger Republic) [30].Some studies based on satellite data vegetation indices [31] [32] [33] have suggested a re-greening of the Sahel after 1994. However, other studies ([34] amongst others) have highlighted the limitations of vegetation indices for the determination of land use and land use changes when using remote sensing at coarse resolution. Land cover studies based on aerial photograph analyses [28] show on the contrary a decrease in vegetation cover in south western Niger.

Although the Sahel is probably re-greening, the western part of Niger and the eastern part of Burkina Faso are still suffering land degradation (see maps in [32] [33]; see figure 24).

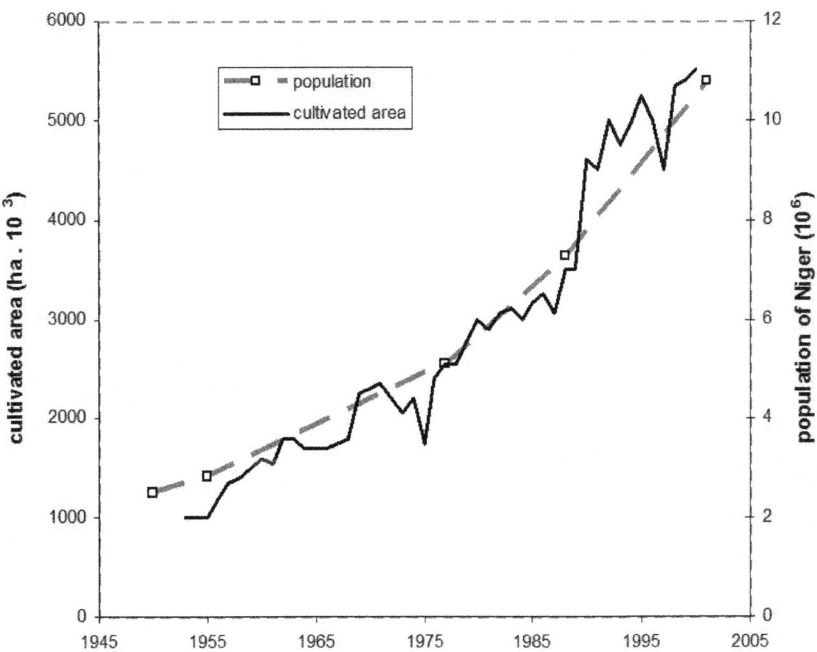

Figure 23: Evolution of population and cultivated area in Niger since 1950 [30].

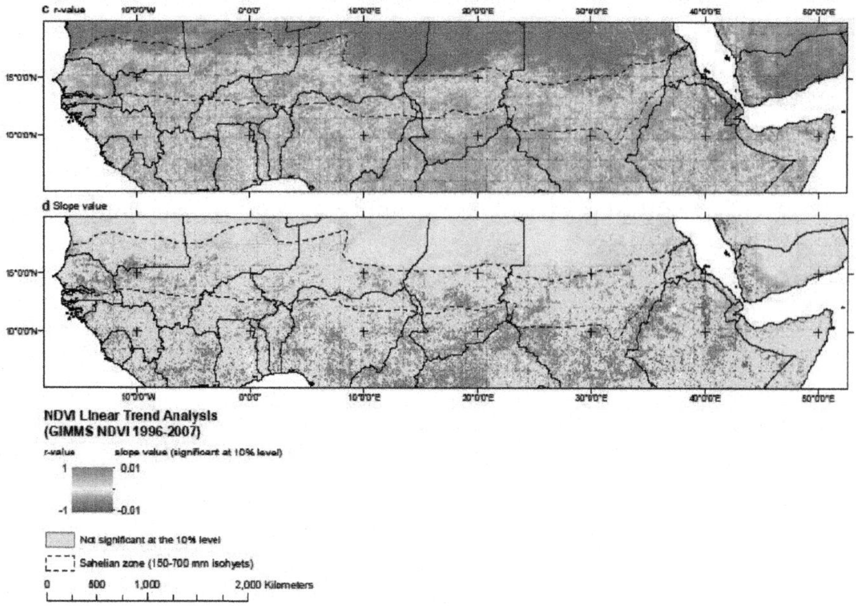

Figure 24: Evolution of NDVI values in the Sahel (area between the two doted lines) (up); down: mapping of the significant pixels (at 10% level) [33].

North of our study area, in eastern Mali, an observed increase in runoff has been attributed to the removal of vegetation during the drought and its inability to recover during recent wetter years due to soil erosion and degradation [35]. In this case, the drought is the cause of the new hydrological behaviour of the area.

Therefore, there is a series of problems to account for in order to improve our knowledge of the Sahelian evolution towards regreening or desertification:

1. The mapping methods to be used: before 1972, there was no satellite data. For this period only aerial photos can be used; therefore, there is no satellite product before the drought which begun in 1968. On the contrary nowadays, there is a large panel of satellite data available;
2. It is not always easy to segregate either the respective role of the drought (climatic, thus "natural" trend) and of the societies (demographic pressure on the resources) within the soil and vegetation degradation (when observed), or the impact of reclamation actions;
3. Studies based on aerial photos are very time-consuming and limited in space: thus, the selected areas for the analyses are not always representative of the entire region.

4. The tools: the NDVI (normalised difference vegetation index) and the RUE (Rain Use Efficiency) are both indexes used to map the desertification trend.
5. Some elements have to be accounted for such as the adaptation of vegetation to drought, the evolution of resilience in a drought context, the weight of graminaceae and herbaceous species in the measurement of NDVI and RUE, particularly in northern Sahelian areas where they represent 99% of the biomass, and then of the NDVI signal; in cultivated areas, the fact that pearl millet crops have a high biomass thus making the NDVI exceed that of the natural vegetation;
6. Overall, there is a very high spatial and temporal variability in the evolution of land use/land cover, making a short - term or small – basin based comparative analysis difficult to generalise.

Finally, at the present time there is no certitude about the re-greening of the Sahel. However, hydrological studies showed that 5 to 10% of the land cover became crusted soils (ERO type, [14]) enough to significantly increase (twice) the runoff coefficient of a basin. Therefore, a general re-greening is consistent with the degradation of 5-10% of the total area, to explain the general increase in runoff in spite of an increase in vegetation cover.

CONCLUSION

A partial recovery of rainfall amount has been observed since the end of 20th century in the eastern and central parts of the Sahel. However, this cannot solely explain the increase in river discharges, since runoff coefficients have strongly increased (twice or more than those observed before the drought for basins of thousands of km^2). Since the other rainfall characteristics did not show any change which could explain the rise in runoff and the earlier occurrence of the yearly flood, this is likely due to the land use change. Instead of deforestation, the land cover evolution is due to the increase in crop areas; since almost all the cultivable areas are now cultivated, the demographic pressure leads to a shortening of fallows. This is the main cause of soil degradation and crusting. Degraded areas are characterised by low infiltration capacity and high runoff coefficients.

The increase in discharge obviously is leading to a rise in flood hazard in Sahelian areas. This is also observed in Sudanian areas as during the monsoon of 2007 when thousands of km^2 where flooded in Burkina Faso, Togo and Ghana [32] [18]. However, this evolution is much more marked in the Sahel because runoff coefficients are rather decreasing in Sudanian areas since the beginning of the drought.

Furthermore, although crops have been destroyed in some areas of the Sahel, the flood hazard is becoming a severe planning problem in urban areas, where extended zones are tarred, leading to a strong reduction in infiltration. Urbanisation in flooding areas (figure 25) is creating a new land management problem. The Sahelian hydrological paradox thus has negative consequences rather than the sole positive opportunity of getting more water to agriculture and grazing. Crops commonly increase infiltration in the Sahel; extension of cropping on unsuitable soils, shortening of fallows and no-use of fertilizers are some of the causes of soil crusting and degradation.

Policy makers should be alerted to the effects of intensive cropping and land clearing, in some areas, on the hydrological regimes of Sahelian rivers. They must consider the flood hazard in downstream urban areas as an emerging severe concern to sub-Saharan populations.

Figure 25: Inundation of a recent housing area in northern Dakar suburbs (Senegal) (left); a street became a temporary river in Ziguinchor (Casamance, Southern Senegal, Sudanian area) (right); pictures taken during the 2012 monsoon.

ACKNOWLEDGEMENTS

We warmly acknowledge the colleagues of the Niger Basin Authority's "Projet Niger-Hycos" (NBA; http://nigerhycos.abn.ne/user-anon/htm/listStationByGroup.php) and the Niger Water Resources Department (DRE) for providing the Niger River discharge data. The Directors, and the colleagues of the Burkina Faso, the Mali and the Niger Meteorological Offices (Direction de la Météorologie Nationale) are also acknowledged. The SIEREM data were provided by A. l'Aour-Crès, N. Rouché and C. Dieulin (IRD-HSM), and the FRIEND animators provided rainfall data. This work was partly funded by French ANR projects ECLIS and ESCAPE, as well as by the AMMA project.

REFERENCES

1. Ali, A.; Lebel, T.. The Sahelian standardized rainfall index revisited. Int. J. Climatol. 2009, 29, 1705-1714.
2. Olivry, J-C. Synthèse des connaissances hydrologiques et potential en resources en eau du Fleuve Niger. Internal report World bank- Niger Basin Authority, 2002, 156 p.
3. Albergel, J. Sécheresse, désertification et ressources en eau de surface : application aux petits bassins du Burkina Faso. In "The Influence of Climate Change and Climatic Variability on the Hydrologic Regime and Water Resources"; IAHS publication N° 168, Wallingford, UK, 1987,pp. 355-365.
4. Mahe, G., Olivry, J.C. Assessment of freshwater yields to the ocean along the intertropical Atlantic coast of Africa". Comptes Rendus de l'Académie des Sciences, Series IIa 1999, 328, 621-626.
5. Amani, A.; Nguetora, M. Evidence d'une modification du régime hydrologique du fleuve Niger à Niamey. In "FRIEND 2002 Regional Hydrology: Bridging the Gap between Research and Practice", Proceedings of the Friend Conference, Cape Town, South Africa, 18-22 March, 2002; Van Lannen, H., Demuth, S., Eds.; IAHS publication N°274, Wallingford, UK, 2002, pp. 449-456.
6. Mahé, G.; Leduc, C.; Amani, A.; Paturel, J-E.; Girard, S.; Servat, E.; Dezetter, A. Augmentation récente du ruissellement de surface en région soudano sahélienne et impact sur les ressources en eau. In "Hydrology of the Mediterranean and Semi-Arid Regions, proceedings of an international symposium. Montpellier (France)", 2003/04/1-4, Servat E., Najem W., Leduc C., Shakeel A. (Ed.); Wallingford, UK, IAHS, 2003, publication n° 278, 2003, p. 215-222..
7. Descroix, L., Mahé, G., Lebel, T., G., Favreau, G., Galle, S., Gautier, E., Olivry, J-C., Albergel, J., Amogu, O., Cappelaere, B., Dessouassi, R., Diedhiou, A., Le Breton, E., Mamadou, I. Sighomnou, D. Spatio-Temporal Variability of Hydrological Regimes Around the Boundaries between Sahelian and Sudanian Areas of West Africa: A Synthesis. Journal of Hydrology, AMMA special issue, 2009,375, 90-102. doi: 10.1016/j.jhydrol.2008.12.012
8. Descroix, L., Laurent, J-P., Vauclin, M., Amogu, O., Boubkraoui, S., Ibrahim, B., Galle, S., Cappelaere, B., Bousquet, S., Mamadou, I., Le Breton, E., Lebel, T., Quantin, G., Ramier, D., Boulain, N. Experimental evidence of deep infiltration under sandy flats and gullies in the Sahel.

Journal of Hydrology, 2012, 424-425, 1-15;. http://dx.doi.org/10.1016/j.jhydrol.2011.11.019

9. Descroix, L., M. Esteves, K. Souley Yéro, J.-L. Rajot, M. Malam Abdou, S. Boubkraoui, J.-M. Lapetite, N. Dessay, I. Zin, O. Amogu, A. Bachir, I. Bouzou Moussa, E. Le Breton, and I. Mamadou. Runoff evolution according to land use change in a small Sahelian catchment. Hydrol. Earth Syst. Sci. Discuss., 2011, 8, 1569-1607, 2011www.hydrol-earth-syst-sci-discuss.net/8/1569/2011/doi:10.5194/hessd-8-1569-2011.
10. Amogu O., Descroix L., Yéro K.S., Le Breton E., Mamadou I., Ali A., Vischel T., Bader J.-C., Moussa I.B., Gautier E., Boubkraoui S., Belleudy P. Increasing River Flows in the Sahel?. Water, 2010, 2(2):170-199.
11. Descroix, L., Genthon, P., Amogu, O., Rajot, J-L., Sighomnou, D., Vauclin, M.. Change in Sahelian Rivers hydrograph: The case of recent red floods of the Niger River in the Niamey region. Global Planetary Change, 2012, 98-99, 18-30.
12. Hulme, M. Climatic perspectives on Sahelian desiccation : 1973–1998. Global Environmental Change 2001, 11, 19–29 http://dx.doi.org/10.1016/S0959-3780(00)00042-X.
13. Nicholson, S.E. On the question of the "recovery" of the rains in the West African Sahel. Journal of Arid Environments, 2005, 63, 615–641. doi:10.1016/j.jaridenv.2005.03.004
14. Casenave, A. & Valentin, C. A runoff capability classification system based on surface features criteria in semi-arid areas of West Africa. Journal of Hydrology, 1992, 130, 231–249 (1992).
15. Asseline, J., Noni, G.D., Chaume, R. Design and use of a low speed, remotely-controlled unmanned aerial vehicle (UAV) for remote sensing [French]. Photo Interprétation, 1999, 37, 3-9. http://www.documentation.ird.fr/hor/fdi:010026172.
16. Sighomnou, D. Evènements de crues du mois d'août 2012 sur le Niger. Projet GIRE 2, Autorité du bassin du Niger,, Niger Basin Authority, Niamey, 8 p. 2012.
17. Karambiri, H. Brève analyse fréquentielle de la pluie du 1er septembre 2009 à Ouagadougou (Burkina Faso). Note technique 2iE, 4 p. 2009.
18. Tschakert, P., Sagoe, R., Ofori-Darko, G. & Codjoe, S.M.. Floods in the Sahel: an analysis of anomalies, memory, and participatory learning. Climatic Change 2010, 103, 471-502. doi: 10.1007/s10584-009-9776-y.
19. Tarhule, A. Damaging rainfall and floodings: the other Sahel hazards. Climatic Change, 2005, 72, 355-377. doi: 10.1007/s10584-005-6792-4.

20. Di Baldassarre, G., A. Montanari, H. Lins, D. Koutsoyiannis, L. Brandimarte, and G. Blöschl. Flood fatalities in Africa: From diagnosis to mitigation, Geophys. Res. Lett., 2010, 37, L22402, doi:10.1029/2010GL045467.

21. Mamadou, I. la dynamique des koris et l'ensablement de leur lit et de celui du fleuve Niger dans la région de Niamey. PhD thesis, Paris 1 Panthéon Sorbonne University and Abdou Moumouni University of Niamey, 260 p. 2012.

22. Mahé, G.; Paturel, J.E.; Servat, E.; Conway, D.; Dezetter, A. Impact of land use change on soil water holding capacity and river modelling of the Nakambe River in Burkina-Faso. J. Hydrol. 2005, 300, 33-43.

23. Mahé, G., Lienou, G., Bamba, F., Paturel, J-E., Adeaga, O., Descroix, L., Mariko, A., Olivry, J-C., Sangaré, S., Ogilvie, A., Clanet, J-C. Le fleuve Niger et le changement climatique au cours des 100 dernières années. « Hydro-climatology variability and change (Proceedings of symposium held during IUGG 2011, Melbourne, Australia)"; IAHS pub. n° 344, 131-137. 2011.

24. Mahé, G.; Paturel, J-E. 1896-2006 Sahelian annual rainfall variability and runoff increase of Sahelian rivers. C.R. Geosciences, 2009, 341, 538-546.

25. Mahé, G. Surface/groundwater interactions in the Bani and Nakambe rivers, tributaries of the Niger and Volta basins. West Africa. Hydrol. Sci. J. 2009, 54, 704-712.

26. Boyer, J. F., Dieulin, C., Rouché, N., Crès, A. Servat, E. Paturel, J. E. & Mahé, G. SIEREM: an environmental information system for water resources. In: Climate Variability and Change – Hydrological Impacts (ed. by S. Demuth, A. Gustard, E. Planos, F. Scatena & E. Servat), 19–25. IAHS Publ. 308., 2006, IAHS Press, Wallingford, UK

27. Leduc, C., Favreau, G. and Shroeter, P. Long term rise in a Sahelian water-table: the Continental terminal in South West Niger. Journal of Hydrology 2001, 243, 43-54.

28. Leblanc, M., Favreau, G., Massuel, S., Tweed, S., Loireau, M., Cappelaere, B. Land clearance and hydrological change in the Sahel: SW Niger. Global and Planetary Change 2008, 61 (1-2), 49-62 (2008).

29. Zaïri, R. Etude géochimique et hydrodynamique de la nappe libre du Bassin du Lac Tchad dans les régions de Diffa (Niger oriental) et du Bornou (nord-est du Nigeria). PhD thesis, Montpellier 2 University, 210 p. 2008.

30. Guengant, J.-P., Banoin, M. Dynamique des populations, disponibilités en terres et adaptation des régimes fonciers: le Niger. FAO-CICRED, Publ., Roma, Paris, 142 p., 2003. http://www.cicred.org/Eng/Publications/pdf/MonoNiger.pdf

31. Anyamba, A. and Tucker, C.J. Analysis of Sahelian vegetation dynamics using NOAA-AVHRR NDVI data from 1981-2003. Journal of Arid Environments 2005, 63(3), 596-614.

32. Prince, S.D., Wessels, K.J., Tucker, C.J., Nicholson, S.E. Desertification in the Sahel: a reinterpretation of a reinterpretation. Global Change Biology 2007,13, 1308–1313. doi:10.1111/j.1365-2486.2007.01356.x.

33. Fensholt, R., Rasmussen, K. Analysis of trends in the Sahelian 'rain-use efficiency' using GIMMS NDVI, RFE and GPCP rainfall data. Remote Sensing of Environment 2011, 115, 438-451. doi:10.1016/j.rse.2010.09.014.

34. Hein, L. & De Ritter, N. Desertification in the Sahel: a reinterpretation. Global Change Biology 2006, 12, 751-758.

35. Gardelle, J., Hiernaux, P., Kergoat, L. & Grippa, M. Less rain, more water in ponds: a remote sensing study of the dynamics of surface waters from 1950 to present in pastoral Sahel. (Gourma region, Mali) Hydro. Earth Syst. Sci. 2010, 14, 309–324. doi:10.5194/hess-14-309-2010.

Chapter 11

CURRENT CHALLENGES IN EXPERIMENTAL WATERSHED HYDROLOGY

Wei-Zu Gu[1], Jiu-Fu Liu[1], Jia-Ju Lu[1] and Jay Frentress[2]

[1]Institute for Hydrology and Water Resources, Nanjing Hydraulic Research Institutes, Nanjing, China

[2]Oregon State University, Corvallis, USA

INTRODUCTION

The river basin, watershed or catchment is central to many of concepts in hydrology [1] including contaminant hydrology. In fact, two different studies in the Seine River basin during the end of seventeenth century, independently made by Pierre Perraut and Edme Mariotte have been identified by A. K. Biswas as the beginning of quantitative hydrology [2], Along with Perraut and Mariotte, Biswas also suggested Edmond Halley as co-founder of experimental hydrology. It was later accepted that scientific hydrology was founded on these two basin studies [3].

However, basin studies developed slowly until the end of nineteenth century when public demands accelerated. The first modern basin studies commenced in Emmental in Switzerland during the 1890s and were focused on hydrological differences between two small, ca 60 ha, basins [1]. A multitude of basin studies have appeared since the early twentieth century from many parts of the world. These include: (1) Wagon Wheel Gap experiment of USA begun in 1910 in two forested basins [4]; (2) Valday Hydrological Laboratory of USSR begun in 1933 and focused on field investigations of multiple hydrological parameters in watersheds with different scales [5]; (3) Coweeta Hydrologic Laboratory of USA set up since 1934 for forest hydrology and, ecological research in two main basins with 32 sub-watersheds under various treatments [6]; (4) Harz Mountains experiment of Germany with a pair of catchments, Wintertal and Large Bramke, begun in 1948 and focusing on hydrological effects of land use [7]; (5) Alrance experiment of France started in 1950 and focused on streamflow patterns[8]; and (6) Bluebrook Runoff Experiment of China established on 1953 and focused on drainage problems of the vast agricultural plain of the North Huai He River [9].

A period of rapid worldwide development of hydrological basin studies resulted from the International Hydrology Decade (IHD) Representative and Experimental Basin Programme 1965- 1974, with an estimated 3000 basin studies conducted during the Decade [1]. "Representative basins (RBs), which are selected as representative of a region of presumed hydrological similarity, are used for intensive investigations of specific problems of the hydrological cycle (or part thereof) under relatively stable, natural conditions" [10]. Experimental basins (EBs) are relatively homogeneous in soil and vegetation, have uniform physical characteristics, and are deliberately modified for study [10]. However, many EBs are just well-instrumented basins of small area [11]. Sometimes both RB and EB are included, as in the fruitful Plynlimon experiment in the UK [12].

If we designate the first phase of basin study (until ca the middle twentieth century) as foundational and the second phase (during/after the IHD) as developmental, it would appear that experimental watershed hydrology is inevitably going into a third phase of transition and innovation. Most field experiments in watershed science to date remain largely descriptive, with results that are difficult to generalize [13]. In effect, "... catchment hydrology is trapped in a dead-end track, a theoretical impasse" [14]. Experimental watershed studies, the core of watershed hydrology, are now confronting tremendous and even more complicated challenges, which are raised mainly from changing environment conditions and partly from the weakness of current watershed studies. This chapter provides an overview of the main challenges facing experimental watershed hydrology. Addressing these challenges will hopefully lead to substantial innovation in the field.

An important and instructive paper addressing a new vision of watershed hydrology [13] asked "What's wrong with the status quo?" If we posed this same question to the experimental hydrologist, the limitations of most existing RBs/EBs might be characterized as mainly twofold. First, study basin designs have been limited by the black box concept and many misconceptions (e.g., the linearity, non-heterogeneity, additivity of hydrologic systems etc.). Second, operation has been substantially bounded by the hydraulic conception of these watersheds as isolated hydrological systems. All of these watershed studies monitor only total runoff at the stream-outlet and the subsurface responses of the watershed are only estimated by hydrograph separation, etc. These characteristics undermine the formulation of a unified theory of watershed hydrology [14] and the development of watershed models [13,15].

There is a clear need to move beyond the status quo and expand from this narrow hydrological perspective to generate hypotheses governing general behavior across places and scales [13], with the ultimate aim "to advance the science of hydrology" [15]. For the third phase, the stage of transition and innovation, instead of classic RBs as described above, another kind of experimental watershed, the Critical Zone Experimental Block (CZEB) is suggested, with its concept and infrastructures very different from that of current EB. Following the concept of Critical Zone defined by the National Research Council [16], the CZEB will have practical boundaries for different geomorphologic regions.

An ongoing trial application of the CZEB approach is presented in this chapter.

CHALLENGES

Numerous challenges face researchers as they move forward to advance hydrology science using experimental watershed studies.

Demands of Advancing Hydrology Science

More than two decades ago, V. Klemes proclaimed that "hydrology is as yet lacking a solid scientific foundation needed for its development as a natural science" [17] and this sentiment has been reiterated by others. Bras and Eagleson said "in the modern science establishment, this niche is vacant" in their paper titled "Hydrology, the fogotten earth science" [18]. Jim Dooge asked "Is hydrology now an established science? Is hydrologic practice now firmly based on scientific principles?"[19]. McDonnell et al concluded that "the critical ideas and positive vision presented in that (Dooge) paper remain just as fresh, relevant and, unfortunately, very much unfulfilled" [13]. Sivapalan also suggested several steps toward a new unified hydrologic theory [14]. Substantial progress in hydrologic science "ultimately depends on new experimental work, new field observations, and new data collection networks" as summarized by Kirchner [15]. But, what shape will these new experiments and networks take?

Couplingprocesses. A watershed is an inherently dynamic ecological system composed of a variety of biotic and abiotic processes. For such systems, it is important not only to focus on hydrological process but also on linked ecological, biogeochemical, pedological and geological processes. Misunderstandings of the interrelatedness of these processes are commonplace (e.g., summarizing the net effect of all biological processes as evapotranspiration). The vegetation root system forms a dynamically complex net for preferential flowpaths. It changes runoff generation mechanisms not only for surface runoff but the subsurface flows, and the root system itself is effected via feedback from the various, changing flow patterns. Vegetation growth also adapts according to climate, infiltration, soil water and even evapotranspiration while feedback mechanisms reinforce the interconnectedness of the watershed system [20].

Innovative measurements. More than twenty years ago, JE Nash suggested that discovery in hydrology sciences has been limited by "a deficiency in our empiricism" and, "our tolerance of poor methods of observations" [21]. Since then, methods of observations have improved significantly. However, our observation techniques still inhibit hydrology and a fundamental change to systematic measurement programs is needed. The current observation programs are aimed mostly at classic watershed hydrological process. Instead we should focus on a holistic description of heterogeneity of not only the watershed hydrological process but all related processes as described above, as well as landscape properties and climate inputs.

Supporting and testing hydrological model. Physically based hydrological models are one of the most promising directions for advancing hydrologic science. Grayson et al concluded that "the models are enabling us to ask more questions, many of which are fundamental to our understanding of the natural systems that are the subject of our models" [22]. However, vital weaknesses in current watershed models exist. First, they are generally based on "well known small-scale theories such as Darcy's law and the Richards equation for coupled balance of mass and momentum" [13] and do not accurately capture the spatial heterogeneity, inherent non-linearity, and non-additive properties of natural watershed systems. Second, a large number of models are heavily over-parameterized, leading to equi-finality, wherein multiple combinations of parameters can yield equivalent results [13]. Grayson et al showed that comparable fits to a hydrograph could be achieved using a saturated overland flow model or Hortonian flow assumptions [22]. Kirchner argued that "parameter-rich models may succeed as mathematical marionettes dancing to match the calibration data even if their underlying premises are unrealistic" [15]. Kirchner concluded that if "the present trend away from physical processes and toward mathematistry ('blackboard hydrology') continues in hydrologic education and practice, hydrology will end up in a dead end as a science and become useless for applications" [17].

To achieve substantive improvements, models should be developed "in conjunction with, and as an integral part of, carefully planned and executed field studies established for the purpose of advancing our knowledge of the natural system"[22]. Key experimental basin requirements for model-generated hypothesis testing include: use of innovative basin designs and inclusion of sufficient variation in physiographic settings, hydroclimates, and watershed scales.

Hydrologic Replumbing and Natural Climate Oscillations

The National Research Council of the US identified anthropogenic perturbation and replumbing of the hydrologic cycle [23] as a fundamental challenge [16] in watershed science. Human influences dominate "the natural cycle of freshwater causing environmental changes that have been argued to move the planet to a new geologic era termed the 'Anthropocene'" [23,24].

Natural climate oscillations with observed large, abrupt events, the widespread millennial scale climate changes of the last glaciation (consisting of Dansgaard-Oeschger oscillations, Heinrich events etc.) are hypothesized to have been forced by North Atlantic atmosphere-ocean-ice interactions [25]. It has been argued that these oscillations are ultimately driven by variations in eccentricity, axial tilt and precession of the Earth's orbit as described in the Milankovitch theory. The decadal and multidecadal climate variations are generally linked with recognized dynamics, such as the El Niño Southern Oscillation (ENSO) [23].

Anthropogenic modification of the water cycle could "push the climate into new regimes" because "climate change has taken the climate system out of the repeated cycle of glacial-interglacial episodes" [23]. These alterations may accelerate the arrival of millennial scale events and trigger a "tipping point" transition resulting in "major climatic perturbations on time scales of decades to centuries"[26]. How might experimentalists, a small proportion of total hydrologists, address these challenges with progressive basin studies?

Long term monitoring. Well-established representative basins and benchmark experimental basins distributed across different physiographic and hydroclimatic conditions are critical for assessing hydrologic changes due to replumbing. Basins instrumented during or before the 1950's may also be used to determine patterns of rainfall redistribution, resultant stream flow and related biological issues.

Consequences of replumbed hydrology. Monitoring and unraveling the hydrological and ecological responses and the feedbacks due to the replumbed hydrological cycle from water conservancy projects, especially those in arid and semi-arid regions, is needed. Dams are perhaps the most dramatic examples of human capacity to transform nature in the name of development [27], but development programs and projects create both winners and losers [28]. Dams in arid and semi-arid basins, especially in endorheic basins, break the natural cycle, looting downstream groundwater and accelerating desertification [29]. Long distance water diversion may reasonably be viewed as an ecological "planning disaster"[30]. There are concerns that development is outpacing scientific understanding, a circumstance that may prove disastrous for the environment for the generations to come.

Fundamental studies. Various hydrological and biogeochemical fluxes through catchments are important to understanding the effects of hydrologic replumbing, but many of these are poorly understood and rarely measured. For example, various subsurface ecological and hydrological components cannot be measured directly and must instead be estimated from a small number of point measurements. Particular emphasis has previously been placed on the downward flux of catchment processes components (i.e., rainfall and discharge) rather than upward flux- land evaporation, transpiration, and upward recharge of groundwater flux, etc – even though "these fluxes serve as important regulators of the dynamics of the cycle"[23]. There are various unknown or poorly understood mechanisms regarding natural catchment behavior that are affected by the hydrologic replumbing occurring at local, regional and global scales. These include: (1) the old water paradox [31]; (2) network-like preferential flow[13], natural symmetry between water, landscape, soil and vegetation and the underlying organizing principles between soil, vegetation and other biotic elements [14]; (3) the combined mechanisms of the geological water-cycle and hydrologic water cycle in the shallow hydrosphere, as well as the deep circulation of groundwater, which go far beyond current hydrogeological boundaries [32]; (4) the puzzle of the missing sink for the remaining 40% of watershed carbon balance, which after years effort a consensus for it still has yet to emerge [33]; (5) the mechanisms involved in low flow hydrology and its relation to shallow and deep groundwater, regolith, land use, water quality and the biodiversity of aquatic ecosystems [34]; and (6) the coupling of basin geomorphologic processes with hydrologic and ecologic processes, including the poorly understood processes and mechanisms of environmental release, transport, and biological transformation of various contaminants in the unsaturated zone, shallow aquifers and deep aquifers.

Historical data on water replumbling. Historical data from sources other than hydrometric stage records will be very helpful to understand large-scale environmental changes, serving as a reference for the natural variation and anthropogenic impact. Figure 1 illustrates desertification in an endorheic basin as the result of human impacts. Four stages of man-earth relationship are illustrated, from the 'stage of natural harmony' until the stage of the environmental disasters, the punishment of nature (seeFigure 1 caption).

Design systematic experimental facilities. The dynamic systems currently facing experimental hydrologists exhibit a wide range of heterogeneity and process complexity across large spatiotemporal scales. There are multiple ways to overcome such scaling problems [13,14], but hydrometric measurement systems in natural basins still face challenges (e.g., the ability of current rain gauges with conventional deployment techniques to obtain precise precipitation inputs for natural basins remains controversial). It is necessary to have systematic experimental facilities consisting of laboratory physical models, hillslope "catchment", small-size natural watershed with uniformly dominant vegetation, small-size natural watershed with more complex vegetation, unchannelized catchments, nested sub-watersheds, and watersheds of large size. Such experimental facilities need to be situated in zones with different physiographic and hydroclimatic conditions. Obviously, this would require collaborative efforts at an international scale.

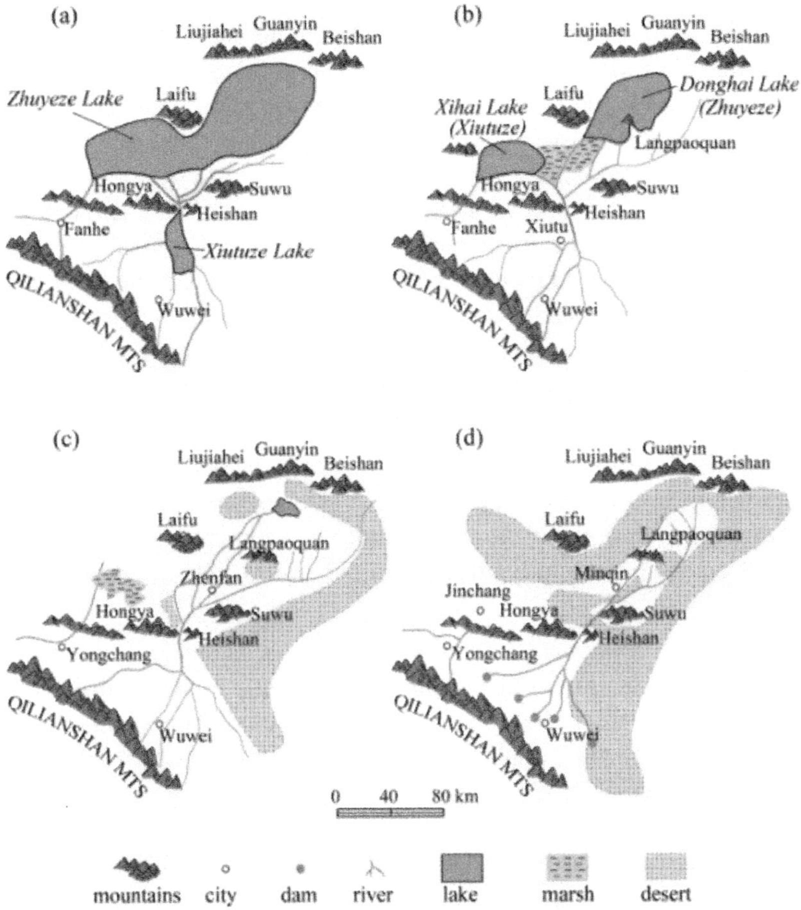

Figure 1: Desertification of the Rocksheep River (Shiyang River) Basin [35], an endorheic basin with total drainage area of 40,690 km^2. (a) prior to 121 BC, the "stage of natural harmony", the area of its terminal lake around 540 km^2; (b) 420- 589 AD, the "stage of relying"; (c) ca 1900AD, the "stage of impact", the cultivated area during Qing Dynasty was developed quickly, which reached to an area about 4/5 of that of modern time. (d) ca 1990's, the "stage of damage", there are 4 dams since 1959, 22 dams until 1972 [35].

THE CRITICAL ZONE EXPERIMENTAL BLOCK (CZEB)

For the future of hydrologic basin study, another kind of experimental basin is suggested: the Critical Zone Experimental *Block* (CZEB) customarily known as Critical Zone Experimental *Basin*.

Deficiencies in The Design and Operation of Current Experimental Basins

Watershed surface laterality. Ignoring the inherent interconnectedness of surface and subsurface watershed components reflects the fact that watershed study is driven by practical rather than scientific requirements. Hydrologists still lack a comprehensive understanding of surface responses to nonvisible and difficult to measure subsurface constraints. A physiological analogy might be studying the skin while ignoring the functions beneath. Focusing on surface responses using surface monitoring will quickly trap the observer with misconceptions. For instance, hydrologists delineate watersheds with surface topography, though actual watershed boundaries may cross topographically determined boundaries depending on subsurface features. Even our most fundamental measurements of the watershed – its shape and size – are strongly influenced by our inability to peer beneath the surface of the catchment.

Hydrologic process laterality. The hydrologic cycle is tightly coupled with other cycles [36]. The hydrologic cycle is also dynamic, not static, in nature. Studying the hydrologic cycle while ignoring the interconnectedness of other cycles is equivalent to studying blood circulation while only looking at blood vessels.

Downward components laterality. There is a tendency to emphasize measurements of downward components (e.g. rainfall, discharge and infiltration) rather than upward components (e.g. evaporation and transpiration aforementioned), even though upward components can be the dominant terms in water balance [37].

Lessons Learned From the Chinese Experimental Basin Studies Over The Last Fifty Years

The Chinese experimental watershed studies experienced a saddle-backed fluctuation with two peaks [9] during their fifty-year history. The lessons learned were summarized in cooperation with J.J. McDonnell, C. Kendall and N.E. Peters, as following [9]:

1. The research facilities need not only the RB and EB of natural conditions, but also those with controlled boundary conditions. All EBs of recent

decades referred only to the natural boundary of the surface watershed, ignoring that of the bedrock. This frequently led to controversial results.

2. The EB should be treated as an integrated system, from surface to bedrock.
3. The runoff response of an EB, including surface and subsurface components, should be monitored hydrometrically. For all EBs in past decades, these components were monitored incompletely or incorrectly, insufficient even for 'black box' evaluations. As a consequence, most of the results can't be explained physically.
4. It is necessary to use multiple tracers to look inside the physical processes of the hydrological cycle and the anthropogenic impacts at the catchment scale.
5. Comparative EBs should be located in different climate and morphological regions and should have comparable monitoring strategies.

It was also suggested that hydrological experimentation should not address "only to the hydrosphere but the interactions between atmosphere, lithosphere, biosphere and the intelligence-sphere." To this end, there is a need to "develop approaches using physical, chemical, isotopic, biogeochemical and hydrometeorological methods for basin studies as a 'hybrid' basin research"[38].

What is CZEB?

The term "critical zone" has been used in many publications to refer to wide-ranging things in areas of geosciences and mineral, etc (e.g., the geological formation, the rhizosphere, the transitional zones in alluvial coastal plain rivers, etc) [36]. In 2001, the National Research Council of US recommended the integrated study of the "Critical Zone" (CZ) [36], defined as "the heterogeneous, near surface environment in which complex interactions involving rock, soil, water, air and living organisms regulate the natural habitat and determine availability of life sustaining resources" [16]. The CZ is one of the most compelling research areas in Earth sciences in the 21st century [36]. This zone was further defined as extending from "the vegetation canopy to the zone of groundwater" in 2005 [39], or "the bedrock to the atmosphere boundary layer" [40] and, "top of the vegetation down to the bottom of the aquifers" [36] by 2010.

The Critical Zone Experimental Block (CZEB) geologically is a monolith-block with its surface, the watershed, bounded by topographical water divides, which define the surface boundary, and which lacks defined subsurface boundaries. A drainage basin conceptually is only its surface part, the visible face (Figure 2). The CZEB is actually an Experimental "Block" within the Critical Zone with a surface drainage basin (the watershed). It is a dynamic ecosystem coupled with various supporting systems but using hydrological processes as the unifying theme. It is a living, breathing, evolving boundary layer where rock, soil, water, air and living organisms interact [41]. The CZEB is the experimental hydrological watershed study in the CZ framework. It follows that the components within the CZEB (including monolith, hillslope, sub-watershed, etc) will be monitored according to the «downward and upward approaches to theory development in catchment hydrology" [14, 42]. "The Critical Zone Observatory network is the only type to integrate biological and geological sciences so tightly"[40], and following this, the CZEB network will perhaps be the only network to integrate the hydrological, biological and geological sciences so tightly as well.

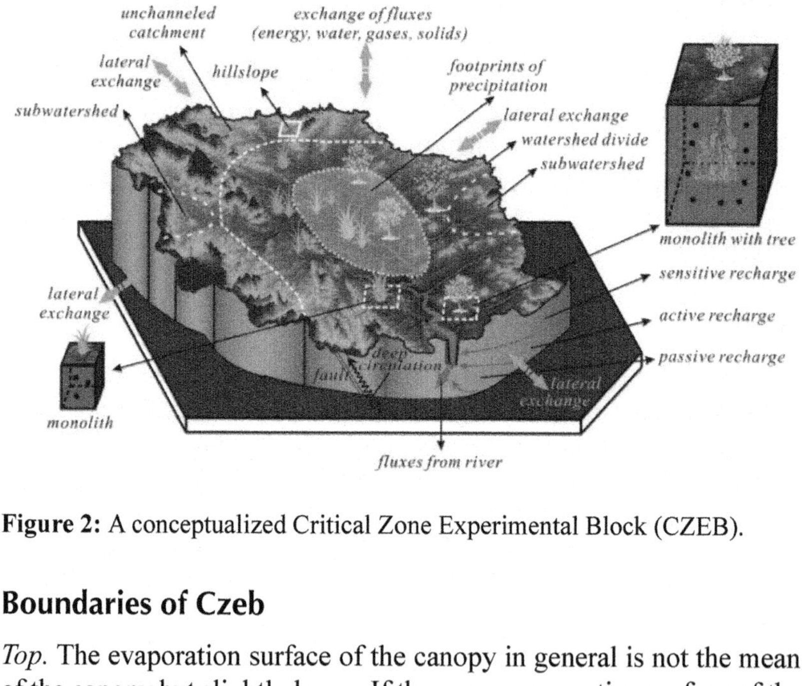

Figure 2: A conceptualized Critical Zone Experimental Block (CZEB).

Boundaries of Czeb

Top. The evaporation surface of the canopy in general is not the mean surface of the canopy but slightly lower. If the mean evaporation surface of the canopy is h *(m)* above the mean ground surface, then the top boundary of a CZEB is defined as $H(m)$, where H=(1.5 to 2.0)×h with coefficients of h varying according to the vegetation. This is mainly for the purpose of energy budget and eddy covariance flux observations. H is just the lower part of the atmosphere boundary layer.

Bottom. There are three cases:

Case I: Bedrock is situated at a relatively shallow depth from the ground surface while the regolith is shallow, too. The bottom boundary is defined as the geological boundary (Figure 3a).

Case II: Bedrock is deep. The bottom boundary is defined as the plane where the tritium content of groundwater approaches zero or the detection limit of ±0.7 TU (the "tritium naught line"(TNL)[43]), which is same as that of Case III (Figure 3b).

Case III: There are stratified alluvia, potentially with multiple aquifers with aquicludes and aquitards, common in flat areas with thick deposit and deep bedrock, up to hundreds of meters or more. Groundwater recharge to the river can be separated into sensitive, active and passive zones. In this case, the bottom boundary of CZEB is defined as the bottom of the active recharge zone, the plane of TNL (Figure 3c).

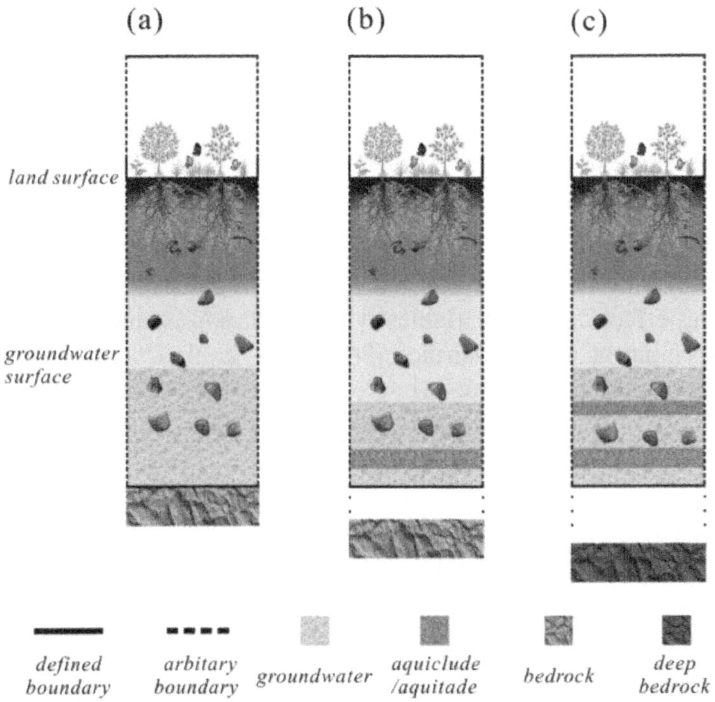

Figure 3: Boundaries of CZEB

Lateral sides

Part I: Above the ground surface up to height *H* as described above, delineated according to the surface topographic watershed boundary (Figure 3).

Part II: Below the ground surface, it is in general defined arbitrarily, except in the case of existing geological boundaries.

Functioning of Czeb

CZEB is an ecological dynamic and evolving system. It also is a natural open system, exchanging mass and energy with its surrounding environment across 'arbitrary' boundaries. It is a dissipative, complex system, however, with some degree of self-organization [36].

Interfaces. CZEB encompasses "the near-surface biosphere and atmosphere, the entire pedosphere, and the surface and near-surface portion of the hydrosphere and lithosphere" [36]. Within the CZEB, various processes including hydrologic, atmospheric, lithospheric, geomorphic and geochemical processes are coupled and dynamically interrelated. To simplify the organization of various interfacing processes throughout the CZEB, compartment zones can be broadly separated as follows:

1. The zone above ground surface, the "aboveground vegetation zone" [36].
2. The unsaturated zone, "the belowground root zone and the deeper vadose zone "[36].
3. The saturated zone, the "saturated aquifer zone" [36].

This layering has a general trend of increasing density with depth, has a dampening effect on state variables with depth, and an increase in distance to energy input at the soil surface [36]. There is also "an overall trend of increasing response time" [36, 44].

Mass – the material aspect of CZEB. The material base of the CZEB is fixed and limited, involving rock, soil, water, air and living organisms [16]. On this point it happens to coincide with the ancient Chinese philosophy, the so-called "*Five-xing*", which holds that five fundamental elements form the universe and the Earth: "Jin"(metal), corresponding with the term "rock"; "*Mu*" (wood), with that of "living organisms"; "Shui"(water) with that of "water"; "Huo"(fire) with that of "air"; and "Tu"(soil).

Energy and force – the driven aspect of CZEB. Continuous energy fluxes and inherent forces drive the CZEB system. External solar energy is certainly the vital component of external energy source. Various inherent forces include: gravity, surface tension, intermolecular forces, capillary force, etc which control its dynamic situation.

Organization and entropy – the philosophical aspect of CZEB. This open system of structural dissipative processes and irreversible evolution tends to increase its entropy spontaneously and go towards disorder [36]. Even the exchange of entropy fluxes across its boundaries is continuous. The second law of thermodynamics, from which the concept of entropy was derived, is one of the fundamental natural laws. However, the role of feedbacks of this nonlinear system will promote "self-organization" as energy dissipates, providing opportunities for dissipating energy to act again within the system towards the direction of order. "Conservation without evolution is death. Evolution without conservation is madness" [45, 46]. Conservation of energy appears in the CZEB at all scales, with observable 'behaviors' of organization, symmetry, genes, etc. Lin suggested including "information" as one of the four general factors (i.e., conservation of energy and mass and, the accumulation of entropy and information, which dictates the evolutionary outcome and the functioning of the CZ) [36]. This is equally applicable to CZEB.

The Reactors of Czeb

The CZEB can be separated into three interdependent zones as mentioned above with each zone consisting of the same base materials, the "*five-xing*" rock, soil, water, air and living organisms as well as non-material bases of energy, force and information. This coincides well with the '*Bagua*' ('*eight-gua'*) of the earliest Chinese philosophical work "*I –Jing"* ("The Book of Chang"). According to I-Jing, the origin of the universe was just a singularity of chaos ('*Tai-ji*'), consisting of '*Yang*' and '*Yin*', the unity of opposites, which produced '*eight-gua,*' the general principles governing development in the material world. Each "*gua*" is a combination of three whole lines ('*Yang*') and broken lines ('*Yin*'). As in the four general characteristics of CZ elaborated by Lin [36]; in the CZEB, the material base refers to the five '*gua*'s (the "*five-xing*") and the non-material base refers to another three '*gua*'s of energy, force and information, which together form a dynamic '*Bagua*'. Each zone has the same five material and three non-material components with different rates, Figure 4.

Each zone will behave as a 'feed-through reactor' according to Anderson et al [47]. It follows then that there are three feed-through reactors coupled together in the CZEB (Figure 4), with probably different rates and residence time of materials.

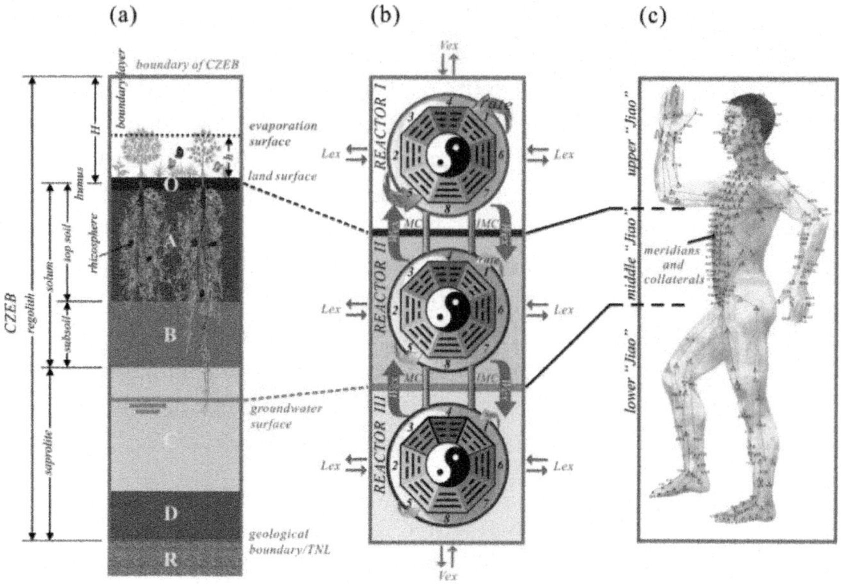

Figure 4: Schematic 'reactors' of the CZEB. The dynamic *'Bagua'* in each 'reactor' operates with its own rate; 1- 5 : the material base (1- air, 2- water, 3- soil, 4- rock, 5- organisms); 6-8: the immaterial base (6- energy, 7- force, 8- information).

Reactor I. The first zone is the above ground surface zone, which also contains above ground vegetation up to its interface with atmosphere (Figure 4a). Evaporation and plant transpiration may account for 50% or more of the total local precipitation and use up to 50% of the total solar energy.

Reactor II. The second zone is the unsaturated zone, extending from the soil surface to the upper surface of the groundwater table. It consists of the whole soil profile: The O horizon (humus), A horizon (topsoil), B horizon (subsoil) and C, and potentially D, horizons (Figure 4a). The unsaturated soil zone has been recognized as "the most complicated biomaterials on the planet" [36].

Reactor III. The third zone is the saturated zone, extending from the groundwater table down to bedrock, including the capillary fringe (Figure 4a). There are two general cases for CZEB, including phreatic groundwater and confined aquifers.

Functioning of reactors

1. Reactor I includes hydrometeorological and ecohydrological processes, Reactor II includes hydropedological and hydroecological processes, while the Reactor III is more exclusively for hydrogeological processes.

2. There is an overall trend of decreasing operation rates from Reactors I to III.
3. Each Reactor has its own lateral flux exchanges *Lex* via the CZEB lateral boundaries (Figure 4b). The current hydrometric runoff data is only that from channel. Reactor I has vertical fluxes exchange *Vex* with atmosphere via the upper boundary of CZEB while Reactor III has *Vex* with deep aquifers and/or deep circulation via the bottom boundary of CZEB (Figure 4b). Fluxes include all material and immaterial components.
4. Reactors are closely coupled within the boundaries of CZEB. There are flux exchanges *Wex*(Figure 5) between Reactor I and II and, and between II and III within the CZEB. There are flux channels *MC* (Figure 4) through I to III for materials, and flux channels *IMC* (Figure 4) through I to III for non-materials.
5. The *MC* appears similar to blood vessel system and *IMC* to meridians and collaterals of the human body from the Chinese art of acupuncture. The three "reactors" are similar to the so-called human "*three Jiaos*" (Figure 4), a central concept of Chinese medicine philosophy.

A TRIAL OF THE CZEB: THE CHUZHOU CZEB EXPERIMENTAL SYSTEM

The Strategy: A Two–Way Multi-Scale Approach

Isolation of the natural system. In order to investigate the natural watershed system, which is filled with complexity and heterogeneity, simplification is incorporated as much as possible (e.g., "*isolation*") [9, 38]. Using detailed observation on ever-decreasing scales allows the investigator to focus and isolate processes from the naturally occurring heterogeneity (i.e., watershed, sub-watershed, unchanneled watershed, hillslope, monolith, a tree) (Figure 5). To some extent it is similar to dissection: splitting "problems into their smallest possible components"[48].

There are limitations in the field, however. Using a single tree as an example, the tree itself can be well instrumented, as well as the subsurface with pits up to meters deep, but still the fluxes from its bottom boundary would be unknown. In this sense, *isolation* refers to a system rather than the physical isolation of a natural unit, which can only be partially isolated in field. Strictly speaking, the field result from such natural systems can be explained only empirically, from a statistic or a stochastic view. Understanding causality mechanisms from observational studies alone is problematic. Thus, McDonnell et al (2007) noted that "most field experiments and observations in watershed science to date remain largely descriptive" [13]. They also discussed a shared 'philosophical path' which claims that "if we characterize enough hillslopes and watersheds around the world through detailed experimentations, some new understanding is bound to emerge eventually", the authors concluded that "what this approach to experimental design has succeeded in doing is to help characterize the idiosyncrasies of more and more watersheds, in different places and at different scales, but with little progress toward realizing the Dooge vision" of "hydrologic laws"[13]. This does not mean, however, that field studies have no place in hydrology. On the contrary, the problem becomes how to "put the pieces back together again" [48]. In our case it is how to put together hydrologic mechanisms for a monolith, a slope, a sub-watershed and, a watershed.

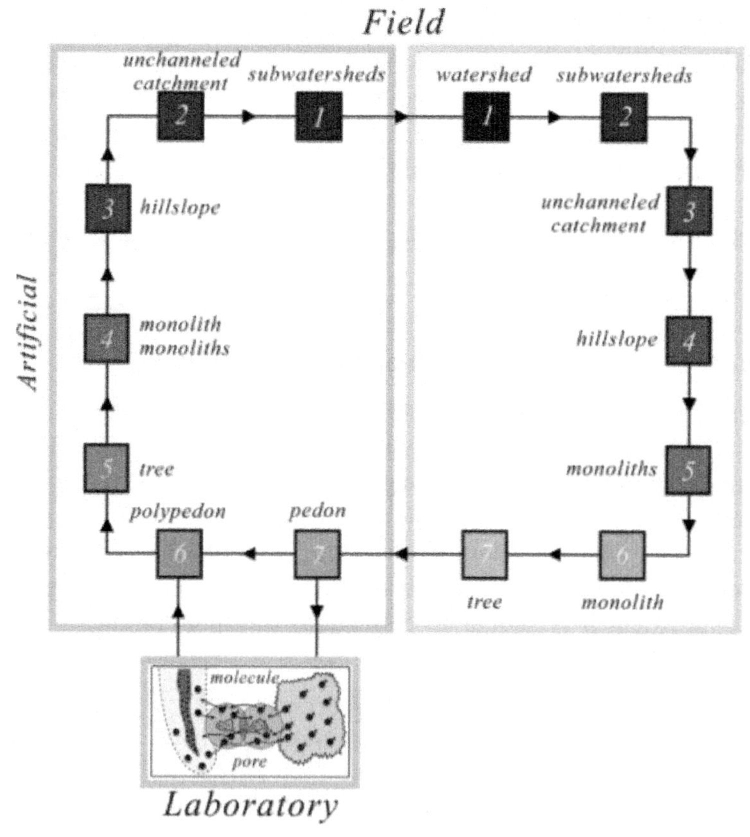

Figure 5: Concepts of the two-way multi-scale approach: natural and artificial.

Control and synthesis in the artificial system. A series of artificial systems are suggested and defined for "control and synthesis" [9, 38]. They encompass the pedon, polypedon, monolith, hillslope, catchment, until subwatershed

(Figure 5). Artificial systems are necessary in order to move beyond problems of boundary and parameter control found in natural systems. It is in fact inspired by the aforementioned empirical impasse of natural systems; artificial systems overcome heterogeneity by controlling all variables and leaving only test variables to fluctuate.

The laboratory system. The two-way multi-scale approach can only be applied at the macroscopic level of watersheds or, maybe the mesoscopic level (e.g., catena monolith). In fact many problems (e.g., the old water paradox) can be trace back to mechanisms occurring on a microscopic level (e.g., the molecular and pore related mechanisms as that of microbial processes of some contaminants transformation). The laboratory system connects with and supplements the artificial system (Figure 5).

Could this be a new vision for watershed hydrology and a unified theory of hydrology? Lin summarized three systems of different levels of complexity and a vast gap between the two extremes [36] as shown in Figure 6. The coupled natural and artificial systems are analogous to '*Yin*' and '*Yang*', which is the successful philosophy of '*Taiji*' (Figure 7). Klmes suggested "a rational search for meaningful conceptualization in hydrology can proceed along two routes: upwards and downwards "[42], or, the "upward or bottom-up approach" and the "downward or top-down approach" [14]. The two-way multi-scale approach perhaps could also be upwards and downwards. Hopefully, it could be a way for a new vision for watershed hydrology [13] and, for the unified theory of hydrology [14].

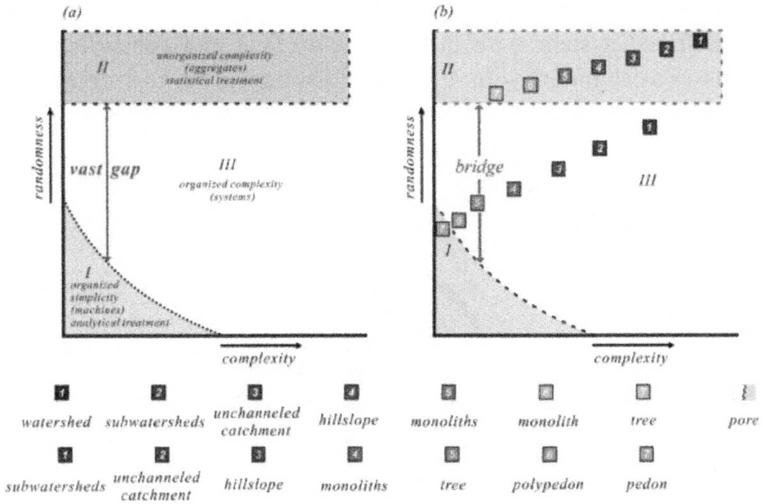

Figure 6: The gap between extremes of complex catchment systems. (a) Three types of complex catchment systems related to different treatment methods [14,19,36]; (b)

The supposed field and artificial two-way multi-scale framework and the gap depending on the level of control. The gap becomes the bridge.

The Trial Practice of The CZEB: The Chuzhou CZEB Experimental System

The Chuzhou CZEB can be attributed to the artistic conception of the *West Brook,* described in a poem of Tang Dynasty where intellectual exiles of Song Dynasty found their Arcadia ca 900 years ago. On the same brook in 1962, a hydrological experimental field station with three watersheds was founded. About fifteen years later, construction of the experimental watershed hydrology had barely survived from the cultural revolution of the 'Road to Perdition'. The Chuzhou CZEB was an adventurous plan for a two-way experimental watershed system including natural catchments and artificial catchments with controlled boundaries. The experiment was, unfortunately, interrupted again during the 'market-economy tides' of China and only partially completed. A turning point emerged a few years ago, with more efforts towards an innovative CZEB system. In the meantime, the components of the natural and artificial systems are: (1) the natural system with main infrastructure of hydrological monitoring and isotope biomonitoring encompasses watersheds with drainage area of 82.1 km^2 (hill 34%, forest 5%, ponds 1.2%), 17.5 km^2 (56%, 10%, 0.6% *idem*), 4.5 km^2 (100%, 2%, 0% *idem*), 3.3 km^2(100%, 100%, 0%*idem*), 2.0 km^2 (5%, 2%, 30% *idem*), 0.06 km^2 (80% paddy field), 4573 m^2 (Morning Glory Catchment, 100% weathered debris), and, a mini CZEB with directly observable surface and subsurface responses with a drainage area of 7897m^2; (2) the artificial system with measurable surface and subsurface responses encompasses a catchment with surficial and bottom area (horizontal projection) of 490m^2, a crop monolith of 150m^2, two grass monoliths (L1, L2) of 32 m^2 each, a saturated monolith (LS) of 1 m^2, and several weighing lysimeters. Here, the focus is on results from the two experimental watersheds - the mini CZEB watershed and the artificial catchment.

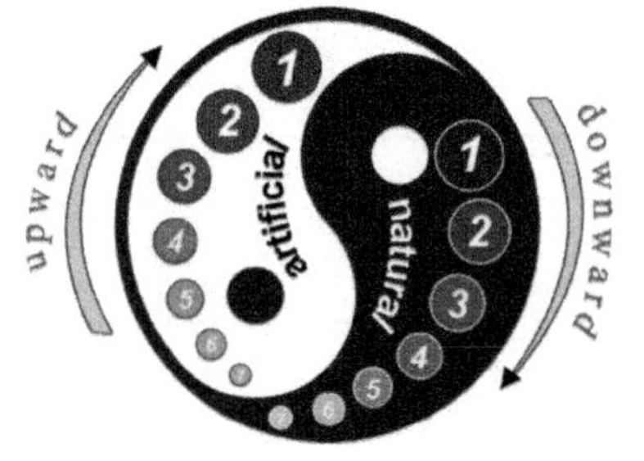

Figure 7: Reconciliation of two-way natural and artificial, *'Yin'* and *'Yang'* [14, 42].

The natural CZEB, Nandadish

Nandadish has a surficial drainage area of 7897 m^2 and rests on the consolidated bedrock of the concordant body of andesitic and tuffaceous facies with a thin weathered layer. The boundary condition of this mini CZEB belongs to the type shown in Figure 3a. The Quaternary regolith overlies the bedrock, its depth ranges from of 1 to 7 m with an average of 2.46 m; its bed rock topography was surveyed via 69 drillings (Figure 8). The regolith consists of brunisolic soil of heavy loam, medium and clay loams; saprolite with prismatic and block structures. Horizontal and vertical fissures and cracks developed in the upper regolith. This watershed has an altitude difference of 12.9 m with a surface slope of 6.7% to 17.1% at different directions. The brook gradient is 6.7%. The coverage during the watershed's construction in 1979 was natural grasses with small shrubs and a few Masson pines aged 5 to 6 years. Since then, coverage has shifted to a dense forest with canopy height ca 8 m.

The main infrastructure includes: (1) measurement for energy flux, water flux, geochemical flux, gas flux; (2) changes of watershed storages of energy, water, gas and that of geochemical ions; (3) CZ tree experiment; and (4) variations of precipitation isotope fingerprints in flux and in storage. The surface and subsurface runoff are monitored directly via a trench, which extends upslope to capture subsurface and surface flow. This CZEB block has deep soils near the divide but only 1-meter depths near the outlet, making the block easy to close via a concrete wall installed to the bedrock at the outlet point. In this way, all surface and subsurface flow drain into discharge measuring structures (Figure 9).

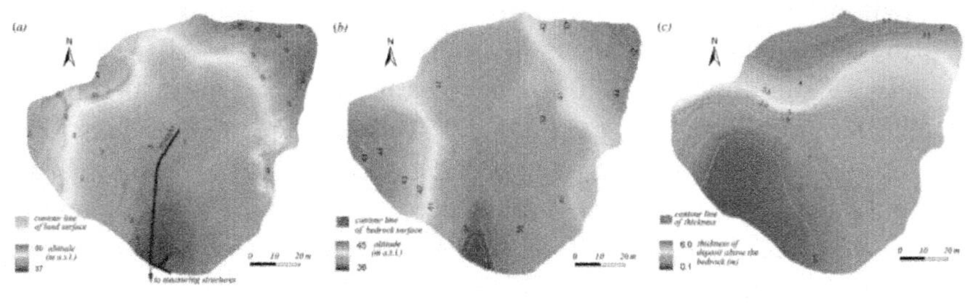

Figure 8: The mini CZEB Nandadish. (a) Its surficial watershed; (b) Its bedrock; (c) The depths of deposit above its bedrock.

Figure 9: Discharge measurement structures for different runoff components from troughs with its location showing in Figure 8(a). 1- for rainfall; 2 and 3- for surface runoff; 4 and 5 – for interflow (30 cm below the soil surface); 6 – for interflow and groundwater flow (down to bedrock); 7 and 8 for total runoff (weirs are not shown). 1,3,5,6 and 8 are V-notch sharp-crested weir; 2, 4 and 7 are the full width rectangular sharp-crested weir; 9 is the tracking water head gauge. The picture design is the original concept, through renovations are now underway for gauge modernization.

The artificial catchment, Hydrohill

The Hydrohill catchment with a drainage area of 490 m^2 (horizontal projection by plane surveying) - 512 m^2 for its inclined surface [49, 50] - is sited on a small andesitic hill. The entire hillslope was first excavated to bedrock to create a bare catchment of ca 4700 m^2. A concrete aquiclude was created above it and consisted of two intersecting slopes dipping towards each other at 10°with an overall downslope gradients of 16.9°(Figure 10a). Impermeable concrete walls enclose the catchment on all sides to prevent any flow of water across the catchment boundaries. Silt-loam soil was removed, layer by layer, from an agricultural field near the Hydrohill site and installed on the artificial aquiclude, layer-by-layer again. Also the bulk density of the soil was adjusted during filling to approximate the natural soil profile of the original agricultural field. Hence, the final 1-m soil 'profile' was identical, at least in composition, to the natural soil profile. Grass was then planted over the surface and soil allowed to settle for 3 years (Figure 10b), after which time a central drainage trench was then excavated at the intersection of the two slopes and the water-sampling instrumentation was installed.

Five fiberglass troughs, each 40 cm wide and 40 m in length, were installed longitudinally in the trench. These troughs were stacked on top of each other to create a set of long zero-tension lysimeters (Figure 11a). Each trough has a 20 cm aluminium lip that extends horizontally into the soil layer to prevent leakage between layers. Water collected in troughs is routed through V-notch based logarithm weirs located in a gauging room (Figure 11b) under the hill where discharge is continuously monitored. Water samples are collected manually above the ponding at the weirs.

As illustrated in Figure 11a, the uppermost trough collects rain; the next lower trough collects surface runoff. The next three troughs collect subsurface flow from soil layers spanning the depths of 0-30, 30-60, and 60-100 cm. These troughs will be referred to as the 30 cm, 60 cm, and 100 cm troughs. The source of the water in these troughs (i.e., whether the water is derived from interflow or saturated flow) varies locally and during storms. The lowermost trough collects either saturated flow or interflow, depending on the height of the water table. When the water table is high, saturated flow may be collected in both of the lower two troughs.

A network of 21 aluminium alloy access tubes for neutron moisture gauges [51], a tensiometer scanner, and 22 wells for water-table measurements and groundwater sampling were installed (Figure 12). The wells were drilled to the aquiclude and are slotted along the lowermost 20 cm. After installation, the spaces around the slotted lengths were packed with sands to allow movement of groundwater to the well, space above the slotted lengths were packed with clay to prevent vertical drainage along the pipes (Figure 11c). The neutron probe access tubes (Figure 11c) and soil water potential tensiometers were positioned adjacent to the wells for soil water monitoring. Catchment evaporation was monitored by methods of water balance, energy budget and, soil water variations above zero flux potential. The plan view of the surface topography, the locations of wells and access tubes, and, the central stacked lysimeter troughs and that of energy budget set are showing in Figure 12.

Figure 10: The artificial catchment Hydrohill: (a) Its concrete aquiclude; (b) Three years waiting for the settling of filled soil to try to match close to the natural soil profile.

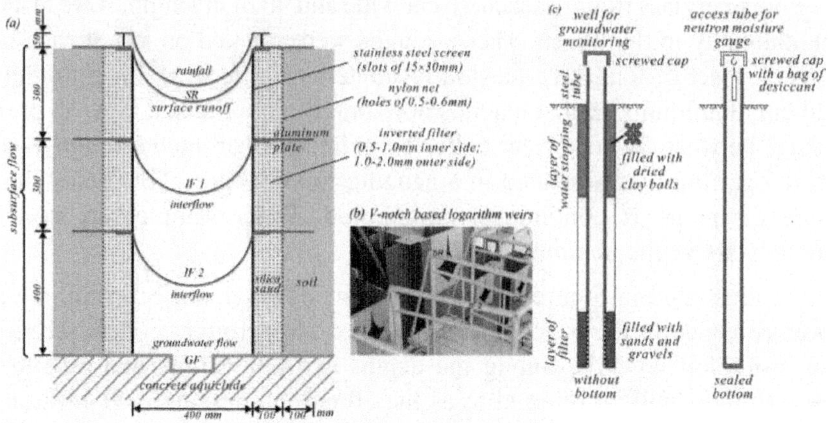

Figure 11: The artificial Hydrohill: (a) Schematic cross-section of rain, surface, and subsurface flow collectors at the catchment; (b) The V-notch based logarithm weirs for discharge measurement for these components; (c) The constructions of wells for groundwater monitoring and access tubes for neutron moisture gauge. (b) and (c) are the original design, now under renovation.

The renovation phase of the main infrastructures for this physically modeled CZEB includes: (1) measurement for energy flux, water flux, geochemical flux, gas flux; (2) changes of watershed storage of energy, water, gas and that of geochemical ions; (3) a small separate laboratory within the gauging room for high-frequency, real time measurement of isotope concentrations in all hydrologic components using of a laser spectrometer LWIA [52]; (4) a removable rainfall simulator on the tracks capable of covering an area of more than 500 m^2, capable of simulating a range of rainfall intensities; and (5) a system for carbon balance of this block.

SELECTED FINDINGS

Runoff Composition

The catchment response to precipitation consists of surface and subsurface components. These components were directly measured at Hydrohill with both isotopic and hydrochemical evidences (Figure 13) [53, 54]. Surface runoff was not always the dominant component during a rainfall-runoff event. The hydrograph composition of surface and subsurface flows was discriminated into four broad categories (Figure 14): 1) S type which is dominated by surface runoff; 2) SS type which is dominated by subsurface flows; 3) Intermediate type M, with largely equal contributions of surface and subsurface flows; and 4) Variation type V, with switching between components during an event.

Figure 12: Hydrometric monitoring of Hydrohill. (a) Plan view of the surface topograph showing the locations of groundwater wells, access tubes for neutron moisture probe with potential scanner, the central stacked lysimeter troughs, the energy budget set. (b) Whole view of its surface watershed in operation during early years (1980's), under renovation now.

Spatial and Temporal Variability in Hydrological Parameters

The physical based distributed-parameter modeling appears to be a promising direction for watershed hydrology. However the scale of model elements, and their parameters, is a lingering problem. This has led to investigations into the establishment of a 'representative elementary area' (REP), 1 km^2[55]. In fact, because of the spatiotemporal heterogeneity of hydrological parameters, little is known regarding hydrologic parameterization in scales less than 1 km^2. Previously unimagined heterogeneities were observed at Hydrohill:

Intrastorm isotopic heterogeneity of shallow groundwater. During a storm in July 1989, the newly developed groundwater showed considerable spatial and temporal variability in the$\delta^{18}O$ (Figure 15a),with values ranging from -12‰ to -6‰ while the groundwater derived from pre-event soil water ranged from 0 to 100% at different times and locations in this catchment [56]. The groundwater samples for $\delta^{18}O$ were collected at three times during the storm from the wells distributed in whole area as shown in Figure 15a. At the end of the storm, the groundwater$\delta^{18}O$ values showed a 4‰ range in composition. This range of compositions during the storm appears to be caused by the combination effects of intrastorm variation in rain$\delta^{18}O$ (Figure 13) and spatial and temporal variability in subsurface flowpaths [56] – including the downward displacement of pre-storm water versus delivery of new water to the aquiclude via macropores [54].

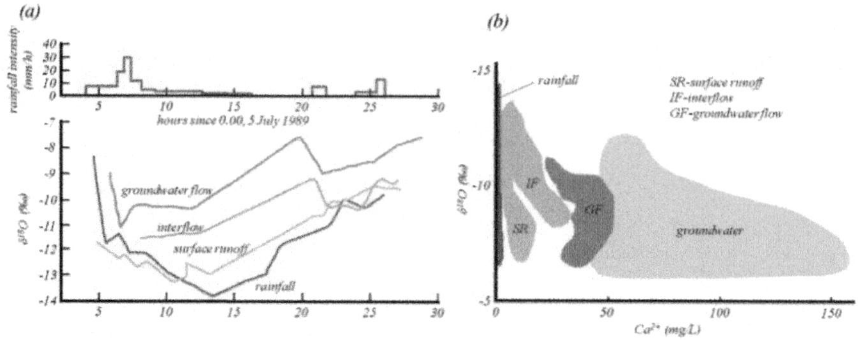

Figure 13: Isotopic and hydrochemical evidences of various runoff components. Samples for isotopes were analysed in Menlo Park laboratory of USGS by C. Kendall (same hereinafter).

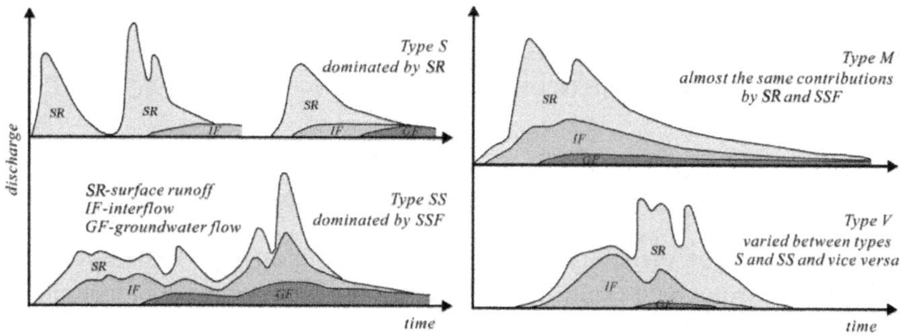

Figure 14: Types of runoff composition summarized from observed data of Hydrohill during 1981- 1992.

Annual heterogeneity of land evaporation rate Land evaporation was measured via the variations of unsaturated soil water at different positions in the soil profile during its matrix potential distribution in that profile showing a zero flux plane. Intensive measurements of volumetric soil moisture by neutron probe at 4 points within each of the 21 access tubes (84 in total) were made during the wet and dry seasons. Representative examples of heterogeneity in the daily evaporation rate are shown in Figure 15b [51].

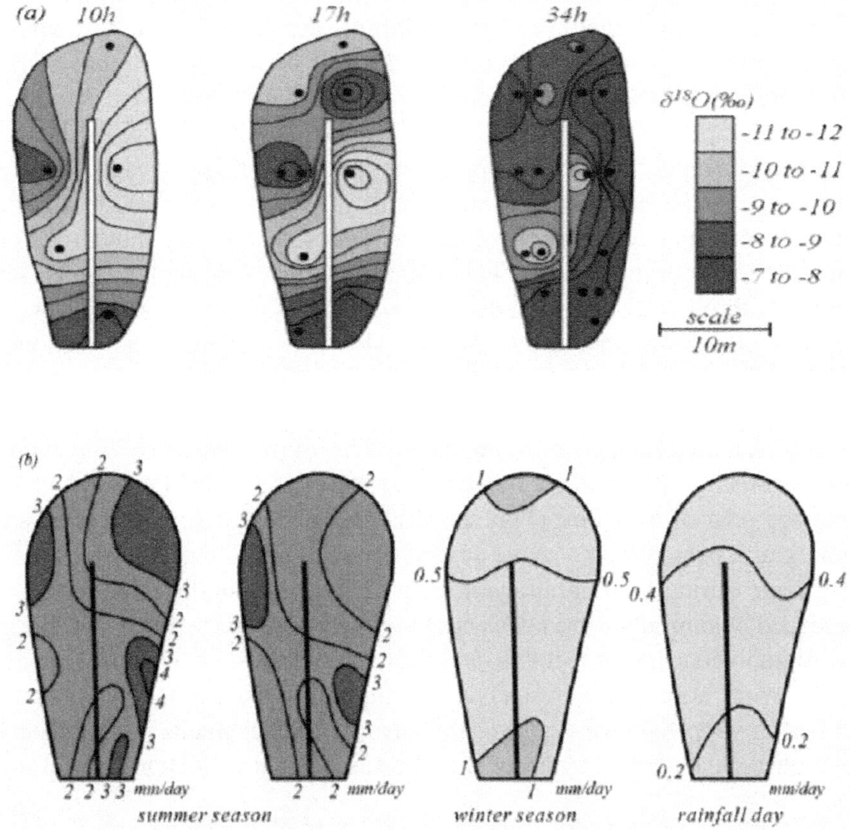

Figure 15: (a) Contour diagrams of $\delta^{18}O$ in groundwater at three sampling times and (b) Typical spatial distribution of the daily land evaporation rate.

The Double Paradox

About ten years ago, Kirchner described a double paradox in catchment hydrology and geochemistry [31]. His objective was to raise questions rather than provide answers and a worthy challenge was then put forward (i.e., "need to take up the search for a unified theory", so that the paradoxes "would no longer seem paradoxes") [31]. For this challenge, perhaps catchments with directly observable surface and subsurface components could be useful in two ways: First, to identify how this double paradox will emerge in the subsurface and surface components individually, instead of in the integrated form at the stream outlet; and secondly, to trace surface and subsurface generation mechanisms using isotopic fingerprints.

The delayed correspondency phenomena. Three isolated rainfall events are presented from May 1987 at Hydrohill, and are ideal for establishing rainfall-runoff relationships as well as the unit hydrograph (Figure 16a). In short, the unit hydrography conceptualization is that rainfall event A produces the hydrograph peak A, rainfall events B and C produce the peaks B and C etc. It can be named as the concept of *'one-to-one correspondency'*. However, using isotope tracers, rainfall-runoff events at Hydrohill showed that rainfall event A corresponded to the hydrography peaks A1 and A2, though A2 only emerged later during event B. This pattern continued throughout subsequent storms with event B corresponding to hydrograph peaks B1 and B2, though B2 only emerged later during event C. This was termed as the phenomena of *delayed correspondency* [57]. In the case of Hydrohill, this delay appears very distinct. It is presumed that such a *delayed correspondency* will tend to diminish as drainage area increases. The diminishment of the *delayed correspondency* phenomena has been verified in a natural watershed with drainage area of 82.1 km^2 (Figure 16b). Rainfall event A produces a runoff hydrograph peak A, some of the constituents of rainfall event A were delayed to emerge during the rainfall event B. It is worth noting that the hydrograph produced by rainfall event A does not have the bell shape referred to in the unit hydrograph concept. In fact this *delayed correspondency* shows the formation of pre-event water ("old" water) during the runoff process. It follows that the current *'one-to-one correspondency',* used to conceptualize rainfall-runoff relationships in applied hydrology, will be associated with large uncertainty.

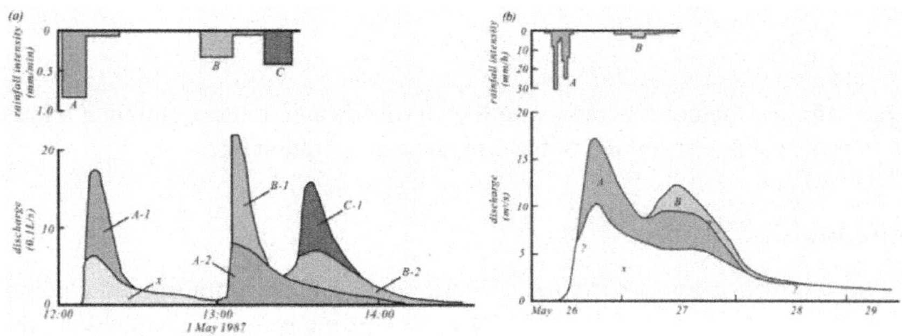

Figure 16: The delayed correspondency phenomena.

"Old" water paradox identified from both surface and subsurface flows. Data from the 'mini CZEB' watershed, Nandadish, suggest that pre-event water ("old water") appeared in all runoff components including surface runoff, interflow and groundwater or saturated flow (Figure 17). Figure 17a and Figure 17b refer to the surface runoff dominated type (type S, from above), and subsurface flow dominated type (type SS) respectively. The surface runoff

and subsurface runoff processes are shown separately with their corresponding proportions of pre-event water. For the type S, the pre-event water accounts for 9% and 24% of the total amount of surface runoff and of subsurface flow respectively while for the type SS, it becomes 11%and 89% respectively [58]. This reveals that even in a catchment with an average soil depth of 2.46 m, large volumes of pre-event water ("old" water) are stored and released promptly by event input.

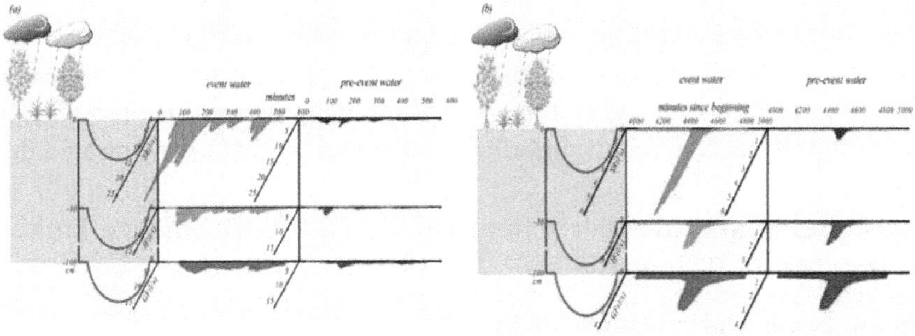

Figure 17: Processes of various runoff components, and that of their pre-event water of Nandadish. (a) for S type; (b) for SS type.

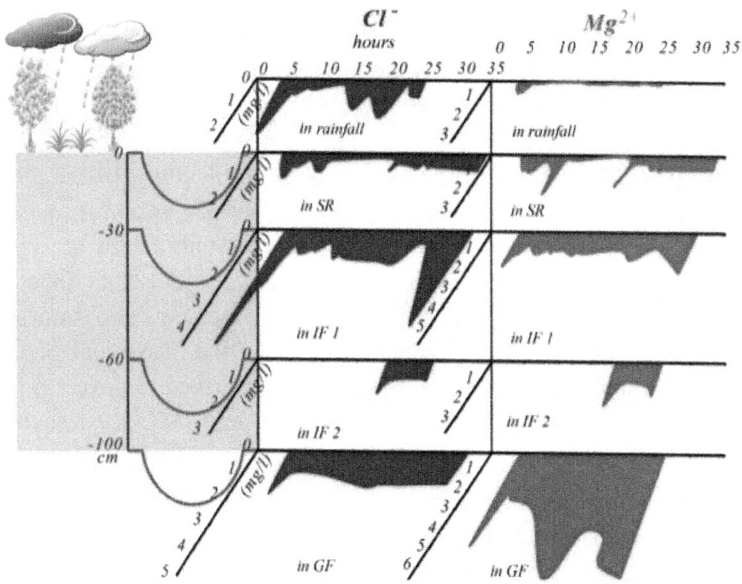

Figure 18: The Mg^{2+} and Cl^- processes in event rainfall and runoff components of Hydrohill. Samples were analysed in Atlanta laboratory of USGS by N.E. Peters (same hereinafter).

Hydrochemistry distribution paradox This was termed by Kirchner as the "'variable chemistry of old water' paradox: although baseflow and stormflow are both composed mostly of 'old' water, they often have very different chemical signatures" [31]. However, the artificial catchment at Hydrohill shows paradoxically the distribution of hydrochemical compositions in different runoff components, including event rainfall. This '*hydrochemistry distribution paradox*' results largely from: (1) inorganic ions in event rainfall input emerge in all runoff components; (2) a strong similarity between rainfall and surface runoff but less similarity in subsurface components; and (3) the fact that the total amount of ions of event rainfall is sometimes much smaller than the sum of all runoff components [59]. Figure 18shows the processes of Mg^{2+} and Cl^- in event rainfall, surface runoff, interflow and groundwater flow (saturated flow).

Does Diel Signal of Hydrochemical Constituents Emerge Linkage Among Multi–Processes?

For the better understanding of the links between contaminants and multi-processes, exploration of diel signals in natural waters may yield "insight into the intricate linkages among hydrological, biological, and geochemical processes [23]". Diel variations of various hydrological constituents in surface and subsurface runoff responses during rainfall events were monitored in artificial catchments and monoliths with examples as following (Figure 19).

Variations of pH. The diel variations of pH in SR, IF, and GF of Hydrohill show their own individuations (solid lines in Figure 19a). Diel variation curve of SR to a large extent appears similar to that of rainfall. However, pH variation curve of IF is contrary to that of rainfall after 10 a.m., while the GF curve appears as the flattened IF curve. The pH variation curve of the interflow of monolith L1 and that of the flow of monolith LS are reasonably similar to that of IF and GF of Hydrohill respectively (dot dash lines in Figure 19a). This reveals that the variation of pH in interflow and saturated flow is not driven only by rainfall input but also by "the biological processes of photosynthesis and respiration [22, 60]." This is ascribed to the coupling of the three "reactors" mentioned above via their MC and IMC for transformation and exchange (Figure 4).

Variation of ionic species. (1) For anion SO_4^{2-} : the diel variation curves of runoff responses of both Hydrohill and monoliths, with few exceptions, are contrary to that of rainfall. Mostly their peak concentrations happened during evening and midnight (Figure 19b). Different from the case of pH, SO_4^{2-} of SR has a strong inversion with the event rainfall curve at night time. (2) For cation Ca^{2+}: the diel curve of rainfall (Fig.19c) shows a small variation after sunrise. However, it triggers variations in runoff responses with their peak concentrations at both a.m. and before midnight. Highest peak happens to LI at afternoon. These diel variations in runoff perhaps are metabolism related. (3) Nimick et al [60] discussed diel cycles of dissolved trace metal concentration in a Rocky Mountain stream. They found that the anionic species "have their highest dissolved concentrations in the late afternoon" while the cationic species "have their highest dissolved concentrations shortly before sunrise"[23,60].

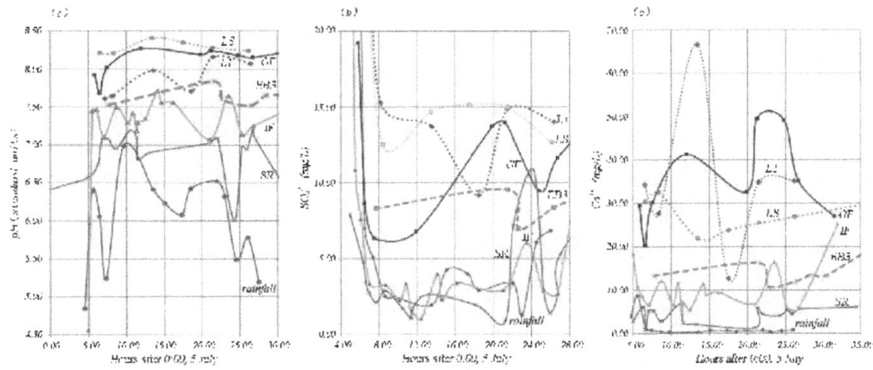

Figure 19: Diel variations of pH, and hydrochemical constituents SO_4^{2-}, Ca^{2+} in event rainfall and runoff responses of Hydrohill and monoliths. SR, IF, GF- runoff responses from Hydrohill; L1- monolith of 32 m^2 serves as an element slope of Hydrohill; LS - saturated monolith of 1m^2 serves as solum in reduction environment; EB3- Morning Glory Catchment of 4573 m^2 serves as a catchment without saturated zone formed by debris only.

Role of soil. Variations of pH, SO_4^{2-}, Ca^{2+} in total runoff of Morning Glory Catchment, a special designed catchment without soil but debris (EB3 in Figure 19 with broken lines), are very similar to that of rainfall curves after 20:00 (EB3 data are not enough before this time). The role of soil in the formation of ionic species in runoff is apparent. Even the variations of pH, SO_4^{2-}, Ca^{2+} in total runoff of EB3 are triggered by the event rainfall, but the resultant variations appear much simpler than various curves of Hydrohill and monoliths. This implies that only a simple process (i.e., mainly the hydrological process is involved in EB3). So, the complex diel variations of the runoff responses from Hydrohill and monoliths (Figure 19) can be reasoned as the results of multi-coupling among hydrological, biological, and geochemical processes. It shows that the reactor II (Figure 4) provides a key operator in contaminant hydrology.

Runoff Generation

The measurements of various runoff components, within artificial systems containing controlled boundaries, provide the possibility to look inside the formation mechanisms of individual components. Isotopic and geochemical tracing can help to investigate these mechanisms but only if significant differences in isotopic compositions of components occur. The general mechanisms for these runoff components (i.e., the surface runoff, interflow and groundwater flow (saturated flow)) are discussed in the following paragraphs [61,62].

Surface runoff (SR). Precipitation input is, of course, the essential condition for the generation of surface runoff. However, in order to actually generate runoff, there must be enough precipitation to form a thin, saturated soil layer (Lsat) at the ground surface. Saturation is key because unsaturated water movement is thought to be too slow to generate runoff. The thickness of Lsat at catchments, monoliths and plots was found to vary between 5 to 50 mm throughout events, increasing downward during rainfall events and receding once rainfall stopped, although these findings were complicated by the irregularities of the soil surface. SR was not generated until a Lsat was established at the surface. Once a Lsat was developed, regardless of its thickness, SR was generated immediately. Overland flowwas only observed on impermeable surface (DO) at artificial plot and, on saturated surface (SO) from a special lysimeter designed for Lsat simulation. Intrastorm variation of isotopic composition of DO andSO can match that of event rainfall. In most cases however, SR is generated within the Lsat, with turbulent mixing of event and pre-event water stored in Lsat, i.e., the saturated mixing surface flow(MS); Alternatively, small amounts of event water act on the surface of Lsat, to force out pre-event water in the Lsat, termed here as the saturated expelled surface flow (ES). The isotopic composition ofMS shows a mixing of event rainfall and pre-event water in Lsat. The isotopic composition of ES is similar to that of Lsat.

Interflow (IF) in the unsaturated zone. Three generation mechanisms are observed. (1) In cases there are soil layers with distinct bulk density and/or hydraulic conductivities, IF can be generated at the interface of soil layers. This was only observed in the 'mini CZEB', Nandadish, where IF occurred at the interface of the layers A (topsoil) and B (subsoil). This is termed as layered interflow (LI). (2) During percolation, soil water moves downward and laterally towards the drainage interface and accumulated until saturation. The saturation will expand if the soil matrix potential $\psi<0$. Once ψ reaches zero, the accumulated saturated soil water will be released immediately at drainage interface. A small amount of infiltration water can trigger this process, to expel the accumulated soil water out as interflow. This process was observed for most IF events in Hydrohill and the 32 m2 monolith. This is termed as expelled interflow (EI). (3) Macropore interflow (MI) was also observed in a special designed catchment without saturated zone with area of 4573 m2.

Groundwater flow (GF) from saturated zone. In most cases, groundwater flow was due to a rising groundwater table resulted from event water recharge via the capillary fringe zone, and was termedrecharged groundwater flow (RG). RG was observed in most cases of both artificial and natural catchments. In addition, the macropore-induced groundwater flow (MG) is the only component not directly observed in this work but inferred from the isotope data in catchment GF during events.

Unreasonableness of Current Two-Component Isotope Hydrograph Separations

Hydrograph separation using conservative, two-component mixing models has been done in various natural basins. However, data from these two catchments, including both the natural and artificial systems indicate that this technique is unacceptable for natural basins and artificial catchments, yielding misleading results due to the unreasonableness of most of the assumptions involved and of the unrealistic operation procedures [63].

Violation of assumptions. In studying natural processes it is inevitable to make simplifying assumptions. However, as Kennedy et al [64] indicated, when simplifying assumptions are made, one runs the risk of drawing misleading conclusions. That may have been the case for the five assumptions [65] summarized for the use of isotopic techniques in precipitation-runoff studies.

The first assumption, that "Groundwater and baseflow are characterized by a single isotopic content" [65], is problematic. Baseflow in natural basins can be recharged by the active zone, passive zone, each with differing flowpaths for recharge resulting in differences in the isotopic composition of baseflow itself, which changes with time and discharge. It is also unreasonable to expect a single isotopic content for groundwater which is subject to both temporal and spatial variations (Figure 16).

The second assumption, that "Rain or snowmelt can be characterized by a single constant isotopic composition or, the variations are documented" [65], is problematic, because, as shown in Figure 14athe intra-storm variability of isotopic composition of rainfall is very significant. High isotopic variability has also been observed in experimental watersheds at Maimai, Panola, etc [56]. Additionally, the spatial variability of isotopic composition of rainfall in a watershed exists and shouldn't be ignored. Isotopic data from a 82.1 km^2 watershed showed that the largest difference in rainfall$\delta^{18}O$ reached 8.2‰.

The third assumption, that "The isotopic composition of rain water is significantly different from that of groundwater/baseflow" [65], can be true. However it must be demonstrated by appropriate sampling [64].

The fourth assumption, that "Contributions from soil water are negligible, or the isotopic composition is identical to that of groundwater" [65] is problematic, because it is a misconception. In a natural watershed, this can only occur if there is no unsaturated zone.

The fifth assumption, that "Contributions from surface water-bodies (such as ponds) are negligible" [65], is reasonable for most upland watersheds [64]. However, in catchments of hilly area with land use including series of ponds and paddy fields, the contributions are large enough to significantly influence results [62].

Based on such problematic assumptions, one of the basic conclusions that "most stormflow is old water"[66], which resulted from two-component hydrograph separations in multiple catchments with different drainage areas, appears suspect.

The physically ambiguous 'old' and 'new' water. This two-component separation model in fact is based on the classic Horton infiltration theory with a very simple runoff generation concept that the soil surface acts as a sieve capable of separating rainfall into two basic components [67]. This 'old' and 'new' water separation model has led to many usages with a variety of defined components in addition to 'old' and 'new' water component of flow, e.g., 'pre-event and event component', 'pre-storm and storm component', 'pre-storm water and rain in storm runoff', 'pre-event and rainfall water', 'groundwater component and event water', 'groundwater and rainwater', 'groundwater discharge and surface runoff'[68], 'surface and subsurface runoff'[69]. The classic two-component method has also been extrapolated to incorporate three components (e.g. that of channel precipitation, soil water and groundwater [70]). Thus, it seems that success can be assured because there are no calibration constraints. As Kirchner [15] noted some models are "often good mathematical marionettes, they can dance to the tune of the calibration data".

The 'old'(O) and 'new'(N) water actually correspond to different generation patterns of runoff components. In fact, multiple runoff mechanisms can result in the same proportions of old versus new water, a problem known as equi-finality. As seen in Hydrohill, surface and subsurface runoff patterns emerge from multiple runoff mechanisms (e.g., macropore flow can result in a large proportion of event water or a large proportion of pre-event water depending on antecedent wetness in the catchment). The runoff components corresponding to old water and new water can be very ambiguous. Applied to surface and subsurface runoff [68, 69], surface runoff is labeled new water [62]. This is a misconception as shown in Figure 20.

The mass balance equations for this model are untenable. The two-component tracer mixing model is stated by two mass balance equations for the composition of stormflow at any time which is used for hydrograph separation: $Q_s=Q_n+Q_o$ and $Q_s\delta_s=Q_n\delta_n+Q_o\delta_o$ where Q is streamflow, δ(Delta) is the D or^{18}O content, the subscripts s, n and o represent the stream, new and old respectively[56]. At any time t, the first equation is correct (i.e., discharge $Q_{s(t)}$ at the outlet of a watershed is equal to the sum of the unknown $Q_{n(t)}$ and $Q_{o(t)}$ right at the same outlet at the time t). During time t, the $Q_{s(t)}$ at the outlet and its isotopic concentrationδ_s are measurable and are the value of known without problem, however both the $Q_{n(t)}$ and $Q_{o(t)}$ at the outlet of time t, are the results of confluence from somewhere within the watershed before time t, i.e., the results of a convolution integration. The $Q_{n(t)}$ and $Q_{o(t)}$ at the outlet of time t are concentrated from the separated isochrone area somewhere within the watershed with different time of concentration. So, the second equation with algebraic sum is physically unrealistic, hinging on the assumptions outlined above. For this linear algebraic sum to be physically realistic, the isotopic concentrations in the reservoirs of new and old water must be constant, without mixing or fractionating during the time interval of event hydrograph. To be physically realistic and representative of the mixing and non-conservative nature of water in natural catchments, it would likely take on the form of an integral or a finite-difference isochrone [63]. No affluxion happens to both the new and old water in the watershed. This can only happen in a small pond, operationally for a watershed, it's "the Procrustean bed [17]".

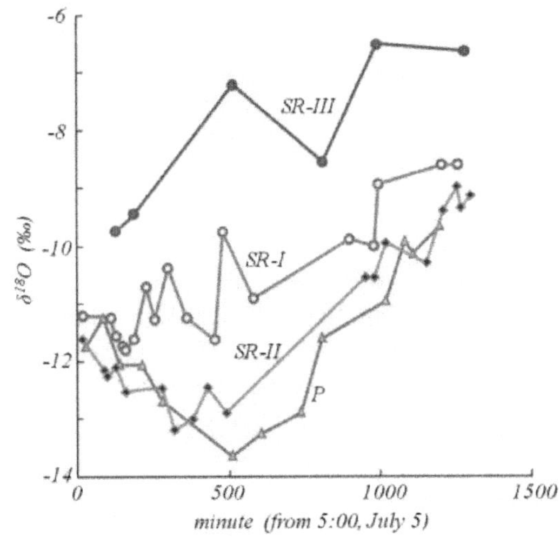

Figure 20: Intrastorm isotopic variations of an event rainfall, and isotopic variations of observed surface runoff in different watersheds. P–rainfall, SRI–surface runoff of the 'mini CZEB', Nandadish, with drainage area of 7897 m^2, SRII–surface runoff of Hydrohill of 490 m^2, SRIII- runoff of Morning Glory Catchment of 4573 m^2.

CONCLUSION

The river basin, watershed or catchment is a central concept in hydrology. Basin studies to assess watershed hydrology are approaching a period of transition and innovation. In fact, experimental watershed studies, the core of watershed hydrology, have tremendous and complicated challenges ahead.

The current challenges in basin experimental hydrology are mainly twofold. Advancing hydrologic science creates a fundamental challenge. Because the watershed system is a dynamic ecological system composed of a variety of biotic and abiotic processes driven by water and climatic processes, experimental watersheds should multi-couple these processes, organizing innovative measurements and approaches while continuing to support and test hydrological models. Anthropogenic hydrologic replumbing and natural climate oscillations are equally challenging. Field studies are the key to understanding and modeling the effect of hydrologic replumbing and climate change. There is a need for long term monitoring, systematic experimental facilities as well as data mining to reclaim historic data useful for determining baseline watershed metrics.

The main problems of the historical experimental basin approach are threefold: watershed surface laterality, hydrologic process laterality and downward components laterality. Lessons from fifty years of Chinese experimental basin studies are: (1) research facilities require natural conditions and also artificial controlled boundaries; (2) address not only the surface watershed, but also downward to the bedrock; (3) surface and subsurface runoff components should be directly monitored hydrometrically; (4) isotopic and hydrochemical tracers are key to understanding runoff generation mechanisms; and (5) account for interactions between the hydrosphere and multiple watershed processes.

Another kind of experimental basin is suggested going forward, namely the Critical Zone Experimental*Block* (CZEB), geologically a monolith-block within the Critical Zone. CZEB is a dynamic ecological and evolving system, coupled with various systems and united by hydrological process. The CZEB is a natural open system; both the energy and mass exchanges exist across its boundaries. It is a dissipative complex system with some degree of self-organization. The function of CZEB is threefold: mass, its material aspect; energy and force, its driven aspect and, organization and entropy, its thermodynamic/philosophical aspect.

To advance watershed hydrology and support development of a unified theory of hydrology, a two-way multi-scale experimental watershed system is suggested, including the natural system and the artificial system. Both have multi-scale subsystems from monolith, slope, sub-watershed to watershed and follow the research idea of upwards and downwards routes. A trial for such a strategy is partly completed at the Chuzhou CZEB Experimental System.

To advance contaminant hydrology, the suggested two-way multi-scale experimental watershed system may provide a key to unravel the complex mechanisms coupling hydrological, biological, and geochemical processes. Contaminant transformation and fate and their effects on regional degradation of groundwater basin involve highly complex mechanisms. This two-way multi-scale system calls for new models using both natural and artificial experimental basin results and using upwards and downwards approaches to multidisciplinary techniques, including isotope tracing.

All hydrological knowledge ultimately comes from observations, experiments, and measurements [15]. Progress in hydrology results mainly from challenges to prevailing approaches and concepts [71]. Hydrological experimentation, including CZEB experimental watershed studies, is the building block for development of a unified theory of hydrology including contaminant hydrology. However, it is important to remember Werner Heisenberg's warning that, "what we observe is not nature herself, but nature exposed

ACKNOWLEDGEMENTS

The renovation of the Chuzhou CZEB Experimental System is supported by the Hydrology Bureau of the Chinese Ministry of Water Resources and, the Nanjing Hydraulic Research Institutes. The adventurous plan for a two-way multi-scale hydrological experimental watershed system and its realization are led by Acadimician Jian-Yun Zhang.

Gu is deeply grateful to Jeffrey McDonnell and the group from USGS, Vance Kennedy, Carol Kendall, Norman (Jake) Peters for their kind help and support - the ever green cedars they planted by Hydrohill are now rooted and full of luxuriant foliage – as well as the wonderful water tracing methodology they taught, extending associations to new generation of colleagues and students. The authors are also deeply grateful to those who have visited Hydrohill for their teaching, help and support, they are: Sklash M. from Canada; Geyh M.A., Plate E. and Seiler K-P from Germany; Gat J. from Israel; Shiklomanov I.A. from Russia; Verhagen B. from South Africa, Littlewood I. from UK, Kinzelbach W. from Switzerland. Thanks to post-doctoral researchers Ma Tao, Xu J-T for their figures. Many thanks to the Editor of this book, Paul Bradley, for his kind help and, encouragements.

REFERENCES

1. J. C Rodda, Basin Studies. In: Rodda JC (ed.) Facets of Hydrology. London: John Wiley & Sons; 1976257297
2. A. K Biswas, History of Hydrology. North-Holland Publishing Company; 1970
3. Unesco/WMO/IAHSThree Centuries of Scientific Hydrology. Paris: Unesco; 1974
4. C. G Bates, A. J Henry, Forest and Stream-flow Experiment at Wagon Wheel Gap. Monthly Weather Review. Supplement.192830179
5. Урываев ВА. Зкспериментальные Гидрологические Исследования на Валдае. Ленинград: Гидрометеорологическое Иэдательство; 1953
6. J. E Douglass, M. D Hoover, Forest Hydrology and Ecology at Coweeta. New York: Springer-Verlag. 1988
7. J Delfs, W Friedrich, H Kiesekamp, A Wagenhoff, Der Einfluss des Waldes und des Kahlschlages auf den Abflussvorgane den Wasserhaushalt und den Bodenabtrag. Aus dem Walde. 19583223325
8. J Jacquet, Les Etudes d'Hydrologie Analytique sur Bassins Versant Experimentaux. Bull. Du Centre de Recherche et d'Essais de Chatou. 19622325
9. W-Z Gu, C-M Liu, X-F Song, J-J Yu, J Xia, Hydrological experimental system and environmental isotope tracing: a review on the occasion of the 50th Anniversary of Chinese basin studies and the 20th Anniversary of Chuzhou Hydrology Laboratory. In: Xi R-Z, Gu W-Z, Seiler K-P (eds.) Research basins and hydrological planning: proceedings of the International Conference on Research Basins and Hydrological Planning, 2231March 2004, Hefei, China. London: A.A. Balkema Publishers; 2004
10. C Toebes, V Ouryvaev, editors. Representative and Experimental Basins. Paris: Unesco; 1970
11. L. L Kelly, L. M Glymph, Experimental watersheds and hydrological research. In: Representative and Experimental Areas: proceedings of the International Symposium of Budapest, 28 September- 1 October. Paris: IASH; 1965
12. C Kirby, M. D Newson, K Gilman, editors. Plynlimon Research: The First Two Decades. Wallingford: Institute of Hydrology; 1991
13. J. J Mcdonnell, M Sivapalan, K Vache, S Dunn, G Grant, R Haggerty, C Hinz, R Hooper, J Kirchner, M. L Roderick, J Selker, M Weiler, Moving beyond Heterogeneity and Process Complexity: A new Vision for Watershed Hydrology. Water Resources Research 2007W07301

14. M Sivapalan, Pattern, Process and Function: Elements of a Unified Theory of Hydrology at the Catchment Scale. In: Anderson MG (ed.) Encyclopedia of Hydrological Science. John Wiley & Sons; 2005193219

15. J. W Kirchner, Getting the Right Answers for the Right Reasons: Linking Measurements, Analyses, and Models to Advance the Science of Hydrology. Water Resources Research 2006W03S04

16. NRCBasic Research Opportunities in Earth Science. Washington D.C.: The National Academies Press; 2001

17. V Klemes, Dilettantism in Hydrology: Transition or Destiny? Water Resources Research 1986S-188S

18. R Bras, P. S Eagleson, Hydrology, the Forgotten Earth Science. Eos 1987

19. J. C Dooge, Looking for Hydrologic Laws. Water Resources Research 1986S- 58S

20. P. S Eagleson, Climate, Soil, and Vegetation, 1. Introduction to Water Balance Dynamics. Water Resources Research 197814705712

21. J. E Nash, Foreward. Journal of Hydrology 1988v-viii.

22. R. B Grayson, I. D Moore, T. A Mcmahon, Physically Based Hydrologic Modeling. 2. Is the Concept Realistic? Water Resources Research 1992261026592666

23. National Research CouncilChallenges and Opportunities in the Hydrologic Sciences. Prepublication copy. Washington D.C.: The National Academies Press; 2012

24. G Vince, An Epoch Debate. Science 20113343237

25. R. B Alley, P. U Clark, L. D Keigwin, R. S Webb, Making Sense of Millennial-Scale Climate Change. In: Clark PU, Webb RS, Keigwin LD (eds.) Geophys. Monogr. Ser., 112Mechanisms of Global Climate Change at Millennial Time Scale. Washington D.C.: AGU; 1999

26. 26. R. B Alley, P. A Mayewski, T Sowers, M Stuiver, K. C Taylor, P. U Clark, Holocene Climatic Instability: A Prominent, Widespread Event 8200 yr ago. Geology 199725483486

27. W. M Adams, Green Development, Environment and Sustainability in the Third World. London: Routledge; 2001

28. T Scudder, A Sociological Framework for the Analysis of New Land Settlements. In: Cernea M (ed.) Putting People First: Sociological Variables in Rural Development. Oxford: Oxford University Press; 1991148167

29. W-Z Gu, J-J Lu, The Disposition of Water Resources of Arid Area with Special Reference to the Alxa Plateau, Inner Mongolia. In: Maldini D,

Maher DM, Troppoli D, Studer M, Goebel J (eds.) Translating Scientific Results into Conservation Actions: New Roles, Challenges and Solutions for 21st Century Scientists. Boston: Earthwatch Institute; 2007

30. P Hall, Great Planning Disasters. London: Weidenfeld and Nicolson; 1980
31. J. W Kirchner, A Double Paradox in Catchment Hydrology and Geochemistry. Hydrological Processes 200317871874
32. W-Z Gu, Isotope Hydrology (in Chinese). Beijing: Science Press; 2011
33. W Steffen, P. J Crutzen, J. R Mcneill, The Anthropocene: Are Humans Now Overwhelming thr Great Forces of Nature? Ambio 200736614642
34. V. U Smakhtin, Low Flow Hydrology: A Review. Journal of Hydrology 2001
35. F-X Li, J-H Yao, Synthetical Research on the Economic Development and the Environmental Rectification for the He-Xi Corridor area (in Chinese). Beijing: Environmental Science Press; 1998
36. H Lin, Earth's Critical Zone and Hydropedology: Concepts, Characteristics, and Advances. Hydrology and Earth System Sciences 2010142545
37. M Falkenmark, J Rockstrom, Balancing Water for Humans and Nature: The New Approach in Ecohydrology. London: Earthscan; 2004
38. W-Z Gu, Experimental and representative basin stuides in China. In: Hooghart JC, Posthumus CWS, Warmerdam PMM. (eds.) Hydrological research basins and the environment: proceedings of the international conference, 2428September 1990, Wageningen, The Netherland. Hague: CHO; 1990
39. S. L Brantley, T. S White, A. F White, D Sparks, D Richter, K Pregitzer, L Derry, J Chorover, O Chadwick, R April, S Anderson, R Amundson, Frontiers in exploration of the Critical Zone: Report of a workshop sponsored by the National Science Foundation. 2005
40. Report prepared by the CZO CommunityFuture directions for Critical Zone observatory (CZO) science. 2010
41. National Critical Zone Observatory Program: http://criticalzoneorg/index.html
42. V Klemes, Conceptualization and Scale in Hydrology. Journal of Hydrology 1983
43. K-P Seiler, W Lindner, Near Surface and Deep Groundwater. Journal of Hydrology 19951653344

44. R. W Arnold, I Szabolcs, V. O Targulian, Global Soil Change. Laxenburg: International Institute for Applied Systems Analysis; 1990

45. G Bateson, Mind and Nature: A Necessary Unity. New York: Dutton; 1979

46. S Weinberg, Dreams of a Final Theory: The Scientist's Search for the Ultimate Laws of Nature. London: Vintage Books; 1994

47. S. P Anderson, F Von Blanckenburg, A. F White, Physical and Chemical Controls on the Critical Zone. Elements 20073315319

48. A Toffler, Forward: Science and Change. In: Prigogine I, Stengers I (eds.). Order out of Chaos. Toronto: Bantam Books; 1984

49. W-Z Gu, Field research on surface water and subsurface water relationships in an artificial experimental catchment. In: Dahlblom P, Lindh G. (eds.) Interaction between groundwater and surface water: proceedings of the international symposium, 30 May- 3 June, Ystad, Sweden. Lund: Lund University; 1988

50. W-Z Gu, J Freer, Patterns of surface and subsurface runoff generation. In: Leibundgut C (ed.) Tracer technologies for hydrological systems: proceedings of symposium H4 at the XXI General Assembly of the International Union of Geodesy and Geophysics, July 1995, Boulder, USA. IAHS Publ. 22919951995265273

51. W-Z Gu, Measurements of spatial evapotranspiration characteristics of an experimental basin using a neutron probe. In: Isotope techniques in water resources development 1987: proceedings of the International Symposium on the Use of Isotope Techniques in Water Resources Development, 30 March- 3 April, Vienna. Vienna: IAEA; 19871987789793

52. J Mcdonnell, Personal communications.

53. C Kendall, J. J Mcdonnell, W-Z Gu, A Look Inside 'Black Box' Hydrograph Separation Models: A Study at the Hydrohill Catchment. Hydrological Processes 20011518771902

54. C Kendall, W-Z Gu, Development of isotopically heterogeneous infiltration waters in an artificial catchment in Chuzhou, China. In: Isotope techniques in water resources development 1991: proceedings of the International Symposium on Isotope Techniques in Water Resources Development, 11-15 March, Vienna. Vienna: IAEA; 199219926173

55. E. F Wood, M Sivapalan, K. J Beven, L Band, Effects of Spatial Variability and Scale with Implications to Hydrologic Modeling. Journal of Hydrology 19881022947

56. C Kendall, J. J Mcdonnell, Effect of intrastorm isotopic heterogeneities of rainfall, soil water, and groundwater on runoff modeling. In: Tracers in hydrology: proceedings of the International Symposium on Tracers in Hydrology, 11-23 July 1993, Yokohama, Japan. IAHS Publ. 215199319934148

57. W-Z Gu, Experimental Research on Catchment Runoff Responses Traced by Environmental Isotopes. Advances in Water Science 1992in Chinese with English abstracts); 34246254

58. W-Z Gu, M-T Shang, S-Y Zhai, J-J Lu, F Jason, J. J Mcdonnell, C Kendall, Rainfall-runoff Paradox from A Natural Experimental Catchment. Advances in Water Science 2010in Chinese with English abstracts); 214471478

59. W-Z Gu, J-J Lu, X Zhao, N. E Peters, Responses of Hydrochemical Inorganic Ions in the Rainfall-runoff Processes of the Experimental Catchments and Its Significance for Tracing. Advances in Water Science 2007in Chinese with English abstracts); 18117

60. D. A Nimick, T. E Cleasby, R. B Mccleskey, Seasonality of Diel Cycles of Dissolved Trace Metal Concentrations in a Rocky Mountain Stream. Environmental Geology 2005475603614

61. W-Z Gu, Challenge on some rainfall-runoff conceptions traced by environmental isotopes in experimental catchments. In: Hotzl H, Werner A. (eds.) Tracer hydrology: proceedings of the 6th International Symposium on Water Tracing, 21-26 September 1992, Karlsruhe, Germany. Rotterdam: A.A. Balkema; 19921992397403

62. W-Z Gu, Various Patterns of Basin Runoff Generation Identified by Hydrological Experimentation and Water Tracing using Environmental Isotopes. Journal of Hydraulic Engineering 1995in Chinese with English abstracts); 5917

63. W-Z Gu, Unreasonableness of current two-component isotopic hydrograph separation for natural basins. In: Isotope techniques in water resources development 1995: proceedings of the International Symposium on Isotopes in Water Resources Development, 20-24 March, 1995, Vienna. Vienna: IAEA; 19961996261264

64. V. C Kennedy, C Kendall, G. W Zellweger, T. A Wyerman, R. J Avanzino, Determination of the Components of Stormflow using Water Chemistry and Environmental Isotopes, Mattole River Basin, California. Journal of Hydrology 198684107140

65. M. G Sklash, R. N Farvolden, The Use of Environmental Isotopes in the Study of High-runoff Episodes in Streams. In: Perry EC, Montgomery

CW. (eds.) Isotope Studies of Hydrologic Processes. Dekalb: N. Illinois University Press; 19826573

66. K. H Bishop, Episodic increases in stream acidity, catchment flow pathways and hydrograph separation. PhD thesis. University of Cambridge, Dep. Geol., Jesus College, Cambridge; 1991
67. R. J Chorley, The Hillslope Hydrological Cycle. In: Kirkby MJ. (ed.) Hillslope Hydrology. Chicheter: John Wiley & Sons; 1979142
68. E Crouzet, P Hubert, Oliver Ph, Siwertz E. Le Tritium dans les Mesures d'Hydrologie de Surface. Determination Experimentale du Coefficient de Ruissellement. Journal of Hydrology 197011217229
69. C Wels, R. J Cornett, B. D Lazerte, Hydrograph Separation: A Comparison of Geochemical and Isotope Tracers. Journal of Hydrology 1991122253274
70. D. R Dewalle, B. R Swistock, W. E Sharpe, Three Component Tracer Model for Stormflow on a Small Appalachian Forested Catchment. Journal of Hydrology 1988104301310
71. V Yevjevich, Misconceptions in Hydrology and Their Consequences. Water Resources Research 196842225232

Chapter 12

EVALUATION OF THE INFLUENCE CAUSED BY TUNNEL CONSTRUCTION ON GROUNDWATER ENVIRONMENT: A CASE STUDY OF TONGLUOSHAN TUNNEL, CHINA

Jian Liu, Dan Liu, and Kai Song

Faculty of Geosciences and Environmental Engineering, Southwest Jiaotong University, Chengdu, Sichuan 610031, China

ABSTRACT

Problems related to water inflow during tunnel construction are challenging to designers, workers, and management departments, as they can threaten tunneling project from safety, time, and economic aspects. Identifying the impacts on groundwater environment resulting from tunnel drainage and making a correct assessment before tunnel construction is essential to better understand troubles that would be encountered during tunnel excavation and helpful to adopt appropriate countermeasures to minimize the influences. This study presents an indicator system and quantifies each indicator of Tongluoshan tunnel, which is located in southwest China with a length of 5.2 km and mainly passes through carbonate rocks and sandstones, based on field investigation and related technological reports. Then, an evaluation is made using fuzzy comprehensive assessment method, with a result showing that it had influenced the local groundwater environment at a moderate degree. Information fed back from environmental investigation and hydrologic monitoring carried out during the main construction period proves the evaluation, as the flow of some springs and streams located beside the tunnel route was found experiencing an apparent decline.

INTRODUCTION

Nowadays, more and more tunnels are constructed for the reason that efficient transport strongly relies on road and railway tunnel, both in long-distance traffic and in metropolitan areas [1, 2]. However, when a tunnel interferes with groundwater in complex geological media, especially in carbonate karstic rocks, serious problems can arise during the excavation because of groundwater inflow, which is known as one of the most common but challenging problems faced by tunnel designers and constructors, leading to unsafe conditions, high construction costs, and delays, not to mention the risk to life and damage to property [1, 3–7]. Yuanliangshan tunnel on Chongqing-Huaihua railway in China came across three large filled caves at Maoba syncline carbonate strata, and a volume of approximately 4200 m^3 of clay erupted within 30 seconds in one cave of them, causing casualties and severe damage to the equipment in tunnel [8]. During the construction of Pinglin tunnel in Taiwan, the major difficulties encountered are caused by sudden high-pressure groundwater inflow, with a yield of approximately 180 L/s in the pilot tunnel as an example, leading to the TBM being trapped and damaged and construction progress being greatly impacted [9]. Due to fluid drainage and pore pressure changes following tunnel construction, vertical settlements with magnitudes reaching 12 cm were measured in fractured crystalline rock several hundred meters above the Gotthard highway tunnel in central Switzerland [10]. A series of hydrogeological problems with geotechnical and environmental impacts, causing spring discharges drying up and leading to a public protest, occurred during the construction of one of the high-speed railway tunnels between Malaga and Córdoba in south Spain [11]. Similar phenomena can also be observed during the drilling of the Firenzuola tunnel in Italy, water inrushes resulting in water table dropping below the level of the valleys and the gaining streams being transformed into losing streams or running completely dry, as did many springs, leading to severe damage to the aquatic fauna and other elements of the ecosystem [12].

Historically, groundwater studies associated with the design and construction of large underground structures have focused primarily on methods for control of groundwater inflows during excavation and for keeping the completed structure free of water. However, within the last several decades, the impacts of such activities on environment have become a major consideration. Though the impacts on groundwater resources of an area by underground excavation may be minimized by such planned constraints as preexcavation grouting and installation of an impermeable lining of the final excavation, such measures may not be efficient for avoiding claims of environmental impacts, particularly in areas of existing water shortages and/or marginal supplies [13]. As any groundwater drawdown alters the natural hydrogeological flow system and can consequently impact groundwater-dependent vegetation, surface streams, lakes, wetlands, and associated aquatic ecosystems, as well as springs and wells, the wise approach is to assess the potential hydrological and ecological impact of a tunnel before building it and take appropriate measures to minimize the impact [12].

As mentioned above, groundwater inflow is one of the most complicated problems which can pose a serious risk and induce impacts on groundwater environment during the tunnel excavation. Accordingly, a number of measures must be adopted to minimize these effects. Appropriate measures can only be taken once the impacts are correctly identified, but it is a difficult task to obtain the correct identification [14]. Fortunately, some relative researches are trying to do this. For instance, a quantitative evaluation of hydrogeological impacts produced by a tunnel of 3 m in diameter and over 7 km in length in the surroundings of the city of Ferrol, NW of Spain, was assessed by means of calculating the hydrogeological behaviors before and after the tunnel excavation using water balance models, and then a comparison was made to allow for the quantitative evaluation of the changes in groundwater flow and the variation in the amount of water corresponding to each component of the model [14]; Tracer tests using uranine and sulforhodamine G and hydrological observations consisting of springs and streams were adopted to evaluate the effects of tunnel drainage on groundwater and surface waters in the Northern Apennines, Italy [12]; Yang et al. [15] utilized a numerical method and MODFLOW codes to simulate groundwater flow pattern in the tunnel area and determine the impact of tunneling excavation on hydrogeological environment in a regional area around the tunnel and a local hot springs area, at the "Tseng-Wen Reservoir Transbasin Diversion Project," in Taiwan.

Since the hydrologic and geologic system are very rare to be completely understood due to their complexity and heterogeneity, there are many difficulties and uncertainties existing in identifying the impacts caused by tunnel excavation on groundwater environment. Compared to problems, such as time delay and increase in costs, and losses to ecosystem and society, induced by drainage from tunnel, it is very essential to make an assessment of the negative effects caused by tunnel excavation prior to constructing it. Despite the fact that some efforts have been made to try achieving this goal, more works should be done to enrich the related researches.

This study presented in this paper focuses on evaluating the influence resulting from tunnel excavation on groundwater environment, by means of employing an indicator system proposed by Liu [16]. The procedure presented in this paper is applied to a case study of the Tongluoshan tunnel constructed in southwest China, where the karstification is well developed [17]. In order to completely understand the impacts caused by the excavation of Tongluoshan tunnel, some representative hydrological points and drainage from tunnel as well as precipitation in the tunnel area have been observed during most of the construction period.

METHODOLOGY

Indicator System for Assessment of the Influence Caused by Tunnel Construction

Water inflow is known as one of the most challenging problems during the tunnel construction, and many other hydrological and geological troubles such as regional water table drawdown, surface subsidence, and wells and springs drying up are induced by it. In order to assess the impacts caused by tunnel excavation on groundwater environment, indicators that closely related to water inflow into tunnel should be firstly taken into consideration. As both mining and tunneling are subsurface activities encountering the risk and damage produced by water inflow [1, 18], the same attention should be paid to the factors controlling water inrush to mine.

As concluded by Liu [16], the factors affecting water inflow during tunnel excavation can be classified into three categories including physical geography, geology and hydrogeology, and tunnel engineering. Each category can also be subdivided into several indicators which extremely explain how this category impacts tunnel inflow. In the category of physical geography, seven indicators such as average annual rainfall, average annual evaporation, area of catchment zone, coefficient of rainfall infiltration, relationship between the tunnel and geomorphology, capacities of reservoirs and lakes on the ground, and flow of surface rivers are included. It is necessary to note that catchment area does not always refer to the whole area of the hydrogeological unit that tunnel is located in but means a zone which collects water from precipitation and contributes to water inflow. Another source of water inflow is surface water, which should not be ignored because of its powerful ability to supply tunnel with abundant water. All the surface water contributing to water inflow should be taken into consideration when quantifying the indicators including capacities of reservoirs and lakes on the ground and flow of surface rivers. The category of geology and hydrogeology is composed of seven indicators too, which are carbonate rocks exposure ratio, water yield properties of the aquifers, water pressure on the tunnel, development of folds, development of fracture zones, formation lithology, and location of tunnel in horizontal and vertical hydrodynamic zoning of groundwater. The last category, tunnel engineering, consists of length of tunnel, area of disturbed range, construction method, burial depth of tunnel, and measures for prevention of groundwater flowing into tunnel. It is worth noting that water inflow does not always increase or decrease over burial depth of tunnel, but a special range of depths may be suitable for groundwater flowing toward tunnel.

The structure of the indicator system constructed by Liu [16] aiming at assessing the negative effects caused by tunnel excavation can be seen in Table 1.

Table 1: Indicator system for assessment of the negative effects caused by tunnel excavation on groundwater environment [16].

Objective layer	Rule layer	Indicator layer	Definition and explanation
[A] assessment of the negative effects caused by tunnel excavation on groundwater environment	[B_1] physical geography	[C_{11}] average annual rainfall (mm)	Average value of annual precipitation in the previous five to ten years
		[C_{12}] average annual evaporation (mm)	Average value of annual evaporation in the previous five to ten years
		[C_{13}] area of catchment zone (km^2)	Area of the catchment zone that collects water contributing to water inrush into tunnel
		[C_{14}] coefficient of rainfall infiltration	The proportion of atmospheric precipitation contributing to groundwater recharge
		[C_{15}] spatial relationship between the tunnel and geomorphology	Spatial relationships between tunnel and geomorphology on cross and longitudinal section
		[C_{16}] capacities of reservoirs and lakes on the ground (m^3)	Capacities of reservoirs and lakes, located on the ground, which may become water sources of tunnel inflow
		[C_{17}] flow of surface rivers (m^3/s)	Flow of surface rivers which may supply tunnel with water
	[B_2] geology and hydrogeology	[C_{21}] carbonate rocks exposure ratio (%)	Areal ratio of outcropping carbonate rocks to the catchment zone contributing to water inflow in the plane
		[C_{22}] water yield property of aquifers	Water yield property of aquifers that may provide tunnel with water
		[C_{23}] water pressure on the tunnel (Mpa)	Hydrostatic pressure on tunnel
		[C_{24}] development of folds	Characteristics and scale of folds, as well as the development of water passages formed during folds formation
		[C_{25}] development of fracture zones	Development of fracture zones which may become water channels primarily including faults-fracture zone, joints concentrated zone, and contact zone of different lithology
		[C_{26}] formation lithology	Strata lithologic and its proportion
		[C_{27}] location of tunnel in horizontal and vertical hydrodynamic zoning of groundwater	Location of tunnel in horizontal hydrodynamic zone of groundwater including recharge zone, runoff zone and discharge zone, and in vertical hydrodynamic zone of groundwater, consisting of epikarst zone, aeration zone, seasonal fluctuation zone, shallow saturation zone, stressful saturation zone and deep circulation zone
	[B_3] tunnel engineers	[C_{31}] length of tunnel (km)	Length along the tunnel axis
		[C_{32}] area of disturbed range (m^2)	Area of the zone that may be disturbed by tunnel excavation
		[C_{33}] construction method	Methods used to excavate tunnel, mainly including drilling and blasting method, New Austrian Tunneling Method, and tunnel boring machine method
		[C_{34}] burial depth of tunnel (m)	Vertical distance from ceiling of the tunnel to ground surface
		[C_{35}] measures for prevention of groundwater flowing into tunnel	Ideas and technologies adopted to prevent and treat groundwater flowing into tunnel

Fuzzy Comprehensive Assessment

The concept of fuzzy sets describing imprecision or vagueness was introduced by Zadeh [19]. Fuzzy logic where an element can belong partially to several subsets may be regarded as an extension of classical Boolean logic where belonging or not to a set is mutually exclusive. It simplifies the process of taking decisions by simulating the way of reasoning of a human expert in environments characterized by uncertainty and imprecision. Fuzzy evaluation methods process all the components according to predetermined weights and decrease the fuzziness by using the membership function; therefore, the sensitivity of fuzzy evaluation is quite high compared to other index evaluation techniques [20–22].

The following procedure describes fuzzy comprehensive assessment [21].

(a) Selection of Factor Set U. Consider

$$U = \{u_i\}, \quad i = 1, 2, \ldots, n, \tag{1}$$

where n is the number of selected evaluation factors. In this study, 19 indicators listed in Table 1 are selected to build the factor set; in other words, $n = 19$.

(b) Construction of Evaluation Criteria Set V. Consider

$$V = \{v_j\}, \quad j = 1, 2, \ldots, m, \tag{2}$$

where m is the number of evaluation criteria categories and V_j is the threshold of the jth criteria category. In the present study, outputs of the assessment are classified into five grades shown in Table 2, along with corresponding evaluation criteria of each indicator. From grade 1 to grade 5, the extent of groundwater environment influenced by tunnel construction ranges from very weak to very strong.

Table 2: Criteria for assessment of the negative effects caused by tunnel excavation on groundwater environment [16].

Indicator	Criteria Very weak	Weak	Moderate	Strong	Very strong
[C_{11}] average annual rainfall (mm)	<600	600~800	800~1000	1000~1600	>1600
[C_{12}] average annual evaporation (mm)	>800	600~800	500~600	400~500	<400
[C_{13}] area of catchment zone (km^2)	<5	5~10	10~30	30~50	>50
[C_{14}] coefficient of rainfall infiltration	<0.05	0.05~0.15	0.15~0.30	0.30~0.50	>0.50
[C_{15}] spatial relationship between the tunnel and geomorphology	Other (such as flat, protruding)	Flat and basin-shaped	Angular space and river crossing	Side below the valley and river crossing	Right below the valley and river crossing
[C_{16}] capacities of reservoirs and lakes on the ground (m^3)	<1	1~10	10~50	50~300	>300
[C_{17}] flow of surface rivers (m^3/s)	<0.1	0.1~0.5	0.5~2.0	2.0~10.0	>10.0
[C_{21}] carbonate rocks exposure ratio (%)	<30	30~50	50~70	70~90	>90
[C_{22}] water yield property of aquifers	<5	5~10	10~15	15~20	>20
[C_{23}] water pressure on the tunnel (Mpa)	<0.5	0.5~1.0	1.0~3.0	3.0~5.0	>5.0
[C_{24}] development of folds	No folds	Folds with undeveloped fissure	Folds with moderately developed fissure	Folds with developed fissure	Folds with developed faults
[C_{25}] development of fracture zones	Rarely developed	Poorly developed	Moderately developed	Developed	Well developed
[C_{26}] formation lithology	Mudstone, shale, or clay	Sandstone or fine sandstone	Granite or igneous rock	Metamorphic rock	Soluble rocks including limestone, dolomite, and so forth
[C_{27}] location of tunnel in horizontal and vertical hydrodynamic zoning of groundwater	Recharge area in horizontal and unsaturated zone in vertical zoning	Recharge area in horizontal and seasonal fluctuation zone in vertical zoning	Runoff area in horizontal and shallow saturation zone or deep circulation zone in vertical zoning	Runoff area in horizontal and stressful saturation zone in vertical zoning	Discharge area in horizontal zoning
[C_{31}] length of tunnel (km)	<1.0	1.0~3.0	3.0~10.0	10.0~30.0	>30.0
[C_{32}] area of disturbed range (m^2)	<50	50~120	120~250	250~350	>350
[C_{33}] construction method	Tunnel boring machine method	New Austrian Tunneling Method	Partial excavation using drilling and blasting method	Benching tunneling using drilling and blasting method	Full face excavation using drilling and blasting method
[C_{34}] burial depth of tunnel (m)	Extremely bad for water inflow	Bad for water inflow	Moderate for water inflow	Good for water inflow	Extremely good for water inflow
[C_{35}] measures for prevention of groundwater flowing into tunnel	Composite lining and pregrouting	Composite lining and exterior waterproof (or postgrouting)	Composite lining	Structural self-waterproof	Drainage

(c) Establishment of Membership Functions. In fuzzy logic, the set A is defined in terms of its membership function by

$$A = \{(f_A(x)),\ x \in X,\ f_A(x) \in [0,1]\}, \tag{3}$$

where X is a domain, with a generic element of X denoted by x, f_A is the membership function of the set A, which maps the domain X onto the interval [0 1], and $f_A(x)$ represents the degree that x belongs to set A. x is a full member of A when $f(x) = 1$, not member of A when $f_A(x) = 0$, and a partial member of A when $f_A(x) = (0, 1)$. The membership function of each factor to the assessment criteria at each grade can be described quantitatively by a set of formulae as follows (4) (Note: if a big value represents a small contribution to water inrush and problems induced by it, then the direction of the inequality in the conditions should be reversed):

$$f_{ij}(x_i) = \begin{cases} 0 & x_i > v_{i(j+1)}, \\ \dfrac{\left(v_{i(j+1)} - x_i\right)}{\left(v_{i(k+1)} - v_{ik}\right)} & v_{ij} \le x_i \le v_{i(j+1)}, \quad j = 1, \\ 1 & x_i < v_{ij}, \end{cases}$$

$$f_{ij}(x_i) = \begin{cases} 0 & x_i > v_{i(j+1)},\ x_i < v_{i(j-1)}, \\ \dfrac{\left(x_i - v_{i(j-1)}\right)}{\left(v_{ij} - v_{i(j-1)}\right)} & v_{i(j-1)} \le x_i \le v_{ij}, \\ \dfrac{\left(v_{i(j+1)} - x_i\right)}{\left(v_{i(j+1)} - v_{ij}\right)} & v_{ij} \le x_i \le v_{i(j+1)}, \end{cases} \tag{4}$$

where i is the number of evaluation factors ($i = 1, 2, 3$), j is the number of assessment criteria levels ($j = 1, 2, 3, 4, 5$), x_i is the actual value of evaluation factor i, V_{ij}, $V_i(j-1)$, $V_i(j+1)$ is the assessment criteria threshold of the ith assessment factor at level $j, j-1, j+1$, respectively, and $f_{ij}(x_i)$ is the membership degree of assessment factor i at level j.

(d) Calculation of Fuzzy Relation Matrix R. Substituting the data of each indicator and the gradation criteria into the membership function listed above, the fuzzy matrix R can be expressed as

$$R = \begin{bmatrix} r_{11} & r_{12} & \cdots & r_{1m} \\ r_{21} & r_{22} & \cdots & r_{2m} \\ \vdots & \vdots & \vdots & \vdots \\ r_{n1} & r_{n2} & \cdots & r_{nm} \end{bmatrix}, \tag{5}$$

where r_{ij} ($i = 1, 2, \ldots, n$; $j = 1, 2, \ldots, m$) is the membership degree of the ith assessment parameter at the jth level.

(e) Determination of Weight Set W. Consider

$$W = \{w_i\}, \quad i = 1, 2, \ldots, n, \tag{6}$$

where n is the number of selected evaluation factors and w_i is the weight of ith factor indicating the relative importance. Determination of the relative weight of each indicator is important for making an appropriate assessment. 10 experts and professors engaged in hydrological and environmental studies coming from home and broad were invited to identify the relative importance of each category and indicator according to their knowledge and experience. After obtaining the sequence and its score representing the relative importance of each factor, weight set (Table 3) is calculated using G1-rank correlation analysis method proposed by Guo [23].

Table 3: Comprehensive weights of the indicators included in the indicator system [16].

Indicator	C_{11}	C_{12}	C_{13}	C_{14}	C_{15}	C_{16}	C_{17}	C_{21}	C_{22}	C_{23}
Weight	0.0535	0.0251	0.0387	0.0386	0.0539	0.0419	0.0419	0.0679	0.0705	0.0393
Indicator	C_{24}	C_{25}	C_{26}	C_{27}	C_{31}	C_{32}	C_{33}	C_{34}	C_{35}	
Weight	0.0617	0.0834	0.0781	0.0681	0.0554	0.0362	0.0477	0.0445	0.0538	

(f) Fuzzy Matrix Composition and Determination of the Final Evaluation Result. Fuzzy composition evaluation can be performed as follows:

$$E = W * R = (w_1, w_2, \ldots, w_n) \begin{bmatrix} r_{11} & \cdots & r_{15} \\ \vdots & \ddots & \vdots \\ r_{n1} & \cdots & r_{n5} \end{bmatrix} = (e_1, e_2, \ldots, e_5), \tag{7}$$

where W and R are the weight set and the fuzzy relationship matrix determined above, respectively, and $*$ is a fuzzy composite operator which is very critical to final evaluation results. In this study, average fuzzy composite operator is chosen.

CASE STUDY OF TONGLUOSHAN TUNNEL

General Description

Project Profile

Tongluoshan tunnel, with a length about 5.2 km, located in Guang'an, Southwest China (Figure 1), is a key project of Dianjiang-Linshui expressway. This project consists of two parallel tunnels, between which the distance is 30 m (from the adjacent walls). The thickest overburden of Tongluoshan tunnel is approximately 280 m, while the minimum value is less than 40 m. In order to save time, excavations were executed simultaneously from the northwest portal and southeast portal to the center during 2005 to 2008.

Figure 1: Location map of Tongluoshan tunnel.

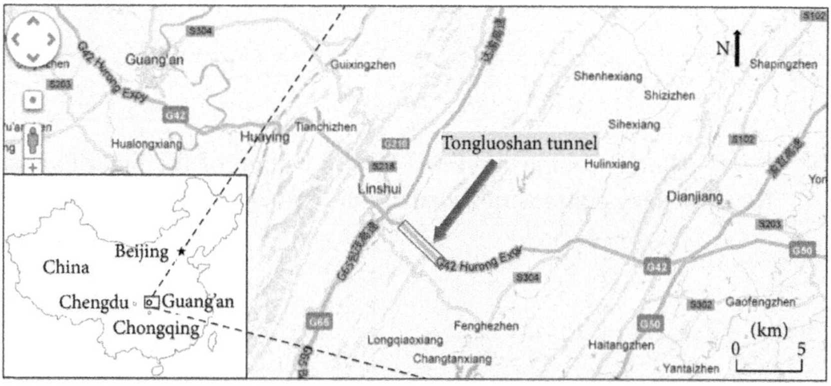

Climatic and Hydrological Features of the Tunnel Area

The tunnel area belongs to subtropics monsoon climate region, where precipitation is abundant but distributing unevenly throughout the year with the majority (70%) falling from May to September. Annual precipitation in the tunnel area oscillates between 836.6 mm and 1529.8 mm with an average of 1215.5 mm. Mean annual temperature in the tunnel area is 16.9°C, with the highest and lowest daily temperature 40.4°C and −3.8°C, respectively.

Tongluoshan tunnel is located in the middle of Tongluoshan anticline, where Yulin River and Zhonghe River, belonging to the secondary and first branch of Yangtze River respectively, are the main regional surface water bodies. The major river in the tunnel area is Qingshuixi River, which originates from Jinzhong reservoir northeast of Tongluoshan tunnel and flows approximately

8.8 km along the axis of Tongluoshan anticline and then turns to southeast. After about 2.5 km, it becomes a part of Zhonghe River. As is recorded, the average annual flow rate of Qingshuixi River is about 0.355 m^3/s before tunnel construction.

Geological and Hydrogeological Characteristics of the Tunnel Area

Ground surface elevation ranges from 156 m to 1053 m in a regional scale while the northern part is higher than the southern part which Yangtze River goes through. In the tunnel area, elevation in the valley along the route of Tongluoshan anticline oscillates between 400 m to 550 m, which is much smaller than that in the anticlinal flanks ranging from 600 m to 750 m.

According to the preliminary geological survey undertaken by Sichuan Institute of Coal Field Geological Engineering Exploration and Design [24] and other information about tunnel design, few except a small fault (F_1) and Tongluoshan anticline are the major geological structures developing in the tunnel area. And the stratigraphic sequence consists of carbonate rocks of lower to middle Triassic (T_2, T_1), coal measure strata of upper Triassic (T_3), mudstone interbedded with siltstone of lower to middle Jurassic (J_2, J_1), and unconsolidated sediments in Quaternary (Q). The geological profile along the route is shown in Figure 2. Tongluoshan tunnel is located in the southern part of Qingshuixi secondary hydrogeological unit, which is bounded by Yujiayakou watershed in the south, Shengouzigou watershed in the north, Yulin River in the west, and Zhonghe River in the east. This hydrogeological unit has an area about 95 km^2 , stretching 11.9 km from the southern boundary to the northern boundary, and 7 to 9 km in the west-east direction. According to the topography, geological structure, water-bearing medium and recharge, runoff, and discharge condition of groundwater in this hydrogeological unit, it can be divided into four aquifer systems, composed of pore water in the unconsolidated formation of Quaternary, pore and fissure water in the clasolite of Jurassic, pore and fissure water in the clasolite of upper Triassic, and carbonate water in the carbonate rocks of lower to middle Triassic. Precipitation is believed to be the primary source of recharge to the aquifers. Groundwater recharge occurs mainly from infiltration of precipitation into outcropping carbonate rocks, including stratum of lower to middle Triassic principally distributed in the core area of Tongluoshan anticline. Controlled by the coal-bearing strata and mudstone interbedded with siltstone located in the anticlinal flanks, karst groundwater generally flows towards the south and mainly discharges in terms of springs along Qingshuixi River, which is viewed as local base level of erosion.

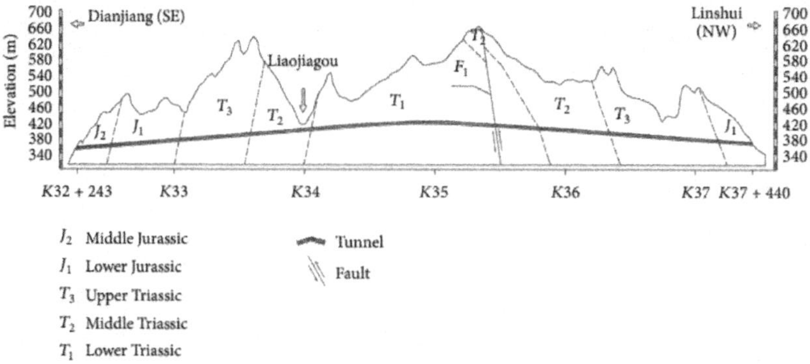

Figure 2: Geological profile of Tongluoshan tunnel.

Fuzzy Comprehensive Evaluation of the Influence Caused by Tunnel Excavation on Groundwater Environment

Quantification of the Indicators

In order to assess the negative effects caused by Tongluoshan tunnel construction on groundwater environment, the value of each indicator included in the indicator system should be previously quantified based on field investigation and technological reports related to the project such as geological survey undertaken by Sichuan Institute of Coal Field Geological Engineering Exploration and Design [24] and so on. As some indicators, for example, lithology of the formation and relation between tunnel and topography and burial depth of tunnel, are unique and of great importance to the evaluation result that may play a vital role in decision making, the work of quantification of the indicators should be done as carefully as it can be, aiming at assessing the influence as accurate as possible. Specific value representing quantification of each indicator is listed in Table 4.

Table 4: Quantification of each factor included in the indicator system.

Category	Indicator	Value of Tonluoshan tunnel
[B_1] physical geography	[C_{11}] average annual rainfall (mm)	According to rainfall records from local meteorological station, the average annual precipitation is 1215.5 mm in the tunnel area.
	[C_{12}] average annual evaporation (mm)	According to evaporation records from local meteorological station, the average annual evaporation is 959.6 mm in the tunnel area.
	[C_{13}] area of catchment zone (km^2)	Tongluoshan tunnel passes through strata consisting of carbonate rocks of high permeability in the core and clastic rocks of low permeability in the flanks of Tongluoshan anticline. When calculating the catchment zone of tunnel inflow, a region composed of local watershed around the route of tunnel in the flanks and the entire karst valley are included. Total area of the catchment zone of Tongluoshan tunnel then is determined to 38.4 km^2.
	[C_{14}] coefficient of rainfall infiltration	Based on the geological investigation report, infiltration coefficient of rainfall in the outcropping carbonate rocks (T_1 and T_2) reaches 0.55, with 0.20 in the coal measure strata (T_3), while it is only 0.054 in the Jurassic clastic rocks (J_1 and J_2). Due to statistics of outcrop in terms of different lithology, 44% of them are carbonate rocks, while 32% are coal bearing strata, and 24% are clastic rocks.
	[C_{15}] spatial relationship between the tunnel and geomorphology	Length consisting of the section between 32 K + 700 and 34 K + 200 and that between 35 K + 800 and 36 K + 300 is about 2 km, belonging to the type of side below the valley and river crossing, while the rest belongs to other type.
	[C_{16}] capacities of reservoirs and lakes on the ground (m^3)	There are no reservoirs and lakes within 2 km from the tunnel axis except Jinzhong reservoir, which is about 9 km far from the tunnel axis, storing approximately 300 thousand cubic meters. Since karstification is well developed in the tunnel area and Jinzhong reservoir is the origin of Qingshuixi River, it is taken into consideration from a safe point of view.
	[C_{17}] flow of surface rivers (m^3/s)	Qingshuixi River is the main surface river which may have hydraulic connection with the water inflow into tunnel.
[B_2] geology and hydrogeology	[C_{21}] carbonate rocks exposure ratio (%)	As mentioned above, this ratio is about 44% in the tunnel area.
	[C_{22}] water yield property of aquifers	Based on the longitudinal profile of Tongluoshan tunnel, area of rocks with poor water yield property occupies 62.37%, with middle standing 30.14%, while the left (7.5%) is considered as aquifer with very good water yield property.
	[C_{23}] water pressure on the tunnel (Mpa)	Average water pressure on the tunnel is estimated as 1.0 MPa.
	[C_{24}] development of folds	Tongluoshan anticline is the main fold developed in the tunnel area with fractures coming from geological and karstic process.
	[C_{25}] development of fracture zones	Fracture zone passed through by tunnel excavation is found moderately developed.
	[C_{26}] formation lithology	Based on outcrops in the tunnel area, mudstone, shale, and clay stand for 21% and sandstone and siltstone stand for 35%, while carbonate rocks occupy 44%.
	[C_{27}] location of tunnel in horizontal and vertical hydrodynamic zoning of groundwater	From a regional scale, Tongluoshan tunnel is located in runoff area in horizontal and stressful saturation zone in vertical zoning.
[B_3] tunnel engineers	[C_{31}] length of tunnel (km)	About 5.2 km.
	[C_{32}] area of disturbed range (m^2)	About 185 m^2 including two tunnels.
	[C_{33}] construction method	About 80% of the tunnel excavated by full face excavation using drilling and blasting method, while the others adopt benching tunneling using drilling and blasting method.
	[C_{34}] burial depth of tunnel (m)	Burial depth between 100 m and 300 m stands for 50%, while the others have a value less than 100 m. From some statistic data and hydrogeological condition in the tunnel area, moderate level is given.
	[C_{35}] measures for prevention of groundwater flowing into tunnel	Composite lining is the prevailing waterproof used by Tongluoshan tunnel, while 20% of which adopts external pregrouting.

Data Preprocessing

Data preprocessing for comprehensive evaluation is to choose an appropriate membership function which determines the membership of each indicator according to given criteria. In this case study, trapezoidal function and semitrapezoidal function (4) are chosen, because of their advantage in simple processing and smooth linking. It is noted that the standard of an indicator at each level can be described as an interval like $[a, b]$, which may have existed or will be built. And it is reasonable for determining the membership of x as 1.0 when $x = (a + b)/2$, so V_{ij} in the equations (e.g., (4)) may be valued as $(a + b)/2$. If difficulties existing in finding an upper or lower limit of some indicators, $\eta = 3\sim5$ or $\eta=1/5\sim1/3$ can be multiplied to achieve this goal. After substitution of actual value quantified in Table 4 and criteria listed in Table 2 into the membership function, membership of each indictor at every level (Table 5) can be gotten, which will directly be transferred into a fuzzy matrix R.

Table 5: Membership of each indictor at every level calculated from quantification of Tongluoshan tunnel and the evaluation criteria.

Indicator	Grades				
	Very weak	Weak	Moderate	Strong	Very strong
[C_{11}] average annual rainfall (mm)	0	0	0.211	0.789	0
[C_{12}] average annual evaporation (mm)	0.288	0.712	0	0	0
[C_{13}] area of catchment zone (km^2)	0	0	0.080	0.920	0
[C_{14}] coefficient of rainfall infiltration	0.158	0.146	0.256	0.330	0.110
[C_{15}] spatial relationship between the tunnel and geomorphology	0.600	0	0	0.400	0
[C_{16}] capacities of reservoirs and lakes on the ground (m^3)	0	0	1.000	0	0
[C_{17}] flow of surface rivers (m^3/s)	0	1.000	0	0	0
[C_{21}] carbonate rocks exposure ratio (%)	0	0.800	0.200	0	0
[C_{22}] water yield property of aquifers	0.624	0	0.301	0	0.075
[C_{23}] water pressure on the tunnel (Mpa)	0	0.800	0.200	0	0
[C_{24}] development of folds	0	0	0	1.000	0
[C_{25}] development of fracture zones	0	0	1.000	0	0
[C_{26}] formation lithology	0.210	0.350	0	0	0.440
[C_{27}] location of tunnel in horizontal and vertical hydrodynamic zoning of groundwater	0	0	0	1.000	0
[C_{31}] length of tunnel (km)	0	0.289	0.711	0	0
[C_{32}] area of disturbed range (m^2)	0	0	1.000	0	0
[C_{33}] construction method	0	0	0	0.200	0.800
[C_{34}] burial depth of tunnel (m)	0	0	1.000	0	0
[C_{35}] measures for prevention of groundwater flowing into tunnel	0.200	0	0.800	0	0

Fuzzy Comprehensive Evaluation

When prepared, evaluation set W and fuzzy matrix R can be composed using average fuzzy composite operator according to (7):

$$E = W * R = (w_1, w_2, \ldots, w_{19}) \begin{bmatrix} r_{1,1} & \cdots & r_{1,5} \\ \vdots & \ddots & \vdots \\ r_{19,1} & \cdots & r_{19,5} \end{bmatrix} \tag{8}$$

$= (0.117, 0.195, 0.355, 0.251, 0.082)$.

Maximum membership principle [25, 26] is a simple and widely used principle on the membership degree matrix. The elements in the vector of evaluation result stand for the membership degree to assessing level. According to this principle, it can be determined that the assessment result of Tongluoshan tunnel is moderate, as 0.355 is the maximum member in the evaluation result vector. Meanwhile, another weighted mean method [27] which allocates a relative rating (1, 2, 3, 4, and 5, resp.) to the five levels and uses accelerations composition method to calculate the total value is employed to perform calculations:

$$\begin{aligned} T &= 1\times e_1 + 2\times e_2 + 3\times e_3 + 4\times e_4 + 5\times e_5 \\ &= 1\times 0.117 + 2\times 0.195 + 3\times 0.355 + 4\times 0.251 + 5 \\ &\quad \times 0.082 \\ &= 2.99. \end{aligned} \tag{9}$$

Based on the result calculated by weighted mean method, middle level tends to be accepted because T is very close to 3. Therefore, from arithmetically speaking, moderate grade is recommended as the influence level of Tongluoshan tunnel, indicating that the local groundwater environment might have suffered a medium degree of impact from the tunnel excavation. Although the indicator system supplying us with a convenient way to evaluate the influence resulted from tunnel construction on groundwater environment, field work such as environmental investigation and hydrological monitoring is important and essential to help judge whether the assessment is appropriate or not. Some details about the procedure and result of environmental investigation and hydrological monitoring carried out in Tongluoshan tunnel area are described in the next section.

Environmental Investigation and Hydrologic Monitoring

Environmental investigation and hydrologic monitoring in the project area that may be affected was implemented for the early detection of environmental impacts caused by tunnel excavation. More than fifty springs, wells, and streams were found in the tunnel area during the first investigation carried in July 2005, and fifteen of them including twelve springs, one well, and two streams, which either supply people with drinking water or flow at a notable rate, were finally selected to perform a long term monitoring program. In addition, discharge from tunnel and rainfall in the project area were monitored together. Figure 3 shows the location of the observation points included in the frequent sampling protocol.

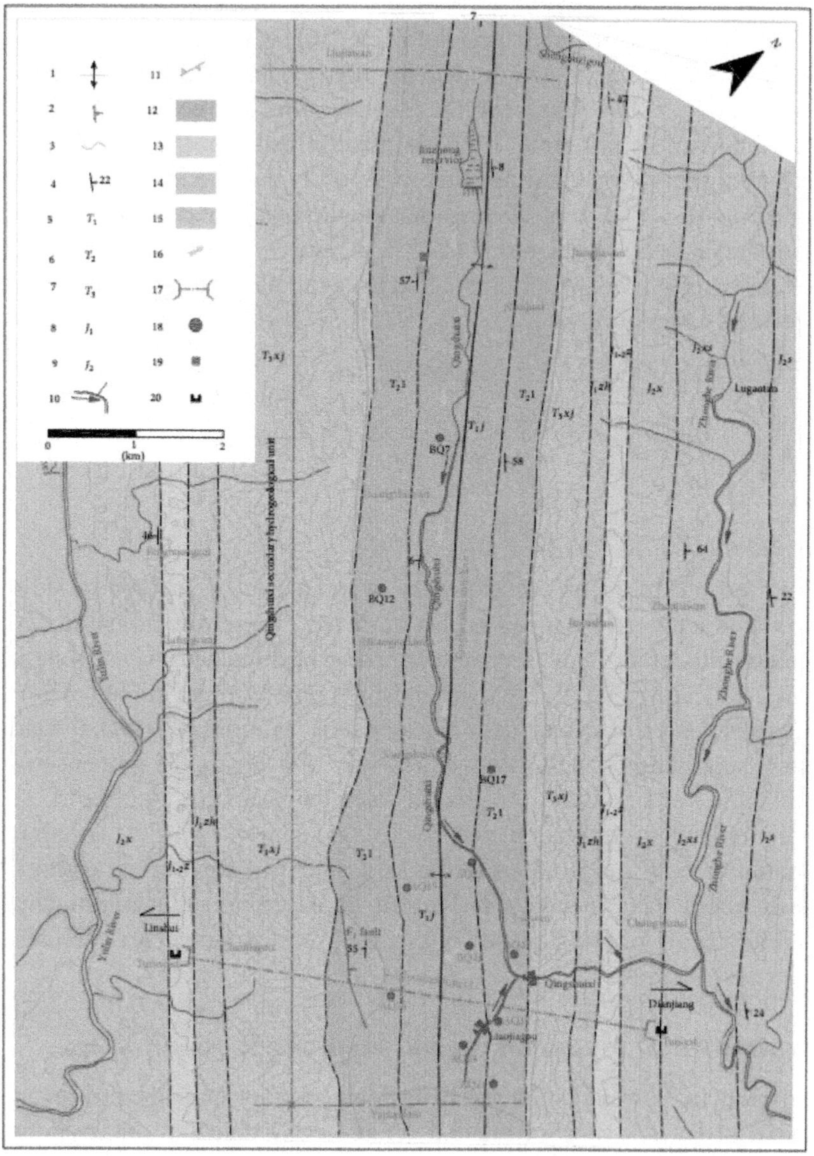

Figure 3: Map showing hydrogeological features and observation points in the project area. 1: anticline; 2: fault; 3: stratigraphic boundary; 4: attitude of strata; 5: lower Triassic; 6: middle Triassic; 7: upper Triassic; 8: lower Jurassic; 9: upper Jurassic; 10: river and its flow direction; 11: watershed; 12: limestone and dolomite; 13: carbonate rocks interbedded with mudstone; 14: sandstone; 15: mudstone interbedded with siltstone; 16: flow direction of groundwater; 17: tunnel; 18: spring; 19: well; 20: drainage from tunnel.

The monitoring process, lasting about 15 months from May 2006 to July 2007, was executed within the main construction period, during which daily discharge rate of tunnel and springs, flow of streams, and water table in well BJ1 were observed. In general, almost all the springs and streams as well as BJ1 respond to rainfall, with more responses from BJ1, BQ7, BQ27, AQ23, AQ37, Qingshuixi, and Liaojiagou, followed by BQ12, BQ25, AQ9, and AQ26. Another type including AQ21 and AQ35 shows evident time lag between discharge of springs and the precipitation, indicating that the fracture network recharging from rainfall and transferring groundwater to the emergence place does not work efficiently. More surprising, there are some springs which become nearly of no interest in precipitation, such as BQ17 and BQ26.

From a water balance point of view, it is believed that groundwater discharge from tunnel acting as a new flow out pattern will reduce the amount of other flow-out patterns such as springs and wells, supposing no changes happening to the flow in patterns. Consequently, hydrograph of surface water and groundwater can help us to identify the influence caused by tunnel construction.

AQ23, BQ25, and BQ27 are karst springs which have never dried up in the past according to local residents, appearing in stratum. It is obviously described in Figure 4(a) that these springs respond quickly to rainfall, especially in some periods with sufficient recharge. However, unfortunate things happen to AQ23 and BQ25 during the monitoring period. AQ23 completely dried up during June 27 to 30, 2006, July 29 to September 4, 2006, September 16 to 27, 2006, December 22, 2006, to January 11, 2007, and March 15 to April 1, 2007. BQ25 had no discharge during July 23 to September 27, 2006, October 15 to November 28, 2006, December 10, 2006, to January 10, 2007, January 18 to February 12, 2007, February 23 to April 1, 2007, and May 1 to 31, 2007. These phenomena indicate that Tongluoshan tunnel may drain groundwater which could have come to surface from AQ23 and BQ25. Compared to them, BQ27 has more luck because it had never dried up despite the fact that the minimum flow rate was about 0.1 L/s, far below the normal average 0.8 L/s.

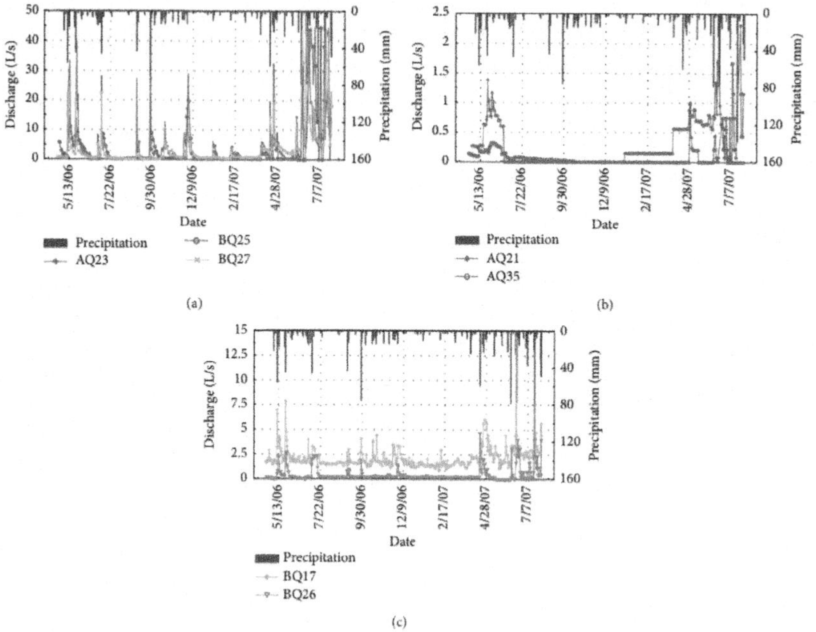

Figure 4: Hydrographs of the principal springs in the project area. (a) Discharge of AQ23, BQ25, and BQ27 versus time. (b) Discharge of AQ21 and AQ35 versus time. (c) Discharge of BQ26 and BQ27 versus time.

AQ21 flowed out from an abandoned mine, which settled in coal bearing layer. It continuously supplied people around with groundwater for drinking prior to tunnel excavation. According to the records shown in Figure 4(b), the flow came to zero during June 22, 2006, to January 9 and July 1 to 31, 2007, in spite of adequate rainfall during January 10 to June 30, 2007. AQ35 is an epikarst spring situated in higher elevation, as is recorded in Figure 4(b); its discharge declined from 0.3 L/s to zero using approximately 4 months and maintains this status for a long time regardless of whether rainfall occurred. Based on the situation described above, it is doubted that AQ21 and AQ35 may had been linked to tunnel construction. According to Figure4(c), BQ17 flowed throughout the period recorded with modest discharge, even during rainy periods. This phenomenon can be attributed to the abundant storage capacity of fractures that feeding BQ17.

Liaojiagou is a branch of Qingshuixi, which is the major stream in the project area. According to Figures 5 and4(a), both Qingshuixi and Liaojiagou bear much resemblance to AQ23, BQ25, and BQ27 in response to rainfall. But Liaojiagou, as with AQ23, did not escape influences from both local meteorological condition and tunnel excavation. They dried up quickly after being recharged from precipitation and continued this state for a long period. However, Qingshuixi showed strong vitality during the whole monitoring period, flowing in quite accordance with precipitation recorded in the project area without any appearance of external leakage, demonstrating that Qingshuixi has not been affected by construction of Tongluoshan tunnel.

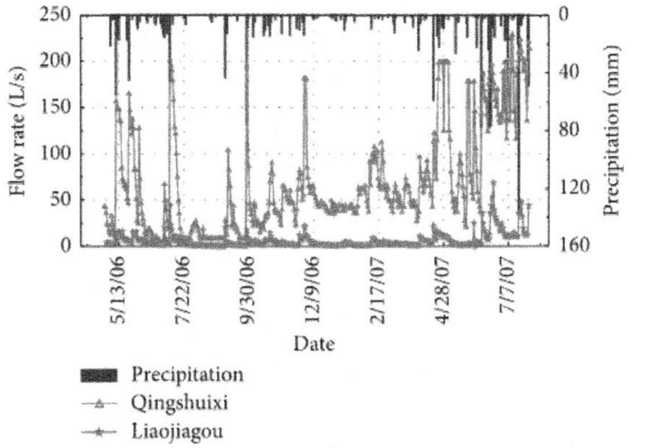

Figure 5: Hydrographs of Qingshuixi and Liaojiagou.

Drainage from Tongluoshan tunnel is depicted in Figure 6, from which we can find a quick increase from 20 L/s to 64 L/s before June 28, 2006, and drastic oscillation between 32 L/s and 69 L/s in the next three months, followed by pediocratic changes between 30 L/s and 40 L/s. In general, tunnel inflow had an obvious ascending tendency when encountering carbonate rocks and performed a slow downtrend after drainage of groundwater stored in the fracture network around tunnel and grouting to prevent water inrush.

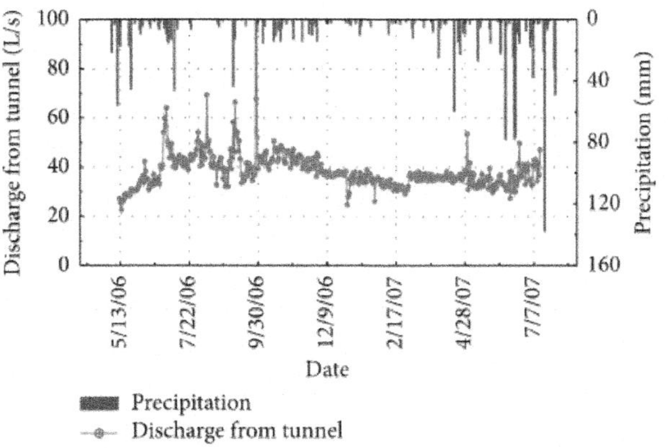

Figure 6: Drainage from tunnel versus time.

Determination of an Impact

While employing the indicator system to assess the negative effects caused by Tongluoshan tunnel construction on groundwater environment, maximum membership principle method makes an assessment result of medium level as 0.355 is the maximum member in the evaluation result vector, and weighted mean method gives nearly the same result because T is very close to 3. From a mathematic point of view, moderate grade is recommended as the influence level of Tongluoshan tunnel.

In order to verify whether the assessment is reasonable or not, environmental investigation and hydrological monitoring program was carried out in the tunnel area. Twelve springs, one well, and two streams, which either supply people with drinking water or flow at a notable rate, were selected to perform a 15-month-long monitoring process and so were the discharge from tunnel and rainfall in the project area. During the main construction period, some of the monitoring points near the tunnel axis, such as AQ21, AQ23, AQ35, and BQ27, have been strongly impacted by tunnel construction and lasting drainage of groundwater. Most of them have never dried up and maintained base flow even in the dry season, but an obvious decline was found throughout the monitoring period. Though strong relation exists between most of the monitoring points and precipitation, it seems that groundwater drainage from tunnel plays a nonnegligible role because some of the monitoring points repeatedly ran dry during the main construction period despite being recharged by precipitation. Based on the information fed back from environmental investigation and hydrologic monitoring carried out during the main construction period, approximately 1 km from the tunnel axis (not including BQ25 which is partly fed by upstream surface water) is inferred as the sphere of influence in the middle karst section of Tongluoshan tunnel, which is less than other typical karst tunnel, such as Yuanliangshan tunnel and Geleshan tunnel [8, 16].

As is analyzed above, moderate grade influence on groundwater system is considered to be an appropriate result for Tongluoshan tunnel construction for the reason that (a) assessment made from the indicator system and fuzzy comprehensive method falls into middle level; (b) some of the monitoring points near the tunnel axis which have never dried up and maintained base flow even in the dry season experienced an obvious decline throughout the monitoring period, revealing that the local groundwater environment might have been impacted by the tunnel excavation; (c) the sphere of influence in the middle karst section of Tongluoshan tunnel approaches approximately 1 km from the tunnel axis, which is less than other typical karst tunnels. So, there is no doubt that the local groundwater environment had been impacted by the excavation of Tongluoshan tunnel, but the influence level is moderate.

CONCLUSION AND OUTLOOK

Making an assessment on how the construction would affect local groundwater environment before excavation is of vital importance to line selection of tunnel. This study adopted an indicator system and fuzzy comprehensive evaluation method to identify the influence caused by Tongluoshan tunnel construction on groundwater environment. It was shown that maximum membership stood in the middle of the evaluation result set and 3.2 was scored by relative rating weighted calculation, indicating that excavation of Tongluoshan tunnel had influenced the local groundwater environment at a moderate degree. Based on the information fed back from environmental investigation and hydrologic monitoring carried out during the main construction period, flow of some springs and Liaojiagou stream located beside the tunnel route was found experiencing an apparent decline; even worse, AQ21, AQ23, AQ35, and BQ25 had dried up for a relatively long time because of lower precipitation during 2006 than that in the past and drainage from tunnel resulting in reduction of other flow-out patterns in water balance.

Through this practice, it can be concluded that making an assessment using the indicator system and fuzzy comprehensive evaluation method would supply us with a simple way to understand the situation that may happen to groundwater environment during tunnel construction and can help us to choose an optimized channel through which less disruption related to water inflow may occur. Once the tunnel location is determined and prepared to construction, we can also use the indicator system to evaluate the potential groundwater environmental impact before building it and take appropriate measures to minimize the impact. The major differences between a building tunnel and a built tunnel focus on two indicators including "construction method" and "measures for prevention of groundwater flowing into tunnel." Taking Tongluoshan tunnel as an example and setting "boring machine method" and "drainage with composite lining" as the initial condition, a result vector of 0.106, 0.195, 0.312, 0.242, and 0.145 indicates that the influence level partly tends to be strong, because the gap between the maximum member and the second maximum member falls to 0.07, smaller than 0.1. In order to minish the potential impact, measurements such as previous and postgrouting from tunnel, pregrouting from surface, and partial excavation using drilling and blasting method were taken at some sections with high risk of water irruption, reducing the influence level and impact on groundwater environment to some extent.

ACKNOWLEDGMENTS

This research was supported by Communications Department of Sichuan Province, Sichuan Dayu Expressway Construction and Development Company Limited (2006-G-02), and the Fundamental Research Funds for the Central Universities (2682015CX020). Many thanks should be given to the team members related to this program, for their help with field investigation and data collection and processing.

REFERENCES

1. H. R. Zarei, A. Uromeihy, and M. Sharifzadeh, "Identifying geological hazards related to tunneling in carbonate karstic rocks-Zagros, Iran," Arabian Journal of Geosciences, vol. 5, no. 3, pp. 457–464, 2012.
2. C. Butscher, P. Huggenberger, and E. Zechner, "Impact of tunneling on regional groundwater flow and implications for swelling of clay-sulfate rocks," Engineering Geology, vol. 117, no. 3-4, pp. 198–206, 2011.
3. U. Chiocchini and F. Castaldi, "The impact of groundwater on the excavation of tunnels in two different hydrogeological settings in central Italy," Hydrogeology Journal, vol. 19, no. 3, pp. 651–669, 2011.
4. J.-H. Hwang and C.-C. Lu, "A semi-analytical method for analyzing the tunnel water inflow," Tunnelling and Underground Space Technology, vol. 22, no. 1, pp. 39–46, 2007.
5. S. Alija, F. J. Torrijo, and M. Quinta-Ferreira, "Geological engineering problems associated with tunnel construction in karst rock masses: the case of Gavarres tunnel (Spain)," Engineering Geology, vol. 157, pp. 103–111, 2013.
6. M. J. Day, "Karstic problems in the construction of Milwaukee's Deep Tunnels," Environmental Geology, vol. 45, no. 6, pp. 859–863, 2004.
7. L. Schwarz, I. Reichl, H. Kirschner, and K. P. Robl, "Risks and hazards caused by groundwater during tunnelling: geotechnical solutions used as demonstrated by recent examples from Tyrol, Austria,"Environmental Geology, vol. 49, no. 6, pp. 858–864, 2006. · ·
8. L. W. Jiang, Y. J. Yi, and Z. M. Jia, "Research on characteristics and formation mechanism of great deep buried infilled caves at Maoba syncline in the Yuanliangshan railway tunnel," Journal of Railway Engineering Society, no. 4, pp. 53–60, 2007 (Chinese).
9. D.-J. Tseng, B.-R. Tsai, and L.-C. Chang, "A case study on ground treatment for a rock tunnel with high groundwater ingression in Taiwan," Tunnelling and Underground Space Technology, vol. 16, no. 3,

pp. 175–183, 2001.

10. C. Zangerl, E. Eberhardt, and S. Loew, "Ground settlements above tunnels in fractured crystalline rock: numerical analysis of coupled hydromechanical mechanisms," Hydrogeology Journal, vol. 11, no. 1, pp. 162–173, 2003.

11. J. Gisbert, A. Vallejos, A. González, and A. Pulido-Bosch, "Environmental and hydrogeological problems in karstic terrains crossed by tunnels: a case study," Environmental Geology, vol. 58, no. 2, pp. 347–357, 2009.

12. V. Vincenzi, A. Gargini, and N. Goldscheider, "Using tracer tests and hydrological observations to evaluate effects of tunnel drainage on groundwater and surface waters in the Northern Apennines (Italy)," Hydrogeology Journal, vol. 17, no. 1, pp. 135–150, 2009.

13. P. M. Attanayake and M. K. Waterman, "Identifying environmental impacts of underground construction," Hydrogeology Journal, vol. 14, no. 7, pp. 1160–1170, 2006.

14. J. R. Raposo, J. Molinero, and J. Dafonte, "Quantitative evaluation of hydrogeological impact produced by tunnel construction using water balance models," Engineering Geology, vol. 116, no. 3-4, pp. 323–332, 2010.

15. F.-R. Yang, C.-H. Lee, W.-J. Kung, and H.-F. Yeh, "The impact of tunneling construction on the hydrogeological environment of 'Tseng-Wen Reservoir Transbasin Diversion Project' in Taiwan,"Engineering Geology, vol. 103, no. 1-2, pp. 39–58, 2009.

16. J. Liu, Study on the evaluation system of negative effects on groundwater environment resulted by tunnel construction in karst area [Ph.D. thesis], Southwest Jiaotong University, Chengdu, China, 2011, (Chinese).

17. D. X. Yuan, Karstology of China, Geological Publishing House, Beijing, China, 1994, (Chinese).

18. J. F. Durand, "The impact of gold mining on the Witwatersrand on the rivers and karst system of Gauteng and North West Province, South Africa," Journal of African Earth Sciences, vol. 68, no. 15, pp. 24–43, 2012.

19. L. A. Zadeh, "Fuzzy sets," Information and Control, vol. 8, pp. 338–353, 1965.

20. G. Onkal-Engin, I. Demir, and H. Hiz, "Assessment of urban air quality in Istanbul using fuzzy synthetic evaluation," Atmospheric Environment, vol. 38, no. 23, pp. 3809–3815, 2004

21. C. Q. Mi, X. D. Zhang, S. M. Li, J. Yang, D. Zhu, and Y. Yang,

"Assessment of environment lodging stress for maize using fuzzy synthetic evaluation," Mathematical and Computer Modelling, vol. 54, no. 3-4, pp. 1053–1060, 2011.

22. S. Dahiya, B. Singh, S. Gaur, V. K. Garg, and H. S. Kushwaha, "Analysis of groundwater quality using fuzzy synthetic evaluation," Journal of Hazardous Materials, vol. 147, no. 3, pp. 938–946, 2007.
23. Y. J. Guo, "New theory and method of dynamic comprehensive evaluation," Journal of Management Sciences, vol. 5, no. 2, pp. 52–57, 2002 (Chinese).
24. Sichuan Institute of Coalfield Geological Engineering Exploration and Designing, Preliminary Geological Survey of Tongluoshan Tunnel, Sichuan Institute of Coalfield Geological Engineering Exploration and Designing, Chengdu, China, 2003.
25. Y. Wang, W. F. Yang, M. Li, and X. Liu, "Risk assessment of floor water inrush in coal mines based on secondary fuzzy comprehensive evaluation," International Journal of Rock Mechanics & Mining Sciences, vol. 52, pp. 50–55, 2012.
26. Q. Zhang, X. Yang, Y. Zhang, and M. Zhong, "Risk assessment of groundwater contamination: a multilevel fuzzy comprehensive evaluation approach based on DRASTIC model," The Scientific World Journal, vol. 2013, Article ID 610390, 9 pages, 2013.
27. H. Y. Wu, K. L. Chen, Z. H. Chen et al., "Evaluation for the ecological quality status of coastal waters in East China Sea using fuzzy integrated assessment method," Marine Pollution Bulletin, vol. 64, no. 3, pp. 546–555, 2012.

CITATION

CHAPTER 1

Shirdel, B. (2015) Analysis of Engineering Geology Indices in Three Units of Atamir Formation in Kope Dagh Zone. Open Journal of Geology, 5, 847-875. doi: 10.4236/ojg.2015.512072.

CHAPTER 2

Selim, S. , Hamdan, A. and Rady, A. (2014) Groundwater Rising as Environmental Problem, Causes and Solutions: Case Study from Aswan City, Upper Egypt. *Open Journal of Geology*, **4**, 324-341. doi: 10.4236/ojg.2014.47025.

CHAPTER 3

Merayyan, S. and Safi, S. (2014) Feasibility of Groundwater Banking under Various Hydrologic Conditions in California, USA. Computational Water, Energy, and Environmental Engineering, 3, 79-92. doi: 10.4236/cweee.2014.33009.

CHAPTER 4

S. Mangala PRAVEENA, M. Harun ABDULLAH, A. Zaharin ARIS and K. BIDIN, "Groundwater Solution Techniques: Environmental Applications," *Journal of Water Resource and Protection*, Vol. 2 No. 1, 2010, pp. 8-13. doi:10.4236/jwarp.2010.21002.

CHAPTER 5

Djémin, J. , Kouamé, J. , Deh, K. , Abinan, A. and Jourda, J. (2016) Contribution of the Sensitivity Analysis in Groundwater Vulnerability Assessing Using the DRASTIC Method: Application to Groundwater in Dabou Region (Southern of Côte d'Ivoire). *Journal of Environmental Protection*, **7**, 129-143. doi: 10.4236/jep.2016.71012.

CHAPTER 6

Rebelo, M. , Santos, G. and Silva, R. (2014) A Methodology to Develop the Integration of the Environmental Management System with Other Standardized Management Systems. Computational Water, Energy, and Environmental Engineering, 3, 170-181. doi: 10.4236/cweee.2014.34018.

CHAPTER 7

Yazdi, A. , Arian, M. and Tabari, M. (2014) Geological and Geotourism Study of Iran Geology Natural Museum, Hormoz Island. *Open Journal of Ecology*, **4**, 703-714. doi: 10.4236/oje.2014.411060.

CHAPTER 8

L. Khodapanah, W. Sulaiman and H. Nassery, "Hydrogeological Framework and Groundwater Balance of a Semi-arid Aquifer, a Case Study from Iran," *Journal of Water Resource and Protection*, Vol. 3 No. 7, 2011, pp. 513-521. doi: 10.4236/jwarp.2011.37061.

CHAPTER 9

Arshad Ashraf (2013). Changing Hydrology of the Himalayan Watershed, Current Perspectives in Contaminant Hydrology and Water Resources Sustainability, Dr. Paul Bradley (Ed.), ISBN: 978-953-51-1046-0, InTech, DOI: 10.5772/54492.

CHAPTER 10

Luc Descroix, Ibrahim Bouzou Moussa, Pierre Genthon, Daniel Sighomnou, Gil Mahé, Ibrahim Mamadou, Jean-Pierre Vandervaere, Emmanuèle Gautier, Oumarou Faran Maiga, Jean-Louis Rajot, Moussa Malam Abdou, Nadine Dessay, Aghali Ingatan, Ibrahim Noma, Kadidiatou Souley Yéro, Harouna Karambiri, Rasmus Fensholt, Jean Albergel and Jean-Claude Olivry (2013). Impact of Drought and Land – Use Changes on Surface – Water Quality and Quantity: The Sahelian Paradox, Current Perspectives in Contaminant

Hydrology and Water Resources Sustainability, Dr. Paul Bradley (Ed.), ISBN: 978-953-51-1046-0, InTech, DOI: 10.5772/54536.

CHAPTER 11

Wei-Zu Gu, Jiu-Fu Liu, Jia-Ju Lu and Jay Frentress (2013). Current Challenges in Experimental Watershed Hydrology, Current Perspectives in Contaminant Hydrology and Water Resources Sustainability, Dr. Paul Bradley (Ed.), ISBN: 978-953-51-1046-0, InTech, DOI: 10.5772/55087.

CHAPTER 12

Jian Liu, Dan Liu, and Kai Song, "Evaluation of the Influence Caused by Tunnel Construction on Groundwater Environment: A Case Study of Tongluoshan Tunnel, China," Advances in Materials Science and Engineering, Article ID 149265, in press

INDEX

K

M

N

O

P

Q

R

S

T

U

W

Encyclopaedia of
APPLIED GEO ENVIRONMENTAL STUDIES

Encyclopaedia of APPLIED GEO ENVIRONMENTAL STUDIES

Volume II: Geoenvironmental Mapping: Methods, Theory and Practice

Contributors

Mohammad Ali Zare Chahouki, Paulo Cesar Fernandes da Silva, and Bizhan Shirdel et al.

AURIS Reference

www.aurisreference.com

Encyclopaedia of Applied Geo Environmental Studies

Volume II: Geoenvironmental Mapping: Methods,Theory and Practice

Contributors: Mohammad Ali Zare Chahouki, Paulo Cesar Fernandes da Silva, and Bizhan Shirdel et al.

Published by Auris Reference Limited

www.aurisreference.com

United Kingdom

Printed in 2017 for Sale in the Indian Subcontinent

Geoenvironmental Mapping: Methods,Theory and Practice

ISBN: 978-1-78154-980-3 (Set of 3 Volumes)

British Library Cataloguing in Publication Data
A CIP record for this book is available from the British Library

Printed in the United Kingdom

Exclusively distributed by CBS Publishers & Distributors Pvt. Ltd.

Sales & Distribution Rights only for India, Pakistan, Bangladesh, Sri Lanka, Nepal and Bhutan.This book is not to be sold outside these territories.

Contents

List of Abbreviations

AHP	Analytic Hierarchy Process
APIs	Application program interfaces
CBR	case-based reasoning
CLR	common language runtime
DEM	Digital elevation model
DEM	Digital elevation model
EROS	Earth Resources Observation and Science
GIS	geographic information systems
GWPs	global warming potentials
IDE	Integrated development environment
IPTA	Interferometric Point Target Analysis
KTF	Kephalonia Transform Fault
MMF	Morgan-Morgan-Finney
ORDBMS	Object-relational database management system
PSI	Persistent Scatterers Interferometry
SLC	Single Look Complex
TM	Thematic Mapper sensorU.S. Geological SurveyUSGS
USLE	Universal Soil Loss Equation
UTM	Universal Transverse MercatorUTM
VES	Vertical electrical soundings
WOE	weights-of-evidence

List of Contributors

Paulo Cesar Fernandes da Silva
Geological Institute - São Paulo State Secretariat of Environment,Brazil

John Canning Cripps
Department of Civil and Structural Engineering, University of Sheffield,United Kingdom

Gebreyesus Brhane Tesfahunegn
College of Agriculture, Aksum University, Aksum, Ethiopia
Centre for Development Research, University of Bonn, Walter-Flex-Street 3, 53113 Bonn, Germany

Lulseged Tamene
International Centre for Tropical Agriculture (CIAT), Chitedze Agricultural Research Station, Lilongwe, Malawi

Paul L. G. Vlek
College of Agriculture, Aksum University, Aksum, Ethiopia

Binbin He
School of Resources and Environment, University of Electronic Science and Technology of China, Chengdu, China

Jianhua Chen
College of Geophysics, Chengdu University of Technology, Chengdu, China

Cuihua Chen
College of Geosciences, Chengdu University of Technology, Chengdu, China

Yue Liu
College of Geosciences, Chengdu University of Technology, Chengdu, China

Sunil Pratap Singh
Department of Physics and Computer Science, Dayalbagh Educational Institute, Agra, India

Preetvanti Singh
Department of Physics and Computer Science, Dayalbagh Educational Institute, Agra, India

Ali Suhardiman
Department of Global Agricultural Sciences, The University of Tokyo, Tokyo, Japan

Satoshi Tsuyuki
Department of Global Agricultural Sciences, The University of Tokyo, Tokyo, Japan

Muhammad Sumaryono
Department of Forest Science, University of Mulawarman, Samarinda, Indonesia

Yohanes Budi Sulistioadi
Department of Forest Science, University of Mulawarman, Samarinda, Indonesia
Division of Geodetic Science, School of Earth Science, The Ohio State University, Columbus, USA

Emmanuel Vassilakis
Department of Dynamics, Tectonics and Applied Geology, Faculty of Geology & Geoenvironment, National & Kapodistrian University of Athens, Panepistimioupoli Zografou, 15784, Athens, Greece

Issaak Parcharidis
Harokopio University of Athens, Department of Geography, El. Venizelou 70, 17671 Athens, Greece

Sotiris Kokkalas
University of Patras, Department of Geology, Division of Physical Geology, Marine Geology and Geodynamics, 265 00 Patras, Greece

Ioannis Fountoulis
National and Kappodistrian University of Athens, Faculty of Geology and Geoenvironment, Department of Dynamic Tectonic and Applied Geology, Panepistimioupolis Zografou, 157 84 Athens, Greece

Michael Foumelis
National and Kappodistrian University of Athens, Faculty of Geology and Geoenvironment, Department of Geophysics and Geothermics, Panepistimioupolis Zografou, 157 84 Athens, Greece

Slavoljub Dragicevic
University of Belgrade, Faculty of Geography, Belgrade, Serbia

Nenad Zivkovic
University of Belgrade, Faculty of Geography, Belgrade, Serbia

Mirjana Roksandic
University of Belgrade, Faculty of Geography, Belgrade, Serbia

Ivan Novkovic
University of Belgrade, Faculty of Geography, Belgrade, Serbia

Stanimir Kostadinov
University of Belgrade, Faculty of Forestry, Belgrade, Serbia

Radislav Tosic
Faculty of Natural Sciences, Banja Luka, Republic of Srpska

Milomir Stepic
Institute for Political Studies,Belgrade, Serbia

Marija Dragicevic
First Elementary School in Obrenovac, Obrenovac, Serbia

Borislava Blagojevic
University of Nis, Faculty of Civil Engineering and Architecture, Serbia

Seth Appiah-Opoku
Geography Department, University of Alabama, Tuscaloosa, USA

Crystal Taylor
Florida State University, USA

Dirk Loehr
Trier University of Applied Sciences, Environmental Campus Birkenfeld, , Germany

Alireza Gharagozlou
Geomatics College of National Cartographic Center of Iran (NCC), Tehran, Iran

Hassan Nazari
Geomatics College of National Cartographic Center of Iran (NCC), Tehran, Iran

Mohammadjavad Seddighi
Geomatics College of National Cartographic Center of Iran (NCC), Tehran, Iran

Giulliana Mondelli,
Institute for Technological Research of São Paulo State, São Paulo State University, University of São Paulo, São Paulo, Brazil

Heraldo Luiz Giacheti and
Institute for Technological Research of São Paulo State, São Paulo State University, University of São Paulo, São Paulo, Brazil

Vagner Roberto Elis
Institute for Technological Research of São Paulo State, São Paulo State University, University of São Paulo, São Paulo, Brazil

Preface

Encyclopaedia of Applied Geo Environmental Studies reaffirms the global need for mapping geoscientific data. The text *Geoenvironmental Mapping: Methods, Theory and Practice* illustrates the range of environmental geoscience mapping presently carried out around the world. Specialists in several countries have contributed a number of sub disciplinary and thematic topics including landslides, dolines, tsunamis, radon potential, medical geology, rainfall erosion, engineering geology, borehole stratigraphy, lake sediment geochemistry, aggregate resources and remote sensing. In first chapter, a number of case studies are presented to demonstrate the methodological as well as the predictive and preventative aspects of geo-environmental management, with a particular view to regional- and semi-detailed scale, satellite image based terrain classification. Second chapter describes a case-based reasoning (CBR) method for mineral prospectivity mapping that takes spatial features of geology data into account and offers an intelligent approach. Third chapter presents the integration of open-source geospatial tools and web technology to visualize and interact with spatial data using web browser. In fourth chapter, we apply geostatistical and geographical information system (GIS) to generate site-suitability mapping for mangrove rehabilitation guidance in the Mahakam delta. Fifth chapter highlights the important role that analysis of multispectral satellite data can play in the identification of surface alterations related to human activity and natural processes. Sixth chapter concerns the monitoring of the potential ground deformation caused by the active tectonism in the cities of Patras and Pyrgos in Western Greece. A case study of the Kolubara river basin in Serbia has been presented in seventh chapter. Eighth chapter introduces the idea that institutions of higher learning leave ecological footprints on the landscape. Ninth chapter focuses on the role of tradable planning permits in environmental land use planning. Tenth chapter develops a framework for flood control and begins with some general comments on the importance of land use planning and outlines some current environmental issues and then presenting environmental models to use in disaster management plan by using GIS and remote sensing results. The aim of eleventh chapter is to derive spatial distribution of hydrophysical parameters and apply them in the Morgan-Morgan-Finney (MMF) model for estimating soil erosion in the Mai-Negus catchment, northern Ethiopia. Last chapter aims to present and discuss the different tests and steps of a geo-environmental site investigation program proposed for municipal solid waste disposal sites.

Chapter 1

GEO-ENVIRONMENTAL TERRAIN ASSESSMENTS BASED ON REMOTE SENSING TOOLS: A REVIEW OF APPLICATIONS TO HAZARD MAPPING AND CONTROL

Paulo Cesar Fernandes[1] da Silva and John Canning Cripps[2]

[1]Geological Institute - São Paulo State Secretariat of Environment,Brazil

[2]Department of Civil and Structural Engineering, University of Sheffield,United Kingdom

INTRODUCTION

The responses of public authorities to natural or induced geological hazards, such as land instability and flooding, vary according to different factors including frequency of occurrence, severity of damage, magnitude of hazardous processes, awareness, predictability, political willingness and availability of financial and technological resources. The responses will also depend upon whether the hazard is 1) known to be already present thus giving rise to risk situations involving people and/or economic loss; or 2) there is a latent or potential hazard that is not yet present so that development and land uses need to be controlled in order to avoid creating risk situations. In this regard, geo-environmental management can take the form of either planning responses and mid- to long-term public policy based territorial zoning tools, or immediate interventions that may involve a number of approaches including preventative and mitigation works, civil defence actions such as hazard warnings, community preparedness, and implementation of contingency and emergency programmes. In most of cases, regional- and local-scale terrain assessments and classification accompanied by susceptibility and/or hazard maps delineating potential problem areas will be used as practical instruments in efforts to tackle problems and their consequences. In terms of planning, such assessments usually provide advice about the types of development that would be acceptable in certain areas but should be precluded in others. Standards for new construction and the upgrading of existing buildings may also be implemented through legally enforceable building codes based on the risks associated with the particular terrain assessment or classification. The response of public authorities also varies depending upon the information available to make decisions. In some areas sufficient geological information and knowledge about the causes of a hazard may be available to enable an area likely to be susceptible to hazardous processes to be predicted with reasonable certainty. In other places a lack of suitable data may result in considerable uncertainty. In this chapter, a number of case studies are presented to demonstrate the methodological as well as the predictive and preventative aspects of geo-environmental management, with a particular view to regional- and semi-detailed scale, satellite image based terrain classification. If available, information on the geology, geomorphology, covering material characteristics and land uses may be used with remotely sensed data to enhance these terrain classification outputs. In addition, examples provided in this chapter demonstrate the identification and delineation of zones or terrain units in terms of the likelihood and consequences of land instability and flooding hazards in different situations. Further applications of these methods include the ranking of abandoned and/or derelict mined sites and other despoiled areas in support of land reclamation and socio-economic regeneration policies. The discussion extends into policy formulation, implementation of environmental management strategies and enforcement regulations.

USE OF REMOTE DENSING TOOLS FOR TERRAIN ASSESSMENTS AND TERRITORIAL ZONING

Engineering and geo-environmental terrain assessments began to play an important role in the planning process as a consequence of changing demands for larger urban areas and related infra-structure, especially housing, industrial development and the services network. In this regard, the inadequacy of conventional agriculturally-orientated land mapping methods prompted the development of terrain classification systems completely based on the properties and characteristics of the land that provide data useful to engineers and urban planners. Such schemes were then adopted and widely used to provide territorial zoning for general and specific purposes. The process of dividing a country or region into area parcels or zones, is generally called land or terrain classification. Such a scheme is illustrated in Table 1. The zones should possess a certain homogeneity of characteristics, properties, and in some cases, conditions and expected behaviour in response to human activities. What is meant by homogeneous will depend on the purpose of the exercise, but generally each zone will contain a mixture of environmental elements such as rocks, soils, relief, vegetation, and other features. The feasibility and practicability of delineating land areas with similar attributes have been demonstrated throughout the world over a long period of time (e.g. Bowman, 1911; Bourne, 1931; Christian, 1958; Mabbutt, 1968; amongst others), and encompass a wide range of specialisms such as earth, biological and agricultural sciences; hydrology and water resources management; military activities; urban and rural planning; civil engineering; nature and wildlife conservation; and even archaeology. According to Cendrero et al. (1979) and Bennett and Doyle (1997), there are two main approaches to geo-environmental terrain assessments and territorial zoning, as follows. 1) The analytical or parametric approach deals with environmental features or components individually.

Table 1: Hierarchical classification of terrain, soil and ecological units [after Mitchell, 1991]

Terrain unit	*Definition*	*Soil unit*	*Vegetation unit*	*Mapping scale (approx.)*	*Remote sensing platform*
Land zone	Major climatic region	Order	-	< 1:50,000,000	
Land division	Gross continental structure	Suborder	Plant panformation Ecological zone	1:20,000,000 to 1:50,000,000	Meteorological satellites
Land province	Second-order structure or large lithological association	Great group	-	1:20,000,000 to 1:50,000,000	
Land region	Lithological unit or association having undergone comparable geomorphic evolution	Subgroup	Sub-province	1:1,000,000 to 1:5,000,000	Landsat SPOT ERS
Land system *	Recurrent pattern of genetically linked land facets	Family	Ecological region	1: 200,000 to 1:1,000,000	Landsat SPOT, ERS, and small scale aerial photographs
Land catena	Major repetitive component of a land system	Association	Ecological sector	1:80,000 to 1:200,000	
Land facet	Reasonably homogeneous tract of landscape distinct from surrounding areas and containing a practical grouping of land elements	Series	Sub-formation; Ecological station	1:10,000 to 1: 80,000	Medium scale aerial photographs, Landsat, and SPOT in some cases
Land clump	A patterned repetition of two or more land elements too contrasting to be a land facet	Complex	Sub-formation; Ecological station	1:10,000 to 1: 80,000	
Land subfacet	Constituent part of a land facet where the main formative processes give material or form subdivisions	Type	-	Not mapped	Large-scale aerial photographs
Land element	Simplest homogeneous part of the landscape, indivisible in form	Pedon	Ecological station element		

The terrain units usually result from the intersection or cartographic summation of several layers of information [thus expressing the probability limits of findings] and their extent may not corresponding directly with ground features. Examples of the parametric approach for urban planning, hazard mapping and engineering purposes are given by Kiefer (1967), Porcher & Guillope (1979), Alonso Herrero et al. (1990), and Dai et al. (2001). 2) In the synthetic approach, also termed integrated, landscape or physiographic approach, the form and spatial distribution of ground features are analysed in an integrated manner relating recurrent landscape patterns expressed by an interaction of environmental components thus allowing the partitioning of the land into units. Since the advent of airborne and orbital sensors, the integrated analysis is based in the first instance, on the interpretation of remotely sensed images and/or aerial photography. In most cases, the content and spatial boundaries of terrain units would directly correspond with ground features. Assumptions that units possessing similar recurrent landscape patterns may be expected to be similar in character are required for valid predictions to be made by extrapolation from known areas. Thus, terrain classification schemes offer rational means of correlating known and unknown areas so that the ground conditions and potential uses of unknown areas can be reasonably predicted (Finlayson, 1984; Bell, 1993). Examples of the applications of the landscape or physiographic approach include ones given by Christian & Stewart (1952, 1968), Vinogradov et al. (1962), Beckett & Webster (1969); Meijerink (1988), and Miliaresis (2001). Griffiths and Edwards (2001) refer to Land Surface Evaluation as a procedure of providing data relevant to the assessment of the sites of proposed engineering work. The sources of data include remotely sensed data and data acquired by the mapping of geomorphological features. Although originally viewed as a process usually undertaken at the reconnaissance or feasibility stages of projects, the authors point out its utility at the constructional and post-construction stages of certain projects and also that it is commonly applied during the planning of engineering development. They also explain that although more reliance on this methodology for deriving the conceptual or predictive ground model on which engineering design and construction are based, was anticipated in the early 1980s, in fact the use of the methods has been more limited. Geo-environmental terrain assessments and territorial zoning generally involve three main stages (IG/SMA 2003; Fernandes da Silva et al. 2005b, 2010): 1) delimitation of terrain units; 2) characterisation of units (e.g. in bio-geographical, engineering geological or geotechnical terms); and 3) evaluation and classification of units. The delimitation stage consists of dividing the territory into zones according to a set of pre-determined physical and environmental characteristics and properties. Regions, zones or units are regarded as distinguishable entities depending upon their internal homogeneity or the internal interrelationships of their parts. The characterisation stage consists of attributing appropriate properties and characteristics to terrain components. Such properties and characterisitics are designed to reflect the ground conditions relevant to the particular application. The characterisation of the units can be achieved either directly or indirectly, for instance by means of: (a) ground observations and measurements, including in-situ tests (e.g. boring, sampling, infiltration tests etc); (b) laboratory tests (e.g. grain size, strength, porosity, permeability etc); (c) inferences derived from existing correlations between relevant parameters and other data such as those obtained from previous mapping, remote sensing, geophysical surveys and geochemical records. The final stage (evaluation and classification) consists of evaluating and classifying the terrain units in a manner relevant to the purposes of the particular application (e.g. regional planning, transportation, hazard mapping). This is based on the analysis and interpretation of properties and characteristics of terrain - identified as relevant - and their potential effects in terms of ground behaviour, particularly in response to human activities. A key issue to be considered is sourcing suitable data on which to base the characterisation, as in many cases derivation by standard mapping techniques may not be feasible. The large size of areas and lack of accessibility, in particular, may pose major technical, operational, and economic constraints. Furthermore, as indicated by Nedovic-Budic (2000), data collection and integration into useful databases are liable to be costly and time-consuming operations. Such problems are particularly prevalent in developing countries in which suitably trained staff, and scarce organizational resources can inhibit public authorities from properly benefiting from geo-environmental terrain assessment outputs in planning and environmental management instruments. In this regard, consideration has been given to increased reliance on remote sensing tools, particularly satellite imagery. The advantages include: (a) the generation of new data in areas where existing data are sparse, discontinuous or non-existent, and (b) the economical coverage of large areas, availability of a variety of spatial resolutions, relatively frequent and periodic updating of images (Lillesand and Kiefer 2000; Latifovic et al. 2005; Akiwumi and Butler 2008). It has also been proposed that developing countries should ensure that options for using low-cost technology, methods and products that fit their specific needs and capabilities are properly considered (Barton et al. 2002, Câmara and Fonseca 2007). Some examples are provided here to demonstrate the feasibility of a low-cost technique based on the analysis of texture of satellite imagery that can be used for delimitation of terrain units. The delimited units may be further analysed for different purposes such as regional and urban planning, hazard mapping, and land reclamation. The physiographic compartmentalisation technique (Vedovello 1993, 2000) utilises the spatial information contained in images and the principles of convergence of evidence (see Sabins 1987) in a systematic deductive process of image interpretation. The technique evolved from engineering applications of the synthetic land classification approach (e.g. Grant, 1968, 1974, 1975; TRRL 1978), by incorporating and advancing the logic and procedures of geological-geomorphological photo-interpretation (see Guy 1966, Howard 1967, Soares and Fiori 1976), which were then converted to monoscopic imagery (as elucidated by Beaumont and Beaven 1977; Verstappen 1977; Soares et al. 1981; Beaumont, 1985; and others). Image interpretation is performed by identifying and delineating textural zones on images according to properties that take into account coarseness, roughness, direction and regularity of texture elements (Table 2). The key assumption proposed by Vedovello (1993, 2000) is that zones with relatively homogeneous textural characteristics in satellite images (or air-photos) correspond with specific combinations of geo-environmental components (such as bedrock, topography and landforms, soils and covering materials) which share a common tectonic history and land surface evolution. The particular combinations of geo-environmental components are expected to be associated with specific ground responses to engineering and other land-use actions. The process of image interpretation (whether or not supported by additional information) leads to a cartographic product in which textural zones constitute comprehensive terrain units delimited by fixed spatial boundaries The latter correspond with ground features. The units are referred to as physiographic compartments or basic compartmentalisation units (BCUs), which are the smallest units for analysis of geo-environmental components at the chosen cartographic scale (Vedovello and Mattos 1998). The spatial resolution of the satellite image or air-photos being used for the analysis and interpretation is assumed to govern the correlation between image texture and terrain characteristics. This correlation is expressed at different scales and levels of compartmentalisation. Figure 1 presents an example of the identification of basic compartmentalisation units (BCUs) based on textural differences on Landsat TM5 images. In this case the features on images are expressions of differences in the distribution and spatial organisation of textural elements related to drainage network and relief. The example shows the contrast between drainage networks of areas consisting of crystalline rocks with those formed on areas of sedimentary rocks, and the resulting BCUs.

TERRAIN SUSCEPTIBILITY MAPS: APPLICATIONS TO REGIONAL AND URBAN PLANNING

Terrain susceptibility maps are designed to depict ground characteristics (e.g. slope steepness, landforms) and observed and potential geodynamic phenomena, such as erosion, instability and flooding, which may entail hazard and potential damage. These maps are useful for a number of applications including development and land use planning, environmental protection, watershed management as well as in initial stages of hazard mapping applications.

Table 2: Description of elements and properties used for recognition and delineation of distinctive textural zones on satellite imagery [after Vedovello 1993, 2000]

Textural entities and properties	Description
Image texture element	The smallest continuous and uniform surface liable to be distinguishable in terms of shape and dimensions, and likely to be repetitive throughout an image. Usual types of image texture elements taken for analysis include: segments of drainage or relief (e.g. crestlines, slope breaks) and grey tones.
Texture density	The quantity of textural elements occurring within an area on image. Texture density is defined as the inverse of the mean distance between texture elements. Although it reflects a quantitative property, textural density is frequently described in qualitative and relative terms such as high, moderate, low etc. Size of texture elements combined with texture density determine features such as coarseness and roughness.
Textural arrangement	The form (ordered or not) by which textural elements occur and are spatially distributed on an image. Texture elements of similar characteristics may be contiguous thus defining alignments or linear features on the image. The spatial distribution may be repetitive and it is usually expressed by 'patterns' that tend to be recurrent (regularity). For example, forms defined by texture elements due to drainage expressed in rectangular, dendritic, or radial patterns.
Structuring (Degree of spatial organisation)	The greater or lesser organisation underlying the spatial distribution of textural elements and defined by repetition of texture elements within a certain rule of placement. Such organisation is usually expressed in terms of regular or systematic spatial relations, such as length, angularity, asymmetry, and especially prevailing orientations (tropy or directionality). Tropy reflects the anisotropic (existence of one, two, or three preferred directions), or the isotropic (multi-directional or no predominant direction) character of textural features. Asymmetry refers to length and angularity of linear features (rows of contiguous texture elements) in relation to a main feature identified on image. The degree of organisation can also be expressed by qualitative terms such as high, moderate, low, or yet as well- or poorly-defined.
Structuring order	Complexity in the organisation of textural elements, mainly reflecting superposition of image structuring. For example, a regional directional trend of textural elements that can be extremely pervasive, distinctive and superimposed on other orientations also observed on imagery. Another example is drainage networks that display different orders with respect to main stream lines and tributaries (1st, 2nd, 3rd orders)

Early multipurpose and comprehensive terrain susceptibility maps include examples by Dearman & Matula, (1977), Matula (1979), and Matula & Letko (1980). These authors described the application of engineering geology zoning methods to the urban planning process in the former Republic of Czechoslovakia. The studies in this and other countries focused on engineering geology problems related to geomorphology and geodynamic processes, seismicity, hydrogeology, and foundation conditions.

Culshaw and Price (2011) point out that in the UK, a major initiative on urban geology began in the mid-1970s with obtaining geological information relevant to aggregates and other industrial minerals together with investigations relating to the planning of the proposed 3rd London Airport. In the latter case, a very wide range of map types was produced, including one that could be viewed in 3D, using green and red anaglyph spectacles. Of particular interest was the summary "Engineering Planning Map which showed areas that were generally suitable for different types of construction and, also, detailed suggested site investigation procedures (Culshaw and Northmore 2002). As Griffiths and Hearn (2001) explain, subsequently about 50 experimental 'environmental geological mapping, 'thematic'geological mapping' and 'applied geological mapping' projects were carried out between 1980 and 1996 Culshaw and Price (2011) explain that this was to investigate the best means of collecting, collating, interpreting and presenting geological data that would be of direct applicability in land-use planning (Brook and Marker 1987). Maps of a variety of geological and terrain types, including industrially despoiled and potentially unstable areas, with mapping at scales between 1:2500 and 1:25000 were produced. The derivation and potential applications of these sets of maps and reports are described by Culshaw et al. (1990) who explain that they include basic data maps, derived maps and environmental potential maps. Typically such thematic map reports comprise a series of maps showing the bedrock and superficial geology, thickness of superficial deposits, groundwater conditions and areas of mining, fill, compressible, or other forms of potentially unstable ground. Maps showing factual information include the positions of boreholes or the positions of known mine workings. Derived maps include areas in which geological and / or environmental information has been deduced, and therefore is subject to some uncertainty. The thematic sets include planning advice maps showing the constraints on, and potential for, development and mineral extraction. Culshaw et al. (1990) also explained that these thematic maps were intended to assist with the formulation of both local (town or city), regional (metropolis or county) structure plans and policies, provide a context for the consideration of development proposals and facilitate access to relevant geological data by engineers and geologists. It was also recognised that these is a need for national (or state) policies and planning to be properly informed about geological conditions, not least to provide a sound basis for planning legislation and the issuing of advice and circulars. Examples of such advice include planning guidance notes concerning the granting of planning permission for development on potentially unstable land which were published (DOE, 1990, 1995) by the UK government. A further series of reports which were intended to assist planners and promote the consideration of geological information in land-use planning decision making were compiled between 1994 and 1998 by consultants on behalf of the UK government. Griffiths (2001) provides details of a selection of land evaluation techniques and relevant case studies.

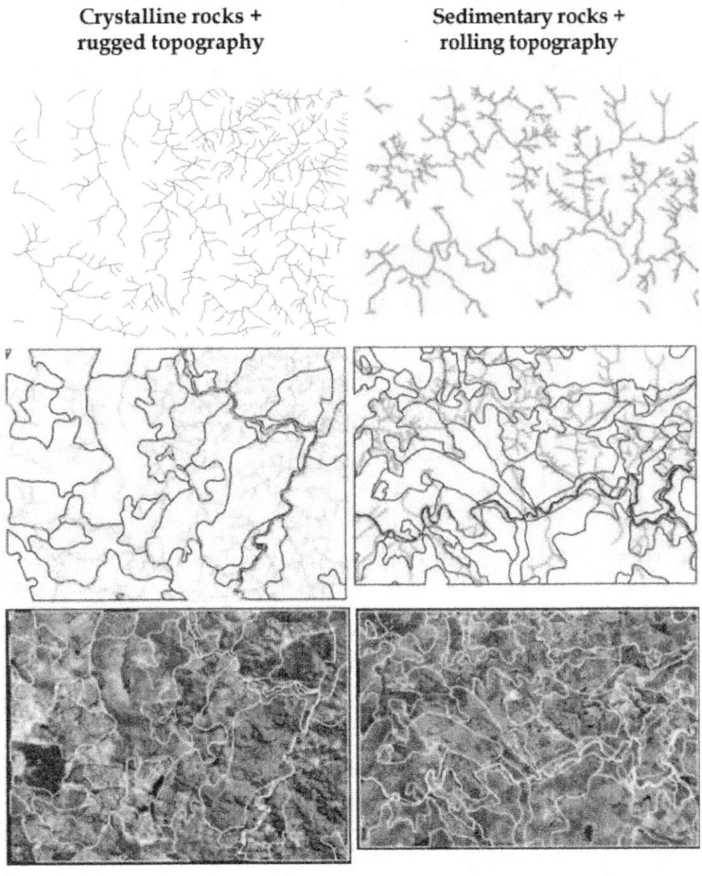

Figure. 1: Identification of basic compartmentalisation units (BCUs) based on textural differences on image. The image for crystalline rocks with rugged topography contrasts with sedimentary rocks with rolling topography. Top: Drainage network. Mid Row: Drainage network and delineated BCUs. Bottom: Composite Landsat TM5 image and delineated BCUs [after Fernandes da Silva et al. 2005b, 2010]

These covered the following themes:

- Environmental Geology in Land Use Planning: Advice for planners and developers (Thompson et al., 1998a)
- Environmental Geology in Land Use Planning: A guide to good practice (Thompson et al., 1998b)
- Environmental Geology in Land Use Planning: Emerging issues (Thompson et al., 1998c)
- Environmental Geology in Land Use Planning: Guide to the sources of earth science information for planning and development (Ellsion and Smith, 1998)

For an extensive review of world-wide examples of geological data outputs intended to assist with urban geology interpretation, land-use planning and utilisation and geological hazard avoidance, reference should be made to Culshaw and Price (2001). Three examples of terrain susceptibility mapping are briefly described and presented in this Section. The physiographic compartmentalisation technique for regional terrain evaluation was explored in these cases, and then terrain units were further characterised in geoenvironmental terms.

Multipurpose Planning

The first example concerns the production of a geohazard prevention map for the City of São Sebastião (IG/SMA 1996), where urban and industrial expansion in the mountainous coastal zone of São Paulo State, Southeast Brazil (Figure 2) led to conflicts in land use as well as to high risks to life and property. Particular land use conflicts arose from the combinations of landscape and economic characteristics of the region, in which a large nature and wildlife park co-exists with popular tourist and leisure encroached bays and beaches, a busy harbour with major oil storage facilities and associated pipelines that cross the area. Physiographic compartmentalisation was utilised to provide a regional terrain classification of the area, and then interpretations were applied in two ways: (i) to provide a territorial zoning based on terrain susceptibility in order to enable mid- to long-term land use planning; and (ii) to identify areas for semi-detailed hazard mapping and risk assessment (Fernandes da Silva et al. 1997a, Vedovello et al., 1997; Cripps et al., 2002). Figure 2 presents the main stages of the study undertaken in response to regional and urban planning needs of local authorities In the Land Susceptibility Map, the units were qualitatively ranked in terms of ground evidence and estimated susceptibility to geodynamic processes including gravitational mass movements, erosion, and flooding. Criteria for terrain unit classification in relation to erosion and mass movements (landslides, creep, slab failure, rock fall, block tilt and glide, mud and debris flow) were the following: a) soil weathering profile (thickness, textural and mineral constituency); b) hillslope profile; c) slope steepness; and d) bedrock structures (fracturing and discontinuities in general). Criteria in relation to flooding included: a) type of sediments; b) slope steepness; and c) hydrography (density and morphology of water courses). The resulting classes of terrain susceptibility can be summarised as follows: Low susceptibility: Areas where mass movements are unlikely. Low restrictions to excavations and man-made cuttings Some units may not be suitable for deep foundations or other engineering works due to possible high soil compressibility and presence of geological structures. In flat areas, such as coastal plains, flooding and river erosion are unlikely. Moderate susceptibility: Areas of moderate to high steep slope (10 to 30%) with little evidence of land instability (small-scale erosional processes may be present) but with potential for occurrence of mass movements. In lowland areas, reported flooding events were associated with the main drainage stream in relevant zones. Terrain units would possess moderate restrictions for land-use with minor engineering solutions and protection measures needed to reduce or avoid potential risks. High susceptibility: Areas of moderate (10 to 20%) and high steep slope (20 to 30%) situated in escarpment and footslope sectors, respectively, with evidence of one or more active land instability phenomena (e.g. erosion + rock falls + landslide) of moderate magnitude. Unfavourable zones for construction work wherein engineering projects would require accurate studies of structural stability, and consequently higher costs. In lowland sectors, recurrent flooding events were reported at intervals of 5 to 10 yrs, associated with main drainage streams and tributaries. Most zones then in use required immediate remedial action including major engineering solutions and protection measures.

Very high susceptibility: Areas of steeper slopes (> 30%) situated in the escarpment and footslope sectors that mainly comprised colluvium and talus deposits. There was evidence of one or more land instability phenomena of significant magnitude requiring full restriction on construction work. In lowland sectors, widespread and frequent flooding events at intervals of less than 5 years were reported and most land-used needed to be avoided in these zones. Units or areas identified as having a moderate to high susceptibility to geodynamic phenomena, and potential conflicts in land use, were selected for detailed engineering geological mapping in a subsequent stage of the study. The outcomes of the further stage of hazard mapping are described and discussed in Section 4.

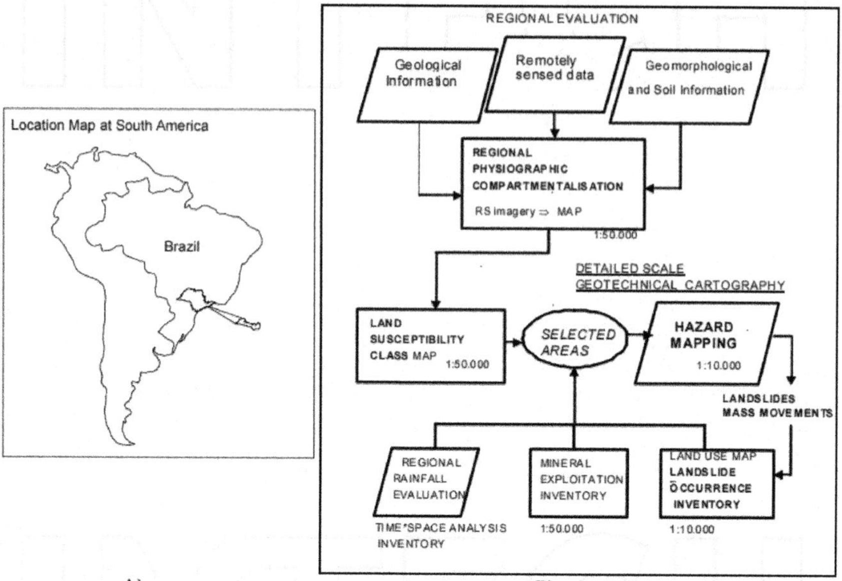

A) B)

Figure. 2: A) Location map for the City of São Sebastião, north shore of São Paulo State, Southeast Brazil. B) Schematic flow diagram for the derivation of the geohazard prevention chart and structural plan (after IG/SMA, 1996).

Watershed Planning and Waste Disposal

The physiographic compartmentalisation technique was also applied in combination with GIS tools in support of watershed planning in the Metropolitan District of Campinas, central-eastern São Paulo State (Figure 3). This regional screening study was performed at 1:50,000 scale to indicate fragilities, restrictions and potentialities of the area for siting waste disposal facilities (IG/SMA, 1999). A set of common characteristics and properties (also referred to as attributes) facilitated the assessment of each BCU (or terrain unit) in terms of susceptibility to the occurrence of geodynamic phenomena (soil erosion and land instability) and the potential for soil and groundwater contamination. As described by Brollo et al. (2000), the terrain units were mostly derived on the basis of qualitative and semi-quantitative inferences from satellite and air-photo images in conjunction with existing information (maps and well logs – digital and papers records) and field checks. The set of attributes included: (1) bedrock lithology; (2) density of lineaments (surrogate expression of underlying fractures and terrain discontinuities); (3) angular relation between rock structures and hillslope; (4) geometry and shape of hillslope (plan view and profile); (5) soil and covering material: type, thickness, profile; (6) water table depth; and (7) estimated permeability. These attributes were cross-referenced with other specific factors, including hydrogeological (groundwater production, number of wells per unit area), climatic (rainfall, prevailing winds), and socio-political data (land use, environmental restrictions). These data were considered to be significant in terms of the selection of potential sites for waste disposal.

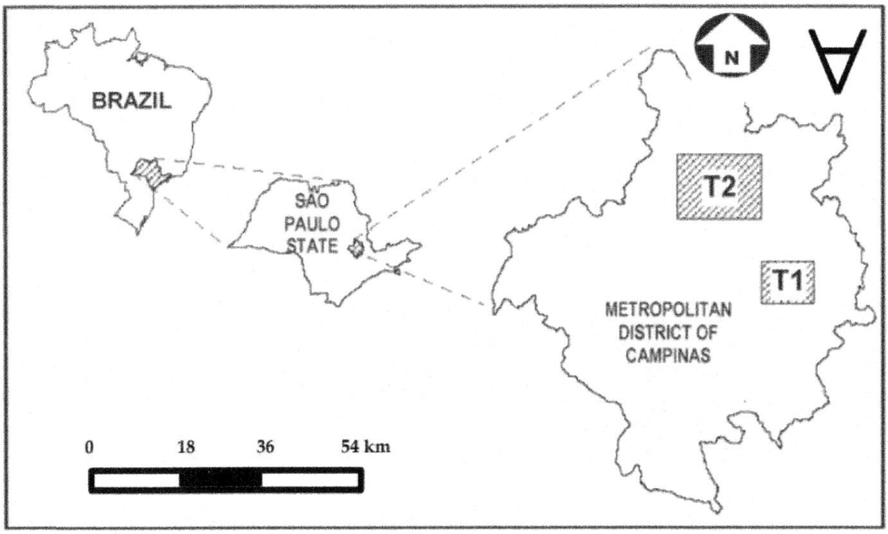

Figure. 3: Location map of the Metropolitan District of Campinas (MDC), central-eastern São Paulo State, Southeast Brazil (see Section 3.2). Detail map depicts Test Areas T1 and T2 within the MDC (see Section 3.3). Scale bar applies to detail map.

Figure 4 displays the study area in detail together with BCUs, and an example of a pop-up window (text box) containing key attribute information, as follows: 1st row - BCU code (COC1), 2nd - bedrock lithology, 3rd - relief (landforms), 4th – textural soil profile constituency, 5th - soil thickness, 6th - water table depth (not show in the example), 7th - bedrock structures in terms of density of fracturing and directionality), 8th - morphometry (degree of dissection of terrain). The BCU coding scheme expresses three levels of compartmentalisation, as follows: 1st letter – major physiographic or landscape domain, 2nd– predominant bedrock lithology, 3rd - predominant landforms, 4th– differential characteristics of the unit such as estimated soil profile and underlying structures. Using the example given in Figure 4, COC1 means: C = crystalline rock basement, O = equigranular gneiss, C = undulating and rolling hills, 1 = estimated soil profile (3 textural horizons and thickness of 5 to 10 m), underlying structures (low to moderate degree of fracturing, multidirectional). In terms of general interpretations for the intended purposes of the study, certain ground characteristics, such as broad valleys filled with alluvial sediments potentially indicate the presence water table level at less than 5 m below ground surface. Flood plains or concave hillside slopes that may indicate convergent surface water flows leading to potentially high susceptibility to erosion, were considered as restrictive factors for the siting of waste disposal facilities (Vedovello et al. 1998).

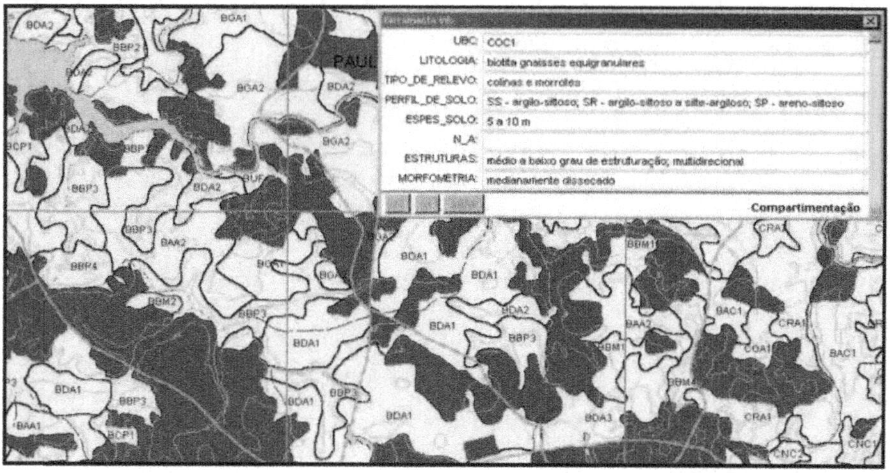

Figure. 4: Basic compartmentalisation units (BCUs) and pop-up window showing key attribute information relevant to BCUs. See text for details. [Not to scale] [after IG/ SMA, 1999]

Regional Development Planning

The third example is a territorial zoning exercise, in which terrain units delimited through physiographic compartmentalisation were further assessed in terms of susceptibility to land instability processes and groundwater vulnerability (Fernandes da Silva et al. 2005b). The study was conducted in two test areas situated in the Metropolitan District of Campinas (Figure 3) in order to assist State of São Paulo authorities in the formulation of regional development policies. It incorporated procedures for inferring the presence and characteristics of underlying geological structures, such as fractures and other discontinuities, then evaluating potential implications to ground stability and the flow of groundwater.

Details of image interpretation procedures for the delimitation of BCUs are described by Fernandes da Silva et al. (2010). The main image properties and image feature characteristics considered were as follows: (a) density of texture elements related to drainage and relief lines; (b) spatial arrangement of drainage and relief lines in terms of form and degree of organisation (direction, regularity and pattern); (c) length of lines and their angular relationships, (d) linearity of mainstream channel and asymmetry of tributaries, (e) density of interfluves, (f) hillside length, and (g) slope forms. These factors were mostly derived by visual interpretation of images, but external ancillary data were also used to assist with the determination of relief-related characteristics, such as slope forms and interfluve dimensions. The example given in Figure 1 shows sub-set images (Landsat TM5) and the basic compartmentalisation units (BCUs) delineated for Test Areas T1 and T2. Based on the principle that image texture correlates with properties and characteristics of the imaged target, deductions can be made about geotechnical-engineering aspects of the terrain (Beaumont and Beaven 1977, Beaumont 1985). The following attributes were firstly considered in the geo-environmental characterisation of BCUs: (a) bedrock lithology and respective weathered materials, (b) tectonic discontinuities (generically referred to as fracturing), (c) soil profile (thickness, texture and mineralogy), (d) slope steepness (as an expression of local topography), and (e) water table depth (estimated). Terrain attributes such as degree of fracturing, bedrock lithology and presence and type of weathered materials were also investigated as indicators of ground properties. For instance, the mineralogy, grain size and fabric of the bedrock and related weathered materials would control properties such as shear strength, pore water suction, infiltration capacity and natural attenuation of contaminants (Vrba and Civita 1994, Hudec 1998, Hill and Rosenbaum 1998, Thornton et al. 2001, Fernandes 2003). Geological structures, such as faults and joints within the rock mass, as well as relict structures in saprolitic soils, are also liable to exert significant influences on shear strength and hydraulic properties of geomaterials (Aydin 2002, Pine and Harrison 2003). In this particular case study, analysis of lineaments extracted from satellite images combined with tectonic modelling underpinned inferences about major and small-scale faults and joints. The approach followed studies by Fernandes and Rudolph (2001) and Fernandes da Silva et al. (2005b) who asserted that empirical models of tectonic history, based on outcrop scale palaeostress regime determinations, can be integrated with lineament analysis to identify areas: i) of greater density and interconnectivity of fractures; and ii) greater probability of open fractures; also to iii) deduce angular relationships between rock structures (strike and dip) and between these and hill slope directions. These procedures facilitated 3-dimensional interpretations and up-scaling from regional up to semi-detailed assessments which were particularly useful for assessments of local ground stability and groundwater flow. The BCUs were then classified into four classes (very high, high, moderate, and low) in terms of susceptibility to land instability and groundwater vulnerability according to qualitative and semi-quantitative rules devised from a mixture of empirical knowledge and statistical approaches. A spreadsheet-based approach that used nominal, interval and numerical average values assigned in attribute tables was used for this. A two-step procedure was adopted to produce the required estimates where, at stage one, selected attributes were analysed and grouped into three score categories (A - high, M - moderate, B - low B) according to their potential influence on groundwater vulnerability and land instability processes. In the second step, all attributes were considered to have the same relative influence and the final classification for each BCU was the sum of the scores A, B, M. The possible combinations of these are illustrated in Table 3. Figure 5 shows overall terrain classifications for susceptibility to land instability.

Table 3: Possible combinations of scores "A" (high), "M" (moderate), and "B" (low) respective to the four attributes (bedrock lithology and weathered materials, fracturing, soil type, and slope steepness) used for classification of units (BCUs) in terms of susceptibility to land instability and groundwater vulnerability.

Combinations of scores	Classification
AAAA	Very high
AAAM, AAAB, AAMM	High
AAMB, AABB, AMMM, AMMB, MMMM	Medium
AMBB, ABBB, MMMB, MMBB, MBBB, BBBB	Low

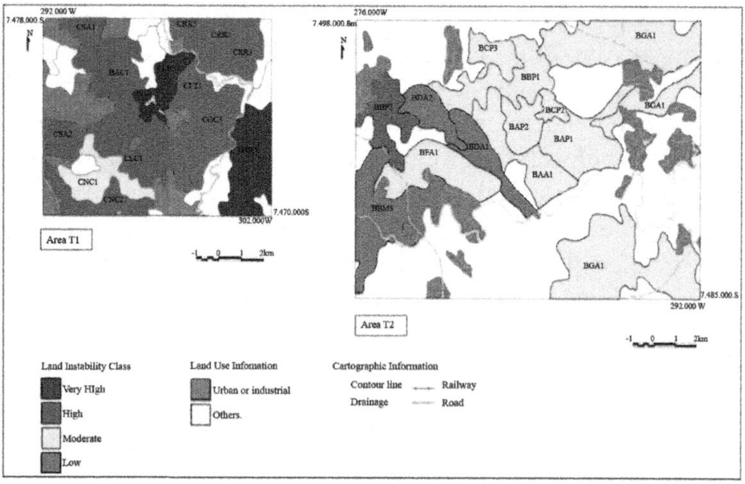

Figure. 5: Maps of susceptibility to land instability processes. Test Areas T1 and T2. UTM projection and coordinates [After Fernandes da Silva et al., 2010]

HAZARD MAPPING: LAND INSTABILITY AND FLOODING

In order to prevent damage to structures and facilities, disruption to production, injury and loss of life, public authorities have a responsibility to assess hazard mitigation and controls that may require remedial engineering work, or emergency and contingency actions. In order to accommodate these different demands, information about the nature of the hazard, and the consequences and likelihood of occurrence, are needed. Hazard maps aim to reduce adverse environmental impacts, prevent disasters, as well as to reconcile conflicting influences on land use. The examples given in this Section demonstrate the identification and zonation in terms of the likelihood and consequences of land instability and flooding hazards. There are several reasons for undertaking such work, for instance to provide public authorities with data on which to base structural plans and building codes as well as civil defence and emergency response programmes.

Application to Local Structural Plans

As indicated in Section 3.1, the BCUs (terrain units) classified as having a moderate to high susceptibility to geodynamic processes (mass movements and flood) were selected for further detailed engineering geological mapping. This was to provide data and supporting information to the structure plan of the City of São Sebastião. The attributes of the selected units were cross-referenced with other data sets, such as regional rainfall distribution, landuse inventory, and mineral exploitation records to estimate the magnitude and frequency of hazards and adverse impacts. Risk assessment was based on the estimated probability of failure occurrence and the potential damage thus caused (security of life, destruction of property, disruption of production). Both the triggering and the predisposing factors were investigated, and, so far as was possible, identified. It is worth noting the great need to consider socio-economic factors in hazard mapping and risk analysis. For instance, areas of consolidated housing and building according to construction patterns and reasonable economic standards were distinguished from areas of unconsolidated/expanding urban occupation. Temporal analysis of imagery and aerial photos, such as densities of vegetation and exposed soil in non-built-up areas, were utilised to supplement the land use inventory. The mineral exploration inventory included the locations of active and abandoned mineral exploitation sites (quarries and open pit mining for aggregates) and certain geotechnical conditions. Besides slope steepness and inappropriate occupancy and land use, the presence of major and minor geological structures was considered to be one of the main predisposing factors to land instability in the region studied. Figure 6 depicts a detail of the hazard map for the City of São Sebastião. Zones of land instability were delimited and identified by code letters that correspond with geodynamic processes as follows: A - landslides, B - creep, C - block tilt/glide, and D - slab failure/rock fall. Within these zones, landsliding and other mass movement hazards were further differentiated according to structural geological predisposing factors as follows: r – occurrence of major tectonic features such as regional faults or brittle-ductile shear zones; f – coincidence of spatial orientations between rock foliation, hillslope, and man-made cuttings; t – high density of fracturing (particularly jointing) in combination with coincidence of spatial orientations between fracture and foliation planes, hillslope, and manmade cuttings (Moura-Fujimoto et al., 1996; Fernandes da Silva et al. 1997b).

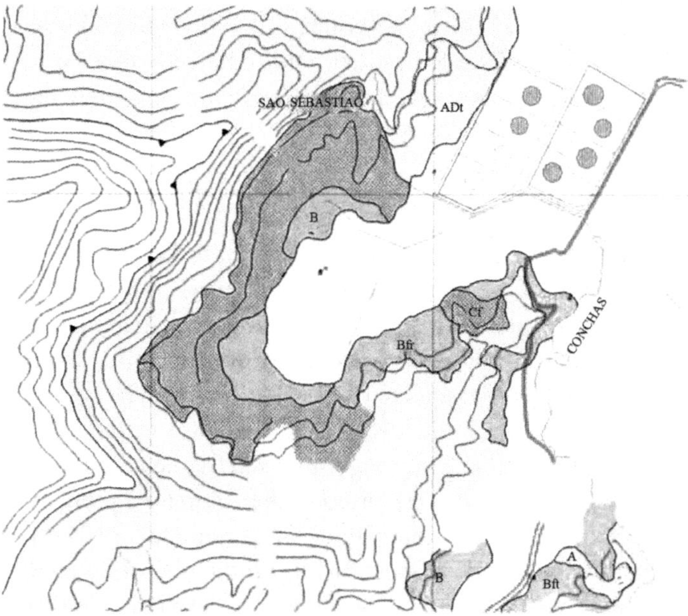

Figure. 6: Example of hazard map from the City of São Sebastião, north shore of São Paulo State, Southeast Brazil. Key for unit classification: Light red = very high susceptibility; Blue = high susceptibility; Light orange = moderate susceptibility; Yellow = low susceptibility. See Section 4.1 for code letters on geodynamic processes and predisposing factors. [after Fernandes da Silva et al. 1997b] (not to scale).

Application to Civil Defence and Emergency Response Programmes

Methods of hazard mapping can be grouped into three main approaches: empirical, probabilistic, and deterministic (Savage et al. 2004, as cited in

Tominaga, 2009b). Empirical approaches are based on terrain characteristics and previous occurrence of geodynamic phenomena in order to estimate both the potential for, and the spatial and temporal distribution of, future phenomena and their effects. Probabilistic approaches employ statistical methods to reduce subjectivity of interpretations. However, the outcomes depend very much on measured patterns defined through site tests and observations, but it is not always feasible to perform this acquisition of data in developing regions and countries. Deterministic approaches focus on mathematical modelling that aims quantitatively to describe certain parameters and rules thought to control physical processes such as slope stability and surface water flow. Their application tends to be restricted to small areas and detailed studies. In the State of Sao Paulo (Southeast Brazil), high rates of population influx and poorly planned land occupation have led to concentration of dwellings in unsuitable areas, thus leading to increasing exposure of the community to risk and impact of hazard events. In addition, over the last 20 years, landsliding and flooding events have been affecting an increasingly large geographical area, so bringing about damage to people and properties (Tominaga et al. 2009a). To deal with this situation, Civil Defence actions including preventive, mitigation, contingency (preparedness), and emergency response programmes have been implemented. The assessment of the potential for the occurrence of landslides, floods and other geodynamic processes, besides the identification and management of associated risks in urban areas has played a key role in Civil Defence programmes. To date, systematic hazard mapping has covered 61 cities in the State of São Paulo, and nine other cities are currently being mapped (Pressinotti et al., 2009). Examples that mix empirical and probabilistic approaches are briefly presented in this Section. The concepts of hazard mapping and risk analysis adopted for these studies followed definitions provided in Varnes (1984) and UN-ISDR (2004), who described risk as an interaction between natural or human induced hazards and vulnerable conditions. According to Tominaga (2009b), a semi-quantitative assessment of risk, R, can be derived from the product R = [H x (V x D)], where: H is the estimated hazard or likelihood of occurrence of a geodynamic and potentially hazardous phenomenon; V is the vulnerability determined by a number of physical, environmental, and socio-economic factors that expose a community and/or facilities to adverse impacts; and D is the potential damage that includes people, properties, and economic activities to be affected. The resulting risk, R, attempts to rate the damage to structures and facilities, injury and loss of lives, and disruption to production. The first example relates to hazard mapping and risk zoning applied to housing urban areas in the City of Diadema (Marchiori-Faria et al. 2006), a densely populated region (around 12,000 inhab. per km2) of only 31.8 km2, situated within the Metropolitan Region of the State Capital – São Paulo (Figure 7). The approach combined the use of high-resolution satellite imagery (Ikonos sensor) and ortho-rectified aerial photographs with ground checks. The aim was to provide civil defence authorities and decision-makers with information about land occupation and ground conditions as well as technical advice on the potential magnitude of instability and flooding, severity of damage, likelihood of hazard, and possible mitigating and remedial measures. Driving factors included the need to produce outcomes in an updateable and reliable manner, and in suitable formats to be conveyed to non-specialists. The outcomes needed to meet preventive and contingency requirements, including terrain accessibility, linear infrastructure conditions (roads and railways in particular), as well as estimations of the number of people who would need to be removed from risk areas and logistics for these actions. Risk zones were firstly identified through field work guided by local authorities. Site observations concentrated on relevant terrain characteristics and ground conditions that included: slope steepness and hillslope geometry, type of slope (natural, cut or fill), soil weathering profile, groundwater and surface water conditions, and land instability features (e.g. erosion rills, landslide scars, river undercutting). In addition, information about periodicity, magnitude, and effects of previous landsliding and flooding events as well as perceptions of potential and future problems were gathered through interviewing of residents. Satellite images were further used to assist with the identification of buildings and houses liable to be affected and the delineation of risk zone boundaries. Risk assessment was based on a qualitative ranking scheme with four levels of risk: R1 (low); R2 (moderate); R3 (high); R4 (very-high). Low risk (R1) zones, for example, comprised only predisposing factors to instability (e.g. informal housing and cuttings in steep slope areas) or to flooding (e.g. informal housing in lowland areas and close to watercourses but no reported flood within the last 5 years). Very-high risk (R4) zones were characterized by significant evidence of land instability (e.g. presence of cracks in soil and walls, subsidence steps, leaning of trees and electricity poles, erosion rills and ravines, landslide scars) or flooding hazards (e.g. flooding height marks on walls, riverbank erosion, proximity of dwellings to river channel, severe floods reported within the last 5 years). The outcomes, including basic and derived data and interpretations, were integrated and then presented on a geo-referenced computational system designed to respond the needs of data displaying and information management of the State of São Paulo Civil Defence authorities (CEDEC). As described by Pressinotti et al. (2007), such system and database, called Map-Risk, includes cartographic data, interpretative maps (risk zoning), imagery, and layers of cadastral information (e.g. urban street network). The system also enabled generation and manipulation of outputs in a varied set of text (reports), tabular (tables), and graphic information including photographic inventories for risk zones. The system was fully conceived and implemented at low cost, utilizing commercial software available that were customized in this visualisation system through target-script programming designed to achieve user functionalities (e.g. ESRI/MapObjects, Delphi, Visual Basic, OCX MapObjects). Examples of delineated risk zones for the City of Diadema and a display of the Map-Risk functionalities are presented in Figure 7. The second example refers to a flooding hazard mapping performed at regional and local scales in the Paraiba do Sul River Watershed, Eastern São Paulo State (Figure 8), in order to provide a rapid and comprehensive understanding of hazard phenomena and their impacts, as well as to enable application of procedures of data integration and mapping in different socio-economic contexts (Andrade et al. 2010). The information was systematised and processed to allow the build-up of a geo-referenced database capable of providing information for both environmental regional planning (economic-ecological zoning) and local scale hazard mapping for civil defence purposes. The regional evaluation covered all the 34 municipalities located in the watershed, and comprised the following stages of work: 1) survey of previous flooding events reported in newspaper and historical archives; 2) data systematisation and consolidation to translate gathered news into useful pieces of technical information; 3) identification of flooding occurrence locations using Google Earth tools; 4) cartographic auditing, geo-referencing and spatial data analysis using a freeware GIS package called SPRING (see Section 5); 5) exploratory statistical analysis of data; 6) preliminary flooding hazard classification on the basis of statistical results. Such preliminary classification used geopolitical (municipality) and hydrographical sub-basin boundaries as units for the analysis.

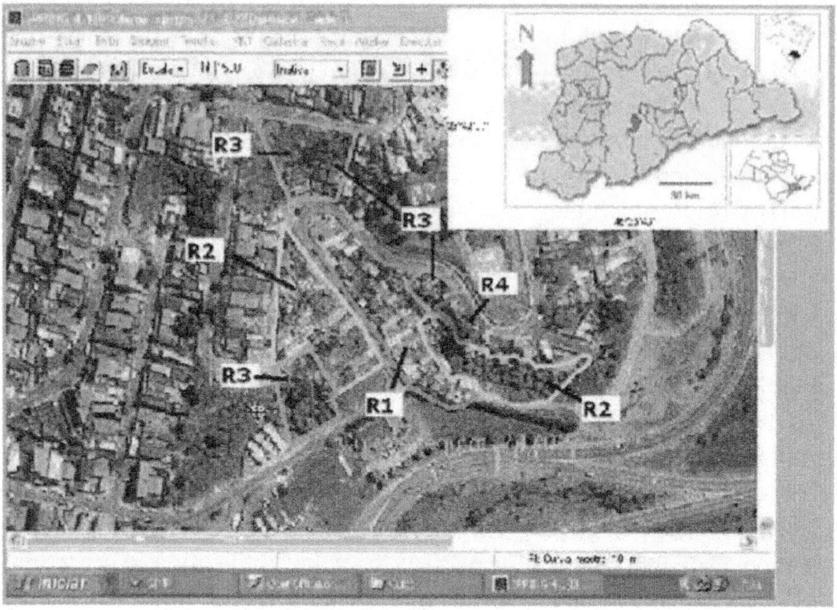

A)

B)

Figure. 7: A) Location of the City of Diadema in the Metropolitan District of São Paulo (State capital), Southeast Brazil and example of delineated risk zones over a high-resolution satellite image (Ikonos). B) Example of Map-Risk system display. See Section 4.2 for details. [after Marchiori-Faria et al., 2006; Pressinotti et al., 2007]

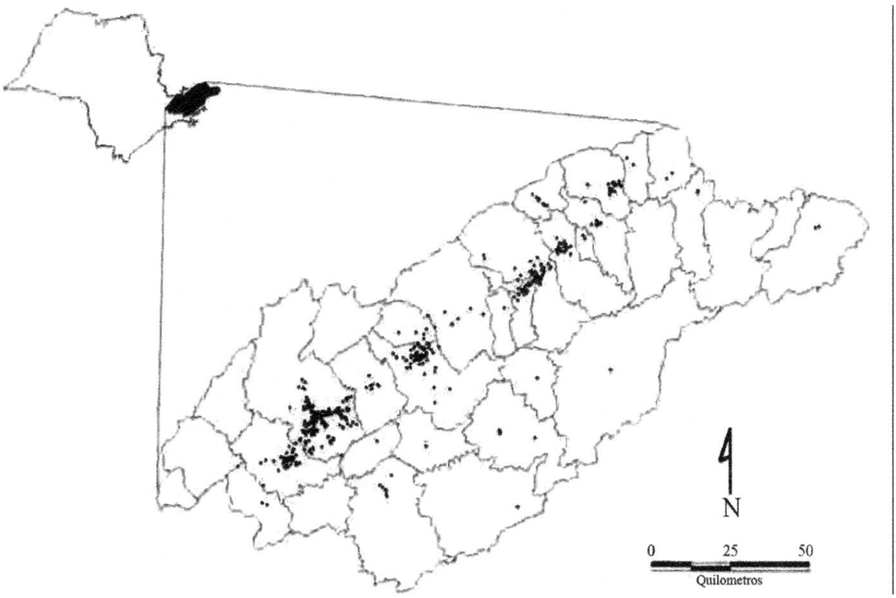

Figure. 8: Location of Paraiba do Sul River Watershed in Eastern São Paulo State and distribution of flooding occurrences. Internal sub-divisions correspond to geopolitical boundaries (municipalities). [After Andrade et al., 2010]

The regional evaluation was followed-up with detailed flooding hazard mapping (1:3,000 scale) in 7 municipalities, which included: a) ground observations - where previous occurrence was reported – to measure and record information on flooding height marks, land occupation, and local terrain, riverbank and water course characteristics; b) georeferencing and spatial data analysis, with generation of interpolated numerical grids on flooding heights and local topography; c) data interpretation and delimitation of flooding hazard zones; d) cross-referencing of hazard zones with land use and economic information leading to delimitation of flooding risk zones. Numerical scoring schemes were devised for ranking hazard and risk zones, thus allowing relative comparisons between different areas. Hazard zone scores were based on intervals of flooding height (observed and interpolated) and temporal recurrence of flooding events. Flooding risk scores were quantified as follows: R = [H x (V x D)], in which potential damage and vulnerability were considered (housing areas, urban infrastructure, facilities and services to be affected) on the basis of image interpretation and cross-referencing with land use maps and information. A detail map (yet unpublished) showing the interpolated grid of flooding heights and delineated hazard zones is presented in Figure 9.

A)

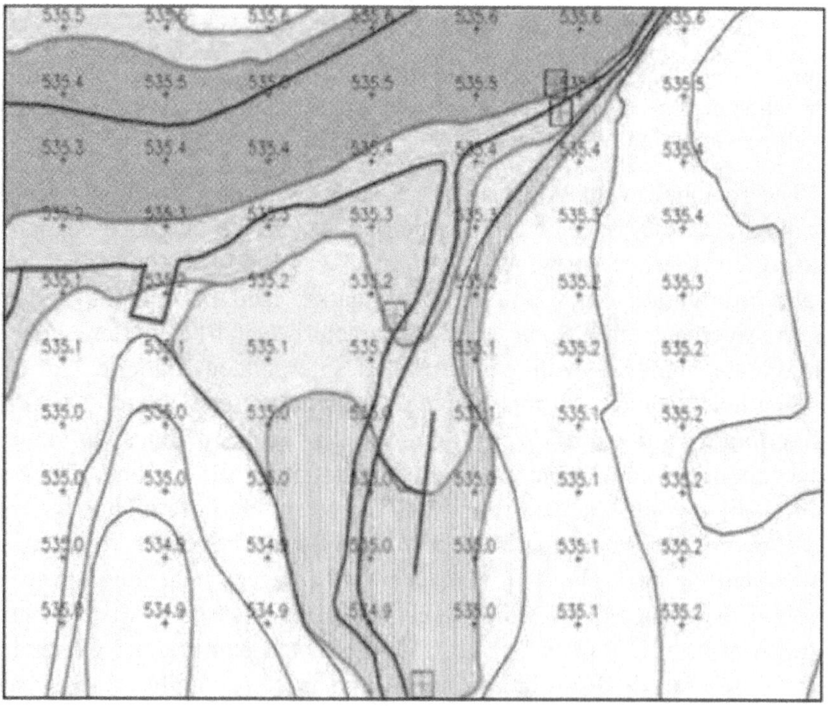

B)

Figure. 9: A) Measurement of maximum flood height for recent flooding event. B) Numerical interpolated grid of flooding heights and delineated flooding hazard zones.

Green = Low probability of occurrence, Estimated flooding heights (Efh) < 0.40 m. Yellow = Moderate probability, 0.40 < Efh < 0.80 m. Light Brown = High probability, 0.80 < Efh < 1.20 m. Red = Very high probability, Efh > 1.20 m. Ground observations and measurements: cross and rectangle. Continuous lines: black = topographic contour lines, blue = main river channel boundaries. Not to scale.

GEO-ENVIRONMENTAL ASSESSMENT: APPLICATIONS TO LAND RECLAMATION POLICIES

Land reclamation of sites of previous mineral exploitation frequently involve actions to minimize environmental damage and aim at re-establish conditions for natural balance and sustainability so reconciling former mined/quarried sites with their surroundings (Brollo et al., 2002). Strategies and programmes for land reclamation need to consider physical and biological characteristics of the local environment as well as socio-economic factors. Socioenvironmental regeneration, involves not only revegetation and land stabilisation engineering, but also rehabilitation or introduction of a new function for the area.

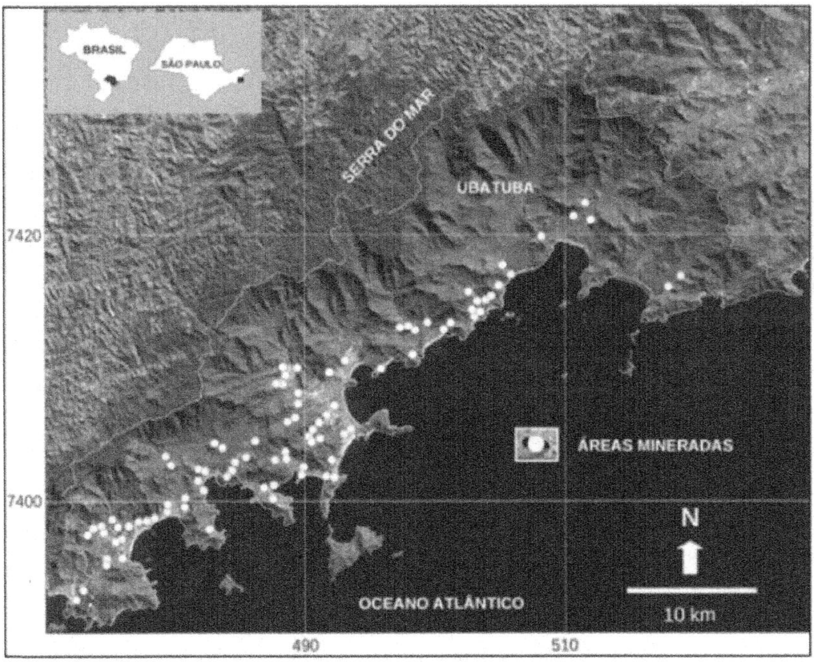

A)

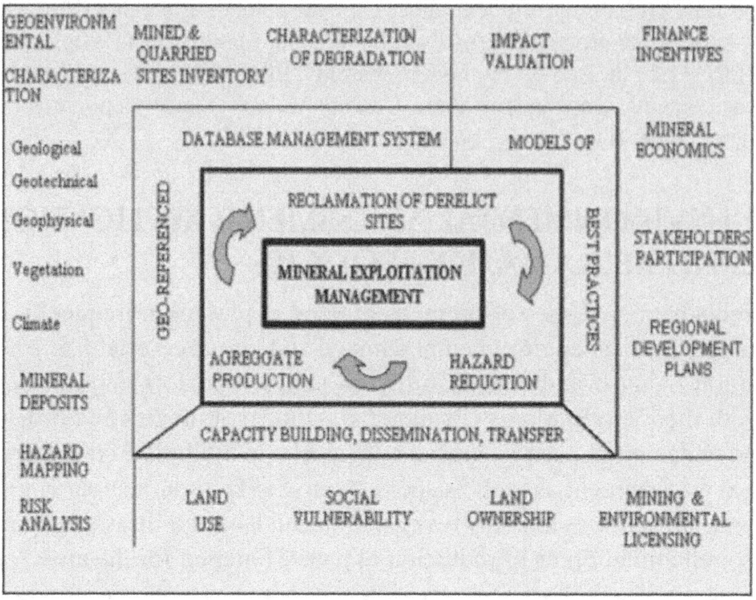

B)

Figure. 10: A) Location map and satellite image of the Municipality of Ubatuba (North Shore São Paulo State, Southeast Brazil). Dots on image represent quarried/mined sites. B) Schematic display of the integrated approach taken to reconcile mineral exploitation management and land reclamation. The scheme shows the three main issues to be addressed (centre) and topics of interest to be studied. [after Ferreira et al., 2006; Ferreira & Fernandes da Silva, 2008]

The case study concerns a GIS-based geo-environmental management scheme to reconcile sustainable mineral exploration of aggregates and construction materials with regeneration of abandoned and/or derelict mined sites in the municipality of Ubatuba (North Shore of São Paulo State, Brazil). Until the early 1990's intensive exploitation of residual soil and ornamental stone (for fill and civil construction) took place in an unplanned and unregulated manner. This led to highly adverse environmental impacts, including the creation of 114 derelict and abandoned sites which resulted in State and Federal authorities enforcing a virtual halt to mining activity in the region. Besides this, the municipality of Ubatuba is highly regarded for its attractive setting and landscape, including encroached coastline with sandy beaches and bays with growing leisure and tourism activities. The area encompasses the Serra do Mar Mountain Range covered by large remnants of Atlantic Forest so that approximately 80% of the municipal territory lies within a nature and wildlife reserve (Figure 10A). As described by Ferreira et al. (2005, 2006), the devised strategy required an integrated approach (Figure 10B) in order to address three key issues: 1) environmental recovery of a number of derelict (abandoned, unsightly) sites; 2) reduction of hazards (land instability, erosion, flooded areas etc), particularly at those sites informally occupied by low income populations; and 3) rational exploitation of materials for local building materials corresponding to local needs. The study was implemented using a freeware GIS and image processing package called SPRING (Câmara et al. 1996, INPE 2009) and ortho-rectified air photos (1-metre resolution, taken in 2001, leading to an approximate scale of 1:3,000). The key output of the land management strategy was a prioritisation ranking scheme based on a comprehensive site critical condition (ICR) score, which synthesised the significance of each factor or issue to addressed (IG/SMA, 2008; Ferreira et al., 2009). Accordingly, the score system consisted of three numerical indicators: 1) environmental degradation indicator (IDE), 2) mineral potential indicator (IPM), and 3) hazard/risk indicator (IRI). Each indicator was normalised to a scalar range (0 to 1), and the ICR was the sum of the three indicators. The ICR was then used to set up directives and recommendations to advise local and State authorities about the possible measures to be taken, through mid- and long-term policies and/or immediate remedial and mitigating actions. According to Ferreira et al. (2008), the IDE comprised four component criteria to estimate the degree of adverse environmental impact (or degradation) of the individual mineral extraction sites: erosional features, terrain irregularity, herbaceous and bushy vegetation, and exposed soil. Information on these factors was acquired from imagery and ground checks. Tracing of linear features on images was investigated as an indicator of the frequency and distribution of erosional processes (rills, ravines, piping scars) as well as for terrain irregularity. In the first case, the sum of linear features representative of erosional processes was ratioed by the area of each site to quantify the estimate. Similarly, linear features related to the contour of cutting berms, rill marks, and slope breaks caused by mining/quarrying activity, were also measured to quantify terrain irregularity. The areal extent of herbaceous and bushy vegetation as well as exposed soil were also delimited on images. The IPM, as described by Ferreira & Fernandes da Silva (2008), was achieved by means of the following procedures: 1) identification and delimitation of quarried/mined sites (polygons) on geo-referenced imagery; 2) derivation of local DEM (digital elevation model) from topographic contour lines to each delimited site; 3) calculation of local volume to material (V1) based on the original geometry of the quarried/mined sites; 4) calculation of volume of material already taken (V2) and exploitable volume of material (V3), so that [V3] = [V1 – V2]; 5) application of classification rules based on legal environmental and land use restrictions. The calculation of volumes of material (residual soil and ornamental stone) was performed by means of GIS operations involving polygons (areas) and numerical grids of topographic heights generated with nearest neighbour interpolator in the SPRING package (Figure 11). The IRI was derived from R = [H x (V x D)] – see Section 4.2 – focussing on mass movement and flood hazards and their consequences to people, property and economic activity. According to Rossini-Penteado (2007) and Tominaga et al. (2008), Hazard, H, was quantified according to the spatial and temporal probability of occurrence of each phenomenon and then weighted in relation to areal distribution of such probabilities (percentage of sq. km). The vulnerability, V, was computed by means of scores assigned to socio-economic aspects such as nature of built structures, spatial regularity of land occupation, presence of urban infrastructure (e.g. water supply, sanitation, health services, refuse collection and disposal method), road/street network, educational and income patterns. Similarly, in order to estimate the extent of potential damage, D, numerical scores were devised and attributed to the estimated number of people and buildings per unit area, and to the proportion of built area in relation to total area of the site.

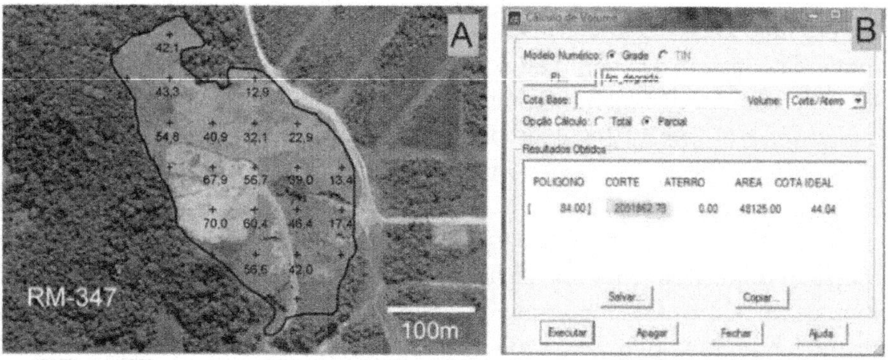

Figure. 11: Example of the mineral potential indicator (IPM) for the Municipality of Ubatuba (North Shore São Paulo State, Southeast Brazil). A) Abandoned/quarried site delimited on image and numerical grid of topographic heights. B) Screen display from GIS-based computation of volumes of exploitable material [after Ferreira & Fernandes da Silva, 2008].

Figure 12 illustrates the application of the ICR scoring scheme to mined/quarried sites in Ubatuba. In summary, 47% of sites were classified as very low priority, 12% as low priority, 19% as moderate priority, 15% as high priority (18 sites), and 7% as very high priority (8 sites). The priorities represent a combination of availability of exploitable volumes of building materials and the need for measures to tackle adverse environmental impacts and high risk situations (Figure 12). Based on the application of the ICR scores and current land use, directives and recommendations for land reclamation and socio-regeneration of mined/quarried sites were consolidated into ten main groups (IG/SMA, 2008; Ferreira et al., 2009). Such directives and recommendations ranged from simple measures such as routine maintenance and cleaning, revegetation with grass and control of water surface flow (including run-off) to the implementation of leisure and multi-purpose public facilities, major land stabilisation projects combined with mineral exploitation, and monitoring.

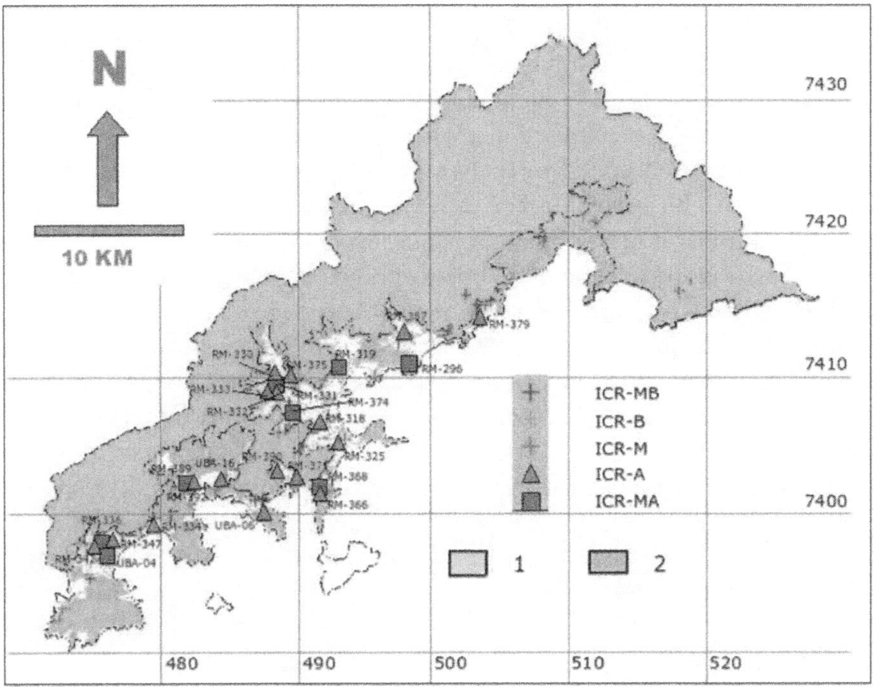

Figure. 12: Spatial distribution of mined/quarried sites classified according to critical condition score scheme (ICR). Sites classified as High (triangles) and Very High (squares) priority are highlighted. Remarks: 1- Serra do Mar Nature and Wildlife State Park, 2- Environmentally sensitive protected areas. ICR-MB = Very Low priority, ICR-B = Low priority, ICR-M = Moderate priority, ICR-A = High priority, ICR-MA = Very high priority [after IG/SMA, 2008].

CONCLUSIONS

Geo-environmental terrain classification may be used as part of the land-use planning decision making and may also provide the basis of responses to emergency situations. In most examples presented here, classification schemes were based on knowledge of the bedrock geology, topography, landforms, superficial geology (soil and weathered materials), groundwater conditions and land-uses. Information for the classification has been variously derived from remote sensing and fieldwork rather than specific site investigations. A framework for carrying out a terrain classification at different scales has been presented. In practice, the effectiveness of land zoning system requires the implementation of planning controls. To do this the Local Authority needs adequate resources and an appropriate legal or planning guidance policy framework. Preferably, the control process should be based on the principle that permission will be given unless there is a good reason for refusal. In granting permission, conditions may then be applied to ensure the safety of the development with regard to landsliding, flooding and other potential problems. However it must be recognised that where practical control over development cannot be exercised, other preventive, mitigative or advisory measures may be all that can be used. Marker (1996) explains that the rate and style of development has a major impact on the information requirements. Without a rigorously enforced planning framework based on accurate information about the ground conditions very rapid urban development will generally lead to construction on areas of less stable land or land which may be subject to hazards such as flooding or pollution. This type of development may also result in the sterilisation of geological resources which it would be expedient to exploit as part of the development process. On the other hand restricting development to designated areas will generally require detailed information about the ground conditions and likelihood and potential impact of hazards to be available at the planning stage. It also assumes existence of the resources and will to enforce the plan. Such models can severely constrain the social and economic development of an area, lead to excessively high population densities and give rise to problems associated with the re-use of previously developed land. Policy formulation may incorporate incentives (e.g. subsidies and reduced taxes) to be provided by local and state governments to encourage such measures and good practice, which can be viewed as kinds of voluntary or induced control. In some cases financial controls exerted by public funding as well as mortgage and insurance providers are the means by which some types of development may be curtailed but such controls may not prevent informal occupancy of hazardous areas. In this regard, some of the examples presented here, from a regional and local perspective, have also demonstrated environmental management regulations may have little meaning in some urban areas subject to rapid expansion where, because of population influx, informal housing is virtually an inevitable consequence. Due to the latter hazard mapping updating and post-episode monitoring [failure episodes] are absolutely vital as these procedures facilitate a contemporary understanding of ground conditions and risk circumstances, which can be essential to provide timely and efficient advice for mitigation and control of hazards as well as to design effective contingency actions and engineering solutions.

ACKNOWLEDGEMENTS

The authors are very much indebted to the Sao Paulo State Geological Institute and its staff for most of the examples provided here.

REFERENCES

1. Akiwumi, F.A., Butler, D.R. (2008). Mining and environmental change in Sierra Leone, West Africa: a remote sensing and hydrogeomorphological study. Environmental Monitoring Assessment, 142: 309–318. ISSN 0167-6369
2. Alonso Herrero, E.; Frances, E., & Cendrero, A. (1990). Environmental geological mapping in Cantabrian Mountains, Spain. In: Proceedings of the 6th Congress of the Intl. Assoc. Engineering Geology and Environment, Amsterdam. Balkema, Rotterdam, 1: 31-35.
3. Andrade, E.; Danna, L.C.; Santos, M.L., & Fernandes da Silva, P.C. (2010). Survey of flooding occurrence in newspaper records as a support for regional planning and hazardmapping. Proceedings of 7th Brazilian Symposium on Geotechnical and Geoenvironmental Cartography. ISSN 2178-1834. Maringa, October 2010. 16p. [in Portuguese]
4. Aydin, F. (2002). Heterogeneity and behaviour of saprolitic slopes. Proceedings of the 9thInternational Congress of the Intl. Assoc. Engineering Geology and Environment, ISBN 0- 620-28559-1, Durban, September 2002. p. 846-856.
5. Barton, J., Alexander, D., Correa, C., Mashelkar, R., Samuels, G., & Thomas S. (2002). Integrating intellectual property rights and development policy. London: UK Department for International Development, Commission on Intellectual Property Rights.
6. Beaumont, T.E. (1985). An application of satellite imagery for highway maintenance andrehabilitation in Niger. International Journal of Remote Sensing, 6 (7): 1263-1267, ISSN 0143-1161.
7. Beaumont, T.E. & Beaven, P.J. (1977). The use of satellite imagery for highway engineering in overseas countries. (T.R.R.L. - Transport & Road Research Laboratory) England, Supplementary Report 279, 19 p.
8. Beckett, P.H.T. & Webster, R. (1969). A Review of Studies on Terrain Evaluation by the Oxford-MEXE-Cambridge Group,1960-1969, MEXE

Report 1123, Christchurch Bell, F.G. (1993). Engineering Geology. Blackwell, Oxford. 269p.

9. Bennett, M.R & Doyle, P. (1997). Environmental Geology. John Wiley and Sons Ltd, Chicester. Bourne, R. (1931). Regional Survey and its Relation to Stocktaking of the Agricultural Resources of the British Empire, Oxford Forestry Memoirs No.13.

10. Bowman, I. (1911). Forest Physiography, Physiography of the US and Principal Soils in Relation to Forestry, John Wiley and Sons, New York.

11. Brollo M.J., Barbosa, J.M., Rocha, F.T., & Martins, S.E. (2002). Research Programme on Characterisation and Reclamation of Environmentally Degraded Areas. Proceedings of the 5th Meeting on Environmental Management Research. São Paulo: CINP/SMA,2002, p. 74-82. [in Portuguese]

12. Brollo, M.J., Vedovello, R., Gutjahr, M.R., Hassuda, S., Iritani, M.A., Fernandes da Silva, P.C., & Holl, MC. (2000). Criteria for selection of areas for waste disposal in regional scale (1:100,000). Application area: Metropolitan Region of Campinas, SaoPaulo State, Brazil. Proceedings of the 8th Congress of the Intl. Assoc. Engineering Geology and Environment. Vancouver, September 1998, 6: 4281-4286.

13. Brook D. & Marker B.R. (1987) Thematic geological mapping as an essential tool in land-use planning. In: Culshaw MG, Bell FG, Cripps JC, O'Hara M (eds) "Planning and Engineering Geology", Engineering Geology Special Publication No. 4. Geological Society, London, pp 211–214.

14. Câmara, G. & Fonseca, F. (2007) Information Policies and Open Source Software in Developing Countries. Journal of the American Society for Information Science and Technology, 58(1): 121–132, ISSN 1532-2890.

15. Câmara, G., Souza, R.C.M., Freitas, U.M., & Garrido, J. (1996). SPRING: Integrating remote sensing and GIS by object-oriented data modelling. Computers & Graphics, 20 (3): 395-403, ISSN 0097-8493.

16. Cendrero, A., Flor, G., Gancedo, R., González-Lastra, J.R., Omenaca, J.S., & Salinas, J.M.(1979). Integrated Assessment and Evaluation of the Coastal Environment of the Province of Vizcaya, Bay of Biscay, Spain. Environmental Geology, 2 (6): 321-331, ISSN 0943-0105.

17. Christian, C.S. (1958). The concepts of land units and land systems. In: Proceedings of the 9th Pacific Science Congress. Vol. 20, pp. 74-81. Christian, C.S. & Stewart, GA. (1952). Summary of general report on Survey of Katherine - Darwin Region. CSIRO Land Research Series No.1

18. Christian, C.S. & Stewart, GA. (1968). Methodology of integrated surveys. Proceedings of theConference on Aerial Surveys and Integrated Studies. UNESCO, Paris. pp. 233-280

19. Cripps, J.C., Fernandes Da Silva, P.C., Culshaw, M.G., Bell, F.G., Maud, R.R., & Foster, A. (2002). The planning response to landslide hazard in São Paulo State - Brazil,

20. Durban - South Africa and Antrim - Northern Ireland. Proceedings of 9thInternational Congress of the Intl. Assoc. Engineering Geology and Environment, ISBN 0- 620-28559-1, Durban, September 2002. p. 1841-1852.

21. Culshaw M.G., Foster, A., Cripps, J.C., & Bell FG (1990). Applied maps for land use planning in Great Britain. Proceedings of the 6th International Congress of the Intl.

22. Assoc. Engineering Geology and Environment. AA Balkema, Rotterdam, 1: 85-93. Culshaw M.G., & Northmore K.J. (2002) An engineering geological map for site investigation planning and construction type identification. In: Van Rooy JL, Jermy

23. CA (eds) Proceedings of the 9th International Association for Engineering Geology and the Environment Congress, Durban, 423–431. South African Institute of Engineering and Environmental Geologists, Pretoria. On CD-ROM.

24. Culshaw M.G. & Price S.J. (2011) The 2010 Hans Cloos lecture: the contribution of urban geology to the development, regeneration and conservation of cities. Bulletin of Engineering Geology and the Environment, 70, Part 3. [In press]

25. Dai, F.C., Lee, C.F., & Zhang, X.H. (2001). GIS-based environmental evaluation for urbanland-use planning: a case study. Engineering Geology, 61 (4): 257 – 271, ISSN 0013- 7952.

26. Dearman, W.R. & Matula, M. (1977). Environmental aspects of engineering geological mapping. Bulletin of the Intl. Assoc. Engineering Geology and Environment, 14: 141-146. ISSN 1435-9529.

27. DOE - Department of the Environment. (1990). Development on unstable ground. Planning Policy Guidance PPG14, HMSO, London.

28. DOE - Department of the Environment. (1995) Development on unstable ground: Landslides and Planning. Planning Policy Guidance PPG14 Annex 1, HMSO, London.

29. Ellison, R. & Smith, A. (1998). Environmental Geology in Land Use Planning: Guide to the sources of earth science information for planning

and development. Published by British Geological Survey (Nottingham, UK) on behalf of Department of Environment, Transport and the Regions, Great Britain.

30. Fernandes, A.J. (2003) The influence of Cenozoic Tectonics on the groundwater vulnerability in fractured rocks: a case study in Sao Paulo, Brazil. Proceedings of the International Conference on Groundwater in Fractured Rocks, Prague, IAH/UNESCO, p.81-82.
31. Fernandes, A.J. & Rudolph, D.L. (2001) The influence of Cenozoic tectonics on the groundwater-production capacity of fractured zones: a case study in Sao Paulo, Brazil. Hydrogeology Journal, 9: 151-167, ISSN 1431-2174.
32. Fernandes Da Silva, P.C., Cripps, J.C., & Wise, S.M. (2005a). The use of Remote Sensing techniques and empirical tectonic models for inference of geological structures: bridging from regional to local scales. Remote Sensing of Environment, 96 (1): 18 – 36. ISSN 0034-4257.
33. Fernandes da Silva, P.C., Ferreira, C.J., Fernandes, A.J., Brollo, M.J., Vedovello, R., Tominaga, L.K., Iritani, M.A., & Cripps, J.C. (2005b)
34. Evaluation of land instability and groundwater pollution hazards applying the physiographic compartmentalisation technique: case study in the Metropolitan District of Campinas, Brazil. Proceedings of the 11th Brazilian Congress on Environmental and Engineering Geology, ISBN 85-7270-017-X. Florianopolis, November 2005. Pp. 383 –402. [In Portuguese]
35. Fernandes da Silva, P.C., Maffra, C.Q.T., Tominaga, L.K., & Vedovello, R. (1997a). Mapping units on São Sebastião Geohazards Prevention Chart, Northshore of São Paulo State, Brazil. In: Environmental Geology, Proceedings of the 30th International Geological Congress. Daoxian, Y. (Ed.), 24: 266-281, VSP Scientific Publisher, ISBN 90-6764-239-8. Utrecht.
36. Fernandes da Silva, P.C., Moura-Fujimoto, N.S.V., Vedovello, R., Holl, M.C., & Maffra, C.Q.T. (1997b). The application of structural geological data in mass movement hazard zoning at the municipality of São Sebastião, Northshore of São Paulo State, Brazil. Proceedings of the 6th Brazilian Symposium on Tectonic Studies, Goiania, May 1997. p. 136-138. [in Portuguese]
37. Fernandes da Silva, P.C., Vedovello, R., Ferreira, C.J., Cripps, J.C., Brollo, M.J., & Fernandes, A.J. (2010) Geo-environmental mapping using physiographic analysis: constraints on the evaluation of land instability and groundwater pollution hazards in the Metropolitan District

of Campinas, Brazil. Environmental Earth Sciences, 61: 1657–1675. ISSN 1866-6280.

38. Ferreira, C.J.; Brollo, M.J.; Ummus, M.E.; & Nery, T.D. (2008). Indicators and quantification of environmental degradation in mined sites, Ubatuba (SP). Revista Brasileira de Geociências, 38(1): 141-152. ISSN 0375-7536 [In Portuguese]

39. Ferreira, C.J. & Fernandes da Silva, P.C. (2008). The use of GIS for prioritising mineral exploitation of building materials in derelict sites: the Ubatuba – SP case. Revista do Instituto Geológico, 29 (1/2): 19-31. ISSN 0100-929X [In Portuguese] Ferreira, C.J., Fernandes da Silva, P.C., & Brollo, M.J. (2009). Directives for socioregeneration of sites affected by mining activity, Ubatuba, Sao Paulo State.

40. Proceedings of the 11th Southeast Brazil Geological Symposia, ISSN 2175-697X, São Pedro, November 2009. pp. 123-124.

41. Ferreira, C.J., Fernandes da Silva, P.C., Brollo, M.J., & Cripps, J.C. (2006). Dereliction problems from exploitation of residual soil and ornamental stone at Sao Paulo State, Brazil. Proceedings of the 10th International Congress of the International Association of Engineering Geology and Environment. Nottingham , September 2006. CD-ROM. 10p.

42. Ferreira, C.J., Fernandes da Silva, P.C, Furlan, S.A., Brollo, M.J., Tominaga, L.K., Vedovello, R., Guedes, A.C.M., Ferreira, D.F., Cripps, J.C., & Perez, F. (2005). Devising strategies for reclamation of derelict sites due to mining of residual soil (Saibro) at Ubatuba, North coast of Sao Paulo State, Brazil: the views and roles of stakeholders. Sociedade e Natureza , Special Issue on Land Degradation, p.643 - 660, ISSN 0103-1570.

43. Finlayson, A. (1984). Land surface evaluation for engineering practices: Applications of the Australian P.U.C.E. system for terrain analysis. Quarterly Journal of Engineering Geology, 17(2): 149-158, ISSN 1470-9236

44. Grant, K. (1968). A Terrain Evaluation System for Engineering. CSIRO, Div. Soil. Mech. Technical Paper No. 2, Melbourne. 27p.

45. Grant, K. (1974). The PUCE Programme for Terrain Evaluation System for Engineering Purposes. II. Procedures for Terrain Classification. CSIRO, Div. Soil. Mech. Technical Paper No. 15, Melbourne. 68p.

46. Grant, K. (1975). The PUCE Programme for Terrain Evaluation System for Engineering Purposes. I. Principles. CSIRO, Div. Soil. Mech. Technical Paper No. 19, Melbourne. 32p.

47. Griffiths, J.S. & Edwards, R.J.G. (2001). The development of land surface

evaluation for engineering practice. In Griffiths, J S (Ed.) Land Surface Evaluation for Engineering Practice. Geological Society, London, Engineering Geological Special Publication, 18, 3 – 9.

48. Griffiths, J S (Ed.) 2001 Land Surface Evaluation for Engineering Practice. Geological Society, London, Engineering Geological Special Publication, 18.
49. Guy, M. (1966). Quelques principles et quelques experiences sur la metodhologie de la photointerpretation. Proceedings of the 2nd International Symposium on Photointerpretation. Paris, 1966. v.1: 2-41.
50. Hill, S.E. & Rosenbaum, M.S. (1998). Assessing the significant factors in a rock weathering system. Quarterly Journal of Engineering Geology, 31: 85-94, ISSN 0013-7952.
51. Howard, A.D. (1967) Drainage analysis in geologic interpretation: a summation. Bulletin of the American Association of Petroleum Geologists, 51: 2246-2254, ISSN 0149-1423.
52. Hudec, P.P. (1998) Rock properties and physical processes of rapid weathering and deterioration. Proceedings of the 8th International Congress of the Intl. Assoc. Engineering Geology and Environment, Vancouver, September 1998. p.335-341.
53. IG/SMA – Instituto Geológico/ São Paulo State Secreariat of Environmet. (1996). São Sebastião Geohazards Prevention Chart, Northshore of São Paulo State, Brazil. São Paulo, Technical Report, 77p. [in Portuguese] IG/SMA - Instituto Geológico/ São Paulo State Secreariat of Environmet. (1999).
54. Methodology for Selection of Suitable Areas for Treatment and Disposal of Solid Waste: Regional Approach. São Paulo, IG/SMA Technical Report, 98p. [in Portuguese]
55. IG/SMA - Instituto Geológico/ São Paulo State Secreariat of Environmet. (2003). Evaluation of land instability and groundwater pollution hazards in the Metropolitan District of Campinas, Brazil. São Paulo, Foreign Commonwealth Office Intl. Cooperation Programme. Technical Report, 57p.
56. IG/SMA - Instituto Geológico/ São Paulo State Secreariat of Environmet. (2008). Directives for Socio-Environmental Regeneration of Derelict Sites Due to Mining of Building Materials, Ubatuba, Northshore of São Paulo State, Brazil. São Paulo, FAPESP – Public Policies Research Programme, Funding Grant n. 03/07182-5. Final Tech. Sci. Report, 168p. [in Portuguese]
57. INPE - Brazilian National Institute for Space Research (2009) SPRING

5.0. http://www.dpi.inpe.br/spring/english/download.php. Cited in 11 December 2009. Kiefer, R.W. (1967). Terrain analysis for metropolitan fringe area planning, Journal of the Urban Planning and Development Division, Proceedings of the American Society of Civil Engineers, UP4, 93, 119-39.

58. Latifovic R., Fytas K., Chen J., & Paraszczak J. (2005) Assessing land cover change resulting from large surface mining development. International Journal of Applied Earth Observation and Geoinformation, 7: 29–48, ISSN 0303-2434.

59. Lillesand, T. & Kiefer, R.W. (2000) Remote Sensing and Image Interpretation. 4th Ed., John Wiley, New York. Mabbut, J.A. 1968. Review of concepts on land classification. In: Land Evaluation. Stewart, G.A. (ed.), p. 11-28, Macmillan, Melbourne.

60. Marchiori-Faria, D.G., Ferreira, C.J., Fernandes da Silva, P.C., Rossini-Pentado, D., & Cripps, J.C. (2006). Hazard mapping as part of Civil Defence preventative and contingency actions: A case study from Diadema, Brazil. Proceedings of the 10th International Congress of the International Association of Engineering Geology and Environment. Nottingham , September 2006. CD-ROM. 10p.

61. Marker, B.R. (1996). Role of the earth sciences in addressing urban resources and constraints In: McCall, G J H ; de Mulder, E F J and Marker, B R (Eds). Urban Geoscience. AGID Special Publication Series No 20 in association with Cogeoenvironment. Balkema (Rotterdam), pp 163-180.

62. Matula, M. (1979). Regional engineering geological evaluation for planning purposes. Bulletin of the Intl. Assoc. Engineering Geology and Environment, 18: 18-24. ISSN 1435-9529

63. Matula, M. & Letko, V. (1980). Engineering geology in planning the metropolitan region of Bratislava. IAEG Bull.,. Bulletin of the Intl. Assoc. Engineering Geology and Environment, 19: 47-52. ISSN 1435-9529

64. Meijerink, A.M. (1988). Data acquisition and data capture through terrain mapping units. ITC Journal, International Journal of Applied Earth Observation and Geoinformation 1: 23-44. ISSN 03032434

65. Miliaresis, G.Ch. (2001). Geomorphometric mapping of Zagros Range at regional scale. Computer and Geosciences, 27 (7): 775 – 786. ISSN 0098-3004. Mitchell, C.W. (1991). Terrain Evaluation. 2nd Ed. Longman, Essex. 497p.

66. Moura-Fujimoto, N.S.V., Holl, M.C., Vedovello, R., Fernandes da Silva, P.C, & Maffra, C.Q.T. (1996). The identification of mass movement

hazard zones at the municipality of São Sebastião, Northshore of São Paulo State, Brazil. Proceedings of the 2nd Brazilian Symposium on Geotechnical Cartography. São Paulo, November 1996. p. 129-137. [in Portuguese]

67. Nedovic-Budic, Z. (2000). Geographic Information Science Implications for Urban and Regional Planning. URISA Journal, 12(2): 81-93. ISSN 1045-8077.

68. Pine, R.J. & Harrison, J.P. (2003). Rock mass properties for engineering design. Quarterly Journal of Engineering Geology and Hydrogeology, 36: 5–16. ISSN1470-9236.

69. Porcher, M. & Guillope, P. (1979). Cartographie des risques ZERMOS appliquées a des plans d'occupation des sols en Normandie. Bull. Mason Lab. des Ponts et Chausses, 99.

70. Pressinotti, M.M.N., Guedes, A.C.M, Fernandes da Silva, P.C, Sultanum, H.J., & Guimarães, R.G. (2007). Automated system for displaying of gravitational mass movement and flooding hazard mapping of São Paulo State. Proceedings of the 6th Brazilian Symposium on Geotechnical and Geo-environmental Cartography. Uberlândia, November 2007. p. 324 – 333. [in Portuguese]

71. Pressinotti, M.M.N, Fernandes da Silva, P.C.; Marchiori-Faria, D.G., & Mendes, R.M. (2009). The experience of the Geological Institute in hazard maping and risk analysis in urban environments. Proceedings of the 11th Southeast Brazil Geological Symposia. São Pedro, November 2009. ISSN 2175-697X. P. 156-157. [in Portuguese] Rossini-Penteado, D., Ferreira, C.J., & Giberti, P.P.C. (2007). Quantification of vulnerability and potential damage applied to hazard mapping and risk analysis, 1:10,000,

72. Ubatuba-SP. Proceedings of the 2nd Brazilian Symposium on Natural and Technological Disasters. Santos, December 2007. [in Portuguese]

73. Sabins Jr., F.F. (1987). Remote Sensing. Principles and Interpretation. Freeman, New York. Soares, P.C. & Fiori, A.P. (1976). Systematic procedures for geological interpretation of aerial photographs. Notícia Geomorlógica 16 (32): 71-104. [In Portuguese] Soares, P.C., Mattos, J.T., Barcellos, P.E., Meneses, P.R., & Guerra, S.M.S. (1981). Regionalmorpho-structural analysis on Radar and Landsat images at the Parana Basin. Proceedings of the 3rd South Brazilian Regional Symposium on Geology, Curitiba. p. 5-23.[In Portuguese]

74. Thompson, A., Hine, P.D., Poole J.S. & Greig, J.R. (1998a) Environmental Geology in LandUse Planning: Advice for planners and developers.

Published by Symonds Travers Morgan (East Grinstead, UK) on behalf of Department of Environment, Transport and the Regions, Great Britain.

75. Thompson, A., Hine, P.D., Poole J.S. & Greig, J.R. (1998b) Environmental Geology in Land Use Planning: A guide to good practice. Published by Symonds Travers Morgan (East Grinstead, UK) on behalf of Department of Environment, Transport and the Regions, Great Britain.

76. Thompson, A., Hine, P.D., Greig, J.R. & Poole J.S. (1998c) Environmental Geology in Land Use Planning: Emerging issues. Published by Symonds Travers Morgan (East Grinstead, UK) on behalf of Department of Environment, Transport and the Regions, Great Britain.

77. Thornton, S.F., Lerner, D.N., & Banwart, S.A. (2001) Assessing the natural attenuation of organic contaminants in aquifers using plume-scale electron and carbon balances: model development with analysis of uncertainty and parameter sensitivity. Journal of Contaminant Hydrology, 53(3-4): 199-232. ISSN 0169-7722.

78. Tominaga, L.K., Rossini-Penteado, D., Ferreira, C.J., Vedovello, R. & Armani, G. (2008). Landsliding hazard assessment through the analysis of multiple geo-environmental factors. Proceedings of the 12th Brazilian Congress on Environmental and Engineering Geology. ABGE, ISBN 978-85-7270-052-8. Porto de Galinhas, November 2008. 15p. [In Portuguese]

79. Tominaga, L.K. (2009a). Natural Disasters: Why they happen ? In: Disasters: Knowing to prevent. Tominaga, L.K., Santoro, J., & Amaral, R (eds). Pp. 147-160, Geological Institute, ISBN 978-85-87235-09-1, São Paulo. [In Portuguese]

80. Tominaga, L.K. (2009b). Risk Analysis and Mapping. In: Disasters: Knowing to prevent. Tominaga, L.K., Santoro, J., & Amaral, R (eds). Pp. 13-23, Geological Institute, ISBN 978-85-87235-09-1, São Paulo. [In Portuguese]

81. TRRL - Transport & Road Research Laboratory. (1978). Terrain evaluation for highway engineering and transport planning: a technique with particular value for developing countries. TRRL Supplementary Report No. 448, 21p.

82. UN-ISDR – United Nations – International Strategy for Disaster Reduction. (2004). Living with Risk: A Global Review of Disaster Reduction Initiatives. In: Inter-Agency Secretariat International Strategy for Disaster Reduction. Geneve, 152p. Accessed in Aug, 2009. Available at http://www.unisddr.org.

83. Varnes, D.J. (1984). Landslide Hazard Zonation: Review of Principles and Practice. Unesco Press, Paris, 56p.

84. Vedovello, R. (2000). Geotechnical Zoning for Environmental Management through Basic Physiographic Units. (unpublished PhD Thesis, Sao Paulo State University at Rio Claro). [In Portuguese]

85. Vedovello, R. (1993). Geotechnical Zoning based on Remote Sensing Techniques for Urban and Regional Planning. São José dos Campos, 186p. (unpublished M.Sc. Dissertation, National Institute for Space Research) [In Portuguese]

86. Vedovello, R. & Mattos, J.T. (1998). The use of basic physiographic units for definition of geotechnical units: a remote sensing approach. Proceedings of the 3rd Brazilian Symposium on Geotechnical Cartography. São Paulo, November 1998, 11p. [In Portuguese]

87. Vedovello, R., Brollo, M.J., & Fernandes da Silva, P.C. (1998). Assessment of erosion as a conditioning factor on the selection of sites for industrial waste disposal: an approach based on regional physiographic compartmentalisation obtained from satellite imagery. Proceedings of the 3rd Brazilian Symposium on Erosion Control. Presidente Prudente, March 1998. 9p. [In Portuguese]

88. Vedovello, R., Tominaga, L.K., Moura-Fujimoto, N.S.V., Fernandes da Silva, P.C., Holl, M.C., & Maffra, C.Q.T. (1997). A mass movement hazard analysis approach for public authority responses. In: Proceedings of the 1st Latin American Forum on Applied Geography. Curitiba, September 1997. 9p. [In Portuguese]

89. Verstappen HTh (1977) Remote Sensing in Geomorphology. Elsevier Sci. Publ., Amsterdam. 214p.

90. Vinogradov, B.V., Gerenchuk, K.I., Isachenko, A.G., Raman, K.G., & Teselchuk, Y.N. (1962). Basic principles of landscape mapping, Soviet Geography: Review and Translation, 3 (6), 15-20.

91. Vrba J. & Civita, M. (1994) Assessment of groundwater vulnerability. In: Guidebook on Mapping Groundwater Vulnerability. Vrba, J. & Zaporozec, A. (eds.). 16: 31-48, International Association of Hydrogeologists (IAH).

Chapter 2

MINERAL PROSPECTIVITY MAPPING METHOD INTEGRATING MULTI-SOURCES GEOLOGY SPATIAL DATA SETS AND CASE-BASED REASONING

Binbin He[1], Jianhua Chen[2], Cuihua Chen[3], Yue Liu[3]

[1]School of Resources and Environment, University of Electronic Science and Technology of China, Chengdu, China

[2]College of Geophysics, Chengdu University of Technology, Chengdu, China

[3]College of Geosciences, Chengdu University of Technology, Chengdu, China

ABSTRACT

Extracting and synthesizing information from existing and massive amounts of geology spatial data sets is of great scientific significance and has considerable value in its applications. To make mineral exploration less expensive, more efficient, and more accurate, it is important to move beyond traditional concepts and establish a rapid, efficient, and intelligent method of predicting the existence and location of minerals. This paper describes a case-based reasoning (CBR) method for mineral prospectivity mapping that takes spatial features of geology data into account and offers an intelligent approach. This method include a metallogenic case representation that combines spatial and attribute features, metallogenic case-based storage organization, and a metallogenic case similarity retrieval model. The experiments were performed in the eastern Kunlun Mountains, China using CBR and weights-of-evidence (WOE), respectively. The results show that the prediction accuracy of the CBR is higher than that of the WOE.

INTRODUCTION

Mineral prospectivity analysis and quantitative resource estimation have been recognized as important when integrating multi-source geology spatial data in recent years [1]. The statistical and mathematical approaches developed recently for multi-resources geological spatial data integration include weights-of-evidence (WOE) [2-8], and the logistic regression [9,10]. The fuzzy logic [11, 12], artificial neural networks [13,14] and the Fractal method [15] have been applied in the assessment of mineral resources potential. Although these methods promote the efficiency and effectiveness of mineral resource prospecting, their algorithms are unable to accumulate knowledge, and lack intelligent reasoning. Meanwhile, similar deposit types occur in similar geological conditions and spatial distributions. The metallogenic geological conditions and spatial distribution of discovered and typical deposits can be used to construct a historical case-base for mineral prospectivity analysis. Traditional analysis methods cannot mine the depth of information or make intelligent inferences. In recent years, some researchers have begun applying case-based reasoning (CBR) to the environment, urban planning, and land use.

Lekkas et al. [16] suggested using CBR to predict air pollution levels; Holt and Benwell [17] tried using CBR to classify soil; Ye et al. [18] integrated CBR and GIS for urban planning approval; and Du et al. [19] applied CBR for land use change prediction. CBR is a branch of artificial intelligence that began in the research of Schank and Abelson [20]. CBR does not require a precise domain model, and it solves new cases by using historical knowledge. Its application is based on two assumptions about the objective world: 1) similar problems have similar solutions; and 2) similar problems may recur. CBR uses the principle of similarity to find strategies for new cases; it also offers a method that resembles the human problem-solving approach of extracting and storing expertise. From a methodological point of view, CBR proposes a comprehensive, problem-oriented approach to analysis that is more adaptable than rule-based and model-based reasoning. CBR is particularly suitable for areas in which it is difficult to summarize, abstract, and express expertise; this makes CBR useful for solving ambiguous problems. CBR can do quantitative analysis and prediction without a careful mechanism study, and it has advantages in the simplification of knowledge acquisition, the improvement of efficiency and quality, and the accumulation of knowledge. Additionally, CBR and the identification process are highly automated and reusable. CBR is an effective method in cases in which prior knowledge is lacking or for constructing complex issues in quantitative models.

In this paper, a method for mineral prospectivity mapping was proposed integrating multi-sources geology spatial data sets and case-based reasoning, including a metallogenic case representation model that combines spatial and attribute features, the metallogenic case-feature weights-determining model, metallogenic case-based storage organisation, and a metallogenic case-similarity retrieval model. The experiment was performed in the eastern Kunlun Mountains, China to predict the existence of potential iron deposits using case-based reasoning and weights-of-evidence, respectively.

METHODOLOGY

The mineral prospectivity mapping method using casebased reasoning include three main components: a metallogenic case representation model, metallogenic case storage, and a metallogenic case retrieval model. Figure 1 describes the flow of mineral prospectivity mapping method using CBR.

Metallogenic Case Representation Model

Generally, a case in a traditional CBR model is composed of both attribute and goal features. Because of the spatial distribution and regional laws of geological entities, the case representation is different from a traditional one. The features of a metallogenic case include both spatial and attribute features, which are selected or extracted from metallogenic entities.

During the construction of a metallogenic case representation model, each grid of a certain size is taken as a representative object. First, typical feature attributes related to ore control that are contained in vector grids of existing mineral points are extracted. Then, the corresponding names of mines in vector grids and relevant result values are determined. The extracted-features attribute, the corresponding names, and the relevant results are all described by the rules of case expression. To extract spatial features, the orientation relations, the metric relations, and the topology relations related to ore control in each vector grid are extracted, and spatial relations are transformed to attribute mode. Therefore, a metallogenic case consists of general attributes and spatial-relation property items. The basic expression is as follows:

$$C=\left(A_{a1},A_{a2},\cdots,A_{ak},A_{s1},A_{s2},\cdots A_{sm},\text{Result}\right) \tag{1}$$

where A_{ai} is the general feature property item, A_{sj} is the spatial-relation feature property item, and Result represents the result of the case. To solve a new case, existing cases can be extracted by spatial relations under certain rules (e.g., spatial coding). After that, candidates for a historical case set are obtained.

Metallogenic Case Storage

After a typical metallogenic case is constructed, it is stored in a spatial database in database tables or into document systems in a text file. The stored cases are then indexed to improve the efficiency of the metallogenic case-similarity retrieval model.

Metallogenic Case Retrieval Model

Because a metallogenic case has spatial features, it is different than a traditional CBR model. First, during the construction of a metallogenic case retrieval model, all vector grids are set as unsolved cases under the metallogenic case representation model. In other words, each case describes typical attribute and spatial features, and the results description (i.e., the case-determining attribute) is set to blank. Second, a similarity-measure threshold is set, and each unsolved case is retrieved for similarity. After a similar case is found, its result is assigned to the unsolved case according to the threshold and the strategy given. If the case obtained is unsatisfactory, it can be modified by expertise. Its result can then be assigned to the unsolved case. The retrieval unsolved cases in all vector grids are then completed. Third, the typical cases obtained or modified can be stored into the case base for expansion and update.

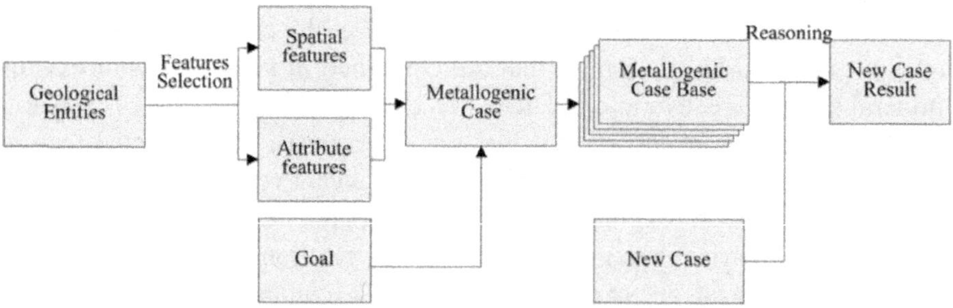

Figure 1: The flow of mineral prospectivity mapping method using CBR.

After a metallogenic case base is constructed, the metallogenic case retrieval model (Figure 2) can use it to compare existing metallogenic cases with new ones. The similarity measurement formulas for existing and new cases are as follows:

$$S_{\%} = \left(100 * \left(1 - \text{sqrt}\left(\frac{\text{distance}}{\text{sum(weights)}}\right)\right)\right) * \left(\frac{\text{searchedWeightsSum}}{\text{totalWeightsSum}}\right) \quad (2)$$

$$\text{distance} = \text{weight}_1 * \text{dist}_1^2 + \text{weight}_2 * \text{dist}_2^2 + \cdots + \text{weight}_n * \text{dist}_n^2 \quad (3)$$

$$\text{dist} = \min\left(1, \frac{\text{diff}(\text{newCaseValue, caseValue})}{(\text{maxValue} - \text{minValue}) * \text{inf initiyCons tan} t}\right) \quad (4)$$

where "$S_{\%}$" is similarity ranged between 0% and 100%; "distance" is the weighted sum of the squares of "$dist_i$" ranged between 0 and 1; "searchedWeightsSum" is the sum of the weights, with the new case feature and the actual case feature both being non-empty; "totalWeightsSum" is the sum of the weights of all case features; "$dist_i$" is the distance between the new case feature and the actual case feature, in which the value is the smaller of either 1 or the Euclid distance between the new case feature and the actual case feature; "newCaseValue" is the new case feature value; "caseValue" is the actual case feature value; "maxValue" and "minValue" are the case corresponding feature's maximum and minimum values; and "infinityConstant" is a large constant.

To measure similarity, each new case is compared with all cases in the case base. The return value is based on the selection strategies of the maximum, threshold, or K nearest neighbors. If the value is unsatisfactory, it can be modified by the return value and relevant expertise. The typical cases obtained and the cases modified can be stored into the case base for expansion.

EXPERIMENTS

To verify the effectiveness of the proposed method, the experiments of mineral potential prediction for iron deposits were performed in the eastern Kunlun Mountains, using the metallogenic CBR model and the weights-ofevidence model, respectively. All of the data sets used in this paper were derived from our established multisource geology spatial database, which contains geological, geophysical, geochemical, and remote-sensing data. The metallogenic CBR model was implemented with C# based on ArcEngine GIS components. The weights-of-evidence model was performed with Arc-SDM [21].

Geological Setting of Study Area

The eastern Kunlun Mountains are within Qinghai Province, China, and are shown as an insertion from left to right to the provincial map (Figure 3). The Mountains are within latitudes 34°57′ and 37°56′N, and longitudes 90°31′ and 100°04′E. Of the study area, the eastern Kunlun orogenic belt is attached to the southern margin of the Qaidam Basin.

The area consists of three major deep crustal-scale faults that divide the area roughly from north to south into subtectonic belts (Figure 4). Kunbei ("Kun" is short for Kunlun. "bei" means north in Chinese) belt is in the north. It belongs to the Kunbei Caledonian back-arc basin situated mainly in the northwestern part of the Kunlun Mountains. The belt is made of early Palaeozoic folding belts dominated by the Ordovician marine sediments

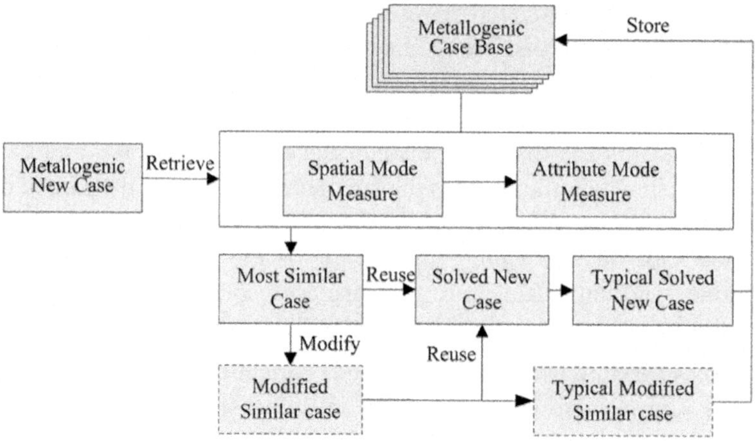

Figure 2: Metallogenic case similarity retrieval model.

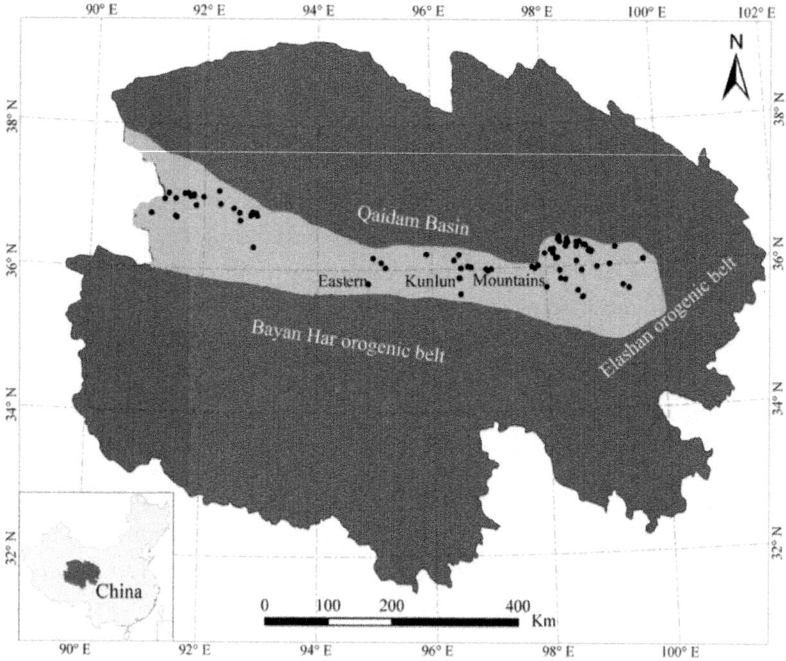

Figure 3: Eastern Kunlun Mountains within Qinghai Province, China.

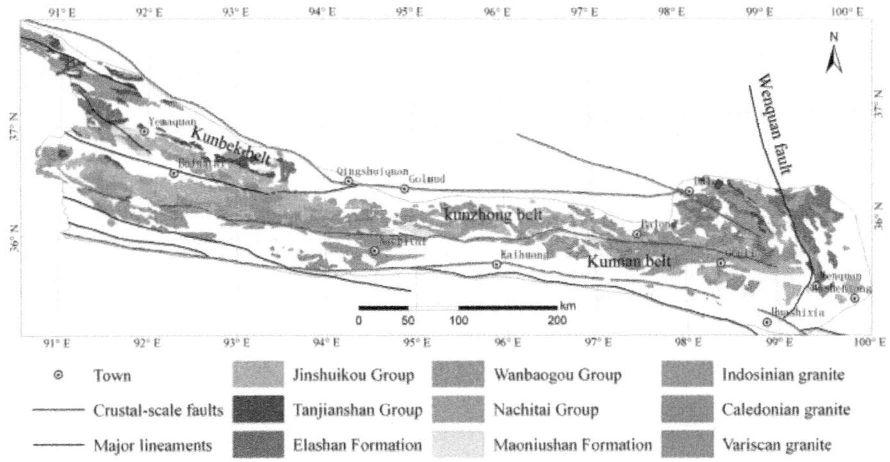

Figure 4: A simplified geological map of the eastern Kunlun Mountains, China.

Note: Showing major lithologic units, stratigraphic units, and crustal-scale faults. The east Kunlun orogenic belt is subdivided into the Kunbei belt, and the Kunnan belt. and low-grade metamorphic rock. Kunzhong ("zhong" is middle in Chinese) belt is the basement of an uplift belt and a granitic belt. It is made predominantly of the middle to late Proterozoic metamorphic sequences, and Palaeozoic and Mesozoic granitic rock. The Devonian continental sandstones, conglomerate, and volcanic rock, and Carboniferous marine limestone and sedimentary rock lie over the metamorphic and plutonic basement. The composition of Kunnan ("nan" means south in Chinese) belt is geologically similar to that of the Kunzhong belt, but it consists of numerous Triassic successions. As of today, there are 81 known sites of iron formation within the area. Their locations are shown as black dots in Figure 3.

Within the study area, regions exposed mainly by lithologic and stratigraphic units are displayed inFigure 4. The Jinshuikou Group is the oldest crystalline basement that comprises gneiss, amphibolite rock, migmatite, and marble. It belongs to a suite of middle-to-high grade of metamorphic rock [22]. The Tanjianshan Group of the Ordovician-Cambrian period is composed of intermediate-mafic volcanic rock, and phyllite crystalline limestone and sandstone. The Elashan Formation of the Triassic time consists basically of volcanic rock that is intermediate-acid. The rock is with sandstone intercalation. The Wanbaogou Group of the New Mesoproterozoic period is subdivided into an upper unit and a lower one, comprising mainly carbonate rock and intermediate-mafic volcanic rock. Both types of rock belong to the pre-Cambrian folding basement with a low-grade of metamorphism [23].

The volcanic rock and carbonaceous slate of the Wanbaogou Group serve as important ore beds of precious metal (e.g., gold) and non-ferrous metals (copper, cobalt, and nickel) in the Kunnan belt [24]. The Nachitai Group of the Ordovician period consists largely of schist, mafic volcanic rock, chert, and crystalline limestone. The Maoniushan Formation of the Devonian time is composed of an intermediate-acid volcanic rock underlain by clastic rock. The Variscan-Indosinian granite is closely associated with the metalliferous mineralization in the region when the granite occurs extensively, diversely and permanently [25]. Known iron mineralization occurred mainly in the Yemaquan metallogenic belt located in the western part of the Mountains, whereas the Dulan-Elashan metallogenic belt lies in the east.

Data Preprocessing and Metallogenic Case Construction

The best ore-controlling variable and threshold were determined using proximity analysis of the weights-ofevidence model. On the basis of a correlation analysis among evidence variables, the authors selected vector ore-controlling data of stratum, unconformity, fault, regional geochemical data, remote-sensing mineralization information, Bouguer gravity data, aeromagnetic dataand mineral occurrence for this experiment. Before constructing the specific metallogenic case, the region was partitioned into 96,576 grids, each one being 1 km by 1 km. All of the evidence-variable data were spatially joined to a grid polygon, and each grid had corresponding feature-attribute values. The unconformity, fault, and mineral occurrences were buffered by with distances of 3000 m, 300 m, and 1000 m, respectively. To extract the spatial features of the metallogenic case, the fault's direction (orientation relationship), the shortest distance between mineral occurrence and faults (metric relationship), the disjoint relationship between mineral occurrence and faults, and the unconformities (topology relationship) were computed and extracted in each grid polygon. The spatial relationships were then transformed into attributes and stored in the attribute tables of each grid polygon. This process paved the way for metallogenic case retrieval. In this way, the metallogenic representation model combined with spatial and attribute features was constructed. Each grid polygon became a case representation object. By analyzing each grid layer's attribute table in tandem with ore-controlling factors, the authors established the metallogenic case's attribute features by using lithological characters, chronostratigraphy, unconformity, fault, regional chemical anomaly, remote-sensing mineralisation anomaly, Bouguer gravity anomaly, and aeromagnetic anomaly. In this way, specific genetic types became object attributes. The metallogenic case representation model in this research is as follows:

C = (unconformity, regional geochemical anomaly, Bouguer gravity anomaly, aeromagnetic anomaly, chronostratigraphy, lithological characters, remote-sensing mineralisation anomaly, fault characters, fault directions, short distance to fault, distance to unconformity, disjoint fault, disjoint unconformity, genetic type).

Prior to analysis, the attribute and spatial features of the above case are set corresponding weights, which are determined and assigned based on the Analytic Hierarchy Process (AHP) [26]. On the basis of expert knowledge, the importance of AHP case features is as follows: regional geochemical anomaly > fault directions > short distance to fault = disjoint fault > fault characters = remote-sensing mineralisation anomaly > chronostratigraphy = lithological characters > distance to unconformity = disjoint unconformity > unconformity characters = Bouguer gravity anomaly = aeromagnetic anomaly. Table 1 shows the comparison matrix of metallogenic CBR features by AHP. The matrix is equalised and simplified to seven features. After calculating, uniformity has been passed and each feature weight determined; identical, important features have the same weights (Table 2).

To grid the polygonal layers that are overlapped by attribute and spatial features, the authors analysed each grid

Table 1: The CBR features comparison matrix by AHP

	B1	B2	B3	B4	B5	B6	B7	B8	B9	B10	B11	B12	B13
B1	1	1/7	1	1	1/3	1/3	1/4	1/4	1/6	1/5	½	1/5	1/2
B2		1	7	7	7/3	7/3	7/4	7/4	7/6	7/5	7/2	7/5	7/2
B3			1	1	1/3	1/3	1/4	1/4	1/6	1/5	½	1/5	1/2
B4				1	1/3	1/3	1/4	1/4	1/6	1/5	½	1/5	1/2
B5					1	1	3/4	3/4	3/6	3/5	3/2	3/5	3/2
B6						1	3/4	3/4	3/6	3/5	3/2	3/5	3/2
B7							1	1	4/6	4/5	4/2	4/5	4/2
B8								1	4/6	4/5	4/2	4/5	4/2
B9									1	6/5	6/2	6/5	6/2
B10										1	5/2	1	5/2
B11											1	2/5	1
B12												1	5/2
B13													1

B1: unconformity; B2: regional geochemical anomaly; B3: Bouguer gravity anomaly; B4: aeromagnetic anomaly; B5: chronostratigraphy; B6: lithological characters; B7: remote-sensing mineralisation anomaly; B8: fault characters; B9: fault directions; B10: short distance to fault; B11: distance to unconformity; B12:disjoint fault; B13:disjoint unconformity.

Table 2: The CBR equivalent property features comparison matrix and weights determined by AHP

	B1	B2	B3	B4	B5	B6	B7	Weights
B1	1	7	7/3	7/4	7/6	7/5	7/2	0.250
B2		1	1/3	¼	1/6	1/5	1/2	0.036
B3			1	¾	3/6	3/5	3/2	0.107
B4				1	4/6	4/5	4/2	0.143
B5					1	6/5	6/2	0.214
B6						1	5/2	0.179
B7							1	0.071

B1: regional geochemical anomaly; B2: Bouguer anomaly; B3: chronostratigraphy; B4: remote-sensing mineralization anomaly; B5: fault directions; B6: short distance to fault; B7: distance to unconformity.

layer's attribute table, selected all the records in which the field showing the genetic type of mineral occurrence was non-empty, and exported those records for further analysis. The final records were stored in a text file in which all attribute values were separated by tabs. The corresponding genetic type case base was then constructed. The attribute tables of relevant grid polygon layers were exported and stored in a text file, and each unsolved case set was constructed (each grid represents an unsolved case object). Each grid in the polygon layers corresponds to an unsolved metallogenic case. After a similarity measurement, each grid was assigned a genetic type, and the similarities were assigned values between 0 and 100%. In this way, the classification strategy automatically outlined a regional metallogenic prediction map showing high, medium, and low potentials.

Mineral Potential Prediction Results and Analysis

Based on the data-processing and metallogenic CBR model described above, an experiment regarding mineral potential prediction for iron deposits was performed in the eastern Kunlun Mountains, China. Figure 5 reports, respectively, the curves representing the relationships between 1) posterior probability based on the WOE and cumulative mineral occurrence; and 2) posterior probability and cumulative areas. Table 3 and Figure 6 show the favorable metallogenic potential regions (i.e., areas of high and medium potential) extracted using weights-ofevidence model. Highand medium-potential areas occupy 21% of the study area and contain 62 points of 81 known deposit points (i.e., 77% of known deposit points). High-potential areas occupy 11% of the total area and include 45 known deposit points (i.e., 56% of known deposit points). Medium-potential areas occupy 10% of the total area and include 17 known deposit points (i.e., 21% of known deposit points).

Table 3 and Figure 7 present the potential prediction results for iron deposits using the proposed metallogenic CBR method. Favorable metallogenic regions (i.e., highand medium-potential areas) account for 21% of the

Table 3: The contrast of prediction results using CBR and WOE

Method	high potential (number, percent)	medium-potential (number, percent)	low-potential (number, percent)
WOE	45, 56%	17, 21%	19, 23%
CBR	68, 84%	5, 6%	8, 10%

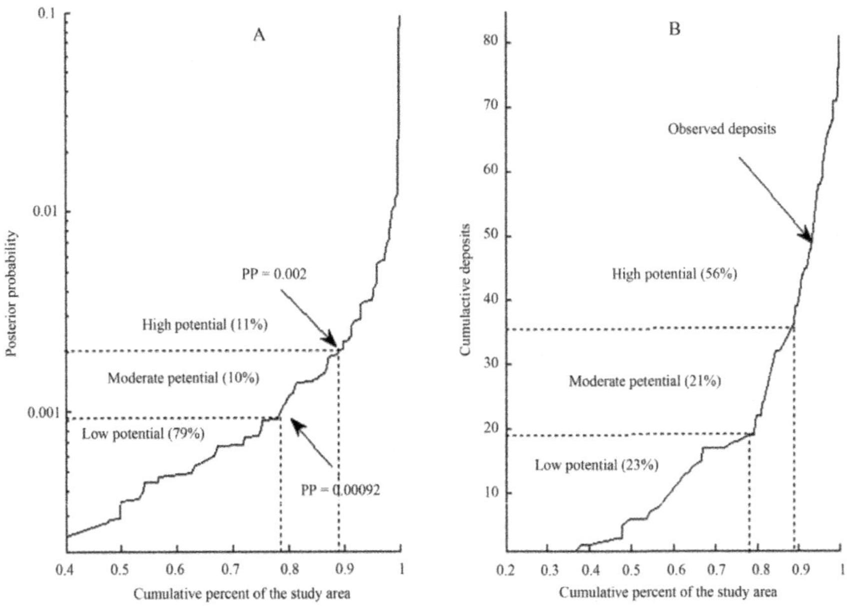

Figure 5: Variation of cumulative area with sum of weights and cumulative deposits using WOE.

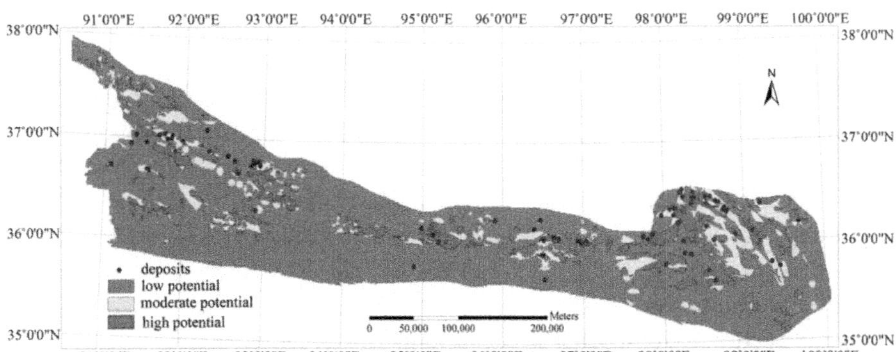

Figure 6: Potential prediction map for iron deposits using WOE in eastern Kunlun Moutains, China.

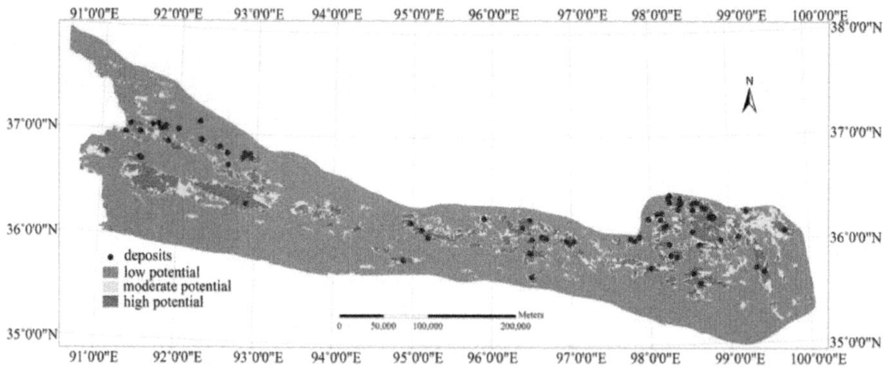

Figure 7: Potential prediction map for iron deposits using CBR in eastern Kunlun Mountains, China.

study area, with high-potential areas accounting for 10% of the total area and the medium-potential areas accounting for 11% of the total area. The prediction results show that known mineral occurrence is highly consistent with the high-potential areas, as analysis predicts that 68 of 81 known mineral occurrences fall into the high-potential areas (84%), 5 fall into the medium-potential areas (6%), and 8 fall into the low-potential areas (10%). Overall prediction accuracy (highand medium-potential areas account for 90%) is significantly higher than the accuracy of the traditional weights-of-evidence model (i.e., 77%).

CONCLUSION

The metallogenic CBR method for regional mineral prospectivity mapping is a new and intelligent prediction method. It makes full use of multisource massive geology spatial data. It also surpasses traditional mineralprediction approaches to improve the intelligence, efficiency, and accuracy of mineral prediction. This paper takes spatial features of geology data into account and proposes an integral metallogenic CBR method, which includes the metallogenic case representation model, metallogenic case storage, and the metallogenic case similarity retrieval model. Finally, an application of mineral potential prediction for iron deposits was performed in the eastern Kunlun Mountains, China, using a metallogenic CBR and WOE, respectively. The results indicated that the prediction accuracy of the metallogenic CBR is significantly higher than the accuracy of the traditional weights-of-evidence model.

ACKNOWLEDGEMENTS

This study was supported by the National Natural Science Foundation of China (Grant No. 41171302 & 40701146) and the National High-Tech R&D Program of China (Grant No. 2007AA12Z227). We express our gratitude to Zhuang Yongcheng from the Qinghai Institute of Geology Survey for his directions.

REFERENCES

1. K. Porwal and O. P. Kreuzer, "Introduction to the Special Issue: Mineral Prospectivity Analysis and Quantitative Resource Estimation," Ore Geology Reviews, Vol. 38, No. 3, 2010, pp. 121-127. doi:10.1016/j.oregeorev.2010.06.002
2. G. F. Bonham-Carter, F. P. Agterberg and D. F. Wright, "Weights of Evidence Modeling: A New Approach to Mapping Mineral Potential," In: F. P. Agterberg, et al., Eds., Statistical Applications in the Earth Sciences, Canadian Government Publishing Centre, Ottawa, 1989, pp. 171-183.
3. F. P. Agterberg, "Combining Indicator Patterns in Weights of Evidence Modeling for Resource Evaluation," Natural Resources Research, Vol. 1, No. 1, 1992, pp. 39-50.doi:10.1007/BF01782111
4. J. R. Harris, L. Wilkinson and E. C. Grunsky, "Effective Use and Interpretation of Lithogeochemical Data in Regional Mineral Exploration Programs: Application of Geographic Information Systems (GIS) Technology," Ore Geology Reviews, Vol. 16, No. 3-4, 2000, pp. 107-143. doi:10.1016/S0169-1368(99)00027-X
5. E. J. M. Carranza, "Weights of Evidence Model of Mineral Potential: A Case Study Using Small Number of Prospects, Abra, Philippines," Natural Resources Research, Vol. 13, No. 3, 2004, pp. 173-187. doi:10.1023/B:NARR.0000046919.87758.f5
6. Daneshfar, A. Desrochers and P. Budkewitsch, "Mineral-Potential Mapping for MVT Deposits with Limited Data Sets Using Landsat Data and Geological Evidence in the Borden Basin, Northern Baffin Island, Nunavut, Canada," Natural Resources Research, Vol. 15, No. 3, 2006, pp. 129-149. doi:10.1007/s11053-006-9020-7
7. Porwal, I. González-Álvarez, V. Markwitz, T. C. McCuaig and A. Mamuse, "Weights-of-Evidence and Logistic Regression Modeling of Magmatic Nickel Sulfide Prospectivity in the Yilgarn Craton, Western Australia," Ore Geology Reviews, Vol. 38, No. 3, 2010, pp. 184-196. doi:10.1016/j.oregeorev.2010.04.002
8. B. He, C. H. Chen and Y. Liu, "Gold Resources Potential Assessment in Eastern Kunlun Mountains of China Combining Weights-of-Evidence Model with GIS Spatial Analysis Technique," Chinese Geographical Science, Vol. 20, No. 5, 2010, pp. 461-470.doi:10.1007/s11769-010-0420-6

9. F. P. Agterberg, G. F. Bonham-Carter, Q. M. Cheng and D. F. Wright, "Weights of Evidence Model and Weighted Logistic Regression in Mineral Potential Mapping," In: J. C. Davis, et al., Eds., Computers in Geology, Oxford University Press, New York, 1993, pp. 13-32.

10. E. J. M. Carranza and M. Hale, "Logistic Regression for Geologically Constrained Mapping of Gold Potential, Baguio District, Philippines," Exploration and Mining Geology, Vol. 10, No. 3, 2001, pp. 165-175. doi:10.2113/0100165

11. F. Aminzadeh, "Applications of Fuzzy Expert Systems in Integrated Oil Exploration," Computers & Electrical Engineering, Vol. 20, No. 2, 1994, pp. 89-97. doi:10.1016/0045-7906(94)90023-X

12. X. Luo and R. Dimitrakopoulos, "Data-Driven Fuzzy Analysis in Quantitative Mineral Resource Assessment," Computers & Geosciences, Vol. 29, No. 1, 2003, pp. 3-13.doi:10.1016/S0098-3004(02)00078-X

13. K. Koike, S. Matsuda, T. Suzuki and M. Ohmi, "Neural Network-Based Estimation of Principal Metal Contents in the Hokuroku District, Northern Japan, for Exploring Kuroko-Type Deposits," Natural Resources Research, Vol. 11, No. 2, 2002, pp. 135-156.doi:10.1023/A:1015520204066

14. J. P. Rigol-Sanchez, M. Chica-Olmo and F. Abarca-Hernandez, "Artificial Neural Networks as a Tool for Mineral Potential Mapping with GIS," International Journal of Remote Sensing, Vol. 24, No. 5, 2003, pp. 1151-1156. doi:10.1080/0143116021000031791

15. P. Gumiel, D. J. Sanderson, M. Arias, S. Roberts and A. Martín-Izard, "Analysis of the Fractal Clustering of Ore Deposits in the Spanish Iberian Pyrite Belt," Ore Geology Reviews, Vol. 38, No. 4, 2010, pp. 307-318. doi:10.1016/j.oregeorev.2010.08.001

16. G. P. Lekkas, N. M. Avouris and L. G. Viras, "CaseBased Reasoning in Environmental Monitoring Applications," Applied Artificial Intelligence, Vol. 8, No. 3, 1994, pp. 359-376.doi:10.1080/08839519408945448

17. Holt and G. L. Benwell, "Applying Case-Based Reasoning Techniques in GIS," International Journal of Geographical Information Science, Vol. 13, No. 1, 1999, pp. 9-25.doi:10.1080/136588199241436

18. J. A. Ye and X. Shi, "Integrating Case-Based Reasoning and GIS for Handling Planning Applications," Journal of Urban Planning Department, No. 3, 2001, pp. 34-40.

19. Y. Y. Du, W. Wen, F. Cao and J. Min, "A Case-Based Reasoning Approach for Land Use Change Prediction," Expert Systems with Applications, Vol. 37, 2010, pp. 5745-5750.doi:10.1016/j.eswa.2010.02.035

20. R. C. Schank and P. A. Robert, "Scripts, Plans, Goals, and Understanding: An Inquiry into Human Knowledge Structures," Lawrence Erlbaum, Hillsdale, 1977.

21. L. D. Kemp, G. F. Bonham-Carter and G. L. Raines, "ArcWofE: ArcView Extension for Weights of Evidence Mapping [EB/OL]," 1999. http://gis.nrcan.gc.ca/software/arcview/wofe

22. G. C. Wang, Q. H. Wang, P. Jian and Y. H. Zhu, "Zircon SHRIMP Ages of Precambrian Metamorphic Basement Rocks and Their Tectonic Significance in the Eastern Kunlun Mountains, Qinghai province, China," Earth Science Frontiers , Vol. 11, No. 4, 2004, pp. 481-490 .

23. Y. S. Pan, W. M. Zhou and R. H. Xu, "The Early Palaozoic Geological Features and Evolutions of the Kunlun Mountain," Science in China (Series D), Vol. 26, No. 4, 1996, pp. 302-307.

24. X. D. Guo, Y. J. Zhang, G. G. Liu, A. J. Pan and F. Zhang, "Metallogenic Regularities and Prospecting Direction of Gold and Copper in Eastern Kunlun," Gold Geology, Vol. 10, No. 4, 2004, pp. 16-22.

25. Z. Z. Qian, Z. G. Hu, J. Q. Liu and H. M. Li, "Active Continental Margin and Regional Metallogenesis of the Palaeo-Tethys in the East Kunlun Mountains," Geotectonica et Metallogenia, Vol. 24, No. 2, 2000, pp. 134-139.

26. T. L. Saaty and L. G. Vargas, "Uncertainty and Rank Order in the Analytic Hierarchy Process," European Journal of Operational Research, Vol. 32, No. 1, 1987, pp. 107- 117.doi:10.1016/0377-2217(87)90275-X

Chapter 3

MAPPING SPATIAL DATA ON THE WEB USING FREE AND OPEN-SOURCE TOOLS: A PROTOTYPE IMPLEMENTATION

Sunil Pratap Singh, Preetvanti Singh

Department of Physics and Computer Science, Dayalbagh Educational Institute, Agra, India

ABSTRACT

There is a growing need for web-based geographic information systems for easy and fast dissemination, sharing, displaying and processing of spatial information. The tremendous growth in the use of web and open-source geospatial resources has sparked development of web-based spatial applications to address multidisciplinary issues with spatial dimensions. This paper presents the integration of open-source geospatial tools and web technology to visualize and interact with spatial data using web browser. The goal of this paper is to implement a prototype system for web-based mapping by providing step-by-step instructions in order to encourage the eager developers and interested readers to publish their maps on the web with no prior technical experience in map servers. The implementation of mapping prototype shows the utilization of open-source geospatial tools which results in a rapid implementation with minimal or no software input cost.

INTRODUCTION

The geographic information systems (GIS) were initially adopted by users in the government sectors, but are now rapidly growing and have profound implications in a large number of application domains including planning [1], marketing [2], retail [3], transportation [4], traffic [5], tourism [6], natural resources [7], real estate [8], agriculture [9] and healthcare [10]. These systems can be valuable planning tools to assist in the development and administration of scientific programs based on geographically identified needs [11]. Linking location to information is a process that applies to many aspects of business such as choosing a site, targeting a market, planning a distribution network or delivery route, drawing up sales territory and allocating resources [12]. The companies, communities and individuals are realizing that they must have geographical information to understand their world.

The tremendous growth in popularity of the internet and the growing public interests in accessing online geo-spatial information have given a move to develop web mapping applications. The web is playing a huge role in spatial application development mainly because of advantages such as platform independency, reduction in distribution costs and maintenance problems, ease of use and widespread access. The most important power of web-based spatial application is the capability to publish and share geo-spatial information on the web which jointly allows information to be exchanged in a rapid and efficient manner thereby helping individuals make important decision quicker. The geo-spatial information not only includes maps or locations of landmarks, but multiple attribute data, socioeconomic data, aerial photographs, satellite images, etc., which may have static or dynamic characteristics [13].

This paper is aimed to provide a method for building a new web-based spatial visualization prototype by using a combination of open-source geospatial packages and Microsoft .NET platform. The following sections specifically document the development and implementation of a prototype system by adopting web-based client-server computing technology.

MATERIALS AND METHODS

Open-Source Geospatial Tools

The commercial software are predominantly "closed source" which means the code cannot be accessed or modified and they may, in some cases, come at a relatively high price with ongoing licensing fees to maintain their use. These systems are thus often priced out of the reach of the resource-constrained areas, particularly within developing countries, which lead them into seeking open-source alternatives [14]. The wide spread free of cost availability, freedom to read, redistribution and modification facilities have made the open-source packages as feature rich alternatives to proprietary software packages [15]. These packages can help the implementation of mapping and spatial analysis tools in a large number of private and public firms, especially in those small organizations that cannot afford proprietary software's cost, complexity, training cost and special requirements. The market of open-source geospatial packages is growing [16] and the quality of these packages is improving with the strong and free collaboration between a large numbers of developers from all around the world.

The open-source project offerings cover the spectrum of tools that interact with geospatial data. The interaction occurs on several fronts: creating, converting, manipulating, analysing or visualizing the geospatial data. These are the common categories of tools or the functional components of open-source geospatial tools. Figure 1 represents the organization of these tools around the geospatial data and some of the well-known geospatial tools are listed in Table 1.

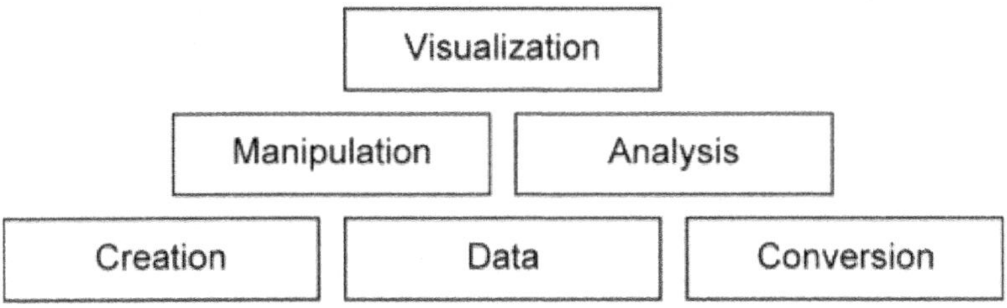

Figure 1: Functional components of open-source tools [17].

Table 1: List of open-source geospatial tools [18]

Category	Open-Source Geospatial Tools
Data Creation	Quantum GIS, OpenEV
Data Conversion	GDAL/OGR, AVCEOO, shp2pgsql (PostGIS)
Data Manipulation	GDAL/OGR, PostGIS, GEOS
Data Analysis	GRASS-GIS, PostGIS
Data Visualization	UMN MapServer, GeoServer, Degree, MapGuide

Microsoft .NET Platform

A range of development technologies is available to develop web based mapping applications. Microsoft's .NET is a revolutionary software programming architecture aimed to create web based applications and offers objectoriented support. The use of object orientation in web applications has increased in recent years since applications are well structured, scalable to new changes and offer better operating performance. The Microsoft .NET Platform provides all the tools and technologies that are needed to build web applications. It exposes a language independent and unified programming model. Its main tools are .NET Framework and Visual Studio .NET (Figure 2).

NET Framework

The .NET Framework is a technology that supports building and running the next generation of web services and applications. It is a managed execution environment that provides a variety of services (memory management, common type system, language interoperability, version compatibility) to its running applications. Its main components include [19]: the common language runtime (CLR), which is the managed execution environment that handles memory allocation, error trapping, and interacting with the operating-system services; the .NET Framework Class Library, which is an extensive collection of programming components and application program interfaces (APIs) for all major areas of application development; ASP.NET, a server side web technology which supports creation of dynamic web applications, websites and web services.

The advantages offered by the .NET Framework include shorter development cycles (code reuse, support for multiple programming languages), easier deployment, fewer data type related bugs due to integral type safety, reduced memory leaks, and, in general more scalable, reliable applications.

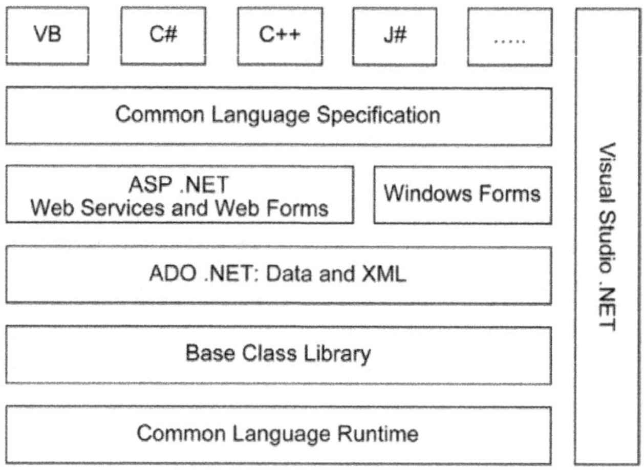

Figure 2: Microsoft .NET platform architecture [19].

Visual Studio .NET

The Microsoft Visual Studio .NET is an integrated development environment (IDE) to develop console and graphical user interface applications along with web applications, web services, websites and windows forms. It supports different programming languages (C#, C++, Visual Basic, J#, etc.) by means of language services, which allow the code editor and debugger to support nearly any programming language, provided a languagespecific service exist. The built-in languages include Visual C++, Visual Basic, Visual C# and Visual J#. It also supports XML/XSTL, HTML/XHTML, JavaScript and CSS. It has comprehensive tools (drag-and-drop form designer, IntelliSense, dynamic help) for rapidly building and integrating applications which increases developer's productivity. The Microsoft Express set of free software applications offer a way around budget limits, as all of the software are free, albeit somewhat restricted. Visual Web Developer 2008 Express is a .NET developer's integrated development environment (IDE) provided for free by Microsoft. It is a lightweight version of Microsoft Visual Studio product line. Figure 3 shows the starting screen of Visual Web Developer 2008 Express. The idea of Express edition is to provide streamlined, easy-to-use and easy-to-learn IDEs for users other than professional software developers. It allows web development with ASP.NET using drag-and-drop user interface designer, enhanced code editor, support for other web technologies (e.g., CSS, JavaScript, XML) and integrated design time validation.

Prototype Development

Preparation of Geospatial Data

The city of Taj, Agra in India is selected as a pilot area to create the prototype spatial mapping system. The geographic location of the scope lies within 27.11° latitude north and 78.20° longitude east and is covered in 1:50,000 scaled sheets 54E/16–54I/4 of Survey of India.

The data that is stored in spatial information system can be thought of as a digital representation of real world objects. These objects can be abstracted into either discrete objects such as houses or continuous fields such as elevation and are stored as vector or raster data structures respectively. The raster model is a representation of the world as a surface that is divided into a regular grid of cells. The vector model represents space as a series of discrete entity-defined point, line or polygon units which are geographically referenced. The vector data models are more compact and predominant across the diverse range of spatial information systems [6,7,20,21] and therefore we have focused on vector model for mapping the geographic data in web-based environment.

The geographic objects together with their attribute data constitute spatial data (known locations on the earth with statistical and non-location data associated with spatial entity). The collection of spatial data under vector model is organized in a thematic approach that categorizes data in layers. Usually, each data layer represents only one type of spatial entities, i.e., point, line or polygon.

There are many sources for obtaining spatial data including hard copy maps; aerial photographs; remotelysensed imagery and existing digital data files. For the scope of this prototype application, a geo-referenced color raster map is obtained in digital format from the Survey of India (government organization). This raster map is processed using Quantum GIS (an open-source desktop mapping engine) to create separate vector layers for different geometry types. Table 2 lists some of the common layers created by extracting geo-objects from the raster map and are used in the prototype application. These layers are saved in ESRI's (Environmental Research System Institute, Redlands, California) shapefile format.

System Components

Database

The analysis on spatial data can be performed well if the data of geographic objects is stored in relational database. For this reason, all the layers (shapefiles) of the study area are converted to their corresponding tables in a PostgreSQL database. PostgreSQL is one of the most popular object-relational database management system (ORDBMS) on the open-source platform. It is not easy to store the spatial data in a standard RDBMS, thus spatial extensions have been developed and standardized by the OGC (Open Geo Consortium).

PostGIS is an open-source and OGC compliant spatial database extender for PostgreSQL [22]. It adds spatial function and specialty geometry data types to the database. It is an excellent way to bring tabular and spatial data together into a common management environment. The conversion of shapefiles to PostgreSQL database tables is achieved using the shp2pgsql utility included as part of the PostGIS extension. This utility takes a shapefile and outputs a series of SQL statements to create a table in PostgreSQL database. The resulting table contains all the attributes of the shapefile including the

Table 2: Layers extracted from base map (raster map)

Sr. No.	Layer	Geometry Type
1	Roads	Line
2	Administrative Buildings	Polygon
3	Points of Interests	Point
4	Railway Lines	Line
5	Public Services (Amenities)	Point

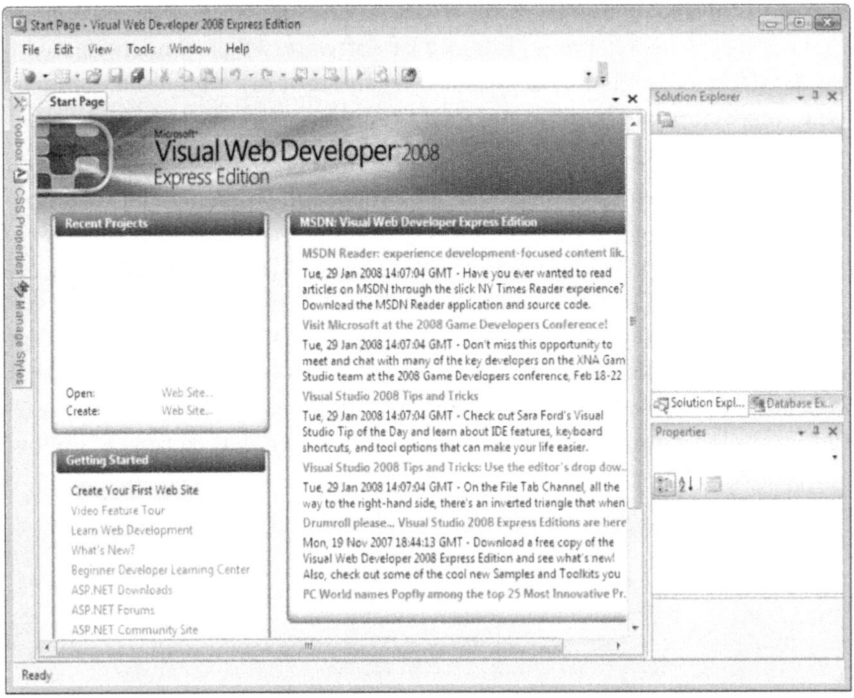

Figure 3: Microsoft Visual Web Developer 2008 Express Edition—start page.

coordinates that define each feature. This process was repeated for each of the vector layers created for the scope of prototype. Figure 4 shows the flow diagram of necessary steps to create the spatial database.

Table 3 represents the open-source geospatial tools used to implement web-based mapping application.

Mapping Server

Internet map server applications easily make spatial data accessible through a web browser interface to end users. The map servers usually serve spatial data based on three OGC standards: WMS (Web Mapping Service) for the display of maps as image; WFS (Web Feature Service) for vector data; and WCS (Web Coverage Service) for raster data. The two best-known open-source map servers are Minnesota MapServer and GeoServer. Both of these are viable alternatives to proprietary map servers.

We have used MapServer for development of current prototype, because it has an excellent performance of functionality and speed on processing large volume dataset. It is considered as the world's leading open-source web mapping tool and complies with WMS, WFS and WCS web specifications of OGC [23]. It is a server-side piece of software which renders spatial data source into cartographic map products on-the-fly [24]. The main components of the MapServer are mapfile and CGI program.

A mapfile defines how data is used and sets the display and query parameters for the map. It needs to set cartographic parameters, cartographic objects, data load- ing, classification, displaying and querying, and graphic elements definition. It is composed of several objects in which each object starts by keyword OBJECT and finishes by END. Each layer object inside the mapfile starts by keyword LAYER, defines the path and connection type to load the specified data, specifies the style and other parameters, and finishes by END. A mapfile allows direct connections for shapefiles and raster files while more complex connections are required for other data formats. Following code represents the part of a mapfile created for the prototype application with connection to PostGIS layer.

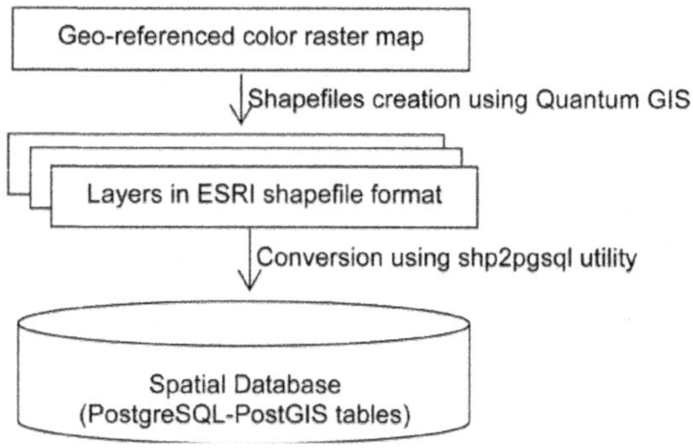

Figure 4: Flow diagram of spatial database development.

Table 3: List of open-source geospatial tools used in study

Category	Open-Source Geospatial Tool
Data Creation	Quantum GIS
Data Conversion	shp2pgsql (PostGIS)
Data Manipulation	PostGIS
Data Analysis	PostGIS
Data Visualization (Internet Mapping)	UMN MapServer

LAYER CONNECTIONTYPE PostGIS CONNECTION "Host = Localhost DbName = dbDemo User = postgresql Password = pgAdmin Port = 5432"

DATA "the_geom FROM POI"

NAME "POI"

TYPE POINT STATUS ON CLASS STYLE COLOR 255 0 0 OUTLINECOLOR 0 0 0 SYMBOL "Circle"

END END END

....

A CGI program is the real data processing engine: started up by the web server, it reads the map file settings and returns processed spatial data as maps, cartographic objects, variable values and query results.

MapServer can be used in two ways, with the CGI interface or with the MapScript. MapScript provides a scripting interface for MapServer for the development of web based stand-alone applications. It supports popular scripting languages such as PHP, Java, Perl, C# etc. The prototype application in this study is implemented with MapScript API, accessed with C# MapScript under ASP.NET environment. To gain access to PostgreSQL database server, Npgsql is used which is an open-source ASP.NET data provider. The description on how to configure the MapServer and taking reference of precompiled binaries is provided in Appendix I.

Development Environment and User Interface

The user interface design makes the user's interaction simple and efficient in terms of accomplishing user goals. It allows the users to select and input the query criteria in order to view the required data. In the present prototype application, the user interface is made up of ".aspx" files which are populated with the appropriate controls using Visual Web Developer 2008 Express as the .NET development environment. Figure 5 demonstrates the running

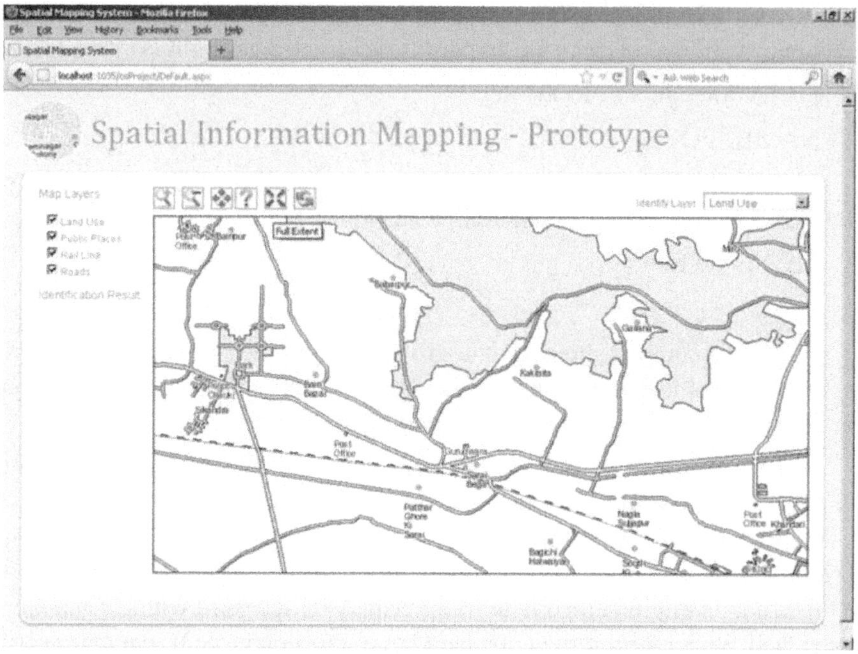

Figure 5: User interface (running view).

view of user interface implemented with ASP.NET by which user interact with the navigation tools such as zoom in, zoom out, zoom to full extent to view the map at different scales and identification of different map features.

System Configuration and Architecture

Hardware Platform Configuration

The server machine used for prototype application development has the following specifications:

CPU: Pentium Dual 3.00 GHz with 2 MB Cache Main Memory: 2048 MB (1024 × 2)

HDD: 80 GB Operating System: Windows XP (SP3)

Web Server: Internet Information Server (IIS) 6.0 DBMS: PostgreSQL with PostGIS

System Architecture

The prototype system adopts client-server architecture where most of the processing takes place on the server side while the client is used for gathering input from the user and displaying map and other associated data based on query parameters. The server-side approach only requires a browser as the spatial mapping and processing; modelling, database and other components will be located on the server. The architecture of the prototype application is shown in the Figure 6, in which the users using any of the popular internet browsers requests geospatial information from the web server.

The web server forwards the request to the ASP.NET application server in which C# Mapscript is called for geospatial data. C# Mapscript forwards the map request to MapServer and the MapServer using the mapfile generates a map based on the geospatial data supplied by the PostGIS. The generated map is sent back to the web server, which forwards it to the web client.

Explanation and Code

The prototype application contains two web forms named Default.aspx and MapStreaming.aspx by following the flow of [25]. The Default.aspx form contains all the controls required to perform various operations on the map and geospatial data. This form is populated with following controls:

- cblLayers (CheckBoxList) for selection of layers to be displayed
- ddlLayers (DropDownList) for selection of layer to be identified
- litIdentifyResult (Literal) for display the result of identified map features
- ibtnMap (ImageButton) for display the map
- ibtnZoomIn (ImageButton) to perform zoom in
- ibtnZoomOut (ImageButton) to perform zoom out
- ibtnPan (ImageButton) to perform panning
- ibtnIdentify (ImageButton) to identify the features of selected layer
- ibtnFullExtent (ImageButton) to display the full extent map
- ibtnRefresh (ImageButton) to refresh the map
- rblAction (RadioButtonList) to set the map action with item values Zoom In, Zoom Out, Pan and Identify The MapStreaming.aspx form does not contain any control and is used to send the image stream to the ibtnMap control. The following Figure 7 represents the flow of code among the components of prototype application.

The next section describes the code to implement the essential functions of mapping (detailed code is available on request from authors).

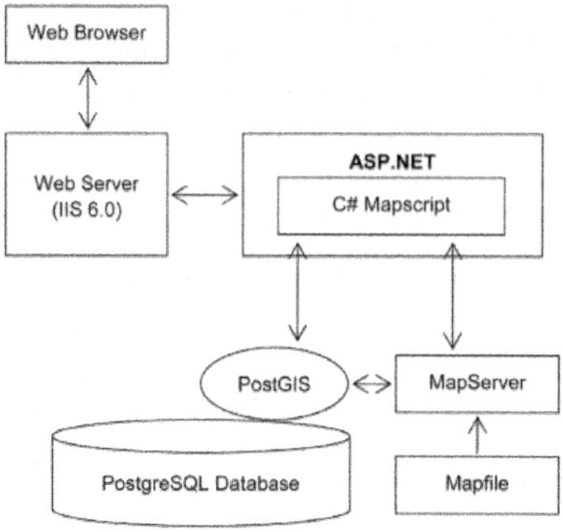

Figure 6: Architecture of prototype mapping application.

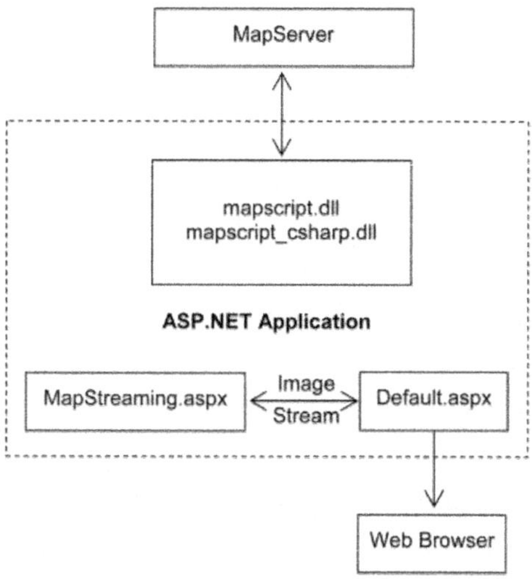

Figure 7: Flow of code [25].

C#.NET Mapscript for "Default.aspx"

```
using System;
using System.Web.UI;
using OSGeo.MapServer; //For MapScript public partial class _Default : System.Web.UI.Page
{
```

```
mapObj map;
protected void Page_Load(object sender, EventArgs e)
{
if (!Page.IsPostBack)//First Access to Map
{
//Image Stream from MapServer ibtnMap.ImageUrl = "MapStreaming.
aspx?MAPACTION =INITMAP";
mapObj map = new mapObj(Server.MapPath("App_Data/ MapFile.map"));
for (int i = 0; i < map.numlayers; i++)
{
layerObj layer = map.getLayer(i);
if (layer.status == (int)mapscript.MS_ON) {
cblLayers.Items[i].Selected = true;
}
}
}
else { //Next Access to Map (By Session)
map = (mapObj)Session["MAP"];
}
}
protected void ibtnMap_Click(object sender, ImageClick EventArgs e)
{
lblInfo.Text = "";
string mapAction = "";
string activeLayer = ddlLayers.SelectedValue;
//Check and Set Map Action switch (rblAction.SelectedItem.Text.ToUpper())
{
case "ZOOM IN":
mapAction = "ZOOMIN";
break;
case "ZOOM OUT":
mapAction = "ZOOMOUT";
```

```
break;
case "PAN":
mapAction = "PAN";
break;
case "IDENTIFY":
mapAction = "IDENTIFY";
break;
}
//Image Stream from MapServer to ibtnMap according to required Map Action
ibtnMap.ImageUrl = "MapStreaming.aspx?MAPACTION =" + mapAction +
"&X=" + e.X + "&Y=" + e.Y + "&ACTIVELAYER=" + activeLayer;
}
//Setting rblAction for different Map Operations protected void ibtnZoomIn_
Click(object sender, Image ClickEventArgs e) {
rblAction.SelectedValue = "Zoom In";
}
protected void ibtnZoomOut_Click(object sender, Image ClickEventArgs e) {
rblAction.SelectedValue = "Zoom Out";
}
protected void ibtnPan_Click(object sender, ImageClick EventArgs e) {
rblAction.SelectedValue = "Pan";
}
protected void ibtnIdentify_Click(object sender, Image ClickEventArgs e) {
rblAction.SelectedValue = "Identify";
}
//For Full Extent Map protected void ibtnFullExtent_Click(object sender,
Image ClickEventArgs e) {
ibtnMap.ImageUrl = "MapStreaming.aspx?MAPACTION =FULLEXTENT";
}
//For Refreshing the Map according to Selected (cblLayers Checked) Map
Layers protected void ibtnRefresh_Click(object sender, Image ClickEventArgs
e)
{
```

```
for (int i = 0; i < cblLayers.Items.Count; i++)
{
layerObj layer = map.getLayerByName(cblLayers.Items[i]. Value);
if (cblLayers.Items[i].Selected) {
layer.status = (int)mapscript.MS_ON;
}
else {
layer.status = (int)mapscript.MS_OFF;
}
}
ibtnMap.ImageUrl = "MapStreaming.aspx?MAPACTION =REFRESHMAP";
}
}
```

C#.NET Mapscript for "MapStreaming.aspx"

```
using System;
using System.IO;
using System.Web.UI;
using System.Drawing;
using OSGeo.MapServer; //For MapScript public partial class MapStreaming
: System.Web.UI.Page
{
mapObj map;
rectObj originalExtent;
private enum zoomMode {zoomIn = 0, zoomOut=1} //Enumerator for
Zooming protected void Page_Load(object sender, EventArgs e)
{
//Read Map from Session (if exist), Otherwhise Create new using Mapfile map
= (mapObj)Session["MAP"];
if (map == null)
{
map = new mapObj(Server.MapPath("App_Data/Map File. map"));
originalExtent = new rectObj(map.extent.minx, map.extent.miny, map.extent.
maxx, map.extent.maxy, 0);
```

```
Session["ORIGINALEXTENT"] = originalExtent;
}
//Variables to read the required Map Action, (x,y) position and active Layer String mapAction;
Double x = 0; //Longitude Double y = 0; //Latitude String activeLayer;
//Read Map Action from QueryString mapAction = Request. QueryString["MAPACTION"]. ToString().ToUpper();
//Read (x,y) position from QueryString if (Request.QueryString["X"] != null && Request.Query String["Y"] != null) {
x = Double.Parse(Request.QueryString["X"].ToString());
y = Double.Parse(Request.QueryString["Y"].ToString());
}
//Read active Layer from QueryString if (Request. QueryString["ACTIVELAYER"] != null) {
activeLayer = Request.QueryString["ACTIVELAYER"]. ToString();
}
//Call for different methods based on Map Action switch (mapAction)
{
case "ZOOMIN":
doZoom(zoomMode.zoomIn, x, y); // Zoom In break;
case "ZOOMOUT":
doZoom(zoomMode.zoomOut, x, y); // Zoom Out break;
case "PAN":
doPan(x,y); // Panning break;
case "IDENTIFY":
doIdentify(x, y, activeLayer); // Feature Identification break;
case "FULLEXTENT":
doFullExtent(); //Full Extent break;
case "REFRESHMAP":
break;
}
doRefresh();
Session["MAP"] = map;
```

```
}
private void doZoom(zoomMode zMode, Double x, Double y) { }
private void doPan(Double x, Double y) {  }
private void doIdentify(Double x, Double y, String active Layer) { }
private void doFullExtent() {  }
private void doRefresh() {  }
```

CONCLUSION

In any project, for the developing countries like India, cost is the primary consideration for the adaptability of spatial technology. Under these circumstances, free and open-source geospatial tools provide all the data storage, analysis and information visualization for free. This paper provides a foundation in web-based mapping using open-source geospatial tools and web technology. It highlights building a prototype mapping application by providing step-by-step instructions which can encourage the eager developers and interested readers to publish their maps on the web with no prior technical experience in map servers. The whole prototype lives in a common web-based application implemented in ASP.NET with C#.NET MapScript exploiting MapServer for mapping functionalities and PostGIS for its connection with the PostgreSQL databse. The techniques and source code of this research work can be applied as the laboratory exercise or a course curriculum and can help a large number of organizations to publish their geographic information on the web, especially those small organizations that cannot afford proprietary software's cost, complexity, training cost and special requirements.

ACKNOWLEDGEMENTS

The corresponding author is thankful to University Grants Commission, India for providing financial support in the form of research fellowship under RFSMS scheme.

REFERENCES

1. R. Mari, L. Bottai, C. Busillo, F. Calastrini, B. Gozzini and G. Gualtieri, "A GIS-Based Interactive Web Decision Support System for Planning Wind Farms in Tuscany (Italy)," Renewable Energy, Vol. 36, No. 2, 2011, pp. 754- 763.http://dx.doi.org/10.1016/j.renene.2010.07.005
2. M. Ozimec, M. Natter and T. Reutterer, "Geographical Information Systems-Based Marketing Decisions: Effects of Alternative Visualizations on Decision Quality," Journal of Marketing, Vol. 74, No. 6, 2010, pp. 94-

110. http://dx.doi.org/10.1509/jmkg.74.6.94

3. Okabe and K. Okunuki, "A Computational Method for Estimating the Demand of Retail Stores on a Street Network and Its Implementation in GIS," Transactions in GIS, Vol. 5, No. 3, 2001, pp. 209-220. http://dx.doi.org/10.1111/1467-9671.00078
4. K. Ziliaskopoulos and S. T. Waller, "An InternetBased Geographic Information System That Integrates Data, Models and Users for Transportation Applications," Transportation Research Part C, Vol. 8, No. 1-6, 2000, pp. 427-444.
5. F. Xie, "Design and Implementation of Highway Management System Based WebGIS," Journal of Networks, Vol. 5, No. 12, 2010, pp. 1389-1392.http://dx.doi.org/10.4304/jnw.5.12.1389-1392
6. S. P. Singh, J. Sharma and P. Singh, "A Geo-Referenced Information System for Tourism (GeoRIST)," International Journal of Geomatics and Geosciences, Vol. 2, No. 2, 2011, pp. 456-464.
7. P. S. Singh, D. Chutiya and S. Sudhakar, "Development of a Web Based GIS Application for Spatial Natural Resources Information System Using Effective Open Source Software and Standards," Journal of Geographic Information System, Vol. 4, No. 3, 2012, pp. 261-266. http://dx.doi.org/10.4236/jgis.2012.43031
8. T. Q. Zeng and Q. Zhou, "Optimal Spatial Decision Making Using GIS: A Prototype of a Real Estate Geographical Information System (REGIS)," International Journal of Geographical Information Science, Vol. 15, No. 4, 2001, pp. 307-321.http://dx.doi.org/10.1080/136588101300304034
9. T. Oswari, E. S. Suhendra, E. Haryatmi and F. Agustina, "Prototype Geographic Information Systems Mapping of Crop Products Featured Local," Journal of Geographic Information System, Vol. 5, No. 3, 2013, pp. 193-197.http://dx.doi.org/10.4236/jgis.2013.53018
10. Evans and C. E. Sabel, "Open-Source Web-Based Geographical Information System for Health Exposure Assessment," International Journal of Health Geographics, Vol. 11, 2012, p. 2.
11. R. M. Mullner, K. Chung, K. G. Croke and E. K. Menash, "Geographic Information Systems in Public Health and Medicine," Journal of Medical Systems, Vol. 28, No. 3, 2004, pp. 215-221. http://dx.doi.org/10.1023/B:JOMS.0000032972.29060.dd
12. Boyles, "GIS Means Business," ESRI, Redlands, 2002.
13. M. N. K. Boulos and K. Honda, "Web GIS in Practice IV: Publishing Your Health Maps and Connecting to Remote WMS Sources Using the Open

Source UMN MapServer and DM Solutions MapLab," International Journal of Health Geographics, Vol. 5, 2006, p. 6.

14. Q. Yi, R. E. Hoskins, E. A. Hillringhouse, S. S. Sorensen, M. W. Oberle, S. S. Fuller and J. C. Wallace, "Integrating Open-Source Technologies to Build Low-Cost Information Systems for Improved Access to Public Health Data," International Journal of Health Geographics, Vol. 7, 2008, p. 29.

15. S. Steiniger and E. Bocher, "An Overview on Current Free and Open Source Desktop GIS Developments," International Journal of Geographical Information Science, Vol. 23, No. 10, 2009, pp. 1345-1370. http://dx.doi.org/10.1080/13658810802634956

16. J. Lowe, "Spatial on a Shoestring: Leveraging free Open Source Software," Geospatial Solutions, Vol. 12, No. 6, 2002, pp. 42-45.

17. T. Mitchell, "Open Source Geo Tools," O'Reilly Where Where 2.0 Conference, San Francisco, 29-30 June 2005.

18. T. Mitchell, "An Introduction to Open Source Geospatial Tools," 2005. http://www.oreillynet.com/pub/a/network/2005/06/10/osgeospatial.html

19. W. M. Lee, "C#. NET Web Developers Guide," Syngress, Rockland, 2002.

20. Mansourian, A. Fasihi and M. Taleai, "A Web-Based Spatial Decision Support System to Enhance Public Participation in Urban Planning Processes," Journal of Spatial Science, Vol. 56, No. 2, 2001, pp. 269-282. http://dx.doi.org/10.1080/14498596.2011.623347

21. G. Othman, B. Mohamed, A. Bahauddin, A. P. M. Som and S. I. Omar, "A Geographic Information System Based Approach for Mapping Tourist Accommodations in the East Coast States of Malaysia," World Applied Sciences Journal, Vol. 10, 2010, pp. 14-23.

22. PostGIS. http://postgis.net/

23. S. Steiniger and A. J. Hunter, "Free and Open Source GIS Software for Building a Spatial Data Infrastructure," In: E. Bocher and M. Neteler, Eds., Geospatial Free and Open Source Software in the 21st Century, Springer-Verlag Berlin Heidelberg, 2012, pp. 247-261.

24. P. Ramsey, "The State of Open Source GIS," 2007.http://www.refractions.net/expertise/whitepapers/opensourcesurvey/survey-open-source-2007-12.pdf

25. P. Corti, "Thinking in GIS," 2006. http://www.paolocorti.net/2006/09/20/mapserver-tutorial-for-c-mapscript-asp-net/

Chapter 4

GEOSTATISTICAL APPROACH FOR SITE SUITABILITY MAPPING OF DEGRADED MANGROVE FOREST IN THE MAHAKAM DELTA, INDONESIA

Ali Suhardiman[1,2], Satoshi Tsuyuki[1], Muhammad Sumaryono[2], Yohanes Budi Sulistioadi[2,3]

[1]Department of Global Agricultural Sciences, The University of Tokyo, Tokyo, Japan

[2]Department of Forest Science, University of Mulawarman, Samarinda, Indonesia

[3]Division of Geodetic Science, School of Earth Science, The Ohio State University, Columbus, USA

ABSTRACT

As part of operational guidance of mangrove forest rehabilitation in the Mahakam delta, Indonesia, site suitability mapping for 14 species of mangrove was modelled by combining 4 underlying factors—clay, sand, salinity and tidal inundation. Semivariogram analysis and a geographic information system (GIS) were used to apply a site-suitability model, while kriging interpolation generated surface layers, based on sample point data collection. The tidal inundation map was derived from a tide table and a digital elevation model from topographic maps. The final site-suitability maps were produced using spatial analysis technique, by overlaying all surface layers. We used a Gaussian model to adjust a semivariogram graph in order to help to understand the variation of sample data values, and create a natural surface layer of data distribution over the area of study. By examining the statistical value and the visual inspection of surface layers, we saw that the models were consistent with the expected data behavior; therefore, we assumed that interpolation has been carried out appropriately. Our site-suitability map showed that Avicennia species was the most suitable species and matched with 50% of the study area, followed by Nypa fruticans, which occupied about 42%. These results were actually consistent with the mangrove zoning pattern in the region prior to deforestation and conversion.

INTRODUCTION

The Mahakam delta is located at the mouth of the Mahakam river on the east coast of Kalimantan, Indonesia. Including the distributary and interdistributary channels, the delta plain covers an area of 1500 km^2. It was originally covered by dense mangrove, composed of mainly Avicennia, Sonneratia, Rhizophora, and Nypa species, and fresh water mangroves [1]. This ecosystem was disturbed since the introduction of shrimp farming. Human settlements also expanded because of the oil and gas exploration and exploitation, which attracted both workers and land speculators [2]. This ecological disturbance reached its peak during Asian economic crisis (1997- 2000) when shrimp price on the global market inflated to 4 - 5 times the normal level due to devaluation of Indonesian rupiah against international currency. This triggered mangrove forest conversion to ponds, involving heavy equipment, on a massive scale [3].

In 2001, 75% of the delta had been changed to ponds [4], which subsequently reduced to 52% in 2007, as reported by the Fishery and Oceanic Affairs Office of the Kutai Kartanegara District government [5]. Almost 90% of the Mahakam delta region belongs to the Indonesian Ministry of Forestry and is classified to be a production forest, however three other different government sectors lay claim to the management of the Mahakam delta: the Fishery Department, Interior Affairs Department, and the Environment Department [3]. Due to unclear management, laws and regulations, land use control of this area failed to be enforced.

Massive degradation of mangrove ecosystems has prompted a world-wide movement to plant new areas of mangroves [6]. Unfortunately, mangrove rehabilitation has often been carried out simply by planting mangrove seedlings, without adequate site assessment, or subsequent evaluation of the success of planting at the ecosystem level [7,8]. In this paper, we applied geostatistical and geographical information system (GIS) to generate site-suitability mapping for mangrove rehabilitation guidance in the Mahakam delta.

MATERIALS AND METHODS

Study Area

The Mahakam delta is located between 117.30°E - 117.62°E and 0.49°S - 0.82°S (Figure 1). It comprises 46 sedimentation-formed islands [9]. For this study, we focused on areas (approximately 40,636 ha) where sample points were distributed relatively balance in term of spatial representation and therefore, data interpolation would give the best prediction.

Model Criteria and Flow Chart of the Study

In this study, we employed the technical guidance on mangrove plantation issued by Indonesian Ministry of Forestry in 2004. Geographical information system model was developed using the criteria presented in Table 1. ArcGIS 10 embedded geostatistical analysis tool was used to apply the model. Data normality were examined prior to geostatistics analysis. Geostatistical analysis tool was used to perform semivariogram analysis, ordinary kriging interpolation and subsequent cross-validation. Ordinary kriging interpolation produced a surface layer (continuous data) of all the data from our sample points. By superimposing all related surface layers, we can produce map of site suitability for mangrove rehabilitation over the entire area of the study. The major analysis steps used in this study were shown in Figure 2.

The physical data collected were soil texture (percentage of clay and sand fraction) and salinity. Tidal inundation data was derived from combination of a digital elevation model (DEM) and the 2009 tide table, obtained from the hydro-oceanographic division of the Indonesian navy. The digital elevation model was derived from the height points of the Mahakam delta topographic maps.

The tide table contained 12 months of hourly sea level rise predictions for a specific area. January, April, July and October were selected to represent the 2009 tidal prediction. We assumed each selected month would capture the seasonal effect of tidal dynamics in the area of interest. The East Kalimantan province is characterized by a tropical rain forest climate with a dry (May to September) and a wet (October to April) season [10]. The dry and wet seasons were represented by the months of July and January, respectively, while April and October represented the transitional months.

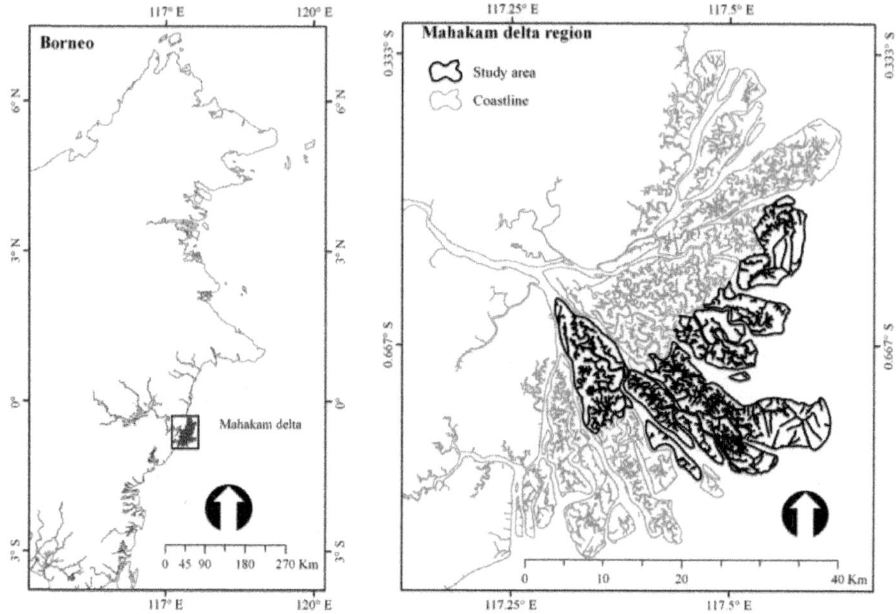

Figure 1: Location of Mahakam delta in the east coast of Borneo Island. The area sat the focus of this study are outlined by bold lines.

Table 1: Technical guidance of mangrove plantation in the forested areas of Indonesia

Zones	Species	Salinity (ppt)	Sandy fraction tolerance	Mud tolerance	Frequency of tidal inundation
A	*Rhizophora mucronata* Lamk. *Rhizophora apiculata* Bl.	10 - 30	Moderate	Suitable	<20 days/month
B	*Rhizophora stylosa* Griff *Sonneratia alba* J.Smith.	10 - 30	Suitable	Suitable	<20 days/month
C	*Bruguiera parviflora* (Roxb.) Wight & Arn. ex Griffith. *Bruguiera sexangula* (Lour.) Poir.	10 - 30	Moderate	Suitable	10 - 20 days/month
D	*Bruguiera gymnorrhiza* (L.) Savigny.	10 - 30	Not suitable	Moderate	10 - 20 days/month
E	*Sonneratia caseolaris* (L.) Engl.	10 - 30	Moderate	Moderate	<20 days/month
F	*Xylocarpus granatum* J.Koenig. *Heritiera littoralis* Dryand.	10 - 30	Moderate	Moderate	>10 days/month
G	*Lumnitzera racemosa* Willd.	10 - 30	Suitable	Moderate	>10 days/month
H	*Cerbera manghas* L.	0 - 10	Moderate	Moderate	>10 days/month
I	*Nypa fruticans* Wurmb.	0 - 10	Not suitable	Suitable	<20 days/month
J	*Avicennia* sp.	10 - 30	Not suitable	Suitable	<20 days/month

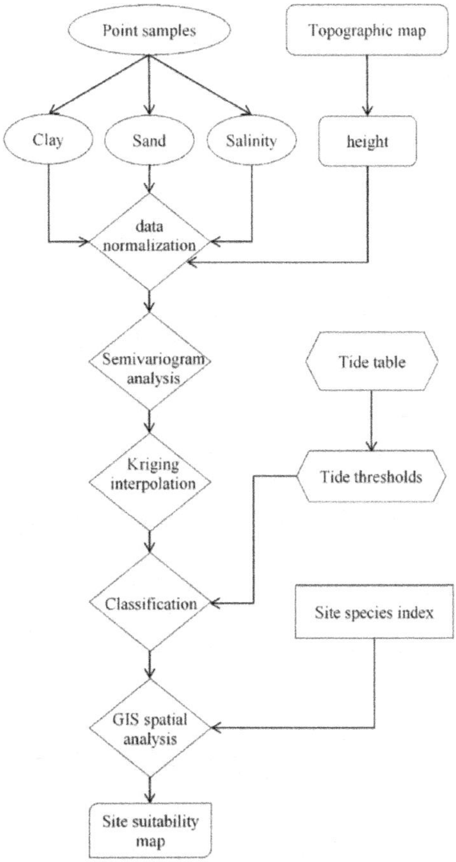

Figure 2: The flow chart of the study, showing the major steps and the site-suitability map as the final result.

Semivariogram

A semivariogram is based on an assumption that properties of regions in proximity to one another are more alike than those farther away. In order to get a better interpolation result based on semivariogram analysis, we needed normally distributed data. The Shapiro-Wilk test was used to quantify the normality of the data. We used the BoxCox and logarithmic transformation to normalize the data. After that, we used the semivariogram analysis to quantify the spatial relationship between the sample points. The semivariogram symbolized by g(h) is the average variance between observations separated by a distance h. It was computed using formula:

$$\gamma(h)=\frac{1}{2N(h)}\sum_{i=1}^{N(h)}\{z(x_i)-z(x_i+h)\}^2 \tag{1}$$

where γ(h) is the experimental semivariogram value at a distance h; N(h) is number of sample value pairs within distance h; and $z(x_1)$, $z(x_1 + h)$ are the sample values at two points separated by distance h. All pairs of points separated by distance h (lagh) were used to calculate the experimental semivariogram.

The semivariogram parameters are the range, total sill and nugget ratio. The total sill is the sum of the sill and nugget, while the nugget ratio is the ratio of the nugget to the total sill [11]. Lag is the distance between two pairs data; the sill is the maximum level of γ(h); the range is the point on the h axis where γ(h) reaches a maximum [12]. The experimental semivariogram will be fitted using a mathematical model used for interpolation. There are many models available, but the most acceptable ones for continuous variables would be spherical, exponential, Gaussian or combinations of these models [12]. In this study, we used a Gaussian model with a small range and sill value, which means this model predicted smaller data variability compared to the spherical and exponential models.

Kriging Interpolation

We applied Kriging interpolation to clay, sand, salinity and height point data following the semivariogram investigation. Reference [13] discovered that kriging interpolation was better suited for prediction and mapping of the distribution of chemical soil properties while reference [14] found that kriging interpolation had successfully explained soil properties. We chose widely used ordinary kriging interpolation to generate the model, assuming the absence of non-linear trends and focusing only on the spatially correlated component [15].

The characteristics of an interpolated surface layer can be controlled by limiting the input points used in the calculation of the output cell values. Specifying the maximum and minimum number of points to be sampled will return the points closest to the output cell location, until the maximum number is reached. This number is then used in calculating output values. In this study, we used a combination of maximum and minimum number of neighboring points to be 5 and 2, respectively. Another method to control the interpolated surface is by defining either a circular or elliptical model to enclose the points that are used to predict output cell values. Furthermore, these circular or elliptical models could be divided into sectors. We used a circular model split by a diagonal cross section (4 sectors with a 45° offset) for calculating output cell values.

Surface layer-based interpolation data were classified according to criteria shown in Table 1. To divide the surface layer into classes (suitable, moderate or not suitable) especially for clay and sand data, we used Jenks natural breaks provided in most of GIS softwares. Salinity data was divided into two classes: 0 - 10 ppt (part per thousand) and 10 - 30 ppt. From DEM data, we defined three classes of tidal inundation based on the number of inudation days in a month: 1) more than 20 days in a month; 2) between 10 - 20 days; and 3) less than 10 days in a month.

Validation

Cross validation of sample points was used to measure the model accuracy. The following statistics were used:

root mean square (RMS), which measured error size sensitive to outliers [16], and standardized RMS [15]. RMS was calculated using the following formula:

$$\mathrm{RMS} = \sqrt{\frac{1}{n}\sum_{i=1}^{n}\left(Z_{i,act} - Z_{i,est}\right)^2} \qquad (2)$$

By dividing RMS with standard error then standardized RMS would be obtained. Here the variables are defined as follows: $Z_{i,act}$ = the known value at point i$Z_{i,est}$ = the estimated value at point in = the number of points.

RESULTS

Normalizing Data

The normality test using Shapiro-Wilk method indicated that our data was not normally distributed (Table 2). Therefore, data normalization was imperative prior to kriging interpolation. Only clay data appeared to be normally distributed, since its p-value was close to 0.05. We performed a standard statistical calculation and found that sand, salinity, and height data exhibited higher variation than the clay data. Coincidently, the high variation was related to the p-value of Shapiro-Wilk. We used the Box-Cox transformation for all data, except clay, which was transformed using a logarithmic method. The Box-Cox improved the efficacy of normalizing and variance, equalizing for both positivelyand negatively-skewed variables [17].

Producing the Tidal Inundation Map

We analyzed 4 monthly tide tables and found that the tide rose somewhere between 0 - 3 m. The lowest altitude area was assumed to be inundated in longer period than higher altitude area. We summed up days exceeding a certain tidal height interval (0.1 m) and determined as inundation period starts from 0 m above sea level. Following tidal inundation classes (Table 1), we obtained specific height thresholds to separate the classes that is 1.1 and 1.6 m of altitude (Figure 3).

Semivariogram and Kriging Interpolation Analysis

Lag size and the number of lags are two parameters that contribute to the number of pair points generated in a semivariogram graph. The number of lags was set to 12, which is the default value set by the geostatistical analysis tool. From Figure 4, we found that variation in the sand data was confirmed to be the highest one, as indicated by the nugget value equal to 16.024, as shown in Table 3. However, we also found that salinity data need a longer distance to reach the sill value (nearly 45 km).

Table 2: Descriptive statistics of the four types of data

No.	Properties	Nuinber	Mean	Coefficient of Variation	Skewness	Kurtosis	Shapiro-Wilks (p-value)
1.	Clay	66	34.97	0.57	0.44	−0.61	0.045
2.	Sand	66	35.01	0.83	0.53	−1.05	7.561×10^{-5}
3.	Salinity	67	12.64	0.92	0.46	−1.14	1.939×10^{-5}
4.	Height	641	1.434	0.83	0.58	0.95	3.763×10^{-24}

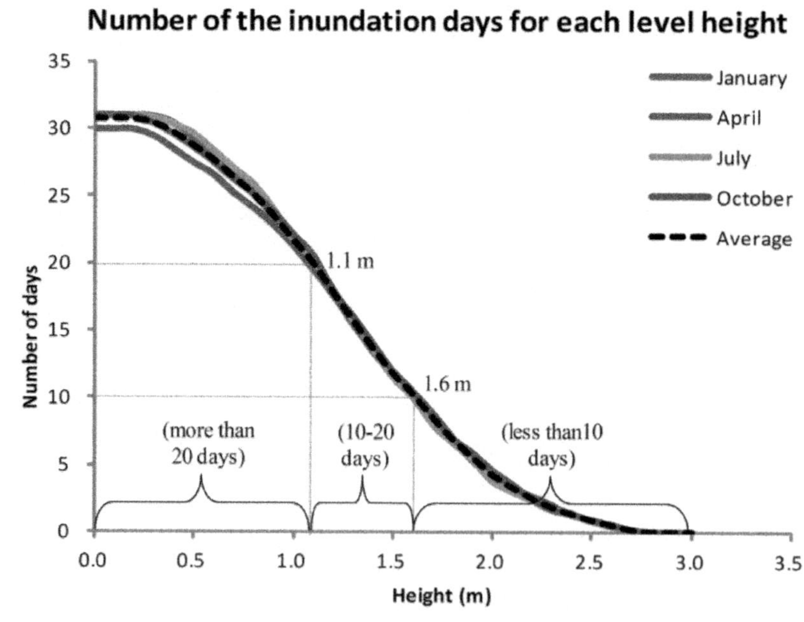

Figure 3: The number of inundation days for each of the height levels summarized from tide table.

Kriging interpolation accuracy was inspected through RMS and standardized RMS. However we could not compared RMS values, since the units were different. From the standardized RMS value, we found that clay and height data underestimated the variability of prediction (standardized RMS < 1), while sand and salinity would be overestimated (standardized RMS > 1).

Surface Classification Using GIS

Kriging interpolation produced surface layers, as shown in Figure 5. Clay fraction data after interpolation ranged between 7.4% and 52.8%. Two clay classes were defined using the Jenks natural break. Moderate class for clay was within the range 7.4% - 28.2%, while the suitable class was 28.2% - 52.8%. Sand data after interpolation ranged between 5.5% and 76.0%. Using similar method, three classes were generated and classified as: not suitable (5.5% - 25.6%), moderate (25.6% - 46.8%) and suitable (46.8% - 76.0%). Digital elevation model was classified as 1) elevation less than 1.1 m, which would be inundated more than 20 days a month; 2) elevation higher than 1.6 m, which would be inundated less than 10 days in a month; and 3) elevation ranging between 1.1 and 1.6 m, which would be inundated within 10 - 20 days in a month.

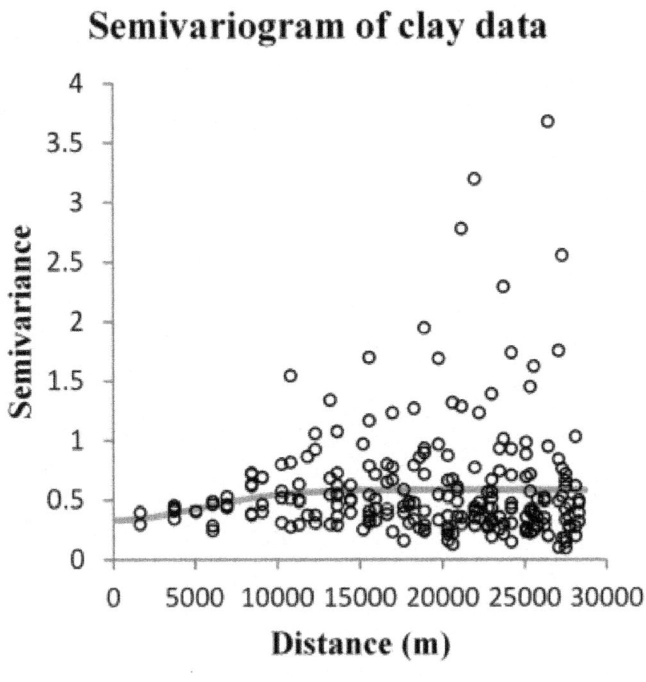

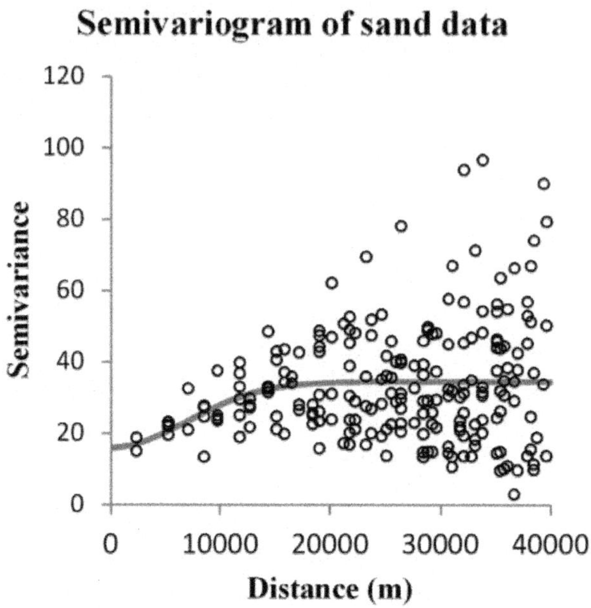
Semivariogram of sand data
Semivariance
120
100
80
60
40
20
0
0
10000
20000
30000
40000
Distance (m)

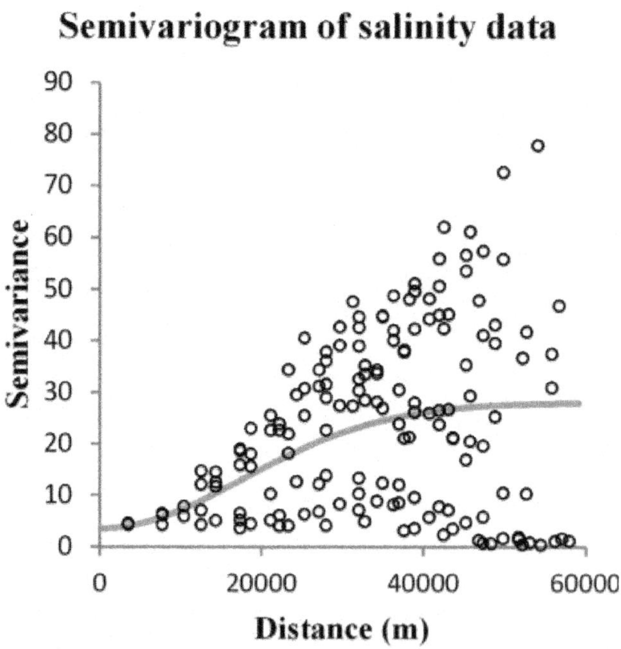
Semivariogram of salinity data
Semivariance
90
80
70
60
50
40
30
20
10
0
0
20000
40000
60000
Distance (m)

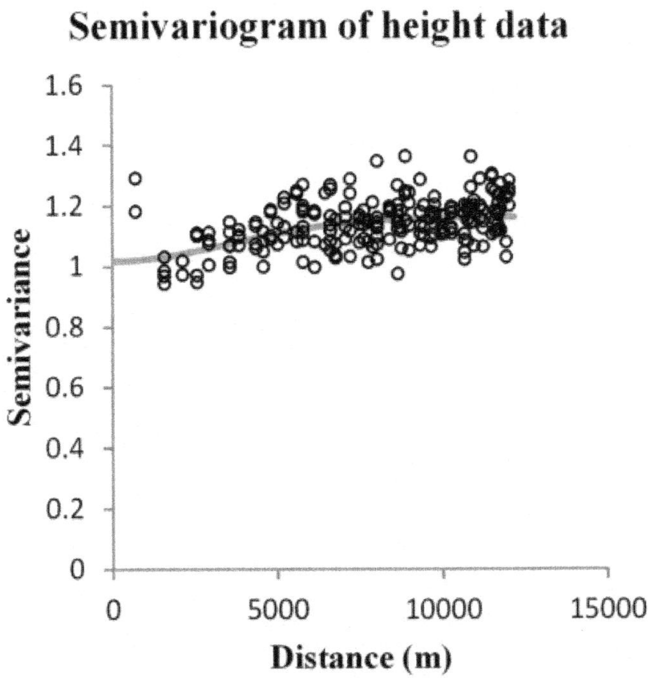

Figure 4: Omnidirectional semivariogram fitted by Gaussian model for all data.

Table 3: Semivariogram parameters for each data with RMS and standardized RMS values after kriging interpolation

Parameters	Semivariogram				Kriging interpolation	
	Range	Nugget	Total sill	Nugget ratio	RMS error	Standardized RMS
Clay	12693.267	0.331	0.584	0.567	21.821	0.779
Sand	17035.230	16.024	34.510	0.464	27.010	1.186
Salinity	43432.550	3.579	28.034	0.021	6.284	1.047
Height	8703.099	1.017	1.163	0.874	1.086	0.878

According to Table 1, there are ten unique zones of mangrove species. Some of the species shared similar site characteristics, such as Rhizophora mucronata and R. apiculata or Bruguiera parviflora and B. sexangula. The GIS-based overlay of whole surface layers produced 12 unique sites with different combination of all data. A site-suitability map of mangrove species was generated corresponding to criteria in Table 1. The results showed that three site zones were not represented in any GISunique sites, i.e. zone B (Rhizophora stylosa and Sonneratia alba), G (Lumnitzera racemosa), and H (Cerbera manghas). The final map of site suitability of mangrove species is presented in Figure 6.

DISCUSSION

Geostatistical Analysis

Reference [18] explained that non-normal data is actually common in geostatistical data, especially if the observed data is discrete, i.e. ppm, percentages of grade, etc. Moreover, reference [19] showed that in the case of soils, they naturally form in different depositional environments; therefore, their physical properties vary from point to point (in the horizontal as well as vertical plane). Thus, we assumed the existing variation of our data were natural. Semivariogram graphs shown in Figure 3 describe the relation of semivariance value (y-axis) and distances (xaxis). All data (clay, sand, salinity and height) seemed to follow the theoretical assumption of semivariogram, where points that are close to each other are expected to have small variance, and points that area farther apart are expected to have higher variance [15].

As shown in Table 3, the highest value for the sill element was that of sand, which means that the variance among the sand sample point data is greater than for other data. However, semivariogram measuring salinity showed that more than 43 km were needed to get variance to decrease, while clay had the smallest such distance. Sand data had the highest nugget value, which might indicate the highest measurement errors. Nevertheless, all semivariogram graphs showed a natural experimental omnivariogram.

Clay data had RMS error equal to 21.82%, which was smaller than sand with 27.01%. While salinity had RMS error of 6.28 ppt and height error 1.09 m. Standardized RMS value had juxtaposed our data into 2 groups (clay with height and sand with salinity) respecting to the estimation of the variability of the predictions. However, according to reference [20], statistical value is not the only source of assessing the interpolation quality; we also need to carefully study the visual quality of the surface model generated by kriging interpolation to detect distinct spatial patterns and visual faithfulness.

Analysis of Spatial Data Interpolation

We compiled 66 sample points for soil texture and 67 sample points for salinity, both onand off-shore. Samples were not spread evenly across the area but rather clumped in the southern part and missing in the central part. As a result, the study only focused on some parts of the Mahakam delta, which would have the best interpolation results. The interpolation process ignored the distributary channel and assumed that the topographic features in the delta plain were negligible [21]. Soil type in the Mahakam delta is loamy, varying from silt loam to clay loam, except for the coastline which is loam with sand [2]. According to Figure 5, we found consistency in this fact, whereas higher clay concentration dominate over the area of study. The sand fraction concentration is located at the outer delta, while clay mostly deposits in the center of delta, straight to the Mahakam river flow discharge direction. Although pond construction altered land, however, soil texture was assumed to remain fairly constant and was unchanged by management practices [22,23]. A suitable class for clay and sand means that both fractions have a high percentage. A high concentration of clay dominated 95% of the study area. A species preference for high percentage of clay implies that species was intolerant to the high percentage of sand and vice versa. Salinity also showed a natural distribution, where higher concentration situated in the outer delta rather than the inner delta. Salinity data is one of the sea water environmental factors that would spread gently and smoothly, especially in the estuary region. The total salinity range of the open ocean is about 30 - 40 parts per thousand (or ‰) while salinity in coastal estuaries can be much lower [24]. Altitude ranged from 0 - 4 m above sea level, with an average 2.26 m. A digital elevation model was generated

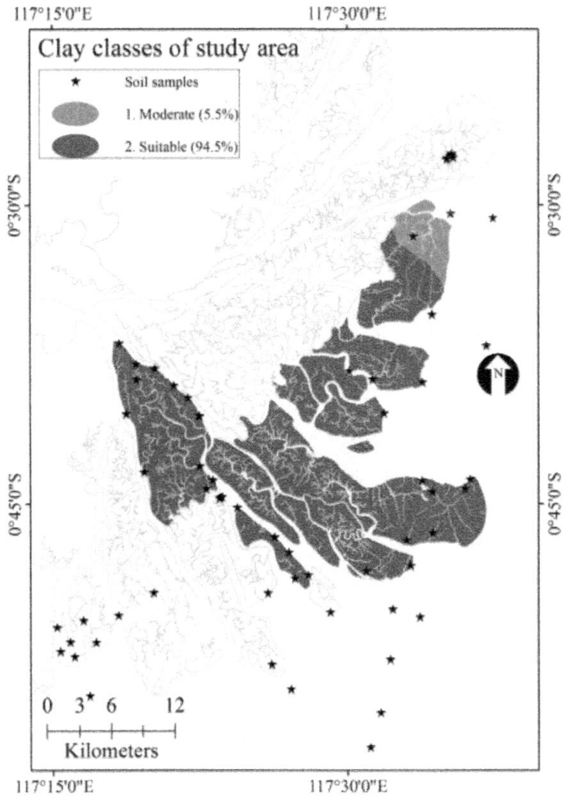

(a)

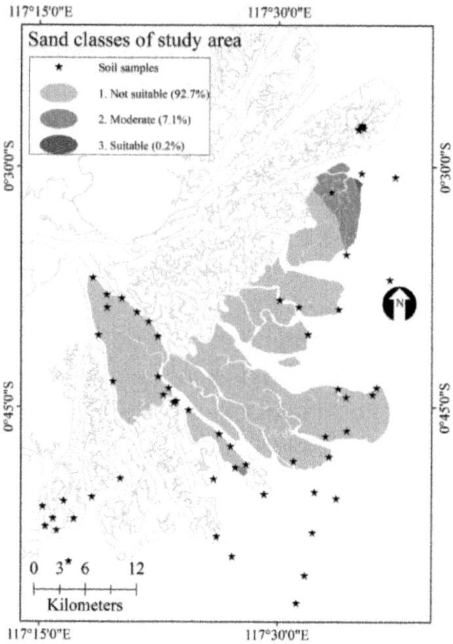
117°15'0"E
117°30'0"E
Sand classes of study area
Soil samples
1. Not suitable (92.7%)
2. Moderate (7.1%)
3. Suitable (0.2%)
0°30'0"S
0°45'0"S
0 3 6 12
Kilometers

(b)

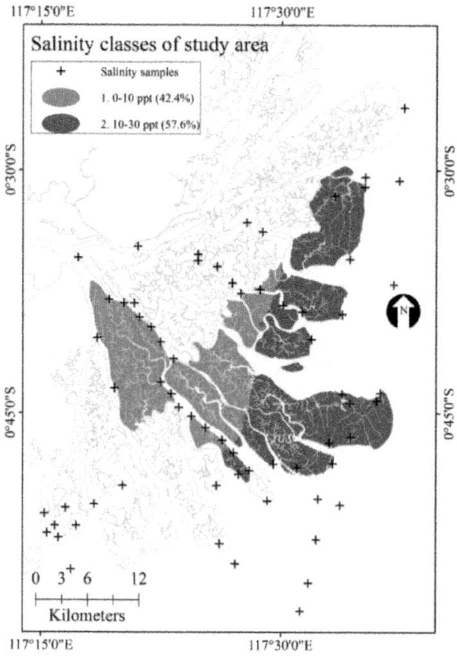
117°15'0"E
117°30'0"E
Salinity classes of study area
Salinity samples
1. 0-10 ppt (42.4%)
2. 10-30 ppt (57.6%)
0°30'0"S
0°45'0"S
0 3 6 12
Kilometers

(c)

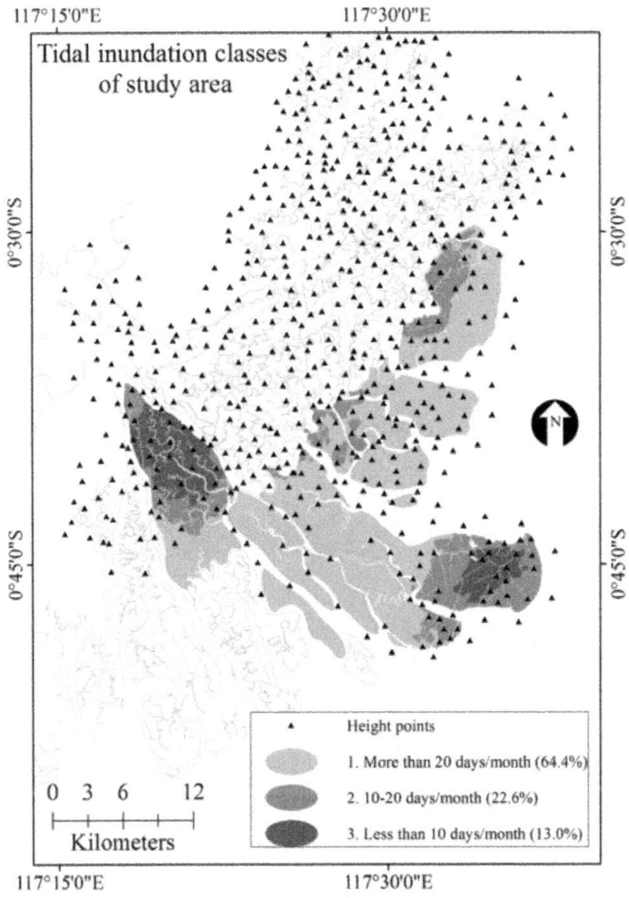

(d)

Figure 5: Classification of surface layers: (a) clay; (b) sand; (c) salinity; and (d) tidal inundation.

by kriging interpolation. The interpolation produced continuous elevation, ranging from 0 to 4.64 m. Normally higher elevation area are located in the inner delta, but interpolation found another high elevation area in the outer delta, near the coastline (Figure 5(d)). This area used to be an agricultural spot planted with coconut trees. However, during the introduction of shrimp pond, this area was also converted. The construction of ponds changed the terrain to flat.

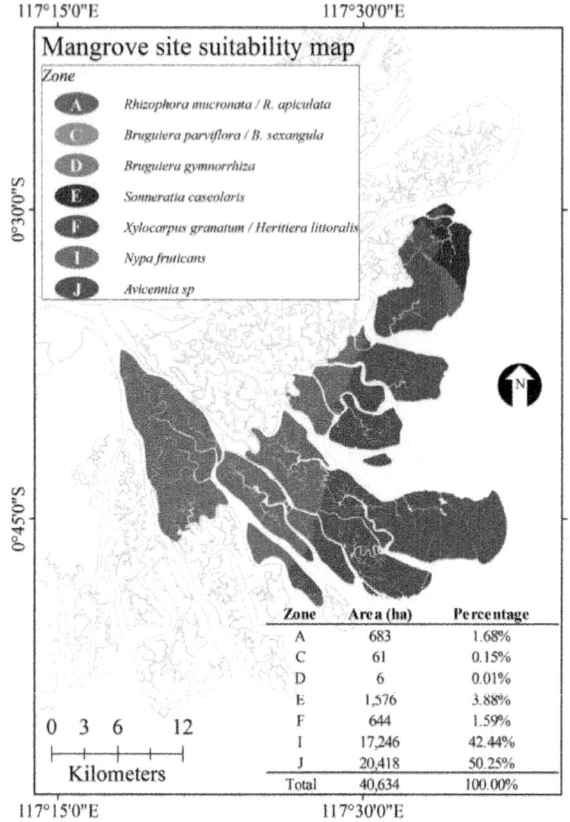

Figure 6: Site suitability map of our study sites generated using geostatistical analysis and GIS operations.

Site Suitability Mapping

Site suitability map as an output of spatial combination of four physical characteristics showed that theoretically 10 mangrove species were suitable for planting in the barren area or abandoned ponds. We calculated that 50.25% of the study area was suitable for Avicennia sp. and the other 42.22% suitable for Nypa fruticans. These two species occupied 92.69% of the study area. Another 7.14% was shared among Sonneratia caseolaris, Rhizophora mucronata, R. apiculata, Xylocarpus granatum and Heritiera littoralis. Only 0.16% would be suitable for Bruguiera species, which is known to prefer to grow at the boundary between mangroves and the coastal ecosystem, in the middle and upper inundation areas [25]. These results were consistent with the mangrove zoning, where Avicennia sp. is naturally found adjacent to the sea, together with Sonneratia species [26]. Meanwhile, Nypa fruticans seemed suitable to areas behind Avicennia sp., where brackish to almost freshwater streams exist and tidal effects decline.

Regarding to the rehabilitation efforts, site suitability map will assist the authority to prepare and implement rehabilitation strategy in this area. However, further studies are needed to validate this conceptual model by establishing small-scale demonstration plots and planting various mangrove species, to periodically compare their growth and mortality. Even though Avicennia sp. and Nypa fruticans seem suitable for about 92% of the total study area; it does not mean that other species will not survive if planted. Reference [26] emphasized that overlapping between species or zones occurred at many sites.

Site Suitability Mapping

Reference [27] reported that Total E&P Indonesia (company developing oil and gas) have planted more than 3 million mangrove trees on cleared areas, mainly along the pipeline distributional network. However, on a broader scale, mangrove rehabilitation in the Mahakam delta is still a minor project, due to unclear land ownership. Most of the ponds are effectively "owned" by the people operating them, so that plantation programs cannot be implemented in all necessary locations. Rehabilitation projects needs to be prepared carefully otherwise they will trigger social conflict. As a top priority, reference [2] suggested restoring mangroves in the areas that are incompatible with pond production due to poor soil quality or required for coastal protection.

In the case of the Mahakam delta region, a social approach for rehabilitation of the "occupied" land is a prerequisite to any such rehabilitation program. It is complicated by a number of factors, including the lack of clarity in land status, which, in theory, mostly belongs to the state, as well as the claims of the local people. Although these claims are illegal, it is impossible to restore the land without compensation to its current users; this is the main obstacle for rehabilitation programs in the region. Spending government fund for compensation is currently only permitted for private, not public, land. Recently, some small-scale silvofishery projects funded by a private company and an NGO had been introduced to this site in the last 3 - 4 years. However, most of the degraded areas remain barren and need immediate action to bring back the ecological function of mangrove forests for future generations.

CONCLUSION

Our study demonstrated an original approach to producing a site-suitability map of mangrove for rehabilitation purposes by combining 4 factors (clay, sand, salinity and height) using geostatistics. We proposed a new approach to making tidal inundation maps. Based on this study, we suggest height points linked with tide tables to produce tidal inundation maps. Statistical values and visual inspection of surface layers as a result of kriging interpolation yielded consistent results therefore, we assumed that interpolation processes have been carried out appropriately.

ACKNOWLEDGEMENTS

We would like to appreciate and thank Total E&P Indonesia, Dr. Iwan Suyatna from the Faculty of Fishery and Marine Sciences, University of Mulawarman, and the Faculty of Forestry, University of Mulawarman, for kind support in providing various data related to this study

REFERENCES

1. P. Sandjatmiko, A. M. Rony, H. Tarumadevyanto, I. Suyatna, Y. B. Sulistioadi, I. Tjitradjaja, L. Adrianto and D. G. Bengen, "Delta Mahakam Dalam Ruang dan Waktu. Ekosistem, Sumberdaya dan Pengelolaannya," BPMIGAS—Total E&P Indonesie dan Institute of Natural & Regional Resources (INRR), Jakarta, 2006. (In Indonesian).
2. R. Bosma, A. S. Sidik, P. Van Zwieten, A. Aditya and L. Visser, "Challenges of a Transition to a Sustainably Managed Shrimp Culture Agro-ecosystem in the Mahakam Delta, East Kalimantan, Indonesia," Wetlands Ecology and Management, Vol. 20, No. 2, 2012, pp. 89-99. http://dx.doi.org/10.1007/s11273-011-9244-0
3. R. A. Bourgeois, A. Gouyon, F. Jesus, P. Levang, W. Langeraar, F. Rahmadani, E. Sudiono and Y. B. Sulistioadi, "A Socio Economic and Institutional Analysis of Mahakam Delta Stakeholders," Total-Fina Elf, Belgium, Final Report to Total-Fina Elf, Contract No. 501125/DKI/204, 2002, pp. 108.
4. P. A. M. Van Zwieten, A. S. Sidik, Noryadi and I. S. Abdunnur, "Aquatic Food Production in the Coastal Zone: Data-based Perceptions on the Trade-off between Mariculture and Fisheries Production of the Mahakam delta and Estuary, East Kalimantan, Indonesia," In: C.T. Hoanh, T. P. Tuong, J. W. Gowing and B. Hardy, Eds., Environmental and Livelihoods in Tropical Coastal Zones: Managing Agriculture-Fishery-Aquaculture Conflicts, CABI, Oxfordshire, 2006, pp. 219-236. http://dx.doi.org/10.1079/9781845931070.0219
5. Anonym, "A Report on Detailed Ponds Mapping of Mahakam delta Using High Resolution Satellite Imagery," Fishery and Oceanic office of Kutai Kartanegara district, East Kalimantan province, Tenggarong, 2007.

(In Indonesian).

6. C. D. Field, "Rehabilitation of Mangrove Ecosystems: An Overview," Marine Pollution Bulletin, Vol.37, No. 8-12, 1999, pp. 383-392. http://dx.doi.org/10.1016/S0025-326X(99)00106-X
7. J. O Bosire, F. Dahdouh-Guebas, M. Walton, B. I. Crona, R. R. Lewis III, C. Field, J. G. Kairo and N. Koedam, "Functionality of Restored Mangroves: A Review," Aquatic Botany, Vol. 89, No. 2, 2008, pp. 251-259. http://dx.doi.org/10.1016/j.aquabot.2008.03.010
8. H. Ren, S. Jian, H. Lu, Q. Zhang, W. Shen, W. Han, Z. Yin and Q. Guo, "Restoration of Mangrove Plantations and Colonization by Native Species in Leizhou Bay, South China," Ecol Res, Vol. 23, No. 2, 2008, pp. 401- 407. http://dx.doi.org/10.1007/s11284-007-0393-9
9. C. Caratini and C. Tissot, "Paleogeographical Evolution of the Mahakam Delta in Kalimantan, Indonesia During the Quaternary and Late Pliocene," Review of Palaeobotany and Palynology, Vol. 55, No. 1-3, 1988, pp. 217-228. http://dx.doi.org/10.1016/0034-6667(88)90087-5
10. M. G. Sassi, A. J. F. Hoitink, B. de Brye, B. Vermeulen and E. Deleersnijder, "Tidal Impact on the Division of River Discharge over Distributary Channels in the Mahakam Delta," Ocean Dynamics, Vol. 61, No. 12, 2011, pp. 2211-2228.http://dx.doi.org/10.1007/s10236-011-0473-9
11. G. Lamorey and E. Jacobson, "Estimation of Semivariogram Parameters and Evaluation of the Effects of Data Sparsity," Mathematical Geology, Vol. 27, No. 3, 1995, pp. 327-358.http://dx.doi.org/10.1007/BF02084606
12. P. J. Curran, "The Semivariogram in Remote Sensing: An Introduction," Remote sensing of Environment, Vol. 24, No. 3, 1998, pp. 493-507. http://dx.doi.org/10.1016/0034-4257(88)90021-1
13. J. Yasrebi, M. Saffari, H. Fathi, N. Karimian, M. Moazallahi and R. Gazni, "Evaluation and Comparison of Ordinary Kriging and Inverse Distance Weighting Methods for Prediction of Spatial Variability of Some Soil Chemical Properties," Research Journal of Biological Sciences, Vol. 4, No. 1, 2009, pp. 93-102.
14. Govaerts and A. Vervoort, "Geostatistical Interpolation of Soil Properties in Boom Clay in Flanders," In: P.M. Atkinson, C.D. Lloyd, Eds., GeoENV VII—Geostatistics for Environmental Applications, Quantitative Geology and geostatistics, Vol. 16, Springer, Netherlands, 2010, pp. 219-230. http://dx.doi.org/10.1007/978-90-481-2322-3_20
15. K. Chang, "Introduction to Geographic Information System," 3rd edition, McGraw-Hill Publishing, New York, 2006, p. 432.

16. J. H. Zar, "Biostatistical Analysis," 4th edition, PrenticeHall, New Jersey, 1999, p. 469.

17. J. W. Osborne, "Improving Your Data Transformation: Applying the Box-Cox Transformation," Practical Assesment, Research & Evaluation, Vol. 15, No. 12, 2010, pp. 1-9. http://pareonline.net/genpare.asp?wh=0&abt=15

18. Ploner, "The Use of the Variogram Cloud in Geostatistical Modeling," Environmetrics, Vol. 10, No. 4, 1999, pp. 413-437.

19. A.L Jones, S.L. Kramer and P. Arduino, "Estimation of Uncertainty in Geotechnical Properties for Performancebased Earthquake Engineering, Pacific Earthquake Engineering Research Center, University of California, Barkeley, 2002, p. 23.http://peer.berkeley.edu/publications/peer_reports/reports_2002/0216.pdf

20. X. Yang and T. Hodler, "Visual and Statistical Comparison of Surface Modelling Techniques for Point-based Environmental Data," Cartography and Geographic Information Science, Vol. 27, No. 2, 2000, pp. 165-75. http://dx.doi.org/10.1559/152304000783547911

21. J. E. A. Storms, R. M. Hoogendoorn, R. A. C. Dam, A. J. F. Hoitink and S. B. Kroonenberg, "Late-Holocene Evolution of the Mahakam Delta, East Kalimantan, Indonesia," Sedimentary Geology, Vol. 180, No. 3-4, 2005, pp. 149-166.http://dx.doi.org/10.1016/j.sedgeo.2005.08.003

22. N. G. Juma, "The Pedosphere and It's Dynamics. A System Approach to Soil Science. Volume 1. Intoduction to Soil Science and Soil Resources," Salman Production, Canada, 2001. http://www.pedosphere.ca/volume01/pdf

23. McCauley, C. Jones and J. Jacobsen, "Soil and Water Management Module 1: Basic Soil Properties," Montana State University Extension Service, 2005.http://landresources.montana.edu/SWM/PDF/Final_proof_SW1.pdf

24. L. D. Talley, "Salinity Patterns in the Ocean," In: M. C. MacCracken and J. S. Perry, Eds., Encyclopedia of Global Environmental Change. Volume 1: The Earth System: Physical and Chemical Dimensions of Global Environmental Change, John Wiley & Sons, 2002, pp. 629- 640.

25. D. Setyawan and Y. I Ulumuddin, 2012,"Species Diversity and Distribution of Bruguiera in Tambelan Islands, Natuna Sea, Indonesia," Proceedings of the Society for Indonesian Biodiversity International Conference, Society for Indonesian Biodiversity, Solo, 22-23 July 2011, Vol. 1. 2012, p. 290. http://biosains.mipa.uns.ac.id/P/index.htm

26. W. Giesen, S. Wulffraat, M. Zieren and L. Scholten, "Mangrove Guidebook for Southeast Asia," FAO and Wetlands International, 2006.

27. C. H. Chaineau, J. Mine and Suripno, "The Integration of Biodiversity Conservation with Oil and Gas Exploration in Sensitive Tropical Environments," Biodiversity Conservation, Vol. 19, No. 2, 2010, pp. 587-600. http://dx.doi.org/10.1007/s10531-009-9733-0

Chapter 5

REMOTE SENSING OF ENVIRONMENTAL CHANGE IN THE ANTIRIO DELTAIC FAN REGION, WESTERN GREECE

Emmanuel Vassilakis

Department of Dynamics, Tectonics and Applied Geology, Faculty of Geology & Geoenvironment, National & Kapodistrian University of Athens, Panepistimioupoli Zografou, 15784, Athens, Greece

ABSTRACT

In the westernmost region of the rapidly widening Corinth rift, Greece, extensive development of roads, bridges and other human infrastructure has caused continuous environmental change over the past twenty years. River networks, the land surface and the coastal environment, have been altered, especially in the areas corresponding to deltaic fans. In this paper we use earth observation systems that have captured these environmental changes, particularly medium (Landsat TM and ETM+) and high (Quickbird) resolution satellite images, to identify environmental changes between the periods 1992, 2000, 2002, and 2005. Six pseudo-color multi-temporal images in different spectral areas were created in order to detect changes to the terrestrial and coastal environment caused mainly by direct or indirect human impact. This methodology provided new data for quantifying significant alterations in the environment on different scales. In many cases this revealed their sequence during the time of observation.

INTRODUCTION

Over the last two decades the Greek government has proceeded with infrastructure development within the western part of Greece. Formerly there was no uninterrupted highway that connected mainland Greece in the north with the Peloponnesus in the south, and the Gulf of Corinth interrupted the land continuity. Until recently, the only direct modern road connecting these large parts of the country was on the easternmost side of the gulf where the natural strait is located. Transportation across the western part of the gulf consisted of small ferries completing the route between Rio (mainland Greece) and Antirio (Peloponnesus). This distance, of not longer than two kilometers, was covered by a ferry in about 15 minutes, not including the time for embarkation and disembarkation, but only during adequate weather conditions.

The Gulf of Corinth is one of the most rapidly spreading rifts in the world, generating large and disastrous earthquakes [1,2,3,4,5]. Thus, one of the main arguments against construction of a bridge across the gulf was the high seismic potential of the area, which is related to its rapid expansion, determined by GPS techniques to be an average of 18 ± 2 mm/yr in an approximately north-south direction across the gulf [6,7,8,9]. New construction techniques allowed for design of the Rio-Antirio Bridge, which was structurally supported by different fault blocks. Its design allowed the bridge to remain intact during future differential movements of the fault blocks [10]. The bridge was finally completed in 2004, providing a modern alternative highway along the western part of Greece. The anticipated increase in traffic along the highways leading to the bridge encouraged the local authorities to extend and widen the existing road network to the north-east and the west, and especially north of the bridge in the mainland sector. Accordingly, several construction sites were developed in the area surrounding the Antirio deltaic fan (Figure 1). Some of these are still ongoing as planned (quarries, highways, junctions, tunnels, *etc.*) and a high quality highway will be completed during the next decade, providing easier, safer and faster transportation in western Greece.

Figure 1: The study area on a natural color (3,2,1—R,G,B) reference image acquired by Landsat 5 on September 2, 1992. The Quickbird image used in this study (1,4,3—R,G,B) is shown as a partially overlapping layer. The white rectangle in the inset shows the location of the study area.

In addition to these sites supporting new highway construction, a new river dam was created to supply fresh water to the eastern part of mainland Greece from the mountainous and densely forested area of western Greece where precipitation is much higher. The dam was constructed on the Evinos River beginning in 1992 and was completed in June 2001. It was filled in October 2002. Water from this dam flows through an artificial tunnel to an older dam (on the Mornos River), before it reaches the network that supplies water to Athens and the surrounding cities. Therefore, in the valley downstream from the dam on the Evinos River, there has been a reduction in the average volume of water that reaches the coastal area just west of the new Rio-Antirio Bridge, along with a reduction in the sediment load carried by the river. Moreover, in the forested area between the Evinos valley and Antirio deltaic fan many wildfires—whether set intentionally or naturally occurring—have occurred over the last several decades. This has had serious consequences for the natural environment because most of the fire damage has not been repaired. Due to the high precipitation in this area, high rates of erosion have been recorded during wintertime, resulting in a large amount of transported material being deposited in a violent way along the Antirio deltaic coastal area [11,12].

This paper concerns the results of a first effort to document the surface changes of this tectonically and geomorphologically active region and to categorize them systematically, for further interpretation. The initial idea was to define whether the changes detected on the earth's surface are caused by the natural procedures of the evolving landscape or due to human interference, or both. The most efficient way to observe environmental alterations on such a wide scale, is the use of various remote sensing data, especially medium to high resolution and in the next section the methodology is described in detail.

DATA PROCESSING AND ANALYSIS

In order to detect environmental changes and the implications caused by the large construction sites along the Antirio fan and surrounding areas, several series of earth observation datasets were created. The satellite images used for this purpose were captured during several time periods beginning in 1992 and ending in 2005, and covering most of the area that was affected by the development of new infrastructure, as described in the introduction. The satellite image processing employed in this study includes atmospheric corrections, co-registration of images and creation of multi-temporal data constructed by creating new datasets for every spectral channel. The resulting false color composites reveal environmental changes not only where the construction sites are located but also in the coastal areas where erosion and deposition occur. The higher resolution images were used to increase the spatial resolution of the color composites, accomplished by merging high resolution images with lower resolution images [13,14] for better evaluation and quantification of the environmental alterations.

The first scene was acquired by the Thematic Mapper sensor (TM), placed on board the Landsat 5 satellite, on September 2, 1992, and consists of path 184 and row 33. Processing begins with geometric ortho-rectification in Universal Transverse Mercator (UTM) projection (zone 34N) by including, in the correction procedure, a highly detailed DEM with 25 meters pixel size [15]. The latter was produced by digitizing topographic map contours at a scale of 1:50,000. Subsequently, a dataset with 1123 columns and 577 rows was created; this covered only a part of the total scene and corresponds to the study area. The same procedure and dimensions were also applied to two other datasets, one derived from a Landsat 7-ETM+ (Enhanced Thematic Mapper) acquired on June 28, 2000 and another from a Landsat 7-ETM+ acquisition on August 13, 2005. All Landsat data were downloaded from U.S. Geological Survey (USGS) Earth Resources Observation and Science (EROS) Center. The last two datasets were ortho-rectified by using the image to image methodology and taking as reference dataset, the oldest one having the same DEM for altitude reference. A high resolution Quickbird 2 image acquired on June 12, 2002 and covering the central part of the study area was also used for larger scale observations with the highest possible accuracy and was imported in a geodatabase for further study along with the other data. These data were ortho-rectified, pan-sharpened and registered to the Landsat 5 ortho-corrected dataset, which was used primarily as the referencing image (Figure 1). The co-registration procedure of all datasets was successful since the RMS error was calculated less than 0.5 pixels either in X or Y geographic axis.

Effects of the atmosphere on remotely sensed data are not considered "errors" because they are part of the signal received by the sensing device. However, it is very important to remove atmospheric effects, especially on visible channels, as it has been noted that the atmospheric contribution to the radiance received by a satellite forms a much greater percentage of the radiance leaving the target area, than in the case of infrared [16]. Because this study is focused on detecting changes in a specific region through time, the main analytical tool relies on ground measurements made over time. Therefore it is very important to correct the radiance values recorded by the sensor for the effects of the atmosphere [17]. The technique that was applied on the Landsat series of datasets was based on the minimum digital number of the middle infrared band 7 for each dataset [18].

In general, when images contain areas of low reflectance, such as clear water bodies and deep shadows caused by extreme relief discontinuities, the corresponding pixels have values very close to zero, especially in the middle infrared band 7 of the Landsat TM and ETM+ sensors. Considering that a water body has virtually no reflection, even at shorter wavelengths such as band 4 or band 5, the digital number of a pixel representing deep sea water in band 7 should not be significantly larger than zero [19]. If it is larger than zero, then the non-zero digital number is due to atmospheric effects. This enables one to identify the effects of the atmosphere over large water bodies. The existence of the large water bodies in the Gulf of Patras west of the study area, and the Gulf of Corinth east of the study area, are thus very useful for calculating atmospheric corrections. After computing the statistics over each band of every dataset, the histogram minimum value for band 7 was calculated. This value was subtracted from the digital number arrays of the other bands of the same dataset and the result was a new array with digital numbers shifted towards the zero value, and presumably representing data with atmospheric corrections. The statistics for every single band were updated and the histograms were saved.

MULTI-TEMPORAL INTERPRETATION

The acquisition dates of the three Landsat images used for the multi-temporal interpretation were almost ideal for the purpose of this study because nearly all of them were captured during the summertime (late June to early September) when the local weather conditions are similar.

The multi-temporal analysis procedure begins with the compilation of six datasets, each one containing a single spectral channel from the three Landsat images. Each of these datasets contains three bands with the same Landsat spectral channel for each acquisition date. The thermal infrared channels of the Landsat 5 and Landsat 7 were not used because the spatial resolution varies for the TM and ETM+ sensors and the thermal reflectance cannot be considered identical throughout long time periods. Thus, the interpretation continues separately for each spectral channel in order to detect changes in the absorption and reflection spectra for specific bandwidths. Subsequently, six pseudochromatic images were produced by using the earliest data in the red channel (R-1992), the latest data in the blue channel (B-2005) and the intermediate data in the green channel (G-2000).

The resultant color composite images for every channel are presented in such a way as to locate the areas with high pixel values for each of the red, green or blue colors. In these cases, the earth's surface changed significantly from one acquisition date to the next. These observations proved to be quite difficult to obtain from images collected during the intermediate observation period (2000), as restoration works for the altered land coverage (river flow, forest fire, quarry operations *etc.*) could have taken place before the capturing date of the last satellite image (2005).

In detail, areas colored nearly pure red in the pseudo-chromatic images reveal areas that suffered significant change after September 2, 1992 and remained unchanged until August 13, 2005. The almost pure green areas represent changes on the earth's surface which happened in the time window between November 3, 1992 and June 28, 2000; with the provision that before and after these dates the land coverage should be more or less similar, or at least the spectral attitude of the area should be the same. If full and successful environmental restorations of an altered area have occurred, these areas would appear as green in each of the six datasets. Otherwise the restoration could not be characterized as completely successful, because the spectral attitude of the rehabilitated area is not the same before and after the environmental alteration. Finally, nearly pure blue areas represent change that happened after June 28, 2000 and before August 13, 2005.

CHANGE DETECTION RESULTS

In the pseudo-color image produced from band 1, the most spectacular change is represented by a blue linear feature connecting mainland Greece to the Peloponnesus; this is the Rio-Antirio cable bridge completed in 2004 and inaugurated just before the Athens Olympic Games (Figure 2). The blue color indicates that this bridge did not exist when the first two satellite images were acquired but did exist when the third satellite image was acquired. Thus it is clearly depicted on the 2005 Landsat image, which has been assigned to the blue channel. Its structural towers can be clearly seen in the southern part of the Quickbird image, which was captured before 2004 and apparently during construction (Figure 1). Additionally, the new road junctions on both sides of the gulf next to the bridge are also blue.

In the same image, just northeast of the town of Nafpaktos, a number of blue linear features indicate parts of the new highway detour designed to minimize traffic increase in the town itself. North of Nafpaktos, the lighter blue features with reddish parts are probably several road widenings with roadside banks constructed before 2005 (blue) or trenches constructed after 1992 (red). A new road constructed after 2000 in the forested area (21°43′E, 38°26′N) is also visible as a blue polyline feature (Figure 2).

Figure 2: Multi-temporal pseudo-color image produced from layer stacking of Landsat satellite data (Band 1, 0.45 to 0.52 μm) with three different acquisition dates. The diachronic changes in onshore and offshore areas, largely due to infrastructure development after 2000, are clearly shown as blue linear features. The black arrows point to alterations of the surface described in the text. Below, two single band-1 enlargements of the black rectangle area show the changes caused by the bridge construction during the period from 1992 (left) to 2005 (right).

The flysch rocktype that makes up most of the bedrock in the study area has great potential for generating large-scale landslides. Frequent heavy rainfall in this area, combined with surface denudation from road construction and river erosion, cause major and minor land movements, which expose bare soil prior to rehabilitation and the renewal of vegetation (Figure 3) Such phenomena can be identified along the road that connects Nafpaktos with Vlachomandra, especially along the western side of the road where red pixels indicate landslides that happened shortly before 1992 and that were successfully restored before 2000. Larger-scale land movements seem to have occurred before 2005 on the eastern bank of the Mornos river (21°53′E, 38°25′N) and smaller scale changes along both banks of the Evinos river, especially when flysch lithologies were exposed along the river (21°36′E, 38°25′N).

Figure 3: Multi-temporal pseudo-color image produced from layer stacking of Landsat satellite data (Band 2, 0.52 to 0.60 μm) with three different acquisition dates. Land surface destruction due to mass landsliding along roads and river banks, as well as other changes in bare soil and vegetation, can be identified. The black arrows point to alterations of the surface described in the text.

A spectacular change in the land surface resulting from restoration of a burnt area can be observed over a large area in the north central portion of the study area (21°44′E, 38°24′N). In the oldest image (1992), this large area represents a region of bare soil in the middle of forest-type vegetation; such features normally correspond to tree destruction. In the multi-temporal pseudo-color images this area is colored reddish, whereas a smaller area in between the large ones is colored green (Figure 3and Figure 4). In the high resolution Quickbird image (Figure 1) it is clear that in the same area there is a densely vegetated area with striped cultivation corresponding to vineyards. This interpretation is based on the reflectance contrast between bare soil and vegetation. Thus bare soil existed during 1992, a large part of which was cultivated before 2000 (red) and the vineyard expanded between 2000 and 2003 (green); as seen in the Quickbird image, the vegetation seems to be quite homogenous throughout this whole area.

Just west of this vineyard another major environmental alteration is indicated by the multi-temporal data analysis. A big quarry was developed sometime after 1992 but before 1996 when air photographs show its operation at a very early stage. This quarry is located on flysch, mainly consisting of shale and sandstone. Such materials are widely used in local housing settlements for decorative reasons. The quarry works have expanded, gradually causing a decrease in the forest vegetation that used to cover the surrounding area. Restoration of the quarry site must have begun between 2003 and 2005 because there is no blue in the false color image at this location, indicating that the reflectance was not that of bare soil when the latest Landsat image was acquired (Figure 4). There is also a smoothing of the relief as shown in the Quickbird image; this is not consistent with the exposure of the steep quarry terraces that are clearly visible in the 1996 high resolution air photographs.

Figure 4: Multi-temporal pseudo-color image produced from layer stacking of Landsat satellite data (Band 3, 0.63 to 0.69 μm) with three different acquisition dates. High contrast spots in the forested area reveal changes in reflectance due to human activity or forest fires. The time at which the reflectance changed can be estimated by the colors on the image, as described in the text (see black arrows).

For many years now, the reduction in water flow along rivers emptying into the Gulf of Corinth has caused coastal erosion and a regression of the shoreline. This could be related to construction of several dams along upstream portions of the rivers or perhaps to global climate change. Multi-temporal interpretation of Landsat images reveals that the northern coastlines of the two gulfs, on either side of Antirio, have suffered serious regression. In the near infrared datasets a discontinuous red stripe of a few pixels width is visible, beginning just east of Nafpaktos through the west of its waterfront, indicating that the beach which existed there in 1992—and possibly for many years before—has eroded (Figure 4 and Figure 5).

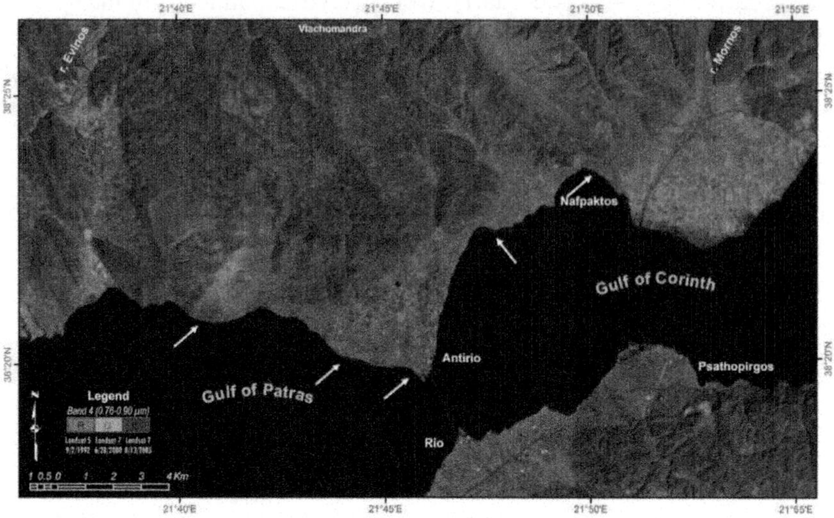

Figure 5: Multi-temporal pseudo-color image produced from layer stacking of Landsat satellite data (Band 4, 0.76 to 0.90 μm) with three different acquisition dates. The high absorbance of water creates a high reflectance contrast along the coastline so that shoreline regression is easily observed and quantified. The white arrows point to alterations of the coastal areas described in the text.

Along the eastern coast of the Antirio fan, no significant changes in the shoreline can be observed. A blue linear feature corresponds to a small pier constructed before 2005 but after 2000. In contrast, along the western part of the coastal deltaic area, an almost continuous red stripe corresponds to a regression of the 1992 shoreline, which in places has regressed by as much as 50 meters. Reddish pixels can also be observed along the western part of the coastline in the study area, where the main outcrops onshore consist of limestone, conglomerate, shale and sandstone, which make up parts of the flysch sequence. The steep topographic slopes present here, combined with the high frequency of planes of weakness in the rock (caused by bedding, joints, *etc.*) have the potential to produce rock fall into the sea; thus the red pixels along the coast line might represent missing land surfaces destroyed by rock fall, with rock fall probably occurring after 2000 and definitely after 1992. The epicenter of the large Aigion earthquake on June 15, 1995 (M=6.1) is located just 35 km eastward of this area, suggesting that rock fall could have been triggered directly by the earthquake or be related to post seismic surface deformation [20].

The Evinos dam, which is not located in the study area, has apparently altered the natural flow regime of the river. This has created temporary changes in the pattern of river incision, especially in the low relief segments of the valley at the northwestern end of the study area. In the multi-temporal pseudo-color images these changes are visible as linear features of different colors and widths; these linear features cross each other in a wide part of the river valley (Figure 6). These shifts in the river bed and incision pattern could have been the result of short-term meteorological changes, but because the three multi-temporal images were all acquired during the summer it is most likely that they are caused by controlled water release at the Evinos dam. Controlled water release at the dam would produce a significant change in the rate of discharge and sediment carrying capacity in downstream portions of the river, potentially causing erosion, landsliding and/or rock fall in the lithological formations that outcrop adjacent to the river valley, especially after sudden flooding.

Similar phenomena do not seem to have occurred along the Mornos river valley, perhaps because this dam has been in operation since 1981 and did not alter its operations during the period when the satellite images were acquired, or because a narrow artificial watercourse was constructed downstream of the dam in order to avoid floods at the river outlet in the Mornos deltaic fan area.

Figure 6: Multi-temporal pseudo-color image produced from layer stacking of Landsat satellite data (Band 5, 1.55 to 1.75 μm) with three different acquisition dates. The pattern of water flow in the rivers has shifted throughout the thirteen year observation period as indicated by crosscutting linear features of different colors. The black arrows point to alterations of the surface described in the text.

Because the lithology of the mountainous area consists mainly of flysch [21] and the climate is generally humid, the forest is extremely dense. During the summer whilst temperatures are high, forest fires occur quite frequently, resulting in large burnt areas with consequent environmental effects. By using the multi-temporal interpretation technique of Landsat data images, the areas affected by forest fires can by highlighted and one might be able to determine the extent of the burned area and the approximate time of the fire. In the case of Antirio, several burnt areas are apparent on the image in Figure 7.

A bluish area just east of the Evinos river valley corresponds to a burnt area that appeared between 2000 and 2005; there has been no effective rehabilitation. There are also many spotty areas in the forests that display a mixture of colors; these are in contrast with the densely vegetated areas that were burned and subsequently restored between the three capturing dates. For a more detailed study in terms of time, a more complete time series of images should be used in the future by applying the same methodology but with shorter periods of time between the capturing dates of the acquired remote sensing data.

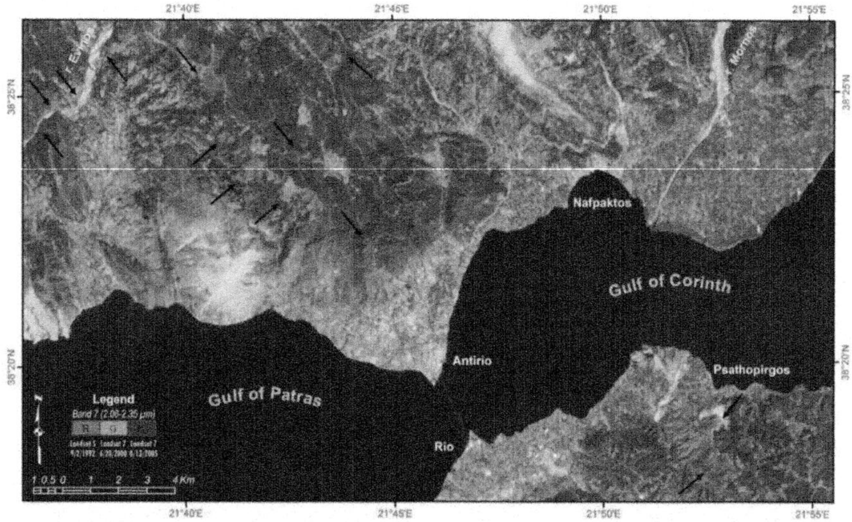

Figure 7: Multi-temporal pseudo-color image produced from layer stacking of Landsat satellite data (Band 7, 2.08 to 2.35 μm) with three different acquisition dates. The reflectance in this band highlights the difference between bare soil and cultivated land, thus showing areas that have been denuded by forest fires. The various colors show surface changes between the three satellite acquisition dates. In particular, large areas of a reddish color appear across the mountainous area where previously burnt areas have been rehabilitated either naturally or artificially before the 2000 Landsat data was acquired. The black arrows point to alterations of the surface described in the text.

CONCLUSIONS

For the last decade, Greece has been making an ongoing investment in its infrastructure in western mainland Greece, including the building of roads, bridges and dams. The human interaction with the natural landscape evolution has a much higher impact than if the human factor was less significant. Almost all the landscape evolution observed for this area is directly or indirectly connected to major ongoing construction activities.

The environmental impact of these activities would not be apparent without the interpretation of remote sensing data such as described in this paper. The region of western Greece is nearly ideal for applying techniques involving multi-temporal remote sensing data because in a very short time period many environmental alterations have occurred due to natural and human activities. The multi-temporal image interpretation technique allows for the identification and, in many cases, the quantification of these environmental effects. The availability of a dense time series of remote sensing multispectral data provides an opportunity to identify changes on the earth's surface. In some cases, these may signal major environmental changes of a serious nature and indicate the need for preventative measures to avoid an environmental disaster.

This paper highlights the important role that analysis of multispectral satellite data can play in the identification of surface alterations related to human activity and natural processes. In some cases high resolution satellite images will also be needed to quantify the spatial extent of surface alterations, but multispectral data, at either medium or high spatial resolution, will remain the key in identifying the existence and timing of changes to the earth's surface environment.

ACKNOWLEDGEMENTS

The author would like to thank L. H. Royden for her constructive comments on an early version of this manuscript, as well as two anonymous reviewers whose suggestions and comments highly improved the text and the figure settings. Special thanks are due to D. J. Papanikolaou for encouraging the publication of this study and E. L. Lekkas for kindly offering the Quickbird raw data.

REFERENCES

1. Jackson, J.A.; Gagnepain, J.; Houseman, G.; King, G.; Papadimitriou, P.; Soufleris, C.; Virieux, J. Seismicity, normal faulting, and the geomorphological development of the Gulf of Corinth (Greece) the Corinth earthquakes of February and March 1981. *Earth Planet. Sci. Lett.* 1982, *57*, 377–397.

2. Bernard, P.; Briole, P.; Meyer, B.; Lyon-Caen, H.; Gomez, J.-M.; Tiberi, C.; Berge, C.; Cattin, R.; Hatzfeld, D.; Lachet, C.; Lebrun, B.; Deschamps, A.; Courboulex, F.; Larroque, C.; Rigo, A.; Massonnet, D.; Papadimitriou, P.; Kassaras, J.; Diagourtas, D.; Makropoulos, K.; Veis, G.; Papazisi, E.; Mitsakaki, C.; Karakostas, V.; Papadimitriou, E.; Papanastassiou, D.; Chouliaras, M.; Stavrakakis, G. The M_s=6.2, June 15, 1995 Aigion earthquake (Greece): evidence for low angle normal faulting in the Corinth rift. *J. Seismol.* 1997, *1*, 131–150.

3. Evangelidis, C.P.; Konstantinou, K.I.; Melis, N.S.; Charalambakis, M.; Stavrakakis, G.N. Waveform Relocation and Focal Mechanism Analysis of an Earthquake Swarm in Trichonis Lake, Western Greece. *Bull. Seismol. Soc. Amer.* 2008, *98*, 804–811.

4. Gaki-Papanastassiou, K.; Papanastassiou, D.; Maroukian, H. Geomorphic and archaeological-historical evidence for past earthquakes in Greece. *Ann. Geofis.* 1996, *39*, 589–601.

5. Papanikolaou, D.; Chronis, G.; Lykousis, V.; Pavlakis, P. Active Tectonics in the Rion-Antirion Strait, Western Greece. In Proceedings of 5th Meeting European Geological Societies, Dubrovnik, Croatia, October 6–7, 1987; pp. 72–73.

6. Briole, P.; Rigo, A.; Lyon-Caen, H.; Ruegg, J.C.; Papazissi, K.; Mitsakaki, C.; Balodimou, A.; Veis, G.; Hatzfeld, D.; Deschamps, A. Active deformation of the Corinth rift, Greece: Results from repeated Global Positioning System surveys between 1990 and 1995. *J. Geophys. Res.*2000, *105*, 25605–25625.

7. Avallone, A.; Briole, P.; Agatza-Balodimou, A.M.; Billiris, H.; Charade, O.; Mitsakaki, C.; Nercessian, A.; Papazissi, K.; Paradissis, D.; Veis, G. Analysis of eleven years of deformation measured by GPS in the Corinth Rift Laboratory area. *Comptes Rendus Geosciences* 2004, *336*, 301–311.

8. Bernard, P.; Lyon-Caen, H.; Briole, P.; Deschamps, A.; Boudin, F.; Makropoulos, K.; Papadimitriou, P.; Lemeille, F.; Patau, G.; Billiris, H.; Paradissis, D.; Papazissi, K.; Castarθde, H.; Charade, O.; Nercessian, A.; Avallone, A.; Pacchiani, F.; Zahradnik, J.; Sacks, S.; Linde, A. Seismicity, deformation and seismic hazard in the western rift of Corinth: New insights from the Corinth Rift Laboratory (CRL). *Tectonophysics* 2006, *426*, 7–30.

9. Cianetti, S.; Tinti, E.; Giunchi, C.; Cocco, M. Modelling deformation rates in the western Gulf of Corinth: rheological constraints. *Geophys. J. Int.* 2008, *174*, 749–757.

10. Parcharidis, I.; Foumelis, M.; Kourkouli, P.; Wegmuller, U. Persistent Scatterers InSAR to detect ground deformation over Rio-Antirio area

(Western Greece) for the period 1992–2000. *J. Appl. Geophys.* 2009, *68*, 348–355.

11. Zelilidis, A. The geometry of fan-deltas and related turbidites in narrow linear basins. *Geol. J.*2003, *38*, 31–46.
12. Bell, R.E.; McNeill, L.C.; Bull, J.M.; Henstock, T.J. Evolution of the offshore western Gulf of Corinth. *Geol. Soc. Amer. Bullet.* 2008, *120*, 156–178.
13. Welch, R.; Ehlers, W. Merging Multiresolution SPOT HRV and Landsat TM Data. *Photogramm. Eng. Remote Sensing* 1987, *53*, 301–303.
14. Rigol, J.; Chica-Olmo, M. Merging remote-sensing images for geological-environmental mapping: application to the Cabo de Gata-Nvjar Natural Park, Spain. *Environ. Geol.* 1998, *34*, 194–202.
15. Pouncey, R.; Swanson, K.; Hart, K. *ERDAS Field Guide*, 5th ed.; ERDAS Inc: Atlanta, GA, USA, 1999; p. 671.
16. Cracknell, A.; Hayes, L. *Introduction to Remote Sensing*; Taylor and Francis: London, UK, 1991; p. 293.
17. Mather, P. *Computer Processing of Remotely-Sensed Images: An Introduction*; Wiley & Sons: West Sussex, UK, 1994; p. 352.
18. Hadjimitsis, D.G.; Clayton, C.R.I.; Hope, V.S. An assessment of the effectiveness of atmospheric correction algorithms through the remote sensing of some reservoirs. *Int. J. Remote Sens.*2004, *25*, 3651–3674.
19. Lillesand, T.; Kiefer, R. *Remote Sensing and Image Interpretation*, 3th ed.; Wiley & Sons: New York, NY, USA, 1994; p. 750.
20. Parcharidis, I.; Metaxas, C.; Vassilakis, E. Earth observation data and geographical information system (GIS) techniques for earthquake risk assessment in the western Gulf of Corinth, Greece.*Canad. J. Remote Sens.* 2006, *32*, 223–227.
21. Jenkins, D. Structural development of Western Greece. *AAPG Bull.* 1972, *56*, 128–149.

Chapter 6

DETECTION AND MONITORING OF ACTIVE FAULTS IN URBAN ENVIRONMENTS: TIME SERIES INTERFEROMETRY ON THE CITIES OF PATRAS AND PYRGOS (PELOPONNESE, GREECE)

Issaak Parcharidis [1], Sotiris Kokkalas [2], Ioannis Fountoulis [3] and Michael Foumelis [4]

[1]Harokopio University of Athens, Department of Geography, El. Venizelou 70, 17671 Athens, Greece

[2]University of Patras, Department of Geology, Division of Physical Geology, Marine Geology and Geodynamics, 265 00 Patras, Greece

[3]National and Kappodistrian University of Athens, Faculty of Geology and Geoenvironment, Department of Dynamic Tectonic and Applied Geology, Panepistimioupolis Zografou, 157 84 Athens, Greece

[4]National and Kappodistrian University of Athens, Faculty of Geology and Geoenvironment, Department of Geophysics and Geothermics, Panepistimioupolis Zografou, 157 84 Athens, Greece

ABSTRACT

Monitoring of active faults in urban areas is of great importance, providing useful information to assess seismic hazards and risks. The present study concerns the monitoring of the potential ground deformation caused by the active tectonism in the cities of Patras and Pyrgos in Western Greece. A PS interferometric analysis technique was applied using a rich data set of ERS–1 & 2 SLC images. The results of the interferometric analysis were compared with the tectonic maps of the two cities. Patras show clearer uplift–subsidence results due to the more distinct fault pattern and intense deformation compared to the Pyrgos area, where more diffused deformation is observed, with no significant displacements on the surface.

INTRODUCTION

The earthquake cycle of an active fault may include coseismic rupture and interseismic deformation. During the interseismic stage that usually ranges from a few hundreds to thousands of years, crustal tectonic strain may be silently accumulated. The strain is released during the interseismic period, especially along creeping active faults.

Understanding active tectonic processes and related energy release through monitoring of the transient deformation of strain accumulation process has become fundamental for several human activities. Additionally, local deformation type of a fault and the area near the fault may determine the extent of the seismic hazard as well. This allows taking into consideration measures and activities for seismic hazard mitigation.

Monitoring of active faults' interseismic behavior in urban areas is of great importance, as the local exposure (population, infrastructures etc.) increases the risks. Recently, interseismic crustal velocities and strains have been determined for a number of active areas, through repeated measurements using a Global Position System. In some cases the terrain is remote and the accessibility is difficult and thus the density of GPS measurements is relatively sparse, or in the case of urban environments, the operation of GPS receivers may be interrupted due to the frequent blockage of signals [1,2].

During the last two decades the SAR Differential Interferometric (DInSAR) technique based on radar satellite data has become a useful tool for ground deformation detection and monitoring [3,4]. Recent developments (since the end of the 90s) in DInSAR have demonstrated the potential to overcome some of the known limitations of repeat–pass interferometry. By examining interferometric phase from stable, point like targets, it is possible to monitor stability and cover an area that is normally characterized by low coherence. Additionally, millimetric target displacement along the line of sight (LOS) directions can be detected allowing the measurement of slow terrain motion [5]. Permanent or Persistent Scatterers Interferometry (PSI) is a technique used to calculate fine motions of individual ground and structure points over wide areas. These reflectors should remain stable (interferometric phase stability over time). Interferometric Point Target Analysis (IPTA) is a specific method of PSI to exploit temporal and spatial characteristics of interferometric signatures collected from point targets to map scatterer deformation history [6]. Monitoring faults in urban environments using different interferormetric techniques (repeat–pass, stacking and PS) has been widely used in seismically active areas of the globe [1,7,8,9,10,11,12,13,14].

During 1993 catastrophic earthquakes affected northwestern Peloponnese in the southern part of Greece. They caused serious damages to the greater area of two densely populated cities, Patras and Pyrgos. These are the capital cities of the neighboring prefectures of Achaia and Ilia. This study concerns the use of the PS interferometric analysis of ERS–1 and 2 satellite data, over the cities of Pyrgos and Patras in order to monitor and reveal the spatial distribution of creep along active faults.

TECTONIC AND GEOLOGICAL FRAMEWORK—ACTIVE TECTONICS

Western Greece is an area that characterized by high seismicity and a fairly complex three-dimensional setting, with along strike changes in the progress of subduction zone. It comprises continent–continent collision in the north and ocean–continent subduction in the south [15,16,17,18,19]. The change between continent–continent and ocean–continent subduction occurs at the Kephalonia Transform Fault (KTF) [20], which is characterized by a dextral strike–slip sense of shear. GPS data from western Greece show a small amount of motion north of the KTF and southwest-directed rapid motion of the overriding plate south of it that reaches rates of 30–35 mm/yr in the western parts of the Peloponnese along with dextral kinematics of the KTF [21,22].

The NW Peloponnese about 70–80 km east of the present day NNW–SSE trending Hellenic Trench (Figure 1) has a complex geological history of tectonics and erosion. However, the active deformation in western Greece could be described in spatio–temporal continuation with the foreland–propagated fold and thrust belt of the External Hellenides, which can be followed along the Hellenic arc [15,19]. During the Eocene, the Peloponnese was characterized by an Alpine collisional history, which led to assemblage of intra-Tethyan continental fragments (e.g., Apulia and Pelagonian microcontinents) and the formation of the Hellenic mountain range [23]. Within this context, Mesozoic early Cenozoic carbonate rocks, originally deposited on a series of platforms (Pre-Apulian and Gavrovo isopic zones) and basins (Ionian and Pindos isopic zones), were telescoped by a system of N–S striking and east-dipping thrust faults that propagated upward and westward into overlying flysch deposits [24,25,26]. From the late Miocene and onwards thrust processes progressively shifted westward from the Peloponnese area to the Ionian Sea [15,27]. Compressional structures of Upper Miocene to Quaternary age have been recognized in Ionian Islands from field based studies, as well as in deep seismic profiles [15,28].

An extensional stress field, in turn, has prevailed in the northwestern Peloponnese from the early Pliocene up to present day generating three major sets of active normal faults with NE–SW, WNW–ESE and ENE–WSW trends, respectively [27,29,30,31,32]. The complexity of the structure in the westernmost end of the Hellenic subduction is imprinted on the great variety of earthquake focal mechanisms.

Patras

The Patras graben is a 40 km long and 15 km wide fault zone which extends from Cape Araxos in the west to the village of Hellinikon in the east. It comprises thick fluvio–deltaic conglomerates, and floodplain or shallow lacustrine sands and clays of Pleistocene age. Towards its eastern part these deposits pass gradually into a coarse conglomerate sequence that was formed as debris flow facies in a fan environment [27].

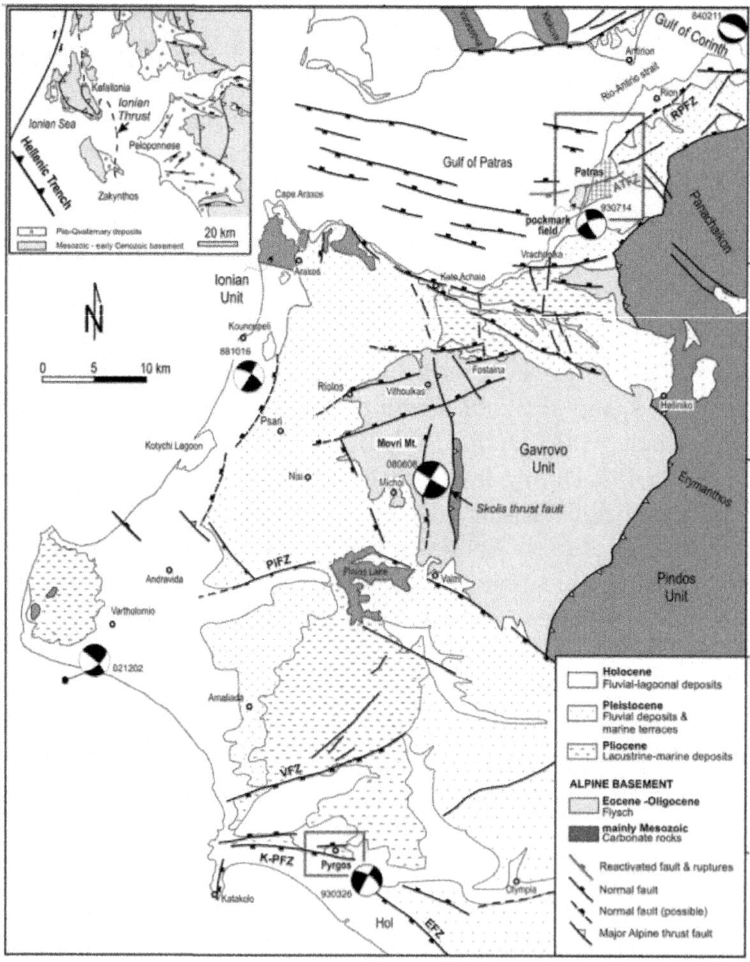

Figure 1: Tectonic setting of the broader study area (ATFZ: Ag. Triada Fault Zone, EFZ: Epitalio fault Zone, RPFZ: Rio–Patra fault Zone, K–PFZ: Katakolo– Pyrgos

Fault Zone, PiFZ: Pinios Fault Zone, VFZ: Vounargo Fault Zone, blue frames indicate the study areas, map modified from Koukouvelas *et al.* [50].

Holocene sediments, with a thickness of 40 m, are arranged in a series of tilted fault blocks bounded by WNW–trending normal faults that accommodate the extension in the area [33]. Generally, WNW trending faults are clearly segmented along strike and most are characterized by variable displacement. Such differences are accommodated by NNE–oriented transfer faults, which permit abrupt changes in depositional conditions along the WNW–trending downthrown blocks and bound small subbasins at the southern margin of the basin.

On the onshore part of the graben, a major WNW–trending fault scarp is formed on the southern part of the graben along the pre-Neogene margin. ENE–directed faults are also present and displace fluviatile terraces. Holocene net vertical slip on the central parts of Patras basin is about 3–5 mm/yr, while on its northern and southern margin is about 1.2 mm/yr [34]. Rifting in the Patras area may have initiated some 1.3–1.5 Ma ago [35].

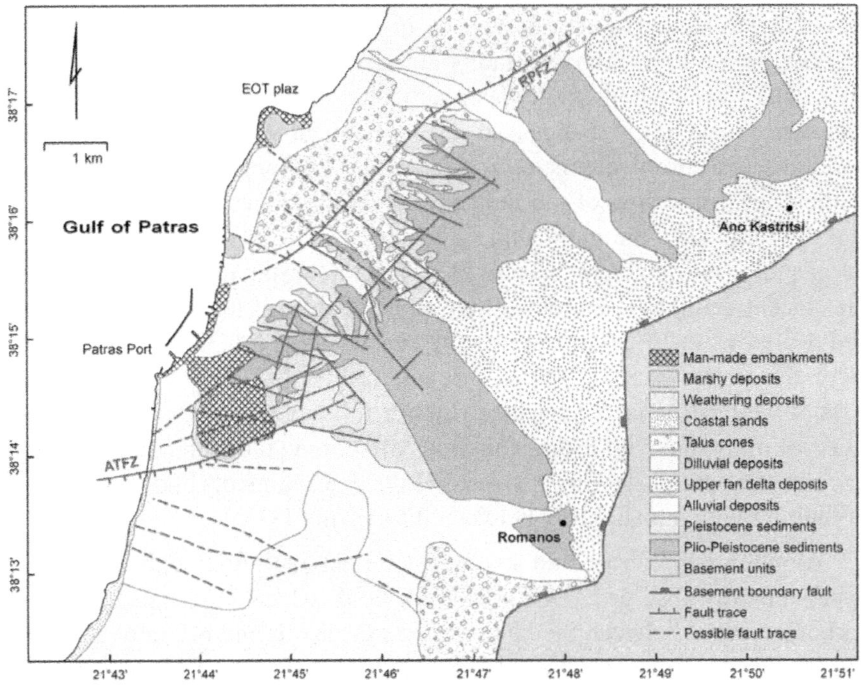

Figure 2: Geological map of Patras City (ATFZ: Ag. Triada Fault Zone, RPFZ: Rio–Patra fault Zone), map modified from Doutsos *et al.* and Koukis *et al.* [27,36].

Further to the east, the Rio graben is developed from the village of Vrahneika to the city of Patras and continues further east through Rio to Nafpaktos town. The Rio graben displays two distinct sedimentary cycles: (a) thick silts and clays deposited during Upper Pliocene into a lacustrine to shallow marine environment and (b) coarse fluvial deposits and alluvial fans of Pleistocene age. Pleistocene Rion series are found uplifted up to 500–600 m high in the hills of Ano Kastritsi. A fairly straight NE–trending fault slope limits this hilly morphology from the narrow Patras littoral plain. Major structural trends in this area vary between NE and ENE (55°–70°). Rio graben acted as a transfer zone between the differently extending Patras and Corinth grabens by reactivation of preexisting NE–SW trending faults in the Pliocene [27].

The city of Patras is founded mainly on Quaternary deposits and Plio–Pleistocene sediments (Figure 2) with a thickness exceeding 300 m, based on borehole data [36]. The fault trace map of the broader residential area of the city, based on fieldwork and air photo interpretation [36] shows a NE–SW main fault trend at the northern part of the city and a more prominent WNW–ESE trend in the southern part. This complex fault interplay in the area is due to the location of Patras city in the junction between Patras and Rio grabens. Some ENE–trending faults are also present, such as the Ag. Triada fault zone (ATFZ). This fault was reactivated during the August 31 (Ms 4.8) 1989 earthquake event and caused serious damage to new multistorey and old buildings, in a narrow elongated zone about 1.5 km long and 50 m wide parallel to the fault. Surveying of the fault motion with geodetic methods for almost eight months after the main shock showed a total subsidence of 25 mm and horizontal displacement of 14 mm [37]. The total estimated throw on the Plio–Pleistocene sediments is on the order of 40 m towards the eastern part of fault and decreases to 15–20 m towards the coastal areas [36,37]. Surface ruptures were also observed in a N70° E orientation for almost 1.5 km. Towards the coastal western parts, the surface rupture follows the course of Diakoniaris river for more than 500 m and continues offshore in the Gulf of Patras dipping towards the south, showing tilting of Holocene sediments and clear evidence of faulting with a total throw between 0.5–5.0 m [38].

Along this fault trace and south of it, a pockmark field with craters of gas expulsion was observed. This pockmark field is located close to Patras new harbor, confined between the 10 m and 45 m isobaths and was activated during another strong earthquake (Ms 5.4) on 1993. The earthquake focal mechanism [39] shows a N058° trending nodal plane. If this nodal plane is the active one, the slip on this plane is dextral with a dip–slip component. Anomalously temperature increase and gas expulsion along an ENE–pockmark string of craters recorded prior the earthquake activity suggesting that this fault system is active and plays an important role in fluid circulation [40].

Pyrgos

The city of Pyrgos is located about 70–80 km east of the NNW–SSE trending Hellenic Trench. Neotectonic structures in the area are not oriented parallel to the arc in the NNW–SSE direction [41]. Instead, there are several WNW–ESE, E–W, and NE–SW trending normal faults forming the margins of the post-Alpine basins developed on top of the thrust sheets of the Hellenides, with alternating horst and graben structures [32,42,43,44].

The broader Pyrgos–Olympia area corresponds to a large 1st order graben structure bounded to the north by the Erymanthos horst, to the east by the Tropea horst and to the south by the Lapithas Mt. horst all of which are built of alpine age sediments. In the Pyrgos–Olympia graben there are more than 3 km of Plio–Quaternary sediments, including some diapiric structures related to the existence of Upper Miocene evaporites at depth, detected from geophysical prospecting and drilling [45]. Continental and lacustrine Pliocene–Lower Pleistocene deposits occur mainly at the eastern part of the Olympia basin whereas marine sediments occurring along the coastal zone of the gulf are mainly of Lower Pleistocene age. The neotectonic evolution of the Pyrgos graben was not the same in all its extent [29]. It was differentiated due to the creation and evolution of smaller size tectonic blocks (2nd, 3rd order macrostructures). One of these smaller order structures is the Pyrgos horst. A great number of faults have been located and studied during neotectonic surveys; some of these were active in previous neotectonic periods (e.g., Late Miocene, Pliocene, Early Pleistocene), whereas others are active Holocene structures [46]. The E–W trending active faults in the Pyrgos area within the Pyrgos–Olympia basin have been analysed after the 1993 destructive earthquake event [30,47].

The outcrops in the broader Pyrgos area (Figure 3) consist of: (i) Alluvial deposits, which crop out in the flat area of the broader Pyrgos area and overlay unconformably the older formations, while their thickness does not exceed 12 m; (ii) Erymanthos Formation, which outcrops over a limited area and consists of red to brown red clays and yellow brown sandy clays of Pleistocene age. It overlies unconformably the Vounargo formation and its thickness varies from 2 m to 8 m; (iii) Vounargo Formation, comprises continuous intercalation of clays, silts, sandstones, sands and marls of Plio–Pleistocene age, with a total thickness up to 600 m [29,46]. The deposits that occur in the area of Pyrgos have undergone neotectonic deformation and are crossed by a number of faults of E–W mean strike that form part of the Katakolo–Pyrgos fault zone, which was responsible for the earthquakes of 26 March 1993 [29].

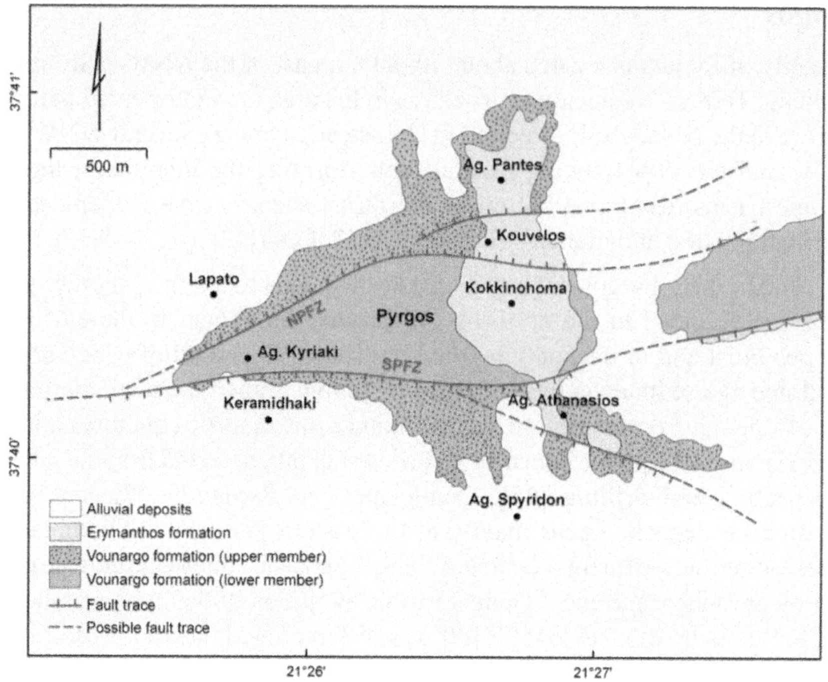

Figure 3: Geological map of Pyrgos City (NPFZ: North Pyrgos Fault Zone; SPFZ: South Pyrgos Fault Zone).

More specifically, the elongated outcrop of the lower member terminates at the South Pyrgos fault, south of which the upper member of the Vounargo formations occurs. It is a normal fault, accompanied by a morphological discontinuity (to the south of the city main square); its throw is at least 50 m (estimated from morphotectonic features) and eastwards it branches into two faults.

To the north there is an identical setting with the North Pyrgos fault (av. strike NE–SW), which also branches into two minor faults to the east. The North Pyrgos fault juxtaposes the outcrops of upper members, while it also crosses some outcrops of Erymanthos formation. Its throw is smaller, (20–30 m) and all along it, seismic fractures caused by the shock of 26 March 1993, were recognized.

The main shock as well as the aftershocks caused extensive damage in Pyrgos and the surrounding area. Several secondary geological effects such as liquefaction, sandblows, landslides and fractures, were observed.

SAR INTERFEROMETRIC PROCESSING AND ANASLYSIS

Interferometric processing for the selected study areas was performed using 42 ERS–1/–2 SAR scenes acquired between 1992 and 2000 along descending track 279 (Table 1) for the Patras case and 39 ERS–1/–2 SAR scenes along descending track 279 for Pyrgos (Table 2).

Table 1: ERS–1 & 2 scenes (Track = 279, Frame = 2835) that have been used (in bold the master reference image)

Count	Master (date)	Satellite	Slave (date)	Orbit	Bp (meters)	dT (days)
1	03/06/1995	ERS–1	12/11/1992	6937	73.9	−933
2	03/06/1995	ERS–1	10/06/1993	9943	−477.4	−723
3	03/06/1995	ERS–1	19/08/1993	10945	−323.7	−653
4	03/06/1995	ERS–1	28/10/1993	11947	569.7	−583
5	03/06/1995	ERS–1	25/03/1995	19305	−1169.9	−70
6	03/06/1995	ERS–1	29/04/1995	19806	−518.5	−35
7	**03/06/1995**	**ERS–1**	**03/06/1995**	**20307**	**0**	**0**
8	03/06/1995	ERS–1	08/07/1995	20808	−583.4	35
9	03/06/1995	ERS–2	13/08/1995	1636	82.9	71
10	03/06/1995	ERS–2	17/09/1995	2137	−360.1	106
11	03/06/1995	ERS–1	21/10/1995	22311	761.3	140
12	03/06/1995	ERS–2	31/12/1995	3640	186.2	211
13	03/06/1995	ERS–2	19/05/1996	5644	71.0	351
14	03/06/1995	ERS–2	23/06/1996	6145	−83.8	386
15	03/06/1995	ERS–2	01/09/1996	7147	−514.6	456
16	03/06/1995	ERS–2	06/10/1996	7648	−392.7	491
17	03/06/1995	ERS–2	10/11/1996	8149	1064.9	526
18	03/06/1995	ERS–2	15/12/1996	8650	−309.2	561
19	03/06/1995	ERS–2	19/01/1997	9151	−28.7	596
20	03/06/1995	ERS–2	23/02/1997	9652	−182.3	631
21	03/06/1995	ERS–2	04/05/1997	10654	−360.3	701
22	03/06/1995	ERS–2	08/06/1997	11155	−166.9	736
23	03/06/1995	ERS–2	13/07/1997	11656	−124.9	771
24	03/06/1995	ERS–2	17/03/1997	12157	111.5	806
25	03/06/1995	ERS–2	21/09/1997	12658	−263.2	841
26	03/06/1995	ERS–2	30/11/1997	13660	177.5	911
27	03/06/1995	ERS–2	04/01/1998	14161	116.4	946
28	03/06/1995	ERS–2	19/04/1998	15664	193.1	1051
29	03/06/1995	ERS–2	24/05/1998	16165	−160.7	1086
30	03/06/1995	ERS–2	28/06/1998	16666	−834.9	1121
31	03/06/1995	ERS–2	02/08/1998	17167	56.7	1156
32	03/06/1995	ERS–2	06/09/1998	17668	47.0	1191
33	03/06/1995	ERS–2	28/02/1999	20173	302.9	1366
34	03/06/1995	ERS–2	13/06/1999	21676	−559.8	1471
35	03/06/1995	ERS–2	18/07/1999	22177	455.9	1506
36	03/06/1995	ERS–2	22/08/1999	22678	1012,1	1541
37	03/06/1995	ERS–2	26/09/1999	23179	448.6	1576
38	03/06/1995	ERS–2	31/10/1999	23680	357.3	1611
39	03/06/1995	ERS–2	05/12/1999	24181	−121.9	1646
40	03/06/1995	ERS–2	09/01/2000	24682	−140.1	1681
41	03/06/1995	ERS–2	23/04/2000	26185	953.7	1786
42	03/06/1995	ERS–2	28/05/2000	26686	806.6	1821

Table 2: ERS–1 & 2 scenes (Track = 279, Frame = 2846) that have been used (in bold the master reference image)

Count	Master (date)	Satellite	Slave (date)	Orbit	Bp (meters)	dT (days)
1	17/8/1997	ERS–1	12/11/1992	6937	−44.7	−1739
2	17/8/1997	ERS–1	10/6/1993	9943	−581.5	−1529
3	17/8/1997	ERS–1	19/8/1993	10945	−432.0	−1459
4	17/8/1997	ERS–1	28/10/1993	11947	455.7	−1389
5	17/8/1997	ERS–1	25/3/1995	19305	−1274.8	−876
6	17/8/1997	ERS–1	29/4/1995	19806	−624.8	−841
7	17/8/1997	ERS–1	3/6/1995	20307	−107.5	−806
8	17/8/1997	ERS–2	13/8/1995	1636	−28.5	−735
9	17/8/1997	ERS–1	21/10/1995	22311	653.7	−666
10	17/8/1997	ERS–1	18/5/1996	25317	39.4	−456
11	17/8/1997	ERS–2	23/6/1996	6145	−196.6	−420
12	17/8/1997	ERS–2	1/9/1996	7147	−622.8	−350
13	17/8/1997	ERS–2	10/11/1996	8149	954.8	−280
14	17/8/1997	ERS–2	15/12/1996	8650	−415.3	−245
15	17/8/1997	ERS–2	19/1/1997	9151	−139.1	−210
16	17/8/1997	ERS–2	23/2/1997	9652	−288.6	−175
17	17/8/1997	ERS–2	4/5/1997	10654	−466.7	−105
18	17/8/1997	ERS–2	8/6/1997	11155	−276.1	−70
19	17/8/1997	ERS–2	13/7/1997	11656	−235.9	−35
20	**17/8/1997**	**ERS–2**	**17/8/1997**	**12157**	**0**	**0**
21	17/8/1997	ERS–2	21/9/1997	12658	−370.4	35
22	17/8/1997	ERS–2	30/11/1997	13660	63.3	105
23	17/8/1997	ERS–2	4/1/1998	14161	9.9	140
24	17/8/1997	ERS–2	19/4/1998	15664	78.8	245
25	17/8/1997	ERS–2	24/5/1998	16165	−271.3	280
26	17/8/1997	ERS–2	28/6/1998	16666	−941.3	315
27	17/8/1997	ERS–2	2/8/1998	17167	−47.8	350
28	17/8/1997	ERS–2	6/9/1998	17668	−60.7	385
29	17/8/1997	ERS–2	28/2/1999	20173	186.5	560
30	17/8/1997	ERS–2	13/6/1999	21676	−674.8	665
31	17/8/1997	ERS–2	18/7/1999	22177	341.6	700
32	17/8/1997	ERS–2	22/8/1999	22678	898.1	735
33	17/8/1997	ERS–2	26/9/1999	23179	345.9	770
34	17/8/1997	ERS–2	31/10/1999	23680	247.4	805
35	17/8/1997	ERS–2	5/12/1999	24181	−236.6	840
36	17/8/1997	ERS–2	9/1/2000	24682	−247.4	875
37	17/8/1997	ERS–2	23/4/2000	26185	838.9	980
38	17/8/1997	ERS–2	19/11/2000	29191	839.0	1190
39	17/8/1997	ERS–2	24/12/2000	29692	−567.1	1225

Initial estimates of the interferometric baselines were calculated from available precise orbit state vectors from Delft Institute (NL) for Earth–Oriented Space Research (DEOS) [48]. The topographic phase was simulated based on SRTM V3 DEM of approximate spatial resolution of 90 m, resampled to 40 m to fit the SAR data resolution.

Time-series analysis was based on the IPTA processing scheme [6]. Temporal and spatial characteristics of interferometric signatures collected from point targets are exploited to accurately map average ground deformation rates and deformation histories.

The selection of the reference points is regarded to be the most critical part of the IPTA, as final deformation rates and histories are greatly affected by this decision. Some criteria for the selection of the reference point are dictated by the applied method, such as the high quality of the point in terms of phase stability overtime. Others are related to the regional tectonic setting of the area and the related pattern of deformation which needs to be extracted.

Interferometric point target analysis

Starting from a stack of coregistered Single Look Complex (SLC) images, the selection of the reference scene was based primarily on the baseline minimization criteria. In addition, the selection of a reference scene acquired near the temporal average of the available SAR acquisitions is also of interest.

The first step of the analysis involves the identification of candidate point targets for which the time-series analysis will be performed. In this case two different approaches were applied. The first approach is based on the spectral properties of each individual SLC image. This is done by identifying point targets of low spectral phase diversity. The second approach involves the identification of candidate point targets based on low intensity variability, since by definition point targets do not show speckle behavior as simple coherent scatterer dominates the echo.

The analysis of the differential interferometric phases in the temporal direction is an important element of an interferometric point target analysis. Point data stack of differential interferograms was generated and analyzed by means of phase regression analysis in the temporal domain using two dimensional bilinear regression model.

Two-dimensional regression analysis is done with the dimensions corresponding to the perpendicular baseline of the interferometric pairs and to the time difference between the two SLC of the interferometric pairs according to:

a0 + a1 × bperp [i] + a2 × delta_t [i]

where a0: phase offset, a1: slope in baseline dimension (can be converted into point height correction), a2: slope in time dimension (can be converted into linear deformation rate), bperp: perpendicular baseline component of interferograms, and delta_t: time interval of interferograms.

The model examines linear dependence of the topographic phase on the perpendicular baseline component as well as linear phase dependence with time, solving respectively for both height correction and constant deformation rate of the point target relative to the reference. The regression analysis of the entire stack of observations was first conducted using multiple patches, within each patch one reference is determined, and then using the selected single reference point as a global reference. This procedure was followed in order to minimize the effect of distance between the two pairs of phase components, as the atmospheric distortion, baseline error (residual orbital phase trends) and higher relative deformation rates result in higher deviations of the individual points from the regression plain. The quality of the preliminary candidate points were then carefully evaluated based on the estimated standard deviation of the differential interferometric phase from the 2-D regression model.

Points with a phase standard deviation larger than the indicated threshold (in this case 1.0) were rejected, significantly reducing the number of scatterers. The majority of the rejected points were located over mountainous areas. A total number of 6,613 and 1,829 targets for Patras and Pyrgos respectively were detected.

The general phase model for IPTA that used is the same as the conventional interferometry. The unwrapped interferometric phase Φunw is expressed as the sum of topographic phase $\Phi topo$, deformation phase Φdef, path delay Φatm (atmospheric phase) and the phase noise $\Phi noise$:

$$\Phi unw = \Phi topo + \Phi def + \Phi atm + \Phi noise$$

Phase terms related to the atmosphere, nonlinear deformation, baseline errors and noise can be discriminated within the residual phases based on their differing spatial and temporal dependencies [6].

Unwrapped phases calculated from the regression analysis described above and the corresponding topographic phases were then used in a least-squares approach for baseline refinement. The analysis was limited over areas exhibiting no deformation as dictated by the linear deformation estimates. Introducing the refined baselines and taking into consideration the early estimated height corrections and linear deformation rates, the interferometric phase model was updated in a second iteration.

Additional processing includes temporal and spatial filtering of newly estimated residual phases to compensate for atmosphere and noise. Atmospheric screen was attributed to large scale nonlinear residuals and subtracted from the model by applying low-pass spatial filtering on the residual phases. Phase noise was treated by spatially filtering of phases around the reference, assuming stability of the area considered.

Further iteration applying the additional corrections results in the final regression model. Results consist of point heights, linear deformation rates, atmospheric phase, refined baselines, quality information (temporal coherence) and nonlinear deformation histories for each point. It is important to mention that the final deformation model, as a consequence of the assumptions made during the estimation of atmospheric phase contribution to the signal, includes nonlinear components of only local scale phase variations.

RESULTS

After transformation of the interferometric results from range–Doppler coordinates into map geometry (geographic coordinates), point targets were imported in a GIS environment and plotted on a panchromatic Landsat–7 ETM+ image and in the Google Earth environment for point target identification (Figure 4 and Figure 5).

Here we should note that, although the reference points are considered stable there are no absolute stable points in the area. In this way, negative velocities do not necessarily represent subsidence, but possibly slower uplift according to the reference point. Thus, when we use the terms subsidence and uplift we refer to relative values with regard to the picked reference point.

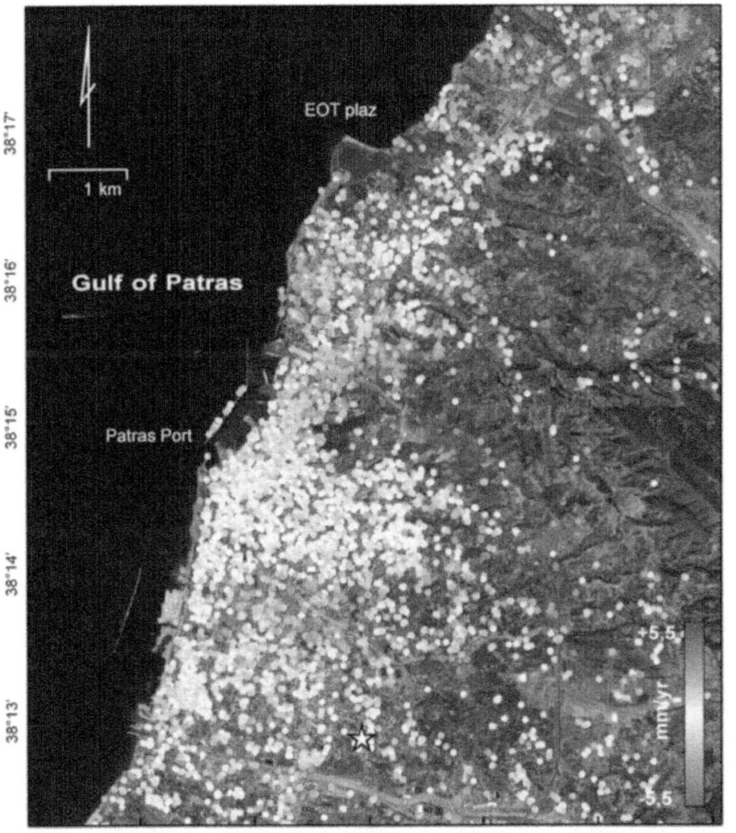

Figure 4: Linear component of ground deformation obtained by IPTA over Patras city. Point targets are plotted on a Panchromatic Landsat–7 ETM+ image. The star on the image refers to the location of the reference point.

Figure 5: Linear component of ground deformation obtained by IPTA over Pyrgos city. Point targets plotted on a QuickBird Pan–Sharpened image. The star on the image refers to the location of the reference point.

Patras

From Figure 6 it is indicated that generally the relative vertical velocities toward and away from the satellite range in the LOS, vary between maximum values of +5 mm/yr and −5 mm/yr, respectively. A remarkable aspect is that there is contrasting subsidence and uplift of PS points along discrete and specific zones, such as the Ag. Triada fault zone in the south of the city and along a W–E trending lineament, north of the Patras port, that doesn't relate with any visible mapped fault trace (Figure 7). Between these two zones only uplift even with low rates is observed. This area represents in a way the relay zone and the hanginwall block of both Ag. Triada and Rio–Patras fault zone, consisting of Plio–Pleistocene sediments. Maximum uplift velocities on the order of +4.0 mm/yr to +5.5 mm/yr are constricted on the footwall block of Pio–Patras fault, while in the relay zone between that fault zone and Ag. Triada fault lower values between +2.5 mm/yr and +4.0 mm/yr are observed. Similar values are also calculated towards west close to the basin bounding Kastritsi fault.

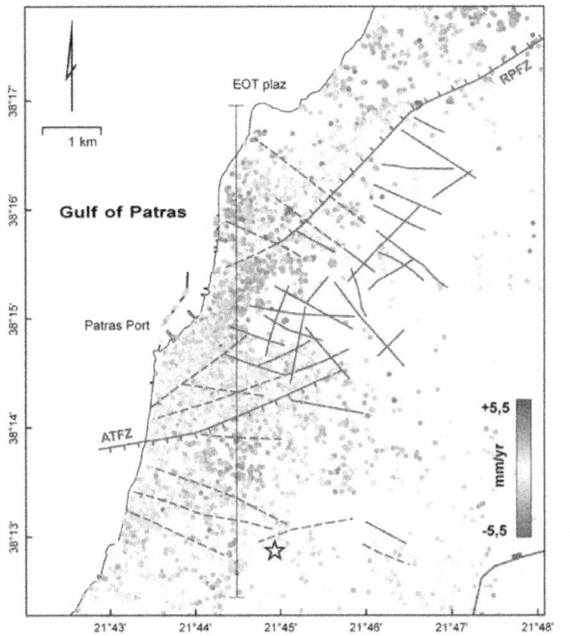

Figure 6: Point targets plotted over the fault map for Patras area (ATFZ: Ag. Triada Fault Zone), solid line corresponds to the section of figure 7, fault pattern modified from Koukis *et al.* [36].

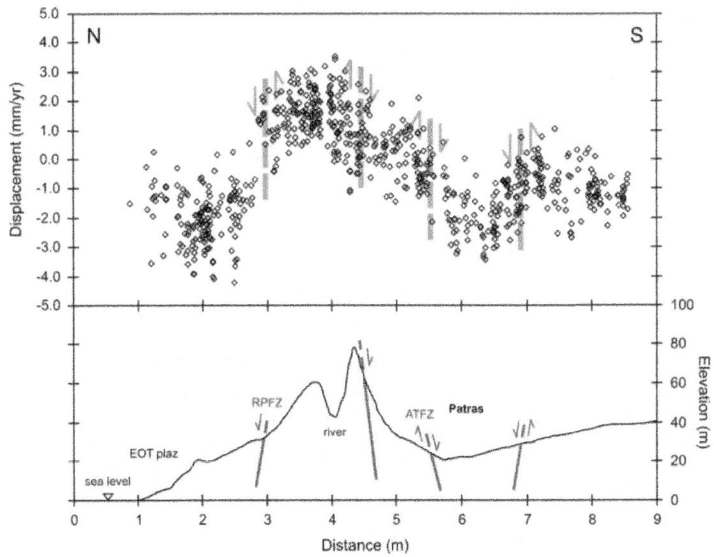

Figure 7: Spatial profile showing the displacement field as observed by IPTA and plotting targets within 150 meters in both sides of the profile (upper), and the related topography (down), red lines correspond to the faults.

Accordingly, maximum subsidence values on the order of −4.5 mm/yr to −5.5 mm/yr are restricted mainly on the hanging wall block of Rio–Patras fault and specifically towards its NE–termination. Lower values between −1.0 mm/yr and −2.8 mm/yr are observed along the downthrown block of Ag. Triada fault, indicating a lower deformation rate compared to the Rio graben. The footwall block of this fault displays low uplift rates between +0.4 and +1.0 mm/yr. Based on the relative uplift and subsidence velocities on footwall and hanginwall blocks of Ag. Triada fault, a vertical velocity between 1.4 mm/yr and 3.8 mm/yr can be estimated for an almost 10 year period of time. Along the main fault trace and in a 50 m width zone, relatively small ground movements, such as cracks on road pavement and buildings, small scale subsidence and fissures are periodically observed. Typical historic deformation diagrams are shown in Figure 8 for two targets from the footwall and hanging wall of Agia Triada fault.

Pyrgos

Based on the spatial distribution of the PS points, the size and the kind of the vertical movement (uplift or subsidence) for the time period 1992–2000, three areas can be distinguished for the Pyrgos city area: the northern, the central and the southern (Figure 9).

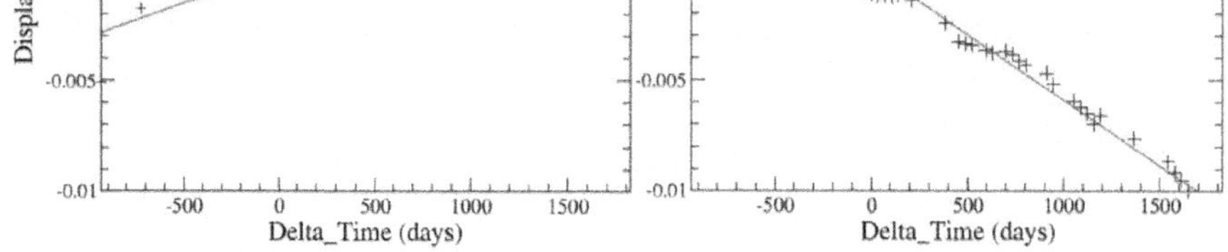

Figure 8: Typical displacement histories obtained by IPTA for specific point targets from the footwall (left diagram) and from the hanginwall (right diagram) of the Agia Triada fault zone.

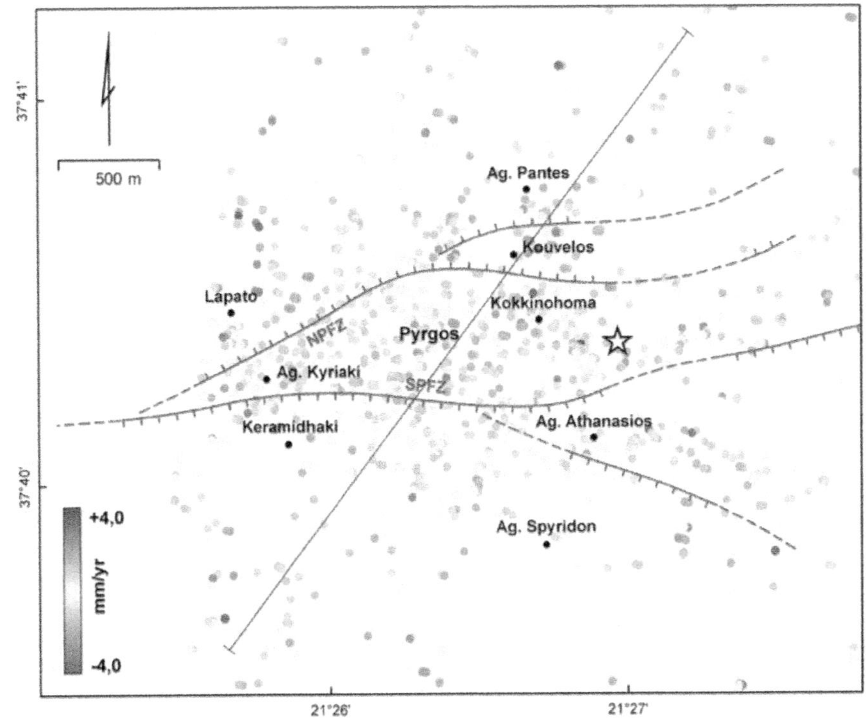

Figure 9: Point targets plotted over the fault map of Pyrgos area, solid line corresponds to the section of Figure 10.

For the northern area the PS points indicate mainly subsidence with values varying from −0.8 to−2.5 mm/yr, while there are few targets, indicating stability. In this area loose alluvial deposits occur, which locally are marshy deposits and the water table of the aquifer is very close to the surface. The area is bounded to the north by the Vounargo fault zone and to the south by the North Pyrgos fault.

The central area, which corresponds to the Pyrgos tectonic horst, is bounded to the north and south by the North and South Pyrgos fault correspondingly [29]. The lower members of the Vounargo formation (thickness >400m, Plio–pleistocene age) occur and the majority of the PS points indicate uplift from +1.5 mm/yr up to +2.5 mm/yr, with very few points presenting subsidence (Figure 10).

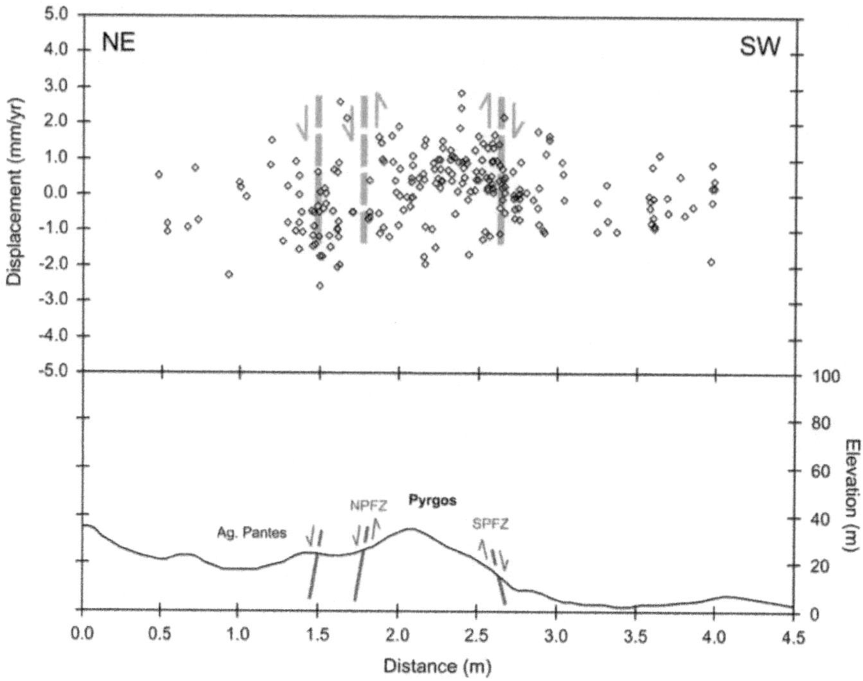

Figure 10: Spatial profile showing the displacement field as observed by IPTA and plotting targets within 150 meters in both sides of the profile (upper), and the related topography (down), red lines correspond to the faults.

At the southern part of Pyrgos, where loose alluvial deposits outcrop, the majority of the targets indicate subsidence, with few being stable and a minority showing uplift.

Typical historic deformation diagram is shown in Figure 11 for a target from the northern part of the city. It has to be mentioned that the measured vertical movements and more specifically the uplift ones are not so intensive. This is in accordance with the low elevation of the Late Pleistocene marine terraces of the broader area, which do not exceed 100 m.

On the contrary, the most destructive earthquakes are related with fault plane solutions of strike slip movements [18,49]. This can explain why the areas of subsidence and uplift are not so clear in the case of Pyrgos being a minor order young neotectonic structure, which has been created after the Middle Pleistocene [29].

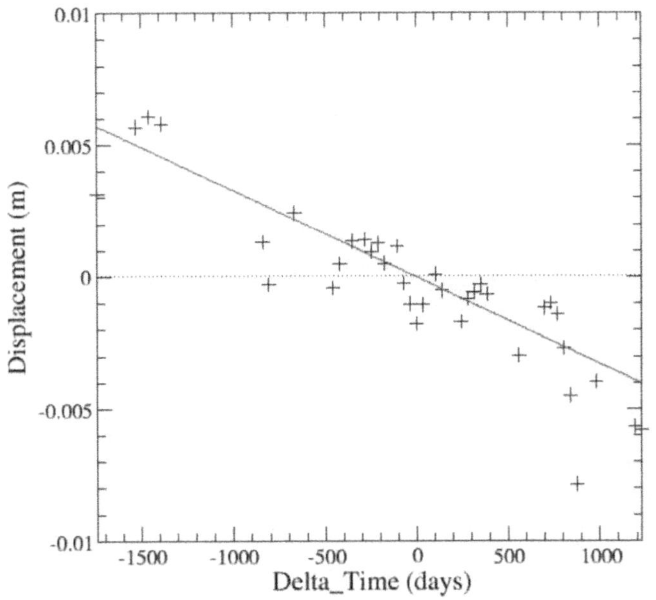

Figure 11: Displacement history obtained by IPTA for specific point target from the northern part of the city.

DISCUSSION

Measurement of ground deformation is a direct and valuable input to models of earthquake risk and for prone areas is of great importance. Given sufficient SAR scenes the submillimeter accuracy of PSI does represent an effective tool for the measurement of tectonic related ground motions.

The PSI technique has proven to be an excellent tool for identifying zones of ground deformation inside urban areas that cannot be easily visible or determined by conventional survey means with unprecedented detail and accuracy. Specifically, the PSI method provides much more information (on both a local and regional scale) on the ground displacements than data obtained by leveling and D–GPS techniques.

Superimposition of several natural and anthropogenic mechanisms that act in different time and scalc and provoking ground deformation should be considered. Finally, a complete multidisciplinary approach (including active tectonic, GPS monitoring and seismology) is the appropriate study scheme.

For this specific study, our results of Patras show clearer uplift–subsidence rates due to the more distinct fault pattern and intense deformation compared to Pyrgos area, where more diffused deformation is observed with no significant displacements on surface. This is maybe due to hidden blind structures activating as strike slip at depth, as it is confirmed by recent earthquakes with such focal mechanisms [49,50].

ACKNOWLEDGEMENTS

Data provided by ESA in the frame of CAT–1 projects. The part of the study concerning Pyrgos city was supported by John S. Latsis Public Benefit Foundation and the Patras city case was supported by ESA in the frame of the 1st Call for Ideas for Greek institutions.

REFERENCES

1. Wright, T.; Parsons, B.; Fielding, E. Measurement of interseismic strain accumulation across the North Anatolian Fault by satellite radar interferometry. *Geophys. Res. Lett.* 2001, *28*, 2117–2120.
2. Tosi, L.; Teatini, P.; Carbognin, L.; Brancolini, G. Using high resolution data to reveal depth–dependent mechanisms that drive land subsidence: The Venice coast, Italy. *Tectonophysics*2009.
3. Massonnet, D.; Rossi, M.; Carmona, C.; Adragna, F.; Peltzer, G.; Feigl, K.; Rabaute, T. The displacement field of the Landers earthquake mapped by radar interferometry. *Nature* 1993,*364*, 138–142.
4. Zebker, H.A.; Rosen, P.A.; Goldstein, R.M.; Gabriel, A.; Werner, C.L. On the derivation of coseismic displacement fields using differential radar interferometry. The Landers earthquake. *J. Geophys. Res.* 1994, *99*, 19617–19634.
5. Ferretti, A.; Prati, C.; Rocca, F. Permanent scatterers in SAR interferometry. *IEEE T. Geosci. Remote* 2001, *39*, 8–20.
6. Werner, C.; Wegmüller, U.; Strozzi, T.; Wiesmann, A. Interferometric point target anaysis for deformation mapping. In Proceedings of the IEEE International Geoscience and Remote Sensing Symposium, Toulouse, France, July 2003; pp. 4362–4364.
7. Rosen, P.; Werner, C.; Fielding, E.; Hensley, S.; Buckley, S.; Vincent, P. Aseismic creep along the San Andreas fault northwest of Parkfield, CA measured by radar interferometry. *Geophys. Res. Lett.* 1998, *25*, 825–828.
8. Colesanti, C.; Ferretti, A.; Ferrucci, F.; Prati, C.; Rocca, F. Monitoring known seismic faults using the Permanent Scatterers (PS) technique. In Proceedings of the IEEE International Geoscience and Remote Sensing Symposium, Honolulu, HI, USA, July 2000; pp. 2221–2223.

9. Lee, J.C.; Angelier, J.; Chu, H.T.; Hu, J.C.; Jeng, F.S.; Rau, R.J. Active fault creep variations at Chihshang, Taiwan, revealed by creep meter monitoring, 1998 – 2001. *J. Geophys. Res-Sol. Ea.*2003, *108*, ETG 4–1–ETG 4–21.

10. Cunha, T.A.; Sarti, F. SAR interferometry as a tool for the detection of active tectonic regions: Preliminary results on the algarve region of the south Portugal. In Proceedings of the FRINGE 2003 Workshop, Frascati, Italy, 1–5 December 2003. ESA SP-550.

11. Taylor, M.; Peltzer, G. Current slip rates on conjugate strike–slip faults in central Tibet using synthetic aperture radar interferometry. *J. Geophys. Res-Sol. Ea.* 2006, *111*, B1240.

12. Biggs, J.; Wright, T.; Lu, Z.; Parsons, B. Multi-interferogram method for measuring interseismic deformation: Denali Fault, Alaska. *Geophys. J. Int.* 2007, *170*, 1165–1179.

13. Cavalié, O.; Lasserre, C.; Doin, M.; Peltzer, G.; Sun, J.; Xu, X.; Shen, Z. Present-day deformation across the Haiyuan fault (Gansu, China), measured by SAR interferometry. In Dragon 1 Programme Final Results 2004–2007, Proceedings of the 2008 Dragon Symposium, Beijing, China, 21–25 April 2008; Lacoste, H., Ouwehand, L., Eds.; ESA Communication Production Office ESTEC: Noordwijk, The Netherlands, 2008; ESA SP-655.

14. Huang, M.; Hu, J.; Ching, K.; Rau, R.; Hsieh, C.; Pathier, E.; Fruneau, B.; Deffontaines, B. Active deformation of Tainan tableland of southwestern Taiwan based on geodetic measurements and SAR interferometry. *Tectonophysics* 2009, *466*, 322–334.

15. Underhill, J.R. Late Cenozoic deformation of the Hellenide foreland, western Greece. *Geo. Soc. Am. Bull.* 1989, *101*, 613–634.

16. Hatzfeld, D.; Pedotti, G.; Hatzidimitriou, P.; Makropoulos, K. The strain pattern in the western Hellenic arc deduced from a microearthquake survey. *Geophys. J. Int.* 1990, *101*, 181–202.

17. Hatzfeld, D.; Kassaras, I.; Panagiotopoulos, D.; Amorese, D.; Makropoulos, K.; Karakaisis, G.; Coutant, O. Microseismicity and strain pattern in northwestern Greece. *Tectonics* 1995, *14*, 773–785.

18. Sachpazi, M.; Hirn, A.; Clément, C.; Haslinger, F.; Laigle, M.; Kissling, E.; Charvis, P.; Hello, Y.; Lépine, J.C.; Sapin, M.; Ansorge, J. Western Hellenic subduction and Cephalonia Transform: local earthquakes and plate transport and strain. *Tectonophysics* 2000, *319*, 301–319.

19. Doutsos, T.; Koukouvelas, I.K.; Xypolias, P. A new orogenic model for the External Hellenides. In *Tectonic Development of the Eastern*

Mediterranean Region; Robertson, A.H.F., Mountrakis, D., Eds.; Geological Society: London, UK, 2006; pp. 507–520.

20. Kokkalas, S.; Xypolias, P.; Koukouvelas, I.K.; Doutsos, T. Post-Ccollisional Contractional and Extensional Deformation in the Aegean Region. Post-Collisional Tectonics and Magmatism in the Mediterranean region and Asia (GSA Special Paper), Dilek, Y., Pavlides, S., Eds.; *Geol. Soc. Am. S.*. 2006, 409, 97–123.
21. Lagios, E.; Sakkas, V.; Papadimitriou, P.; Parcharidis, I.; Damiata, B.N.; Chousianitis, K.; Vassilopoulou, S. Crustal deformation in the Central Ionian Islands (Greece): Results from DGPS and DInSAR analyses (1995–2006). *Tectonophysics* 2007, *444*, 119–145.
22. Hollenstein, Ch.; Müller, M.D.; Geiger, A.; Kahle, H.G. Crustal motion and deformation in Greece from a decade of GPS measurements, 1993–2003. *Tectonophysics* 2008, *449*, 17–40.
23. Doutsos, T.; Pe-Piper, G.; Boronkay, K.; Koukouvelas, I. Kinematics of the Central Hellenides.*Tectonics* 1993, *12*, 936–953.
24. Xypolias, P.; Doutsos, T. Kinematics of rock flow in a crustal scale shear zone: implication for the orogenic evolution of the SW Hellenides. *Geol. Mag.* 2000, *137*, 81–96.
25. Skourlis, K.; Doutsos, T. The Pindos Fold and Thrust Belt (Greece): Inversion kinematics of a passive continental margin. *Int. J. Earth Sci.* 2003, *92*, 891–903.
26. Sotiropoulos, S.; Kamberis, E.; Triantaphyllou, M.; Doutsos, T. Thrust sequences at the central part of the External Hellenides. *Geol. Mag.* 2003, *140*, 661–668.
27. Doutsos, T.; Kontopoulos, N.; Poulimenos, G. The Corinth–Patras rift as the initial stage of continental fragmentation behind an active island arc (Greece). *Basin Res.* 1988, *1*, 177–190.
28. Doutsos, T.; Kontopoulos, N.; Frydas, D. Neotectonic evolution of northwestern continental Greece. *Geol. Rundsch.* 1987, *76*, 433–452.
29. Lekkas, E.; Papanikolaou, D.; Fountoulis, I. The Pyrgos Earthquake: The geological and geotechnical conditions of Pyrgos area (W. Peloponnese, Greece). In Field–guide for the Pre–Congress Excursion of the XV Congress of the Carpatho–Balcan Geological Association, Athens, Greece, September 1995; pp. 42–46.
30. Koukouvelas, I.; Mpresiakas, A.; Sokos, E.; Doutsos, T. The tectonic setting and earthquake ground hazards of the 1993 Pyrgos earthquake, Peloponnese, Greece. *J. Geol. Soc. London*1996, *153*, 39–49.

31. Doutsos, T.; Kokkalas, S. Stress and deformation patterns in the Aegean region. *J. Struct. Geol.*2001, *23*, 455–472.

32. Papanikolaou, D.; Fountoulis, I.; Metaxas, C. Active faults, deformation rates and Quaternary paleogeography at Kyparissiakos Gulf (SW Greece) deduced from on-shore and off-shore data.*Quatern. Int.* 2007, *171–172*, 14–30.

33. Zelilidis, A.; Koukouvelas, I.; Doutsos, T. Neogene paleostress changes behind the forearc fold belt in the Patraikos Gulf area, Western Greece. *Neues Jahrb. Geol. P. M.* 1988, 311–325.

34. Flotte, N.; Sorel, D.; Muller, C.; Tensi, J. Along strike changes in the structural evolution over a brittle detachment fault: example of the Pleistocene Corinth – Patras rift (Greece).*Tectonophysics* 2005, *403*, 77–94.

35. Chronis, G.; Piper, D.W.; Anagnostou, C. Late Quaternary evolution of the Gulf of Patras, Greece: Tectonism, deltaic sedimentation and sea–level change. *Mar. Geol.* 1991, *97*, 191–209.

36. Koukis, G.; Sabatakakis, N.; Tsiambaos, G.; Katrivesis, N. Engineering geological approach to the evaluation of seismic risk in metropolitan regions: Case study of Patras, Greece. *Bull. Eng. Geol. Environ.* 2005, *64*, 219–235.

37. Kalteziotis, N.; Koukis, G.; Tsiambaos, G.; Sabatakakis, N.; Zervogiannis, H. Structural damage in a populated area due to an active fault. In Proceedings of the 2nd International Conference on Recent Advancces in geotechnical Earthquake Engineering and soil Dynamics, Rolla, MO, USA, 11–15 March 1991; pp. 1709–1716.

38. Hasiotis, T.; Papatheodorou, G.; Kastanos, N.; Ferentinos, G. A pockmark field in the Patras Gulf (Greece) and its activation during the 14/7/93 seismic event. *Mar. Geol.* 1996, *130*, 333–344.

39. Papazachos, B.C.; Papazachou, C. *The Earthquakes of Greece*; Ziti Publications: Thessaloniki, Greece, 1997.

40. Papatheodorou, G.; Christodoulou, D.; Geraga, M.; Etiope, G.; Ferentinos, G. The pockmark field of the Gulf of Patras: An ideal natural laboratory for studying seabed fluid flow. In Filed trips guide book "Sedimentology of western and central Greece from Triassic to recent", Proceedings of the 25th IAS Meeting of Sedimentology, Patras, Greece, 4–7 September 2007.

41. Lepichon, X.; Angelier, J. The Hellenic arc and trench system: a key to the neotectonic evolution of the Eastern Mediterranean area. *Tectonophysics* 1979, *60*, 1–42.

42. Mariolakos, I.; Papanikolaou, D. The neogene basins of the Aegean Arc from the Paleogeographic and the Geodynamic point of view. In Proceedings of the Int. Symp. Hell. Arc and Trench (HEAT), Athens, Greece, 1981; Vol. I, pp. 383–399.

43. Mariolakos, I.; Papanikolaou, D. Deformation pattern and relation between deformation and seismicity in the Hellenic arc. *Bull. Geol. Soc. Greece* 1987, *XIX*, 59–76. (in Greek).

44. Mariolakos, I.; Papanikolaou, D.; Lagios, E. A neotectonic geodynamic model of Peloponnesus based on: morphotectonics, repeated gravity measurements and seismicity. *Geol. Jb.* 1985, *50*, 3–17.

45. Kamberis, E. Geology and petroleum geology study of NW Peloponnese, Greece. Ph.D. Thesis, National Technical University of Athens, Athens, Greece, 1987. (in Greek).

46. Lekkas, E.; Papanikolaou, D.; Fountoulis, I. Neotectonic map of Greece, sheet "Pyrgos" – "Tropaia" (scale 1/100.000). Project-University of Athens, Athens, Greece, 1992, 120. (in Greek).

47. Lekkas, E.; Fountoulis, I.; Papanikolaou, D. Intensity distribution and Neotectonic macrostructure Pyrgos Earthquake data (Greece, 26 March 1993). *Nat. Hazards* 2000, *21*, 19–33.

48. Scharoo, R.; Visser, P.N.A.M. Precise orbit determination and gravity field improvement for the ERS satellites. *J. Geophys. Res.* 1998, *103*, 8113–8127.

49. Roumelioti, Z.; Benetatos, Ch.; Kiratzi, A.; Stavrakakis, G.; Melis, N. A study of the 2 December 2002 (M5.5) Vartholomio (western Peloponnese, Greece) earthquake and of its largest aftershocks. *Tectonophysics* 2004, *387*, 65–79.

50. Koukouvelas, I.; Kokkalas, S.; Xypolias, P. Surface deformation during the Mw 6.4 (8 June 2008) Movri earthquake in the Peloponnese and its implications for the seismotectonics of Western Greece. *Int. Geol. Rev.* 2009.

Chapter 7

LAND USE CHANGES AND ENVIRONMENTAL PROBLEMS CAUSED BY BANK EROSION: A CASE STUDY OF THE KOLUBARA RIVER BASIN IN SERBIA

Slavoljub Dragicevic[1], Nenad Zivkovic[1], Mirjana Roksandic[1], Ivan Novkovic[1], Stanimir Kostadinov[2], Radislav Tosic[3], Milomir Stepic[4], Marija Dragicevic[5] and Borislava Blagojevic[6]

[1]University of Belgrade, Faculty of Geography, Belgrade, Serbia

[2]University of Belgrade, Faculty of Forestry, Belgrade, Serbia

[3]Faculty of Natural Sciences, Banja Luka, Republic of Srpska

[4]Institute for Political Studies,Belgrade, Serbia

[5]First Elementary School in Obrenovac, Obrenovac, Serbia

[6]University of Nis, Faculty of Civil Engineering and Architecture, Serbia

INTRODUCTION

Geomorphological analysis of the dominant erosion processes and their intensity quantification were done in the previous researches of the Kolubara River basin [1-3]. The results showed that, the level of the landscape degradation and modification of geomorphologic processes by human activities has been increased in the past decades [4], and it was initiated by very fast demographic, socio-economic and technological changes in Serbia, likewise in the region [5-7], and in the world [8-11].

According to level and type of degradation, the Kolubara River basin belongs to the most endangered areas in Serbia. Due to the lignite exploitation in the Kolubara River basin, human impact led to morphological change of the entire area, as well as to the changes of the intensity of different geomorphologic processes: changes in river course [12,13], the intensity of bank erosion [14,15], sediment deposition [16] and environmental problems [17,18].

Unlike the other rivers with similar hydrological characteristics, the river network in the lower part of the Kolubara River basin were changed rapidly during the XX century because of direct human impact. Anthropogenic influences on the hydrological network in the study area were very intensive since 1959, when the huge river regulation works were done in the lower part of the Kolubara River. Spatial planning of the area, which included diverting of the Kolubara's river bed, had an aim to prepare the site for the lignite exploitation within the Kolubara mining basin. The Kolubara River divides the mining basin in two parts: eastern and western part. The productive area of the basin (geologic contours of lignite deposits) is 520 km^2. Kolubara mining basin is situated about 40 km south-southeast of Belgrade and represents the largest lignite deposits in the central part of Serbia; the annual production is 30 million tons of lignite, and it is the opencast mine. The mine expansion caused the need for technical solutions of diverting and removing river beds in this area. According to "General project of diverting the Kolubara River and its tributaries for the purpose of lignite exploitation", the Kolubara's riverbed was diverted into the Pestan's riverbed (its right tributary). This caused many problems which were not predicted by the General project.

In this way, anthropogenic factor modified existing natural conditions: the process of fluvial erosion was changed; bank erosion became stronger and resulted in soil loss, larger amounts of sediment load deposition, cutting off the meanders and fossilization of certain parts of the riverbed, floods, land use changes, landscape degradation, sediment load pollution, etc.

RESEARCH AREA

Regarding to natural conditions, the Kolubara River basin is similar to the other river basins in the area. Tectonic movements had an influence on a morphological evolution of the river network in the past. During the Paleogene and the Early Neogene a small bay of the Pannonian Sea named the Kolubara's bay existed in the area of the Kolubara River basin. After the sea recession, the fluvial erosion started in this bay and it formed today's hydrological network of the Kolubara River. Tectonic characteristics of this area, more precisely Kolubarsko-pestanski fault and Posavski fault had influenced the orientation of the hydrological network in the Kolubara River basin. But today the Kolubara's hydrological network is influenced by fluvial erosion and anthropogenic factors.

The Kolubara River Basin encompasses the western part of Serbia and covers 4.12% of Serbia's surface area. The highest point of the drainage basin is at 1,346 m, and the lowest has altitude of 73 m. The Kolubara River is the last large right tributary of the Sava River, and according to the flow length (86.4 km) and the basin area (3,641 km^2) it is classified as a middle-sized river on the territory of Serbia [3]. The lower part of the Kolubara River basin is called the Donjokolubarski basin (area of 1,810 km²) and is situated in the municipality of Obrenovac. The Donjokolubarski basin encompasses the catchment area of the Kolubara's confluences (the Pestan River, the Turija River with the Beljanica River, the Tamnava River with the Ub River and the Kladnica River) and the lower part of the Kolubara's valley. The average altitude of the Donjokolubarski basin is 168 m, the highest point is at 695 m, and the lowest has an altitude of 73 m.

According to the nearest meteorological station in Obrenovac, this area is characterized by continental climate, the average temperature was 11°C, and the mean annual precipitation from year 1925 to 2000 was 722 mm [12]. The average annual runoff of the Kolubara River (at Drazevac gauging station) for the period 1961-2005 was 21.8 m³/s.

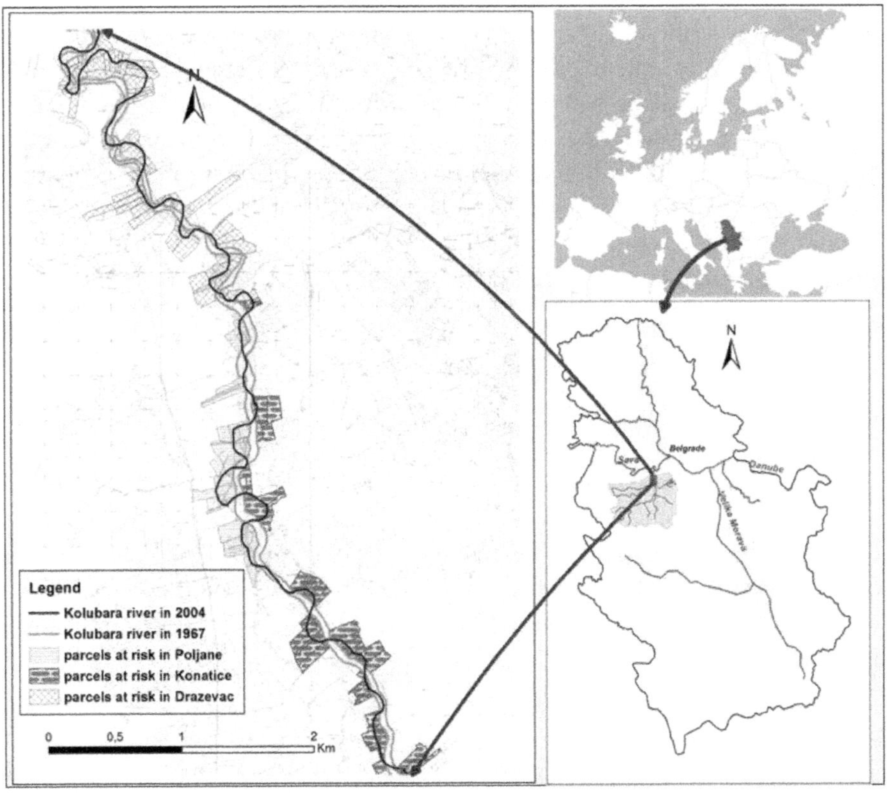

Figure 1: Position of the Kolubara River basin in Serbia (right) and study area (left)

METHODOLOGY

In this research we used diffrent methods that can be devided into the field and lab work methods. The GIS methods were used for the modeling of terrain evolution and landscape changes, which represents the base for bank erosion intensity quantification.

Analysis of topographical maps, aerial photo and orthophoto images were used in the previous researches aiming to determine the evolution of the riverbed [6,15,19-23]. The results showed that the application of GIS has an advantage in quantification of river migration processes.

For the purposes of this study, comparative analyses have been made on the base of Cadastral maps scale 1:2500 from 1967 and orthophoto images from 2004; reconstruction of the hydrological system has been done for the periods from 1967 to 2004. By comparing the data from two periods, we determined the evolution of the Kolubara River course in 37 years. River bank lines were digitized and the extent of bank erosion was calculated under Geomedia professional. The same software was used for the estimation of the Kolubara River lateral migration rate. This rate was estimated using the calculated area between river positions in 1967 and in 2004 (area of river migration), which was divided by the total length of the river course in 1967. The loss of land (*S*) is expressed as the ratio between area of endangered land parcels (ha) in 1967 (P1967) and area of endangered land parcels (ha) in 2004 (P2004) [15]:

$$S = \frac{P_{1967} - P_{2004}}{P_{1967}} * 100$$

River erosion and frequent floods make great material damages to people, villages and economy. The owners of the arable land parcels on the Kolubara River banks loose the parts of the parcels that river carries away. The reduction of parcels on the Kolubara River banks, land loss and land use changes were estimated comparing the cadastral maps from 1967 and orthophoto images from 2004.

Land use structure in the area of villages: Drazevac, Konatice and Poljane are characterized by: arable land (which people used for farming mostly wheat and corn-crop rotation practice), forests (alluvial forests of willows and poplars) and few pastures. The river dynamic is intensive in the Kolubara's alluvial zone, which influenced sandbank formation, mostly on the concave side of the river. By statistical analysis of a land use structure [24] in the three villages with degraded land parcels on the river banks, we obtained the results which show significant reduction in arable land. And by analyses of the questionnaire carried out among the owners of degraded land parcels in the villages Drazevac, Konatice and Poljane, it can be concluded that it was significant decrease in the agricultural production. The risks from the floods and further soil loss influenced the land owners' decision making about farming the degraded land parcels.

The change of fluvial erosion intensity was analyzed regarding to changes in water balance and sediment load transport on two hydrological profiles. The results of water balance that D. Dukic [25] has made in his research for the period of 1925-1960 and the results obtained in this study were analyzed and compared. This comparative analysis appoints to the amount of water which Donjokolubarski basin disposed before regulatory changes of Kolubara in 1959/60 and after them. River flow regimes of different periods were compared because that could be a factor which has a significant influence on the observed process. All these efforts should confirm or eliminate the influence of natural factors on the river banks degradation in the Donjokolubarska valley.

Having data of extreme discharges, in order to estimate the impact of future floods on bank erosion, we have made a probability curve of maximum discharges of the Kolubara River and its tributaries.

Because of intense anthropogenic impacts in the Donjokolubarska valley, we have sampled the suspended sediments from the Kolubara's riverbed and later analyzed the pollution of the accumulated load. Since the processes of bank erosion and sediment accumulation occur close to the villages and that endangered land parcels are used for food production, such approach points to ecological aspect of researched problem.

The sediment samples were taken on two locations in the Kolubara's riverbed. For heavy metals and carbon analysis the soil was milled to a fine powder. Heavy metals were determinate by AAS method.

THE INTENSITY OF BANK EROSION

Natural Conditions Changes as a Factor of Bank Erosion in the Study Area

On the research sector (Fig. 1) the Kolubara River length in 1967 was 8.2 km and 10.6 km in 2004. This fact appoints to the river course evolution through the landscape. In the period between 1967 and 1981 the Kolubara River has migrated 50 m, actually 27 m into left and 23 m into right, and the average migration of the Kolubara River was 3.6 m per year. By further comparison of aerial photo image from 1981 and orthophoto image from 2004 it can be observed that the Kolubara's riverbed was stabilized and during 23 years migrated only 26 m. So, the Kolubara River average migration in this period was 1.1 m per year which is three times less comparing to the previous period of observation (1967-1981) [13].

The rate of the Kolubara river lateral migration along the research sector is 47 m in average for the period of 37 years, which means 1.27 m per year. At the most endangered part (in the area of Drazevac village) the most intensive migration rate of the riverbed was 224 m in 37 years, with the average of 6.05 m per a year [15].

The changes of fluvial erosion intensity may result from changes in climatic-hydrological characteristics of the river basin (which are manifested in discharge regime changes) and various human impacts. Therefore, the natural factors of the Donjokolubarski basin were analyzed to determine whether they have influenced the stronger bank erosion.

The results showed that average mean annual discharge of the Kolubara River measured in Drazevac was 22.3 m^3/s in the observation period 1961-1990, and 21.3 m^3/s in the observation period 1991-2005. Amplitude of average high and low flows in the period 1961-1990 was 77.94 m^3/s, and in the period 1991-2005 it was 64.66 m^3/s [13].

To study water balance of the Donjokolubarski basin we used the following periods: 1925-1960, 1961-1990 and 1991-2005. With this approach it was possible to determine the changes that may be occurred after diverting the Kolubara River into the Pestan's riverbed in 1959. Briefly, precipitation analysis showed that the second period was a bit wetter than the first, actually about 60 mm in the Pestan River basin and 80 mm in the Turija River basin and the Tamnava River basin. Meanwhile, higher air temperatures and higher evaporation caused almost the same specific discharges of these rivers. The last period was in mean values similar to the second, apart from intensified variation of extreme values of all climatic elements. The discharges were influenced by more frequent alternation of wet and dry periods, which could be seen on figure 2.

Monthly coefficients of variation of the period 1991-2005 are higher in all river sub basins except in July and August. These differences are significant, the variation of discharges in eight months are higher than the highest coefficients of variation of the period 1961-1990, which is 1.5. The more important is the fact that the period of appearances of unstable discharges is March-April (over 2.5), which is related to snow melting. That is the period of maximum discharges and any sudden disturbance of soil moisture resulting in serious disorder of river bank stability. Some of the natural factors have been changed in the last two decades, for example, March used to be, in Serbia, the month with the most stable discharges (and the highest). The differences in March discharges during the observed periods are over 20 m^3/s (almost twice reduced), and it was followed by extreme discharge variations. These are the significant changes since the area of the Kolubara River basin is bigger than 3500 km^{2}and maximum discharges are higher than 500 m^3/s. Although the mean values are not of crucial importance, they indicate some disturbances which should be kept in mind; particularly because the last period of observation is twice shorter than the previous one and all analyses in the world indicate that extreme values of natural phenomena are more pronounced and more frequent.

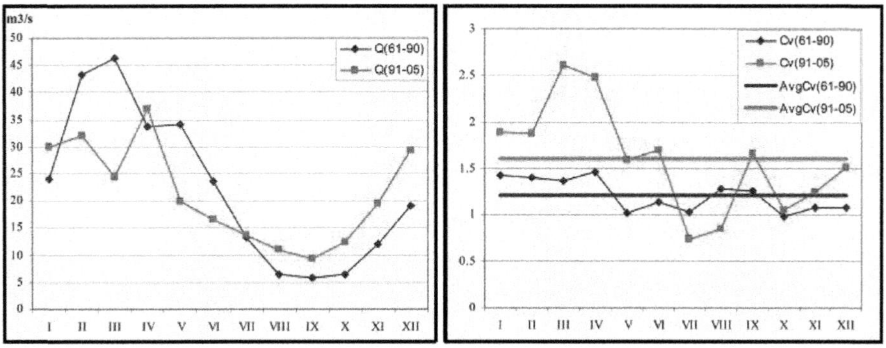

Figure 2: Mean monthly discharges (Q) and coefficients of variation (Cv) of the Kolubara River measured in Drazevac gauging station for the both periods of observation

Preliminary results of the Donjokolubarski basin annual flow variation show discrepancy in spring and summer monthly flow among two studied periods (Figure 2 - right). Further research should examine correlation of monthly flow and bank erosion intensity.

Anthropogenic Influences as a Factor of Bank Erosion in the Study Area

The erosion control works in the river channel can cause changes in river morphology, since they influence the changes in river regimes, river bank characteristics and amount of sediment transport [26]. The consequences of these interventions are numerous, and often lead to riverbed widening and undermining concave sides of the river banks. The processes of river bank collapsing and erosion are complex since they are results of several factors, including sediment transport, ground lithology, stratigraphy, slope, flow geometry and anthropogenic activities [27].

Opencast lignite exploitation in Kolubara mining basin started in 1952 when the mining field "A" was open (it was exploited till 1966). Mining field "B" was opened in 1952, mining field "D" in 1961, "Tamnava-East Mining Field" in 1979, "Tamnava-West Mining Field" in 1994 and mining field "Veliki Crljeni" in 2008 [28].

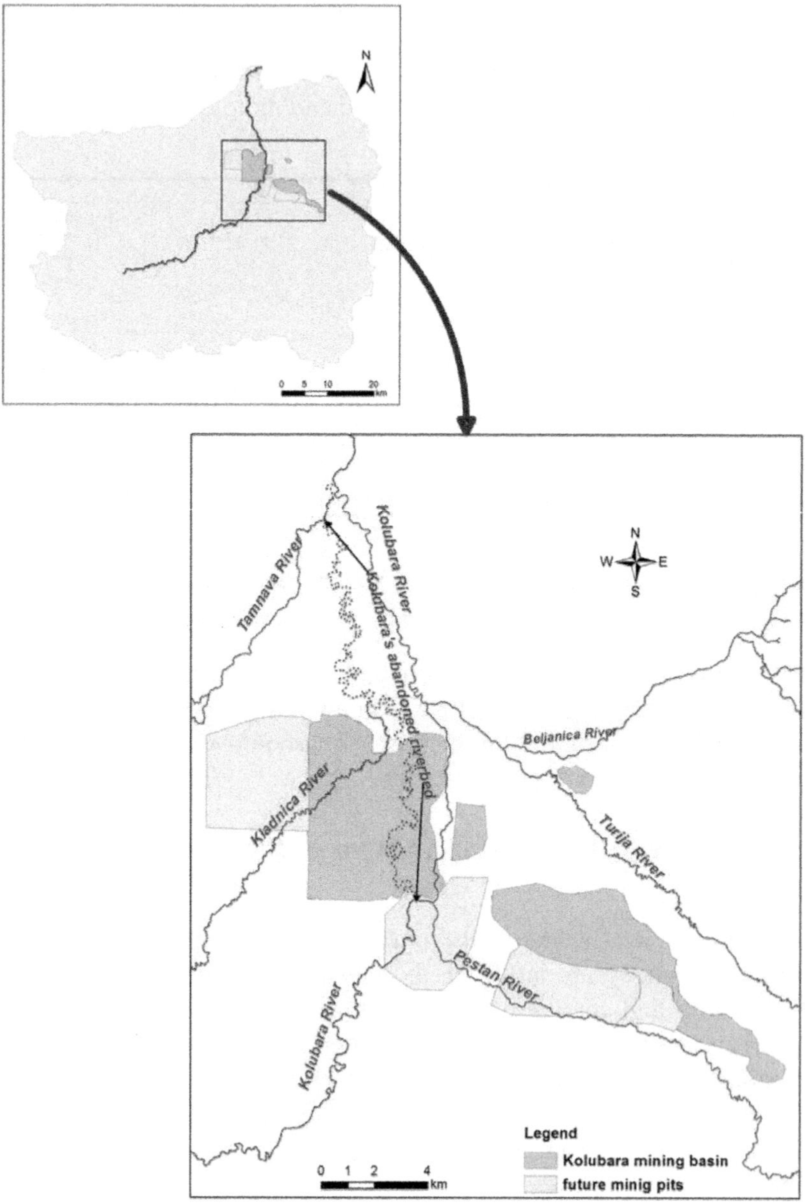

Figure 3: Plan of the mining fields in Kolubara mining basin [28]

Beogradsko-posavska water community “Beograd” has made a project „Regulation of the Kolubara River and its right tributaries from Ćelije to Poljane (km 23+200 – km 55+506)“ in 1957. Regulation works on the Kolubara River and its tributaries had begun in 1959/60. The diverting of the Kolubara River was done to clear the area for lignite exploitation, actually for opening new mining fields.

From that moment the Kolubara River flows through the Pestan›s riverbed (its right tributary), and previous Kolubara›s riverbed is abandoned with the periodic flow. The length of the Kolubara River was shortened by 20 km because of diverting its riverbed, while the length of the Pestan River was also shortened because its confluence was moved to the South. By diverting the Kolubara›s riverbed into the Pestan›s riverbed, which morphologically was not predisposed for kinetic energy of stronger flow, bank erosion became a dominating geomorphological process in the area and initiated processes of digging the riverbanks, transportation and deposition of eroded material. It is obvious that river system changes in lower part of the Kolubara River are demonstrated in domination of fluvial (lateral) erosion on one hand, and in cutting the meanders and fossilization of certain parts of the riverbed, on the other hand.

THE CONSEQUENCES OF BANK EROSION

Forming of Meanders

Map of the Kolubara’s basin first trend of relief energy [2] shows that almost whole area of the Donjokolubarska valley is under tectonic movements of slowly sinking. For this reason the sediments are accumulated in the riverbed, river velocity decreases which cause the riverbed meandering and stronger bank erosion. This natural process became more intensive since the Kolubara River was diverted into the riverbed of Pestan. In the Donjokolubarski basin there are numerous sectors with abandoned riverbeds and cut off meanders.

Forming of meanders and cutting the “necks” are recent geomorphologic-hydrological process, which is dominated in the study area. According to results of the recent researches [13], there are 89 abandoned parts of the riverbeds and cut off meanders in the area of the Donjokolubarski basin. The Kolubara and its tributaries tend to move to the east because of the Kolubarsko-pestanski fault, which indicate more abandoned riverbeds and cut off meanders on the left side of the Kolubara valley (64), compared to the right side (25).

In the study area there are 40 cut off meanders with total length of 20.30 km while the number of abandoned parts of the riverbeds is 49 with total length of 76.03 km. Hence, the total length of all abandoned riverbeds and cut off meanders in the Donjokolubarski basin is 96.33 km, and their total surface is 3.35 km^2. The longest cut off meander is 1.7 km long and the shortest is 185.7 m long. The longest abandoned riverbed is 6.49 km long.

The length of the Kolubara's riverbed is influenced by stronger bank erosion and formation of meanders, which is clearly perceived in the field. According to orthophoto image from 2004 and satellite image (Google Earth) the Kolubara River length (in the Donjokolubarska valley) is 66.52 km, while according to topographical map from 1970 it was 67.5 km, and according to topographical map from 1925 it was 87.6 km.

After cut off meander, the riverbed itself morphologically adjusts to the new state [29]. Morphological changes of the rivers are reflected in digging the concave river banks and sediment accumulation on the convex river banks.

Land Use Changes

As we earlier indicated, river erosion and frequent floods can make great material damages to people, villages and economy. Since the lateral erosion has more intensity, the river banks on concave side of the Kolubara River often collapse and farmers who have arable land parcels on the river bank (in the area of three villages the Kolubara flows through) loose the parts of the parcels which were carried away by the river. Based on Cadastral maps from 1967 and orthophoto images from 2004 we have estimated the area of diminished land parcels and their land loss.

Farmers who have land parcels in three villages (Drazevac, Konatice and Poljane) on the Kolubara river bank cannot farm them in whole, because the river has changed its course and took some parts of the land parcels away. The cadastral maps of the researched area scale 1:2500 from 1967 and orthophoto images from 2004 were compared. Using the results of this comparative analysis, the evolution of the hydrological system in the period from 1967 to 2004 was presented. The previous research showed that 60.37 ha was lost and degraded by the river bank erosion, which means that the land loss is 50.57 % of the land parcels from 1967 [15].

Table 1: Land use structure in Drazevac, Konatice and Poljane

Land use	**Total**	Arable land	**Woods**	**Pastures**	**Mead-ows**	Sand banks	Other (roads...)
Drazevac							
number of end-agered parcels	95	39	16	1	5	34	-
area in 1967 (ha)	57.76	34.96	5.56	0.25	2.93	14.07	-
area in 2004 (ha)	28.23	21.76	2.48	0.02	1.30	2.67	-
loss of land (ha)	29.53	13.20	3.08	0.23	1.63	11.40	-
Konatice							
number of end-agered parcels	86	56	2	2	-	25	1
area in 1967 (ha)	50.44	42.73	0.49	1.02	-	6.12	0.09
area in 2004 (ha)	32.13	29.57	0.34	0.55	-	1.60	0.06
loss of land (ha)	18.31	13.16	0.15	0.47	-	4.52	0.03
Poljane							
number of end-agered parcels	66	41	3	-	-	21	1
area in 1967 (ha)	40.09	33.91	0.82	-	-	5.11	0.25
area in 2004 (ha)	26.26	25.15	0.19	-	-	0.89	0.03
loss of land (ha)	13.83	8.76	0.63	-	-	4.22	0.22

On the basis of the recent and more accurate data from Obrenovac Municipality Cadastre we have determined land use structure of degraded land parcels on the Kolubara River banks. According to these data, total area of all 247 endangered land parcels was 148.3 ha in 1967, and 86.62 ha in 2004. Therefore, 61.68 ha of soil were lost within 37 years [13].

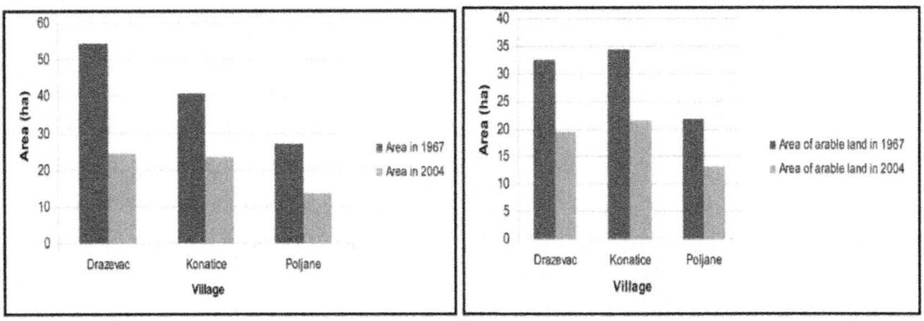

Figure 4: Land use changes in total area (left) and area of arable land (right) in Drazevac, Konatice and Poljane between 1967 and 2004

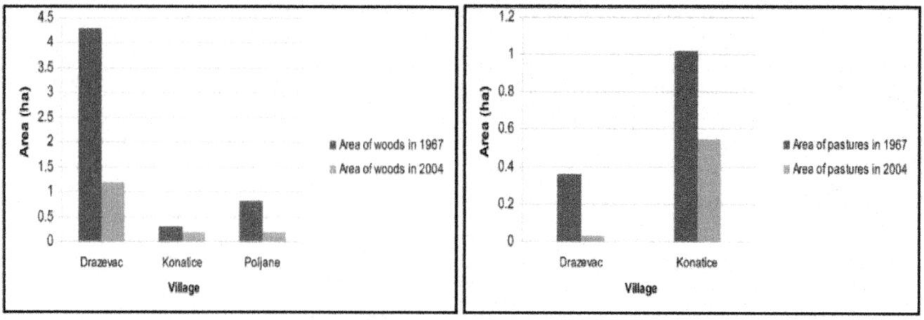

Figure 5: Land use changes in area of woods (left) and area of pastures (right) in Drazevac, Konatice and Poljane between 1967 and 2004

From 247 endangered land parcels, 136 are arable land with the area of 111.6 ha in 1967, and 76.48 ha in 2004, which means that within 37 years 35.12 ha of arable land was lost for farming, and it is 31.5 % of the initial area (in 1967). The woods comprise 21 of all endangered land parcels with area of 6.87 ha in 1967, and 3.01 ha in 2004, which means that it has been lost 3.86 ha of woods. There are only the three endangered land parcels with pastures, and their area was 1.27 ha in 1967, and 0.57 ha in 2004. All five endangered land parcels with meadows are in the area of Drazevac village.

Analyzing the area of endangered parts in the three villages, one can conclude that erosion was the most intensive in the period 1967-1981, when 50.9 ha of soil was lost within 14 years. The riverbed was stabilized later and the erosion decreased. This appoints to the fact that diverting the Kolubara River into the Pestan's riverbed caused more intensive bank erosion since in time erosion was diminished which brought to the riverbed stabilization.

Three villages on the Kolubara River banks (Drazevac, Konatice and Poljane) were characterized by agricultural production and agricultural population. Analyzing the land use structure of endangered parcels one can conclude that arable land parcels are the most endangered and degraded by intensified lateral erosion of Kolubara.

In Serbia there is 4.25 million ha of arable land, and each year 500000 ha (which means 11.74 %) of arable land remain uncultivated [30]. In the above mentioned three villages 33.47 % of arable land (on the Kolubara River banks) remain uncultivated, which is three times more than the average in Republic of Serbia. During the field work, the interviewed owners of endangered arable land parcels pointed that they do not farm their land on the river banks because of flood risks. The Kolubara River floods almost every year and crop is ruined. Therefore, besides the loss of arable land, frequent floods are huge problem in this area.

The economic consequences of bank erosion in the area of the Donjokolubarski basin could be analyzed through losses that the owners of endangered arable land parcels had (because the arable land parcels were reduced). The area of arable land (on the river banks) was diminished by 35.12 ha within 37 years. In the research area the average annual yield is 3-4 t per hectare, so the annual losses of crops (mostly wheat and corn) in recent years are between 100 and 140 t per year.

Sediment Load Discharge

Changes of land use structure and changes in sediment regimes are the direct consequences of the bank erosion [22]. The calculation of one-day sediment load discharge at the monitored hydrological profile includes the values of mean daily flow (Q – m^3/s) and the relevant concentration of the suspended load (C–mg/l). The assessment of sediment deposition rate is based on the results of RHMSS [31] measurements and the results of own daily measurements of suspended load concentration during the period (1985-2004). The results show that 193253.8 tons of material was accumulated between two hydrological profiles. And the riverbed itself was raised for 36 cm, which is nine times enlarged comparing to previous research when it was raised for 4.2 cm (with a shorter time series) [2]. The extraction of the river deposited sediments from the Kolubara's riverbed was stopped. Although the river deposits were hand extracted with low intensity, it certainly had great positive effects from the aspect of maintaining the surface of riverbed profile. Simple solutions, like the river deposit extraction, do not need huge investments for the implementation and they can be carried out without limitation of the natural conditions.

REIK Kolubara has a negative ecological impact on the Donjokolubarska valley. There are lots of waste waters after the ore production. Waste waters from the mine "REIK Kolubara" are discharged without any treatment into the Kolubara River. Therefore, the Kolubara River contains waste waters from the mine and after each flood the soil on the river banks is contaminated by substances from the waste waters. The results of soil analysis in the Kolubara river basin show increased concentration of nickel, arsenic and lead in the area of the Donjokolubarski basin [18].

The eroded material from the river banks is accumulated downstream. The accumulated sediments can contain considerable concentrations of heavy metals and that is threat for the aquatic habitats and for the people [8, 32-34].

Ecological aspects of mechanical water pollution by suspended sediment, chemical water pollution by organic and mineral fertilizers used in plant production in the catchment, nutrients found in the soil as well as chemical pollution of water and sediment by pesticides and heavy metals are very important ecological problem in the study area. On two locations we have sampled the accumulated material from the Kolubara's riverbed to examine the transport of contamination and accumulation of contaminated sediments due to bank erosion processes.

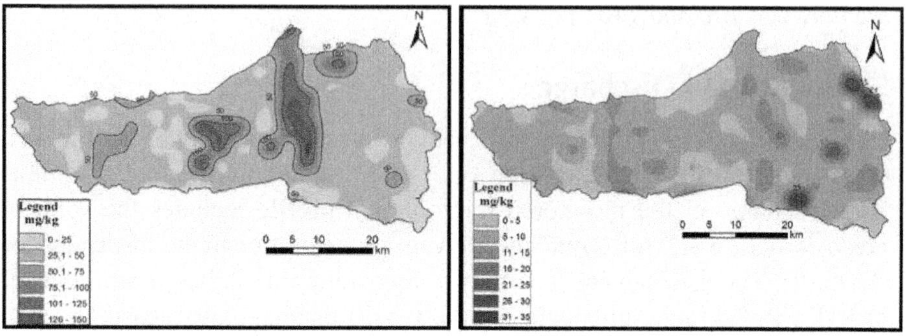

Figure 6: Distribution of Ni (left) and As (right) in the soil of the Donjokolubarski basin [18].

Figure 7: Sampling of deposited sediments on location 1 (left) and 2 (right)

The deposited sediments have sandy-clay texture. Chemical characteristics of deposited sediments from Kolubara's riverbed are: mildly alkaline reaction, high bases saturation degree and low humus content. The average heavy metal concentration in sediments decreased in the order: Ni > Cr > Zn > Pb > Cu > Cd.

In Serbia there is no law defining limitation of heavy metals in suspended sediments. Some European countries have such laws [34], but the differences between countries are significant. In most of the cases the critical values are obtained using the equilibrium method and maximum acceptable concentration

(MAC) for the surface waters with regard to direct and indirect effects on living organisms in the water-sediments systems. According to these data, the range for different elements is as follows: 15 - 100 mg.kg^{-1} for Pb; 0.6 – 2.4 mg.kg^{-1} for Cd; 36 - 120 mg.kg^{-1} for Cu; 123 - 1050 mg.kg^{-1} for Zn; 10 - 180 mg.kg^{-1} for Ni and 37 - 120 mg.kg^{-1} for Cr. These ranges are bigger then estimated ecotoxic criteria [35], which are: 5 - 50 mg.kg^{-1} for Pb, Ni i Cu; 0.1 – 1.0 mg.kg^{-1} for Cd; 50 - 500 mg.kg^{-1} for Zn; 10 - 100 mg.kg^{-1} for Cr.

Table 2: Heavy metal contents in the deposited sediment load

Sample	**Zn**	Cu	Pb	Cd	Cr	Ni
	mg.k^{g-}1					
Location 1	39.3	15.9	23.0	0.0	94.0	198.3
Location 2	41.0	20.2	27.9	0.1	103.0	210.9

Respecting the above mentioned criteria, mean measured concentration of Pb, Zn and Cr are within the limits (after de Vries and Bakker [35]), while concentration of Cu and Cd are below the limits and concentration of Ni are above the limits. According to OSPAR limitation values [36], average concentrations of Pb, Cu and Cd are below the limits, average concentrations of Cr and Zn are within the limits, while average concentrations of Ni are above the limits.

Floods

The Kolubara River is a good example which represents the existence of all conditions for frequent and large scale floods. As an indirect consequence of the anthropogenic influence on the hydrological system in the lower part of the Kolubara valley, once a year (sometimes twice a year) the Kolubara River overflows, and the area of lower part of the Kolubara River basin is endangered by floods. Catastrophic floods of the Kolubara River and its tributaries spread over the area of lower part of the Kolubara River basin during the spring of 1937, and they lasted two months approximately (from March to May). In this area large scale floods also happened in 1965, 1975, 1981, 1996, 1998, 1999, 2001, 2004, 2006, 2008 and 2010.

The highest discharge of the Kolubara River in the period of 1959-2000 was 646 m^3/s and it was registered on Drazevac hydrological station. According to probability curve of high discharges the discharge of 646 m^3/s may occur once in a 46 years. The lowest value of annual maximum discharge would be about 25 m^3/s, the highest discharge in a hundred years would be 740 m^3/s, and the highest discharge in a thousand years would be 960 m^3/s. During the first decade of XXI century almost every two years the flood wave was bigger than the biggest one which occurs once in a fifty years. Huge flood waves were occurred in 2001, 2004, 2006, 2008 and 2010. The last flood in December 2010 had already reached the maximum value which occurs once in a hundred years (according to probability calculation (until and including) year of 2000)). Since the floods are directly and indirectly related to bank erosion these data should be included in bank erosion analysis because their analogy is proved, although there is no quantification of their correlation. Therefore, researches should be focused on causes of floods, and on reduction of bank erosion uncertainties. Many factors that influence the Kolubara River floods are already known. Firstly, there is a difference in flows in the upper and lower part of the Kolubara River basin. The drainage conditions in the upper part are more favorable. The area of hydrological profile Slovac is less than 1/3 of the whole basin, but it drains a half of all waters in the Kolubara River basin. Downstream hydrological profile Beli Brod encompasses a half of the basin, and its discharge is 3/4 of Kolubara's discharge. In the Donjokolubarski basin the drainage conditions are different, and the most significant factor is slope (the slope of the river flow and the slope of the river basin). The distance between Beli Brod and the Kolubara's confluence with the Sava River is 50 km and the altitude difference is 20 m. The present slope of 40 cm/km (0.4 ‰) is declining every year due to intensive sediment accumulation in the riverbed. Relating these processes with the shape of the river basin and rapid concentration of water downstream of the Beli Brod, it does not surprise that Kolubara River "ramp" over its alluvial plain. Moreover, in the last decade there is simultaneity of frequent rains of high intensity with extended duration and sudden snowmelt, and that is the reason for increased concern. Considering that rivers in the sub basins are mostly torrential, this concern is even more enhanced. Additionally, in this area rivers were diverted to bring the economic benefits. Because of all these reasons, the life in the coastal zone of the Kolubara River basin is gloomy but real with lot of uncertainties.

In order to prevent the frequent floods there are a several plans to deal with the actual situation in the area. Construction of several small accumulations on the Kolubara's tributaries is at its first phase, but there is no indication for solving the existing water problems. There is an idea to channelize the Kolubara's riverbed for sailing (i.e. for the transportation of lignite from the Kolubara mine), but it is still in the early phase of planning, although the initiative appeared long time ago. The height difference between the Kolubara's River confluence with Sava River and the location of lignite exploitation in the Kolubara mine basin is 23 m [12]. This height difference and the wideness of the Kolubara's riverbed would facilitate its riverbed training works, enabling cheaper lignite transportation from the Kolubara mine basin. Training works the Kolubara's riverbed and its preparation for lignite transportation could easily be carried out, so the invested means would be economically justified. Thus, the meandering flow would be straightened, and the strong bank erosion in the riverbed would be regulated which means that some factors of flooding would be eliminated.

CONCLUSION

Bank erosion, soil loss, sediment load deposition, changes in the river course, floods, landslides, soil and water pollution are the major environmental problems in the Kolubara River basin which could be aggravated by the land-use changes. The solutions for all mentioned environmental problems demand a complex analysis of the area characteristics and development of the strategy for solving the existing water problems in this area, but in the same time they have to provide necessary conditions for the further lignite exploitation. Some villages are located in the lower part of the Kolubara River basin, in the area which is planned for the expansion of the Kolubara mining basin, so it is an important factor for the future sustainable landscape planning.

Hydrological network of the Donjokolubarski basin is constantly changing due to natural factors and anthropogenic impacts. The damage which is done cannot be compensated, but even worse is the fact that no one feels responsible and that the population in this area is still left to the mercy of torrential river. Numerous calls for helping endangered people and goods were sent to the different addresses, but no one tried to help. Apparently, the problem goes beyond the "values" of a few villages and the state interest (lignite exploitation) has absolute priority, like in the case of neighboring Dubrava and unique sources of Obrenovac Municipality [34]. This situation lasted till the catastrophic floods in June 2010, when the shocking images of flood damage terrified the publicity, and problem could not be ignored anymore. As an attempt to repair the flood consequences, during 2010 two dikes were constructed with the length of 200 m in total. The first location was repaired for bridge protection, and the second one for household protection. The total cost of construction works was 100 000 euros. Since, the total length of all degraded river banks of first category is about 5 km; the economic profitability of this repairing method is questioned. It made sense in the initial phase of degradation, but now it goes beyond the reality of existing situation. It seems that, after the construction of two dikes, somebody tries to justify the negligence, because it is obvious that these two dykes are insufficient to solve the problem. In cases like this one, even not doing anything for protection of degraded areas represents a serious violation of principles of sustainable management of natural resources, actually that is an offence. The responsible for effects of the changes in the Kolubara River basin is still unknown, is it nature or man?

Making the constant pressure on state institutions through various appeals, indicating to unsustainability of current situation and stand by position of constant fear, this paper is one of many attempts to help the endangered population. In this context, the monitoring of the Kolubara River in the Donjokolubarski basin is a logical solution and our contribution, with particular results and recommendations, to fight for the basic human right to live without fear from hazards.

What kind of message can be sent to people living in this area and dealing with above mentioned problems? As they say, finding that the state does not protect them from the problems that come upon them, they give up farming the parcels of endangered area (along the river). The even more irrational, is the fact that they still pay taxes on the parcels, which does not exist anymore or they are significantly reduced, because the taxes calculation is made according to Cadaster from 1967! The estimation of all unnecessary loss of land, land values, personal losses of individuals and damage done to whole community is in the course. At the time when the personal status is far beyond collective responsibility due to difficult economic situation, this scientific approach is the only way to inspire the responsible ones in finding the solution.

This research could be the warning for the future anthropogenic activities on the river system since the new changes on the hydrological network were planned in this area. The four new mining fields should be opened, and if it happens, the hydrological network will be changed again and new problems will appear in the river basin.

ACKNOWLEDGEMENT

This paper was realized as a part of the projects "Studying climate change and its influence on the environment: impacts, adaptation and mitigation" (43007) and "The Democratic and National Capacities of Serbia's Institutions in the Process of International Integrations" (179009) financed by the Ministry of Education and Science of the Republic of Serbia within the framework of integrated and interdisciplinary research for the period 2011-2014. Translation and language correction was performed by Ljiljana Stanarevic.

REFERENCES

1. S. Dragićević, M. Stepić, 2006Changes of the erosion intensity in the Ljig River basin- the influence of the antropogenic factor. Bull. Serbian Geogr. Soc. 85(2): 37-44 (in Serbian with English abstract)
2. S. Dragićević, 2007Dominant Processes of Erosion in the Kolubara Basin. Faculty of Geography, Belgrade: Jantar groupe, 1245in Serbian

with summary in English).

3. S. Dragićević, I. Carević, S. Kostadinov, I. Novković, B. Abolmasov, B. Milojković, D. Simić, 2012Landslide susceptibility zonation in the Kolubara river basin (western Serbia)- analisys of input data. Carpathian Journal of Earth and Environmental Sciences 723747
4. S. Dragićević, I. Milevski, 2010Human Impact on the Landscape-Examples from Serbia and Macedonia. Advances in GeoEcology, 41Global Change- Challenges for soil management (Editor M. Zlatic), CATENA VERLAG GMBH, Germany. 298309
5. R. Tošić, 2006Soil erosion in the catchment Ukrina. Geographic Society of the Republic of Srpska, Banja Luka. Special issue, 13150in Serbian with summary in English)
6. V. Blanka, T. Kiss, 2011Effect of different water stages on bank erosion, case study of river Hernad, Hungary. Carpathian Journal of Earth and Environmental Sciences. 62101108
7. I. Milevski, 2011Factors, Forms, Assessment and Human Impact on Excess Erosion and Deposition in Upper Bregalnica Watershed (Republic of Macedonia). In: Human Impact on Landscape, Eds. S. Harnischmachter and D. Loczy. Zeitschrift für Geomorphologie, 55Suplementary 1Stuttgart, 7794
8. J. Chen, J. Z. Chen, M. Z. Tan, Z. T. Gong, 2002Soil degradation: a global problem endangering sustainable development. Journal of Geographical Sciences, 122243252
9. A. Goudie, 2006The human impact on the natural environment: past, present and future. Blackwell Publishing, USA, sixth edition, 357
10. L. Li, X. Lu, Z. Chen, 2007River channel change during the last 50 years in the middle Yangtze River, the Jianli reach. Geomorpholgy. 85185196
11. L. Denes, 2010Anthropogenic Geomorphology in Environmental Management. In: Anthropogenic Geomorphology- A Guide to Man-Made Landforms. Eds. József S., Lóránt D., Dénes L., Springer, 2538
12. S. Dragićević, N. Živković, V. Ducić, 2007Factors of flooding on the territory of the Obrenovac municipality. Collection of the papers, Faculty of Geography, Belgrade, 553954
13. M. Roksandić, 2012Causes and consequences of changes of hydrographic network in Donjokolubarski basin. Unpublished PhD thesis, University of Belgrade, Faculty of Geography, 197in Serbian with summary in English)
14. S. Dragićević, N. Živković, S. Kostadinov, 2008aChanges of hydrological

system in the lower course of the Kolubara river. In proceedings of the XXIV Conference of the Danubian countries on the hydrological forecasting and hydrological bases of water management, Bled, Slovenia.

15. M. Roksandić, S. Dragićević, N. Živković, S. Kostadinov, M. Zlatić, M. Martinović, 2011Bank erosion as a factor of soil loss and land use changes in the Kolubara river basin, Serbia. African journal of agricultural research, 63266046608DOI:AJAR11.736
16. S. Dragićević, 2002Sediment Load balance in the Kolubara basin. Faculty of Geography, Belgrade. 184in Serbian with summary in English)
17. S. Dragićević, M. Stepić, M. Karić, 2008bNatural potentials and degraded areas of Obrenovac municipality. Jantar groupe, Belgrade. 1180in Serbian with summary in English)
18. S. Dragićević, N. Živković, I. Novković, 2011Preparation of numerical and spatial data basis for the assessment of land and water diffuse pollution in the Kolubara River basin. Ministry of Environment, Mining and Spatial Planning, Environmental Protection Agency, Beograd (in Serbian)
19. J. Hooke, C. E. Redmond, 1989River-channel changes in England and Wales. Journal of Institution of Water and Environmental Management, 3328335
20. Large R G A, Petts E G1996Historical channel-floodplain dynamics along the River Trent, Implications for river rehabilitation. Applied Geography, 163191209
21. Q. Weng, 2002Land use change analysis in the Zhujiang Delta of China using satellite remote sensing, GIS and stochastic modeling. Journal of Environmental Management, 3273284
22. T. Kiss, K. Fiala, G. Sipos, 2008Alterations of channel parameters in response to river regulation works since 1840 on the Lower Tisza River (Hungary). Geomorphology, 9896110
23. R. C. De Rose, L. R. Basher, 2011Measurement of river bank and cliff erosion from sequential LIDAR and historical aerial photography. Geomorphology, 126132147
24. Cadastral register of the Municipality of Obrenovac,1967
25. D. Dukić, 1974Kolubara's regime and water management problems in its river basin. Bulletin of Serbian Academy of Science and Arts, Department of Science, 36Belgrade (in Serbian).
26. N. Surian, M. Rinaldi, 2003Morphological response to river engineering and management in alluvial channels in Italy. Geomorphology, 50307326

27. Youdeowei P O1997Bank collapse and erosion at the upper reaches of the Ekole creek in teh Niger delta area of Nigeria. Bulletin of the International Association of Engineering Geology 55167172
28. ***Diverting of Kolubara River: http://www.neshvyl.com/doc/prica_o_kolubari.pdf www. rbkolubara.rs
29. J. Hooke, 1995River channel adjustment to meander cutoffs on the River Bollin and River Dane, northwest England. Geomorphology, 14235253
30. R. Pejanović, Z. Njegovan, 2009Actual problems of Serbian agriculture and villages. Industry. 37(1): 87-99 (in Serbian)
31. Republic Hydro Meteorological Service of Serbia. Values of precipitation, water discharge, sediment load concentration for Kolubara river basin (1961-2005
32. D. Chen, J. G. Duan, 2006Modeling with adjustment in meandring channels. Journal of hydrology, 3215976
33. S. Dragićević, S. Nenadović, B. Jovanović, M. Milanović, I. Novković, D. Pavić, M. Lješević, 2010Degradation of Topciderska River water quality (Belgrade). Carpathian Journal of Earth and Environmental Sciences, 52177184
34. N. Zivkovic, S. Dragicevic, I. Brceski, R. Ristic, I. Novkovic, S. Jovanovic, M. Djokic, S. Simic, 2012Groundwater Quality Degradation in Obrenovac Municipality, Serbia. In: K. Voudouris and D. Voutsa (Eds.), Water Quality Monitoring and Assessment, Rijeka: InTech. 283300
35. W. De Vries, D. J. Bakker, 1998Manual for calculating critical loads of heavy metals for terestial ecosystems. Guidelines for critical limits, calculation methods and input data. Wageningen, DLO Winand Staring Centre. Report 166. 144
36. OSPAR/ICES2004Workshop on the evaluation and update of background reference concentrations (B/RCs) and ecotoxicological assessment criteria (EACs) and how these assessment tools should be used in assessing contaminants in water, sediment and biota, 913February 2004. The Hague, Final report, OSPAR Commission, 1-90442-652-2

Chapter 8

ENVIRONMENTAL LAND USE AND THE ECOLOGICAL FOOTPRINT OF HIGHER LEARNING

Seth Appiah-Opoku[1] and Crystal Taylor[2]

[1]Geography Department, University of Alabama, Tuscaloosa, USA

[2]Florida State University, USA

INTRODUCTION

The lifestyles of individuals, groups, or nations can be measured by utilizing an accounting tool known as ecological footprint. Ecological footprint refers to the productive land needed to support a given population. As discussed by Wackernagel and Rees (1996), "The ecological footprint concept is simple, yet potentially comprehensive: it accounts for the flows of energy and matter to and from any defined economy and converts these into the corresponding land/water are required from nature to support these flows" (p. 3). A concept known as "overshoot" occurs if demands by humans exceed the supply of a given biologically productive area (Turner et al., 2006). Thus, a larger ecological footprint indicates a less sustainable society.

Research on ecological footprint literature links together the concepts of footprint size and economic development. In other words, footprints represent population size and consumption levels (Wackernagel & Rees, 1996). Furthermore, more-developed countries contain market economies that consume greater levels of natural resources, and environmental degradation is largely driven by the growth and intensification of market economies (Jorgenson, as cited in Jorgenson & Burns, 2006). For example, Americans when compared to the rest of the world exhibit a large ecological footprint due to an intensely consumption-oriented lifestyle. The average ecological footprint for an American is 23.68 acres as compared to the world's average of 5.53 acres (Global Footprint Network, 2003). Further research suggests an economical discrepancy between those who possess large ecological footprints and those who possess small ecological footprints. Wackernagel et al. (2003) found that those contributing most to climate change through their energy intensive lifestyles will most likely be less affected by, and better shielded from, the outfalls of climate change than poor people living on marginal land or in underserved urban conditions.

Though ecological footprint can be used as a useful tool to help measure sustainability, some scientists have criticized ecological footprint calculations for oversimplifying ecosystem processes to numerical values. Assumptions may not be valid as the ecological footprint arbitrarily assumes both zero greenhouse gas emissions, which may not be optimal, and national boundaries, which makes extrapolating from the average ecological footprint problematic (Fiala, 2008). Despite these criticisms, the ecological footprint calculation can serve as a heuristic tool for designing and implementing plans for today as well as for tomorrow. Moreover, plans that take environmental calculations into consideration will have a far greater potential of keeping the Earth as a stakeholder in the planning process than those plans without such calculations.

Colleges and Universities across the world serve as incubators for tomorrow's leaders. In essence, they leave an educational imprint on individuals in an effort to educate and facilitate the development of tomorrow's leaders. These institutions serve as the setting where ideas can take form and this is where ideas can be implemented in a semi contained setting as part of the larger community. Though it is well established that educational institutions leave their imprints on innovative minds, this chapter introduces the idea that institutions of higher learning also leave ecological footprints on the landscape. Universities provide support to environmental issues through policies, programs, and research. The idea of greening campuses has become so popular that the Princeton Review has posted a Green Rating Honor Roll to document the top schools that provide a healthy and sustainable quality of life for the students, environmentally-minded and educational preparations for the future workforce, and environmentally responsible school policies for all to follow (Princeton Review, 2008).

Thinking green has been a hot topic among US Colleges in recent years. To think green is to incorporate environmental impacts into decision-making activities that affect daily lifestyles. The impact that a society imposes on the environment holds importance, as it is a key issue of sustainability. Sustainability refers to the dilemma of how to "meet the needs of the present without compromising the ability of future generations to meet their own needs" (Wackernagel & Rees, 1996, p. 33). Fortunately, one place where sustainable initiatives have spread is on campuses throughout America. Universities have provided support to environmental issues through policies, programs, and research.

The Princeton Review has posted a Green Rating Honor Roll to document the top schools that provide a healthy and sustainable quality of life for the students, environmentally-minded and educational preparations for the future workforce, and environmentally responsible school policies for all to follow (Princeton Review, 2008). The Princeton Review ranked the following eleven colleges throughout the United States as receiving a green rating of ninety-nine points.

- Arizona State University, Tempe
- Bates College
- College of the Atlantic
- Emory University
- Georgia Institute of Technology
- Harvard University
- State University of New York at Binghamton
- University of New Hampshire
- University of Oregon
- University of Washington
- Yale University

Although all the above-listed universities have displayed an extraordinary commitment to green initiatives, Harvard University located in Cambridge, MA; Emory University located near Atlanta, GA; and Bates College in Lewiston, ME, were chosen for closer examination in part due to the accessibility of online information concerning green programs as well as in respect to their diverse financial strategies for integrating sustainable principles. An inventory was performed encompassing a list of similarities and differences concerning green initiatives and strategies. Moreover, this inventory can serve as a framework for other colleges to follow in the future. It is in this context that we discuss the current consumption and environmental awareness levels associated with the use of water and energy resources for dormitory students on The University of Alabama's campus.

The University of Alabama is in the preliminary stages of moving toward a more sustainable campus. Currently, it is difficult to track environmentally friendly progress on campus, as no study has been previously performed to establish where The University of Alabama is concerning environmental initiatives. Thus, if a snapshot of the University were established to include both consumption and environmental awareness levels, then those findings would serve as a benchmark from which the implementation of green strategies may be evaluated in terms of effectiveness. Accordingly, this research documented the environmental awareness and consumption levels of dormitory students concerning energy and water resources on The University of Alabama's campus. Moreover, findings were gathered from the dormitories Ridgecrest East and Lakeside East. The goal was to measure the current ecological footprint of dormitory students on The University of Alabama campus. Specific research objectives were to (a) determine the current state of students' environmental awareness, and (b) determine the current consumption levels in terms of electricity and water usage for specific dormitories on campus.

RESEARCH METHODS

A case study approach was utilized during this research. According to Theodorson and Theodorson (as cited in Punch, 1998) a case study is defined as "a method of studying social phenomena through the thorough analysis of an individual case. Described simply, a case study provides a snapshot of particular social phenomena (Hakim, 1987, p. 61). Thus, the case study approach allows for in-depth research on specific populations, such as the dormitory students that will serve as the focus for this research. This approach also permits the researcher to evaluate subjects in a naturalistic setting as well as conduct research from a wide array of methods such as interviews, observations, numerical data, and questionnaires (Punch, 1998, p. 153). Suitably, interviews, observations, surveys, and data analysis are the primary methods utilized in this research. Even as the case study approach proves to be a viable research tool, a limitation is the inability of the researcher to derive generalizations from specific instances (Punch, 1998, p. 155). In light of this accusation, it is of importance to note that the case study approach warrants merit as this research requires an in-depth inquiry into a particular situation that has yet to be documented.

As mentioned previously, the focus of this research is centered around dormitory students residing on The University of Alabama campus located in Tuscaloosa, Alabama. In the fall of 2008, The University of Alabama reached a record enrollment of 27,052 students (Andreen, 2008). Of the 27,052 students approximately 7,000 students are housed on campus (E. Russell, e-mail, February 24, 2009).[1] -Therefore, on-campus residency accounts for approximately 26% of the student population as illustrated by Figure 1.

For this study two dormitories were chosen for sampling. The selection was done by methods of random sampling. Random sampling allowed every dormitory to have an equal opportunity of being selected. The process entailed writing down the names of all the possible dormitories on campus on individual slips of paper. The dormitory names were mixed up and then drawn out of a hat. The dormitories Lakeside East and Ridgecrest East were selected for an analysis of energy and water usage records. The coed student populations housed within Lakeside East and Ridgecrest East are 238 and 316 students, respectively.

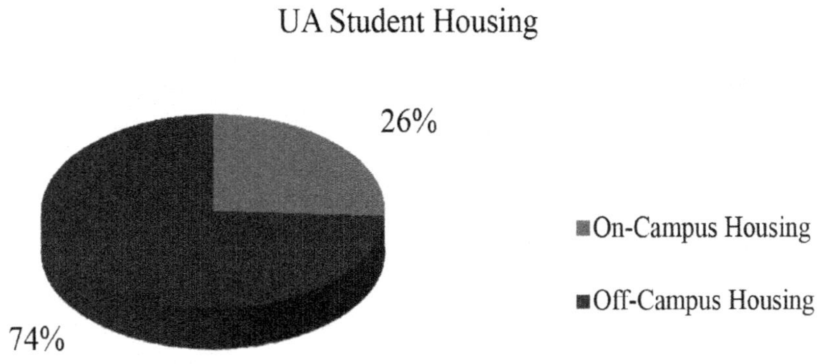

Figure 1: UA student housing.

SURVEY AND DATA ANALYSIS

The University of Alabama's Department of Energy Management aided in providing energy and water consumption records concerning the Ridgecrest East and Lakeside East dormitories. A content analysis of the records was performed to determine applicable

Figure 2: Lakeside East Residential Hall.

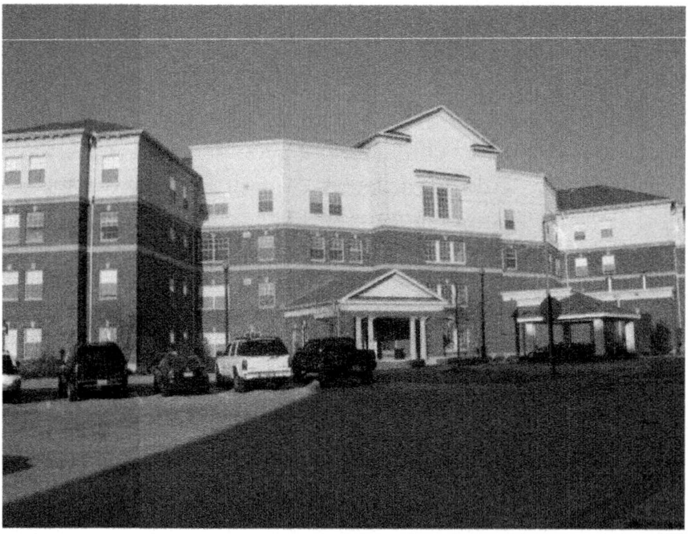

Figure 3: Ridgecrest East Residential Hall.

themes and patterns. Additionally, the records assisted with the calculation of the ecological footprint analysis of energy and water usage in dormitories on campus. The energy records acquired reported monthly electrical and natural gas usage figures for the two dormitories from 2007 and 2008. Due to some technical problems with the water meters, only the last five months of 2008 were available for analysis. However, water usage assumptions were derived for the entire year of 2008. Ecological footprint calculations were projected from estimates of the average water usage in 2008 and from the actual natural gas and electricity usage figures from 2007 and 2008. Even from water approximations, the derived ecological footprint has the ability to serve as a benchmark that can be utilized in future research. During the analysis of water and energy records, the data concerning the population rates for Ridgecrest East and Lakeside East during 2007 were unfortunately unattainable; consequently, the 2008 population numbers were substituted. In addition to the analysis of energy records, an interview with the Director of Energy Management was conducted in an effort to get a proper vision of the campus in terms of resource management.

CALCULATING THE ECOLOGICAL FOOTPRINT

Data from the Department of Energy Management were utilized in the ecological footprint calculation. The following identifies the process for calculating ecological footprints:

- Estimate the average population size.
- Estimate the average annual consumption for a particular item.
- Estimate the land area appropriated per capita for the production of items consumed.
- Estimate the ecological footprint of the average person for all items consumed.
- Multiply the population by the per capita footprint.

The ecological footprint calculation was utilized to determine land use requirements associated with the consumption of resources. The calculation was performed utilizing water, electric, and natural gas records. All the records used in this study were obtained from the Department of Energy Management.

Water

Water usage records were acquired pertaining to Ridgecrest East and Lakeside East Residential Halls from August to December 2008. Due to some technical problems with the water meters, accurate water usage readings prior to August 2008 were unattainable. The trend for water usage at Ridgecrest East showed little variation during the months of August, September, and October as consumption ranged from approximately 72,000 to 80,000 cubic feet or approximately 538,000 gallons to 599,000 gallons. Usage dropped slightly during November followed by a dramatic decrease in December. Lakeside East Residential Hall demonstrated more drastic trends than Ridgecrest East as usage in August peaked at nearly 140,000 cubic feet followed by a marked decline in September as Lakeside levels dropped around 40%. A slight increase occurred during the month of October. In November and December consumption decreased drastically as water usage dipped below Ridgecrest levels.

Figure 11 details water usage in cubic feet consumed. Figure 12 depicts the steps we were utilized to calculate the ecological footprint of water resources consumed in the dormitories Lakeside East and Ridgecrest East. First, the populations of Lakeside East and Ridgecrest East were established. As mentioned previously, 238 students reside within Lakeside East, whereas 316 students live in Ridgecrest East. A full twelve months of records were unavailable, so estimations were used to approximate the yearly water consumption levels within the dormitories. The 2008 yearly estimations for each building were derived from taking the average amount of water used during the five months and then multiplying that average by twelve months. For Ridgecrest East the figure 746,616 cubic feet was used as the 2008 water usage estimate, while the figure 911,496 cubic feet was used for Lakeside East.

Thus, an ecological footprint calculation concerning water resources can be derived by utilizing the water consumption estimates for the two residential halls as indicated above. Initially, the amount of water consumed in cubic feet per dormitory student must be established. The number was calculated by dividing the total water estimates for each

Source: University of Alabama Department of Energy Management

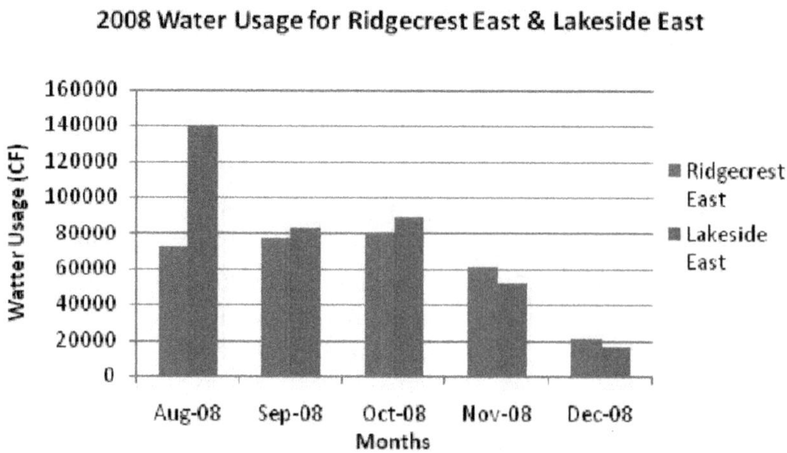

Figure 4: Water Usage for Ridgecrest East and Lakeside East.

Figure 5: Ecological footprint procedure for water.

dormitory by the subsequent student populations residing in each residential hall. Consequently, the average amount of water consumed per student for Lakeside East was 3,830 cubic feet (28,649 gallons) and 2,363 cubic feet (17,674 gallons) for Ridgecrest East.

To obtain a real-world comparison, consumption figures of the individual dormitory student are listed in gallons as well as cubic feet. The individual usage levels can further be broken down into daily usage figures by dividing by 365 to represent the approximate number of days in a year. As a result the daily consumption level for an individual residing in Lakeside East was 10.49 cubic feet or 78.49 gallons and 6.47 cubic feet or 48.42 gallons for those in Ridgecrest East. Daily usage figures are useful as they can be easily compared to the national average of the average American. According to the Environmental Protection Agency (2003), the average American consumes 90 gallons of water daily in the home, as compared to the average European consuming 53 gallons daily, and the typical Sub-Saharan African citizen consuming only 3-5 gallons per day.

After establishing the consumption levels for water resources, it was necessary to determine the amount of land required for the utilization of water resources. Thus, water resources were converted to cubic meters by multiplying by 0.0283 and then divided by 1,500 m^3/ha/yr to accommodate the amount of forested land needed to accommodate the water consumed (Anundson et al., 2001, p. 26). The result was equivalent to

0.0723 hectares (0.1785 acres) per dormitory student in Lakeside East and 0.0446 hectares (0.1101 acres) per dormitory student in Ridgecrest East

Table 1: Ecological Footprint for Water 200

Ecological Footprint for Water 2008	**Lakeside East**	**Ridgecrest East**
Total Water Usage 2008 (cubic ft)	911,496	746,616

Water Usage per Month (cubic ft)	75,958	62,218
Water per Student in 2008 (cubic ft)	3,830	2,363
Total Land Area in Hectares per Dormitory Student	0.0723	0.0446
Total Land Area in Acres per Dormitory Student	0.1785	0.1101

It is germane to keep in mind that all of these figures, concerning hectares/acreage required, only apply to the land required concerning water resources utilized during the consumption of housing. Accordingly, "the ecological footprint concept is based on the idea that for every item of material or energy consumption, a certain amount of land in one or more ecosystem categories is required to provide the consumption-related resource flows and waste sinks" (Wackernagel & Rees, 1996, p. 63). Thus, a complete ecological footprint calculation encompasses many different goods and services as this study looks specifically at water and energy resources associated with housing needs of dormitory students on The University of Alabama's campus.

Electricity

In addition to supplying the water records, as indicated in the findings in the previous section, the Department of Energy Management also provided electric and natural gas records for use in this research. To assist with the analysis of Lakeside East and Ridgecrest East Residential Communities, complete electrical and natural gas records were gathered from January 2007 to December 2008. Energy consumptions records from both 2007 and 2008 show a general trend of Lakeside East utilizing slightly less electricity per month with the exception of a peak on September 2007. During September 2007, Lakeside East Residential Hall experienced a spike in usage as

315,007 kilowatt hours (kWh) were consumed. This consumption stands-out on the electrical records as neither Lakeside East nor Ridgecrest East demonstrated another usage level over 140,000 kilowatt hours during the two-year span.

Despite the September peak for Lakeside East, electricity usage throughout the 2007 year remained somewhat consistent as January through March accounted for a range of approximately 50,000 to 65,000 kWh. April to May experienced a slight increase with consumption hovering near 80,000 kWh. June to July numbers were barely below 70,000 kWh, while August numbers increased back up to nearly 80,000 kWh. October boasted the second highest usage for 2007 at 87,151 kWh. Finally, during the months of November and December consumption ranged from 65,000 to 55,000 kWh. Interestingly, even as Lakeside East consistently consumed less power per month during 2007 with the exception of the September spike, the total 2007 energy consumption figures for Lakeside East (1,067,609 kWh) were slightly higher than Ridgecrest East (1,066,400 kWh).

2007 Electricity Usage for Lakeside East & Ridgecrest East

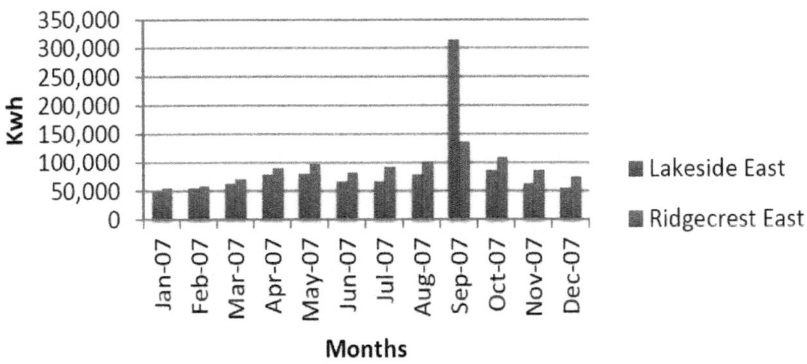

Source: University of Alabama Department of Energy Management

Figure 6: Electricity Usage for Lakeside East and Ridgecrest East.

As mentioned previously, Ridgecrest East has in general consumed a higher amount of electricity in terms of kilowatt hours per month during 2007 when compared to Lakeside East. Those higher consumption rates for Ridgecrest East are indicated as the following approximated percentages above Lakeside East's usage levels: January was 11% higher, February displayed an 8% increase, March had an 11% increase, April saw a 16% increase, May's increase jumped up 22%, June displayed a 24% increase, July had a 35% increase, October displayed a 27% rise, November increased to 35%, and finally December had a 36% increase over Lakeside East's consumption levels. Electricity consumption for Ridgecrest East during September 2007 was only about 44% of what Lakeside East consumed.

During 2008, Lakeside East consumed less total electricity each month than Ridgecrest East. Moreover, when the total consumption figures of 2008 for both dormitories are compared to the 2007 fiscal year, together the buildings show an overall decrease in electrical usage. Lakeside East displayed the following monthly consumption during 2008 recorded in kilowatt hours: January was 39,628 kWh; followed by February with 62,320 kWh; March consumed 58,206 kWh; April used 65,469 kWh; May was 68,613 kWh; June was recorded at 59,222 kWh; July had 62,597 kWh; August consumed 72,264 kWh; September was recorded at 108,040 kWh; October used 81,022 kWh; November had 60,623 kWh of usage; and finally during December 57,632 kWh were utilized. Similar to the methodology utilized to calculate the ecological footprint concerning water resources, Figure 15 depicts the ecological footprint procedure from which the electrical impact of students was derived.

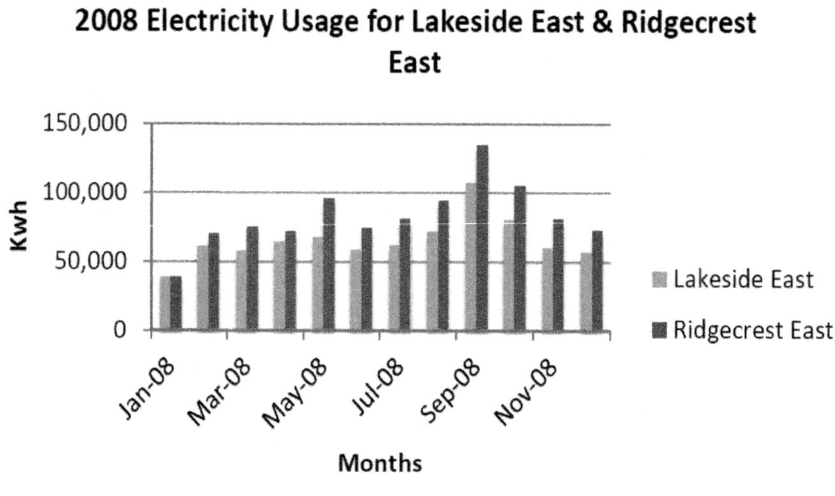

Source: University of Alabama Department of Energy Management

Figure 7: Electricity Usage for Lakeside East and Ridgecrest East.

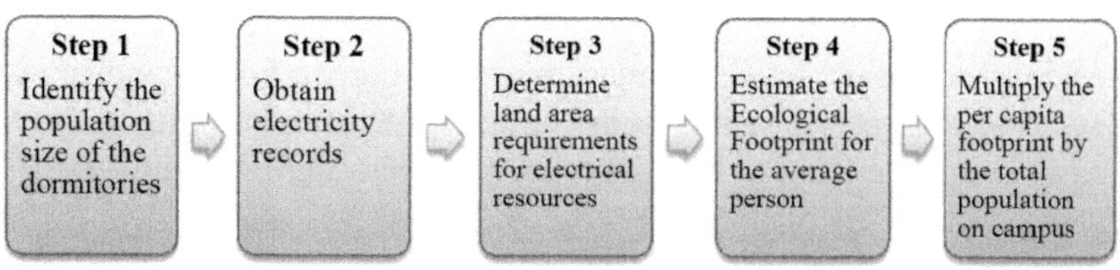

Figure 8: Ecological footprint procedure for electricity

For a more in-depth analysis of electrical usage for the two dormitories, the amount of energy utilized by each dormitory student for the year was calculated as the total electricity consumption numbers were divided by the amount of students residing within each dormitory. This accounted for the amount of electricity utilized per student to be 4,486 kWh at Lakeside East and 3,375 kWh at Ridgecrest East. It is important to note that even though the energy consumption numbers showed little variation during the 2007 fiscal year, the higher population numbers within Ridgecrest East resulted in energy usage per student that was considerably less than those found at Lakeside East. Just as the 2007 electricity records were broken down for analysis, the 2008 electricity records were evaluated for individual usage levels.

To acquire the electricity consumed per dormitory student during 2008, the electrical totals were divided by the amount of the respective residential populations. Thus, the average student consumed 3,343 kWh within Lakeside East and 3,177 kWh for Ridgecrest East. To relate student electricity consumption rates to a real-world example the 2007 and 2008 figures were broken into monthly averages. The 2007 monthly rates per dormitory student were calculated to be approximately 374 kWh for Lakeside East and approximately 281 kWh for Ridgecrest East. For 2008 the monthly averages were approximately 279 kWh for Lakeside East and approximately 265 kWh for Ridgecrest East. According to the Energy Information Administration (2007), the average Alabama household consumes 1,305 kWh per month.

After the consumption levels were successfully calculated for electrical resources, the amount of land could be determined for the usage of electrical resources. To accommodate the carbon emissions from the utilization of electricity the rate of 169 m^2 of forest for every 100 kWh of electricity was used for the following ecological footprint calculations (Anundson et al., 2001, p.11). Thus, the individual amount of electricity per dormitory student was first divided by 100 kWh and then multiplied by 169 m^2. Accordingly during 2007 for Lakeside East, the amount of land needed per dormitory student was 7,581 m^2 (0.758 hectares or 1.873 acres) and for Ridgecrest East 5,703 m^2 (0.570 hectares or 1.409 acres). During 2008, the amount of forested land area necessary per student amounted to 5,650 m2 (0.565 hectares or 1.396 acres) for Lakeside East and 5,369 m^2 (0.537 or 1.327 acres) for Ridgecrest East. In Table 7, meters squared were converted to hectares by dividing by 10,000. Additionally, hectares were converted by multiplying by 2.471.

Table 2: Ecological Footprint for Electricity 2007 and 2008

Ecological Footprint for Electricity 2007	**Lakeside East**	**Ridgecrest East**

Total Electricity 2007 (kWh)	1,067,609	1,066,400
Electricity per Student in 2007 (kWh)	4,486	3,375
2007 Total Land $(m)^2$ per dormitory student	7,581	5,703
2007 Total Land in Hectares per dormitory student	0.758	0.570
2007 Total Land in Acres per dormitory student	1.873	1.409
Ecological Footprint for Electricity 2008	**Lakeside East**	**Ridgecrest East**
Total Electricity 2008 (kWh)	795,636	1,004,000
Electricity per Student in 2008 (kWh)	3,343	3,177
2008 Total Land $(m)^2$ per dormitory student	5,650	5,369
2008 Total Land in Hectares per dormitory student	0.565	0.537
2008 Total Land in Acres per dormitory student	1.396	1.327

As a reminder, it is important to note that all the ecological footprint analysis that has been mentioned in this section pertains only to the electrical energy consumption as related to housing concerns. In reality electricity consumed for housing is only one area of a person's life where electricity is utilized. Therefore, the electrical usage and subsequent land area may in fact be larger than the estimates listed above. In general, ecological footprint calculations encompass a variety of goods and services associated with a person's lifestyle. This research looked specifically at water and energy usage of the footprint equation as related to housing needs.

In addition, each student's consumption of natural gas was calculated in the same way. Thereafter, each students total land area requirement at Lakeside East was calculated as follows: 0.179 acres for water resources in 2008, 1.873 acres for electricity in 2007, 1.396 acres for electricity in 2008, 0.170 acres for natural gas in 2007, and 0.177 acres for natural gas in 2008. Furthermore, Ridgecrest East's numbers were 0.110 acres for water in 2008, 1.409 acres for electricity in 2007, 1.327 acres for electricity in 2008, 0.142 acres for natural gas in 2007, and 0.140 acres for natural gas in 2008. Thus, if the entire student population that resides on-campus of approximately 7,000 individuals adopted the consumption habits of either Lakesides East or Ridgecrest East residents, then the land acreage as illustrated in Table 7 would have been needed.

When evaluating these figures it is important to understand that Lakeside East and Ridgecrest East are both relatively new buildings found on The University of Alabama's campus. As this study represents a sample of consumption levels taken from the new and therefore more efficiently constructed dormitories, the land requirement estimations for the students living on-campus are likely to be a best-case scenario. Overall, from the ecological footprint calculations utilized, Ridgecrest East displayed a lower environmental impact or land requirement than Lakeside East for water, electricity, and natural gas.

Additionally, land requirements decreased for electricity needs for both dormitories from 2007 to 2008. On the other hand, during the two year-span the land requirements for natural gas showed only a slight decrease for Ridgecrest East while Lakeside East showed an increase in demand. Acreage for water resources were not compared from 2007 to 2008 as the required data were unattainable.

Table 3: Ecological Footprint for the On-Campus Population

Ecological Footprint: Land Requirements in Acres for the Dormitory Student Population	**Lakeside East**	**Ridgecrest East**
From 2008 Water Consumed	1,253	770
From 2007 Electricity Consumed	13,111	9,863
From 2008 Electricity Consumed	9,772	9,289
From 2007 Natural Gas Consumed	1,190	994
From 2008 Natural Gas Consumed	1,239	980

CONCLUSION AND POLICY IMPLICATIONS

Although much progress has been made in recent years there is more that The University of Alabama can do in support of sustainable practices, as exemplified by green universities across the country. The first step toward becoming a green campus merely entails setting the goal of wanting to be more sustainable. The President of University of Alabama's message to the student body during fall of 2008 was the initial step required to set the tone for the campus. Now that a goal has been set, a subsequent plan will need to be developed. Objectives will need to be established in order to facilitate progress toward the end goal.

Before any other steps of the plan can be formulated lest carried out, it is essential to stop and take an inventory. The inventory determines where the campus is now so that progress may be more accurately measured. Thus, this research has served as a snapshot of where the campus currently is, during the academic semesters of fall of 2008 to early spring of 2009 in terms of sustainability. The snapshot is a useful tool as it was used to compare The University of Alabama to the top green schools. These prestigious universities were utilized in this analysis to serve as the pinnacle of where The University of Alabama may strive to be concerning environmental initiatives.

Taking the other schools analyzed in this research into consideration, our first recommendation is to formulate an official environmental plan that involves a variety of stakeholders in the planning process. This initiative needs the involvement of students, faculty, staff, alumni, investors, and the community as a whole. During the planning process, objectives must be set that are measurable as well as quantifiable to the overall goal of the plan. If these objectives are to serve as serve as milestones towards the goal of sustainability. Ecological footprint calculations as used in this study will be beneficial for monitoring progress towards this goal.

Our second recommendation is to strive to establish a recognizable environmental office on campus supported by a full-time staff. This ensures availability of knowledgeable staff to assist with inquiries from environmentally-aware students and community members as well as to address sustainability issues in accordance with the campus's environmental plan. According to data gathered on sustainable universities by the Sustainable Endowments Institute (2009), a considerable number of schools have recognized the need for full-time campus sustainability administrators. Currently, 56 percent report having dedicated sustainability staff.

We also recommend the incorporation of green building elements within residential student housing just. Generally speaking, universities are long-term owners of institutions. Hence, looking at the cost of operation over the period of a product's life cycle will help them accept some of the additional costs associated with green building methods. According to Moskow (2008), "Sustainable developments are more cost-effective in the long term and, therefore, ultimately, more valuable" (p.xv). This is especially true as the price of resources such as electricity and natural gas continue to rise. Additionally, green buildings have been noted to promote a healthy, productive work environment that would benefit the welfare and academic status of The University of Alabama.

Fortunately, The University of Alabama has already begun incorporating some green features in buildings such as low flow toilets, low flow faucets, low flow showerheads as well as plans for lighting controls and high efficiency hoods for new projects. Though those efforts are commendable, our recommendation is to use Bates College as an example to strive toward concerning green buildings. Due to cost restrictions, Bates College has not filed for the proper LEED certification for their structures. Despite not having filed, Bates College has used the LEED criteria as a standard in which to construct LEED equivalent buildings. Furthermore, green is marketable and green building designs are a good way to promote The University of Alabama's image.

Our final recommendation is education. Additional educational opportunities may in fact reduce the environmental impact of the University. Due to the fact that the role of academic institutions is to educate and facilitate in the development of tomorrow's leaders, this is a prime environment within which to integrate green technologies. Leaders that are unable to recognize the mismanagement of resources will be incapable of solving environmental problems. If environmentally friendly strategies are to be incorporated into future policies, then exposure to sustainable education is essential.

An expansion of research concerning ecological footprint analysis would be beneficial in an effort to determine the environmental impact of the campus. Though food and recycling strategies were only briefly discussed in this study, a more in-depth analysis may be needed to evaluate whether or not the University should try to promote locally or organically supplied food in the cafeterias and whether or not to participate in the *RecycleMania* competition. Additionally as only dormitory students were analyzed in this study, more sample groups could be evaluated and include both on-campus and off-campus students. Studies on climate change, transportation issues, student led initiatives, and a plethora of other opportunities exist for exploration.

In conclusion, we wish to emphasize that if places of higher learning are able to lessen their ecological footprints, they would ultimately have a greater positive impact on humanity and the dwindling resources of the World.

REFERENCES

1. American College & University Presidents Climate Commitment.2008About Environmental Land Use and the Ecological Footprint of Higher Learningfrom http://www.presidentsclimatecommitment.org/html/about.php
2. C. Andreen, 2008UA Enrollment Reaches Record 27,052 Students; Freshman Class Tops 5,000.Retrieved February 26, 2009, fromhttp://uanews.ua.edu/anewssep08/enrollment091608.htm
3. B. Anundson, J. Crooks, A. Fletcher, M. Frank, et al.2001A Study of the Ecological Footprint of Allegheny College. Unpublished manuscript.
4. 2008, November 20N. Barrella, (200, 20. November, competition. In, seeks. Harvard, "green. to, dorms. up", Harvard seeks to "green up"

dorms. Harvard Law Record. Retrieved December 3, 2008, fromhttp://media.www.hlrecord.org/media/storage/paper609/news/2008/11/20/News/In.Competition.Harvard.Seeks.To.green.Up.Dorms-3554348.shtml

5. Bates College.2008Student Housing at 280 College Street. Retrieved December 2, 2008, from http://www.bates.edu/x175547.xml
6. 2008, September 10B. Bralley, (200, 10. September, U. A. Construction, Green. Goes, White. Crimson, September. 1. Retrieved, 200, fromhttp://www.cw.ua.edu/ua
7. 2009, January 7K. Bursch, (200, 7. January, U. A. Starts, Green. New, Crimson. Campaign, White, Retrieved January 15, 2009, fromhttp://www.cw.ua.edu/ua_starts_newgreen_campaign
8. Emory University.2008aEmory Sustainable Initiative: History. Retrieved December 11, 2008, fromhttp://sustainability.emory.edu/page/1015/History
9. Emory University.2008bEmory Sustainable Initiative: Sustainable Food. Retrieved December 11, 2008, fromhttp://sustainability.emory.edu/page/1008/Sustainable-Food
10. J. Enck, S. Turner, 2003ASHRAE Green Guide: An ASHRAE Publication Addressing Matters of Interest to Those Involved in Green or Sustainable Design of Buildings. Atlanta: American Society of Heating, Refrigerating and Air-Conditioning Engineers, Inc.
11. Energy Information Administration.2001Natural Gas Consumption and Expeditures in U.S. Households by End Uses and Census Region, 2001. Retrieved March 3, 2009, from http://www.eia.doe.gov/emeu/recs/byfuels/2001/byfuel_ng.pdf
12. Energy Information Administration.2007U.S. Average Monthly Bill by Sector, Census Division, and State 2007. Retrieved March 3, 2009, fromhttp://www.eia.doe.gov/cneaf/electricity/esr/table5.html
13. Environmental Protection Agency.2003Water on tap: What you need to know.Retreived March 25, 2009, fromhttp://www.epa.gov/safewater/wot/pdfs/book_waterontap_full.pdf
14. Environmental Protection Agency.2008Wastes- Non-Hazardous Wastes. Retrieved October 12, 2008, fromhttp://www.epa.gov/epawaste/nonhaz/index.htm
15. Environmental Protection Agency.2009aBrownfields and Land Revitalization. Retrieved March 19, 2009, from http://www.epa.gov/brownfields/

16. Environmental Protection Agency.2009bWastes- Resource Conservation-Reduce, Reuse, Recycle. Retrieved April 16, 2009, fromhttp://www.epa.gov/osw/conserve/rrr/reduce.htm
17. N. Fiala, 2008Measuring sustainability: Why the ecological footprint is bad economic and bad environmental science. Ecological Economics. Retrieved October 11, 2008, from ScienceDirect database.
18. T. Glavinich, 2008Contractor's Guide to Green Building Construction: Management, Project Delivery, Documentation, and Risk Reduction. New Jersey: John Wiley & Sons, Inc.
19. Global Footprint Network.2003National Footprints.Global Footprint Network. Retrieved October 1, 2008, fromhttp://www.footprintnetwork.org/gfn_sub.php?content=national_footprints
20. "Global Warming". (n.d.).New York Times. Retrieved October 11,2008fromhttp://topics.nytimes.com/top/news/science/topics/globalwarming/index.html
21. "Graduate Green Living Program Enters Its Second Year".2007, SpringHarvard Green Campus Newsletter. Retrieved December 4, 2008, fromhttp://www.greencampus.harvard.edu/newsletter/archives/2007/05/graduate_green.php
22. 2008, September 19K. Gray, (200, 19. September, Emory Freshman Live 'Green' in New Housing. [News Release]. Retrieved December 3, 2008, from http://www.emory.edu/home/news/releases/2008/09/green-dorms-open.html
23. C. Hakim, 1987Environmental Land Use and the Ecological Footprint of Higher LearningLondon: Allen & Unwin.
24. 2005, March 31B. Handwerk, (200, 31, Earth's Health in Sharp Decline, Massive Study Finds. National Geographic. Retrieved October 10, 2008, from http://news.nationalgeographic.com/news/2005/03/0331_050330_unenvironment.html
25. C. Harper, 2008Environment and Society: Human Perspectives on Environmental 4thed.). New Jersey: Pearson Education, Inc.
26. Harvard University.2008Green Campus Loan Fund: Harvard Office for Sustainability. Retrieved December 11, 2008, fromhttp://www.greencampus.harvard.edu/gclf/
27. 2008, November 13C. Ireland, (200, 13. November, Living in the green zone at "Rock Hall". Harvard University Gazette Online. Retrieved December 3, 2008, from http://www.news.harvard.edu/gazette/2008/11.13/11 -rockefeller.html

28. A. Jorgenson, T. Burns, 2006Environmental Land Use and the Ecological Footprint of Higher LearningSocial Science ResearchRetrieved October 11, 2008, from Science Direct database.

29. J. Kitzes, A. Galli, M. Bagliani, J. Barrett, G. Dige, Ede, S. , et al.2008Environmental Land Use and the Ecological Footprint of Higher LearningEcological EconomicsRetrieved October 11, 2008, from ScienceDirect database.

30. B. Kobet, S. Lee, C. Mondor, 1999Green Buildings: Environmental Land Use and the Ecological Footprint of Higher Learningection.

31. T. Leopard, 2008SGA Memo: University Efforts in Regard to Sustainability. [Memo]. University of Alabama.

32. 2007, AutumnM. Loftus, (200, Autumn, Emory sprouts new green residence halls. Emory Magazine. Retrieved December 3, 2008, fromhttp://www.emory.edu/EMORY_MAGAZINE/2007/autumn/halls.html

33. 2002, January 10H. Mayell, (200, 10. January, Green Group Gives Earth Failing Report Card. National Geographic. Retrieved on October 11, 2008, from http://news.nationalgeographic.com/news/2002/01/0110_020110worldwatch.html

34. K. Moskow, 2008Environmental Land Use and the Ecological Footprint of Higher LearningNew York: McGraw-Hill

35. National Geographic Website.2008Effects of Global Warming. Retrieved October 11, 2008, fromhttp://environment.nationalgeographic.com/environment/global-warming/gw-effects.html

36. J. Pinkse, M. Dommisse, 2008Environmental Land Use and the Ecological Footprint of Higher LearningBusiness Strategy and the EnvironmentRetrieved September 12, 2008, from Wiley InterScience database.

37. Princeton Review.2008Green Rating Honor Roll. Retrieved September 10, 2008, from, http://www.princetonreview.com/green-honor-roll.aspx?uidbadge=07

38. K. Punch, 1998Environmental Land Use and the Ecological Footprint of Higher Learningndon: Sage Publications.

39. RecycleMania2009General Overview. Retrieved January 05, 2009, from http://www.recyclemania.org/

40. R. Spiegel, D. Meadows, 2006Green Building Materials: A Guide to Product Selection and Specification. (2nd ed.). New Jersey: John Wiley & Sons, Inc.

41. Sustainable Endowments Institute.2009Administration- Leaders-Green Report Card 2009. Retrieved March 9, 2009, fromhttp://www.greenreportcard.org/report-card-2009/categories/administration.
42. K. Turner, M. Lenzen, T. Wiedmann, J. Barret, (2006, 2006Examing the global environmental impact of regional consumption activities-Part 1: A technical note on combining input-output and ecological footprint analysis. Ecological EconomicsRetrieved October 10, 2008, from ScienceDirect database.
43. M. Wackernagel, C. Monfreda, N. Schulz, K. Erb, H. Haberl, F. Krausmann, 2003Environmental Land Use and the Ecological Footprint of Higher Learnings. Land Use PolicyRetrieved October 10, 2008, from ScienceDirect database.
44. M. Wackernagel, W. Rees, 1996Our Ecological Footprint: Reducing Human Impact on the Earth.Gabriola Island: New Society Publishers.
45. 2007, JanuaryS. Wallace, (200, January, the. Farming, Last. Amazon, the. of, Amazon, National Geographic. Retrieved October 11, 2008, fromhttp://environment.nationalgeographic.com/environment/habitats/last-of-amazon.html
46. M. Wymer, 2008UA Experts Offer 'Going Green' Advice. Retrieved February 16, 2009, fromhttp://uanews.edu/anews2008/apr08/earthday040808.htm

Chapter 9

THE ROLE OF TRADABLE PLANNING PERMITS IN ENVIRONMENTAL LAND USE PLANNING: A STOCKTAKE OF THE GERMAN DISCUSSION

Dirk Loehr

Trier University of Applied Sciences, Environmental Campus Birkenfeld, , Germany

INTRODUCTION

The idea of tradable planning permits is subject to broad discussion in some developed countries such as Switzerland (for example, see [1]), but particularly in Germany (for example, see [2]).

The German federal government intends to reduce the daily land consumption to 30 ha per day in 2020 [3]. In 13 years between 1993 and 2010, land consumption in Germany was significantly higher than 100 ha per day. In the other 5 years, the undershooting of the 100 ha mark has been mostly due to lower economic growth rates or an economic slump [4]. Particularly rural areas were affected by excessive land consumption. Almost 50% of the converted land is sealed [5].

In order to achieve the 30 ha target, there is a broad consensus about the necessity to support planning by means of economic instruments. In this discussion, tradable planning permits turned out to be the instrument of choice, at least among the scientists. In Germany, a lot of research has been underway on this issue for years now (for example, see [6]). Among others, a pilot project is also in preparation [7,8], as it was planned in the coalition agreement of the current federal government [9].

The idea of tradable planning permits stems from the concept of tradable CO_2 rights, more accurately the cap and trade system. Within the cap and trade regime, pollution rights should be limited in quantity and made tradable. Due to the cap on the pollution possible, the system is considered ecologically effective. If the mechanism is applied to the field of land use planning, the communal development plans are only legally valid if they are backed by planning permits, which have to be held by the communes. The communes

– as the planning authorities – and not the land owners are the holders of the planning permits. This is an important difference from "tradable development rights", where private-sector actors are the sellers and buyers (for example, see [10]).

Due to the trade, the scheme is also regarded as being efficient because only those actors with the lowest marginal abatement costs reduce the emissions. The permits can be bought and sold by the communes on an organized trading platform.

The cap on the permits helps to circumvent rationality traps (game theory) which otherwise would appear. In Germany, for instance, communes competed against each other to attract new inhabitants and industries in order to get more tax revenues and higher shares out of the financial equalization scheme. This competition was a race to the bottom in many cases. Among others, the results in many cases have been almost empty residential or commercial areas and high infrastructure costs. However, if a community waives the preparation of new building areas, the neighbouring municipality takes the chance.

Within the cap and trade scheme, such rationality traps might be broken up [11]: Due to the costs of the permits, only such communes whose benefits of land development exceed the costs of the permits will buy planning permits and carry out land development. If land conversion can be avoided at costs below the costs of the planning permits, communes waive the right to further development. Maybe they can also reconvert the land into a natural state. Hence, if there is no need for holding permits, such communes will sell them to other communes (for example, see [12]). If they do not sell the "free" rights, they will suffer opportunity costs. This means that communes with high marginal abatement costs (tax revenues, jobs etc.) are the buyers of the rights, while communes with low marginal abatement costs are the sellers. In the end, all marginal abatement costs equalize at the price of the tradable permits. In the trading planning permits scheme, the secondary market is the institutional heart of the mechanism.

Although it sounds quite appealing at first glance, we want to show that the application of the cap and trade scheme to land use planning is anything but self-evident and not a promising approach per se.

HYPOTHESIS: NO MAGIC BULLET

Tradable planning permits are considered to be a sort of magic bullet. On the one hand, the cap on development permits makes the system effective. On the other hand, only those communes with the highest benefits (additional taxes and shares from financial equalization schemes) carry out the development.

Communes with low opportunity costs waive the right to development. Hence the scheme is also efficient, because the planning rights are used at the locations with the highest benefits.

However, contrary to what intuition would suggest, we want to show that effectiveness and efficiency don't harmonize if the concept is applied to land use planning. In contrast, the cap and trade approach cannot meet the goals of efficiency and effectiveness at the same time ("incompatibility thesis") [13]. The argument is based on the following two statements:

- In order to be efficient, a cap and trade system needs wide system boundaries. At least in small or medium-sized countries, such wide system boundaries go hand in hand with a unified planning permit and a unit price.
- In contrast to CO_2, effective land use planning doesn't require control of a scale, but of a structure. A land use structure cannot be controlled effectively by a single planning permission with a single price.

THEORETICAL ISSUES AND REVIEW OF LITERATURE

There is a central difference between the cap and trade on CO_2 and the cap and trade on planning permits. Considering the consequences for global warming, it does not matter where the CO_2 is emitted due to the diffusion characteristics of the greenhouse gas. Hence the task is to control a scale (maximal CO_2 emissions anywhere) by capping the quantity of emissions. However, regarding land use planning, not only the scale but also the structure of land use has to be controlled. The quantity of land as a whole can hardly be extended. Instead, the relevant issue relates to changes in the structure of land use, which is for instance forestry, agriculture, industry, settlements etc. It is of central importance where the land use takes place and for which purpose. Hence CO_2 permits are a homogenous good, but land use rights shouldn't be.

Consensus: Primacy of Planning

At present, the structure of land use is controlled by the planning system. According to the proponents of the tradable development rights idea, land use planning should not be substituted but supported by the economic tool ("primacy of planning") [14].

Planning is necessary to break up a possible Nash equilibrium [11] caused by the behaviour of land owners: If, in the absence of any planning, only the willingness to pay decides about land use patterns, a spatial disaster may result and people may run into a rationality trap. If, for example, German people were allowed to realize the favoured model of the detached one-family house in green surroundings, urban sprawl would happen, with negative ecological, economic and social impacts.

At the same time, planning is necessary to protect such forms with weak financial endowments which cause important positive external effects. If no plan provided public spaces e.g. for kindergartens and schools, such forms would have to compete with actors with a high willingness to pay (e.g. banks). Hence they could not be realized. However, without such facilities, the value of the area would often be lower than with them. Good planning should consider the variety of functions of land (e.g. ecological, spiritual). Such forms of land use that move beyond efficiency and profitability are not only important for the cohesion of the social system, but in many cases also for the resilience of the ecological system (for example, see [15]). Planning has to balance the competing demands of various stakeholders, including groups with low budgets and the protection of nature.

Thesis: Efficiency Needs Wide System Boundaries

Although the primacy of planning is wide consensus, in recent debates it has been argued that tradable planning permits may counteract land use planning [13]. The "incompatibility thesis" is based on the required design of a cap and trade regime. A major justification of the system is its efficiency. The efficiency of the cap and trade system is caused by differences in marginal abatement costs:

- Those communes with high marginal abatement costs buy planning permits on the market at the lower market price. The difference is the benefits from the cap and trade system.
- On the other hand, such communes with low marginal abatement costs reduce their harmful activities and sell the free certificates on the market. The difference between market price and marginal abatement costs is the profit from abatement.

In the end, the marginal abatement costs of all the actors equal the market price of the planning permits. The higher the differences of marginal abatement costs of the acting communes, the higher the efficiency potential of the regime will be.

However, high differences in abatement costs can be achieved by a wide design in terms of space, time, participants and the objects of trade:

- In categorical terms, diverse spatial categories (living, commerce, mixed use, traffic etc.) have to be gathered in one single planning permit ("universal" certificate). This is any land for human settlement and transport infrastructure without regard to its different components;

- Regarding the market participants, the discussion is about also including individuals or NGOs instead of only permitting communes as traders;
- In spatial terms, scientists agree that the trade boundaries have to be as wide as possible (e.g. whole of Germany, no single states);
- Considering the time dimension, banking and borrowing is also discussed in order to use differences in the marginal abatement costs over the timeline.

In Germany, in the last decade the preferred design is characterized by

- A country-wide regime (although the pilot project mentioned above will only comprise selected communes) which incorporates the administrative support of the different states of the federation [16].
- A universal certificate which comprises the whole area for settlement and traffic [16]
- Banking should be allowed (at least the transfer into the following provisional period), in order to allow long-term development strategies for the communes. In contrast, there is much scepticism with regard to providing the opportunity to use the rights before owning them (borrowing) [17].
- Regarding the market participants, an extension of participants beyond the communes has not been discussed seriously so far.

Hence the efficiency potential can only be exhausted if the target is "scale" instead of "structure" (which would make sub-markets necessary). The scale target has to fix wide system boundaries (in contrast to other tradable development rights schemes; for example, see [13]). The scale target and the wide system boundaries are mutually dependent.

There is also another reason why a working trade system would not be possible without wide boundaries: Narrow markets cause high price volatility of the permits. The higher the price volatility, the more insecure the economic success of abatement activities and the less abatement activities will take place. Thus the target is to set the condition for organized trade of the permits.

Thesis: Controlling Structure by Economic Tools Needs Tightly Segmented Sub-Markets

Having shown the necessity for wide system boundaries in a cap and trade model, the next question is whether a land use structure – not a scale – can be controlled within such a system. We want to illustrate the problem within figure 1 below. The land use plan sets the allowed land use A (e.g. industry) at the maximum of C_A. The maximal land use B (e.g. housing) is limited by C_B. The marginal abatement costs (MAC) are MAC_A for land use type A and MAC_B for land use type B. For simplification purposes, the illustration doesn't include more land use types.

If a commune waives the right to development of additional sites, it has to suffer marginal abatement costs. Such marginal abatement costs are mainly opportunity costs. If, for instance, a residential area is not realized, a German municipality gets lower shares of the income tax revenues, lower property tax revenues and lower revenues out of the fiscal equalization scheme. If an industrial site cannot be realized, the opportunity costs also comprise lost business tax revenues. Also indirect effects have to be considered, such as income multiplier effects which otherwise would have been initiated by construction activities. All these interrelations and effects are quite complex and include feedbacks within the system. Basically, the scale of opportunity costs is not quite clear. Fiscal impact analysis could provide for more cost transparency, but it is in an early stage. Some fiscal impact tools used so far for residential areas turned out to have quite different performance; for industrial areas no reliable fiscal impact tool is available so far. Also within the above-mentioned pilot model of tradable development rights the development of reliable fiscal impact analysis tools is acknowledged to be important in order to get a better idea about the marginal abatement costs.

However, in the subsequent figure we assume, contrary to the facts, that there is an accurate idea about the volume of the marginal abatement costs of land use type A and B. Hence we can derive a mathematical function of MAC, being dependent on the scale of land use of the different types.

First, let's assume that the caps for land use type A (C_A) and B (C_B) are set according to the land use plans. We assume that the land use planning also properly computes the marginal damages (MD), which are illustrated with the dotted line for both types of land use. With this theoretical "trick" we can take into account that planners care for quantity as well as for quality of land use (for example, see [17]). Therefore, the planning target (C_A and C_B) corresponds perfectly with the intersection of marginal abatement costs and marginal damage of land use type A (E_A) and B (E_B).

Moreover, theoretically the marginal damage of land use type B can be expressed in equivalents of the marginal damage of land use type A. Such equivalents are useful for the definition of a universal cap. Analogous equivalents are also used in the greenhouse gas emission permit schemes. In the Kyoto regime for instance, global warming potentials (GWPs) are used in order to express the global warming potential of the other greenhouse gases in relation to CO_2, whose GWP is standardized to 1. Hence, if caps for different land use types should reflect such equivalents (for simplification purposes a linear function is used below), we get for instance a function such as:

$$C_B = e \times C_A \tag{1}$$

The total cap results by aggregation of the caps of the individual land use types:

$$C_{A+B} = C_A \times e + C_A \tag{2}$$

It is important to note that the equivalents only reflect an average consideration. However, different communes may have different structures of land use; thus the equivalents don't represent their individual situation.

In the subsequent diagram, the added marginal abatement costs (MAC_{A+B}) show the aggregate demand curve, and the aggregated cap (C_{A+B}) shows the aggregate supply curve of all land used for settlement and traffic, set by the planning authorities. The intersection point determines the unit price of the universal development right P* [13]. The illustration holds true for an individual municipality as well as for the aggregation of municipalities.

The figure shows why a unit price causes economic incentives to violate the land use plans:

Regarding land use type A, the mayor in charge will extend the land use until the intersection point (I_A) of the unit price P* and the marginal abatement costs MAC_A. From this point on, the costs for additional planning permits exceed the benefits from additional land use. However, regarding the land use plan (and the marginal damage), more development of land use type A would be possible (up to E_A and C_A respectively). Insofar there is a loss of welfare, indicated by the gap between M_A and C_A. In order to support the land use planning, price P_A would be necessary instead of the unit price P*.

Looking at land use type B, the mayor increases land development until his/her benefits of additional development MAC_B equal the unit price of the development P* (in point I_B or $M_B$respectively). However, this is much more than the land use plans have fixed (C_B). If the caps reflect the intersection point E_B of marginal damage function MD_B and marginal abatement costs MAC_B, this point I_B is also far beyond efficiency (in E_B). In order to get an effective and efficient result, price P_B would have been necessary. From the figure above we may derive two important results:

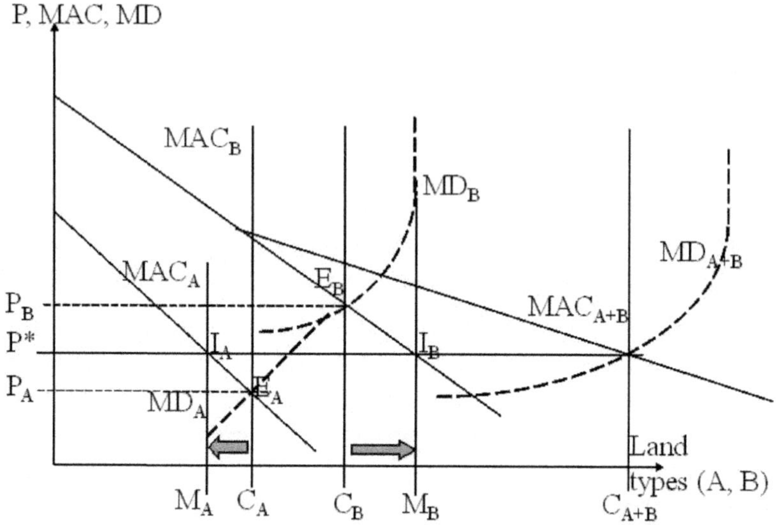

Figure 1: Aberrations with a universal planning permit (adapted from [13])

The trading model, which is based on wide system boundaries and a single tradable planning permit, is not able to support land use planning. With a unit price, it is only able to control scale but not structure.

Nonetheless, the supporters assert effectiveness. Obviously, they refer to the CO_2 blueprint and only consider the control of scale, but ignore the necessity to control structure. There is no assessment of the welfare losses which are caused by overshooting and undershooting of the planning targets so far (differences between $M_{A,B}$ and $C_{A,B}$). Therefore, also no proper statement about the net efficiency gains (efficiency gains minus the welfare losses, due to overshooting and undershooting) of the cap and trade system can be provided.

Sometimes, supporters suppose that the deviations and aberrations are negligible. Depending on the price of the permits, the position of the caps and the marginal abatement costs, the aberrations and welfare losses can be randomly quite high or low. There is no evidence for a correction mechanism, which might be able to reduce such aberrations systematically. Hence, overshootings and undershootings probably don't equalize in aggregation. The system is as effective as a poor marksman who is currently missing the target in different directions. Aggregating the shooting errors doesn't turn the poor marksman into a champion.

The "incompatibility thesis"

Against this background, we can describe the central system conflict as follows:

The rationale of the cap and trade regime is efficiency. However, efficiency can only be achieved within a wide design of the system, among others based on a universal certificate (including all categories of land for human settlement and transport infrastructure);

In contrast, the required primacy of planning can only be maintained within a tightly structured system. In contrast to the controlling of the scale of CO_2 emissions, the settlement structure cannot be controlled by one cap (certificate) and one price, but needs a diversity of prices with a diversity of certificates.

If a structure is "treated" with a single price, welfare losses have to be expected due to overshooting and undershooting of the planning targets. Hence the net efficiency gains and net welfare effects of a cap and trade regime for land use policy are not clear at all.

The supporters of the tradable planning permits claim that planning should impose a correcting action if necessary. However, our argument is that precisely due to the counteracting economic signals, planning cannot impose such an action.

The problem could theoretically be solved by a variety of tightly segmented sub-markets with diverse price settings. Sub-markets which adapt the categories of land use planning (market segmentation in categorical terms) have been discussed e.g. by Henger and Schröter-Schlaack [17]. Spatial boundaries which limit the trading rights to regions with similar protection status (market segmentation in spatial terms) have been addressed e.g. by Williams [18], Walz et al. [16], Henger and Bizer [19]. However, in most countries such markets would be too small and thus inefficient [17]. At least in the German discussion, the conflict of goals which appears in the cap and trade scheme was decided in favour of efficiency and against market segmentation. Within a system with wide boundaries, such an incompatibility could perhaps be avoided in a few countries with high population and centralized land use planning systems, such as China. However, so far there is no sound research about the minimum size of such markets and the requirements for the land use planning system.

Addendum: Initial Distribution

Within a cap and trade system, the initial distribution might be done by auction or by providing the permits to the communes without costs (according to alternative allotment formulas). Simulation experiments with German municipalities related to the cap and trade scheme also showed that the initial distribution of rights is quite a critical issue, which may endanger acceptance [20]. In order to guarantee the acquis of the communes, discussions have so far favoured "grandfathering" schemes, in which the status quo of land consumption is not touched. However, such grandfathering schemes are probably less efficient than auction schemes [13, 17].

RESULTS AND DISCUSSION

Basically, the required primacy of planning is not compatible with the system of tradable planning permits. If the basic idea of capping planning permits should be kept, a redesign of the regime is necessary.

Cap and Auction

In order to support the planning system ("primacy of planning"), the planning permits should best be defined in a tight manner which is in line with the categories of the planning system. For instance, if planners are thinking in categories such as residential, industrial or mixed areas, traffic etc., the planning permits should also follow these categories. This also facilitates the handling of the system. Planning permits should basically be mandatory for all sorts of developed areas. For instance, recreational space may have also negative ecological impacts.

The caps could be administered on different administrative levels, e.g. at the level of the states, even at the level of regions (or in other countries: at county level). However, the administration needs a certain human capacity. In Germany, the 30 ha cap might be broken down into lower administrative levels without problems.

However, within a tight definition of a variety of planning permits at a low administration level (e.g. region), the "markets" for each right would be quite narrow. Hence the allocation mechanism shouldn't be based on the secondary market (trade – "horizontal coordination") but on the primary market, namely auctions ("vertical coordination"). The auction of planning permits to the communes could be done periodically. Giving the focus to auction doesn't mean a complete ban on trade but a reduction of its significance. Due to the tight design of the sub-markets, organized trade wouldn't be possible anyway. Instead, over-the-counter trade would be feasible, regulated by the planning permits administration. Within the proposed design, different types of planning rights (residential, industrial etc.) were auctioned and traded at different prices.

The administration should also guarantee that it will take back the planning permits for a fixed price (based on the auction price). Thus communes could also think about changing the land use plans and the redevelopment of shrinking areas into the natural state. Deconstruction would be encouraged, because the communes could be sure about the compensation.

A system which is based on auction might be designed tightly with regard to space, time, participants and objects of the design. The system may work with only a few participants. It may work even at regional level. Within the auction, the development rights are allocated to those communes that can make the best use of them. Moreover, without the overshooting and undershooting of the cap and trade regime, welfare losses also are avoided.

Despite the segregation of sub-markets, the system is also efficient. However, in terms of efficiency it is not clear if such a cap and auction system will or will not compete with a cap and trade system, which is based on a single, universal development right. Nonetheless, if in doubt, recognizing the primacy of planning within the conflicts of goals means subordination of efficiency. From a system theory point of view, economic efficiency shouldn't be a guiding value [15] of superior significance anyway, at least not in land use management. Instead, the guiding value of efficiency has to be balanced with other guiding values.

Completion By Means Of a Financial Equalization Scheme

The regime sketched out above has to face some serious counterarguments relating to political viability: In an auction, the powerful communes with a high willingness to pay will prevail. Moreover, the financial situation of the communes would be even more strained and the municipalities' acquis would be encroached (see section 3.5.). Hence the question is how to increase the acceptance of a cap and auction model.

On the one hand, certain transition regulations such as free development rights for existing settlements would certainly be helpful, but not nearly enough on their own.

On the other hand, the view of the discussion about climate policy might be promising. Within the Kyoto regime, it has not been possible to put in place effective caps, mainly because the problems of distribution and "climate justice" have not been solved yet [21]. The regime was based on "grandfathering". Hence those countries with the most aggressive occupancy of the atmosphere and most responsibility for the climate problem got most rights. This was considered as being unjust by countries with developing and emerging economies. Basically, the same holds true in terms of land use permits.

However, many of the objections mentioned above could be countered by establishing a redistribution mechanism. Within such a redistribution mechanism, the money paid by the communes in the auction could be firstly collected in a fund ("land trust"), which is administered by the affected communes. Second, the money is redistributed to the communes, preferably according to the number of their inhabitants (other redistribution keys are also possible). In this respect, all inhabitants are considered as "co-owners" of the planning permits, and thus they should participate in equal shares from the revenues of the auction. Considering the CO_2 emission trading schemes, this idea has been popularized by Peter Barnes [22]. Applied to land use planning, a similar redistribution scheme has already been suggested by Krumm [23]. However, his proposal was based on a price-steering basis: Basically, communes should be charged for any new land conversion using a fixed rate per square meter. The money should be pooled in a fund and redistributed to the communes, preferably according to the number of citizens.

If redistribution to the communes were carried out according to the population, the payments into the "land trust" would be according to the land used per capita, whereas the redistribution would be according to the average use per capita. Hence, besides the cap, an additional incentive for a sustainable land use is implemented:

If the actual land use per capita is higher than the average land use, the commune in charge is a net contributor to the "land trust";

If the actual land use per capita is lower than the average, the responsible commune is a net beneficiary;

If the actual land use corresponds to average land use, there is no difference compared with the status quo.

Because every commune tries to get net benefits out of the land trust, there will be a current dynamic incentive to carry out efficient and effective land use management. In terms of microeconomics, the dynamic incentive is pushed by the substitution effect, whereas the income effect is eliminated by the redistribution scheme ("Slutsky equation", see [24]).

Moreover, within this redistribution mechanism, an average access to the planning permits is granted, also for financially weak communes. The redistribution mechanism serves as an ecological financial equalization scheme between the municipalities. Not unlike a lease mechanism, communes with land consumption rates above average pay to communes with land consumption rates below average.

The effects of the redistribution system are far reaching. To mention just some of them:

Currently, for instance, some German communes can take some "fiscal rents" due to their location, at the expense of other municipalities. This holds true particularly for the communes in the wealthy commuter belt of bigger cities ("Speckgürtel"). They benefit from the migration out of the bigger cities (e.g. young families), which are "bleeding". In such peripheral communes, land prices are often lower and the environmental conditions are often better than in the big cities. However, a great deal of the attractiveness is caused by uncompensated spillovers. According to the central locations principle [25] the bigger cities provide a variety of public goods at the commuter belt's benefit. Thus urban sprawl is fuelled, and the financial performance of bigger cities gets weaker and weaker. However, in the proposed regime, the whole fiscal surplus will be skimmed off dependent on the type of auction. The willingness to pay of the commuter belt's communes includes the expected fiscal rents (from spillovers). The fiscal rents are redistributed to all communes according to the number of the people, also to the bigger cities. Due to the higher density of population, the redistribution scheme will compensate the bigger cities for their efforts.

The model is applicable in situations of growth as well as in shrinking areas. In aging societies such as Germany, in particular rural regions are affected by shrinking. However, land conversion and land consumption is highest ex urbia. Urban sprawl turns out to be luxury which is increasingly difficult to finance. The redistribution model may stimulate migration to more compact settlements, with a higher supply of public goods. The system would provide an incentive for renaturation measures in rural communes. This would have positive side effects, considering e.g. vacancy rates and the value of existing properties.

By skimming off the rents from certain types of land use, communes get more indifferent towards land use alternatives. On a regional level, coordination between municipalities and the allotment of certain functions (industry, tourism etc.) towards different communes is easier than today. Thus, integrated approaches of regional development might be put in place without high resistance of the communes affected.

With regard to the technical implementation of the system, some minor problems have to be solved. For instance, in order to create equal conditions in the auction, the communes should pay into the land trust in the same "logical second" as the redistribution happens. This means the communes are only charged or rewarded by the balances (net position of pay-in and pay-out). Moreover, it has to be figured out on which administrative level the system should be applied. Basically, the redistribution mechanism should be tied to the scope of the cap and auction scheme.

One should be clear about the fact that no money for natural protection would be raised within the redistribution scheme. However, modifications are possible: If, for instance, a natural park as a common public good has to be financed, the redistribution could be carried out after first deducting the expenses for covering the park. Such decisions depend on the land trust and the planning authorities. A legal basis for the cooperation arrangement is necessary.

The proposed model may be appealing, but it is not a "silver bullet". The framework has to be completed by other instruments. For instance, the price of real estate may rise due to successful capping of planning rights. Thus, access problems for socially weak groups might be caused. Hence a suitable land taxation system which transfers shares of land rents and land value to the community would be desirable for example.

CONCLUSIONS

The concept of tradable planning permits transforms the idea of the CO_2 cap and trade regime to spatial planning. Analyzing the tool, we have at least to refer to effectiveness (planning, ecology), allocation (economy) and distributional aspects (social).

Regarding effectiveness, there is a broad consensus about the primacy of planning. Any economic tool should support planning instead of substituting it. However, planning land is not the same as planning the maximum permissible load of CO_2 in the atmosphere. The former requires a planning of structure, the latter a planning of scale, since it is irrelevant where the emission takes place.

Planning the structure of land use cannot be supported by a unit price, as a result of a universal certificate for all types of land use (for settlement and traffic). In contrast, a variety of sub-markets are necessary, with a different price setting. Meanwhile, more and more planners are also becoming sceptical about the supporting effects of a cap and trade regime.

Supporting the planning of a structure within a variety of sub-markets may be inferior compared with the efficiency of a cap and trade system with wide system boundaries. On the other hand, efficiency losses due to overshooting or undershooting might be avoided. The efficiency losses might be minimized by auctioning the permits to the needy communes on the primary market ("vertical allocation"). Although trade shouldn't be forbidden, an organized secondary market is dispensable (subordination of a "horizontal allocation mechanism"). Moreover, both systems would have to prevent strategic acquisitions of permits (impediment of development in other communes by an artificial shortage of supply), e.g. by a current devaluation of the permits.

Regarding the blueprint of CO_2 trade, a comprehensive arrangement on a global scale has so far failed due to distribution disputes. Also a cap and auction system for planning permits wouldn't be acceptable particularly for communes with a weak financial endowment if there were no correction. This is the reason why the cap and auction regime should be completed by a redistribution mechanism which is based on equal stakes in the scarce land use opportunities.

However, more research is necessary in order to deal with the details of the counter-proposal outlined in this article. So far, in Germany politics has supported the cap and trade approach; as has the allocation of research funds. Critics who pointed out the incompatibility between effectiveness and efficiency in the cap and trade approach have been pushed aside. This also holds true for the combination of caps, auction and redistribution, which couldn't be assessed so far. However, in experimental simulations the acceptance of the cap and trade regime among the practitioners was obviously not very high. Among others, the results of the cap and trade game turned out to be quite sensitive in terms of an increase of the complexity of the framework [20]. In contrast, at least the redistribution approach of Krumm was highly accepted (here, basically, also no fiscal impact assessment is necessary) [26]. Maybe it is time to widen the scope of the research paradigm to extend beyond the cap and trade regime.

REFERENCES

1. F. Zollinger, I. Seidl, 2005Flächenzertifikate für eine nachhaltige Raumentwicklung?- Ein Konzept für Baden-Württemberg und Erkenntnisse aus der Übertragung auf die Schweiz. Informationen zur Raumentwicklung 4/5273280
2. U. Kriese, 2005Handelbare Flächenfestsetzungskontingente- Anforderungen an ein Mittel zur Beendigung des Landschaftsverbrauchs. Informationen zur Raumentwicklung 4/5297306
3. Federal Ministry for the Environment, Natural Conservation and Nuclear Safety1998Nachhaltige Entwicklung in Deutschland- Entwurf eines umweltpolitischen Schwerpunktprogramms, Berlin, 147 p.
4. G. Penn-Bressel, 2011Flächenneuinanspruchnahme- Wirkungen auf Umwelt, Städtebau und Ökonomie. Wirtschaftsdienst 11800802
5. S. Frerichs, M. Lieber, T. Preuß, 2010Flächen- und Standortinformationen erheben und bewerten- Methoden und Konzepte für ein nachhaltiges Flächenmanagement. In: Frerichs, S, Lieber, M, Preuß, T, editors. Flächen- und Standortbewertung für ein nachhaltiges Flächenmanagement. Berlin: Difu 1127

6. Homepage of the "Refina" research programme, funded by the federal government: http://www.refina-info.de.Accessed: 2012Mar 15.
7. Federal Ministry for the Environment, Natural Conservation and Nuclear Safety2010Environment research plan (Umweltforschungsplan) 2010. Project Z 691Project 3710
8. Homepage of the project "Forum Flächenzertifikate".Available: http://www.ufz.de/index.php?de=21103Accessed 2012Mar 15
9. C. D. U. C. S. U. F. D. P. , 2009Wachstum, Bildung, Zusammenhalt, Coalition Agreement. Berlin.
10. LeJava, J P2009Transfer of Development Rights in New Jersey- A background paper, New Jersey. Available:http://www.dvrpc.org/TDR/pdf/200910 _LeJava_Background_Paper.pdfAccessed 2012 Mar 15.
11. Nash J F1950Non-cooperative Games; Ph.D. Thesis; Princeton University: Princeton, USA. Available:http://www.princeton.edu/mudd/news/faq/topics/Non-Cooperative_Games_Nash.pdf.Accessed 2012 Mar 15.
12. R. Krumm, 2004Nachhaltigkeitskonforme Flächennutzungspolitik- Ökonomische Steuerungsinstrumente und deren gesellschaftliche Akzeptanz, IAW research report, Tübingen: IAW. 136 p.
13. D. Loehr, 2006Handelbare Flächenausweisungskontingente: Eine gute Idee auf Abwegen, in: Zeitschrift für Umweltpolitik und Umweltrecht 4529544
14. J. Bovet, 2006Handelbare Flächenausweisungsrechte aus Steuerungsinstrument zur Reduzierung der Flächeninanspruchnahme. Natur und Recht 6473479
15. H. Bossel, 1998Globale Wende- Wege zu einem gesellschaftlichen und ökologischen Strukturwandel, Munich: Droemer. 464 p.
16. R. Walz, et al.2005Gestaltung eines Modells handelbarer Flächenausweisungskontingente unter Berücksichtigung ökologischer, ökonomischer, rechtlicher und sozialer Aspekte, Final Report. Research Project, funded by the Federal Environmental Office, Project 203Dessau-Roßlau. 170 p. Available: http://opus.kobv.de/zlb/volltexte/2009/7870/pdf/3839.pdf.Accessed 2012 Mar 15.
17. R. Henger, C. Schröter-Schlaack, 2008Designoptionen für den Handel mit Flächenausweisungsrechten in Deutschland, Land Use Economics and Planning- Discussion Paper, 08-020802University of Göttingen, September. Available: http://www.uni-goettingen.de/en/80714.htmlAccessed 2012 Mar 15.

18. Williams R C2003Cost-Effectiveness vs. Hot Spots: Determining the optimal size of emissions permit trading zones. University of Texas at Austin Working Paper.

19. R. Henger, K. Bizer, 2008Tradable Planning Permits for Land-use Control in Germany. Land Use Economics and Planning- Discussion Paper 08-010801

20. C. Küpfer, et al.2010Handelbare Flächenausweisungszertifikate-Experiment Spiel.Raum: Ergebnisse einer Simulation in 14 Kommunen. Naturschutz und Landschaftsplanung 4223947

21. L. Wicke, 2006Das Versagen des Kyoto-Protokolls in seiner jetzigen Form und seine strukturelle Weiterentwicklung. Zeitschrift für Sozialökonomie 15039

22. P. Barnes, R. Pomerance, 2000Pie in the Sky- The Battle for Atmospheric Scarcity Rent, Washington. Available: http://community-wealth.org/_pdfs/articles-publications/commons/paper-barnes-pomerance.pdfAccessed 2012 Mar 20.

23. R. Krumm, 2002Die Baulandausweisungsumlage als ökonomisches Steuerungsinstrument einer nachhaltigen Flächenpolitik. IAW Discussion Papers 7. Available: http://iaw.edu/iaw/De:Publikationen:IAW-Reihen:IAW-Diskussionspapiere:2002Accessed 2012 Mar 20.

24. H. Varian, 1992Microeconomic Analysis, 3rd ed., New York: W.W. Norton & Company Inc. 506 p.

25. W. Christaller, 1933Die zentralen Orte in Süddeutschland. 2nd ed. 1968. Darmstadt: Wissenschaftliche Buchgesellschaft. 331 p.

26. T. Preuß, et al.2007Perspektive Flächenkreislaufwirtschaft, 3Neue Instrumente für neue Ziele. Berlin / Bonn: Difu. 109 p.

Chapter 10

SPATIAL ANALYSIS FOR FLOOD CONTROL BY USING ENVIRONMENTAL MODELING

Alireza Gharagozlou, Hassan Nazari, Mohammadjavad Seddighi

Geomatics College of National Cartographic Center of Iran (NCC), Tehran, Iran

ABSTRACT

To create the final spatial information and analysis, flood hazard maps and land development priority maps and information, data for the flood events to 2009 in north of Iran were incorporated with using Geo-spatial Information System data of physiographic divisions, geologic divisions, land cover classification, elevation, drainage network, administrative districts and population density and environmental parameters modeling. Special analysis also attention was paid to population density for the construction of the land development priority map and using satellite image analysis to determine land use changes and analysis of geo-spatial information, because highly dense populated areas represent the highly important urban and industrial areas. While geo-information technology offers an opportunity to support flood management adequate geo-spatial information is a prerequisite for sustainable development, but many parts of the world lack adequate information on environmental resources. Such information providing, which serves as an important tool for decision-making in land use planning, can help provide effective information to natural disaster management. This paper develops a framework for flood control and begins with some general comments on the importance of land use planning and outlines some current environmental issues and then presenting environmental models to use in disaster management plan by using GIS and remote sensing results. Flood control is a complex problem that requires cooperation of many scientists in different fields. The article also discusses the role that geo-information and environmental planning and GIS and remote sensing technology play in disaster management control to reduce negative impacts of flood and present proper alternatives for developing of Gorganrood in the north of Iran. Advanced high-resolution sensor technology has provided immense scope to the decision makers for analysis of flood and damages details using GIS and remote sensing.

INTRODUCTION

This article begins with some general comments on the importance of land use planning and outlines some current environmental issues. It also highlights the connection between land use planning and sustainable development and the discussion describes several key methods of resource identification, with particular emphasis on existing potential of geo-information technology that offers an opportunity to support disaster management: floods and environmental impacts assessment and natural disaster in national level. The article also discusses the role of geo-information data in promoting geographic information system use. By attention to natural disasters in Iran especially flood in Golestan in the North of Iran proper assessment of flood by using environmental development models and GIS and with attention to Sustainable development approach and disaster management are presented. The article offers proposed models that illustrate how GIS and remote sensing data can be used in land use planning programs that take a sustainable development approach [1] and disaster management (flood). Excessive land use and increased human impacts have imposed significant pressures on the environment worldwide. These effects are increasingly noticeable from a scientific and technical viewpoint. Future development should proceed on the basis of proper land use planning, with minimum destruction of the environment because impacts of human activities results natural disasters in some area [2]. Planning assessments must therefore consider environmental issues and natural disaster (flood) and use environmental and dereferenced information to refine decisions. Gathering information reveals the available potential of the environment; development planning at the nationwide level can help decision-makers identify resources and target their future scientific studies to reach sustainable development. Moreover, each minute, 5.6 hectares of forest are being destroyed and some other human activities cause disaster such as flood in some parts of the world.

MATERIALS AND METHODS

Floods are one of the most common hazards in the world also in the some parts of north of Iran. Flood effects can be local, impacting a neighborhood or community, or very large, affecting entire river basins and multiple states [3]. However, all floods are not alike. Floods themselves average four billion dollars annually in property damage alone. Some floods develop slowly, sometimes over a period of days. But flash floods can develop quickly, sometimes in just a few minutes and without any visible signs of rain. Flash floods often have a dangerous wall of roaring water that carries rocks, mud, and other debris and can sweep away most things in its path. Overland flooding occurs outside a defined river or stream, such as when a levee is breached, but still can be destructive. Some general reasons of flood include: weather related reasons: heavy rainfall, duration of precipitation, sudden snow melting and physical conditions: soil variety, slop of lands, land degradation and human activities: deforestation, misusing of land and transforming to grasslands or agricultural area, misconstruction of roads, bridges, dams and environmental situations. Flooding can also occur when a dam breaks, producing effects similar to flash floods. Be aware of flood hazards no matter where you live, but especially if you live in a low-lying area, near water or downstream from a dam. Even very small streams, gullies, creeks, culverts, dry streambeds, or low-lying ground that appear harmless in dry weather can flood every state is at risk from this hazard [4]. Some scientists think the major problem about natural disaster and flood is in the improper exploitation of land [5]. By using process of plan compilation with a land use planning approach some important negative impacts that cause flood is under our control.

Identification of Land Resources for Planning

Statistics and sampling, conversion of the aerial photos, satellite images and topographic maps, automatic conversion of aerial photos and satellite images and data of remote sensing, geographic information systems are different methods of identification of resources. One of the objectives of this study is to utilize GIS data to construct a set of GIS data, a flood hazard map, and land development priority map to help the responsible authorities develop, design and operate flood control infrastructure and prepared aid and relief operations for high-risk areas during future floods. In recent years the combination of 3D-laser scanning and side-scan can be very beneficial for mapping complicated water side areas; the two systems are complementary [6]. To geo-reference the relative location, GPS positioning required. It should be clear that presenting an environmental development model to be used in a GIS for natural disaster management has a lot of restrictions and limitations [7] whose description would lead too far here. Some factors that have been considered in presenting the model include; industrial sites, transportation networks, weather and climate data, landform, elevation, slop, geology, bedrock, soil, water resources, vegetation, installations and buildings, energy transmission stations, natural resources, gardens, forests, parks, etc. the priority of the mentioned parameters are different in the model[8]. It is clear that north of Iran and Golestan has an environmental development context and is under the interactive effects of the large region. Also it is thus impossible to correctly analyze the environmental conditions for natural disaster management without considering the social and economic activities in this district.

Modeling

Environmental modeling is a complex problem that requires cooperation of many scientists in different areas. In this paper, the architecture and results of environmental modeling and using satellite image processing and GIS for Flood control is presented. Set out below are mathematical linear models for flood management in Golestan. Flood inundation modeling requires distributed model predictions to inform major decisions relating to planning [9]. Present flood model integrates GIS with the environmental modeling and greatest daily of precipitation from1995 to 2009 to determine improper area for development. A linear mathematical model for flood management has been Introduced because proper planning based on environmental potentials cause reduce risk potentials of flood in future and this linear model with attention to planning results sufficient for development impacts on disaster. FF refer to specific model with environmental planning approach for flood management and attention proper land use planning in the region that present location of improper area for development. Predicting the river's flood is one of the important factors for design of dams and hydraulic structures and regional and urban development planning. As geo-information data also used in flood management, many problems occur in flood estimation. One of the methods for planning is determination improper area for development by environmental modeling, statistics and using GIS/RS technology [10].

$$R = S\,(4,5,6,7,8) + AS\,(1,2,3,7,8) + H\,(1,2,3,4) + B\,((x,y) > RR)$$

$$RR = R1\,(V1+M1) + R2\,(V2+M2) + R3\,(V3+M3) + \ldots \quad FF = S(5,6) + H(5,6) + QA(2,3,4) + MA(1,2,3)$$

$$+ WS(5,6) + SO(1,2,4,5,6) + SW(1,2,3) + NI(1,2,3) + HP(1,2,3) + HBU(1,2,3) + HBR(1,2)$$

$$+ So\,(3,4,5,7,8,9,10,11) + Sd\,(4,5) + Prc\,(1,2)$$

where: S is slope, H is height or altitude, A is aspect, QA is fault line, MA is distance from ravine areas, WS is wind speed, SO is soil components, SW is distance from subterranean water resources, NI is distance from industrial sites and HP is historical landmark, HBU is Distance from urban habitat and HBR is distance from rural habitat and Lo, Ma(R-year) is the maximum precipitation based on geographical location. For purposes of the linear models, the terms used have the following definitions: "Slope" (S) includes six classes: 0 to 2% (class 1), 2 to 5% (class 2), 5 to 8% (class 3), 8 to 12% (class 4), 12 to 15% (class 5), and more than 15% (class 6). "Height" (H) includes six altitude classes: less than 1000 meters (class 1), 1000 - 1200 meters (class 2), 1200 - 1400 meters (class 3), 1400 - 1600 meters (class 4), 1600 - 1800 meters (class

5), and more than 1800 meters (class 6). "Distance from ravine areas" (MA) includes four classes: less than 50 meters (class 1), 50 - 300 meters (class 2), 300 - 500 meters (class 3), and more than 500 meters (class 4). "Subterranean water resources" (SW) divides resources into four classes, based on distance to the water resource: less than 100 meters (class 1), 100 - 500 meters (class 2), 500 - 1000 meters (class 3), and more than one kilometer (class 4). "Distance from industrial sites" (NI) includes three classes: less than 5 kilometers (class 1), 5 - 10 kilometers (class 2), and 10 - 20 kilometers (class 3). "Distance from Urban Habitat" (HBU) divides 4 four classes: less than 5 kilometers (class 1), between 5 to 10 kilometers (class 2), between 10 - 20 kilometers (class 3) and more than 20 kilometers (class 4), "Historical landmark" (HP) divides historical places into four classes, based on how far away they are located: less than 5 kilometers (class 1), 5 - 10 kilometers (class 2), 10 - 20 kilometers (class 3), and more than 20 kilometers (class 4), Distance from rural habitat "HBR" divides rural area and around this to 4 classes: less than 2 kilometers (class 1), 2 - 4 kilometers (class 2), 4 - 8 kilometers (class 3) and more than 8 kilometers class 4 and Prc is precipitation in mm in 7 classes more than 2000, 1800 - 2000, 1200 - 1800, 800 - 1200, 500 - 800, 200 - 500, 50 - 200 and less than 200 mm and So is soil construction in 11 classes and Sd is soil depth in 5 classes include more than 180, 120 - 180, 60 - 120, 30 - 60, less than 30 cm. With attention to linear model of flood control and data analysis and using GIS the result of analysis are as maps presented that are presented in Figures 1-4. Satellite data also can be effectively used for mapping and monitoring the flood inundated areas, flood damage assessment, flood hazard zoning and post-flood survey of rivers configuration and protection works. Analyzing the satellite images reveal a noticeable reduction of forestlands in north of Iran due to the expansion of the urban limits misuse from these area. The other fixed natural resources of the region too have been overused resulting in environmental destruction of the area. The amount of residential areas during 1995 and 2009 show an 8% growth while there is no increase in the number of forestlands. There are several definitions of sustainability in the urban forestry sector that we attention in this paper based on analysis results and conditions in urban area. The amount of forestlands declined about 1 hectare every year and open areas have been reduced thus leading to the conclusion that most of the construction activity took place in forestlands and results suiting with other experiences in this filed.

CONCLUSIONS

In this paper Innovations in the filed of environmental modeling which are based on natural disaster management for flood in Iran and using proper models for analysis in GIS are presented. Geo-information technology offers an opportunity to support disaster management and floods as the natural disaster management. Gorganrood rivers waterways trails as a pedestrian pass with a stream of spring and rain which also can hold heavy rain water as a flood control waterway. One of the obvious and prominent aspectsof innovations in this paper, are the models that can integrate between Geo-information technology and environmental modeling and natural disaster management. Analysis positioning of improper locations by using GIS/RS technology for development area determination based on the environmental capacities with a flood control approach. At the same time, by using GIS necessary analysis to find flood risk in the region and impacts of flood on natural and human facilities are presented. Choosing proper linear models based on environmental capacity with a flood management emphasize determining the ecological potentials of the area and using GIS is the important point of this paper. The joint application of GIS and environmental modeling and using remote sensing technology can help land use planners apply optimal development planning guidelines. The other key idea we suggest here is the need to compare the results of these analyses with future development plans. Comparing the natural potential of the territory with predicted development plans can result in better decision making to reduce the cost of flood in rural and urban area. The use of GIS technique during the last decade are increasing being applied for identification of natural resources but the practice of analyzing the development models with the use of GIS/RS in development planning for flood management is a new experience.

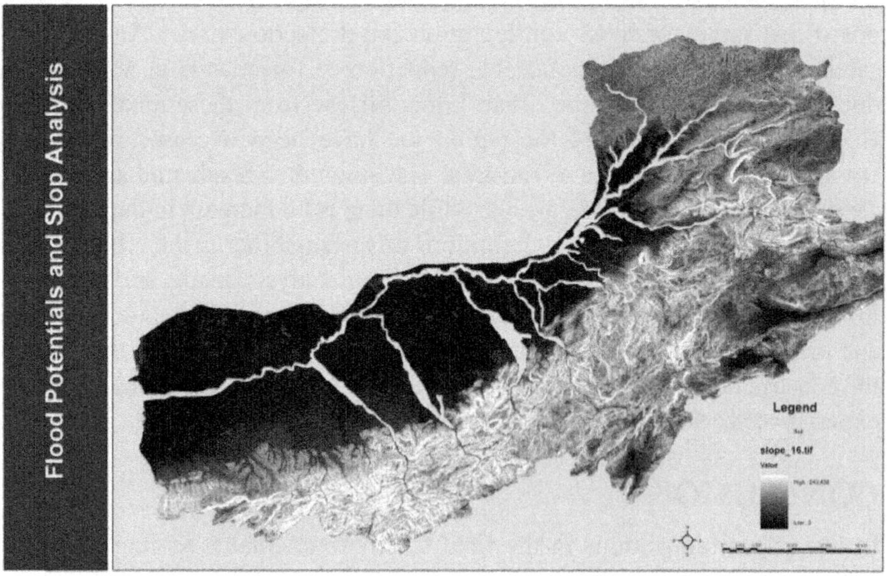

Figure 1: Spatial Analysis for determination flood risk area and improper area for development using flood modeling in north of Iran

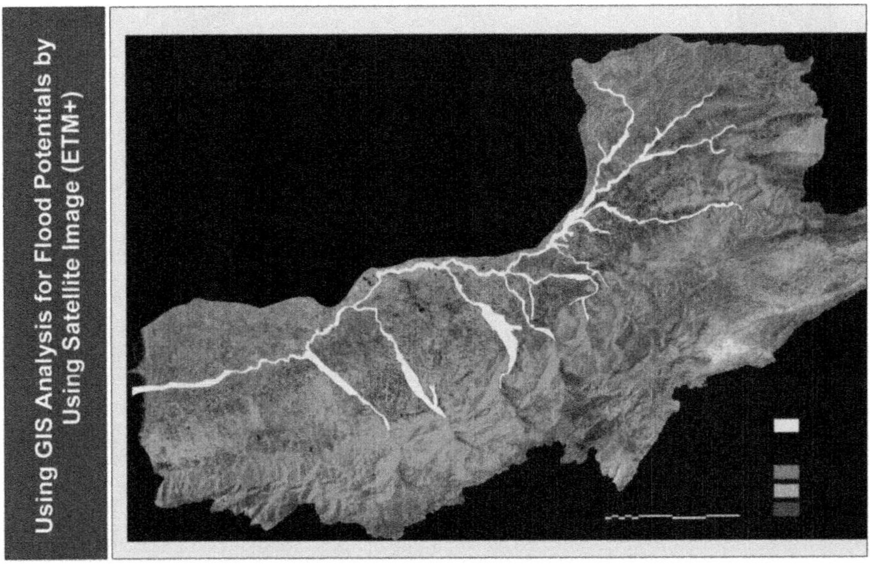

Figure 2: Satellite image processing (ETM+) in Golestan, Gorganrood, and risk assessment and modeling for flood control.

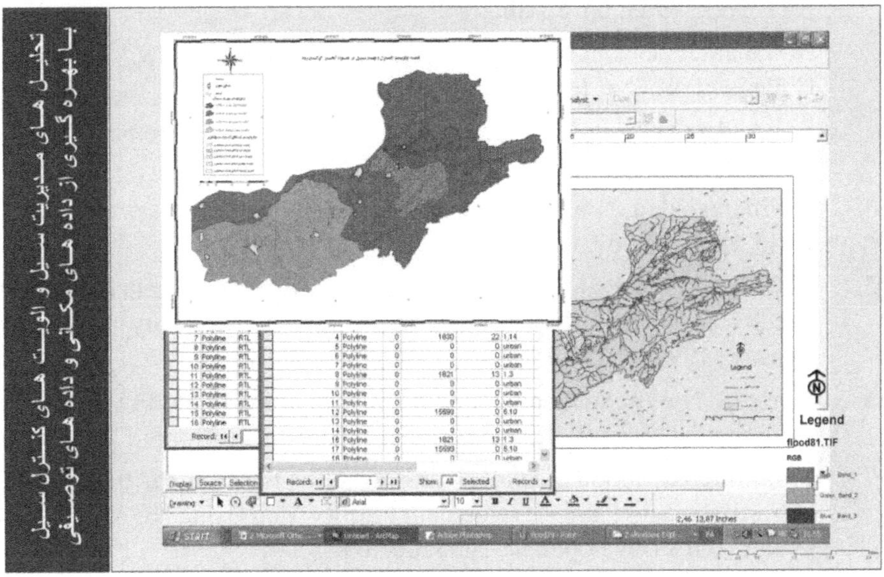

Figure 3: Districts risk classification for flood control by using GIS.

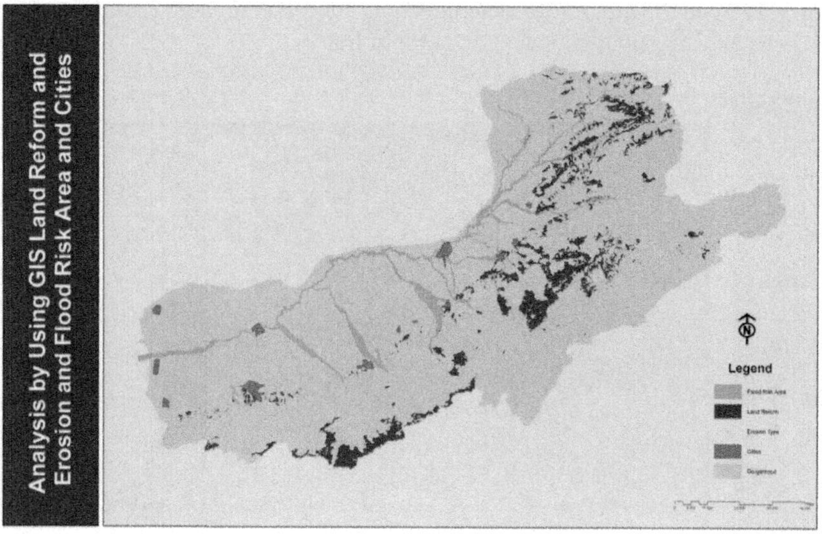

Figure 4: Land reform and erosion and cities under flood risk.

REFERENCES

1. J. Hardisty, D. M. Taylora and S. E. Metcalf, "Computerized Environmental Modeling: A Practical Introduction Using Excel," Wiley, New York, 1993, pp. 80-88.
2. A. Gharagozlou, "Crisis Management (Flood) and GIS," Geomatics College of NCC of Iran, Tehran, 2010, pp. 23-29.
3. S. Andera, "Geo-Information for Disaster Management," GIM International, Amsterdam, 2005, pp. 10-12.
4. A. Gharagozlou, "Environmental Planning for Natural Disaster by Using GIS," ISPRS, 2007, China.
5. E. Wolk and B. Zagajewski, "Remote Sensing of Environment Laboratory, Faculty of Geography and Regional Studies," University of Warsaw, Warsaw, 1999, pp. 44-57.
6. Fairley; "Environmental Planning," Department of Environmental Planning, University of Stratchlyde, Stratchlyde, 2001, pp. 111
7. S. kimitrero, "Flood Hazard Assessment for the Constructionof Flood Hazard Map and Land Development Priority Map Using NOAA/AVHRR DATA," GIS Development, 2006, pp. 3-12.
8. Heywood LAN, S, Cornelius and S. Carver, "Cornelius; -An Introduction to Geographic Information Systems," chap., 1998, pp. 2-5.

9. K. Clayton, “The Land Form Space, Environmental Science for Environmental Management,” Longman, London, 1990, pp. 198-222.
10. F. Ferrini, “Sustainable management techniques for trees in the urban area, Journal of Biodiversity and Ecological Sciences,” IAU University, Tonekabon, 2010, pp. 1-19

Chapter 9

SOIL EROSION PREDICTION USING MORGAN-MORGAN-FINNEY MODEL IN A GIS ENVIRONMENT IN NORTHERN ETHIOPIA CATCHMENT

Gebreyesus Brhane Tesfahunegn,[1,2] Lulseged Tamene,[3] and Paul L. G. Vlek[1]

[1]College of Agriculture, Aksum University, Aksum, Ethiopia

[2]Centre for Development Research, University of Bonn, Walter-Flex-Street 3, 53113 Bonn, Germany

[3]International Centre for Tropical Agriculture (CIAT), Chitedze Agricultural Research Station, Lilongwe, Malawi

ABSTRACT

Even though scientific information on spatial distribution of hydrophysical parameters is critical for understanding erosion processes and designing suitable technologies, little is known in Geographical Information System (GIS) application in developing spatial hydrophysical data inputs and their application in Morgan-Morgan-Finney (MMF) erosion model. This study was aimed to derive spatial distribution of hydrophysical parameters and apply them in the Morgan-Morgan-Finney (MMF) model for estimating soil erosion in the Mai-Negus catchment, northern Ethiopia. Major data input for the model include climate, topography, land use, and soil data. This study demonstrated using MMF model that the rate of soil detachment varied from <20 $t\,ha^{-1}\,y^{-1}$ to >170 $t\,ha^{-1}\,y^{-1}$, whereas the soil transport capacity of overland flow (TC) ranged from 5 $t\,ha^{-1}\,y^{-1}$ to >42 $t\,ha^{-1}\,y^{-1}$. The average soil loss estimated by TC using MMF model at catchment level was 26 $t\,ha^{-1}\,y^{-1}$. In most parts of the catchment (>80%), the model predicted soil loss rates higher than the maximum tolerable rate (18 $t\,ha^{-1}\,y^{-1}$) estimated for Ethiopia. Hence, introducing appropriate interventions based on the erosion severity predicted by MMF model in the catchment is crucial for sustainable natural resources management.

INTRODUCTION

Soil erosion is the dominant cause of soil degradation at a global scale [1–3]. This is accounted for between 70 and 90% of total soil degradation [2, 4]. The adverse influences of soil erosion as a cause for soil degradation have long been recognized as a severe problem for sustainability of economic development [5]. This is because a large portion of fertile soil is lost annually which negatively influences the goal of achieving food security [3]. However, estimation of soil erosion rate is often difficult due to a complex interplay of many factors, besides the differences in scale and methodological components of the studies [6, 7]. In a country like Ethiopia, with an agriculture-based economy for more than 85% of population, having reliable soil loss data is indeed a matter of great concerns and not a matter of choice. Regardless of the great deal of management practices undertaken aggressively in Ethiopia catchments by the government in the past 1-2 decades to reduce soil degradation, soil erosion by water is still recognized to be a severe threat to the national economy [2, 6, 7]. This indicated that the existing literature on the rate of soil erosion in Ethiopia calls for a wise decision supporting tools in order to reduce the degradation level. For instance, past studies on soil erosion in the catchments of Tigray region (northern Ethiopia) showed variability ranging from $7\,t\,ha^{-1}y^{-1}$ [8] to more than $24\,t\,ha^{-1}y^{-1}$ [7] and $80\,t\,ha^{-1}y^{-1}$ [9]. According to the report by FAO [10], erosion rate is estimated up to $130\,t\,ha^{-1}y^{-1}$ from cropland and $35\,t\,ha^{-1}y^{-1}$ averaged over all land use types in the highlands of Ethiopia. Such discrepancies in the rates of erosion by the studies mainly attributed to changes and differences in land use, management practices, and methods employed while developing input data and their respective scale of analysis. Predominantly, previous erosion related input data were developed from simple point observation such as runoff plot, and data were interpolated through conventional method [8, 11, 12]. Such method poses many limitations in terms of cost, representation, and reliability of the resulting data [5]. Recently, to reduce such limitations geostatistic techniques that interpolate data for an entire catchment from appropriately sampled point measurements are readily available [13, 14].

Mapping through conventional methods demands an intensive data collection, which is often difficult to practice in complex terrains like in northern Ethiopia [7, 14]. The Geographic Information System (GIS) techniques can provide easy and time effective tools to map and analyze erosion input data of hydrophysical parameters. These techniques coupled with the concept of catchment priority can help in identifying areas where treatment plans should be first located. Many studies (e.g., Sharad et al. [15]; Sanware et al. [16]) revealed that GIS techniques can have a great role in characterization and prioritization of subcatchments. The catchment level assessment and mapping of hydrophysical resources can support the identification of constraints, ecological problems, and adoption of effective management practices that sustain land and water resources using integrated catchment management strategies [17]. In addition, the availability of hydrophysical parameters in a GIS map format can be used readily for erosion model running in order to understand spatial distribution of ecological problems such as soil erosion. In many environmental studies, data inputs are measured at single points in space, even though classical statistics assume that measured data are independent and thus are not sufficient to analyze spatially dependent variables [13, 14, 18]. However, information is required for the entire catchment space, which necessitates methods that interpolate data to estimate the mean value within an area [13]. Geostatistics provides the basis for interpolation spatial variability of hydrophysical erosion model input parameters that affect runoff and soil loss [13, 14, 19–21].

To estimate soil erosion and suggest appropriate management plans, many erosion models such as Universal Soil Loss Equation (USLE) [22], Morgan-Morgan-Finney (MMF) [23], Water Erosion Prediction Project (WEPP) [24], Soil and Water Assessment Tool (SWAT) [25], European Soil Erosion Model (EUROSEM) [26], and Annualized Agricultural Non-Point Source (AnnAGNPS) [27] have been developed and used data inputs generated through GIS. Among these models, the USLE has remained the most practical method of estimating soil erosion potential for more than 40 years [28, 29], despite the fact that it has many limitations for application at catchment-scale. On the other-hand, process-based erosion models developed afterward have limitations in applicability due to intensive data and computation requirements [30]. The application of process-based models is not always an easy task since these require large amounts of information which is often not available, mainly in data scarce developing regions. The MMF model was selected to estimate annual soil loss, since this model endeavours to retain the simplicity of USLE and also encompasses the understanding of erosion processes into water and sediment phases [23]. Meaning, the MMF model was selected to be applied in this study because of its simplicity and flexibility as compared to the physical-base models and has a stronger physical base than USLE. In addition, since the MMF model is a physically based-empirical model (mix model), it needs less data than most of the other erosion predictive models [23].

Understanding the hydrophysical parameters that can influence erosion rate in a catchment is complex due to the combined nature of the natural processes and man-made features [7, 31]. Therefore, research to obtain quantitative description of hydrology/erosion in a catchment must consider these spatial heterogeneities. In order to tackle against hydrological related problems (runoff, sedimentation), accurate representation or locating the spatial distribution and variability of the influencing parameters using GIS is necessary [7, 32]. This study was aimed to derive and assess the spatial distribution of hydrophysical parameters developed using GIS technique and apply them in the MMF model for estimating the spatial variability of soil loss in the Mai-Negus catchment, northern Ethiopia. The spatial map can be used for prioritizing areas within the catchment that require immediate management measures on the basis of the severity of runoff/soil loss.

MATERIALS AND METHODS

Study Area

The study area, Mai-Negus catchment, is located in the Tigray region, northern Ethiopia (Figure 1). The catchment has an area of 1240 ha and altitude ranges from 2060 to 2650 m above sea level [33]. In the catchment, mean annual temperature of 22°C and annual rainfall of 700 mm have been recorded (Meteorology Agency-Mekelle branch). The highest amount of rainfall (>70%) is received between July and August. Land use is predominantly arable, with teff (Eragrostis tef) being the major crop along with different proportions of pasture land and scattered patches of trees, bushes, and shrubs. The major rock types are lava pyroclastic and metavolcanic. Leptosols are found mainly on the very steep positions, Cambisols on gentle to steep slopes, and Vertisols on the flat areas of the catchment [34].

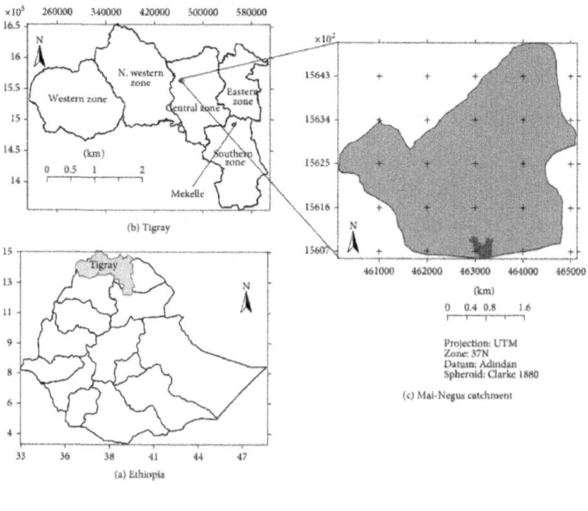

Figure 1: Study area: Ethiopia (a), Tigray (b), and Mai-Negus catchments (c). The blue colour shaded area is the reservoir.

Research Approach

The research approach in this study consisted of five main steps. These are (i) identification of hydrophysical parameters which are inputs of MMF model, (ii) field surveys and informal discussions in order to identify representative soil sampling zones within the study catchment for soil sampling and analysis and also the corresponding vegetation cover conditions, (iii) application of empirical relations which are described by Morgan et al. [23, 35] to calculate intermediate MMF model inputs, (iv) application of geostatistic interpolation technique for spatial model inputs development, and (v) application of MMF model in GIS environment while estimating spatially distributed erosion outputs such as total overland transport capacity and soil detachment rate.

MMF Model Inputs Preparation

Input data for MMF model include rainfall (mm), land use, digital elevation model (DEM) for slope map derivation, soil texture, soil moisture content at field capacity (%w w^{-1}), soil detachability index (g J^{-1}), bulk density of soil (Mg m^{-3}), cohesion of soil surface (KPa), soil moisture storage capacity (*Rc*), effective hydrological top soil depth (EHD), and ratio of actual to potential evapotranspiration (E_t/E_o) [23, 35]. Input data were collected from different sources such as field and laboratory determination, empirical relations, and the literature. Meteorological data such as rainfall and *Et* data were obtained from the meteorology station near the study area in 2009. Slope was derived from DEM developed from the topographic-map available at the Ethiopian Mapping Agency for Aksum area [33]. The map was scanned, and contours and spot heights were digitized and tagged with elevation values in a GIS environment. The vector elevation map was converted to raster and projected using the Universal Transverse Mercator 37 North (UTM-37N) reference system.

Crop and soil parameters were collected from 117 plots scattered throughout the catchment considering major land use and cover types (bush land, protected area, cultivated, abandoned fields, grazing land, mixed-forest, and residential). Supervised classification and visual interpretation of the land satellite image of November 2009 was carried out for general land use and cover mapping. In addition to this, crop covers for the different crop types and their corresponding geographic coordinates were collected using field survey in September 2009. Data related to rainfall such as rainfall intensity, number of rainy days, and total rainfall were assumed to be similar in the study catchment. The reason for having only one weather station in the study catchment is that the Office of Meteorology Agency believed that rainfall variability is negligible within such a small area regardless of the differences in elevation. Rainfall intensity was assumed at 25 mm h^{-1} which is erosive for tropical climates such as Mai-Neguse catchment because no actual intensity data was found for the study catchment. Soil detachability index (K) (g J^{-1}) was determined from the literature that corresponds to the soil texture observed in the study catchment.

Soil Sampling Zones and Sample Collection

In order to prepare MMF model soil related inputs, soil sampling that considered soil variability in the study catchment was executed. Sampling approaches that divided a field into small units (zone sampling) can capture variability and provide more information about soil-test levels compared with one composite sample collected from an entire large sampling area [36]. To reduce the number of samples and sampling costs zone sampling is suggested to provide a way to group the spatial variability of soils while maintaining acceptable information about soil properties [36]. Sampling by zone assumes that sampling areas are likely to remain temporally stable [36, 37].

In this study, the zone sampling technique (divide a field into homogenous units that allow capturing variability and provide more information) was used to collect soil samples based on previous and existing knowledge of the soil and land use systems in the entire study catchment. The natural and management factors across the landscape that influenced soil properties spatial variability were considered while identifying the soil sampling zones. Three soil sampling zones that represented the soil quality (SQ) categories, long-term land use and soil management systems, and different erosion status sites in the catchment were identified using farmers' opinions and researcher and extension experts' judgment. The data that divided the catchment into the soil sampling zones was derived during the field reconnaissance surveys in June 2009. The SQ sampling zone was entirely used for arable land in the catchment whereas the other two sampling zones belonged to all the land use systems in the catchment. The sampling zones were further subdivided into different subsampling zones considering the variability within each zone and analytical costs.

The SQ sampling zone was divided into three subzones as high, medium, and low SQ based on farmers' knowledge. They used indicators such as yield and yield component, soil depth, colour, and fertility conditions to divide into these subzones. The details on how local farmers' classified soil into different SQ categories in the study catchment can be found in Tesfahunegn et al. [38].

Eight representative long-term land use system sampling zones were identified based on farmers' historical and present information acquired in the catchment. These are (i) natural forest; (ii) plantation of protected area;

(iii) grazed land; (iv) teff (Eragrostis tef)-faba bean (Vicia faba) rotation; (v) teff-wheat (Triticum vulgare)/barley(Hordeum vulgare) rotation; (vi) teff monocropping; (vii) maize (Zea mays) monocropping; and (vii) uncultivated marginal land. The age of the systems varied from 5-6 years for teff monocropping and 20–30 years of maize monocropping system. Average age of the other systems was about 10 years except for the plantation, grazed land, and uncultivated marginal land systems with more than 15 years.

The erosion status-based sampling zone was divided into three subzones as stable, eroded, and deposition (aggrading) sites. Information from the local farmers, extension agents, and researcher's (first author) observation on the level of topsoil depth (A-horizon), deposition, rills, pedestals, root and subsoil exposure, and gullies indicators were considered while identifying the three erosion-status sampling subzones. Those areas having A-horizon and minimum erosion indicators were considered as stable sites and the reverse of this as eroded sites. Depositional sites were also easily identified as they are mainly located in depression and flat areas with evidences of recent sediment deposition. In total, there were 14 subsampling zones across the erosion-status sites in the catchment for the soil samples collection. After doing all this identification and division, the soil sampling points in each subzone were located at the centre, considering soils in that point best represent the samples. Each sampling point was georeferenced as their distribution in the catchment is shown in Figure 2. The sampling distance was not regular, ranging from 40 to 180 m.

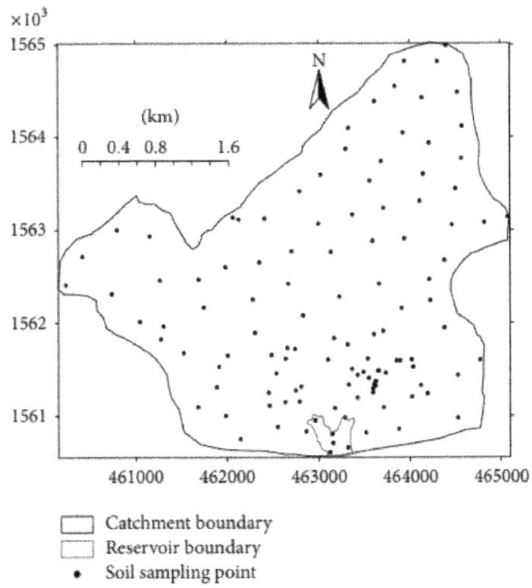

Figure 2: The distribution of soil sampling and vegetation cover points in the study catchment.

Soil samples were collected in June 2009. A total of 51 soil samples (3 subzones × 17 samples) were collected from the SQ based sampling zone. From the long-term land use systems, a total of 24 soil samples (8 subzones × 3 samples) were collected. It was also collected 42 soil/sediment samples (3 subzones × 12 samples in the catchment and 6 sampling points in the reservoir) from the erosion-status sites. The grand total of the composite samples collected across the sampling subzones was 117. Each composite soil sample was collected using 5–8 samples from each representative subsampling zone depending on the size and homogeneity of the sampling area (100–300 m^2). All the composite soil samples were collected at the soil depth of 0–20 cm (the plough depth) since this is where most changes are expected to occur due to erosion, long-term land use, and soil management practices. The composite soil samples were pooled into a bucket and mixed thoroughly to homogenize it. Finally, a subsample of 500 g from the pooled composite samples was taken and soil samples were air dried and sieved to pass 2 mm mesh sieves before analysis for soil textures. On the other hand, two undisturbed soil samples were collected from each soil sampling point for bulk density and soil moisture determination. In addition, field level observation and measurement for parameters such as effective hydrological top soil depth (m), ground cover, and cover factor were carried out from the sampling points and georeferenced.

Soil Analysis

The soil samples collected in the soil sampling zones were determined for soil texture using the Bouyoucos hydrometer method [39], soil bulk density (BD) by the core method [40], and soil moisture content at field capacity ($w\,w^{-1}$) by equilibrating the soil with water through capillary action in KR box [41].

Empirical Relations in Deriving Inputs of MMF Model

Some intermediate input parameters were estimated from observed data in the catchment using the empirical relations described in Morgan et al. [23] as

$$E = R\left(11.9 + 8.7\log_{10}(I)\right),$$

$$R_c = 1000^*\mathrm{MS}^*\mathrm{BD}^*\mathrm{EHD}^*\left(\frac{E_t}{E_o}\right)^{0.5},$$

$$\mathrm{SR} = R\exp\left(-\frac{R_C}{R_o}\right),$$

$$R_o = \frac{R}{R_n}, \qquad (1)$$

where E is annual kinetic energy of rainfall (J m^{-2}), I is intensity of rainfall which is assumed to be 25 mm h^{-1} in tropical conditions, SR is surface runoff/overland flow (mm), R_n is number of rainy days, R is average annual rainfall (mm), R_c is soil moisture storage capacity (mm), R_o is annual rain per rain day, MS is soil moisture content at field capacity (w w^{-1}), BD is bulk density of the topsoil layer (Mg m^{-3}), EHD (m) is effective hydrological topsoil depth defined as the depth of soil from the surface to an impermeable or stony layer to the base of A horizon or to the dominant root base, and E_t/E_o is the ratio of actual (E_t) to potential (E_o) evapotranspiration. EHD is the top soil depth within which the storage of water affects the generation of runoff.

Intermediate maps derived on the basis of land use/cover map (Figure 3) included ratio of actual to potential evapotranspiration (E_t/E_o), permanent rainfall contributing to permanent interception, and stream flow (A) and crop cover management factor (C_f). The C_f combines C and P factors of the Universal Soil Loss Equation to give ratio of soil loss under a given management to that of bare ground with down-slope tillage, other conditions being equal. These were determined in the field. Intermediate layers derived from soil map (soil texture) included soil detachability index (K) and cohesion of topsoil (COH) that were generated using ArcGIS 9.3 software. According to Morgan et al. [23] and Dinka [35], K is defined as the weight of soil detached from the soil mass per unit of rainfall energy. The values of plant and soil related hydrological parameters are shown in Tables 1 and 2, respectively. Inputs such as plant related (e.g., EHD, A, CC) and soil related (e.g., K, COH) parameters were adopted from Morgan et al. [23] and Dinka [35], in which such values corresponded to crop type and cover conditions and soil textures observed in the field.

Table 1: Plant related hydrological parametersfor different land use and land cover in the Mai-Negus catchment

Land use	EHD (cm)	A	CC (%)	PH (m)	GC (—)	E_t/E_o	C_f
Teff	5	0.15	85	0.2	0.85	0.5	0.5
Barley/wheat	12	0.3	60	0.3	0.80	0.58	0.3
Maize	12	0.25	50	1	0.75	0.68	0.25
Pulse	12	0.2	55	0.2	0.80	0.65	0.25
Grazing	12	0.2	80	0.18	0.80	0.8	0.1
Closure (grass/shrubs/bushes)	15	0.4	55	1.5	0.60	0.85	0.1
Mixed forest	20	0.5	65	1.8	0.65	0.9	0.01
Residential	19	0.1	5	0.75	0.55	0.6	0.13

EHD: Effective root depth; A: the percentage of rainfall contributing to permanent interception; CC: canopy cover fraction; PH: plant height; GC: ground cover; E_t/E_o: the ratio of actual to potential evapotranspiration; and C_f: the crop cover management factor.

Table 2: Soil related hydrological parameters values for soil texture in the Mai-Negus catchment

Texture	K (gm J^{-1})	MS (% w w^{-1})	BD (Mg m^{-3})	COH (KPa)
Sandy loam	0.7	18.6	1.87	2
Sandy clay loam	0.3	23.7	1.72	3
Clay loam	0.4	35.38	1.48	10
Sandy clay	0.35	21.21	1.51	1
Silt clay loam	0.3	28.63	1.31	9

K: soil detachability index; MS: soil moisture at field capacity (1/3 bar) tension; BD: bulk density top soil; and COH: cohesion of topsoil.

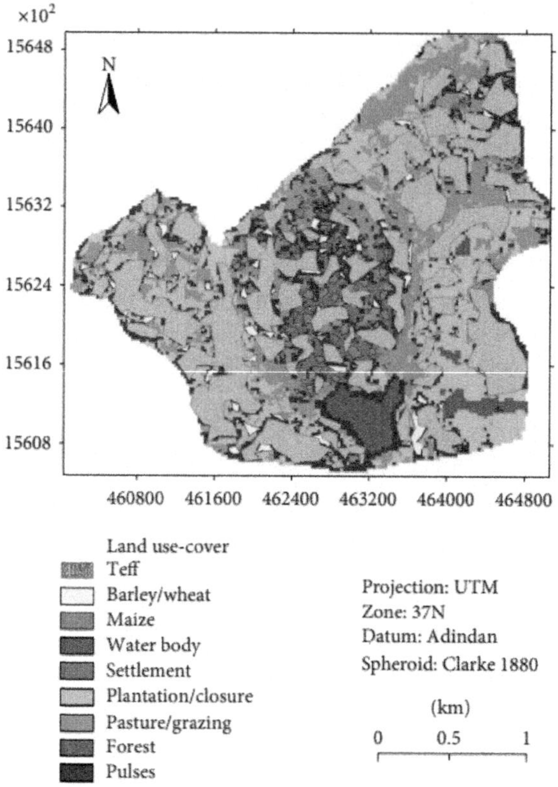

Figure 3: Land use and land cover of the study catchment.

Application of Geostatistical Interpolation Technique

After the point data and their corresponding coordinates were entered into ArcGIS 9.3 software, maps of hydrophysical model input parameters were developed using kriging interpolation technique [42]. Ordinary kriging was selected as the preferred interpolation method for MMF model spatial inputs derivation because it was more reliable than the other interpolation methods based on the mean squared error which compares the measured values with the predicted ones. Moreover, since the spacing of the measured or observed hydro-physical input parameters were relatively sparse and randomly chosen for each subsampling zone, ordinary kriging is the best unbiased predictor for the random process at specific unsampled locations [43]. Ordinary kriging also has an additional advantage of minimizing the influence of outliers [44]. The semivariogram analyses were conducted before the application of ordinary kriging interpolation of the input parameters. This is because the semivariogram model determined the interpolation function [14]. Semivariogram models were chosen by using the cross-validation technique that compares statistical mean square error values estimated from the semivariogram models and actual values.

MMF Model Application

The MMF modelling processes erosion in two phases, that is, the water and sediment phases [23]. The water phase mainly comprises of prediction of soil detachment by rain splash. It thus requires data related to intensity of rainfall (I, mm h^{-1}), number of rainy days (R_n), and average annual rainfall (R, mm). After developing the different input spatial maps (layers), the rate of soil detachment by rain drop impact (F, kg m^{-2}), rate of soil detachment by runoff (H, kg m^{-2}), and transport capacity of overland flow (runoff) (TC, kg m^{-2}) are calculated in the GIS environment as follows:

$$F = 10^{-3} * K * \left(E * e^{-0.05A}\right),$$

$$H = 10^{-3} * (0.5\text{COH})^{-1}(\text{SR})^{1.5} \sin(S)\,(1 - \text{GC}),$$

$$\text{TC} = 10^{-3} * C_f * \text{SR}^2 * \sin(S), \tag{2}$$

where K is soil detachability index (g J^{-1}), E is annual kinetic energy of rainfall (J m^{-1}), A is percentage of rainfall contributing to permanent interception and stream flow (%), COH is cohesion of the soil surface (KPa), GC is fraction of ground (vegetation) cover (0^{-1}), Cf is the crop cover management factor, and S is the steepness of the ground slope expressed in degree. Total particle detachment ($D = F+H$) is finally computed as sum of soil particle detachment by runoff (H) and soil particle detachment by raindrop (F) impacts. The model compares the predicted rate of splash detachment (D) and the transport capacity for overland flow (TC), and the minimum value is taken as the erosion rate (annual soil loss) estimated for fthe study catchment (Figure 4).

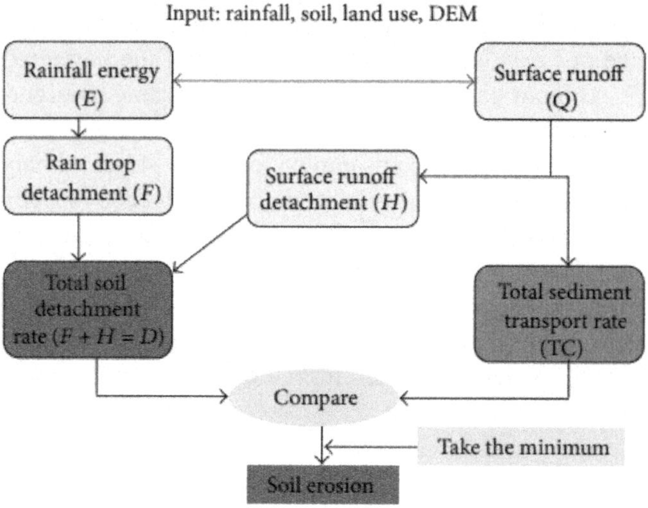

Figure 4: Flow chart depicting the methodology used for MMF modelling (source: Morgan et al. [23]).

Model Evaluation

The percent difference (*D*) was used as methods for goodness-of-fit measure of model prediction. The model estimated soil loss rate in this study was evaluated with respect to the sediment deposited surveyed in the reservoir.The percent difference (*D*) measures the average difference between the simulated and measured values as

$$D = \left(\frac{p-q}{q}\right)100, \tag{3}$$

where p is model simulated value and q is measure value. "*A* value close to 0% is best for *D*; however, higher values of *D* are acceptable if the accuracy in which the observed data have gathered is relatively poor" [45].

In addition, different studies showed that soil profile data such as degree of truncation of the top soil horizon by erosion can be used to assess the performance of models (e.g., Tamene [7]; Desmet and Govers [46]; Mitasova et al. [47]; and Turnage et al. [48]). In this study, the soil profile data was applied to evaluate the estimated erosion result by MMF model. For this purpose, areas with possible erosion processes were selected, and soil profile data related to the truncation level of the A horizon were documented in the study catchment. These data were then compared with the soil loss prediction made by the model. The main purpose was to evaluate whether the spatial pattern of erosion predicted by MMF model correlated well with the depth of soil profile data which semiquantitatively verified the performance of the model. With regard to this, 10 soil profiles (pits) were identified and georeferenced and then compared to the corresponding model spatial erosion results. The MMF model result also compared with the outputs of other models which simulated in the study area and other highlands of Tigray region in northern Ethiopia.

RESULTS AND DISCUSSION

Spatial Distribution of Hydrophysical Parameters Influencing Erosion

Spatial variability in slope, rainfall, vegetation, soil texture, and land use and cover are among the main factors which influence the distribution of erosion risk in a catchment [7]. Since there was only one meteorological station, weather data such as rainfall was assumed to be the same throughout the study catchment. The variability in the kriging interpolated maps of the other erosion influencing factors in the catchment is shown in Figure 5. The slope of the study catchment (Figure 5(a)) showed that the northern and eastern part of the catchment had a very steep slope (up to 77°) and slope steepness decreases in the direction to the reservoir (south of the catchment). This indicated that sources of hydrological losses as runoff and sediment yield can be higher on steep areas as compared to flat to gentle slopes provided that the other factors are similar. Such influence of slope can be explained by soil texture variability; that is, silt and clay soils dominated towards the reservoir and flat areas whereas there was a coarser texture on the steep slopes of the catchment (Figure 5(b)). This could be associated with the selective behaviour of erosion in transporting fine particles [7, 49].

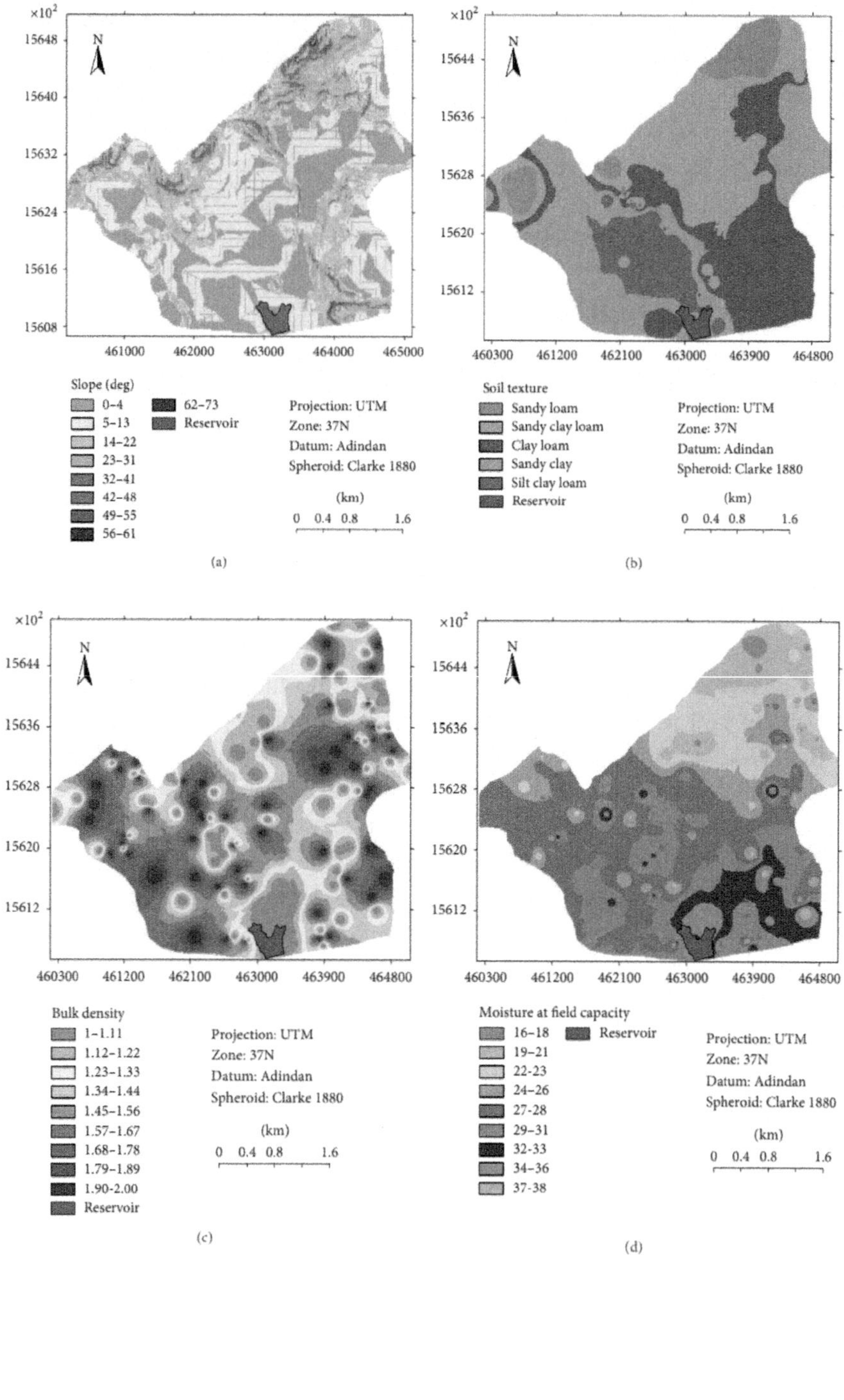

Slope (deg)
0–4
5–13
14–22
23–31
32–41
42–48
49–55
56–61
62–73
Reservoir
Projection: UTM
Zone: 37N
Datum: Adindan
Spheroid: Clarke 1880
(km)
0 0.4 0.8 1.6
(a)
Soil texture
Sandy loam
Sandy clay loam
Clay loam
Sandy clay
Silt clay loam
Reservoir
Projection: UTM
Zone: 37N
Datum: Adindan
Spheroid: Clarke 1880
(km)
0 0.4 0.8 1.6
(b)
Bulk density
1–1.11
1.12–1.22
1.23–1.33
1.34–1.44
1.45–1.56
1.57–1.67
1.68–1.78
1.79–1.89
1.90-2.00
Reservoir
Projection: UTM
Zone: 37N
Datum: Adindan
Spheroid: Clarke 1880
(km)
0 0.4 0.8 1.6
(c)
Moisture at field capacity
16–18
19–21
22–23
24–26
27–28
29–31
32–33
34–36
37–38
Reservoir
Projection: UTM
Zone: 37N
Datum: Adindan
Spheroid: Clarke 1880
(km)
0 0.4 0.8 1.6
(d)

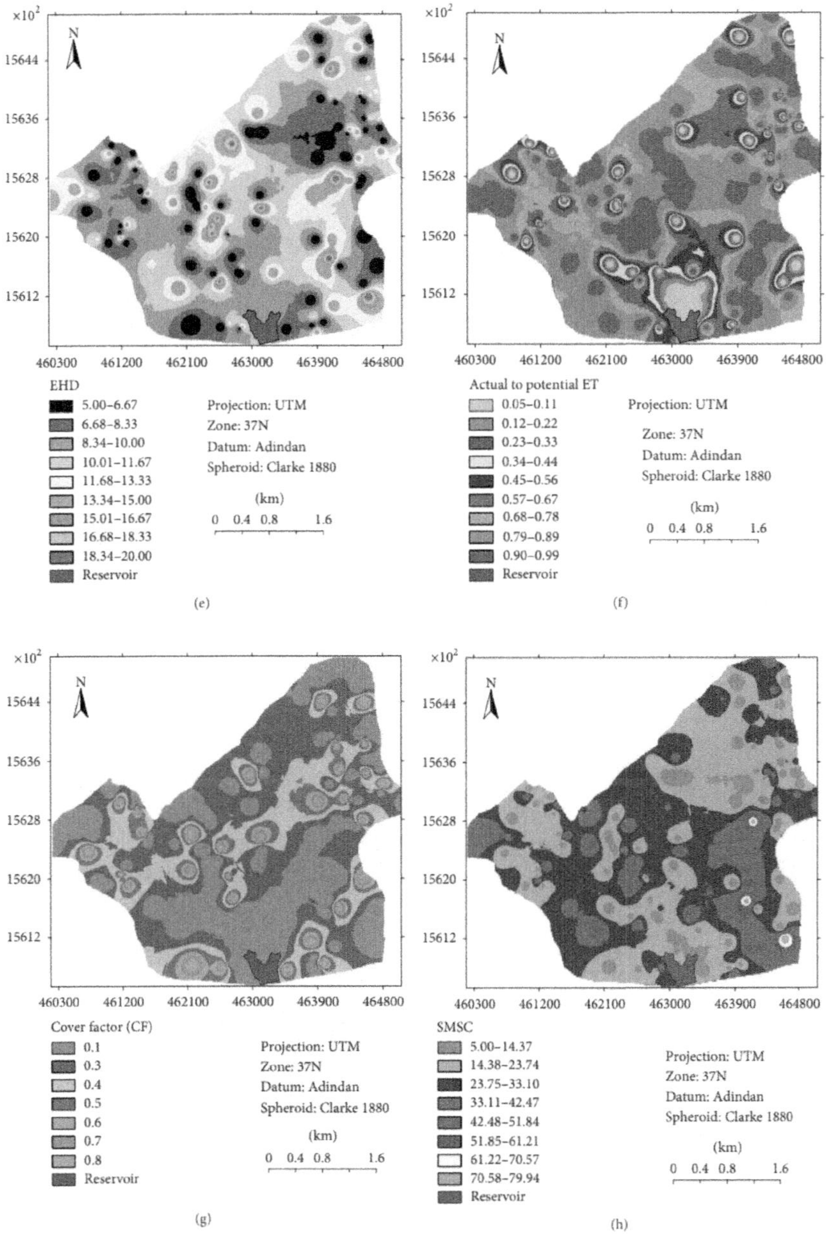
×10²
15644
15636
15628
15620
15612
460300 461200 462100 463000 463900 464800
N
EHD
5.00–6.67
6.68–8.33
8.34–10.00
10.01–11.67
11.68–13.33
13.34–15.00
15.01–16.67
16.68–18.33
18.34–20.00
Reservoir
Projection: UTM
Zone: 37N
Datum: Adindan
Spheroid: Clarke 1880
(km)
0 0.4 0.8 1.6
(e)
Actual to potential ET
0.05–0.11
0.12–0.22
0.23–0.33
0.34–0.44
0.45–0.56
0.57–0.67
0.68–0.78
0.79–0.89
0.90–0.99
Reservoir
(f)
Cover factor (CF)
0.1
0.3
0.4
0.5
0.6
0.7
0.8
Reservoir
(g)
SMSC
5.00–14.37
14.38–23.74
23.75–33.10
33.11–42.47
42.48–51.84
51.85–61.21
61.22–70.57
70.58–79.94
Reservoir
(h)

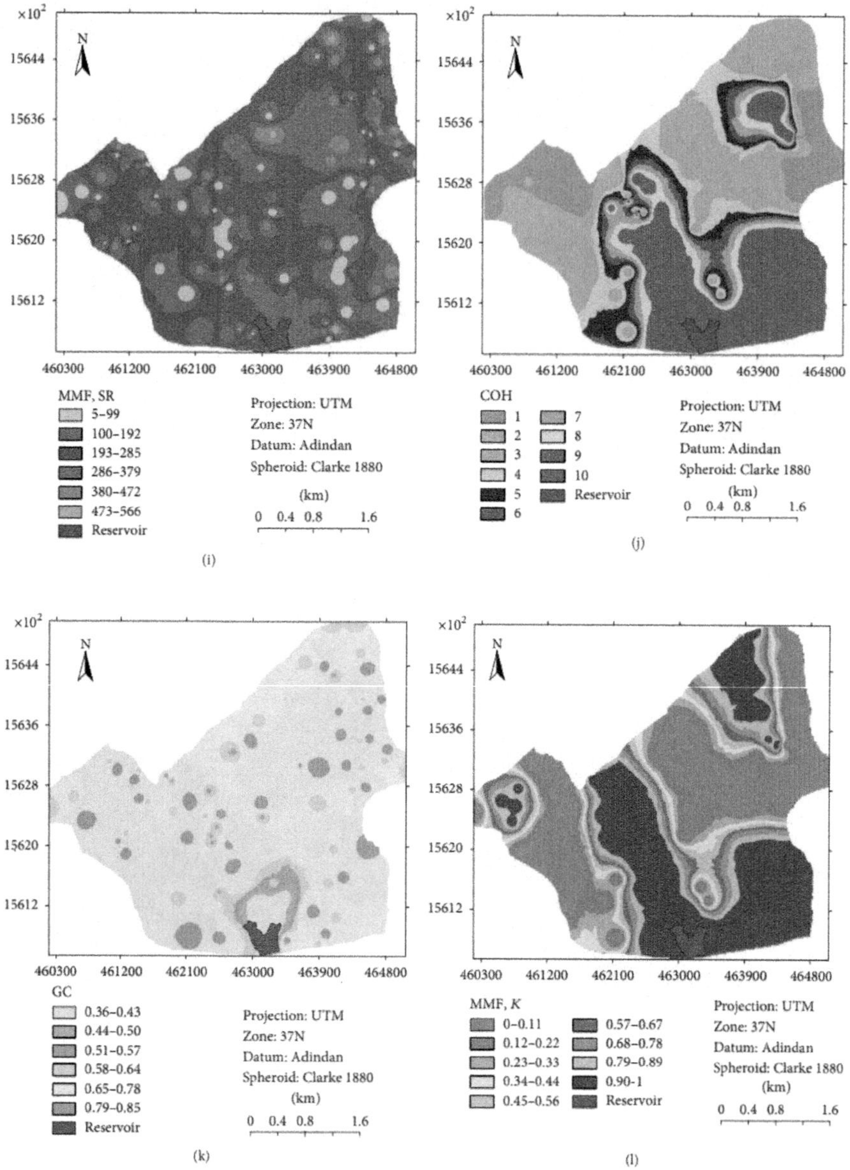

Figure 5: Spatial distribution of parameters influencing the hydrology/erosion of the catchment: (a) slope; (b) soil texture; (c) bulk density (Mg m^{-3}); (d) moisture at field capacity (%); (e) EHD, effective hydrological top soil depth (cm); (f) E_t/E_o, actual to potential evaporation; (g) cover management factor; (h) SMSC, soil moisture storage capacity (mm); (i) SR, surface runoff (Q) (mm); (j) COH, cohesion of the soil surface (KPa); (k) GC, fraction of ground cover (0-1); and (l) *K*, soil detachability index.

Bulk density is higher (up to 1.90 Mg m^{-3}) in the steep slopes with poor vegetation cover and soil management practices and decreased to about 1.10 Mg m^{-3} around the reservoir and valley which are located at the foot-slope of the catchment (Figure 5(c)). From this figure, it is possible to observe higher erosion on areas dominated by higher bulk density as they are located on the steep slopes, low SQ, eroded sites, and marginal land soils and soils with poor vegetation cover. An increase in bulk density can negatively affect the circulation of air, water, and plant nutrients and their root system and in-turn raises rate of soil erosion [41, 50]. Such higher BD increases surface runoff which is the driving force for soil loss by decreasing soil infiltration and soil water holding capacity [51].

The spatial distribution of soil moisture content at field capacity (MS) showed lower value in the hilly part of catchment as compared to that of flat to gentle slopes under similar management and cover conditions (Figure5(d)). This is consistent with the report by Behera et al. [52] who reported lower values of soil parameters such as MS in the topsoil layer of hilly areas. However, regardless of the slope steepness the highest MS was found in some sites of the study catchment such as in the forest and closed pasture lands with relatively higher organic matter (data not shown). The lowest MS in the catchment was associated with low SQ, poor land use, and soil management systems (e.g., marginal land, over grazing land, and eroded sites with shallow soil depth).

The effective hydrological top soil depth (EHD) spatial map (Figure 5(e)) indicated higher values (>15 cm) in relatively better vegetation-cover areas, high SQ and stable sites, and flat to gentle slopes. The EHD values were lower (15 cm) in relatively better vegetation-cover areas, high SQ and stable sites, and flat to gentle slopes. The EHD values were lower (15 cm) in marginalized areas, cultivated and degraded grazing lands. Majority (78%) of the study catchment showed low EHD, indicating that such sites can be the source of higher runoff and soil loss. Similarly, the higher values (0.68 to 0.90) of E_t/E_o corresponded to forest land, whereas lower values ($0.05^{-0}.33$) corresponded to agricultural areas. The intermediate E_t/E_o values can correspond to other areas such as bush land (Figure 5(f)). Such values are thus more influenced by the spatial distribution of crop cover and management factor (C_f) which ranges from $0.2^{-0}.8$ (Figure 5(g)). Areas of the catchment with higher C_f are expected to have lower E_t/E_o values and vice versa [52]. As part of the C_f factor, vegetation maintains high rates of evapotranspiration, rainfall interception, and runoff infiltration [52, 53].

The soil moisture storage capacity (R_c) calculated as a function of MS, BD, EHD, and E_t/E_o varied spatially from 5 to 79 mm (Figure 5(h)). Most of the small R_c values were located on steep slopes with shallow soil depth, poor surface cover, and marginal lands. Farmers who cultivated their land on steep slopes confirmed that they often face crop failure due to moisture stress related to low R_c. Generally, low R_c can be used as an indicator of a source of higher runoff. Such soil could be susceptible to soil detachment by raindrop and runoff impacts as a result of less vegetation cover. Reduction in the EHD (Figure 5(e)) can lead to low soil moisture storage capacity, R_c (Figure 5(h)), which resulted in higher surface runoff (Figure 5(i)).

For estimating soil detachment rate by runoff (*H*), understanding the spatial distribution of slope (*S*) (Figure 5(a)), surface runoff (SR) (Figure 5(i)), cohesion of the soil surface (COH) (Figure 5(j)), and fraction of ground cover (GC) (Figure 5(k)) are important conditions. The soil detachability by raindrop impact is also influenced by the soil detachability index (*K*) which showed higher values (0.79–0.98 g J−1) on flat areas with coarser soils and poor soil cover and lower values (0.11–0.33 g J−1) on steep slope with fine soils and good vegetation cover in the study catchment (Figure 5(l)). The lower *K* values on steep slope could be associated with rainfall drop impact angle which is falling to the ground surface on inclination.

Estimated Soil Loss Using MMF Model

The spatial distribution of the rate of soil detachment by raindrop (*F*) indicated that flat areas (south of the catchment) were exposed more to *F* as compared to that of hilly land (Figure 6(a)). This could be attributed to the perpendicular fall of raindrop energy (strong energy) on flat areas as compared to raindrop impact angle on inclination [2, 3]. This implied that steep slope reduces the impact of rainfall drop energy because rain drop met the soil surface on inclination. Despite this, the net detached and transported soil is almost the same on flat lands, whereas on steep surfaces more soil particles are thrown downslope than upslope during the detachment process, resulting in a net movement of material downslope [3]. Generally, the *F* values estimated by the model ranged from 160 t ha^{-1} y^{-1} in the study catchment. The plateau and valley parts of the catchment showed higher *F* values. Increasing soil cover and practicing zero grazing, for example, can be part of the solution for reducing the amount of soil detached by raindrops. This is because such practices can increase vegetation cover that dissipates rainfall and runoff energy [2, 3].

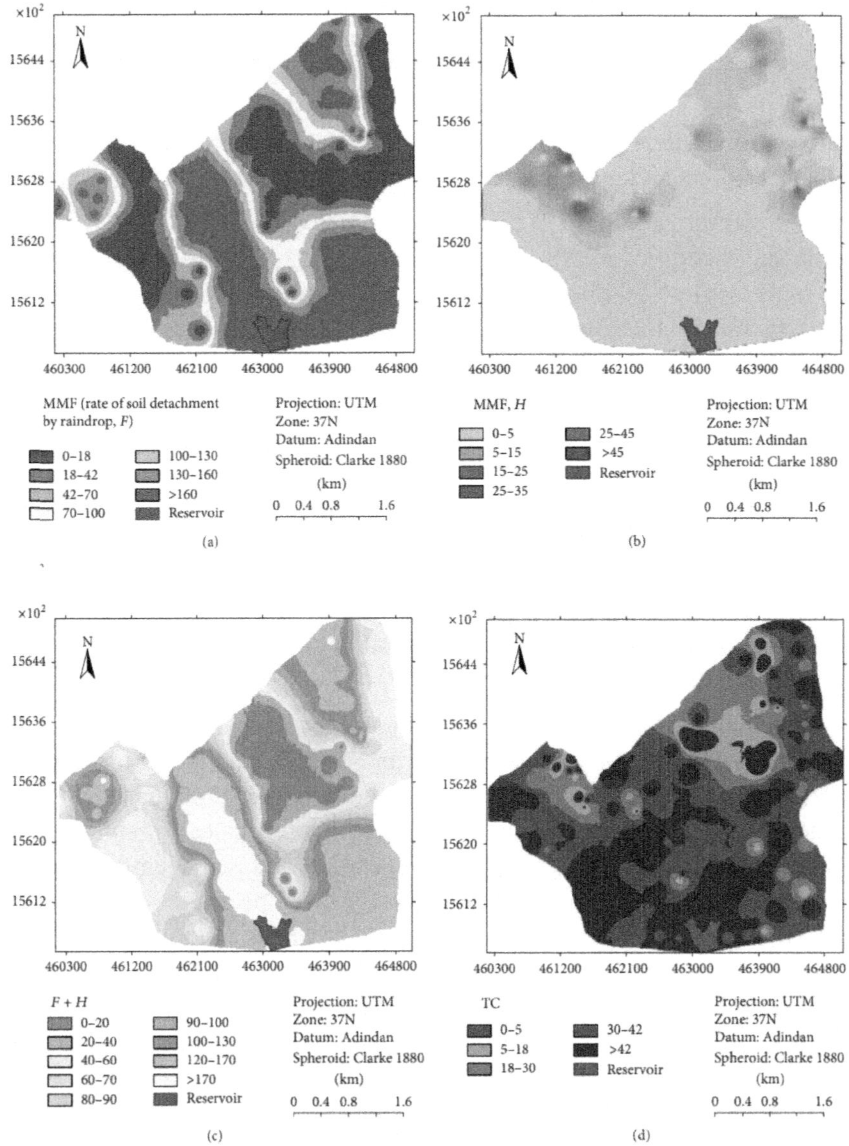

Figure 6: Spatial distribution of MMF erosion estimation. (a) Rate of soil detachment by rain drop impact (*F*) (t ha−1 y−1); (b) rate of soil detachment by runoff impact (*H*) (t ha−1 y−1); (c) total soil detachment (*F*+*H*) (t ha−1 y−1); and (d) soil transport capacity of overland flow (TC) (t ha−1 y−1).

The soil detachment rate by runoff (H) is higher (35 t ha^{-1} y^{-1}) on the steep-hilly parts of the catchment (Figure 6(b)), as slope steepness increases the amount and velocity of overland flow or surface runoff (SR) (Figure 5(i)). In addition, the expansion of cropland and open grazing practices to the steep parts of the catchment could make the soil more vulnerable to soil erosion by runoff. A similar view to this study was reported by Tamene [7] who stated that on steeper slopes the soils are likely to be thinner and the flow velocity is high, which results in high runoff; whereas runoff threshold is high since water is likely to pond and infiltrate, resulting in little or no runoff on flat slope land.

The spatial variability of the total soil detachment rate (*F*+ *H*) as a result of the summation of *F* and *H* for the catchment is shown in Figure 6(c) and this ranged from 170 t ha^{-1} y^{-1}. The highest rates of *F*+*H* occurred in low SQ fields, marginal lands, and subsoil exposed soils having low soil resistance to detaching forces. The lowest was observed in forest land, protected plantation areas, and farm lands with high soil quality regardless of the slope steepness. This study generalized that the rate of soil loss increased with an increase of detaching forces. It was observed from the field that the process of erosion can continue until first the topsoil and finally the subsoil disappear unless suitable controlling measures are implemented. However, it is a matter of fact that all soils detached cannot be reached at the outlet of the catchment because of deposition areas on the way to the outlet. It is therefore important to have information that shows the spatial distribution of soil transport capacity of the overland flow (TC) in the study catchment.

The spatial distribution of TC (Figure 6(d)) indicated higher values (>42 t ha^{-1} y^{-1}) on steep slope, marginal and over grazed lands, and sites with bare soils and intensively cultivated without proper soil management and conservation measures, whereas it indicated lower values (<5 t ha^{-1} y^{-1}) from relatively less disturbed areas (better vegetation and management practices). This indicated that TC is influenced not only by slope but also by cover crop, supporting practices, and soil erodibility and erosivity conditions. These could be the factors that led to the irregular spatial distribution and variability of the soil loss as TC which was estimated by MMF model.

In general, the TC value was lower than that of *F* + *H* which attributed to the transport of fewer amounts of detached soils by rainfall drop and runoff impacts. Thus, erosion is transport limited in the study catchment. However, there were conditions whereby TC could be larger than *F*+*H*, for example, steep areas, compact soils, marginal lands, and farmlands with poor SQ. This could be related to the fact that in steep slope land, the possibility of deposition to take place in the natural flow is low and the time for soil infiltration is also short. The mean and spatial distribution of total soil detachment (*F*+*H*) was higher than the TC, indicating that the value of TC is taken to show the magnitude of annual soil loss from the study catchment. Thus, considering the TC values, the higher erosion risk areas can be identified scattered throughout the catchment. The model generally predicted erosion being limited by transportation, indicating that the amount of soil detachment is very high in the study catchment. This indicated that soil cover is too poor in which soil is exposed to detaching forces and also, the steep slope which generate high runoff can detach and transport large mass of soils.

Results of MMF Model Related to Reservoir Sediment Survey and Soil Profile Data

The model estimated soil loss at catchment level was compared with the survey based measured sediment yield from the reservoir located at the outlet of the catchment. The average catchment level soil loss (TC) estimated by MMF model was $26\,t\,ha^{-1}\,y^{-1}$ whereas the sediment yield measured from the reservoir survey was $20\,t\,ha^{-1}\,y^{-1}$. This resulted in percent difference (D) value of 30%. The similarity between the measured and model simulated value at catchment level was 70%, which is moderately acceptable. The MMF model seems to be overestimating the average soil loss (TC) from the entire catchment as compared to the sediment yield measured from reservoir survey, indicating that the model can pronounce erosion processes observed in the catchment using some extremely higher values of soil loss rate.

In addition, the soil erosion rate predicted by the model was compared with observed soil profile depth data in selected sites of the study catchment. The MMF model was assessed in terms of its capacity to identify areas of soil truncation and/or to predict lower erosion on areas where buried soils and/or alluvial/colluvial deposits or stable soils were observed. The MMF model accurately predicted erosion in about 80% of the pits observed in the catchment whereas the 20% disagreement was located at the upslope position of the catchment where the model predicted slight erosion for sites of a very truncated soil profile. In most sites (80%) of the catchment that characterized by truncated soil profile the model estimated high soil loss rate ($>20\,t\,ha^{-1}\,y^{-1}$).

MMF Model Evaluation in relation to Other Studies (Models)

For most of the catchment (>80%), the MMF model predicted higher soil loss rates than the maximum tolerable soil loss rate of $18\,t\,ha^{-1}\,y^{-1}$ estimated for the country [11]. This could be related to the reason that the study catchment could be characterized by very low soil infiltration, low soil water holding capacity and poor vegetation cover (inappropriate land use system), and low conservation measures. Consistent with this finding and explanation many previous reports indicated that a higher soil loss rate is strongly associated with high runoff resulting from absence of runoff flow obstacles such as vegetation cover, conservation measures, impoundments, and soil with low infiltration rate and low soil water holding capacity [54–60]. If an average annual soil generation rate of 6 t ha^{-1} y^{-1} [61] is considered, the soil loss rates estimated by the model in most parts of the catchment could be beyond this acceptable level. It was less than 1% of the catchment area that experienced a soil erosion rate below 6 t ha^{-1} y^{-1}.

This study showed lower sediment yield (26 t ha^{-1} y^{-1}) as compared to a previous study using Soil and Water Analysis Tool (SWAT) model simulated result (30 t ha^{-1} y^{-1}) in the same catchment [31]. However, such differences can be attributed to model variability in the scale of data requirement. In addition, SWAT model is a continuous and long-time simulation model that might not be represented by the existing land use which can increase model prediction uncertainty [31]. Generally, the variability in soil loss predicted by both models is not higher than 15%, which is an acceptable model uncertainty. Sediment yield of 21 t ha^{-1} y^{-1} based on an in-filled dam with a catchment area of 6.7 km^2 in one of the Tekezze River's tributaries in the Tigray region is also reported by Machado et al. [62].

In addition, a previous study using four catchments in the Tigray region reported an average soil erosion rate of 24 t ha^{-1} y^{-1} [7]. In general, this and the above results are consistent with the trend of erosion estimated using MMF model in this study. Such comparison considers their similarity in the scale of measurement, input data requirement, and the farming system within the catchments. Therefore, the application of MMF model for estimating and identifying the severity of water erosion in order to introduce appropriate interventions is acceptable in conditions similar to the Mai-Negus catchment (study area).

CONCLUSION

This study concluded that GIS is a useful tool to integrate and manage spatially distributed hydrophysical data while assessing the spatial distribution of erosion. In this study, the MMF model results showed a lower rate of erosion for the soil transport capacity of overland flow (TC) when compared to the rate of soil detachment. This indicated that erosion is transport limited, and thus TC can show a realistic image of soil erosion hotspot sites in the study catchment conditions while introducing suitable soil conservation and/or management practices. Availability of such a study can help in making a quick assessment of runoff and soil loss status for the process of appropriate decision making. The average catchment level soil erosion rate (TC) estimated by MMF model was 26 t ha^{-1} y^{-1} but there were sites with erosion rates higher than 42 t ha^{-1} y^{-1}. The sources of such higher soil loss rates were identified mainly to be the eroded sites, low soil quality soils, marginal land, overgrazed lands, and mono-cropping cultivated land system with poor soil management and conservation measures located in the mountainous (north-west) and central-ridge landforms of the study catchment. Therefore, introducing appropriate site specific interventions such as agroforestry, agronomic practices, exclosure of degraded lands, conservation measures based on the model erosion maps produced for the study catchment are suggested to be a practical solution for attaining sustainable environmental management and production services. However, the high soil detachment rates that occurred in many fields of the study catchment as a result of differences in spatial distribution of soil erosion influencing factors should be considered while designing and promoting appropriate practices to improve and maintain soil and water resources.

ACKNOWLEDGMENTS

The authors gratefully acknowledge the financial support of DAAD/GIZ (Germany) through the Centre for Development Research (ZEF), University of Bonn (Germany), and the support of Aksum University (Ethiopia) during the field work. The authors also greatly appreciate the assistance offered by the local farmers and extension workers during the field study.

REFERENCES

1. S. J. Scherr, "Soil degradation: a threat to developing-country food security by 2020?" Food, Agriculture, and Environment Discussion Paper 27, International Food Policy Research Institute, Washington, DC, USA, 1999.
2. R. Lal, "Soil degradation by erosion," Land Degradation and Development, vol. 12, no. 6, pp. 519–539, 2001.
3. R. P. C. Morgan, Soil Erosion and Conservation, Blackwell, Malden, Mass, USA, 3rd edition, 2005.
4. M. A. Zoebisch and E. DePauw, "Degradation and food security on a global scale," in Encyclopedia of Soil Science, R. Lal, Ed., pp. 281–286, Marcel Dekker, 2002.
5. R. Lal, "Soil quality and sustainability," in Methods for Assessment of Soil Degradation, R. Lal, W. H. Blum, C. Valentine, and B. A. Stewart, Eds., pp. 17–30, CRC Press, Boca Raton, Fla, USA, 1998.
6. J. Ananda and G. Herath, "Soil erosion in developing countries: a socio-economic appraisal," Journal of Environmental Management, vol. 68, no. 4, pp. 343–353, 2003.

7. L. Tamene, Reservoir siltation in the drylands of northern Ethiopia: causes, source areas and management options [Ph.D. thesis], University of Bonn, Bonn, Germany, 2005.
8. J. Nyssen, Erosion processes and soil conservation in a tropical mountain catchment under threat of anthropogenic desertification—a case study from Northern Ethiopia [Ph.D. thesis], Katholieke University, Leuven, Belgium, 2001.
9. G. Tekeste and D. S. Paul, "Soil and water conservation in Tigray, Ethiopia," Report of a Consultancy Visit to Tigray, University of Wageningen, Wageningen, The Netherlands, 1989.
10. FAO, "Ethiopian highland reclamation study: Ethiopia," Final Report, FAO, Rome, Italy, 1986.
11. H. Hurni, "Erosion—productivity—conservation systems in Ethiopia," in Soil Conservation and Productivity, Proceeding of the 4th International Conference on Soil Conservation, I. P. Sentis, Ed., pp. 654–674, Maracay, Venezuela, November 1985.
12. G. B. Tesfahunegn, Soil erosion modeling and soil quality evaluation for catchment management strategies in northern Ethiopia [Ph.D. thesis], University of Bonn, Bonn, Germany, 2011.
13. H. Pohlmann, "Geostatistical modelling of environmental data," Catena, vol. 20, no. 1-2, pp. 191–198, 1993.
14. G. B. Tesfahunegn, L. Tamene, and P. L. G. Vlek, "Catchment-scale spatial variability of soil properties and implications on site-specific soil management in northern Ethiopia," Soil and Tillage Research, vol. 117, pp. 124–139, 2011.
15. D. Sharad, M. V. Ravi Kumar, L. Venkataratnam, and T. Mel lerwara Rao, "Watershed prioritization for soil conservation- a GIS approach," GeoCarto International, vol. 1, no. 2, pp. 27–34, 1993.
16. P. G. Sanware, C. P. Singb, and R. L. Karale, "Remote sensing application for prioritization of subwatersheds using sediment yield and runoff indices in the catchment of Marani barrage (Sahibi),"UNDP/ FAO Project 13, Remote Sensing Center, AIS LUS, Government of India, New Delhi, India, 1998.
17. D. K. Das, K. S. S. Sharma, and N. Kalra, "Education and training in remote sensing and GIS for sustainable agricultural development," in Proceeding of the 15th Asian Conference on Remote Sensing, pp. 1–6, Banagalore, India, 1994.
18. E. Özgöz, "Long term conventional tillage effect on spatial variability of

some soil physical properties,"Journal of Sustainable Agriculture, vol. 33, no. 2, pp. 142–160, 2009.

19. R. Webster, "Quantitative spatial analysis of soil in the field," in Advance In Soil Science, B. A. Stewart, Ed., vol. 3, pp. 1–70, Springer, New York, NY, USA, 1985.
20. R. Webster and M. A. Oliver, Statistical Methods in Soil and Land Resource Survey, Oxford University Press, Oxford, UK, 1990.
21. C. A. Cambardella, T. B. Moorman, J. M. Novak et al., "Field-scale variability of soil properties in central Iowa soils," Soil Science Society of America Journal, vol. 58, no. 5, pp. 1501–1511, 1994.
22. W. H. Wischmeier and D. D. Smith, Predicting Rainfall Erosion Losses: A Guide to Conservation Planning, vol. 537, US Department of Agriculture, Washington, DC, USA, 1978.
23. R. P. C. Morgan, D. D. V. Morgan, and H. J. Finney, "A predictive model for the assessment of soil erosion risk," Journal of Agricultural Engineering Research, vol. 30, pp. 245–253, 1984.
24. D. C. Flanagan and M. A. Nearing, "USDA-Water Erosion Prediction Project (WEPP)," Tech. Rep. 10, National Soil Erosion Research Laboratory, USDA-ARS-MWA, West Lafayette, Ind, USA, 1995.
25. J. G. Arnold, R. Srinivasan, R. S. Muttiah, and J. R. Williams, "Large area hydrologic modeling and assessment part I: model development," Journal of the American Water Resources Association, vol. 34, no. 1, pp. 73–89, 1998.
26. R. P. C. Morgan, J. N. Quinton, R. E. Smith et al., "The European Soil Erosion Model (EUROSEM): a dynamic approach for predicting sediment transport from fields and small catchments," Earth Surface Processes and Landforms, vol. 23, no. 6, pp. 527–544, 1998.
27. R. L. Bingner and F. D. Theurer, "AnnAGNPS: estimating sediment yield by particle size for sheet & rill erosion," in Proceedings of Sedimentation: Monitoring, Modeling, and Managing, 7th Federal Interagency Sedimentation Conference, vol. 1, pp. 1–7, Reno, Nev, USA, March 2001.
28. D. M. Fox and R. B. Bryan, "The relationship of soil loss by interrill erosion to slope gradient," Catena, vol. 38, no. 3, pp. 211–222, 2000.
29. P. I. A. Kinnell, "AGNPS-UM: applying the USLE-M within the agricultural non point source pollution model," Environmental Modelling and Software, vol. 15, no. 3, pp. 331–341, 2000.
30. K. J. Lim, M. Sagong, B. A. Engel, Z. Tang, J. Choi, and K.-S. Kim,

"GIS-based sediment assessment tool,"Catena, vol. 64, no. 1, pp. 61–80, 2005.

31. G. B. Tesfahunegn, P. L. G. Vlek, and L. Tamene, "Management strategies for reducing soil degradation through modeling in a GIS environment in northern Ethiopia catchment," Nutrient Cycling in Agroecosystems, vol. 92, no. 3, pp. 255–272, 2012.
32. L. Tamene, S. J. Park, R. Dikau, and P. L. G. Vlek, "Analysis of factors determining sediment yield variability in the highlands of northern Ethiopia," Geomorphology, vol. 76, no. 1-2, pp. 76–91, 2006.
33. EMA, Ethiopian Mapping Agency, "Ethiopia 1:50000 topographic maps: Aksum Sheet," Tech. Rep., Ethiopian Mapping Agency, Addis Ababa, Ethiopia, 1997.
34. FAO, Food and Agriculture Organization of the United Nations, "The Soil and Terrain Database for Northeastern Africa," FAO, Rome, Italy, 1998.
35. T. M. Dinka, Application of the Morgan, Morgan Finney Model in Adulala Mariyam Watershed, Ethiopia: GIS based erosion risk assessment and testing of alternative land management options [M.S. thesis], Wageningen University, Wageningen, The Netherlands, 2007.
36. S. J. Birrell, K. A. Sudduth, and N. R. Kitchen, "Nutrient mapping implications of short-range variability," in Proceedings of the 3rd International Conference on Precision Agriculture, P. C. Robert, R. H. Rust, and W. E. Larson, Eds., pp. 206–216, America Society of Agronomy, Crop Science Society of America, Soil Science Society of America, Madison, Wis, USA, 1996.
37. D. W. Franzen, A. D. Halvorson, J. Krupinsky, V. L. Hofman, and L. J. Cihacek, "Directed sampling using topography as a logical basis," in Proceedings of the 4th International Conference on Precision Agriculture, P. C. Robert, R. H. Rust, and W. E. Larson, Eds., pp. 1559–1568, America Society of Agronomy, Crop Science Society of America, Soil Science Society of America, Madison, Wis, USA, 1998.
38. G. B. Tesfahunegn, L. Tamene, and P. L. G. Vlek, "A participatory soil quality assessment in Northern Ethiopia›s Mai-Negus catchment," Catena, vol. 86, no. 1, pp. 1–13, 2011.
39. G. W. Gee and J. W. Bauder, "Particle-size analysis," in Methods of Soil Analysis, Part 1, A. Klute, Ed., pp. 383–411, America Society of Agronomy, Soil Science Society of America, Madison, Wis, USA, 2nd edition, 1986.
40. G. R. Blake and K. H. Hartge, "Bulk density," in Methods of Soil

Analysis, Part 1, A. Klute, Ed., vol. 9 ofAgronomy Monograph, pp. 363–375, America Society of Agronomy, Madison, Wis, USA, 2nd edition, 1986.

41. T. C. Baruah and H. P. Barthakur, A Text Book of Soil Analysis, Vikas, New Delhi, India, 1999.
42. Utset, T. López, and M. Díaz, "A comparison of soil maps, kriging and a combined method for spatially predicting bulk density and field capacity of ferralsols in the Havana-Matanzas Plain,"Geoderma, vol. 96, no. 3, pp. 199–213, 2000.
43. N. A. C. Cressie, Statistics for Spatial Data, Revised Edition, John Wiley & Sons, New York, NY, USA, 1993.
44. J. Triantafilis, I. O. A. Odeh, and A. B. McBratney, "Five geostatistical models to predict soil salinity from electromagnetic induction data across irrigated cotton," Soil Science Society of America Journal, vol. 65, no. 3, pp. 869–878, 2001.
45. D. Chekol, Modeling of hydrology and soil erosion of upper Awash river basin, Ethiopia [Ph.D. thesis], University of Bonn, Bonn, Germany, 2006.
46. P. J. J. Desmet and G. Govers, "GIS-based simulation of erosion and deposition patterns in an agricultural landscape: a comparison of model results with soil map information," Catena, vol. 25, no. 1–4, pp. 389–401, 1995.
47. H. Mitasova, L. Mitas, W. M. Brown, and D. M. Johnston, GIS Tools for Erosion/Deposition Modelling and Multidimensional Visualization, Geographic Modelling and Systems Laboratory, University of Illinois at Urbana-Champaign, Urbana, Ill, USA, 1997.
48. K. M. Turnage, S. Y. Lee, J. E. Foss, K. H. Kim, and I. L. Larsen, "Comparison of soil erosion and deposition rates using radiocesium, RUSLE, and buried soils in dolines in East Tennessee,"Environmental Geology, vol. 29, no. 1-2, pp. 1–10, 1997. ·
49. N. C. Brady and R. R. Weil, Eds., The Nature and Properties of Soils, Prentice Hall, Upper Saddle River, NJ, USA, 13th edition, 2002.
50. J. W. Doran, "Soil health and global sustainability: translating science into practice," Agriculture, Ecosystems & Environment, vol. 88, no. 2, pp. 119–127, 2002. ·
51. H. A. Ahmed, E. S. Gerald, and H. H. Richard, "Soil bulk density and water infiltration as affected by grazing systems," Journal of Range Management, vol. 40, no. 4, pp. 307–309, 1987.
52. P. Behera, K. H. V. Durga Rao, and K. K. Das, "Soil erosion modeling

using MMF model—a remote sensing and GIS perspective," Journal of Indian Society of Remote Sensing, vol. 33, no. 1, pp. 165–176, 2005.

53. O. T. Ande, Y. Alaga, and G. A. Oluwatosin, "Soil erosion prediction using MMF model on highly dissected hilly terrain of Ekiti environs in southwestern Nigeria," International Journal of Physical Sciences, vol. 4, no. 2, pp. 53–57, 2009.
54. V. Garg and V. Jothiprakash, "Sediment yield assessment of a large basin using PSIAC approach in GIS environment," Water Resources Management, vol. 26, no. 3, pp. 799–840, 2012.
55. Q. B. Le, R. Seidl, and R. W. Scholz, "Feedback loops and types of adaptation in the modelling of land-use decisions in an agent-based simulation," Environmental Modelling and Software, vol. 27-28, pp. 83–96, 2012.
56. L. B. Asmamaw, A. A. Mohammed, and T. D. Lulseged, "Land use/ cover dynamics and their effects in the Gerado catchment, Northeastern Ethiopia," International Journal of Environmental Studies, vol. 68, no. 6, pp. 883–900, 2011.
57. P. Kaini, K. Artita, and J. W. Nicklow, "Optimizing structural best management practices using SWAT and genetic algorithm to improve water quality goals," Water Resources Management, vol. 26, no. 7, pp. 1827–1845, 2012.
58. L. Sun, W. Lu, O. Yang, J. D. Martín, and D. Li, "Ecological compensation estimation of soil and water conservation based on cost-benefit analysis," Water Resources Management, vol. 27, no. 8, pp. 2709–2727, 2013.
59. G. B. Tesfahunegn, P. L. G. Vlek, and L. Tamene, "Application of SWAT model to assess erosion hotspot for sub-catchment management at Mai-Negus catchment in northern Ethiopia," East African Journal of Science and Technology, vol. 2, no. 2, pp. 97–123, 2013.
60. Munodawafa, "The effect of rainfall characteristics and tillage on sheet erosion and maize grain yield in semiarid conditions and granitic sandy soils of Zimbabwe," Applied and Environmental Soil Science, vol. 2012, Article ID 243815, 8 pages, 2012.
61. H. Hurni, "Soil formation rates in Ethiopia," Working Paper 2, Ethiopian Highlands Reclamation Studies, Addis Ababa, Ethiopia, 1983.
62. M. J. Machado, A. Perez-Gonzalez, and G. Benito, "Assessment of soil erosion using a predictive model," in Rehabilitation of DegradIng and Degraded Areas of Tigray, Northern Ethiopia, E. Feoli, Ed., pp. 237–248, Department of Biology, University of Trieste, Trieste, Italy, 1996.

Chapter 12

GEO-ENVIRONMENTAL SITE INVESTIGATION FOR MUNICIPAL SOLID WASTE DISPOSAL SITES

Giulliana Mondelli, Heraldo Luiz Giacheti and Vagner Roberto Elis

Institute for Technological Research of São Paulo State, São Paulo State University, University of São Paulo, São Paulo, Brazil

INTRODUCTION

Deactivated dump sites and inadequate sanitary landfills can be a serious potential source of contamination. Due to the large number of these sites, contaminants can be generated and migrate into the ground. For this reason, site investigation programs which consider different types of soils and contaminants are needed. In many countries, sanitary landfills have been recently built according to the engineering standards with satisfactory procedures and following the requirements imposed by the local environmental agencies, including monitoring and proper operations. There is a great concern with the operation of sanitary landfills and with the future of deactivated dump sites located in small and medium-size cities, since many of them were inappropriately constructed and operated. Consequently, the main surficial streams and groundwater aquifers, subsoil and air become vulnerable to contamination and pollution. A continuous site investigation program, including in-situ and laboratory tests are necessary to identify the typical soil profile, hydrogeological characteristics and background chemical values. The site surroundings and the subsoil contamination plume can be characterized based on geotechnical, geochemical and mineralogical techniques. The demand for the geo-environmental site investigation of contaminated and noncontaminated sites intended for future sanitary landfill planning has substantially increased in the last few years due to the lack of space in metropolitan areas. Environmental agencies from several countries have proposed different site investigation methodologies in order to diagnose and confirm different contamination levels in sites with diverse physical characteristics in order to guide the remediation plan whenever it is necessary. The experience already achieved on site investigations has shown that the best methodology is site specific, and depends on subsoil and chemical contaminants; geotechnical, geological and hydrogeological aspects; evolution of the contamination plume and the possible risks it poses. Several field and laboratory investigation techniques (direct and indirect) have been proposed and used. Sometimes one technique is more suitable than another depending on the physical and natural characteristics of the site. Countries with recent concerns over the environment tend to adapt the experience gained from the more developed countries with the reality of their own environmental laws, economy, industrialization, size, etc. Having outlined this scenario, this chapter aims to present and discuss the different tests and steps of a geo-environmental site investigation program proposed for municipal solid waste disposal sites.

GEO-ENVIRONMENTAL SITE INVESTIGATION STEPS

The various steps that can be included during a geo-environmental site investigation are discussed in this item and can be systematized in the chart shown in Figure 1, which was prepared considering experience gained on several geo-environmental site characterization campaigns at municipal solid waste disposal sites in Brazil. The major focus of these studies was to assess contamination of medium size municipal solid waste disposal sites, installed over typical Brazilian tropical soils (residual soils and sandstones).

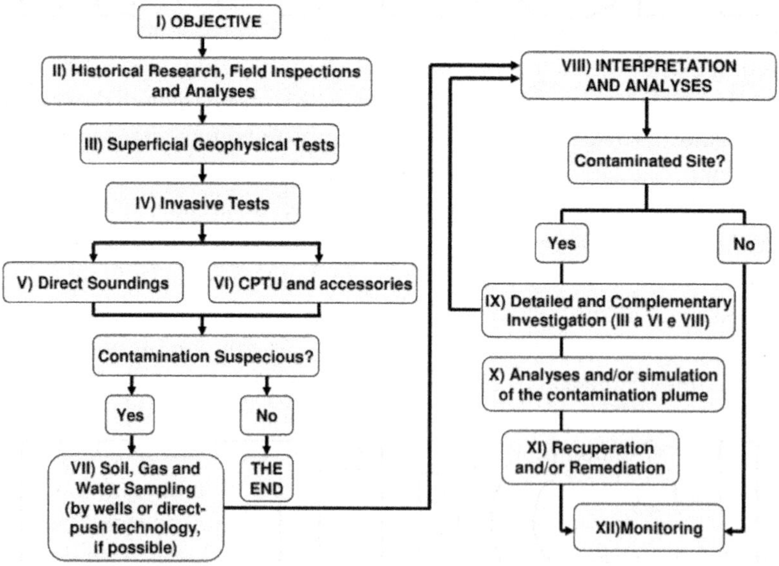

Figure. 1: Proposed geo-environmental site investigation steps for municipal solid waste disposal sites (Mondelli, 2008).

It is important to emphasize that the very first step in any site investigation program is to define the problem. For example, if the objective of the site investigation is just to characterize the local subsoil before the installation of a sanitary landfill, it will probably be necessary to carried out the site investigation until the fifth or sixth step presented in the Figure 1. Depending of the enterprise size or of the natural environmental and anthropological site conditions, the contamination will be assessed in a later step. Therefore, before starting any site investigation test it is fundamental to clearly define the objective of the site investigation program and determine the possible sources and/or targets of contamination. If the sources and/or targets do not exist, there is no sense in continuing the site investigation after the seventh step (Figure 1).

PRELIMINARY INVESTIGATION

Environmental agencies usually recommend gathering historical information and carrying out a site inspection during the preliminary investigation. Compilation of physical and historical data and all the existing information about the site need to be collected before starting any site investigation. This phase (second step in Figure 1) has proved to be very important for the future steps, based on the interpretation of the previous and current topography, hydrology, aerial photographs and historical documents. A good job in this phase can avoid unnecessary tests in the next steps of the site investigation, reducing costs without compromising the quality of the final results. Before installing a new sanitary landfill, closing an old dump or in a regular landfill, or just to investigate the extension of a contamination plume caused by leachate from an irregular municipal solid waste disposal site, it is necessary to contact the responsible company and environmental agencies involved, visit the site and request the following data:

- Location: geographical coordinates, original and current topography, distance from the urban zone, access, surroundings, rivers and airport proximity. Depending on the local regulations, some of these aspects may constrain the construction or extension of new sanitary landfills and may also help the site investigation plan and the final design.
- Municipal waste management: information such as amounts of municipal, hospital, demolition/construction and sweeping/pruning waste generated per person and the locations and manner in which they are disposed of, also needs to be checked. Furthermore, some questions need to be addressed, like: “How long the current landfill will continue to operate? Has it obtained operating and/or commissioning licenses? Is there a plan for hazardous waste management? What percentage of

the waste is recycled by the municipality? Are there any environmental education programs for the local community?"

- Aerial photography: a very complete study of the site to be investigated, its uses and occupational activities and physical surrounds can be done by using aerial photography taken over a period of time up to present day photos.
- Old reports: reports of past site investigations and geological surveys need to be available for consultation and search of the physical media data to be also used or supplemented during the new site investigation phase.
- Weather Aspects: climate data, such as mean annual temperatures, rainfall, radiation, etc for the city or that region where the site investigation will be conducted, should be collected for analysis and calculation of the water balance. The water balance can be carried out using the Thornthwaite and Mather (1955) methodology.

After the analyses and interpretation of all integrated existing data from the site to be investigated, a Preliminary Investigation Report can be submitted. Therefore, the site investigation plan will be planned and conducted based on all information collected on site, facilitating the achievement of the final objective of the site investigation, in order to reduce time and costs. Sometimes, this next phase is called Confirmatory Investigation, when suspicion of contamination occurs. More recently, this phase can integrate the use of different field and laboratory tests. The most well known and widely used tests are surficial geophysics (non-invasive and indirect techniques), followed by piezocone tests (CPTU) (invasive and indirect/direct techniques) with special sensors and samplers for environmental purposes, and groundwater level monitoring wells. Accordingly with Figure 1, some of these techniques, which are more frequently used for geo-environmental site investigation purposes, will be briefly discussed.

GEOPHYSICAL SURVEY

Geophysical methods, particularly electrical techniques, can be used to study those different environmental characteristics which are important during the site characterization for waste disposal and the monitoring of migratory contamination plumes. The major advantage of these methods is that the measurements of soil and contaminant properties are indirect, but it can also be a disadvantage too. It allows a quick investigation of a large area avoiding direct contact with contaminants. Direct techniques based on previous tests, such as simple reconnaissance boring with Standard Penetration Test (SPT) and monitoring through groundwater wells and piezometers, can be used and are considered a traditional practice in developing countries. In this context, much effort has been concentrated on measuring electrical resistivity, a property that is highly dependent on lithology and contamination, which are essential factors to be detected during a geo-environmental site investigation. The major interest in measuring electrical resistivity in a geo-environmental site investigation of contaminated sites is due to the fact that it is an indirect measurement that can be achieved by means of subsurface and/or surface tests, which is also sensitive to temporal variations of the physical environment. This also allows it to be used as a monitoring technique and not only as a method for the investigation and initial reconnaissance of the subsoil. The following geological, geotechnical, hydrogeological and environmental characteristics can be assessed using geophysical methods: (a) rock depth; (b) discontinuities; (c) changes of soil texture; (d) groundwater level; (e) groundwater flow; (f) presence and threedimensional distribution of waste; (g) contaminated soil; (h) contaminated groundwater and plume shape. The first five characteristics are essential for the assessment of any waste disposal sites. All these submitted characteristics are recommended in the geoenvironmental monitoring of a waste disposal facility. The geophysical methods which present good results for these applications are the Resistivity and the Low Frequency Electromagnetics (Ground Conductivity Meters).

Resistivity

The Resistivity Method applies an artificial electrical current I, introducing two electrodes into the ground (A and B), with the objective of measuring the differential potential ΔV generated by two other electrodes (M and N) in the electrical current flow extremities (Figure 2). This arrangement allows the calculation of apparent resistivity ρa in subsurface, using the following equation:

$$\rho_a = K.\frac{\Delta V}{I} \text{ (ohm.m)} \tag{1}$$

where K is a geometrical factor that depends on the electrodes position in the ground and can be calculated for any electrode arrangement using the following general expression:

$$K = \frac{2\pi}{(1/AM)-(1/AN)-(1/BM)+(1/BN)} \text{ (m)} \tag{2}$$

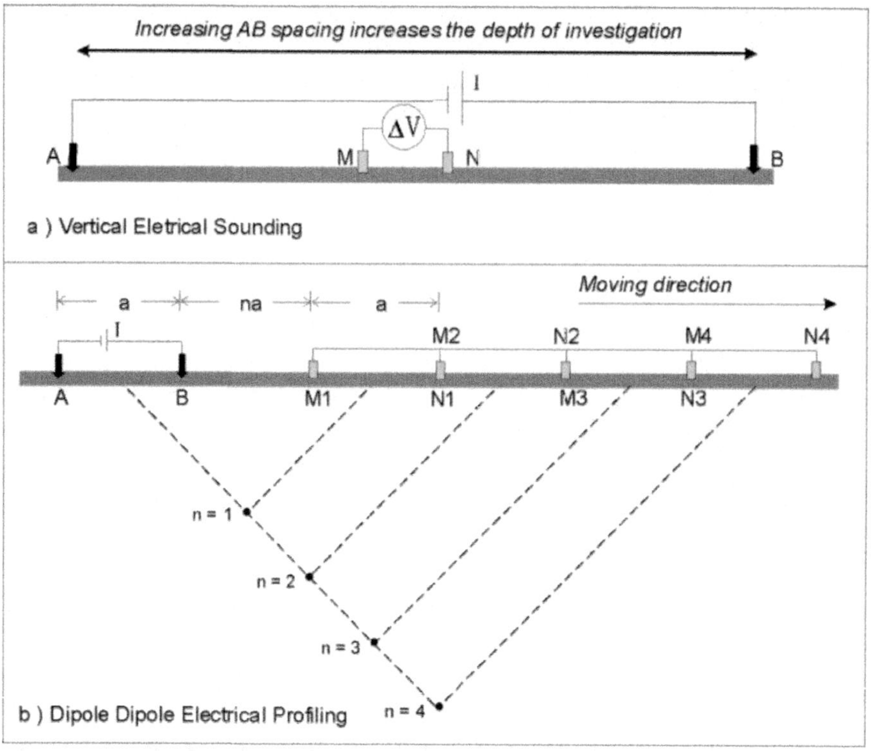

Figure. 2: Configuration of 1D and 2D resistivity surveys: (a) Vertical Electrical Sounding and (b) Dipole Dipole Electrical Profiling.

The soil and rock resistivity values are affected mainly by four factors: mineralogy composition, porosity, moisture, amount and nature of dissolved salts. The most important factors are the porous water content and the salinity content in the porous media. The increase of these two factors tends to decrease the resistivity values. These characteristics allow a good applicability of this geophysical method of application in environmental and hydrogeological studies, where inorganic contaminants occur. In the specific case of municipal solid waste disposal sites, the formation of leachate with elevated concentration of ions, characterizing the polluted areas with very low resistivity values.

The equipment used for the resistivity method consists basically of a controlled electrical current source emission and differential potential measurements. The source potential can range from some kilowatts up to hundreds of watts. This equipment can work using direct or alternating current supply with low frequency, preferably lower than 60 Hz (Telford et al., 1990). The Resistivity Method encompasses various techniques for the application of field tests, which basically consist of vertical electrical sounding (1D resistivity survey) and electrical profiling (2D resistivity imaging) and includes a wide variety of possible electrode configurations (Schlumberger, Wenner, Dipole Dipole, Pole Dipole, Lee and other), rendering the method highly versatile. Currently, 3D tests are possible with the use of multielectrode equipment or a series of 2D profiles grouped and equidistant (Loke & Barker, 1996; Dahlin et al., 2002). Figure 2 presents electrical soundings and dipole dipole electrical profiling field arrangements. As the subsoil is not homogeneous, for any measurement point an apparent resistivity value is obtained. Appropriate software is used for data acquisition and interpretation, for obtaining the electrical stratigraphy of the subsoil (1D survey) or the resistivity cross-section (2D survey). There are several possibilities for carrying out these tests allowing the use of different techniques and arrangements, depending on the site investigation objectives. Usually the most important objective of a geo-environmental characterization is to delineate contamination plumes. The vertical electrical soundings can give important information on site characterization of waste disposal sites. The distribution of different geomaterials on the subsurface profile (several soil layers, waste and rock substrate top) and the depth of the saturated zone can be estimated using vertical electrical soundings. Figure 3 presents the results and the interpretation of a vertical electrical sounding carried out to expand the knowledge of the stratigraphy profile and for defining the position of the saturated zone. The combined results obtained from several vertical electrical soundings allow the construction of a contour map of the groundwater level showing the direction of the groundwater flow. Waste disposal sites can be mapped using several vertical electrical soundings (VES) points or using electrical profiling (continuous way). The VES are more appropriate to vertical subsoil profile definition (waste thickness and landfill bottom). The electrical profiling gives a better delineation of the waste disposal dimensions. 2D inversion surveys of electrical profiling show both lateral and vertical variations of the materials in the cross-section. Figure 4 presents two municipal solid waste trenches excavated on sandy geological substrate delineated by electrical Resistivity Method results (resistivity values lower than 20 ohm.m characterize the places filled with municipal solid waste). Figure 5 shows the results of a survey performed in a contaminated site downstream from a landfill in Brazil. This site was investigated using a 3-D resistivity imaging technique. The purpose of this investigation was to detect and delineate the contamination plume produced by the waste and to acquire detailed information about the affected area. The data set consisted of a series of parallel electrical profile data acquired with the dipole-dipole arrangement, and was inverted as a complete 3-D survey. The resistivity model identifies the disposed waste and the contaminant, marked by the isosurface values lower than 20 ohm.m. The results indicate the presence of a contamination plume and its preferred path. Monitoring wells were installed in the affected area and their chemical analysis confirmed the influence of contaminants.

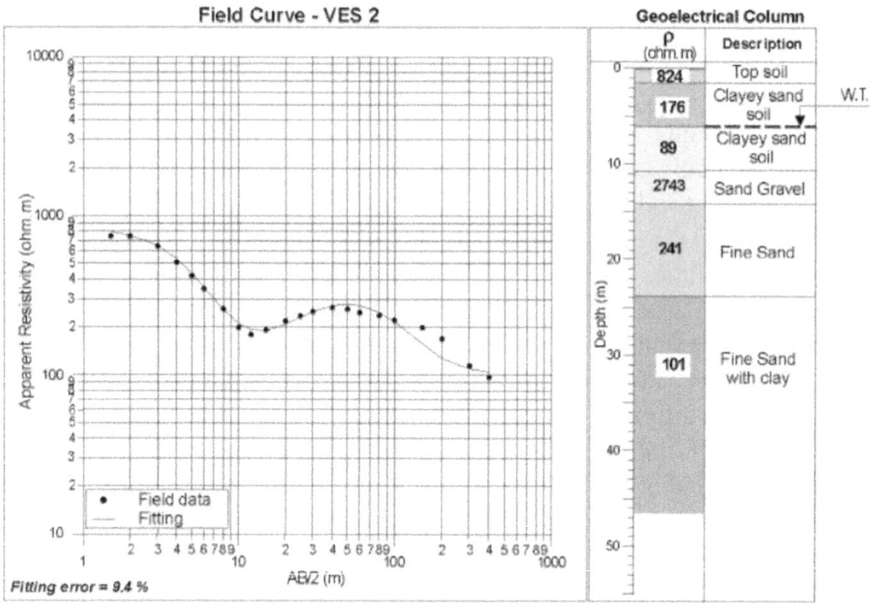

Figure. 3: Results of VES for studying the stratigraphic profile and position of the saturated zone (W.T.).

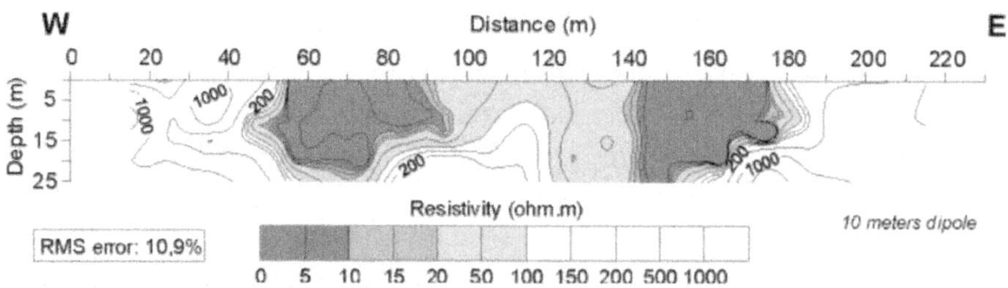

Figure. 4: Resistivity cross-section of waste disposal area. Lower resistivity zones characterize two trenches filled with municipal solid waste.

Low Frequency Electromagnetics (Ground Conductivity Meters)

The Electromagnetic Methods (EM) involves low frequency electromagnetic field propagation. When an alternating current (AC) is established in a wire placed on the ground surface, electrical currents flow in subsurface conductors. This process is known as electromagnetic induction and constitutes the basis of the operation of low frequency electromagnetic methods.

This method is very fast and it uses simple equipment easily operated. It explains the extensive application of this method for geo-environmental studies. The equipment used can be generally called ground conductivity meters. It consists of two coils (transmitter and receiver). The transmitter coil emits a primary magnetic field H_p, which induces electrical currents on the subsurface, generating a secondary magnetic field H_s (Figure 6). The combination of these two fields is measured by the receiver coil. Under certain conditions it is known that there is a linear relationship between the modules of the two fields (McNeil, 1980), technically defined as "low induction number operation". Accordingly, the apparent conductivity can be calculated by σ_a:

$$\sigma_a = \frac{4}{\varpi \mu_0 s^2}\left(\frac{H_s}{H_p}\right) \quad (\mathrm{mS/m}) \tag{3}$$

where ω = angular frequency, $2\pi f$; μ_0 = vacuum magnetic permeability; s = intercoil spacing.

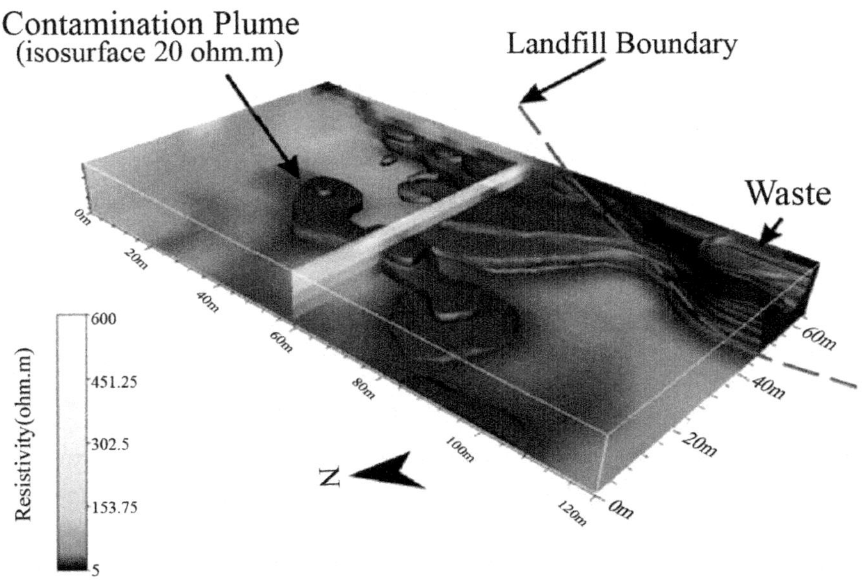

Figure. 5: 3D resistivity model of a contaminant plume generated by a municipal solid waste disposal site. The isosurface of 20 ohm.m delineates the contaminant plume, which is moving in a northwest direction.

The equipment is built to allow a direct reading of the apparent conductivity. The most commonly used equipment at present time was designed to explore pre-defined depths, between 7.5 to 60.0 meters, depending on the coils orientation (vertical loop and horizontal loop modes), the operating frequency and the intercoil spacing. The field tests are usually conducted in profiles, which, due to convenience of operation and transportation of the equipment, are carried out very quickly. The apparent conductivity data can be plotted on several profiles and depending on the distance, a set of profiles can form a site map. The interpretation of these data is basically qualitative, but nowadays there is inversion software designed for quantitative interpretation of conductivity profiling data (Monteiro Santos, 2004).

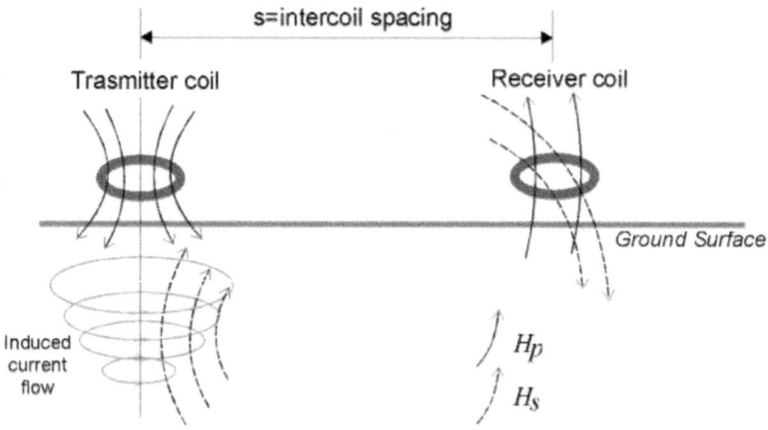

Figure. 6: Operating principle of the Ground Conductivity Meter.

The method is a cost effective and low time-consuming tool for investigating contamination generated by waste disposal sites, and allows delineation of the area affected by contaminants. Figure 7 shows the mapped out area of the contamination plume delineation, which has been created by a municipal solid waste disposal site. The MSW boundaries are marked in a broken white line (black points are measurement stations). The northwestern part of the landfill presents low electrical conductivity, characterizing a granitic base. The site is occupied by waste deposited onto tertiary stream sediments, and the groundwater level appears at around 5 meters depth. The contours of the waste disposed at the site can be identified based on the values of apparent electrical conductivity higher than 50 mS/m. The contamination plume can also be observed in this figure, which flows to the West, marked by high values of apparent conductivity outside of the waste filled area.

INVASIVE TESTS

In some particular sites it is difficult to identify the subsoil properties contrasts based only on geophysical data or index properties, especially in a very heterogeneous municipal solid waste (MSW) disposal site on unsaturated tropical soils. The proper site investigation in these sites relies on invasive tests, including soil, groundwater and direct gas sampling. The modern approach for geo-environmental site characterization using invasive tests relies on the piezocone technology. The piezocone and it accessories are used as a screening tool for stratigraphic logging of geotechnical and chemical measurements. This approach often allows identifying potentially critical zones which may require more specific tests to measure or monitor contaminants based on specific sampling and/or monitoring.

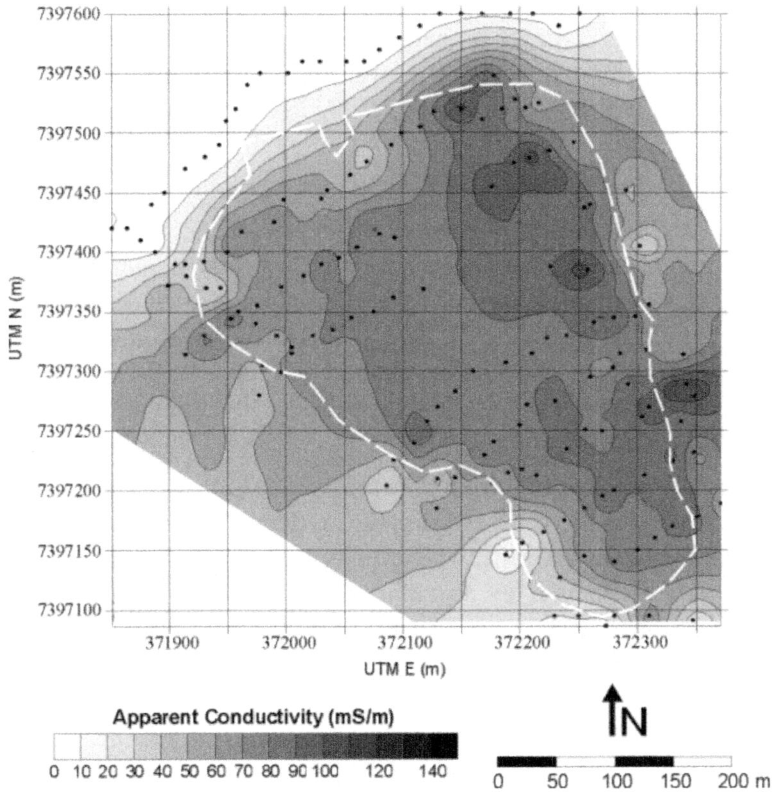

Figure. 7: Apparent conductivity map of a waste disposal site. Theoretical depth of exploration is 7.5 meters. The broken white line marks the limits of the deposit. Note the good match of the boundaries of the deposit with the high apparent conductivity values (> 50 mS/m).

Piezocone Test and Accessories

The Piezocone

Piezocone penetration test (CPTU) is a standard site investigation tool currently used for geotechnical site characterization (Lunne et al, 1997). According to Shinn II & Bratton (1995), the heart of a modern site characterization approach uses the piezocone penetration test. The CPTU test uses a standard instrument probe (ASTM D 5778-07) with a 60° apex and typically 35.68 mm diameter (10 cm2 area) fitted on the end of a series of rods (Figure 8). The probe is pushed into the ground at an approximate constant rate of 2 cm/sec by a hydraulic pushing source, such as a standard drill rig or cone pushing vehicle. As the cone is advanced, the forces measured by the tip and friction sleeve will vary with the material properties of the soil being penetrated. Tip resistance (qc), friction sleeve stress (fs) and dynamic pore-pressure (U) response are measured by calibrated electrical instruments. All channels are continuously monitored and typically digitized at 10, 20, 25 or 50 mm intervals.

Piezocone Data and Interpretation

Piezocone test data is gathered by a computer which allows the user to carry out straightforward post-investigative analysis. The three parameters (q_c, R_f and U) in various combinations, such as friction ratio ($R_f = (f_s/q_c)100\%$), are used to delineate site stratigraphy by using soil classification charts, as proposed by Robertson et al. (1986) and presented in Figure 9. Empirical and semi-theoretical correlations are available in relevant literature in order to estimate mechanical properties of the soil.

Measurement of equilibrium pore pressure at full pore pressure dissipation allows quantifying the vertical hydraulic gradients by using a single sounding and groundwater flow regime, whenever multiple soundings are available. The pore pressure dissipation test also allows estimation of hydraulic conductivity (Campanella, 2008).

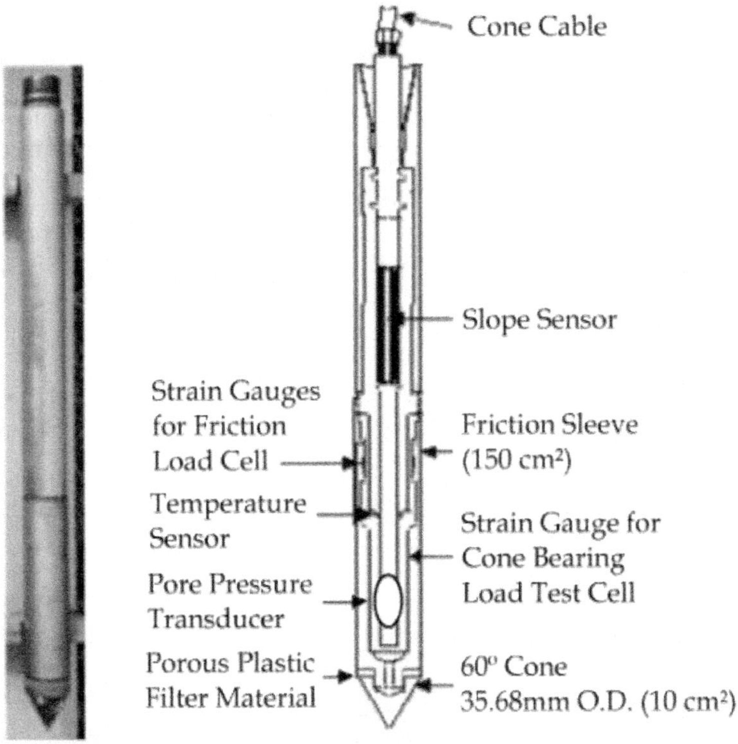

Figure. 8: Schematic drawing and photo of a piezocone probe (modified from Davies & Campanella, 1995).

Accessories

As pointed out by US EPA (1989), a geo-environmental site characterization requires information on the chemical distribution and sources (s)/receptor(s) for potential or existing contaminants. The piezocone technology for geo-environmental application included specific sensors for temperature, resistivity, pH, laser-induced fluorescence, among others. This technology also includes samplers to be used together with the piezocone for sampling soils, water and gas (Lunne et al, 1997). Robertson & Cabal (2008) present in details the main piezocone accessories available for geo-environmental site characterization. The use of the piezocone test and accessories for geo-environmental applications essentially creates no cuttings, produces little disturbance, and reduces contact between field personnel and the contaminants, as the penetrometer push rods can be decontaminated during retrieval (Robertson, 1998). The major limitation of the piezocone test is the impossibility of carrying out the test in gravel.

Resistivity Piezocone

Resistivity is one of the piezocone accessories which is called the resistivity piezocone (RCPTU) test. It measures the electrical resistance of a current flow in the ground. This additional information is extremely useful due to the significant effects that dissolved and free product constituents have on bulk soil resistivity (Campanella, 2008).

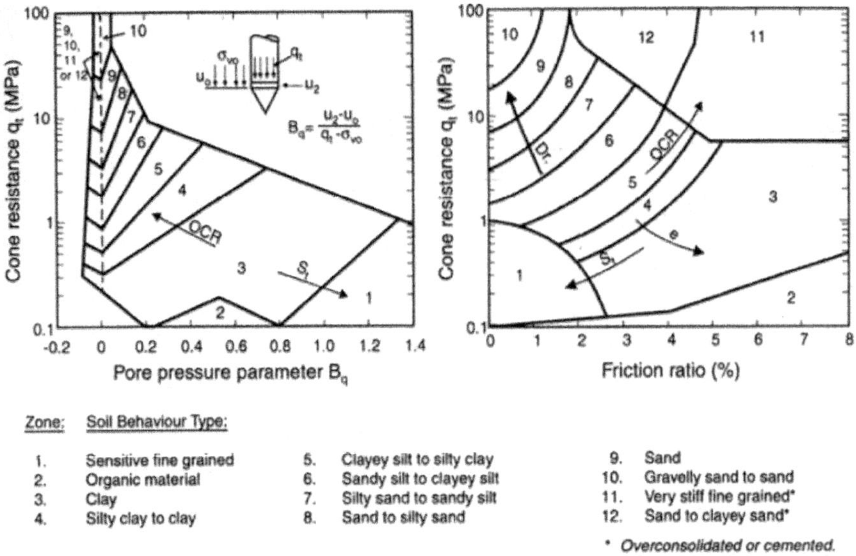

Figure. 9: Piezocone soil behavior type charts proposed by Robertson et al (1986).

The preparations involved in the RCPTU test are similar to those of any other CPT test (Robertson & Campanella, 1988). The only additional procedure is the connection of a signal generator to the data acquisition system for controlling the current level and frequency of the electrical resistance measurements. Weemees (1990) discusses the importance of working with alternating current and sufficiently high frequencies to prevent polarization of the external electrodes. A frequency of 1000 Hz is usually used. Figure 10 presents a Wenner-type resistivity piezocone (RCPTU) with an array of four electrodes. Resistivity measurements are taken with the inner electrodes and the current is applied through outer electrodes. These measurements are digitally recorded at 25 mm intervals, providing essentially continuous in-situ data sampling in addition to all the other standard CPTU measurements. Measurements of bulk resistivity trends indicate whether some dissolved or free product exists below or above the background values. The background values are usually established from RCPTU tests carried out on-site. According to Campanella (2008), the areas where readings are very different (anomalies) from the background values are then further evaluated with appropriate groundwater sampling at discrete depths for detailed chemical analysis. Considerable practical value is gained from the fact that the measured resistivity in saturated soil is almost totally governed by the pore fluid chemistry.

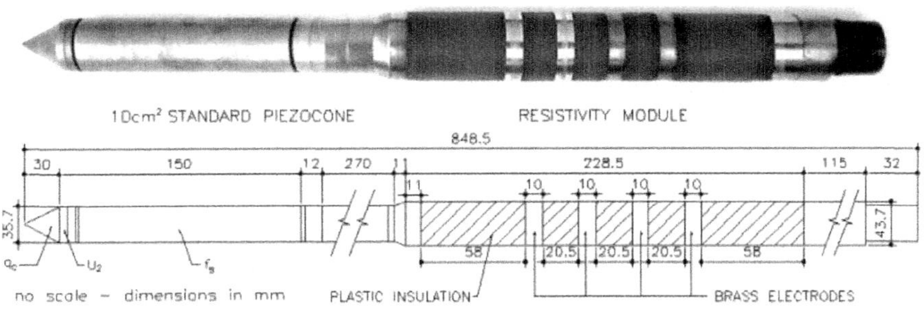

Figure. 10: Schematic representation and photo of a resistivity piezocone probe with a fourelectrode arrangement (Mondelli et al., 2007).

Samplers

A large number of samplers for gas, water and soil sampling have been developed for pushing into the ground using the same equipment that is used for the piezocone test. These samplers are designed for either one-time samples or as monitoring wells. Robertson & Cabal (2008) describe in details some samplers available for geo-environmental site investigation and they will be briefly presented in the follow items.

Soil Sampling

A wide variety of push-in discrete depth soil samplers are available and most of them are based on designs similar to the Gouda soil samplers. Robertson & Cabal (2008) describe some soil samplers including the Gouda type sampler. It is pushed into the ground at the desired depth in a closed position and it has an inner cone tip that is retracted to the locked position leaving a hollow sampler with small diameter (25 mm) stainless steel or brass sample tubes. The hollow sampler is then pushed to collect a sample. The filled sampler and push rods are then retrieved to the ground level. Figure 11 shows a schematic drawing of a typical CPT based soil sampler (Gouda type), as described in detail by Robertson & Cabal (2008).

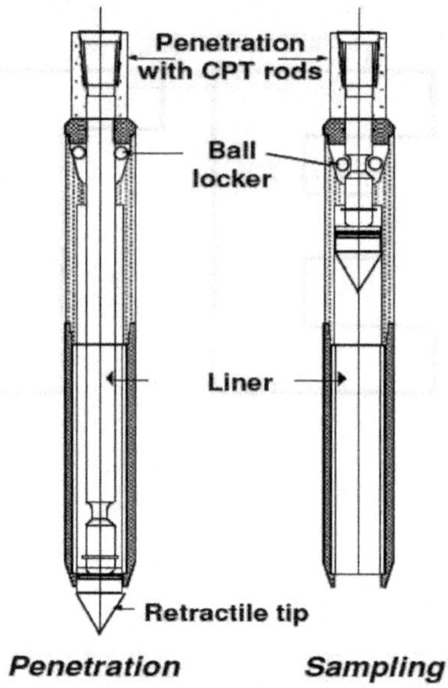

Figure. 11: Schematic drawing of a typical (Gouda type) CPT based soil sampler.

Water Sampling

Robertson & Cabal (2008) affirm that the most common direct-push, discrete depth, in-situ water sampler is the Hydropunch and to a lesser extent, the BAT, Simulprobe and Waterloo Profiler. The Hydropunch sampler and its variations is a simple sampling tool that is pushed down to the desired depth and the push rods withdrawn to expose the filter screen and is described in detail by Robertson & Cabal (2008). A modification of the commercially available BAT groundwater sampler is recommended by Campanella (2008) for obtaining in-situ pore fluid samples and it is presented in Figure 12. This author discribes the original BAT system, which consists of a sampling tip that is accessed through sterile evacuated glass sample tubes and a double-ended hypodermic needle set-up pushed through septum seals. The tube sampler is lowered either by cable or electrical wire depending upon whether a pore fluid sample is taken with or without a pressure test being carried out. Acording to Campanella (2008), the BAT probe is also able to take pore gas samples for collecting volatile contaminants.

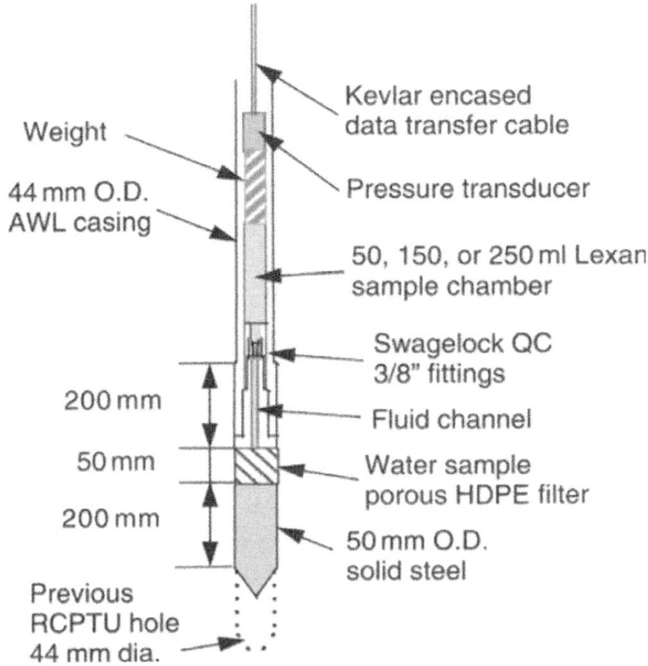

Figure. 12: UBC modified BAT groundwater sampler (Campanella, 2008).

Gas Sampling

Gas samples can be collected in a similar way to the one previously described for groundwater samples. The most common gas sampler that uses the direct-push, discrete depth, in-situ method is the hydropunch type sampler. Robertson & Cabal (2008) also describes this sampler in detail, which is pushed to the required depth, thus the filter element is exposed and a vacuum is applied to draw a vapor sample to the surface. Special disposable plastic tubing is used to draw the sample to the surface.

Example of Using RCPTU Test and Samplers in a MSW Disposal Site

An example of using the piezocone technology to assess contamination of a MSW disposal site from Brazil, by using different techniques is going to be briefly presented based on Mondelli's (2008) research. The site can be considered to be a controlled dump site because it is a planned landfill that incorporates some of the features of a sanitary landfill. According to Mondelli et al. (2007), the site's geology has sandstone from the Adamantina and Marilia Formations, covered by alluvial sandy soils or colluvial clayey sands. Residual soils from sandstone are found underneath these layers. The hydraulic conductivity of the soil was found to vary from 10-7 to 10-6 m/s. The depth of the groundwater level is about 5 meters below the base of the landfill. In order to protect the subsurface and the shallow groundwater table from leakages from the landfill, four 20-cm-thick layers were compacted at the bottom of the landfill using local soil. The upper surface of these layers was coated twice with diluted asphalt emulsion to seal and protect the bottom of the landfill. This procedure was used because it was inexpensive and acceptable for controlled dumps in Brazil when this landfill was established. This site was first investigated using a 3-D resistivity imaging technique, as described by Ustra et al. (2011). Several piezocone and resistivity piezocone tests were carried out on this site. In the first campaign, a four standard piezocone was carried out, measuring cone tip resistance (q_c), sleeve friction (f_s) and pore pressure (u_2). A resistivity piezocone was used in the other campaigns (16 tests). Pore pressure was recorded using a slot-filter filled with automotive grease in all the tests, as suggested by Larsson (1995). A multifunction reaction system equipped with a hydraulic device with a capacity of 200 kN was used to carry out the tests. Details of all the different tests carried out in this site are presented by Mondelli et al. (2007). Two piezocone tests were conducted at the highest parts of the landfill, to obtain a reference profile of the uncontaminated soil. However, resistance to cone penetration was greater than the capacity of the penetration system needed for reaching the groundwater level in this region, where residual soils and sandstone are shallower. Most of the RCPTU tests were performed downstream of the landfill, where sedimentary soils or more homogeneous tropical soils occur. All the RCPTU tests that reached groundwater presented an abrupt reduction in resistivity. This information was useful to help identifying groundwater level (GWL). Two of the RCPTU test results are presented in Figure 13 and will be briefly discussed. The resistivity profiles shown in this figure for the saturated zone, indicate that the resistivity measurements are affected by soil texture and mineralogy. Mondelli et al. (2007) point out that the resistivity values were higher in sandy layers than in clayey layers. At that time the influence of soil type on the resistivity values of the RCPTU-14 and RCPTU-15 tests were not so clear, since the resistivity value found in the RCPTU-14 test was around 50 ohm.m, while that of the RCPTU-15 test was around 20 ohm.m for saturated zone. It indicates a migration of the contamination plume through the sandy layer, based on water sampling using a direct-push sampler at depths of 8.0 to 9.0 m, since it presented low electrical resistivity, which is indicative of the presence of contaminants dissolved in the water. Mondelli et al. (2007) conclude that the RCPTU interpretation for this particular site required collection of soil and groundwater. It was also necessary to adapt a system to measure the electrical resistivity of undisturbed soil samples in the laboratory for better understanding the resistivity variation according to the different saturation conditions and lithological characteristics that occur in tropical regions. This system, as well as the test results, is presented by Mondelli et al. (2010a). The sampling depths were selected based on the interpretation of nearby piezocone and other tests. Mondelli et al. (2010b) present in Figure 14 the relation between the resistivity values of the saturated layers and the results of the characterization tests on soil samples collected using a direct-push sampler, in order to allow identifying and visualizing the layers susceptible to contamination. These results indicate a tendency for increasing values of resistivity with decreasing fines content, clay content or with clay mineral activity. However, layers that do not clearly follow this tendency are also shown in Figure 13, presenting low values of resistivity (between 20 and 40 ohm.m) for the sandier soils. These results demonstrate that values of resistivity ranging from 20 to 40 ohm.m for predominantly sandy layers indicate the presence of leachate in the groundwater. This kind of interpretation goes beyond the merely comparative one of the resistivity values obtained from all the tests discussed by Mondelli et al. (2007), which had diagnosed the presence of contaminants only in tests RCPTU 14 and 15 presented in Figure 13. A certain degree of superposition of values may occur in this type of analysis, since the more clayey layers also show low values of resistivity. Since the flow of contaminants tends to occur between the sandier layers (more permeable), the problem can be simplified in an attempt to identify only these layers.

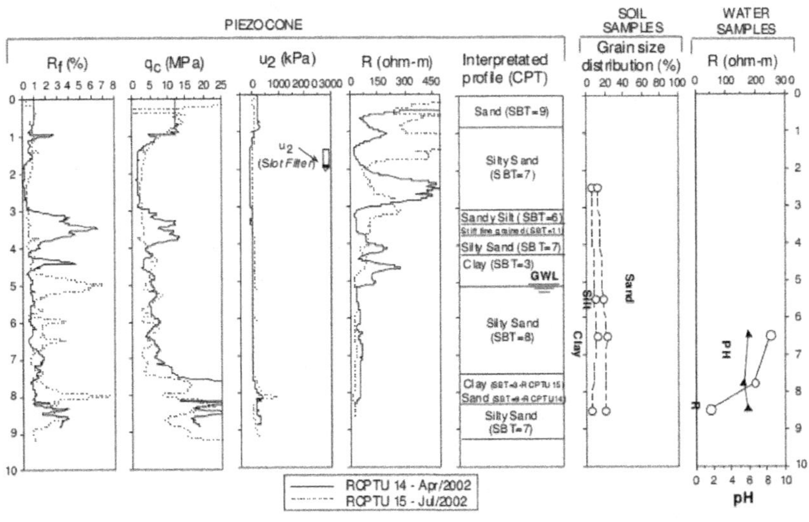

Figure. 13: RCPTU test results, grain size distribution and groundwater analyses for a MSW disposal site, adjacent to the edge of the landfill (Mondelli et al., 2007).

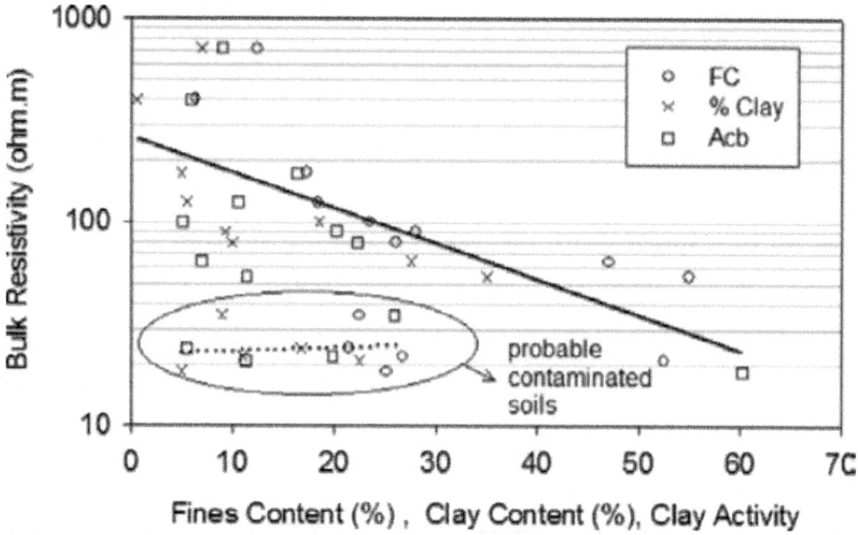

Figure. 14: Electrical resistivity in saturated soil samples with characterization of the fine fraction of soils for a MSW disposal site (Mondelli et al., 2010b).

Mondelli et al. (2010b) studies concluded that the resistivity piezocone test helped identifying contaminated zones. However, this technique presented some limitations for tropical soils, since the groundwater level is sometimes deeper than the penetrable capacity of the cone. The RCPTU tests were useful for detailing stratigraphic soil profile. The relation between corrected point resistance (q_t) and friction ration (R_f) allowed identifying the highly heterogeneous stratigraphic profiles of alluvium and colluvium. This identification for the residual soils was restricted to the behavior of the materials in response to the penetration of the piezocone, requiring soil sampling. The resistivity profiles were useful for the identification of the position of the groundwater level, and were sensitive to variations in the soil's texture and its saturation. A good relationship between the soil behavior index (I_c) and the fines content in the soil samples collected adjacent to the landfill helped the identification of contaminated regions of the saturated zone. The proper interpretation of piezocone tests enabled the identification of those layers more susceptible to contamination, which was confirmed by the chemical analyses of the groundwater samples collected from the monitoring wells. The main conclusion of this study is that the interpretation of RCPTU tests is not as straightforward as it is for sedimentary soils because the tropical soils genesis complexity.

LABORATORY TESTS FOR GEO-ENVIRONMENTAL SITE CHARACTERIZATION

Laboratory tests can improve the interpretation of the in-situ test results, complementing, controlling and detailing the geo-environmental site investigation. A complete geoenvironmental site investigation using current engineering practice is usually difficult to carry out because it is expensive and normally requires authorization and multidisciplinary teams. In addition, an appropriate testing campaign is time consuming and only a few private companies can do it. Limited publications on monitoring and modeling of the contamination plume caused by MSW disposal sites are available. The most commonly known case histories (Cherry et al., 1983; MacFarlane et al., 1983; Mackay et al., 1986; Kjeldsen et al., 1998; Zuquette et al., 2005) have installed more than a hundred monitoring wells, with multilevel soil and groundwater sampling, and are taken from the same sites which have been studied for years. All these studies show that geophysical tests were carried out before the intrusive tests, which were also performed to confirm contamination.

Christensen et al. (2000) explain that, contrary to the specific and known sources of contamination, – such as leaking fuel tanks (hydrocarbons) and chlorinated solvents – the municipal solid waste disposal sites are large, heterogeneous and receive various types of waste over time, have different flow pathways, hydraulic gradients and contamination plumes. Moreover, the size of the landfills will often prohibit removal of the waste and source of the leachate, suggesting that the landfill body and the leachate plume should be considered as a continuum in the context of natural attenuation. In Brazil, most of the laboratory tests for sanitary landfills are preventive, for bottom or cover liner design, using low permeability compacted tropical soils with or without flexible membrane liners (Ritter, 1998; Boscov et al., 2001; Leite & Paraguassú, 2002; Stuermer, 2005; Azevedo et al., 2006). This approach does not consider the older waste deposits, where the contaminants are in direct contact with the natural subsoil. In Brazil, there are some researches on geo-environmental site investigation of industrial and municipal solid wastes disposal sites, using geophysics or boreholes to detect contamination plumes (Grazinolli et al., 1999; Elis & Zuquette, 2002; Lago et al., 2003; Anirban et al., 2004; Bolinelli Jr., 2004; Porsani et al., 2004;). The in-situ techniques permit assessing the natural conditions of the soils, while the laboratory tests provide greater detail, but in minor scale, trying to represent the field conditions based on soil and water/gas samplings. Those difficulties in defining a typical soil profile, hydrogeological parameters and background resistivity (or conductivity) values due to geological complexity (like tropical sites, for example) demand more and varied site investigation techniques. After a better understanding of the site history, topography, hydrogeology and the landfill construction characteristics, laboratory tests can be carried out using disturbed and undisturbed soil samples, in order to investigate a new site for the installation of a sanitary landfill, or to confirm a pre-indicated contamination or to provide support for future remediation actions. The typical soil profile, hydrogeological characteristics and background values can be confirmed and obtained also using laboratory tests. The soil can be characterized using geotechnical, geochemical and mineralogical techniques. Batch sorption and column leaching tests can be carried out using those geomaterials found at the study site. The landfill's leachate can be used to estimate the dispersion and retardation coefficients, for future contamination flow transport modeling. Electrical resistivity values and mineralogical constituents of non-contaminated soils can be determined in the laboratory.

Soil Characterization

The characterization tests are fundamental for the classification of soils, due to their complex geological formation, and even more when performing a geo-environmental site investigation, where the interaction between the contaminants and the physical environment usually occurs. The more usual laboratory characterization tests to investigate a municipal solid waste disposal sites are:

- Geotechnical Properties (Lambe & Whitman, 1969): classical laboratory tests for physical soil characterization are essential for geo-environmental site investigation. Indexes like in-situ moisture content (w), natural unit weight (g), specific gravity (Gs), grain size distribution with and without sodium hexametaphosphate for determination of clay, silt, sand and gravel contents, liquid limit (LL) and plastic limit (PL), shall be determined using disturbed and undisturbed soil samples. Also, permeability / hydraulic conductivity (K) tests are necessary for a good monitoring and pollution prevention of a waste disposal site or for future sanitary landfill, respectively. Depending on the groundwater position and the characteristics of the site, tests to determine the soil retention curve can also be important in a site investigation and monitoring program.
- Organic Matter Content (OMC): OMC is considered an important property of the soil solid phase, as it is responsible for retaining a good part of the organic contaminants and leachate ions, together with clay minerals and hydroxides. OMC can be determined by incineration of the soil sample in a muffle, heating to temperature of 440oC, as described in ABNT NBR 13600 (1996).

- Blue Methyl Adsorption: this test allows the estimative of the cation exchange capacity (CEC), specific soil surface (SE) and activity of clay minerals. This test also indicates the predominant type of clay mineral matrix in fines content of the soil sample. Lan (1977) and Pejon (1992) describe this test method in detail.
- X-Ray Diffraction (XRD), Differential Thermal Analysis (DTA) and Gravimetrical Thermal Analysis (GTA): These tests allow characterization of the soil fines content, by heating up the soil sample to 1000oC, in order to identify and quantify different clay minerals, hydroxides and any solid constituent of the soil. Consequently, the tests can be used to identify the soil mineralogy and its geochemical interaction with the landfill, or for assessing natural attenuation of the contaminants.
- pH: Determination of pH of granular materials in deionized water (usually ratio of 1 : 2.5), by measurement of the effective electrochemical concentration of H+ ions in soil solution or waste using a combined electrode immersed in suspension. The test method is that one recommended by US EPA (1993).
- X-Ray Fluorescence (XRF): XRF spectrometry is used to identify elements in a soil and quantify the amount of those elements (metals) present. An element is identified by its characteristic X-ray emission wavelength (λ) or energy (E). The amount of an element present is quantified by measuring the intensity of its characteristic line. It is known that original rock composition can or cannot present natural high concentrations of some metals. For geo-environmental purposes, these conditions cannot exceed safety, health or potability exigencies for the local population. Therefore, it is noted that the levels of these metals in soils and sediments depend on the composition and proportion of them in their original solid phase. The background values need to be diagnosed before or after the waste disposal. Innov-X Systems Alpha Series (2007) presents handheld EDXRF spectrometers, ideally suited for field and laboratory analysis of soil and waste metals.

As example, Figure 15 presents a typical soil profile, characterized for construction of a future waste disposal site for a very small city in the interior of the São Paulo State, Brazil. IPT (2011) carried out a low cost geo-environmental site investigation campaign with the objective of detecting possible contamination caused by an older controlled dump near to the interest site. The results presented in Figure 15 shows the log of the number of blows from Standard Penetration Test (N_{SPT}) up to 16.5 m depth, when the impenetrable layer was reached. During the SPT tests, deformed soil samples were collected, in order to maintain the natural moisture, and were transported to the laboratory. In the laboratory soil samples were identified based on tactile visual inspection. So, "new"soil samples were composed using the same lythology. The collected soil samples were characterized, determining moisture content (w), organic matter content (OMC), pH, electrical conductivity, grain size distribution and XRF analysis using an Innov-X handheld EDXRF spectrometer. Figure 15 shows that the soil profile is formed by a very loose to loose sand, up to about 9.5 m depth. Between 9.5 and 13.0 m the subsoil has a medium compacity (9 < N_{SPT} 20. Residual soil from sandstone was found below 15.0 m depth. The moisture content was around 7-8 %, which changed just when the groundwater level was reached, at 13.0 m depth, increasing to 17 %.

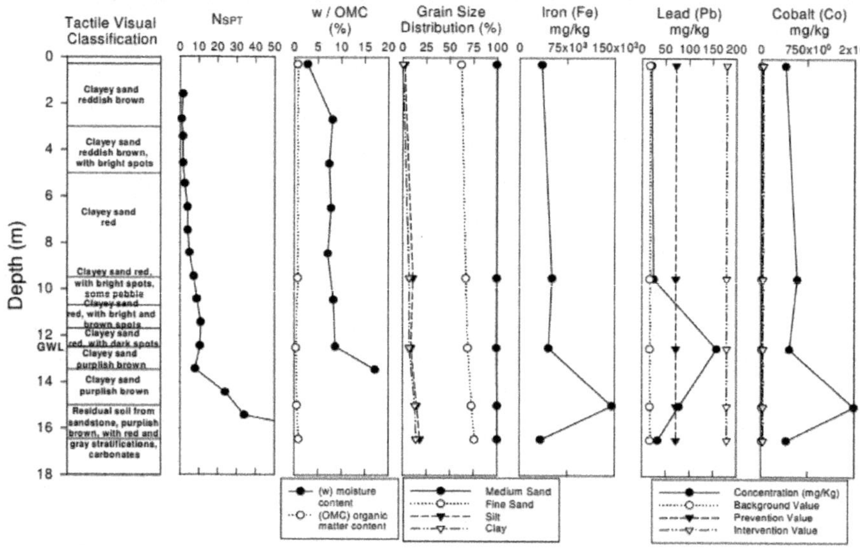

Figure. 15: Soil characterization profiles for a geo-environmental site investigation to install future sanitary trenches in a small city from Brazil.

The local soil is basically constituted of a fine to medium sand. Clay content tends to increase with depth, ranging from 0 to 16%. The organic matter content was found to be very low, ranging between 0.7 and 0.9%, tending to decrease to 0.3 to 0.4%, for the residual soil from sandstone. The geochemical profiles obtained for iron, cobalt and lead are also presented in Figure 15. They have to be compared with the background, prevention and intervention values determined for agricultural use by the Environmental Company from São Paulo State (CETESB, 2005). There is no reference value published for iron in the São Paulo State, Brazil. It is observed that all parameters have reached peak concentrations when residual soil from sandstone is encountered. The cobalt value is greater than the intervention value and lead levels are greater than the reference value throughout all the subsoil profiling. The lead level is also greater than the intervention value for the residual soil. These high levels of Co, Pb and Fe may be associated with the constitution of the natural bedrock and in the case of where iron, iron oxides and hydroxides appear, they work like a cementing constituent for the quartz particles. The electrical conductivity (EC) values measured for the local soil were very low, ranging between 10.4 and 18.0 μS/cm, indicating that the soil is in its natural state, without ever coming into contact with leachate (highly conductive) from the older existing waste dump.

Electrical Resistivity

In order to assess the changes in the resistivity values with the degree of saturation, contamination and mineralogy of some typical geomaterials found at waste disposal sites, undisturbed soil samples and leachate can be carefully selected and sampled around a particular site, for a controlled and detailed study in the laboratory, in order to interpret the in-situ resistivity maps and profiles obtained during geophysical campaigns. Mondelli et al. (2010a) present laboratory electrical resistivity measurements to detect the contamination plumes surrounding and below a waste deposit. These authors measured electrical resistivity (R) of natural undisturbed samples as well as specimens saturated with distilled water, salt solution and original leachate from a MSW disposal site. While the specimens were naturally air dried during a specific period, the resistivity and the degree of saturation were measured. The system consists of an insulating material (PVC) mold, with two stainless-steel cylindrical electrodes, one on the top and another on the bottom of the specimen for electrical current measurement, similar to the one developed by Daniel (1997). Two more 3 mm diameter stainless-steel electrodes were attached into the middle of the specimen, observing the same distance between them (Wenner array) for differential potential transfer. For data acquisition, a Syscal Pro equipment was used. Figure 16 presents the apparatus used.

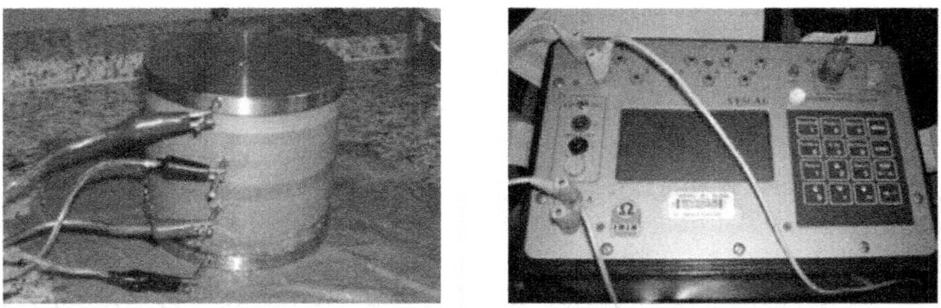

Figure. 16: Instrumented undisturbed specimen and Syscal Pro equipment used for laboratory electrical resistivity measurements (Mondelli et al., 2010a).

Mondelli et al. (2010a) concluded that the laboratory test results supported the interpretation of the in-situ test data, identifying the contamination spots, plume depths, capillarity and saturation zones and the different types of local soil. The results demonstrated the high influence of mineralogy and degree of weathering of the tropical soils on electrical resistivity values, also suggesting reference values for the study site. Figure 17 presents the results obtained for samples percolated with leachate (EC = 25,000 μS/cm and R = 0.4 ohm.m). At the beginning of the test, the resistivity values did not exceed 10 ohm.m until a 50 % degree of saturation was achieved. Below this degree of saturation, resistivity increased exponentially. Due to the high electrical conductivity of the leachate, the resistivity values became very low, with no influence of mineralogy in this case. These results could be interpreted using Archie's Law (Archie, 1942), as studied by Mondelli (2008).

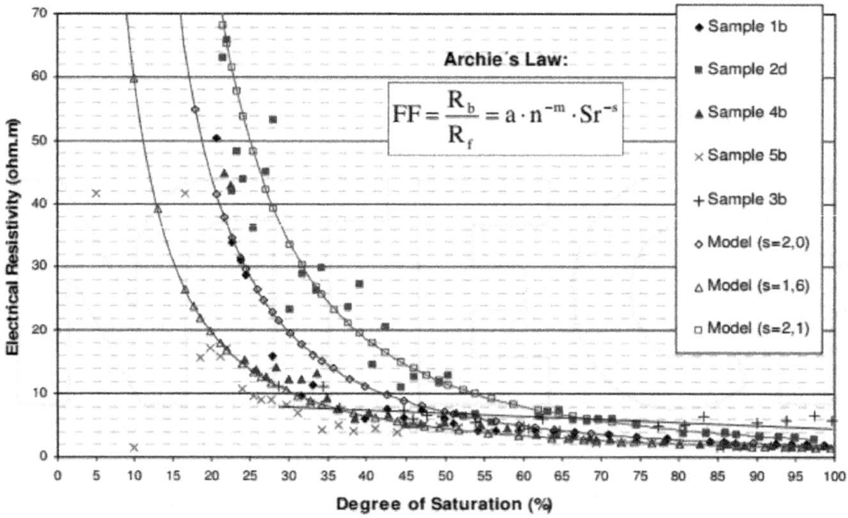

Figure. 17: Electrical resistivity versus degree of saturation for different soil samples percolated with the leachate from a MSW disposal site from Brazil (Mondelli, 2008).

Pollutant Transport Parameters

Estimates of pollutant transport parameters based on laboratory tests such as column, diffusion and batch equilibrium tests are also important in the more detailed phases of geoenvironmental investigations (Freeze & Cherry, 1979; Rowe et al., 1988; Barone et al., 1989; Shackelford & Daniel, 1991; Yong et al., 1992; Shackelford, 1993; ong, 2001). These laboratory tests can be conducted with the objective of a better interpretation of the in-situ test results, and for a numerical modeling performance, using a good conceptual model definition for the problem. Compacted or undisturbed soil samples can be used to assess the pollutant transport parameters, targeting the contamination plume behavior when in contact with the natural environment, using equipment and methods which allows the study of the anisotropy, degree of saturation and chemical extraction. When the determination of pollutant transport parameters is being planned, one of the most important aspects to be defined is the interest pollutants or solutions to be used during the laboratory tests. For geo-environmental investigation of MSW disposal sites, original or artificial leachate can be used. The detection of a concentration of several constituents of leachate (like heavy metals, metals, chloride, organic compounds, chemical oxygen demand and biological oxygen demand), solutions and effluents collected during the tests, need to be chemically analyzed, using emission or atomic absorption spectrometers, flame photometry, chromatography and any other appropriate techniques. The preparation and preservation proceeds to detect any kind of chemical compound need to follow the standards described in APHA, AWWA, WEF (1995).

The laboratory apparatus for column tests also needs to be planned, depending on the sample state (compacted or undisturbed), the hydraulic gradients involved defined in accordance with the field conditions and if the column system will use rigid or flexible wall permeameters. All components of the apparatus were made of non-reactive materials. This test estimates the hydraulic conductivity (k), hydrodynamic dispersion coefficient (D_{hl}) and retardation factor (R_d), due to pollutant solutions or leachate percolation through the soil sample. Once the breakthrough curves (relation between solute concentration and initial concentration – C/C0 in function of the soil pore volumes percolated with the pollutant solution – T) are obtained for each interest solute, considering predominantly advective transport, the hydrodynamic dispersion coefficient (D_{hl}) and the retardation factor (R_d) can be assessed using the following equation proposed by Shackelford (1994):

$$\frac{C}{C_0}=\frac{1}{2}\left\{\operatorname{erfc}\left[\frac{1-(T/R_d)}{2\sqrt{\frac{(t/R_d)}{(uL)/D_h}}}\right]+\exp\left[\frac{vL}{D_h}\right]\operatorname{erfc}\left[\frac{1+(T/R_d)}{2\sqrt{\frac{(t/R_d)}{(uL)/D_h}}}\right]\right\} \tag{4}$$

where: t = test time; T = time factor or soil pore volumes percolated with the pollutant solution; u = Darcy'Law specific velocity; L = height of the specimen; (uL/D_h) = Peclet number.

For non-misceble and reactive solutes, the retardation factor (Rd) can be obtained from batch sorption tests. This test can be carried out following ASTM D 4646-03 and ASTM C 1733-10, standards, using a previously defined soil-solution ratio, as 1:4 (25 g of dry soil sieved through a 40 mesh to 100 mL of leachate or pollutant solution). Rd is given by equation 5:

$$R_d = 1 + \frac{\rho_d}{n} \cdot K_d \tag{5}$$

where: ρ_d = soil specific dried mass; n = soil porosity; K_d = partition coefficient.

Partition coefficient (Kd) represents the angular coefficient of the linear sorption isotherm, defined for an interest solute, during batch sorption tests. For the construction of the sorption isotherms, the amount of soil or solute concentration in the solution can change. When original leachate is used during the test, the solute concentrations remain constant. After a previously determined equilibrium time, the supernatants are filtered, adequatly preserved and stored at 4°C until the chemical analysis, for a limited time, depending on the interest solute. The results are expressed as the adsorption degree, as a function of the concentration. The adsorption degree (S) is defined by the following equation:

$$S = \frac{(C_0 - C_e) \cdot V}{M} \tag{6}$$

where: C_0 = initial concentration of the solution; C_e = equilibrium concentration; V = volume of the solution in the flask; M = dry soil mass.

Frempong & Yanful (2008) and Mondelli (2008) studied the sorption capacity of tropical soils from Ghana (West Africa) and São Paulo State (Brazil), using original leachate from local MSW disposal sites, and batch and column tests. Both these studies showed the following ion sorption selectivity order: $Ca^{2+} < Cl- < Na^+ \leq Br- < Zn^{2+} < Fe2 < K^+$. A significant finding from these studies is the observation that kaolinite and aluminum and iron oxyhydroxides with variable particle surface charges present in the soil allowed sorption of anions, such as C_l- and B_r- , generally considered conservative nonreactive, based on leachate-liner compatibility studies on soils from temperate regions. Frempong & Yanful (2008) found Rd ranging from 1.1 to 47.9 for Na, K, Br and C_l for clayey tropical soils. Mondelli (2008) found major values of Rd, ranging from 1 to 383 for N_a, K, Fe and Z_n for sandy tropical soils, with 20 % of clay content. Ritter (1998) classified the solute as essentially immobile when K_d is higher than 10. On the other hand, R_d values display the same order of magnitude as those obtained by Nascentes (2003) with similar soil and range of tested concentrations. Azevedo et al. (2003) point out that the high Rd values obtained by Nascentes (2003) can be the result of the low concentrations of metals used in the pollutant solutions. Figure 19 presents the sorption isotherm obtained for potassium (K) by Mondelli (2008), indicating that adsorption degree (S) increases along with the increase in equilibrium concentration (C_e). In this case, the linear isotherm was the one that best described the results, presenting a better fit to the testing data and K_d value of 29.8 mL/g.

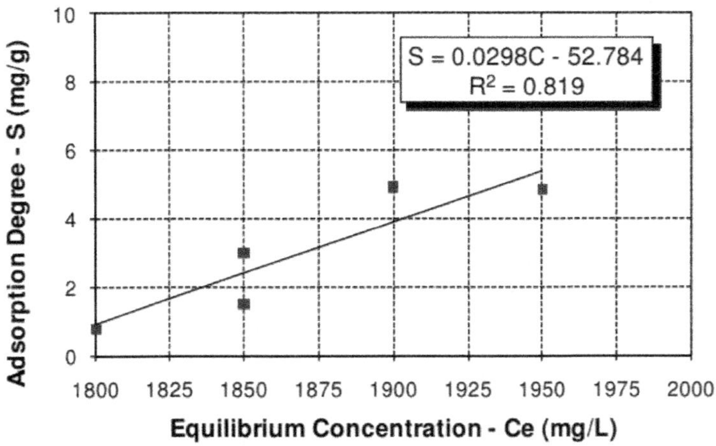

Figure. 18: Linear sorption isotherm obtained for potassium from leachate and a soil sample taken from a MSW disposal site in Brazil (Mondelli, 2008)

CONCLUSION

The chapter shortly presented and discussed the steps of a geo-environmental site investigation program for municipal solid waste disposal sites, and also, different and modern in-situ and laboratory test techniques for this purpose. A literature research was conducted on different areas of knowledge, comprising the current problems on the inadequate disposal of waste, environmental liabilities and techniques for the investigation of contaminated sites. Research results on Environmental Geotechnics were also included, aimed at the appropriate monitoring and proper installation of sanitary landfills. Therefore, it can be concluded that it is fundamental to know exactly all the goals, local interests, background factors and resources involved in each case, for an effective and optimized geoenvironmental site investigation. This approach will provide positive results for the local population and for the physical environment.

ACKNOWLEDGMENTS

The authors gratefully acknowledge the Brazilian research agencies FAPESP (State of São Paulo Research Foundation), CAPES (Brazilian Federal Agency for Support and Evaluation of Graduate Education) and CNPq (National Council for Scientific and Technological Development) for funding their researches.

REFERENCES

1. ABNT NBR – 13600 (1996). Solo – Determinação do Teor de Matéria Orgânica por Queima a 440°C. Brazilian Association for Technical Standards, São Paulo-SP, Brazil.
2. APHA, AWWA, WEF (1995). Standard Methods for the Examination of Water and Wastewater. American Public Health Association, American Water Works Association, Water Environmental Federation. Managing Editor Mary Ann H. Franson. 19th Edition, U. S. A., pp. 3-1 – 3-2.
3. ASTM D 5778-07. Standard Test Method for Electronic Friction Cone and Piezocone Penetration Testing of Soils. ASTM International, Pennsylvania, U.S.A.
4. ASTM D 4646- 03 (Reapproved 2008). Standard Test Method for 24-h Batch-Type Measurement of Contaminant Sorption by Soils and Sediments. ASTM Intern., Pennsylvania, U.S.A.
5. ASTM C 1733-10. Standard Test Method for Distribution Coefficients of Inorganic Species by the Batch Method. ASTM Intern., Pennsylvania, U.S.A.
6. Anirban, D. E.; Matasovic, N. & Dunn, R. J. (2004). Site Characterization of Five Hazardous Waste Landfills. Proceedings of the 2nd International Conference on Site Characterization, Porto, Portugal, September 2004, pp. 1075-1080.
7. Archie, G. E. (1942). The Electrical Resistivity Log as an Aid in Determining Some Reservoir Characteristics. Transactions of American Institute of Mining and Metallurgy Engineering, Vol.146, pp. 54-62.
8. Azevedo, I. C. D.; Nascentes, C. R.; Matos, A. T. & Azevedo, R. F. (2006). Determination of Transport Parameters for Heavy Metal in Residual Compacted Soil Using Two Methodologies. Canadian Journal of Civil Engineering, Vol.33, pp. 912-917.
9. Azevedo, I. C. D. , Nascentes, C. R.; Azevedo, R. F.; Matos, A. T. & Guimarães, L. M. (2003). Coeficiente de Dispersão Hidrodinâmica e Fator de Retardamento de Metais Pesados em Solo Residual Compactado. Solos e Rochas, Vol.26, pp.229-249.
10. Barone, F. S.; Yanful, E. K.; Quigley, R. M. & Rowe, R. K. (1989). Effect

of Multiple Contaminant Migration on Diffusion and Adsorption of Some Domestic Waste Contaminants in a Natural Clayey Soil. Canadian Geotechnical Journal, Vol.26, pp. 189-198.

11. Bolinelli Jr., H. L. (2004). Piezocone de Resistividade: Primeiros Resultados de Investigação Geoambiental em Solos Tropicais. MSc, Department of Civil Engineering, São Paulo State University, Bauru-SP, Brazil.
12. Boscov, M. E. G.; Cunha, I. I. & Saito, R. (2001). Radium Migration Through Clay Liners at Waste Disposal Sites. Science of t he Total Environment, Vol.266, pp. 259-264.
13. Campanella, R. G. (2008). Geo-Environmental Site Characterization. Proceedings of the 3rd International Conference on Geotechnical and Geophysical Site Characterization, Taipei –Taiwan, Taylor & Francis Group, pp. 3-16.
14. CETESB (2005). Relatório de Estabelecimento de Valores Orientadores para Solos e ÁguasSubterâneas no Estado de São Paulo. São Paulo-SP, Brasil.
15. Cherry, J. A.; Gillham, R. W.; Anderson, E. G. & Johnson, P. E. (1983). Migration of Contaminants in Groundwater at a Landfill: a Case Study 2. Groundwater Monitoring Devices. Journal of Hydrology, Vol.63, pp. 31-49.
16. Christensen, T. H., Bjerg, P. L. & Kjeldsen, P. (2000). Natural attenuation: a feasible approach to remediation of ground water pollution at landfills? Ground Water Monitoring & Remediation, Vol.20, pp. 69-77.
17. Dahlin, T.; Bernstone, C. & Loke, M. H. (2002). A 3-D Resistivity Investigation of a Contaminated Site at Lernacken, Sweden. Geophysics, Vol.67, pp. 1692-1700.
18. Daniel, C.R. (1997). An Investigation of the Factors Affecting Bulk Soil Electrical Resistivity. BASc, University of British Columbia, Department of Civil Engineering. Davies, M. P. & Campanella, R. G. (1995). Environmental Site Characterization Using In-Situ Testing Methods, 48th Canadian Geotechnical Conference, Vancouver-BC, Canada.
19. Elis, V. R. & Zuquette, L. V. (2002). Caracterização Geofísica de Áreas Utilizadas para Disposição de Resíduos Sólidos Urbanos. Revista Brasileira de Geociências, Vol.32, pp. 119-134.
20. Frempong, E. M. & Yanful, E. K. (2008). Interactions Between Three Tropical Soils and Municipal Solid Waste Landfill Leachate. Journal of Geotechnical and Geoenvironmental Engineering, Vol.134, pp.379-396.

21. Freeze, R. A. & Cherry, J. A. (1979). Groundwater. Prentice Hall, Inc. Englewood Cliffs, New Jersey, U. S. A.

22. Geoprobe Direct Image (2004). MP 6500 – Membrane Interface Probe (MIP): User Manual. Document N. 3083, Revision 1.0 10-12-2004, Kejr, Inc., Salina, Kansas, U.S.A.

23. Grazinolli, P. L.; Costa, A.; Campos, T. M. P. & Vargas Jr., E. A. (1999). Aplicações do Radar de Penetração no Solo (GPR) e da Eletrorresistividade para a Detecção de Compostos Orgânicos. Proceeding of the 4° Congresso Brasileiro de Geotecnia Ambiental, São José dos Campos, Brazil, December 1999, pp 127-13.

24. Innov-X Systems Alpha Series (2007). X-Ray Fluorescence Spectrometers. P/N 100392, Revision B. Woburn, MA, U. S. A.

25. IPT (2011). Estudo de Viabilidade para Ampliação da Área de Disposição Final de Resíduos Sólidos Urbanos no Município de Anhembi-SP. Institute for Technological Research, Center for Energy and Environmental Technologies. Parecer Técnico 19455-301/11, complementar ao Parecer Técnico 19287-301/11.

26. Kjeldsen, P.; Bjerg, P. L.; Rügge, K.; Christensen, T. H. & Pedersen, J. K. (1998). Characterization of an Old Municipal Landfill (Grindsted, Denmark) as a Groundwater Pollution Source: Landfill Hydrology and Leachate Migration. Waste Management & Research, Vol.16, pp. 14-22.

27. Lago, A. L.; Silva, E. M. A.; Elis, V. R. & Giacheti, H. L. (2003). Aplicação de Eletrorresistividade e Polarização Induzida em Área de Disposição de Resíduos Sólidos Urbanos em Bauru-SP. Proceedings of the 8th International Congress of the Brazilian Geophysical Society, Rio de Janeiro, Brazil, November 2003, CD-ROM.

28. Lambe, T. W. & Whitman, R. V. (1969). Soil Mechanics. New York: J. Wiley, 553p. Lan, T. N. (1977). Um Nouvel Esai D'identification des Sols: L'essai au Bleu de Methyléne. Bulletin Liaison Laboratoire dês Ponts et Chaussée, Paris, pp.136-137.

29. Larsson, R. (1995). Use of a Thin Slot as Filter in Piezocone Tests. Proceedings of International Symposium on Cone Penetration Testing (CPT'95), Linköping, Sweden, pp. 35–40.

30. Leite, A. L. & Paraguassú, A. B. (2002). Diffusion of Inorganic Chemicals in Some Compacted Tropical Soils. Proceedings of the 4th International Congress In Environmental Geotechnics, Rio de Janeiro, Brazil, August 2002, pp 39-45.

31. Loke, M. H. & Barker, R. D. (1996). Practical Techniques for 3D Resistivity Survey and Data Inversion. Geophysical Prospecting, Vol.44,

pp. 499-523.

32. Lunne, T.; Robertson, P. K. & Powell, J. (1997). Cone Penetration Test in Geotechnical Practice. Blackie Academic & Professional, London.

33. Mackay, D. M.; Freyberg, D. L.; Roberts, P. V. & Cherry, J. A. (1986). A Natural Gradient Experiment on Solute Transport in a Sand Aquifer: 1. Approach and Overview of Plume Movement. Water Resources Research, Vol.22, pp. 2017-2029.

34. MacFarlane, D. S.; Cherry, J. A.; Ghillham, R. W. & Sudicky, E. A. (1983). Migration of Contaminants in Groundwater at a Landfill: a Case Study, 1. Groundwater Flow and Plume Delineation. Journal of Hydrology, Vol.63, pp. 1-29.

35. McNeill, J. D. (1980). Electromagnetic Terrain Conductivity Measurement at Low Induction Numbers. Geonics Technical Note, No.6, 15p.

36. Mondelli, G. (2008). Integração de Diferentes Técnicas de Investigação para Avaliação da Poluição e Contaminação de uma Área de Disposição de Resíduos Sólidos Urbanos. Ph.D. Thesis, Department of Geotechnical Engineering, University of São Paulo, São Carlos-SP, Brazil.

37. Mondelli, G.; Giacheti, H. L.; Boscov, M. E. G.; Elis, V. R. & Hamada, J. (2007). Geoenvironmental Site Investigation Using Different Techniques in a Municipal Solid Waste Disposal Site in Brazil. Environmental Geology, Vol.52, pp. 871-887.

38. Mondelli, G.; Giacheti, H. L. & Elis, V. R. (2010a). The Use of Resistivity for Detecting MSW Contamination Plumes in a Tropical Soil Site. Proceedings of 6th International Conference on Environmental Geotechnics, New Delhi, India, pp. 1544-1549.

39. Mondelli, G.; Giacheti, H. L. & Howie, J. A. (2010b). Interpretation of Resistivity Piezocone Tests in a Contaminated Municipal Solid Waste Disposal Site. Geotechnical Testing Journal, Vol. 33, N. 2, pp. 123-136.

40. Monteiro Santos, F. A. (2004). 1-D Laterally Constrained Inversion of EM34 Profiling Data. Journal of Applied Geophysics, Vol.56, pp. 123-134.

41. Nascentes, C. R. (2003). Coeficiente de Dispersão Hidrodinâmica e Fator de Retardamento de Metais Pesados em Solo Residual Compactado. MSc, Department of Civil Engineering, Federal University of Viçosa, Viçosa-MG, Brazil.

42. Pejon, O. J. (1992). Mapeamento Geotécnico da Folha de Piracicaba-SP (Escala 1:100.000): Estudo de Aspectos Metodológicos, de Caracterização e de Apresentação dos Atributos. Ph.D.

43. Thesis, Department of Geotechnical Engineering, University of São Paulo, São Carlos-SP, Brazil.

44. Porsani, J. L.; Malagutti Filho, W.; Elis, V. R.; Fisseha, S., Dourado, J. C. & Moura, H. P. (2004). The Use of GPR and VES in Delineating a Contamination Plume in a Landfill Site: A Case Study in SE Brazil. Journal of Applied Geophysics, Vol.55, pp.199-209.

45. Ritter, E. (1998). Efeito da Salinidade na Difusão e Sorção de Alguns Íons Inorgânicos em um Solo Argiloso Saturado. PhD, Department of Civil Engineering, Federal University of Rio de Janeiro, Rio de Janeiro-RJ, Brazil.

46. Robertson, P. K. (1998). Geo-Environmental Investigation, Characterization and Monitoring Using Penetration Techniques. Simpósio Brasileiro de Geotecnia Ambiental, São PauloSP, Brazil.

47. Robertson, P. K.; Campanella, R. G., Gillespie, D. & Greig, J. (1986). Use of piezometer cone data, Proc. of the In-Situ-86, ASCE Specialty Conference, p. 1263-1280

48. Robertson, P. K. & Cabal, K. L. (2008). Guide to Cone Penetration Testing for Geo-Environmental Engineering. Gregg Drilling & Testing, Inc., 2nd Edition, 84 p.

49. Robertson, P. K. & Campanella, R. G. (1988). Guidelines for Geotechnical Design Using CPT and CPTU Data. Report FHWA, 340p.

50. Rowe, R. K.; Caers, C. J. & Barone, F. S. (1988). Laboratory Determination of Diffusion and Distribution Coefficients of Contaminants Using Undisturbed Clayey Soil. Canadian Geotechnical Journal, Vol.25, No.1, pp. 108-118.

51. Shackelford, C. D. (1993). Contaminant Transport. Geotechnical Practice for Waste Disposal, Chapman & Hall, London, pp. 33-65.

52. Shackelford, C. D. (1994). Critical Concepts for Column Testing. Journal of Geotechnical Engineering, Vol.120, pp.1804-1828.

53. Shackelford, C. D. & Daniel, D. E. (1991). Diffusion in Saturated Soil. I: Background. Journal of Geotechnical Engineering, Vol.117, No.3, pp. 467-484.

54. Shinn II, J.D. & Bratton, W.L. (1995). Innovations with CPT for environmental site characterization», Proceedings of CPT›95, Vol. 2, pp. 93-98.

55. Stuermer, M. M. (2005). Contribuição ao Estudo de um Solo Saprolítico como Revestimento Impermeabilizante de Fundo de Aterros de Resíduos. PhD, Department of Structures and Geotechnical Engineering, University

of São Paulo, São Paulo-SP, Brazil.

56. Telford, W. M.; Geldart, L. P.; Sheriff, R. E. & Keys, D. A. (1990). Applied Geophysics, Cambridge University Press, 860 p.
57. Thornthwaite, C.W. & Mather, J.R. (1955). The water balance. Publications in Climatology. New Jersey: Drexel Institute of Technology, 104p.
58. US EPA (1989). Seminar on Site Characterization for Subsurface Remediations. United States Environmental Protection Agency, Technology Transfer, Report CERI-89-224, 350p.
59. US EPA (1993). SW-846 pH in Liquid and Soil. Method 9040 (Liquid) and SW-846 Method 9045 (Soil).
60. Ustra, A. T.; Elis, V. R.; Mondelli, G.; Zuquette, L. V. & Giacheti, H. L. (2011). Case Study: A 3D Resistivity and Induced Polarization Imaging from Downstream a Waste Disposal Site in Brazil. Environmental Earth Sciences, in press (online).
61. Weemes, I. (1990). A Resistivity Cone Penetrometer for Ground-Water Studies. MASc, University of British Columbia, Department of Civil Engineering.
62. Yong, R. N. (2001). Geoenvironmental Engineering: Contaminated Soils, Pollutant Fate & Mitigation. CRC Press, USA, 307p
63. Yong, R. N.; Mohamed, A. M. O. & Warkentin, B. P. (1992). Principles of Contaminant Transport in Soils. Elsevier Science Publishers B.V., 327 p.
64. Zuquette, L. V.; Palma, J. B. & Pejon, O. J. (2005). Environmental Assessment of anUncontrolled Sanitary Landfill. Bulletin of Engineering Geology and the Environment, Vol.64, pp.257-271.

CITATION

CHAPTER 1

Paulo Cesar Fernandes da Silva and John Canning Cripps (2011). Geo-environmental Terrain Assessments Based on Remote Sensing Tools: A Review of Applications to Hazard Mapping and Control, Environmental Management in Practice, Dr. Elzbieta Broniewicz (Ed.), ISBN: 978-953-307-358-3, InTech, DOI: 10.5772/24487.

CHAPTER 2

B. He, J. Chen, C. Chen and Y. Liu, "Mineral Prospectivity Mapping Method Integrating Multi-Sources Geology Spatial Data Sets and Case-Based Reasoning," Journal of Geographic Information System, Vol. 4 No. 2, 2012, pp. 77-85. doi: 10.4236/jgis.2012.42011.

CHAPTER 3

S. Singh and P. Singh, "Mapping Spatial Data on the Web Using Free and Open-Source Tools: A Prototype Implementation," Journal of Geographic Information System, Vol. 6 No. 1, 2014, pp. 30-39. doi: 10.4236/jgis.2014.61004.

CHAPTER 4

A. Suhardiman, S. Tsuyuki, M. Sumaryono and Y. Sulistioadi, "Geostatistical Approach for Site Suitability Mapping of Degraded Mangrove Forest in the Mahakam Delta, Indonesia," Journal of Geographic Information System, Vol. 5 No. 5, 2013, pp. 419-428. doi: 10.4236/jgis.2013.55040.

CHAPTER 5

Emmanuel Vassilakis, Remote Sensing of Environmental Change in the Antirio Deltaic Fan Region, Western Greece, doi:10.3390/rs2112547.

CHAPTER 6

Issaak Parcharidis Sotiris Kokkalas, Ioannis Fountoulis and Michael Foumelis, Detection and Monitoring of Active Faults in Urban Environments: Time Series Interferometry on the Cities of Patras and Pyrgos (Peloponnese, Greece), doi:10.3390/rs1040676.

CHAPTER 7

Slavoljub Dragicevic, Nenad Zivkovic, Mirjana Roksandic, Stanimir Kostadinov, Ivan Novkovic, Radislav Tosic, Milomir Stepic, Marija Dragicevic and Borislava Blagojevic (2012). Land Use Changes and Environmental Problems Caused by Bank Erosion: A Case Study of the Kolubara River Basin in Serbia, Environmental Land Use Planning, Dr. Seth Appiah-Opoku (Ed.), ISBN: 978-953-51-0832-0, InTech, DOI: 10.5772/50580.

CHAPTER 8

Seth Appiah-Opoku and Crystal Taylor (2012). Environmental Land Use and the Ecological Footprint of Higher Learning, Environmental Land Use Planning, Dr. Seth Appiah-Opoku (Ed.), ISBN: 978-953-51-0832-0, InTech, DOI: 10.5772/48191.

CHAPTER 9

Dirk Loehr (2012). The Role of Tradable Planning Permits in Environmental Land Use Planning: A Stocktake of the German Discussion, Environmental Land Use Planning, Dr. Seth Appiah-Opoku (Ed.), ISBN: 978-953-51-0832-0, InTech, DOI: 10.5772/50469.

CHAPTER 10

A. Gharagozlou, H. Nazari and M. Seddighi, "Spatial Analysis for Flood Control by Using Environmental Modeling," Journal of Geographic Information System, Vol. 3 No. 4, 2011, pp. 367-372. doi: 10.4236/jgis.2011.34035.

CHAPTER 11

Gebreyesus Brhane Tesfahunegn, Lulseged Tamene, and Paul L. G. Vlek, "Soil Erosion Prediction Using Morgan-Morgan-Finney Model in a GIS Environment in Northern Ethiopia Catchment," Applied and Environmental Soil Science, vol. 2014, Article ID 468751, 15 pages, 2014. doi:10.1155/2014/468751.

CHAPTER 12

Giulliana Mondelli, Heraldo Luiz Giacheti and Vagner Roberto Elis (2012). Geo-Environmental Site Investigation for Municipal Solid Waste Disposal Sites, Municipal and Industrial Waste Disposal, Dr. Xiao-Ying Yu (Ed.), ISBN: 978-953-51-0501-5, InTech, DOI: 10.5772/28835.

INDEX